T0295245

GALAXY EVOLUTION AND FEEDBACK ACROSS DIFFERENT ENVIRONMENTS

IAU SYMPOSIUM 359

COVER ILLUSTRATION:

Background: Gas column density in the cosmic web, z = 2, comoving 15 Mpc side (credit: Rainer Weinberger, IllustrisTNG Collaboration). From this cosmic web, galaxies and clusters have been formed, as shown with the foreground images, revealing different forms of feedback.

Top left: Jellyfish galaxy JO201(credit: GASP collaboration, 2019) suffering feedback within a galaxy cluster.

Top right: Composite image of galaxy cluster MS0735.6 + 7421 showing feedback effects in the X-ray gas (Credit: X-ray: NASA/CXC/Univ. Waterloo/ B.McNamara; Optical: NASA/ESA/STScI/Univ. Waterloo/B.McNamara; Radio: NRAO/Ohio Univ./L.Birzan et al., 2006, revised 2018).

Bottom left: HST [OIII] narrow-band image of the QSO J135251 + 654113 showing the ionized gas extending by 30 kpc (orange) ovelaid on a continuum-band image of the host galaxy (grey) (Credit: Storchi-Bergmann et al. 2018, ApJ 868, 14; figure by Fausto Barbosa), revealing Supermassive Black Hole feedback beyond the host galaxy.

Bottom right: Composite image of the Starburst galaxy M82 showing feedback from star formation (Credit: X-ray: NASA/CXC/JHU/D. Strickland; Optical: NASA/ESA/STScI/AURA/The Hubble Heritage Team; Infrared: NASA/ JPL-Caltech/Univ. of Arizona/C. Engelbracht).

IAU SYMPOSIUM PROCEEDINGS SERIES

INTERNATIONAL ASTRONOMICAL UNION
UNION ASTRONOMIQUE INTERNATIONALE

GALAXY EVOLUTION AND FEEDBACK ACROSS DIFFERENT ENVIRONMENTS

PROCEEDINGS OF THE 359th SYMPOSIUM OF THE INTERNATIONAL ASTRONOMICAL UNION HELD IN BENTO GONÇALVES, BRAZIL 2–6 MARCH, 2020

Edited by

THAISA STORCHI-BERGMANN
Departamento de Astronomia, Instituto de Física, UFRGS, Brazil

WILLIAM FORMAN
Harvard-Smithsonian Center for Astrophysics, USA

RODERIK OVERZIER
Observatório Nacional, Brazil

and

ROGÉRIO RIFFEL
Departamento de Astronomia, Instituto de Física, UFRGS, Brazil

CAMBRIDGE UNIVERSITY PRESS
University Printing House, Cambridge CB2 8BS, United Kingdom
1 Liberty Plaza, Floor 20, New York, NY 10006, USA
10 Stamford Road, Oakleigh, Melbourne 3166, Australia

First published 2021

Printed in the UK by Bell & Bain, Glasgow, UK

Typeset in System LaTeX 2ε

A catalogue record for this book is available from the British Library Library of Congress Cataloguing in Publication data

This journal issue has been printed on FSC™-certified paper and cover board. FSC is an independent, non-governmental, not-for-profit organization established to promote the responsible management of the world's forests. Please see www.fsc.org for information.

ISBN 9781108490689 hardback
ISSN 1743-9213

Table of Contents

Session 4: Secular evolution and internal processes: Mass quenching, stellar and AGN feedback over different z's

Session 7: Discussion - questions and answers

Preface

The history of IAU Symposium 359 – Galaxy Evolution and Feedback across Different Environments, GALFEED for short, began with a meeting, by the end of 2014, between two of us at the Harvard/Smithsonian CfA: William and Thaisa, who was visiting the CfA. Due to our common interest in Active Galactic Nuclei (AGN) feedback, at that opportunity, we decided to create the IAU Commission X1 – Supermassive Black Holes, Feedback and Galaxy Evolution. And shortly afterwards, we began to discuss the promotion of an IAU Symposium on this topic in the south of Brazil. But these Symposia are very competitive, and the first proposal, focusing on feedback from AGN, did not succeed. There had been too many recent Symposia on this topic. Discussing this fact with our young colleagues from the astronomy group at the Instituto de Física, UFRGS (Universidade Federal do Rio Grande do Sul), Porto Alegre, Brazil, and with their support, we changed the proposal to include the broader topic of galaxy evolution and a strong community outreach component, that led to its selection by the IAU.

Our goal with this Symposium was, thus, to bring together the AGN and galaxy evolution science communities, that usually propose separate Symposia. However, we now know that AGN is a phase occuring in most galaxies that critically influences their evolution. Thus it is important to study the two processes together and to enable researchers, in both topics, to learn from each other. Key questions we have discussed during the Symposium include: How do galaxies acquire their gas and how efficiently is it transformed into stars? How is the SMBH in the galaxy center fuelled to become an AGN? What is the main physical mechanism that quenches star formation? How powerful are the stellar and AGN feedback processes in regulating galaxy evolution? What is the role of the environment on galaxy evolution and AGN triggering?

During our Symposium, the above questions have been discussed in 118 scientific contributions, grouped in six sessions:

(1) The first session addresses the formation and early evolution of galaxies and supermassive black holes via models, simulations and observations, targeting massive galaxies, starburst galaxies and QSOs;

(2) The second session focuses on the epoch of Cosmic Noon, addressing black hole masses, galaxy mergers, star-forming galaxies, distant quasar host galaxies, including the claim that there is no quenching of star formation in quasars;

(3) In the third session we discuss high density environments, from protoclusters, through the effect of black hole outbursts on galaxy clusters, to the effect of the environment on starburst and AGN activity;

(4) The fourth session addresses feedback associated with secular evolution, including the effect of AGN jets and outflows on galaxies, via models and observations, and a discussion of the important role of the adopted gas density on the calculation of outflow power from ionised gas kinematics;

(5) The fifth session concentrates on the fueling of star formation and nuclear activity, including the observation of gas flows with ALMA down to the inner few parsecs around the supermassive black hole and models of the flows in the innermost parsec;

(6) In the sixth session, the primary topic is the state of the present day Universe, via the study of resolved stellar populations of galaxies to derive their star formation history and the investigation of resolved excitation and kinematics of the gas in individual nearby active and non-active galaxies.

The seventh and last session comprises a compilation of 25 specific questions proposed by the participants and discussed on the last day of the Symposium together with the proposed answers given both at the conference or afterwards by the participants.

It was a pleasure to receive the approximately 200 colleagues and students who came to the Symposium in Bento Gonçalves during the week of March 2 to 6 of 2020, making the meeting a success, just before the cancelation of all similar meetings due to the COVID-19 pandemic. We were pleased, in particular, by the enthusiastic participation of Brazilian students and post-docs who helped with the reception of our foreign colleagues as well as with the outreach activities that engaged about 2000 people during the 5 days of the Symposium.

The SOC, from left to right: William Forman, Françoise Combes, Thaisa Storchi-Bergmann, Richard Davies, Roderik Overzier and Keiichi Wada. Missing from the picture: Raffaella Morganti and Sebastian Sanchez.

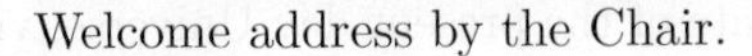
Welcome address by the Chair.

The audience of the Symposium.

Thaisa Storchi-Bergmann, *Editor and Chair of the SOC*
William Forman, *Co-editor and Co-chair*
Roderik Overzier, *Co-editor and Co-chair*
Rogério Riffel, *Co-editor and Chair of the LOC*

Dedication

We dedicate this volume to the memory of Dr. João Evangelista Steiner, who prematurely passed away on September 10, 2020, by the time we were finishing these Proceedings. As co-chair Roderik Overzier posed, "the great oak of Brazilian Astronomy has fallen, and we will be lost without its shadow".

Left: João Steiner. **Right - From left to right:** Daniel May, Roberto Menezes, Patricia da Silva, João Steiner, Tiago Ricci, Catarina Aydar, Pedro H. Cezar

João was a Professor of Astrophysics at IAG-USP (Instituto de Astronomia, Geofísica e Ciências Atmosféricas - Universidade de São Paulo) and a member of the Brazilian Academy of Sciences. He was an active astrophysicist who made important scientific contributions in the area of compact stars (in particular cataclysmic variables) and Active Galactic Nuclei. A 1983 paper by him and Jules Halpern, written when both were at the Harvard/Smithsonian Center for Astrophyiscs, Cambridge, US, is an important reference to this date on the nature of LINER galactic nuclei.

Besides being a productive researcher, and having advised more than 20 students and post-docs, João was restless to promote scientific research not only in astrophysics but also in other areas of knowledge, having been secretary general of the Brazilian Society for the Progress of Science (SBPC), Secretary for the Coordination of Research Units of the Ministry of Science, Technology and Innovation (MCTI), directed the Institute for Advanced Studies of University of São Paulo (IES/USP) and was also director of INPE (Instituto Nacional de Pesquisas Espaciais). João was particularly concerned that brazilian astrophysics had the necessary infrastructure to be at the forefront of astrophysical research. He coordinated the efforts of modernization of the Pico dos Dias Observatory, the establishment of the National Astrophysics Laboratory (LNA) and played a fundamental role in the Brazilian participation in the Gemini and SOAR observatories consortia. More recently, he had been the lead scientist corrdinating the Brazilian participation in the GMT (Giant Magellan Telescope).

João came with his 6 students and post-docs (see figure above) to participate in the Symposium, happily reporting on the results of their recent survey of nearby southern galaxies with the Gemini Multi-Object Integral Field Spectrograph called "Diving 3D", described along 9 contributions to the present volume. His contributions, guidance and leadership will be missed.

Thaisa Storchi-Bergmann, William Forman, Roderik Overzier and Rogério Riffel
The Editors

Editors

Thaisa Storchi-Bergmann
Departamento de Astronomia, Instituto de Física, UFRGS, Brazil

William Forman
Harvard-Smithsonian Center for Astrophysics, USA

Roderik Overzier
Observatório Nacional, Brazil

Rogério Riffel
Departamento de Astronomia, Instituto de Física, UFRGS, Brazil

Organizing Committee

Scientific Organizing Committee

Thaisa Storchi-Bergmann (Chair)	Instituto de Física, UFRGS, Porto Alegre, Brazil
Roderik Overzier (Co-Chair)	Observatório Nacional, Rio de Janeiro, Brazil
William Forman (Co-Chair)	Harvard-Smithsonian CfA, Cambridge, US
Raffaella Morganti (Co-Chair)	ASTRON, Dwingeloo, Netherlands
Francesco Massaro	University of Turin, Italy
Maria Victoria Alonso	IATE, Cordoba, Argentina
Francoise Combes	Paris Observatory, Paris, France
Luis Colina Robledo	Centro de Astrobiologia, Madrid, Spain
Richard Davies	MPE, Garching, Germany
Luis Ho	KIAA, Peking University, China
Stefanie Komossa	Max Planck Inst. Radio Ast., Bonn, Germany
Paulina Lira	University of Chile, Santiago, Chile
Alessandro Marconi	University of Florence, Italy
Nicole Nesvadba	Inst. d'Astroph. Spatiale, Orsay, France
Sebastian Sanchez	UNAM, Ciudad de Mexico, Mexico
Keiichi Wada	Kagoshima University, Japan

Local Organizing Committee

Rogério Riffel (Chair)	Instituto de Física, UFRGS, Porto Alegre, Brazil
Ana L. Chies-Santos (Co-Chair)	Instituto de Física, UFRGS, Porto Alegre, Brazil
Cristina Furlanetto (Co-Chair)	Instituto de Física, UFRGS, Porto Alegre, Brazil
Allan Schnorr-Müller (Co-Chair)	Instituto de Física, UFRGS, Porto Alegre, Brazil
Marina Trevisan (Co-Chair)	Instituto de Física, UFRGS, Porto Alegre, Brazil
Carpes Hekatelyne	Instituto de Física, UFRGS, Porto Alegre, Brazil
Juliana Motter	Instituto de Física, UFRGS, Porto Alegre, Brazil
Rogemar Riffel	Universidade Federal de Santa Maria, Brasil
Tiago Ricci	Universidade Federal da Fronteira Sul, Cerro Largo, Brazil
Jaderson Schimoia	Universidade Federal de Santa Maria, Brazil

Members of the LOC - from left to right Rodrigo Flores, Tiago Ricci, Hekatelyne Carpes, Cristina Furlanetto, Juliana Motter, Ana L. Chies-Santos, Eduardo Brock, Marina Trevisan and Jaderson Schimoia (not in the picture: Allan Schnorr-Müller, Rogério Riffel and Rogemar Riffel).

Acknowledgements

IAU 359 was a very special and successful meeting because of the remarkable efforts of many people. First, it gathered a superb cast of scientists from around the world for a superb set of talks. Second, it provided a showcase for South American postdocs and students to present their exciting work. The posters and the flash talks were dramatic and interesting – too many to absorb in too short a time – but the symposium proceedings are now before you! Read, enjoy, and reflect. Third, our public outreach event was the highlight of the season for the young and old of Bento Gonçalves and surrounding towns that packed the lecture hall until it overflowed and waited for over an hour after the public lecture in order to ask questions and have their photos taken with the speakers. It was inspiring to experience the enthusiasm of this audience for astrophysics!

We would like to thank the support from the IAU, SAB (Sociedade Astronômica Brasileira), Instituto Serrapilheira and FAPERGS (RS foundation for research support). We thank also the SOC for the guidance; the LOC for the logistic support, the creation of the "Universe at your feet" sandals, bags and notebooks; the astronomy students from our Physics Institute of Universidade Federal do Rio Grande do Sul (IF-UFRGS) and from other institutes for the support during the sessions and outreach activities; our University UFRGS for its support and the provost of research Rafael Roisler, for his speech in the opening ceremony; our partner Universities Universidade Federal da Fronteira Sul and Universidade Federal de Santa Maria for their support and the town of Bento Gonçalves, represented in the opening ceremony by the Secretary of Tourism, Rodrigo Parisotto, who welcomed the participants.

Thanks to the students, for their support throughout all activities.

The partnership with the town administration allowed us to use the space Casa das Artes, adjacent to the hotel, to develop our many educational and outreach activities, under the leadership of the director of the UFRGS Planetarium Daniela Pavani. We thank her and her team for the promoted activities for young school students during the days, and for the general public and high school students during the nights.

Kids waiting for the Planetarium session.

Audience of the nighttime outreach talks.

About 2,000 people attended these activities that included shows in an inflatable Planetarium, nighttime observations and talks by renowned international participants of the Symposium.

A word from Henrique Roberto Schmitt, Thaisa's first PhD student

In the name of Thaisa's students, I would like to point out that this symposium also marks a major point in Dr. Thaisa Storchi-Bergmann's career. After a long tenure as a professor at the Astronomy Department at UFRGS, Thaisa is retiring from teaching and administrative duties, which will allow her to dedicate more time to research and other interests. We would like to congratulate her on this major achievement, as well as on her highly accomplished professional career, spanning more than three decades, with seminal contributions to several areas of extragalactic astrophysics. Furthermore, her current and former students and post-docs would like to thank her for the guidance, friendship and encouragement that she provided through the different stages of their careers. Thanks to her energetic and gregarious nature, they were brought into a vibrant group, pursuing leading edge research in her many areas of interest. By incorporating these students into the daily activities of the group, not only were they taught the basics of the field, and guided on how to navigate the arduous and sometimes convoluted process of obtaining a degree, they were also encouraged to pursue their own interests and explore their own ideas. This independence was essential, not only for the professional development of the students and their future careers, but also for the overall success of the group, keeping it in the forefront of the field, with collaborations throughout the globe.

Special thanks to (from left to right): Daniela Pavani for coordinating the outreach activities; Grazyna Stasinska, Christopher Harrison and Christine Jones, for giving the nighttime talks.

Thaisa's present and former students and post-docs.Left: top, from left to right, Bruna Arajo, Natalie Schreiber, Henrique Schmitt, Jaderson Shimoia, [Thaisa], Gabriel Roier, Cristina Furlanetto, Julio Cesar, Hekatelyne Carpes, Daniel Ruschel Dutra, Oli Dors Jr.. Bottom: Edwin David, Bruno Dall'Agnol de Oliveira, Guilherme Couto. **Right:** Thaisa with former student Rodrigo Nemmen

A word from the Chair

Being in a phase of my career when I am retiring from teaching and administrative duties at the University, I would like to use this opportunity to thank all the participants and in particular my students, post-docs, my PhD advisor, Miriani Pastoriza, my colleagues and collaborators, from whom I learned so much during the years we worked together. Special thanks go to my former students Henrique Schmitt and Rodrigo Nemmen for their kind words during the Symposium. I also use this opportunity to pay tribute to Astronomy and the Universe with the following poem. My cousin Roberto Niederauer has composed a melody to go with it and played it at the Symposium. The song is available on Youtube:

https://www.youtube.com/watch?v=I-5j6Nb-3mg.

The Universe Song

When you look far, you see our Universe's past
Took shape at the Big Bang,
Then the first atoms began to circulate, away
And where they stayed, galaxies they created
Inside all galaxies, stars took form and Black Holes were born
Inside the stars, new heavier elements were forged
Then Supernovae and Black Holes
Spread the elements around the Universe
And where they stayed, new stars they created
And from star dust, this rich star dust, planets formed around
And in one planet, a little blue planet we were born,
We were finally born.

Roberto Niederauer and Thaisa: the authors of the Universe song

Thaisa Storchi-Bergmann
Chair of the SOC

CONFERENCE PHOTOGRAPH

List of Participants

Name & Organization	Email
Adam Thomas, Australian National University, Australia	adam.thomas@anu.edu.au
Aitor Carlos Robleto Orús, Universidad de Guanajuato, Mexico	arobleto@astro.ugto.mx
Alberto Nigoche-Netro, Universidad de Guadalajara, Mexico	anigoche@gmail.com
Alberto Rodríguez-Ardila, Laboratório Nacional de Astrofísica, Brazil	aardila@lna.br
Alessandro Marconi, Physics & Astronomy Dept., University of Florence, Italy	alessandro.marconi@unifi.it
Alexandre Vazdekis, Instituto de Astrofísica de Canarias, Spain	vazdekis@iac.es
Allan Schnorr Müller, Universidade Federal do Rio Grande do Sul, Brazil	allanschnorr@gmail.com
Amirnezam Amiri, University of Florence (UniFi) & Inistitute for research in fundamental (IPM), Italy and IRAN	amirnezamamiri@gmail.com
Ana Carolina Pichel, IAFE, Argentina	ana.pichel@gmail.com
Ana L. Chies-Santos, Universidade Federal do Rio Grande do Sul, Brazil	ana.chies@ufrgs.br
Anderson Caproni, NAT - Universidade Cidade de São Paulo, Brazil	anderson.caproni@cruzeirodosul.edu.br
André Luiz de Amorim, UFSC, Brazil	andre@astro.ufsc.br
Andrew Newman, Carnegie Observatories, USA	anewman@carnegiescience.edu
Andrey Vayner, Johns Hopkins University, USA	avayner1@jhu.edu
Anelise Audibert, National Observatory of Athens, Greece	anelise.audibert@obspm.fr
Anna D. Kapinska, NRAO, USA	akapinska@nrao.edu
Anna Luiza Trindade Falcao, The Catholic University of America, USA	anna.trindade04@gmail.com
Annagrazia Puglisi, CEA Saclay, Durham University, United Kingdom	anna.gp27@gmail.com
Ariane Trudeau, University of Victoria, Canada	ariane.trudeau@videotron.ca
Arianna Cortesi, Observatorio do Valongo, Brazil	aricorte@gmail.com
Ariel Werle, IAG - USP, Brazil	ariel.werle@gmail.com
Artemi Camps Fariña, Instituto de Astronomia, UNAM, Mexico	acamps@astro.unam.mx
Augusto Lassen, Universidade Federal do Rio Grande do Sul - UFRGS, Brazil	augusto.lassen@gmail.com
Beena Meena, Georgia State University, USA	bmeena@astro.gsu.edu
Benjamin Lee Davis, Swinburne University of Technology, Australia	benjamindavis@swin.edu.au
Bianca Poggianti, INAF-OAPd, Italy	bianca.poggianti@inaf.it
Bruna Lorrany de Castro Araujo, Universidade Federal do Piauí, Brazil	araujo.brunalc@gmail.com
Bruno Dall'Agnol de Oliveira, Universidade Federal do Rio Grande do Sul, Brazil	bruno.ddeo@gmail.com
Bruno Rodríguez Del Pino, Centro de Astrobiología, CSIC-INTA, Spain	brodriguez@cab.inta-csic.es
Carlo Cannarozzo, Alma Mater Studiorum Università di Bologna - Dipartimento di Fisica e Astronomia, Italy	carlo.cannarozzo3@unibo.it
Carlos Gomez-Guijarro, CEA Saclay, France	carlos.gomezguijarro@cea.fr
Carlos López-Cobá, Instituto de Astronomía - Universidad Nacional Autónoma de México, Mexico	clopez@astro.unam.mx
Carlos Roberto de Melo Carneiro, Universidade Federal do Rio Grande do Sul, Brazil	carlos.melo@ufrgs.br

Name & Organization	Email
Carolina Queiroz de Abreu Silva, Instituto de Física, Universidade de São Paulo, Brazil	cqueiroz@if.usp.br
Catarina Pasta Aydar, Universidade de São Paulo, Brazil	catarina.aydar@gmail.com
César Augusto Caretta, Universidad de Guanajuato, Mexico	caretta@astro.ugto.mx
Chris Harrison, Newcastle University, United Kingdom	c.m.harrison@mail.com
Christine Jones, Harvard-Smithsonian Center for Astrophysics, USA	cjones@cfa.harvard.edu
Clare Wethers, University of Turku (FINCA), Finland	clare.wethers@utu.fi
Cody Carr, University of Minnesota, USA	CodyCarr24@gmail.com
Cristina Furlanetto, Universidade Federal do Rio Grande do Sul, Brazil	cristina.furlanetto@ufrgs.br
Cristina Ramos Almeida, Instituto de Astrofísica de Canarias, Spain	cra@iac.es
Damien Spérone-Longin, LASTRO - EPFL, Switzerland	damien.sperone-longin@epfl.ch
Daniel Alberto Marostica, Universidade Tecnológica Federal do Paraná, Brazil	danielmarostica@live.com
Daniel May Nicolazzi, IAG-USP, Brazil	dmay.astro@gmail.com
Daniel Michael Crenshaw, Georgia State University, USA	crenshaw@astro.gsu.edu
Daniel Ruschel Dutra, Universidade Federal de Santa Catarina, Brazil	daniel@astro.ufsc.br
Daniela Borges Pavani, Universidade Federal do Rio Grande do Sul, Brazil	dpavani@if.ufrgs.br
Daniela Hiromi Okido, Universidade Federal do Rio Grande do Sul, Brazil	dhiromiokido@gmail.com
Darshan Kakkad, European Southern Observatory, Chile	dkakkad@eso.org
Davi Cabral Rodrigues, Universidade Federal do Espírito Santo, Brazil	davi.rodrigues@cosmo-ufes.org
David Williamson, University of Southampton, United Kingdom	d.j.williamson@soton.ac.uk
Denimara Dias dos Santos, National Institute for Space Research (INPE), Brazil	denimaradds@id.uff.br
Diogo Henrique Francis de Souza, Universidade Federal do Rio Grande do Sul, Brazil	diogo.henrique@ufrgs.br
Dominika Wylezalek, ESO, Germany	dwylezal@eso.org
Eduardo Alberto Duarte Lacerda, Instituto de Astronomía - Universidade Nacional Autónoma de México, México	lacerda@astro.unam.mx
Eduardo Telles, Observatorio Nacional, Brazil	etelles@on.br
Edwin Gouveia David, Universidade Federal do Rio Grande do Sul, Brazil	edwin.godavid@gmail.com
Elismar Lösch, Universidade Federal de Santa Catarina, Brazil	elismar@astro.ufsc.br
Emilio Zanatta, Universidade Federal do Rio Grande do Sul, Brazil	emiliojbzanatta@gmail.com
Fabio Cafardo, IAG - USP, Brazil	fabio.cafardo@usp.br
Felícia Palacios, Universidade Federal do Rio Grande do Sul, Brazil	felicia.palacios@gmail.com
Felipe Schmidt Lohmann, Universidade Federal do Rio Grande do Sul, Brazil	felipeslohmann@gmail.com
Fernanda Roman Oliveira, Universidade Federal do Rio Grande do Sul, Brazil	fernanda.oliveira@ufrgs.br
Filippo Maccagni, INAF - OAC, Italy	filippo.maccagni@inaf.it
Fiorella Lucia Polles, Observatiore de Paris, France	fiorella.polles@obspm.fr
Flávio Benevenuto da Silva Junior, NAT-UNICSUL, Brazil	flavio.ben@outlook.com

Name & Organization	Email
Francoise Combes, Observatoire de Paris, France	francoise.combes@obspm.fr
Frederick Hamann, University of California, USA	fhamann@ucr.edu
Gabriel Maciel Azevedo, Universidade Federal do Rio Grande do Sul, Brazil	gabriel.maciel.azevedo@gmail.com
Gabriel Roberto Hauschild Roier, Universidade Federal do Rio Grande do Sul, Brazil	gabrielrhroier@gmail.com
Gabriele Ilha, Universidade Federal de Santa Maria, Brazil	gabrieleilha1994@gmail.com
Gaia Gaspar, Observatorio Astronomico de Cordoba, Argentina	gaiagaspar@gmail.com
Giacomo Venturi, Pontificia Universidad Católica de Chile, Instituto de Astrofísica, Chile	gventuri@astro.puc.cl
Grazyna Stasinska, Observatoire de Paris, France	grazyna.stasinska@obspm.fr
Guilherme Couto, Universidad de Antofagasta, Chile	guilherme.couto@uamail.cl
Gustavo A. Lanfranchi, NAT - Universidade Cidade de São Paulo/Universidade Cruzeiro do Sul, Brazil	gustavo.lanfranchi@cruzeirodosul.edu.br
Gustavo Soares, IAG-USP, Brazil	gustavo.9891@gmail.com
Carpes Hekatelyne, Universidade Federal do Rio Grande do Sul, Brazil	hekatelyne.carpes@gmail.com
Henrique Schmitt, Naval Research Laboratory, USA	henrique.schmitt@nrl.navy.mil
Horacio Dottori, Universidade Federal do Rio Grande do Sul, Brazil	hdottori@gmail.com
Ilaria Pagotto, Leibniz-Institut für Astrophysik Potsdam, Germany	ilaria.pagotto@gmail.com
Itziar Aretxaga, INAOE, Mexico	itziar@inaoep.mx
Ivan Almeida, IAG-USP, Brazil	ivan.almeida@usp.br
Izabel Cristina Freitas dos Santos, Universidade Federal de Santa Maria, Brazil	izabelfisica@gmail.com
Jaderson da Silva Schimoia, Universidade Federal de Santa Maria, Brazil	jaderfisico@gmail.com
Jari Kotilainen, University of Turku, Finland	jarkot@utu.fi
Jason K. Chu, Gemini Observatory, USA	jchu@gemini.edu
João Evangelista Steiner, Universidade de São Paulo, Brazil	joao.steiner@iag.usp.br
João Pedro Verardo Benedetti, Universidade Federal do Rio Grande do Sul, Brazil	jpvbene@gmail.com
Johan Matheus Marques, Universidade Federal de Santa Maria, Brazil	johanmatheusmarques@gmail.com
Jonathan Cohn, Texas A&M University, USA	joncohn@tamu.edu
Jorge Moreno, Pomona College & Harvard CfA, USA	jorge.moreno@pomona.edu
Juliana Cristina Motter, Universidade Federal do Rio Grande do Sul, Brazil	jujucmotter@gmail.com
Júlio César Matte Figueiró, Universidade Federal do Rio Grande do Sul, Brazil	jcmf1998@outlook.com
Jullian Henrique Barbosa dos Santos, Universidade de São Paulo, Brazil	jullian.santos@usp.br
Karin Menendez-Delmestre, Valongo Observatory, Federal University of Rio de Janeiro, Brazil	kmd@astro.ufrj.br
Karla Alejandra Cutiva Alvarez, Universidad de Guanajuato, Mexico	kacutivaa@unal.edu.co
Katia Slodkowski Clerici, Universidade Federal de Santa Catarina, Brazil	clericikatia@gmail.com
Kei Ito, NAOJ/SOKENDAI, Japan	kei.ito@nao.ac.jp
Keiichi Wada, Kagoshima University, Japan	wada@astrophysics.jp
Kelen Tonet, Universidade Federal do Rio Grande do Sul, Brazil	kelenkaty@gmail.com
Kirsty May Butler, Sterrewacht Leiden, Netherlands	butler@strw.leidenuniv.nl
Konrad Tristram, European Southern Observatory, Chile	konrad.tristram@eso.org

Name & Organization	Email
Konstantinos Kolokythas, Centre for Space Research, North-West University, South Africa	K.Kolok@nwu.ac.za
Kristina Nyland, NRC fellow, resident at NRL, United States	knyland@nrao.edu
Kshama Sara Kurian, European Southern Observatory, Germany	kshama.sara@gmail.com
Lilianne Mariko Izuti Nakazono, Universidade de São Paulo, Brazil	lilianne.nakazono@usp.br
Lucimara Pires Martins, Universidade Cruzeiro do Sul, Brazil	lucimara.martins@cruzeirodosul.edu.br
Luidhy Santana da Silva, Federal University of Rio de Janeiro/Valongo, Brazil	luidhy@astro.ufrj.br
Luigi Spinoglio, Instituto di Astrofisica e Planetologia Spaziali - INAF, Italy	luigi.spinoglio@inaf.it
Luis Gabriel Dahmer Hahn, Laboratório Nacional de Astrofísica, Brazil	luisgdh@gmail.com
Lyndsay Old, European Space Agency (ESA), France	lyndsay.old@esa.int
Makoto Ando, The University of Tokyo, Japan	mando@astron.s.u-tokyo.ac.jp
Mar Mezcua, Institute of Space Sciences, Spain	marmezcua.astro@gmail.com
Marcela Yasmín González Paillalef, Universidad de Concepción, Chile	marcegonzalezp@udec.cl
Marco Antonio Canossa Gosteinski, Universidade Federal do Rio Grande do Sul, Brazil	canossa.marco@gmail.com
Marcos Fonseca Faria, Instituto Nacional de Pesquisas Espaciais, Brazil	marcos.faria@inpe.br
Maria Luisa Gomes Buzzo, Universidade de São Paulo, Brazil	luisa.buzzo@gmail.com
Maria Paz Aguero, Cordoba Observatory, Argentina	mpaguero@unc.edu.ar
Mariela Martinez Paredes, Korea Astronomy and Space Science Institute, South Korea	mariellauriga@gmail.com
Marina Bianchin, Universidade Federal de Santa Maria, Brazil	marinabianchin17@gmail.com
Marina Dal Ponte, Universidade Federal do Rio Grande do Sul, Brazil	mari.dalponte@gmail.com
Marina Trevisan, Universidade Federal do Rio Grande do Sul, Brazil	marina.trevisan@gmail.com
Martin Hardcastle, University of Hertfordshire, United Kingdom	m.j.hardcastle@herts.ac.uk
Marzena Sniegowska, Nicolaus Copernicus Astronomical Center Polish Academy of Sciences, Poland	msniegowska@camk.edu.pl
Matilde Mingozzi, Astronomical Observatory of Padova, Italy	matilde.mingozzi@inaf.it
Maximilien Franco, University of Hertfordshire, United Kingdom	m.franco@herts.ac.uk
Michael Beasley ,Instituto de Astrofísica de Canarias, Spain	beasley@iac.es
Michael Brotherton, University of Wyoming, USA	mbrother@uwyo.edu
Minju Lee, Max Planck Institute for Extraterrestrial Physics, Germany	minju@mpe.mpg.de
Miriani G. Pastoriza, Universidade Federal do Rio Grande do Sul, Brazil	miriani.pastoriza@gmail.com
Mônica Tergolina, Universidade Federal do Rio Grande do Sul, Brazil	monica.tergolina@ufrgs.br
Montserrat Villar Martin, Centro de Astrobiologia, Spain	villarmm@cab.inta-csic.es
Muryel Guolo Pereira, University of Santa Catarina, Brazil	muryel@astro.ufsc.br

Name & Organization	Email
Natacha Zanon Dametto, Universidad de Antofagasta, Chile	natacha.dametto@uamail.cl
Natalia Vale Asari, Universidade Federal de Santa Catarina, Brazil	natalia@astro.ufsc.br
Natalie Nicole Schreiber Bensley, Universidade Federal do Rio Grande do Sul, Brazil	nataliemaggori@gmail.com
Nícolas Dullius Mallmann, Universidade Federal do Rio Grande do Sul, Brazil	nicolas.mallmann@ufrgs.br
Nicolas Peschken, University Nicolaus Copernicus, Poland	npeschken@umk.pl
Oli Luiz Dors Junior, Universidada do Vale do Paraiba, Brazil	olidors@univap.br
Pablo G. Pérez-González, Centro de Astrobiología (CAB/CSICINTA), Spain	pgperez@cab.inta-csic.es
Paramita Barai, Universidade Cruzeiro do Sul, Brazil	paramita.barai@iag.usp.br
Patrícia da Silva, Universidade de São Paulo, Brazil	p.silva2201@gmail.com
Pedro Henrique Cezar Remião de Macedo, Universidade de São Paulo, Brazil	pedrocezar@usp.br
Peter Laursen, Institute of theoretical astrophysics, University of Oslo, Norway	pela@astro.uio.no
Rafael S. de Souza, Shanghai Astronomical Observatory, China	drsouza@ad.unc.edu
Raffaella Morganti, ASTRON/Kapteyn Institute, The Netherlands	morganti@astron.nl
Rainer Simon Weinberger, Center for Astrophysics — Harvard & Smithsonian, USA	rainerweinberger@googlemail.com
Raniere de Menezes, Universidade de São Paulo, Brazil	raniere.m.menezes@gmail.com
Raquel Santiago Nascimento, Laboratório Nacional de Astrofísica, Brazil	rnascimento@lna.br
Richard Bower, Durham University, United Kingdom	r.g.bower@durham.ac.uk
Richard Davies, MPE, Germany	davies@mpe.mpg.de
Roberto Bertoldo Menezes, Instituto Mauá de Tecnologia, Brazil	lindeterob@gmail.com
Roberto Hazenfratz Marks, Universidade de São Paulo, Brazil	robertohm@usp.br
Roderik Overzier, Observatório Nacional, Brazil	roderikoverzier@gmail.com
Rodolfo Brumel Cardoso Spindler, Universidade Federal do Rio Grande do Sul, Brazil	rodolfo.spindler@ufrgs.br
Rodrigo Flores de Freitas, Universidade Federal do Rio Grande do Sul, Brazil	rodrigofloresdefreitas@gmail.com
Rodrigo Nemmen, Universidade de São Paulo, Brazil	rodrigo.nemmen@iag.usp.br
Rogemar A. Riffel, Universidade Federal de Santa Maria, Brazil	rogemar@ufsm.br
Rogério Riffel, Universidade Federal do Rio Grande do Sul, Brazil	riffel@ufrgs.br
Rosemary Coogan, Max Planck Institute for Extraterrestrial Physics, Germany	rcoogan@mpe.mpg.de
Sandra Isabel Raimundo, DARK, Niels Bohr Institute, University of Copenhagen, Denmark	sandra.raimundo@nbi.ku.dk
Santiago GARCIABURILLO, OAN-IGN, Spain	s.gburillo@oan.es
Saqib Hussain, Universidade de São Paulo, Brazil	s.hussain2907@gmail.com
Satish Shripati Sonkamble, NCRA-TIFR, India	ssonkamble@ncra.tifr.res.in
Sebastián Francisco Sanchez, IA-UNAM, México	sfsanchez@astro.unam.mx
Sedona Price, Max-Planck-Institut für extraterrestrische Physik (MPE), Germany	sedona@mpe.mpg.de
Stela Adduci Faria, Instituto de Astronomia, Geofísica e Ciências Atmosféricas, Brazil	stela.faria@usp.br

Name & Organization	Email
Steve Kraemer, Catholic University of America, USA	kraemer@cua.edu
Takuma Izumi, NAOJ, Japan	takuma.izumi@nao.ac.jp
Taro Shimizu, MPE, Germany	shimizu@mpe.mpg.de
Thaísa Storchi Bergmann, Universidade Federal do Rio Grande do Sul, Brazil	thaisa@ufrgs.br
Tiago Vecchi Ricci, UFFS - Campus Cerro Largo, Brazil	tiago.ricci@uffs.edu.br
Timothy Rawle, ESA/STScI, USA	tim.rawle@esa.int
Travis Fischer, United States Naval Observatory/Space Telescope Science Institute, USA	fischertc13@gmail.com
Valeria Olivares, Observatoire de Paris, France	valeria.olivares@obspm.fr
Vanessa Lorenzoni, Universidade Federal de Santa Maria, Brazil	vanessalorenzonii@gmail.com
Venkatessh Ramakrishnan, Universidad de Concepcion, Chile	vramakrishnan@udec.cl
Victoria Reynaldi, Facultad de Ciencias Astronómicas y Geofísicas - UNLP, Argentina	victoria.reynaldi@gmail.com
Vincenzo Mainieri, ESO, Germany	vmainier@eso.org
Vitor Eduardo Buss Bootz, Universidade Federal do Rio Grande do Sul, Brazil	vitorbootz@gmail.com
William Forman, CfA/SAO, USA	wforman@cfa.harvard.edu
Yiqing Song, University of Virginia, USA	ys7jf@virginia.edu

Session 1: Formation and early growth of galaxies and SMBHs

Galaxy Evolution and Feedback across Different Environments
Proceedings IAU Symposium No. 359, 2020
T. Storchi-Bergmann, W. Forman, R. Overzier & R. Riffel, eds.
doi:10.1017/S1743921320002306

INVITED LECTURES

Recent insights into massive galaxy formation from observing structural evolution (Review)

Andrew B. Newman

Observatories of the Carnegie Institution for Science, 813 Santa Barbara St., Pasadena, CA, USA 91101

Abstract. New observations are probing the structures and kinematics of massive galaxies at a much greater level of detail than previously possible, especially during the first half of cosmic history. ALMA data now resolve the distribution of dust and molecular gas in massive galaxies to $z \sim 5$. The stellar kinematics of several massive galaxies at $z \sim 2-3$ have been spatially resolved using gravitational lensing, providing new information on the connection between quenching and morphological transformation. Star formation histories have been reconstructed for growing samples at $z \sim 0.8-2$, revealing a wide range of timescales that correlate with galaxies' sizes and environments, providing evidence for multiple paths to quiescence. I review these and other developments and summarize the insights they have provided into massive galaxies' evolution.

Keywords. galaxies: high-redshift, galaxies: kinematics and dynamics, galaxies: structure

---

1. Introduction

The evolution of galactic structures and kinematics offers many insights into massive galaxies' histories. First, galaxy structures are indicative of the formation physics, although the connection is sometimes direct and uncontroversial (e.g., classical ellipticals experienced dry mergers, high central densities require dissipation) and sometimes not (major mergers may not always destroy disks). Second, particularly in massive galaxies, the evolution of surface brightness profiles is intimately linked to mergers and accretion and so can be used to study these processes. Third, galactic structures (e.g., bulge fraction, central density) are strongly correlated with the quenching of star formation empirically, although the origin of this correlation, and even whether there is a causal relationship, is a subject of debate.

Since massive galaxies formed most of their stars very early, understanding their formation history requires observations at high redshifts, when much of the action occurred. This poses a number of observational difficulties: high-z galaxies are faint, small in angular size, and dusty (particularly high-mass galaxies). This short review will consider recent observations that address these challenges using various techniques, including deep near-infrared spectroscopy, radio interferometry, and gravitational lensing. I will attempt to highlight areas where recent observations have been particularly illuminating, but naturally the scope of this review only permits a small subset of results in this developing area to be discussed. We will proceed chronologically, starting with observations of very early massive galaxies at $z = 3-6$; we will then turn to some observations that resolve distribution, kinematics, and chemistry of gas and stars in the "cosmic noon" era at $z \sim 2-3$;

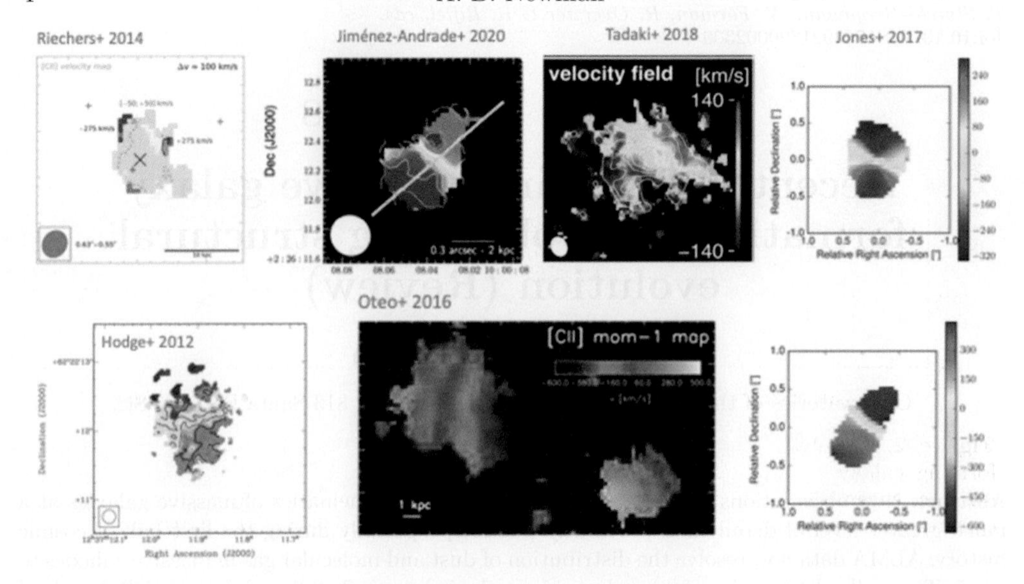

Figure 1. Velocity fields of massive, highly star-forming galaxies at $z > 4$ derived from radio interferometry, reproduced from Hodge *et al.* (2012); Riechers *et al.* (2014); Oteo *et al.* (2016); Jones *et al.* (2017); Tadaki *et al.* (2018); Jiménez-Andrade *et al.* (2020). The majority are remarkably kinematically mature with high V and V/σ as discussed in the text.

and finally we will review evidence for multiple evolutionary tracks at $z \sim 0.8-2$ that has come from connecting galaxies' star formation histories and structures.

2. Early massive galaxies and quenching ($z = 3-6$)

At the highest redshifts most structural information on massive galaxies comes from radio interferometry due to the high angular resolution it affords and its insensitivity to dust obscuration. For a handful of luminous galaxies at $z > 4$ with extremely high star-formation rates (>1000 $M_\odot$ yr^{-1}), gas velocity fields have been resolved. While the first few objects seemed to have rather mixed properties, it appears now that these early massive galaxies are remarkably dense yet kinematically mature. The ratio of rotational to random motion, $V/\sigma = 3 - 5$ or more (see references in Fig. 1), a range typical of Milky Way-mass galaxies at much later epochs $z \sim 1.5$ (e.g., Simons *et al.* 2016). However, these $z > 4$ galaxies have reported rotation speeds often exceeding 400 km s^{-1}, indicating that they have reached very high densities.

The high densities and short gas depletion times of these highly star-forming galaxies beyond $z = 4$ make them good candidates for progenitors of the first population of quenched galaxies (e.g., Toft *et al.* 2014) at $z > 3$. Deep near-infrared spectroscopic observations are now confirming the first samples of galaxies that had already quenched by $z = 3$ and have begun to characterize their star formation histories (e.g,. Glazebrook *et al.* 2017; Schreiber *et al.* 2018a,b; Tanaka *et al.* 2019; Valentino *et al.* 2020). The number densities of early quiescent galaxies are remarkably high, comprising perhaps 35% of massive galaxies by some estimates (Straatman *et al.* 2014), in tension with many models. Reconstructions of the past star-formation in these galaxies suggest past averages of 300–1000 $M_\odot$ yr^{-1}, broadly consistent with sub-mm galaxies at $z > 4$. Kinematic or structural data are needed to help evaluate this connection, but so far there is very little of such information for quiescent galaxies beyond $z > 3$.

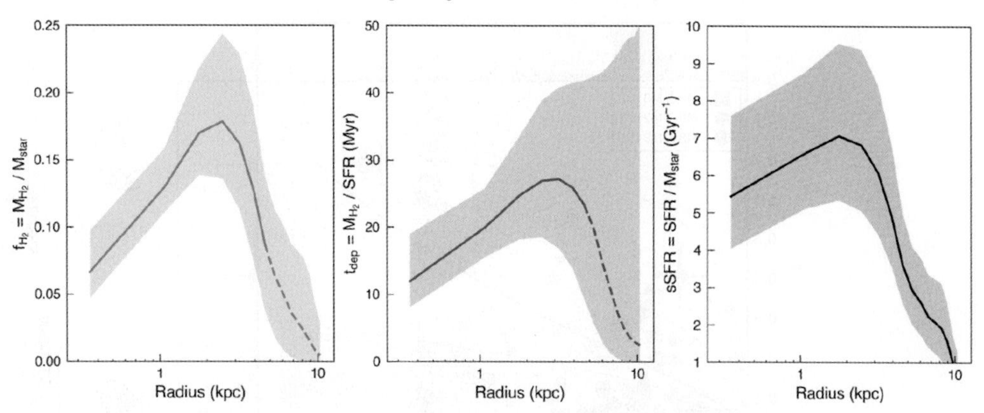

Figure 2. Resolved profiles of H_2, gas depletion time, and specific SFR in a compact star-forming galaxy at $z = 2.2$. Note the decline in gas fraction and depletion time in the inner ~2 kpc, suggesting that star formation will soon cease in the inner galaxy due to rapid gas consumption coupled with likely outflows. Reproduced from Spilker *et al.* (2019).

3. Resolving gas and Stellar structure at cosmic noon ($z = 2-3$)

The study of galaxy structure and kinematics at $z \sim 2$ now comprises a large and rich literature. This short review will focus on ALMA and JVLA observations of massive galaxies and systems thought to be on the cusp of quenching.

Many massive ($> 10^{11}$ $M_\odot$) galaxies at $z \sim 2.5$ are in the process of building bulges. In ALMA observations of 25 massive star-forming galaxies at this epoch, Tadaki *et al.* (2017a) detected compact dust emission in 9 cases with a half-light radius more than 2× smaller than the stellar light. The high star formation rate density implies that these nuclear starbursts will assemble a dense stellar bulge with a surface density $\Sigma_{*,1\text{kpc}} > 10^{10}$ $M_\odot$ kpc^{-2} within a few hundred Myr. This is comparable to the gas depletion time, suggesting that gas exhaustion could end star formation after the bulge is formed. These forming bulges remain rotation-dominated with $V/\sigma \approx 4-7$ (Tadaki *et al.* 2017b).

Mapping the star-forming gas in galaxies that are thought to be on the cusp of quenching is particularly interesting. Candidates observed with ALMA or JVLA include compact star-forming galaxies that are already structurally similar to quiescent galaxies (Barro *et al.* 2013, 2014, 2016, 2017; Talia *et al.* 2018) and sub-mm–selected galaxies (Lang *et al.* 2019). The general trends are that (1) the gas, dust, and star formation are centrally concentrated, just as was seen by Tadaki et al. in general samples of massive star-forming galaxies; (2) molecular gas fractions are often low and gas depletion times are short, on the order of 100 Myr or less, suggesting that the central star formation is nearing its end; (3) the galaxies remain rotationally supported. Spilker *et al.* (2019) resolved a compact star-forming galaxy using JVLA (Fig. 2) and showed that the molecular gas fraction declines significantly in the inner 2 kpc. This is perhaps the most direct evidence of star formation ending first in the inner regions of early massive galaxies ("inside out"), as the molecular gas supply is rapidly removed by star formation likely coupled with outflows.

These radio observations have illuminated some key questions concerning massive galaxies' evolution and suggested new ones. *Are quiescent galaxies so compact because they have "shrunk"?* In some systems there is clearly some "shrinking"—as defined by a decline in the half-light radius—that must be produced given that the star formation is more compact than the existing stars. What triggers this nuclear star formation is hard to pin down observationally, but theoretical models suggest that gas-rich high-z disks are very unstable and are susceptible to perturbations from mergers, interactions, accretion flows, etc. (e.g. Dekel & Burkert 2014; Zolotov *et al.* 2015). *Is feedback needed to finish*

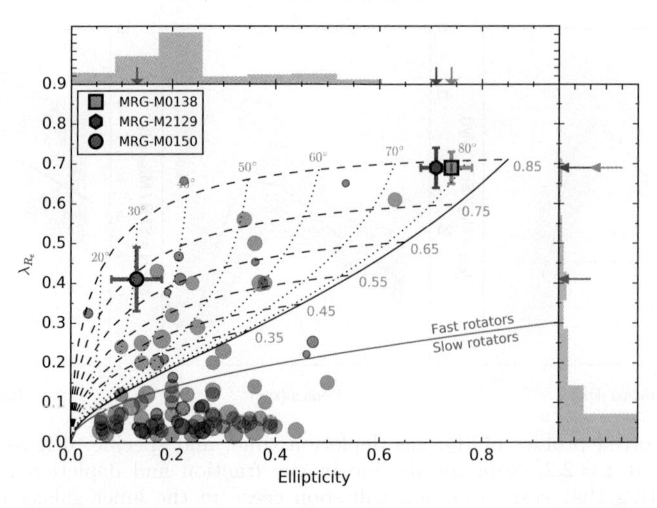

Figure 3. The projected ellipticities and angular momentum parameters λ_{R_e} of 3 lensed quiescent galaxies at $z = 2.0-2.6$ (points with error bars) are compared to local early-type galaxies (circles). The $z = 2$ quiescent galaxies have similar and very flat intrinsic shapes ($e_{\rm intr} \approx 0.75-0.85$) and much more specific angular momentum than a typical early-type galaxy of equal or higher mass in the local universe, represented here by data from the ATLAS$^{\rm 3D}$ (red circles; Cappellari *et al.* 2013) and MASSIVE (blue circles; Veale *et al.* 2017) surveys. The grid is a family of models in which dashed lines have constant intrinsic ellipticity $e_{\rm intr}$ (labeled at right) and dotted lines have constant inclination angle (labeled at top). Reproduced from Newman *et al.* (2018b).

off star formation? Simple consumption by star formation will quickly exhaust the fuel in many of the galaxies discussed in this section that are thought to be in the process quenching. Provided there is a "maintenance" mechanism to block fresh fuel from the disk, must we invoke feedback? Perhaps not, but it is worthwhile to recall evidence that nuclear outflows appear to be quite common in massive $z \sim 2$ galaxies (e.g., Genzel *et al.* 2014). They may well provide an important additional "sink" for star-forming gas. As usual, the difficulty is estimating the mass outflow rate and thus the consequences for modulating the star formation. *When do elliptical galaxies form?* So far all observations discussed have indicated that massive galaxies remain rotation dominated right up until quenching. Yet we know that quiescent galaxies today are very structurally distinct from star-forming systems. When and how did this emerge?

Answering this question requires measuring the structure and kinematics of quenched galaxies all the way back to their formation. Although the morphologies, sizes, and ellipticities have now been measured for large samples of galaxies, arguably the angular momentum is the most fundamental parameter underlying morphological differences. Its measurement has remained elusive for quiescent galaxies much beyond $z \sim 1$ due to their faintness and small angular sizes. The most practical way to circumvent these difficulties is to use gravitational lensing to gain angular resolution. The difficulty is that magnified high-z quiescent galaxies are very rare. Nevertheless, searches have succeeded in turning up modest but very valuable samples (Newman *et al.* 2015; Toft *et al.* 2017; Newman *et al.* 2018a,b).

Newman *et al.* (2018b) spatially resolved the stellar kinematics in 4 lensed quiescent galaxies spanning the redshift range $z = 2-2.6$. These galaxies are viewed typically $\sim$1 Gyr after quenching. Remarkably, all show significant rotation and would be classified as "fast rotators" based on criteria used to classify low-z early-type galaxies (see Fig. 3). For the 3 galaxies with a lens model that permits the source to be reconstructed, the inferred dynamical masses exceed $\gtrsim 2 \times 10^{11} M_{\odot}$, placing them already in the mass range where "slow rotators" (classical ellipticals) are dominant in the local universe. Considering that some mass growth is expected, these galaxies are very likely to evolve into giant ellipticals. Yet just after quenching, they are rotating at 290–352 km s^{-1} and are primarily rotation supported with $V/\sigma \approx 2$ (c.f. $V/\sigma \lesssim 0.1$ for slow rotators). This is smaller than the typical V/σ reported for massive, coeval star-forming galaxies, which might indicate that quenching is accompanied by partial erosion of rotational support, although more data on the kinematics of the $z > 3$ progenitors of $z \sim 2$ quiescent galaxies is needed to ascertain this. However, it seems clear that most of the decline in angular momentum—by a factor of 5–10×—comes after quenching. Rather than a single event that simultaneously transforms a galaxy's morphology and kills off star formation, as envisioned in classical major mergers models, the morphological transformation appears to be separate. Simulations indicate that this is likely due to the gas-rich character of high-z mergers, which make disks more robust, while the transformation to an elliptical post-quenching probably arises from a series of mergers that increase the galaxy's size and mass while reducing its net angular momentum.

4. Star-formation histories and galaxy structure ($z \approx 0.8-2$)

A growing number of spectroscopic observations at $z \approx 0.8-2$ reach depths sufficient to reconstruct the past star-formation histories of moderately sized samples of massive galaxies. Even galaxies which are quiescent at the epoch of observation appear to have experienced a wide range of star formation histories over the prior few Gyr, and these star formation histories are correlated with galaxies' structures.

At all epochs since at least $z \sim 3$, there is evidence for a population of recently quenched or "post-starburst" galaxies in which star formation has shut down recently and rapidly. In the local universe, these are rare and typically low-mass galaxies, but beyond $z \sim 1$ they are more frequent and occur at higher masses. Since much of the red sequence was in place at $z \sim 1$, this raises the question of whether most galaxies joined it rapidly or more gradually.

Post-starbursts can confidently be identified by their spectroscopic signatures, but they also present distinctive broad-band colors that are useful to estimate their evolving number density using imaging surveys, particularly at $z > 1$ where continuum spectroscopy is difficult (Fig. 4). A key uncertainty in this approach is the duration of the post-starburst phase, i.e., the length of time galaxies spend traversing the post-starburst box outlined in Fig. 4. Belli *et al.* (2019) used a library of deep Keck/MOSFIRE spectroscopy to calibrate this timescale. They then compared the flux of galaxies through the post-starburst box with the rate that galaxies appear in the quiescent region of Fig. 4.

Belli *et al.* (2019) found that most galaxies at $z \sim 1.5$ are not "quenching" particularly quickly: perhaps 20% pass through a post-starburst phase. Wild *et al.* (2020) inferred a fraction 25–50% that is smaller but still implies that rapidly quenched galaxies are a minority. By necessity, the post-starburst fraction increases with redshift (galaxies cannot end star formation early *and* slowly), but even at $z \sim 2.2$ Belli *et al.* (2019) find that post-starbursts account for half of the growth of the red sequence. Interestingly, the recently, rapidly quenched galaxies tend to be smaller in size and located in overdense environments. This suggests that there are distinct and independent physical mechanisms that produce the "rapidly" and "slowly" quenched galaxies.

Wu *et al.* (2018) investigated similar questions at $z \sim 0.8$ using very deep spectra of massive galaxies collected as part of the LEGA-C survey. They find that, in general, more recently quenched galaxies are larger than older ones. This is expected due to the extra growth they had time to undergo while star-forming. But post-starburst galaxies buck the trend: despite their young ages, they are the smallest galaxy population. Among the

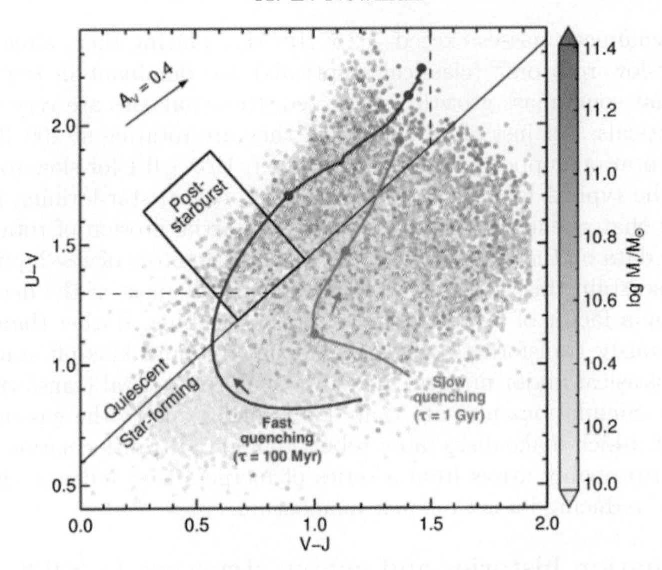

Figure 4. Color-color diagram of $z = 1.5–2$ galaxies separating star-forming (lower-right) and quiescent (upper-left) systems. Galaxies that experience "fast quenching" pass through the blue end of the red sequence, depicted here as the box labeled "Post-starburst". Galaxies that quench slowly join the red sequence without passing through the post-starburst box. Comparing the flux of galaxies onto the red sequence and through the post-starburst box shows that most galaxies quench rather slowly. Reproduced from Belli *et al.* (2019).

post-starbusts, the color correlates with size. This trend can successfully be understood using a simple model in which a brief nuclear starburst occurs within an extended disk, just before the galaxy as a whole is quenched (Wu *et al.* 2020).

These observations strongly suggest that post-starbursts at $z \approx 1–2$ are generally produced when gas is funneled into galactic centers, producing a brief nuclear star-burst that increases the central stellar density (thus decreasing the half-light radius) and extinguishes star formation in the galaxy (presumably by gas consumption and feedback). The trigger for this phase is less clear. In the theoretical "compaction" paradigm, instabilities in gas-rich disks lead to nuclear star formation, but the timescales are rather long: $0.35t_H \approx 1.6$ Gyr at $z = 1.4$ according to Zolotov *et al.* (2015), more akin to what observational studies typically classify as "slow" quenching. Bursty star-formation timescales of a few hundred Myr are reported in major merger simulations (e.g., Di Matteo *et al.* 2008), but it is not clear whether this is a plausible route for all post-starbusts.

5. Summary

At the highest redshifts probed ($z > 4$), radio observations of intensely star-forming massive galaxies often show them to be dominated by remarkably regular disks. If these galaxies are about to quench, they will produce quiescent galaxies at $z > 3$, a population that is perhaps larger than expected and is just beginning to be studied in detail.

At "cosmic noon", many massive star-forming galaxies are actively assembling bulges, as shown by observations of centrally concentrated, dusty star formation in many systems. The short gas depletion times (∼100 Myr) are comparable to the time required to assemble a stellar surface density approaching the observed maximum, suggesting that the nuclear star formation galaxies will soon end as envisioned in "inside-out" quenching scenarios. Star-forming galaxies that may be on the cusp of quenching remain rotation supported. Initial observations resolving the kinematics of $z \sim 2$ quiescent galaxies show that they, too, are quite disky and rotation supported. This implies that the transformation to an elliptical morphology is not likely to be coincident with quenching in most cases; rather, it probably occurs later through a series of dry mergers.

At $z \approx 0.8-2$, star-formation histories of quiescent galaxies, reconstructed from deep spectra, show a wide range of timescales. Relatively "slow" quenching (multiple Gyr) seems to be dominant and may involve relatively little structural transformation. "Fast" quenching (hundreds of Myr, post-starbursts) is observed in a minority of galaxies that are more compact and are located in denser environments than average. These and other observations imply that "fast" quenching is associated with a short nuclear starburst.

The observations of massive galaxies' evolving structures, kinematics, and star-formation activity covered in this review have provided several insights: quenching at high-z and formation of ellipticals are probably rather disconnected; quenching occurs over a range of timescales likely through multiple independent processes; rapid quenching requires a dissipative process, etc. But they do not uniquely identify physical mechanisms that, for instance, trigger a starburst or terminate star formation. Making such identifications, if possible, will require a synthesis of many observational and theoretical approaches.

References

Barro, G., Faber, S. M., Pérez-González, P. G., *et al.* 2013, *ApJ*, 765, 104
Barro, G., Faber, S. M., Pérez-González, P. G., *et al.* 2014, *ApJ*, 791, 52
Barro, G., Kriek, M., Pérez-González, P. G., *et al.* 2016, *ApJL*, 827, L32
Barro, G., Kriek, M., Pérez-González, P. G., *et al.* 2017, *ApJL*, 851, L40
Belli, S., Newman, A. B., & Ellis, R. S. 2019, *ApJ*, 874, 17
Cappellari, M., Scott, N., Alatalo, K., *et al.* 2013, *MNRAS*, 432, 1709
Dekel, A. & Burkert, A. 2014, *MNRAS*, 438, 1870
Di Matteo, P., Bournaud, F., Martig, M., *et al.* 2008, *A&A*, 492, 31
Genzel, R., Förster Schreiber, N. M., Rosario, D., *et al.* 2014, *ApJ*, 796, 7
Glazebrook, K., Schreiber, C., Labbé, I., *et al.* 2017, *Nature*, 544, 71
Hodge, J. A., Carilli, C. L., Walter, F., *et al.* *ApJ*, 760, 11
Jiménez-Andrade, E. F., Zavala, J. A., Magnelli, B., *et al.* 2020, *ApJ*, 890, 171
Jones, G. C., Carilli, C. L., Shao, Y., *et al.* 2017, *ApJ*, 850, 180
Lang, P., Schinnerer, E., Smail, I., *et al.* 2019, *ApJ*, 879, 54
Newman, A. B., Belli, S., & Ellis, R. S. 2015, *ApJL*, 813, L7
Newman, A. B., Belli, S., Ellis, R. S., *et al.* 2018a, *ApJ*, 862, 125
—. 2018b, *ApJ*, 862, 126
Oteo, I., Ivison, R. J., Dunne, L., *et al.* 2016, *ApJ*, 827, 34
Riechers, D. A., Carilli, C. L., Capak, P. L., *et al.* 2014, *ApJ*, 796, 84
Schreiber, C., Labbé, I., Glazebrook, K., *et al.* 2018a, *A&A*, 611, A22
Schreiber, C., Glazebrook, K., Nanayakkara, T., *et al.* 2018b, *A&A*, 618, A85
Simons, R. C., Kassin, S. A., Trump, J. R., *et al.* 2016, *ApJ*, 830, 14
Spilker, J. S., Bezanson, R., Weiner, B. J., *et al.* 2019, *ApJ*, 883, 81
Straatman, C. M. S., Labbé, I., Spitler, L. R., *et al.* 2014, *ApJL*, 783, L14
Tadaki, K., Iono, D., Yun, M. S. *et al.* 2018, *Nature*, 560, 613
Tadaki, K.-i., Genzel, R., Kodama, T., *et al.* 2017a, *ApJ*, 834, 135
Tadaki, K.-i., Kodama, T., Nelson, E. J., *et al.* 2017b, *ApJL*, 841, L25
Talia, M., Pozzi, F., Vallini, L., *et al.* 2018, *MNRAS*, 476, 3956
Tanaka, M., Valentino, F., Toft, S., *et al.* 2019, *ApJL*, 885, L34
Toft, S., Smolčić, V., Magnelli, B. *et al.* 2014, *ApJ*, 782, 68
Toft, S., Zabl, J., Richard, J., *et al.* 2017, *Nature*, 546, 510

Valentino, F., Tanaka, M., Davidzon, I., *et al.* 2020, *ApJ*, 889, 93
Veale, M., Ma, C.-P., Thomas, J., *et al.* 2017, *MNRAS*, 464, 356
Wild, V., Taj Aldeen, L., Carnall, A., *et al.* 2020, *MNRAS*, 494, 529
Wu, P.-F., van der Wel, A., Bezanson, R., *et al.* 2018, *ApJ*, 868, 37
—. 2020, *ApJ*, 888, 77
Zolotov, A., Dekel, A., Mandelker, N., *et al.* 2015, *MNRAS*, 450, 2327

Galaxy Evolution and Feedback across Different Environments
Proceedings IAU Symposium No. 359, 2020
T. Storchi-Bergmann, W. Forman, R. Overzier & R. Riffel, eds.
doi:10.1017/S1743921320001751

Models for galaxy and massive black hole formation and early evolution

Rainer Weinberger

Center for Astrophysics | Harvard & Smithsonian, 60 Garden Street, MS-51,
Cambridge, MA 02138, USA
email: rainer.weinberger@cfa.harvard.edu

Abstract. Models for massive black holes are a key ingredient for modern cosmological simulations of galaxy formation. The necessity of efficient AGN feedback in these simulations makes it essential to model the formation, growth and evolution of massive black holes, and parameterize these complex processes in a simplified fashion. While the exact formation mechanism is secondary for most galaxy formation purposes, accretion modeling turns out to be crucial. It can be informed by the properties of the high redshift quasars, accreting close to their Eddington limit, by the quasar luminosity function at peak activity and by low-redshift scaling relations. The need for halo-wide feedback implies a feedback-induced reduction of the accretion rate towards low redshift, amplifying the cosmological trend towards lower accretion rates at low redshift.

Keywords. galaxies: formation, galaxies: nuclei, galaxies: active, methods: numerical, black hole physics, accretion

---

1. Introduction

Massive black holes (MBHs) are an essential part of cosmological structure formation. Modern simulations of galaxy or galaxy cluster formation rely on models for MBH formation and growth, since feedback effects from these objects have been shown to potentially explain some properties of massive galaxies and galaxy clusters (Somerville & Davé 2015). Two examples are the bimodal distribution of central galaxy colors (Trayford *et al.* 2016; Nelson *et al.* 2018), with more massive galaxies being redder and less star-forming, and the so-called cooling-flow problem, where gas cooling can be commonly observed in galaxy clusters, yet, does not lead to expected levels of star formation (e.g. Fabian 2012).

Cosmological simulations of galaxy formation model the formation of structure of the Universe as an initial value problem, numerically evolving the dark matter and gas distribution over most of cosmic time to redshift zero (for a detailed review, see Vogelsberger *et al.* 2020). While non-radiative, or generally non-dissipative simulations can be readily run and result in virialized halos, the introduction of dissipative terms in the form of radiative cooling of the gas will lead to a runaway collapse which will prohibit a numerical time-integration over cosmic timescales. The introduction of a simple closure at small scales, in which gas exceeding a certain density threshold is transformed to a collisionless 'star-particle' will alleviate this computational problem, however, yield vastly different results when compared to the observed galaxy population.

Significant progress has been made in recent years, showing that a more multi-facetted closure, including the effects of stellar feedback (e.g. Springel & Hernquist 2003; Dalla Vecchia & Schaye 2008) as well as a feedback component from active galactic nuclei (e.g. Sijacki *et al.* 2007; Dubois *et al.* 2012), can produce broad agreement between the simulated and observed galaxy population (e.g. Vogelsberger *et al.* 2013; Crain *et al.* 2015;

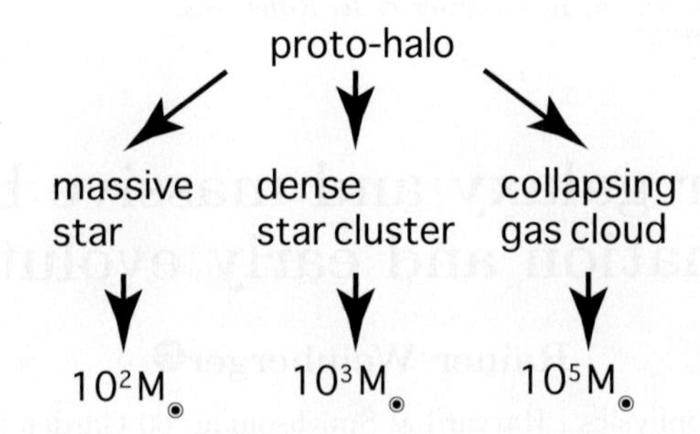

Figure 1. Broad classification of massive black hole seeding channels with expected seed mass.

Pillepich *et al.* 2018). Since these closure or sub-grid models are inspired by a simplified physical understanding, they do not cover the full complexity of the underlying process, therefore limiting the predictive power of the simulation. Yet, these models can be used as a guiding line and a test of plausibility of specific ideas which astrophysical processes are responsible for specific observational signatures.

I will review models for MBHs in cosmological simulations from seeding at high redshift until the onset of quenching around redshifts $z = 1-2$, discuss some of the difficulties related to these models and present open questions about the evolution of MBHs over cosmic time.

2. Different phases

The evolution of MBHs can be divided into 4 phases. First, MBH seed formation; second, early growth and the highest redshift quasars; third, the peak of the cosmic accretion rate density, and finally the epoch in which active galactic nucleus (AGN) feedback impacts the evolution of the entire host galaxy.

2.1. *Seeding*

One of the most uncertain aspects of MBHs is their formation. Since structure formation of non-dissipative components such as dark matter stops at virialization, further gravitational collapse is only possible via dissipative processes, i.e. radiative cooling of gas. It is therefore unsurprising that the formation of MBHs depends critically on the physics of radiative cooling, which in itself depends on chemical composition as well as external radiation fields. Theories for different channels of high redshift MBH formation have been around for some time (see Rees 1984 and Volonteri 2010 for reviews on this topic), yet the precise mechanism and possible observational evidence for it are subject to active research. Figure 1 shows a broad categorization into three different channels.

The first is the stellar-remnant channel, in which black holes of mass of order 10^2 M$_\odot$ are produced as remnants of massive, so-called population III stars (Carr *et al.* 1984). These short-lived stars are the first stars in the Universe, and form only in the absence of chemically enriched gas in so-called mini-halos.

The second channel is operating when cooling is slightly more efficient. Cool gas can form in the halo center, fragment and collapse into individual stars, thus forming a dense star cluster. In this star cluster, through collisional n-body dynamics, core-collapse can occur (Begelman & Rees 1978), which leads to the formation of a MBH in the center, with masses of order 10^3 M$_\odot$ (Devecchi & Volonteri 2009).

A third channel is possible if the formation of molecular hydrogen and consequently cooling to low temperatures is inhibited, e.g. by a sufficiently strong UV radiation flux. This leads to a larger Jeans length, i.e. prevents fragmentation, and a direct collapse of an entire massive gas cloud into a MBH seed (Bromm & Loeb 2003). The expected mass of a black hole forming via this channel is of the order of 10^5 M$_\odot$.

While none of these channels is fully understood, there are only a comparably small number of properties of crucial importance for cosmological simulations of galaxy formation: when these seed MBHs form, which mass they have and how frequent they are.

Given the typical mass resolution of a cosmological volume simulation (targeted towards a $z = 0$ galaxy population) is around 10^5 to 10^6 M$_\odot$, it is often omitted to distinguish between different seed scenarios. Instead, a commonly used way to seed MBHs in simulations is to simply assume that they are present in every halo exceeding a specific mass, typically around 10^{10}–10^{11} M$_\odot$. While the exact numbers are somewhat arbitrary, this is a numerically very robust way to introduce MBHs in the simulation. However, implicitly assumes that low-mass MBHs grow in the same way as halos, which leads to a relatively flat distribution of seed times. A metallicity and gas density based seeding, which is an alternative and used in some simulations, leads to a peak of seeding at high redshift, with practically no seeding events at lower redshift, reflecting more the theoretical expectation that MBH seeds require a low metallicity environment to form (Tremmel *et al.* 2017). While these vastly different ways to introduce MBHs in the simulation likely lead to very different predictions about the early and the low-mass MBH population, it is important to keep in mind that the properties of the high-mass population is strongly influenced by gas accretion and hierarchical merging of halos (Weinberger *et al.* 2018), which leads to similar properties at low redshift independent of the details of the seeding.

2.2. *Early growth*

Once formed, MBHs grow in two different ways: via mergers with other black holes, and via gas accretion. While mergers will contribute, the initial growth is dominated by rapid accretion of gas. Evidence for this is provided by the existence of high redshift quasars with associated MBH masses of order 10^9 M$_\odot$ at redshifts $\gtrsim 7$ (Bañados *et al.* 2018). These high mass MBHs at these redshifts place strict constraints on the combination of seed redshift, seed mass and maximum accretion rate at which a MBH can accrete. Assuming this maximum accretion rate is the Eddington limit, it becomes very hard to explain these black holes from population III remnants. Viable solutions are high-mass seeds from direct collapse or accretion rates that exceed the Eddington limit (see Smith *et al.* 2017 for a more detailed discussion).

Simulations of these high redshift quasars are very challenging, since the low number density of these objects requires to simulate a significant fraction of the visible universe to obtain a meaningful sample. An illustration of the scales involved is shown in Figure 2. The mean inter-object separation of high redshift quasars is of the order 10^9 pc, beyond the reach of most cosmological simulations targeting galaxy formation. Simply increasing the simulated volume is not possible, since there are resolution requirements to consider for modeling MBH seeds, even the direct collapse ones, as well as for black hole accretion. Fully satisfying these two opposing requirements is not possible at present day, yet some studies exist trying to address this problem in cosmological volume simulations that stop at high redshift (Di Matteo *et al.* 2017) as well as dedicated zoom simulations focusing on single halos (Smidt *et al.* 2018). In the latter case it has recently become possible to include the effects of radiation self-consistently in the simulation, which, by definition is crucial for objects accreting at the Eddington limit.

Cosmological volume simulations to $z = 0$ focusing on galaxy formation to-date do neither include radiation-hydrodynamics nor have the volume to produce these rare, high redshift quasars, which implies on the one hand that dedicated simulations are required to study them, on the other hand that their presence and abundance cannot be used as a

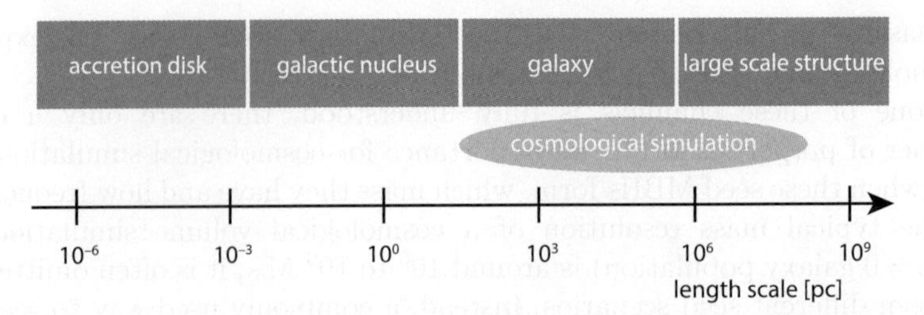

Figure 2. Spatial scales relevant for massive black holes. Cosmological simulations typically cover spatial scales from 300 Mpc to 30 pc. Therefore the covered volume is too small to contain rare objects such as the most luminous quasars, while the resolution limits require to marginalize over 6 orders of magnitude in spatial scales and related small-scale processes around massive black holes.

direct constraint for simulations. Yet, the general notion that MBH at high redshift seem to be able to accrete at or close to their Eddington limit is reassuring that the commonly employed assumption in simulations that accretion is limited to the Eddington rate is not unreasonable (note that due to the lack of radiation in these simulations, the simulated MBHs could have super-Eddington accretion rates at high redshift if not limited by the accretion model).

2.3. *Peak of activity*

Towards lower redshift, at the peak of the MBH accretion rate and star formation rate density at $z = 2{-}3$, the quasar luminosity function can be determined since a significant fraction of active galactic nuclei are observable. This quasar luminosity function is an important constraint on gas accretion onto MBHs. From a modeling perspective, the estimate of the accretion rate for the bulk of the of MBH population is very uncertain. While for very high redshifts and for the most luminous objects the assumption of accretion at the Eddington limit, i.e. a radiation pressure limited accretion, is a reasonable one, this ceases to be the case for the less extreme cases (Weinberger *et al.* 2018). For these less extreme cases it is hard to determine the limiting factor for accretion (possible factors are angular momentum, cooling, interactions with small-scale outflows, ...), let alone to estimate the accretion rate accurately from properties at galactic scales. Many models used in simulations are based on the Bondi accretion, frequently with some modifications. More recently, simulations using other prescriptions such as relations based on a torque, i.e. angular momentum limited accretion, have been performed and have shown to yield orders of magnitude different results for the same large scale conditions (Anglés-Alcázar *et al.* 2013). Considering the large range of unresolved scales (see Figure 2), and the fact that these modes assume different limiting factors for accretion, this discrepancy is not entirely surprising. But considering this uncertainty, it is rather surprising that these models are at all able to produce reasonable agreement with both low redshift MBH scaling relations, as well as high redshift quasar luminosity function constraints.

The main reason for this, in most simulations, is the self-termination of rapid accretion due to AGN feedback towards late times (however, see Anglés-Alcázar *et al.* 2013). Feedback at late times, unlike accretion, acts on resolved, galactic scales, which makes modeling easier.

2.4. *Downsizing and quenching*

Towards low redshift, both the star formation rate density as well as the MBH accretion rate density decrease towards redshift zero. Cosmological simulations reveal that this trend with redshift is not solely caused by feedback, but also visible in the global gas accretion rate density onto halos (van de Voort *et al.* 2011). The drastic decrease in number density of very luminous AGN can consequently, at least in part, be caused by a decrease in luminosity of individual AGNs due to decreased fueling, in combination with a steep negative slope of the quasar luminosity function at high luminosities. Therefore, the existence of these observed global trends cannot be interpreted as evidence for AGN feedback (however the trends might be enhanced due to AGN feedback).

The need for AGN feedback in cosmological simulations is more evident when trying to reproduce a population of massive, central galaxies with sustained low star formation rates, and consequently red intrinsic colors, as well as the X-ray properties of galaxy clusters. Matching observations in this respect has so far only been possible by invoking efficient AGN feedback in galaxies more massive than the Milky Way (e.g. Weinberger *et al.* 2017), a feature all cosmological simulations aiming to reproduce these high mass objects have in common (e.g. Khandai *et al.* 2015; Beckmann *et al.* 2017; Davé *et al.* 2019).

One of the key remaining questions in which simulations differ is how AGN feedback comes to be efficient in massive galaxies, while not being that relevant for the less massive galaxy population that remains star forming to the present day. Different simulations overcome this problem in different ways, some pointing towards the properties of stellar feedback (e.g. Bower *et al.* 2017), others achieve a similar effect by a change in mode of AGN feedback (Weinberger *et al.* 2018), possibly induced by small-scale accretion disk physics or a change in black hole spin (Bustamante & Springel 2019). Future observations, for example of the hot, soft X-ray emitting halo gas might be able to rule out certain models (Oppenheimer *et al.* 2020; Truong *et al.* 2020).

Another major aspect that requires more detailed study is the coupling of the energy released by the AGN with the host galaxy. In cosmological simulations, this is implicitly assumed when constructing a model on kpc scales. Studying this in more detail on smaller scales (e.g. Cielo *et al.* 2018), will be necessary to make a convincing case that whatever, for now, is assumed and required in cosmological simulations, is actually realistic.

3. Summary

Cosmological simulations require AGN feedback to reproduce the properties of massive galaxies. This need for AGN feedback makes it necessary to parameterize the rich and complex physics of MBH formation and evolution in a simplified fashion, but at the same time also allows to investigate the evolution of MBHs over cosmic time in a realistic environment. Some important takeaways are:

- There might be different seeding channels, however they are not modeled in most cosmological simulations, and the high-mass MBHs have likely lost all information about seeding. However, information about seeding can be obtained from low-mass MBHs and gravitational wave events.
- High redshift quasars indicate that early growth in the most extreme environments is close to Eddington-limited. This is very informative for accretion rate estimates at high redshift.
- Accretion rates of less extreme MBHs are significantly more difficult to estimate due to poorly understood physics at unresolved scales.
- Towards low redshift, AGN feedback is required to produce massive, quiescent central galaxies. This trend likely amplifies the general reduction of star formation rate density and downsizing, however is not the sole cause for it.
- The physical cause of the transition from a growth dominated to a feedback dominated regime is still debated, with upcoming observations having the potential to rule out some scenarios.

While cosmological simulations have been remarkably successful over the past years, presenting plausible scenarios of MBH evolution (and rule out a number of alternative ones), future studies that connect modeling of individual processes from first principle with the cosmological evolution are needed to gain further understanding about MBH formation and evolution.

References

Anglés-Alcázar, D., Özel, F., & Davé, R. 2013, *ApJ*, 770, 5
Bañados, E., Venemans, B. P., Mazzucchelli, C., *et al.* 2018, *Nature*, 553, 473
Beckmann, R. S., Devriendt, J., Slyz, A., *et al.* 2017, *MNRAS*, 472, 949
Begelman, M. C. & Rees, M. J. 1978, *MNRAS*, 185, 847
Bower, R. G., Benson, A. J., Malbon, R., *et al.* 2006, *MNRAS*, 370, 645
Bower, R. G., Schaye, J., Frenk, C. S., *et al.* 2017, *MNRAS*, 465, 32
Bromm, V. & Loeb, A. 2003, *ApJ*, 596, 34
Bustamante, S. & Springel, V. 2019, *MNRAS*, 490, 4133
Carr, B. J., Bond, J. R., & Arnett, W. D. 1984, *ApJ*, 277, 445
Cielo, S., Bieri, R., Volonteri, M., *et al.* 2018, *MNRAS*, 477, 1336
Crain, R. A., Schaye, J., Bower, R. G., *et al.* 2015, *MNRAS*, 450, 1937
Dalla Vecchia, C. & Schaye, J. 2008, *MNRAS*, 387, 1431
Davé, R., Anglés-Alcázar, D., Narayanan, D., *et al.* 2019, *MNRAS*, 486, 2827
Devecchi, B. & Volonteri, M. 2009, *ApJ*, 694, 302
Di Matteo, T., Croft, R. A. C., Feng, Y., *et al.* 2017, *MNRAS*, 467, 4243
Dubois, Y., Devriendt, J., Slyz, A., *et al.* 2012, *MNRAS*, 420, 2662
Fabian, A. C. 2012, *ARAA*, 50, 455
Khandai, N., Di Matteo, T., Croft, R., *et al.* 2015, *MNRAS*, 450, 1349
Nelson, D., Pillepich, A., Springel, V., *et al.* 2018, *MNRAS*, 475, 624
Oppenheimer, B. D., Bogdán, Á., Crain, R. A., *et al.* 2020, *ApJL*, 893, L24
Pillepich, A., Springel, V., Nelson, D., *et al.* 2018, *MNRAS*, 473, 4077
Rees, M. J. 1984, *ARAA*, 22, 471
Sijacki, D., Springel, V., Di Matteo, T., *et al.* 2007, *MNRAS*, 380, 877
Smidt, J., Whalen, D. J., Johnson, J. L., *et al.* 2018, *ApJ*, 865, 126
Smith, A., Bromm, V., & Loeb, A. 2017, *Astronomy and Geophysics*, 58, 3.22
Somerville, R. S., & Davé, R. 2015, *ARAA*, 53, 51
Springel, V. & Hernquist, L. 2003, *MNRAS*, 339, 289
Trayford, J. W., Theuns, T., Bower, R. G., *et al.* 2016, *MNRAS*, 460, 3925
Tremmel, M., Karcher, M., Governato, F., *et al.* 2017, *MNRAS*, 470, 1121
Truong, N., Pillepich, A., Werner, N., *et al.* 2020, *MNRAS*, 494, 549
van de Voort, F., Schaye, J., Booth, C. M., *et al.* 2011, *MNRAS*, 415, 2782
Vogelsberger, M., Genel, S., Sijacki, D., *et al.* 2013, *MNRAS*, 436, 3031
Vogelsberger, M., Marinacci, F., Torrey, P., *et al.* 2020, *Nature Reviews Physics*, 2, 42
Volonteri, M. 2010, *A&ARv*, 18, 279
Weinberger, R., Springel, V., Hernquist, L., *et al.* 2017, *MNRAS*, 465, 3291
Weinberger, R., Springel, V., Pakmor, R., *et al.* 2018, *MNRAS*, 479, 4056

Galaxy Evolution and Feedback across Different Environments
Proceedings IAU Symposium No. 359, 2020
T. Storchi-Bergmann, W. Forman, R. Overzier & R. Riffel, eds.
doi:10.1017/S1743921320004032

ORAL CONTRIBUTIONS

Tracing young SMBHs in the dusty distant universe – a Chandra view of DOGs

Karín Menéndez-Delmestre[1], Laurie Riguccini[1] and Ezequiel Treister[2]

[1]Valongo Observatory, Federal University of Rio de Janeiro, Ladeira Pedro Antônio 43, Centro, Rio de Janeiro, RJ, Brazil
email: kmd@astro.ufrj.br

[2]Pontificia Universidad Católica, Instituto de Astrofísica, Vicuña Mackenna 4860, Macul, Santiago, Chile

Abstract. The coexistence of star formation and AGN activity has geared much attention to dusty galaxies at high redshifts, in the interest of understanding the origin of the Magorrian relation observed locally, where the mass of the stellar bulk in a galaxy appears to be tied to the mass of the underlying supermassive black hole. We exploit the combined use of far-infrared (IR) Herschel data and deep Chandra ∼160 ksec depth X-ray imaging of the COSMOS field to probe for AGN signatures in a large sample of >100 Dust-Obscured Galaxies (DOGs). Only a handful (∼20%) present individual X-ray detections pointing to the presence of significant AGN activity, while X-ray stacking analysis on the X-ray undetected DOGs points to a mix between AGN activity and star formation. Together, they are typically found on the main sequence of star-forming galaxies or below it, suggesting that they are either still undergoing significant build up of the stellar bulk or have started quenching. We find only ∼30% (6) Compton-thick AGN candidates ($N_H > 10^{24}$ cm^{-2}), which is the same frequency found within other soft- and hard-X-ray selected AGN populations. This suggests that the large column densities responsible for the obscuration in Compton-thick AGNs must be nuclear and have little to do with the dust obscuration of the host galaxy. We find that DOGs identified to have an AGN share similar near-IR and mid-to-far-IR colors, independently of whether they are individually detected or not in the X-ray. The main difference between the X-ray detected and the X-ray undetected populations appears to be in their redshift distributions, with the X-ray undetected ones being typically found at larger distances. This strongly underlines the critical need for multiwavelength studies in order to obtain a more complete census of the obscured AGN population out to higher redshifts. For more details, we refer the reader to Riguccini *et al.* (2019).

Keywords. galaxies: active, galaxies: evolution, galaxies: high-redshift

---

1. Introduction

Massive galaxies continue to pose important challenges to our current understanding of galaxy formation. Stellar population studies and the presence of an underlying supermassive black hole (SMBH) point to the idea that these galaxies are the result of mergers between gas-rich galaxies. Within this formation scenario, a dust-obscured phase – where starburst episodes coexist with activity from a galactic nucleus (AGN) associated to the growth of a young SMBH – is traced back to the so-called ultra-luminous infrared (IR) galaxies (ULIRGs: $L_{IR} > 10^{12} L_{\odot}$; Sanders *et al.* 1988). ULIRGs are locally rare, but appear to dominate the co-moving energy density at higher redshifts ($z > 2$; e.g., Casey *et al.* 2014). Many of these galaxies have been identified by the detection of their thermal dust emission at submillimeter wavelengths (the so-called submillimeter galaxies or SMGs; Blain *et al.* 1999). Detailed ground-based follow-up in the optical/near-IR and mm/radio, as well as space-based studies with Chandra, HST and Spitzer have revealed intricate morphologies reminiscent of major mergers (e.g., Swinbank *et al.* 2010; Engel *et al.* 2010; Alaghband-Zadeh *et al.* 2012; Menéndez-Delmestre *et al.* 2013), and the predominance of (weak) AGN, establishing that star formation and AGN activity coexist in these objects (e.g., Chapman *et al.* 2005; Alexander *et al.* 2005; Pope *et al.* 2008; Menéndez-Delmestre *et al.* 2009). Other dusty galaxies have been selected by their high dust obscuration in optical bands (F24/FR>1000) and named 'dust-obscured galaxies' or DOGs (Dey *et al.* 2008). They are typically characterized by a rising power-law continuum of hot dust (~200–1000 K) in the near-IR, indicating that their mid-IR luminosity is dominated by an AGN and that they likely trace a later phase in the merger-ULIRG scenario (e.g., Riguccini *et al.* 2015).

Riguccini *et al.* (2015) showed that a sub-sample of DOGs with far-IR (100–500μm) detection have a significant contribution from AGN activity at higher IR luminosities. Although X-ray surveys are a powerful tool to select unobscured and mildly-obscured AGNs, the current census of actively-growing SMBHs still remains far from complete (e.g, Treister *et al.* 2004; Worsley *et al.* 2005; Tozzi *et al.* 2006; Page *et al.* 2006; Fiore *et al.* 2009; Juneau *et al.* 2013). Because DOGs are selected based on their far-IR output, at longer wavelengths than the AGNs selected by near-through-mid IR surveys, far-IR selected DOGs can potentially represent a distinctly-defined population of AGN candidates. In this work we exploit the Chandra COSMOS Legacy Survey (Civano *et al.* 2016) to assess the AGN fraction in DOGs using the most recent and exquisite combination of far-IR and X-ray data. We here focus in the main results and refer the reader to Riguccini *et al.* (2019) for further details and discussion.

2. Far-IR DOGs and the search for X-ray counterparts in the Chandra COSMOS Legacy Survey

We build our "far-IR DOG" parent sample using the catalogues provided by the PEP and HerMES Herschel surveys (Berta *et al.* 2011; Roseboom *et al.* 2010) to identify DOGs in the COSMOS field, all detected in at least 3 of the 5 Herschel bands. We identified a total of 108 far-IR DOGs, 22 with spectroscopic redshifts from Salvato *et al.* (in prep.) and the rest with photometric redshifts determined by Riguccini *et al.* (2015).

The Chandra COSMOS Legacy Survey (Civano *et al.* 2016) covers a total area of ~2.2 deg^2, uniformly covering the ~1.7 deg^2 COSMOS/HST field at a ~160 ksec depth, expanding on the deep C-COSMOS area (1.45 vs 0.44 deg^2) by a factor of ~3 at $\sim 3 \times 10^{16}$ erg cm^{-2} s^{-1}. The deeper and wider coverage of the Chandra COSMOS Legacy survey compared to previous X-ray observations of the COSMOS field (e.g., Brusa *et al.* 2007, 2010; Salvato *et al.* 2009) allows us to detect new X-ray DOGs that have been missed by previous X-ray surveys. We identify individual X-ray counterparts for 22 of the far-IR DOGs. From these 22, 9 are detected in X-rays for the first time thanks to the increased field coverage of the Chandra COSMOS Legacy Survey.

Stacking of the DOGs with no individual X-ray detections suggests a mixture of star-formation and AGN activity. Stacking also showed that X-ray fluxes increase with 24 μm flux (see also Dey *et al.* 2008 and Fiore *et al.* 2009), pointing to an increase of the total AGN fraction in the brightest 24 μm bins. This indicates that the combined population of X-ray detected and far-IR DOGs is effective at selecting AGNs, compared to the 24 μm population as a whole.

3. Obscured hosts and the search for Compton-thick AGNs

Because of low number counts, detailed fitting to the observed X-ray spectrum is not possible for the majority of the X-ray detected DOGs. However, we were able to estimate the neutral hydrogen column density along the line of sight (N_H) based on the methodology described in Treister *et al.* (2009) which assumes an intrinsic power-law spectrum with spectral index $\Gamma = 1.9$ (corresponding to the observed average AGN spectrum; Nandra & Pounds 1994) and computes the expected hardness ratio (HR) for the source (more details in Riguccini *et al.* 2019). Using this approach we find that 6 out of the 22 X-ray detected DOGs (i.e., 27%) are plausible Compton-thick AGNs ($N_H > 10^{24} cm^{-2}$), 15 are moderately-obscured AGNs (10^{22} cm$^{-2} < N_H < 10^{24}$ cm^{-2}), and only one has a low hardness-ratio consistent with being unobscured ($N_H < 10^{22} cm^{-2}$).

The fraction of Compton-thick AGNs that we find in our sample is consistent with previous local and low-redshift reports (e.g., Georgakakis *et al.* 2010; Ricci *et al.* 2015; Aird *et al.* 2015; Burlon *et al.* 2011). In particular, based on NuSTAR observations at 5-80 keV – with a sensitivity peak at 10-30 keV – the observed fraction is 11-20% (Civano *et al.* 2015; Masini *et al.* 2018), while Lansbury *et al.* (2017) report a fraction of ~30% based on the NuSTAR Serendipitous survey. It would have been reasonable to expect DOGs to have a higher fraction of Compton-thick sources because by definition they have dustier host galaxies. However, our results appear to indicate that there is no significant difference with the general AGN population. Thus, we can speculate that the obscuration, at least in the most extreme cases has to be nuclear and roughly independent of the properties of the host galaxy (see also Ricci *et al.* 2015; Buchner & Bauer 2017).

4. DOGs with AGN signatures – a step closer to quenching

In Figure 1 (left panel) we show the evolution of the specific star-formation rate (sSFR = SFR/M*) of DOGs with cosmic time. The SFR rate is calculated from the IR luminosity obtained from the SED-fitting analysis of Riguccini *et al.* (2015), after the AGN contribution has been removed. The stellar masses are from Ilbert *et al.* (2009), based on the SED fits of 30 bands from the COSMOS survey. We immediately note that DOGs with no AGN signatures based on the far-IR lie on the Main Sequence of star-forming galaxies and within the starburst regime (Riguccini *et al.* 2015). The stacking of all X-ray undetected DOGs also places this population within the main-sequence region. On the other hand, the DOGs with AGN signatures (both in the X-ray and the far-IR) predominantly appear to present lower sSFRs, occupying the region of the main-sequence or below it. Considering that these dust-enshrouded galaxies, where both star formation and AGN coexist, likely trace stages in the merger-ULIRG-quasar scenario called upon to explain the formation of massive galaxies (e.g., Hopkins *et al.* 2008; Toft *et al.* 2014), Figure 1 suggests that DOGs with an AGN signature may be a subpopulation that is further ahead in the ULIRG-quasar scenario, already quenching the star-formation activity.

Only two DOG sources with AGN signatures lie well above the main sequence: one Compton-thick candidate and a moderately obscured AGN. According to the evolutionary scenario of Treister *et al.* (2010), the highly obscured Compton-thick phase corresponds to the early, very dust-enshrouded phase of a major merger where a SMBH is rapidly growing, while the moderately obscured AGN phase traces a later stage in this evolutionary scenario, when feedback from the AGN has already started heating up the dust and gas of the galaxy, shutting down star formation activity. In this picture, we would expect Compton-thick candidates to lie slightly above the moderately-obscured AGNs in the sSFR-redshift diagram. We do not observe this in Figure 1. This further

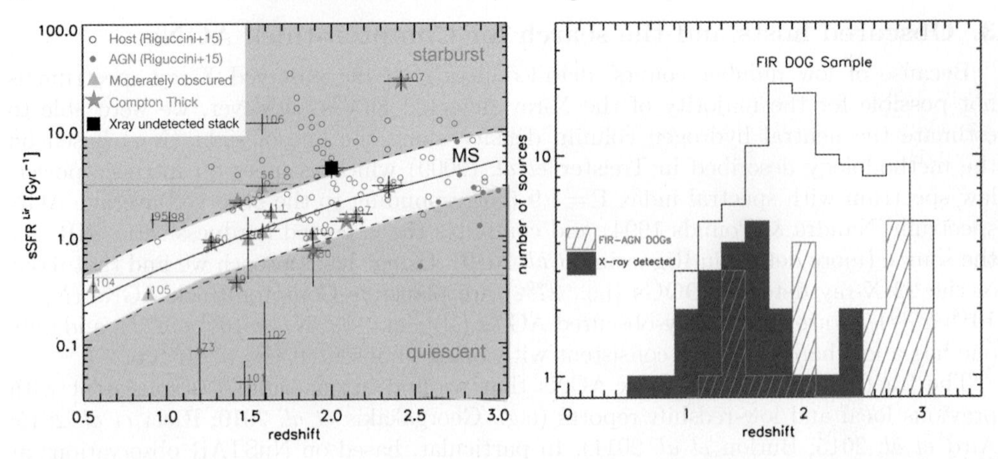

Figure 1. [Left Panel]: Evolution of the specific SFR (sSFR) of DOGs with cosmic time. DOGs with no AGN signatures (empty circles, identified as "host") based on the far-IR analysis of Riguccini *et al.* (2015) lie on the Main Sequence of star-forming galaxies and within the starburst regime. DOGs identified to have an AGN either based on the far-IR (red circles) or X-ray (green triangles and blue stars; this work) predominantly lie on the Main Sequence or below. This suggests that this AGN sub-population of DOGs may already be quenching star formation and be in a latter phase of the ULIRG-quasar scenario. [Right Panel]: Redshift distribution of the DOG parent sample (black histogram), which includes 108 DOGs detected in at least 3 of the five Herschel bands (100–500 μm). The filled and hatched histograms highlight DOGs with AGN signatures in the X-ray (based on the present work) and mid-to-far-IR SED fitting analysis from Riguccini *et al.* (2015), respectively. The two populations only seem to differ in their redshift distributions. This emphasizes the need for a multi-band approach to get a full census of the obscured AGN population out to higher redshifts. Figure from Riguccini *et al.* (2019).

supports that, at least for the most extreme sources, the obscuration is nuclear and thus not directly connected to the evolutionary stage of the host galaxy.

We also bring to the attention of the reader that both DOGs with AGN signatures detected in the X-ray and those detected in the far-IR share the same behaviour vis-a-vis their position with respect to the main sequence of star-forming galaxies: they are typically found on the main-sequence or below it. They also typically display similar near-IR and mid-to-far IR colors. The main difference between these populations appears to be in their redshift distribution (Figure 1, right panel). The DOGs with AGN signatures in the far-IR are typically found at higher redshifts than those with X-ray signatures. Together, these results suggest that the two populations share most of their physical properties and that the lack of detection in the X-ray band for the bulk of far-IR AGN DOGs is explained by the difference in redshift distributions. This emphasizes the crucial need for a multi-wavelength approach to obtain a more complete census of the obscured AGN population out to higher redshifts.

We report here the X-ray and far-IR properties of 108 DOGs from the COSMOS field. Considering that selection criteria of the DOG sample (i.e., F24/FR>1000) selects z$\sim$2 highly dust-enshrouded galaxies in the LIRG/ULIRG regime, there was an expectation to uncover privileged sites of highly-obscured AGNs with a higher frequency of Compton-thick AGNs. However, ~70% are moderately obscured AGNs ($N_H \sim 10^{22-24}$ cm^{-2}) and only ~30% (6) Compton-thick AGN candidates ($N_H > 10^{24}$ cm^{-2}). This is the same as within other AGN populations, so the fact that by looking at DOGs we are indeed going after more heavily-obscured AGNs seems to make no difference. That is, the large N_H of Compton-thick AGNs must be nuclear and have little to do with the dust obscuration from the host galaxy. However, with a higher fraction of AGNs (based on X-ray and far-IR analysis) than the whole 24μm population, DOGs present an interesting population to select AGN candidates at higher redshifts (moderately, highly-obscured). This work emphasizes the important role that the DOG population, in particular the combined X-ray and far-IR detected DOG population, plays in the effort to get a more complete census of the AGN population at high redshift, particularly for the highly obscured population.

References

Alaghband-Zadeh, S., Chapman, S. C., Swinbank, A. M., *et al.* 2012, *MNRAS*, 424, 2232
Alexander, D. M., Bauer, F. E., Chapman, S. C., *et al.* 2005, *ApJ*, 632, 736
Aird, J., Coil, A. L., Georgakakis, A., *et al.* 2015, *MNRAS*, 451, 1892
Berta, S., Magnelli, B., Nordon, R., *et al.* 2011, *A&A*, 532, A49
Blain, A. W., Small, I., Ivison, R. J., *et al.* 1999, "The Hy-Redshift Universe: Galaxy Formation and Evolution at High Redshift", ASP Conference Proceedings, vol. 193, eds. A. J. Bunker and W. J. M. van Breugel, p. 425
Brusa, M., Zamorani, G., Comastri, A., *et al.* 2007, *ApJ Supplement Series*, 172, 353
Brusa, M., Civano, F., Comastri, A., *et al.* 2010, *ApJ*, 716, 348
Buchner, J. & Bauer, F. E. 2017, *MNRAS*, 465, 4348
Burlon, D., Ajello, M., Greiner, J., *et al.* 2011, *ApJ*, 728, 58
Casey, C. M., Narayanan, D., & Cooray, A. 2014, *Physics Reports*, 541, 45
Chapman, S. C., Blain, A. W., Smail, I., *et al.* 2005, *ApJ*, 622, 772
Civano, F., Hickox, R. C., Puccetti, S., *et al.* 2015, *ApJ*, 808, 185
Civano, F., Marchesi, S., Comastri, A., *et al.* 2016, *ApJ*, 819, 62
Dey, A., Soifer, B. T., Desai, V., *et al.* 2008, *ApJ*, 677, 943
Engel, H., Tacconi, L. J., Davies, R. I., *et al.* 2010, *ApJ*, 724, 233
Fiore, F., Puccetti, S., Brusa, M., *et al.* 2009, *ApJ*, 693, 447
Georgakakis, A., Rowan-Robinson, M., Nandra, K., *et al.* 2010, *MNRAS*, 406, 420
Hopkins, P. F., Hernquist, L., Cox, T. J., *et al.* 2008, *ApJ Supplement Series*, 175, 356
Ilbert, O., Capak, P., Salvato, M., *et al.* 2009, *ApJ*, 690, 1236
Juneau, S., Dickinson, M., Bournaud, F., *et al.* 2013, *ApJ*, 764, 176
Lansbury, G. B., Alexander, D. M., Aird, J., *et al.* 2017, *ApJ*, 846, 20
Masini, A., Civano, F., Comastri, A., *et al.* 2018, *ApJ Supplement Series*, 235, 17
Menéndez-Delmestre, K., Blain, A. W., Smail, I., *et al.* 2009, *ApJ*, 699, 667
Menéndez-Delmestre, K., Blain, A. W., Swinbank, M., *et al.* 2013, *ApJ*, 767, 151
Nandra, K. & Pounds, K. A. 1994, *MNRAS*, 268, 405
Page, M. J., Loaring, N. S., Dwelly, T., *et al.* 2006, *MNRAS*, 369, 156
Pope, A., Bussmann, R. S., Dey, A., *et al.* 2008, *ApJ*, 689, 127
Ricci, C., Ueda, Y., Koss, M. J., *et al.* 2015, *ApJ Letters*, 815, L13
Ricci, C., Bauer, F. E., Treister, E., *et al.* 2017, *MNRAS*, 468, 1273
Riguccini, L., Le Floc'h, E., Mullaney, J. R., *et al.* 2015, *MNRAS*, 452, 470
Riguccini, L. A., Treister, E., Menéndez-Delmestre, K., *et al.* 2019, *AJ*, 157, 233
Roseboom, I. G., Oliver, S. J., Kunz, M., *et al.* 2010, *MNRAS*, 409, 48
Salvato, M., Hasinger, G., Ilbert, O., *et al.* 2009, *ApJ*, 690, 1250
Sanders, D. B., Soifer, B. T., Elias, J. H., *et al.* 1988, *ApJ*, 325, 74
Swinbank, A. M., Smail, I., Chapman, S. C., *et al.* 2010, *MNRAS*, 405, 234
Toft, S., Smolčić, V., Magnelli, B., *et al.* 2014, *ApJ*, 782, 68
Tozzi, P., Gilli, R., Mainieri, V., *et al.* 2006, *A&A*, 451, 457
Treister, E., Urry, C. M., Chatzichristou, E., *et al.* 2004, *ApJ*, 616, 123
Treister, E., Cardamone, C. N., Schawinski, K., *et al.* 2009, *ApJ*, 706, 535
Treister, E., Urry, C. M., Schawinski, K., *et al.* 2010, *ApJ Letters*, 722, L238
Worsley, M. A., Fabian, A. C., Bauer, F. E., *et al.* 2005, *MNRAS*, 357, 1281

Galaxy Evolution and Feedback across Different Environments
Proceedings IAU Symposium No. 359, 2020
T. Storchi-Bergmann, W. Forman, R. Overzier & R. Riffel, eds.
doi:10.1017/S1743921320002380

High-redshift starbursts as progenitors of massive galaxies

Carlos Gómez-Guijarro

AIM, CEA, CNRS, Université Paris-Saclay, Université Paris Diderot, Sorbonne Paris Cité, F-91191 Gif-sur-Yvette, France
email: carlos.gomezguijarro@cea.fr

Abstract. Starbursting dust-rich galaxies are capable of assembling large amounts of stellar mass very quickly. They have been proposed as progenitors of the population of compact massive quiescent galaxies at $z \sim 2$. To test this connection, we present a detailed spatially-resolved study of the stars, dust, and stellar mass in a sample of six submillimeter-bright starburst galaxies at $z \sim 4.5$. We found that the systems are undergoing minor mergers and the bulk star formation is located in extremely compact regions. On the other hand, optically-compact star forming galaxies have also been proposed as immediate progenitors of compact massive quiescent galaxies. Were they formed in slow secular processes or in rapid merger-driven starbursts? We explored the location of galaxies with respect to star-forming and structural relations and study the burstiness of star formation. Our results suggest that compact star-forming galaxies could be starbursts winding down and eventually becoming quiescent.

Keywords. galaxies: bulges, galaxies: evolution, galaxies: formation, galaxies: fundamental parameters, galaxies: high-redshift, galaxies: interactions, galaxies: ISM, galaxies: starburst

1. Introduction

Local elliptical galaxies are the largest, oldest, and most massive galaxies in the local universe. How did they form? Already at $z \sim 2$ half of the most massive galaxies appear red and dead, but they are also extremely compact. These are the so-called compact quiescent galaxies (cQGs; e.g., Toft *et al.* 2007; van Dokkum *et al.* 2008), also commonly known in the literature as "red nuggets". The evolutionary pathways of these galaxies to become local ellipticals are thought to be dominated by passive aging and dry minor mergers (e.g., Bezanson *et al.* 2009). How did cQGs at $z \sim 2$ form? What are their progenitors?

Gas-rich major mergers are mechanisms capable of igniting strong nuclear starbursts that create large amounts of stars in compact regions very rapidly. Dusty star-forming galaxies (DSFGs), with the best-studied example being the so-called submillimeter galaxies (SMGs), are the most intense starbursts known with a typical star formation rate (SFR) that goes from hundreds up to thousands of solar masses per year, making them easily observable at far-infrared/submillimeter (FIR/sub-mm) wavelengths at high redshifts (see Casey *et al.* 2014, for a review). Toft *et al.* (2014) proposed a direct evolutionary connection between $z \gtrsim 3$ SMGs and $z \sim 2$ cQGs.

In this text the main results from Gómez-Guijarro *et al.* (2018) and Gómez-Guijarro *et al.* (2019) are outlined, to which the reader is referred for further details.

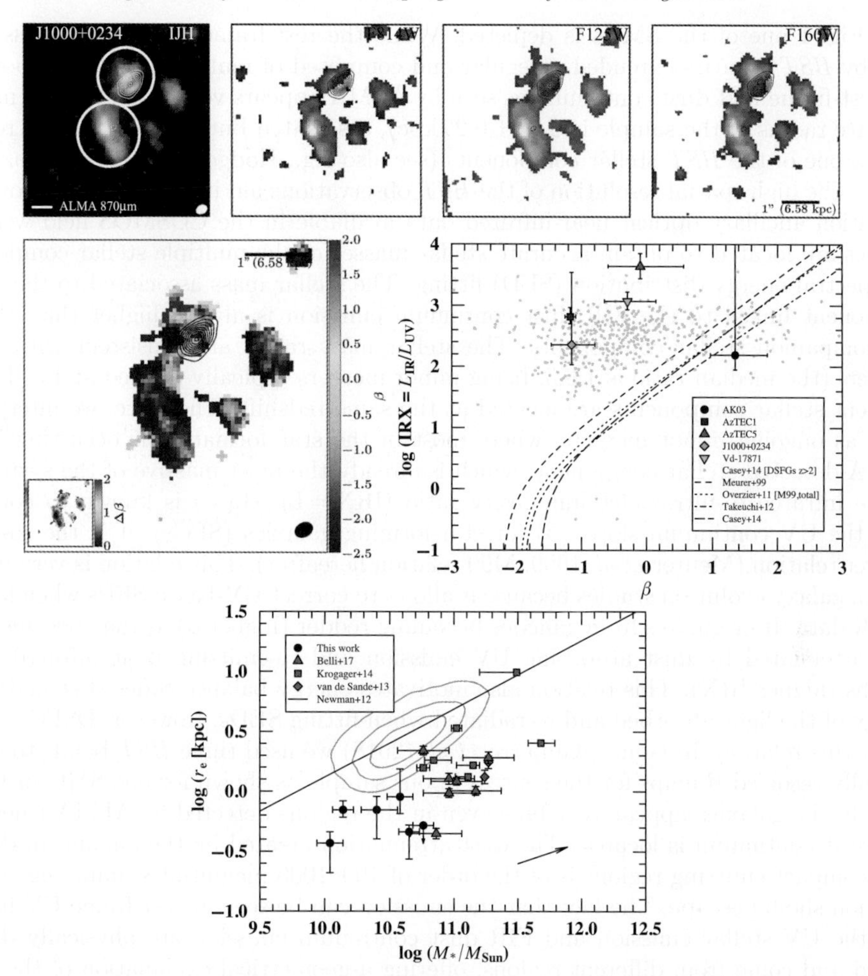

Figure 1. From Gómez-Guijarro *et al.* (2018). Top panel: *HST* images for one of the observed SMGs. ALMA band 7 ($\sim$870 μm) contours are overlayed. Middle left panel: UV continuum slope map for the same SMG. The error map is shown in the bottom left corner. Middle right panel: IRX-β plane. Symbols indicate *HST* photometry performed over the region above the 3σ contour in the ALMA image. Bottom panel: Stellar mass-size plane. SMG sample along with $z\sim2$ cQGs. The bottom-right black arrow represents the expected evolution of the SMG sample growing in stellar mass via in their current starburst episode and minor merger contribution, and in size via minor mergers.

2. Connecting submillimeter and quiescent galaxies

A very important piece in making such evolutionary connection between $z\gtrsim3$ SMGs and $z\sim2$ cQGs comes from the stellar and dust structure of SMGs and how it compares with the proposed quiescent descendants. In the pre-ALMA era, lacking of sufficient spatial resolution very little was known about the dust continuum structure of SMGs and how it related with the stellar structure. In Gómez-Guijarro *et al.* (2018) we tackled this problem and test the evolutionary sequence by targeting a sample of six SMGs, mostly spectroscopically confirmed at $z\sim4.5$, with both *HST* and ALMA. This study provided a spatially-resolved view of the stars, dust, and stellar mass to have a complete picture of both the unobscured and obscured star formation processes.

In Fig. 1 one of the SMGs is depicted. While the rest-frame UV stellar emission as seen by *HST* appears extended, irregular and composed of multiple stellar components, the rest-frame FIR dust continuum as seen by ALMA appears very compact (the median effective radius of the sample is 0.70 ± 0.29 kpc), associated but misaligned with respect to just one of the *HST* stellar components (see also e.g., Hodge *et al.* 2016; Elbaz *et al.* 2018). The high spatial resolution of the *HST* observations aid in the deblending of lower resolution ancillary optical–near-infrared data available in the COSMOS field were the sources are located to obtain accurate stellar masses of the multiple stellar components via spectral energy distribution (SED) fitting. The stellar mass associated to the stellar component linked to the FIR dust continuum emission is always higher than that of the companion stellar components. The stellar mass ratios are consistent with minor mergers (the median ratio is 1:6.5, being minor mergers typically defined at 1:3–4). The different stellar components are located at the same redshift. Therefore, we interpreted them as ongoing minor mergers, where most of the star formation is occurring in the ALMA-detected stellar component, which is already the most massive of the system.

The infrared-to-ultraviolet luminosity ratio ($\mathrm{IRX} = \mathrm{L_{IR}}/\mathrm{L_{UV}}$) is known to correlate with the UV continuum slope (β) for star-forming galaxies (SFGs), it is the so-called Meurer relation (Meurer *et al.* 1999, M99 relation hereafter). This relation is very important in galaxy evolution studies because it allows to correct UV-based SFRs when lacking of FIR data. It originates from galaxies becoming redder (higher β) as they become more dust attenuated by dust absorbing UV emission and re-emitting it at infrared wavelengths (higher IRX). This relation also motivates energy balance codes that match the energy of the light absorbed and re-radiated when fitting SEDs. However, DSFGs do not follow this relation. In Gómez-Guijarro *et al.* (2018) we used three *HST* bands to create spatially-resolved β maps for the sources in our sample as shown for one SMG in Fig. 1. Overall, the galaxies appear very blue even in the regions detected by ALMA where the FIR dust continuum is located. The dust attenuation created by the amount of dust in such compact emitting regions is of the order of 100–1000 magnitudes, implying that no emission should escape. The fact that the emission can be seen at rest-frame UV implies that the UV stellar emission and FIR dust continuum emission are physically disconnected and come from different regions, offering a geometrical explanation of the offset of DSFGs with respect to the M99 relation.

In terms of the evolutionary sequence between $z \gtrsim 3$ SMGs and $z \sim 2$ cQGs, thanks to the high spatial resolution observations from *HST* and ALMA that provided a more complete picture of the structure of SMGs, we revisited the stellar mass-size plane in Gómez-Guijarro *et al.* (2018) as shown in Fig. 1. The sample of SMGs and their proposed cQGs descendants at $z \sim 2$ are located in distinct regions of the diagram, but the separation is what is expected given that the SMGs will grow in stellar mass in their current starburst episode and in size due to the observed minor mergers, bringing the location of the populations into agreement and furthering the strength of their evolutionary connection. This evolutionary sequence is also strengthened by recent analysis showing evidence for gaseous rotationally-support disks in some of the SMGs in the sample (e.g., Jones *et al.* 2017; Tadaki *et al.* 2020) and other results on $z \sim 2$ cQGs exhibiting rotationally-supported stellar disks (Newman *et al.* 2015; Toft *et al.* 2017).

3. Investigating the transition from star-forming to quiescent galaxies

SFGs are known to form a tight correlation between SFR and stellar mass, the so-called main sequence (MS) of star formation (e.g., Elbaz *et al.* 2007; Noeske *et al.* 2007), which exhibits a small scatter implying that secular growth must be the dominant mode of galaxy evolution. On the contrary, quiescent galaxies (QGs) are located below the MS. In terms of structure, at fixed stellar mass and redshift SFGs are larger than QGs (e.g.,

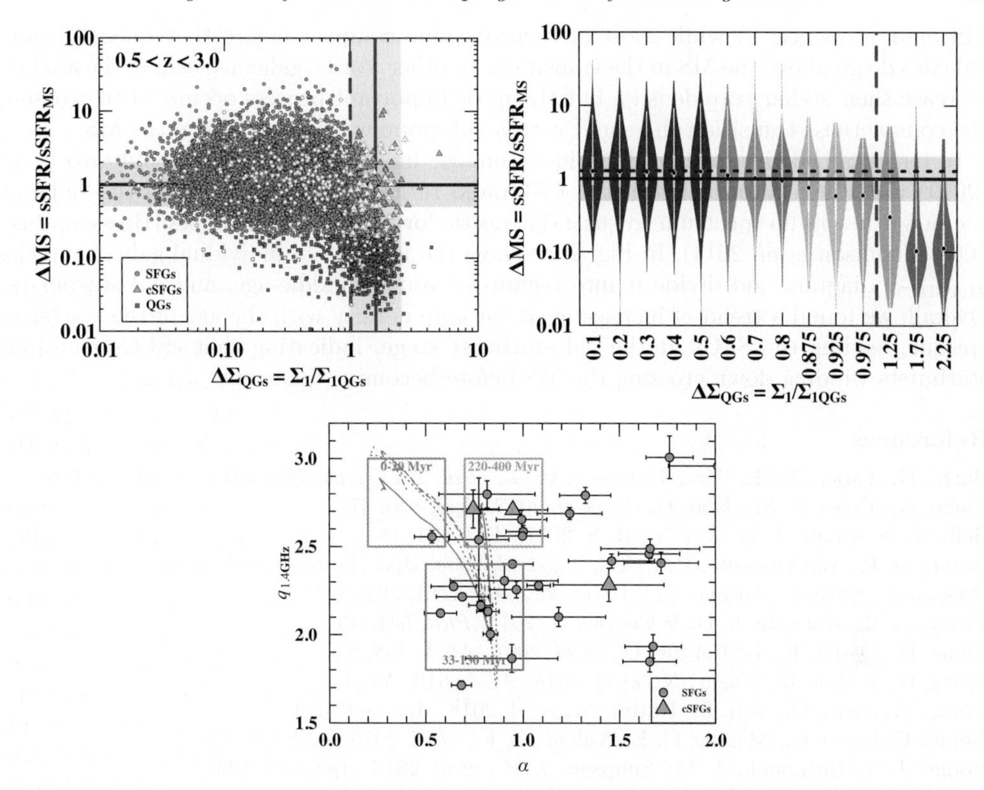

Figure 2. From Gómez-Guijarro *et al.* (2019). Top left panel: ΔMS-$\Delta\Sigma_{\rm QGs}$ plane. Top right panel: Same panel binned in the X axis. Bottom panel: $q_{1.4\rm GHz}$-α plane. (Bressan *et al.* 2002) evolutionary tracks for the age of starbursts are plotted as purple lines.

van der Wel *et al.* 2014). These two fundamental relations of galaxy properties based on star formation and structure put together has lead to several authors to point out that the quenching of star formation and the departure from the MS implies the build up of a central stellar core (e.g., Kauffmann *et al.* 2003; Barro *et al.* 2017). In this context, a population of galaxies arises as the immediate link between the more extended SFG and the more compact QGs, the so-called compact star-forming galaxies (cSFGs; e.g., Barro *et al.* 2013; van Dokkum *et al.* 2015), also commonly refered in the literature as "blue nuggets". cSFG are galaxies that follow SFG in terms of their location within the fundamental MS of star formation, but that follow the QGs in terms of their location with respect to fundamental structural relations of QGs. Did the build up of the stellar core, formation of cSFGs, and quenching happen slowly as the product of the secular evolution of SFGs, or rapidly requiring a merger-driven starburst episode? In Gómez-Guijarro *et al.* (2019) we studied SFGs and QGs with respect to the MS and structural relations and investigate diagnostic of the burstiness of star formation to address this question.

In Fig. 2 we show individual galaxies in the plane formed by the distance to the MS (ΔMS) at its stellar mass and redshift versus the distance to the compactness selection threshold in the core density based on the structural relation of QGs ($\Delta\Sigma_{\rm QGs} = \Sigma_1/\Sigma_{1\rm QGs}$) at its stellar mass and redshift (see Barro *et al.* 2017, for details about the compactness threshold). Overall, galaxies become compact before they quench reproducing the L-shape reported by previous studies (e.g., Barro *et al.* 2017; Lee *et al.* 2018). Additionally, we reported two other distinct behaviours in this diagram: 1) as galaxies become progressively more compact (centrally concentrated) they smoothly abandon the

MS main sequence; 2) while most of them become compact before they quench, some galaxies do go above the MS in the transition. In other words, galaxies seem to die as they increase their stellar core density, but the most important process occurs when crossing the compactness transition threshold with a sub-population going above the MS.

In terms of the diagnostic of burstiness, one we investigated in Gómez-Guijarro *et al.* (2019) was the relation between the FIR/radio ratio ($q \propto L_{\rm IR}/L_{\rm radio}$) and the slope of the power law radio spectrum (α) as a diagnostic for the age of starbursts (Bressan *et al.* 2002; Thomson *et al.* 2014). In Fig. 2 we show the location of individual galaxies in the $q_{\rm 1.4GHz}$-α diagram and divide it into regions of young, middle-age, and old starbursts. Overall, we found a trend of increasing stellar core density with the age of the starburst episode, leading to cSFGs at the old starburst stage, indicating that cSFGs could be starbursts winding down crossing the MS before becoming quiescent.

References

Barro, G., Faber, S. M., Pérez-González, P. G., *et al.* 2013, *ApJ*, 765, 104
Barro, G., Faber, S. M., Koo, D. C., *et al.* 2017, *ApJ*, 840, 47
Belli, S., Newman, A. B., & Ellis, R. S. 2017, *ApJ*, 834, 18
Bezanson, R., van Dokkum, P.G., Tal, T., *et al.* 2009, *ApJ* (Letters), 697, 1290
Bressan A., Silva, L., Granato, G. L., *et al.* 2002, *A&A*, 392, 377
Casey, C. M., Narayanan, D. & Cooray, A. 2014, *PhR*, 541, 45
Elbaz, D., Daddi, E., Le Borgne, D., *et al.* 2007, *A&A*, 468, 33
Elbaz, D., Leiton, R., Nagar, N., *et al.* 2018, *A&A*, 616, A110
Gómez-Guijarro, C., Toft, S., Karim, A., *et al.* 2018, *ApJ*, 856, 121
Gómez-Guijarro, C., Magdis, G. E., Valentino, F., *et al.* 2019, *ApJ*, 886, 88
Hodge, J. A., Swinbank, A. M., Simpson, J. M., *et al.* 2016, *ApJ*, 833, 103
Jones, G. C., Carilli, C. L., Shao, Y., *et al.* 2017, *ApJ*, 850, 180
Kauffmann, G., Heckman, T. M., Tremonti, C., *et al.* 2003, *MNRAS*, 346, 1055
Krogager, J. K., Zirm, A. W., Toft, S., *et al.* 2014, *ApJ*, 797, 17
Meurer, G. R., Heckman, T. M., Calzetti, D., *et al.* 1999, *ApJ*, 521, 64
Newman, A. B., Ellis, R. S., Bundy, K., *et al.* 2012, *ApJ*, 746, 162
Newman, A. B., Belli, S., Ellis, R. S., *et al.* 2015, *ApJ* (Letters), 813, L7
Noeske, K. G., Weiner, B. J., Faber, S. M., *et al.* 2007, *ApJ* (Letters), 660, L43
Lee B., Giavalisco, M., Whitaker, K., *et al.* 2018, *ApJ*, 853, 131
Tadaki, K., Iono, D., Yun, M., *et al.* 2020, *ApJ*, 889, 141
Thomson A. P., Ivison, R. J., Simpson, J. M., *et al.* 2014, *MNRAS*, 442, 577
Toft, S., van Dokkum, P., Franx, M., *et al.* 2007, *ApJ*, 671, 285
Toft, S., Smolčić, V., Magnelli, B., *et al.* 2014, *ApJ*, 782, 68
Toft, S., Zabl, J., Richard, J., *et al.* 2017, *Nature*, 546, 510
van Dokkum, P. G., Franx, M., Kriek, M., *et al.* 2008, *ApJL* (Letters), 677, L5
van Dokkum, P. G., Nelson, E. J., Franx, M., *et al.* 2015, *ApJ*, 813, 23
van de Sande, J., Kriek, M., Franx, M., *et al.* 2013, *ApJ*, 771, 85
van der Wel, A., Franx, M., van Dokkum, P. G., *et al.* 2014, *ApJ*, 788, 28

Galaxy Evolution and Feedback across Different Environments
Proceedings IAU Symposium No. 359, 2020
T. Storchi-Bergmann, W. Forman, R. Overzier & R. Riffel, eds.
doi:10.1017/S1743921320001921

Variable radio AGN at high redshift identified in the VLA Sky Survey

Kristina Nyland[1], Dillon Dong[2], Pallavi Patil[3,4], Mark Lacy[4], Amy Kimball[5], Gregg Hallinan[2], Sumit Sarbadhicary[6], Emil Polisensky[7], Namir Kassim[7], Wendy Peters[7], Tracy Clarke[7], Dipanjan Mukherjee[8], Sjoert van Velzen[9,10] and Vivienne Baldassare[11,12]

[1]National Research Council fellow, resident at NRL, Washington, DC, USA
[2]California Institute of Technology, Pasadena, CA, USA
[3]University of Virginia, Charlottesville, VA, USA
[4]NRAO, Charlottesville, VA, USA
[5]NRAO, Socorro, NM, USA
[6]Michigan State University, Lansing, MI, USA
[7]NRL, Washington, DC, USA
[8]IUCAA, Pune, India
[9]New York University, New York, NY, USA
[10]University of Maryland, College Park, MD, USA
[11]Yale University, New Haven, CT, USA
[12]Einstein Fellow

Abstract. As part of an on-going study of radio transients in Epoch 1 (2017–2019) of the Very Large Array Sky Survey (VLASS), we have discovered a sample of $0.2 < z < 3.2$ active galactic nuclei (AGN) selected in the optical/infrared that have recently brightened dramatically in the radio. These sources would have previously been classified as radio-quiet based on upper limits from the Faint Images of the Radio Sky at Twenty-centimeters (FIRST; 1993-2011) survey; however, they are now consistent with radio-loud quasars. We present a quasi-simultaneous, multi-band (1–18 GHz) VLA follow-up campaign of our sample of AGN with extreme radio variability. We conclude that the radio properties are most consistent with AGN that have recently launched jets within the past few decades, potentially making them among the youngest radio AGN known.

Keywords. galaxies: active, galaxies: jets, galaxies: evolution

---

1. Introduction

The slow (timescales of seconds to years) radio transient population is dominated by active galactic nuclei (AGN), the majority of which are associated with variability (Mooley *et al.* 2016). AGN variability in the radio may arise from extrinsic propagation effects (e.g. interstellar scattering) or intrinsic mechanisms directly related to the AGN itself (e.g. Thyagarajan *et al.* 2011). Whether driven by extrinsic or intrinsic effects, variable radio AGN beyond the low-z universe are inherently compact in nature. AGN with compact, sub-galactic jets hosted by gas-rich galaxies, especially at $1 \lesssim z \lesssim 3$, are an important, yet still poorly studied, class of objects for understanding the link between jet-ISM feedback and galaxy evolution. In particular, *the prevalence of compact radio jets as a*

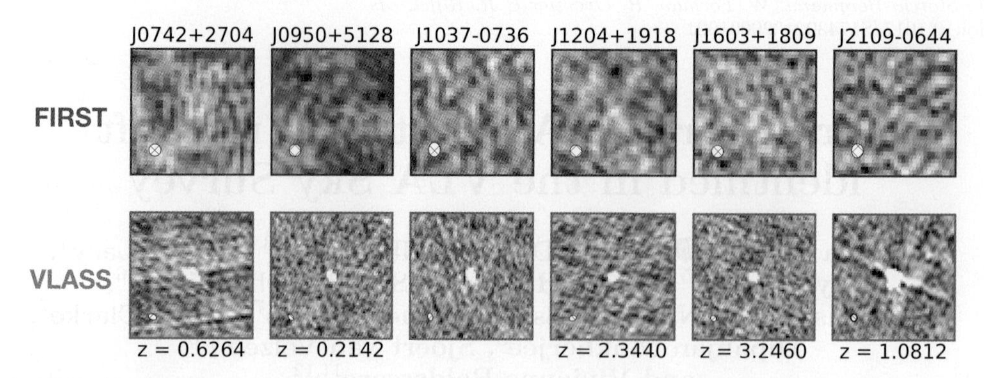

Figure 1. Cut-out images ($1' \times 1'$) from FIRST (1.4 GHz; 1993–2011) and VLASS Epoch 1 (3 GHz; 2017–2019) of a subset of our sample highlighting their variability on timescales of 1–2 decades. The synthesized beam is shown in the lower left corner of each cut-out image.

function of redshift and host galaxy properties remains unknown. AGN with compact jets are challenging to identify in single-epoch radio surveys due to observational limitations (e.g. resolution and sensitivity), as well as inherent limitations due to variability. Recently, multi-epoch radio surveys have begun to make major advancements identifying AGN with compact jets on the basis of radio variability.

2. Sample

As part of our ongoing search for radio transients in Epoch 1 (2017–2019) of the Very Large Array Sky Survey (VLASS; Lacy *et al.* 2020), we have identified a preliminary sample of ~2000 candidate "transients" that are compact and >1 mJy in VLASS, but below the 5σ detection threshold (< 0.75 mJy) of the 1.4 GHz FIRST survey (Becker *et al.* 1995) observed 1–2 decades earlier. VLASS is a synoptic survey at S-band (2–4 GHz) of the entire northern sky visible to the VLA ($\delta > -40°$). A unique feature of VLASS is its synoptic design consisting of 3 epochs with a cadence of 32 months.

Selection of Variable Radio AGN. We considered optical and IR AGN selection diagnostics to capture both the unobscured and obscured AGN populations. In the optical, we searched for spectroscopically verified quasars from the Sloan Digital Sky Survey (SDSS; York *et al.* 2000) DR14 (Pâris *et al.* 2018) and found 52 candidates within $1''$ of the VLASS position. Using data from the Widefield Infrared Survey Explorer (WISE; Wright *et al.* 2010) and the AGN diagnostic criteria from Assef *et al.* (2018), we found 144 obscured AGN with variable radio emission. We further required VLASS fluxes >3 mJy to rule-out steady but optically-thick sources with spectral indices up to $\alpha = 2.5$ (where $S \sim \nu^{\alpha}$). Our final sample contains 26 radio-variable AGN in the 25% of VLASS Epoch 1 analyzed thus far.

Source Properties. Spectroscopic redshifts ($0.2 \lesssim z \lesssim 3.2$) are available for the optically-selected subset of our sample (13/26 sources). These sources are classified as broad-line quasars and have radio luminosities of $\log(L_{3\,\mathrm{GHz}}/\mathrm{erg\,s^{-1}}) \approx 40{-}42$. Shen *et al.* 2011 report bolometric luminosities of $\log(L_{\mathrm{bol}}/\mathrm{erg\,s^{-1}}) \approx 45.2{-}46.8$ and virial supermassive black hole (SMBH) masses of $\log(M/\mathrm{M}_{\odot}) \approx 8.0{-}9.7$. We are in the process of obtaining spectroscopic redshifts for the IR-selected, obscured AGN portion of our sample. We note that some of the IR-selected AGN have extreme colors in WISE that are consistent with heavily obscured, hyperluminous quasars (Lonsdale *et al.* 2015, Patil *et al.* 2020). Example cut-out images illustrating the remarkable increase in radio flux between FIRST and VLASS are shown in Fig. 1.

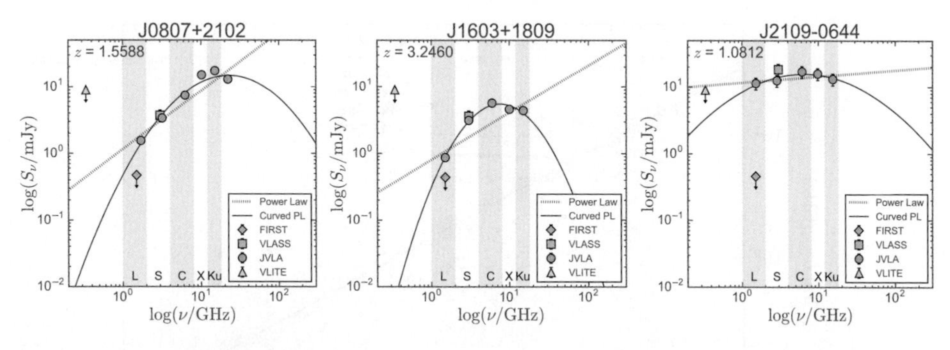

Figure 2. Example radio SEDs showing our new quasi-simultaneous VLA data (purple circles), the VLASS detections (orange squares), and the FIRST 3σ upper limits (green diamonds). The yellow triangles denote 340 MHz upper limits from the commensal VLA Low-band Ionosphere and Transient Experiment (VLITE; Clarke *et al.* 2016; Polisensky *et al.* 2016) data that were recorded during our new VLA observations. Two model fits to our quasi-simultaneous new VLA data are shown: 1) a standard non-thermal power-model and 2) a curved power-law model.

3. Observations

We obtained 1–18 GHz VLA/A-configuration data (project 19A-422; PI: Gregg Hallinan) from July 23 - October 13, 2019 for 14/26 variable radio AGN. For each source, the multi-band VLA observations were performed "quasi-simultaneously" within a single scheduling block $\lesssim$3 hr. This enabled a consistent and robust analysis of the radio spectral energy distribution (SED) shapes, thereby mitigating the confounding influence of intrinsic variability on timescales longer than a few hours. Further details will be presented in a forthcoming paper (Nyland *et al.* in prep.).

Variability Constraints. Comparison between VLASS and our follow-up VLA S-band data taken a few months later reveals good agreement within the current $\sim$20% flux uncertainties of the VLASS quicklook images (Lacy *et al.* 2020). This suggests a typical variability timescale $>$ a few months. At L-band, the current peak flux values range from 1–11 mJy, which corresponds to large variability amplitudes of $\sim$2$-$20$\times$ in the 1–2 decades since FIRST. To put this into context, blazar variability amplitudes in the cm-wave regime are typically at the 10-20% level (Pietka *et al.* 2015).

Broadband Radio SEDs. In Fig. 2, we show examples of the broadband radio SEDs. We perform a least squares fit to the quasi-simultaneous data using two basic synchrotron emission models (Callingham *et al.* 2017, and references therein): 1) a standard non-thermal power-law model of the form $S_\nu = a\nu^\alpha$, where a is the amplitude and α is the spectral index, and 2) a curved power-law model defined as $S_\nu = a\nu^\alpha e^{q(\ln \nu)^2}$, where q is the degree of spectral curvature (i.e. the breadth of the model at the peak flux value) and is defined by $\nu_{\rm peak} = e^{-\alpha/2q}$, where $\nu_{\rm peak}$ is the turnover frequency. The presence of substantial spectral curvature is typically defined as $|q| \geqslant 0.2$ (Duffy & Blundell 2012). Our sources have significant spectral curvature $(0.18 < |q| < 1.09)$ with turnover frequencies of $\sim$6.6 GHz (2.5–22.7 GHz). The quasi-simultaneous VLA data thus demonstrate that the SEDs of our sources are best modeled by curved power-law fits.

Intrinsic Sizes. Our sources have compact radio emission over the full frequency range of our VLA study, placing an upper limit to their intrinsic sizes of $\theta_{\rm max} < 0.1''$ (<1 kpc). We show the turnover-size relation in Fig. 3. This relationship arises from the expansion and subsequent energy loss of young jets, which causes the peak frequency (or turnover) of the relativistic electron energy distribution to shift to lower frequencies. Assuming our sources follow the turnover-size relation, we can obtain rough estimates of their sizes and ages. For example, the source J0807+2102 at $z = 1.5588$ has a turnover frequency of $\sim$22.7 GHz ($\sim$58.1 GHz in the rest frame). We thus estimate an intrinsic jet extent of $<$1 pc and (assuming a jet speed of $0.1c$) an age of $\lesssim$30 yr for this source.

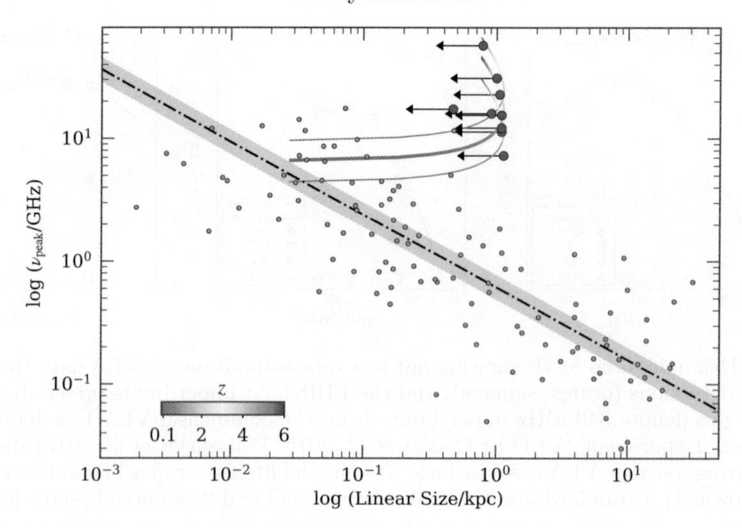

Figure 3. Spectral turnover as a function of linear size. The small gray circles are literature measurements of young radio AGN (Jeyakumar 2016). The black dot-dashed line shows the empirical fit to the turnover-size relation (O'Dea 1998), and the dark gray shaded region indicates the uncertainty. Linear source size upper limits are shown by the large purple circles. For sources lacking redshifts, the linear size upper limits are shown over a range of possible redshifts as indicated by the rainbow-colored arcs.

4. Origin of the radio variability?

Extrinsic vs. Intrinsic Variability. The radio SED shapes, variability timescale and amplitude constraints, high radio luminosities, and high SMBH masses rule-out variability driven by extrinsic propagation effects or transient phenomena (including on-axis GRBs, radio supernovae afterglows, and tidal disruption events). We thus conclude that the increase in radio flux is caused by *intrinsic* AGN variability. Intrinsic radio AGN variability may arise from blazar-like behavior (e.g. shocks propagating along the jet due to magnetic field turbulence; Marscher & Gear 1985), the rapid re-orientation of a compact jet towards our line of sight (e.g. Bodo *et al.* 2013), and newborn radio jets that have been recently launched (possibly following a state transition of the accreting SMBH; Wójtowicz *et al.* 2020).

Young Radio Jets. Given the radio properties of our sources (in particular variability amplitude and timescale constraints), we favor the jet youth scenario in which the radio variability is the result of parsec-scale jets that were launched in the last 1–2 decades. Confirmation of the jet youth scenario will require continued monitoring of the evolution of radio fluxes and SED shapes, as well as tighter size constraints from higher-resolution radio observations.

Young radio AGN, such as Gigahertz peaked spectrum (GPS) sources, are characterized by compact morphologies and inverted radio SEDs below their turnover frequencies, which are typically in the GHz regime (O'Dea 1998), consistent with the morphologies and radio SEDs of our sources. After a jet is launched, models predict a rapid increase in luminosity ($P_{\rm radio} \sim t^{2/5}$) as the dominant energy loss mechanism transitions from adiabatic to synchrotron losses (An & Baan 2012), making the identification of young radio AGN in VLASS that have emerged in the time since FIRST (~20 yr) plausible. The identification of such young radio AGN is not unprecedented; the youngest known sources have kinematic ages as low as 20 yr (Gugliucci *et al.* 2005). For a nascent radio jet that has been triggered within the last 20 yr, the model of An (2012) suggests a $> 3\times$ increase in radio luminosity, which is consistent with the observed radio brightening at 1.4 GHz between FIRST and our 2019 L-band observations.

5. Discussion

Radio-changing-state Quasars? While our sources would have previously been classified as radio-quiet based on their upper limits in FIRST, VLASS has revealed that they are now consistent with radio-loud quasars (e.g. Kellermann *et al.* 2016). Other multi-epoch radio surveys have recently identified AGN with similar radio variability amplitudes and timescales and concluded that jets with lifetimes of 10^{4-5} yr may be common (Jarvis *et al.* 2019; Wołowska *et al.* 2017). If this is indeed the case, jet-ISM feedback may play a more important role in the regulation of SMBH growth and star formation than typically assumed. Large statistical studies of compact radio AGN will ultimately enable key improvements to prescriptions for sub-grid models of AGN feedback in cosmological simulations.

Future Prospects. As part of our continued effort to study AGN that have recently transitioned from radio-quiet to radio-loud, new multi-band radio observations with the Very Long Baseline Array, which will provide milliarcsecond-scale resolution, are currently in progress. Over the next several years, the completion of the remaining epochs of VLASS, as well as future multi-epoch, high-resolution studies with prospective instruments such as the next-generation Very Large Array (Nyland *et al.* 2018), will enable major advancements in our understanding of the triggering of radio jets and their connection to galaxy evolution.

References

An, T. & Baan, W. A. 2012, *ApJ*, 760, 77
Assef, R. J., Stern, D., Noirot, G., *et al.* 2018, *ApJS*, 234, 23
Becker, R. H., White, R. L., & Helfand, D. J. 1995, *ApJ*, 450, 559
Bodo, G., Mamatsashvili, G., Rossi, P., *et al.* 2013, *MNRAS*, 434, 3030
Callingham, J. R., Ekers, R. D., Gaensler, B. M., *et al.* 2017, *ApJ*, 836, 174
Clarke, T. E., Kassim, N. E., Brisken, W., *et al.* 2016, Society of Photo-Optical Instrumentation Engineers (SPIE) Conference Series, Vol. 9906, Commensal low frequency observing on the NRAO VLA: VLITE status and future plans, 99065B
Duffy, P. & Blundell, K. M. 2012, *MNRAS*, 421, 108
Gugliucci, N. E., Taylor, G. B., Peck, A. B., *et al.* 2005, *ApJ*, 622, 136
Jarvis, M. E., Harrison, C. M., Thomson, A. P., *et al.* 2019, *MNRAS*, 485, 2710
Jeyakumar, S. 2016, *MNRAS*, 458, 3786
Kellermann, K. I., Condon, J. J., Kimball, A. E., *et al.* 2016, *ApJ*, 831, 168
Lacy, M., Baum, S. A., Chandler, C. J., *et al.* 2020, *PASP*, 132, 035001
Lonsdale, C. J., Lacy, M., Kimball, A. E., *et al.* 2015, *ApJ*, 813, 45
Marscher, A. P. & Gear, W. K. 1985, *ApJ*, 298, 114
Mooley, K. P., Hallinan, G., Bourke, S., *et al.* 2016, *ApJ*, 818, 105
Nyland, K., Harwood, J. J., Mukherjee, D., *et al.* 2018, *ApJ*, 859, 23
O'Dea, C. 1998, *PASP*, 110, 493
Pâris, I., Petitjean, P., Aubourg, É., *et al.* 2018, *A&A*, 613, A51
Patil, P., Nyland, K., Whittle, M., *et al.* 2020, arXiv e-prints, arXiv:2004.07914
Pietka, M., Fender, R. P., & Keane, E. F. 2015, *MNRAS*, 446, 3687
Polisensky, E., Lane, W. M., Hyman, S. D., *et al.* 2016, *ApJ*, 832, 60
Shen, Y., Richards, G. T., Strauss, M. A., *et al.* 2011, *ApJS*, 194, 45
Thyagarajan, N., Helfand, D. J., White, R. L., *et al.* 2011, *ApJ*, 742, 49
Wójtowicz, A., Staarz, ł, & Cheung, C. C. 2020, *ApJ* 892, 116
Wołowska, A., Kunert-Bajraszewska, M., Mooley, K., *et al.* 2017, *Frontiers in Astronomy and Space Sciences*, 4, 38
Wright, E. L., *et al.* 2010, *AJ*, 140, 1868
York, D. G., *et al.* 2000, *AJ*, 120, 1579

Rainer Weinberger

Andrew Newman

Kristina Nyland

Galaxy Evolution and Feedback across Different Environments
Proceedings IAU Symposium No. 359, 2020
T. Storchi-Bergmann, W. Forman, R. Overzier & R. Riffel, eds.
doi:10.1017/S1743921320001659

POSTERS

The evolution of star formation in QSOs according to WISE

K. A. Cutiva-Alvarez, R. Coziol, J. P. Torres-Papaqui, H. Andernach and A. C. Robleto-Orús

Depto. de Astronomía, DCNE, Universidad de Guanajuato, CP 36023, Gto., México
email: kacutivaa@unal.edu.co

Abstract. Using WISE data, we calibrated the W2-W3 colors in terms of star formation rates (SFRs) and applied this calibration to a sample of 1285 QSOs with the highest flux quality, covering a range in redshift from $z \sim 0.3$ to $z \sim 3.8$. According to our calibration, the SFR increases continuously, reaching a value at $z \sim 3.8$ about 3 times higher on average than at lower redshift. This increase in SFR is accompanied by an increase of the BH mass by a factor 100 and a gradual increase of the mean Eddington ratio from 0.1 to 0.3 up to $z \sim 1.5$–2.0, above which the ratio stays constant, despite a significant increase in BH mass. Therefore, QSOs at high redshifts have both more active BHs and higher levels of star formation activity.

Keywords. (galaxies:) quasars: general; infrared: galaxies; galaxies: evolution

1. Introduction

In Coziol *et al.* (2014), a new mid-infrared (MIR) diagram (MIRDD) based on WISE (Wright *et al.* 2010 was devised that allows to separate type II AGNs from star forming galaxies (SFGs) in a way similar to the BPT-VO diagram in the optical (Baldwin *et al.* 1981; Veilleux & Osterbrock 1987). The key of the MIRDD was shown to be the sensitivity of the W2-W3 color to the intensity of star formation: as the star formation (SF) intensity increases in the host galaxies, W2-W3 becomes significantly redder (Donoso *et al.* 2012; Jarrett *et al.* 2013). In Coziol *et al.* (2015) the authors observed the same phenomenon in type I AGNs, suggesting an increase of SF with the redshift (Leipski *et al.* 2014; Delvecchio *et al.* 2014). Using the data in Coziol *et al.* (2015), we calibrated the W2-W3 color in terms of star formation rate (SFR) and applied this calibration to a sample of 1285 QSOs from the SDSS DR12Q sample of Paris *et al.* (2017), with high quality MIR fluxes (A in all the four fluxes), redshifts between $z \sim 0.3$ and $z \sim 3.8$, and continuum luminosity and BH mass previously determined by Kozlowski (2017). Depending on the emission line used to determine the BH mass (MgIIλ2798 at low redshift and CIVλ1549 at higher redshifts), our sample was separated in two (identified simply as MgII and CIV).

2. Results

In Fig. 1 we compare in the two samples the SFR, the BH mass (M_{BH}), and Eddington ratio, $n_{\rm Edd} = \mathrm{Log}(L_{\rm bol}/L_{\rm Edd})$, in different redshift bins. In the MgII sample (upper panels), SFR and M_{BH} increase with the redshift (up to $z \sim 2.5$) by a factor 3 and 100 respectively, while the Eddington ratio increases from 0.1 to 0.3. In the CIV sample the SFR and M_{BH} increase by a factor 10 from $z \sim 1.0$ to $z \sim 3.8$, while the mean Eddington ratio stays almost constant at its highest value (~0.3).

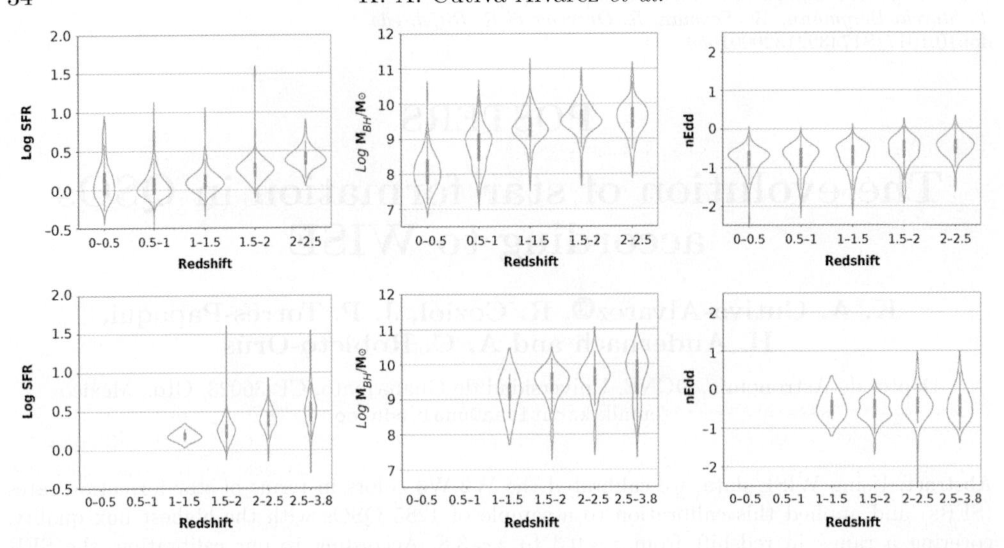

Figure 1. In the upper panel, the MgII sample was divided in 5 redshift bins, to highlight the evolution in SFR (left), BH mass (middle) and Eddington ratio (right). In the lower panel the same comparison is done for the CIV sample divided in 4 redshift bins. The core of the violin shows the mean (white dot), first and third quartiles (central vertical bar), while the body of the violin traces the (rotated kernel) density distribution.

3. Conclusions

The observation that the BH mass decreases at low redshift more rapidly than the SFR suggests that the timescales of the mass accretion by the BHs and the growth in mass of the galaxies through SF are not directly correlated. What is remarkable, in particular, is the high values of n_{Edd} at high redshifts. Since $n_{Edd} \propto L_{AGN}/M_{BH}$, the fact that this ratio stays high despite the increase in M_{BH} suggests the mass accretion onto the BH and the SF activity in the host galaxies are high at the same time. Taken at face value, this might imply a lack of influence of the BHs on their host galaxies (no BH feedback). Note however that our sample is biased towards QSOs with high quality fluxes in MIR, which is a small fraction of QSOs. Note also that there seem to be QSOs with unusually high SFRs at low redshift (the neck of the violin with lowest redshift bin in Fig. 1 left upper panel), and QSOs with unusually low BH masses and high SFRs at high redshifts (the end pin of the violin with highest redshift bin in Fig. 1 central lower panel). These cases could be consistent with narrow line QSOs, and the fact they have extreme SFRs and n_{Edd} might suggest they are examples of QSOs in an early phase of formation.

References

Baldwin, J. A., Phillips, M. M., & Terlevich, R. 1995, *PASP*, 93, 5
Coziol, R, Torres-Papaqui, J. P., Plauchu-Frayn, I. *et al.* 2014, *RMxAA*, 50, 255
Coziol, R., Torres-Papaqui, J. P., & Andernach, H. 2015, *AJ*, 142, 192
Delvecchio, I., Gruppioni, C., Pozzi, F., *et al.* 2014, *MNRAS*, 439, 2736
Donoso, E., Yan, L., Tsai, C., *et al.* 2012, *ApJ*, 145, 6
Jarrett, T. H., Masci, F., Tsai, C. W., *et al.* 2013, *AJ*, 145, 6
Kozłowski, S. 2017, *ApJS*, 228, 9
Leipski, C., Meisenheimer, K., Walter, F., *et al.* 2014, *ApJ*, 785, 154
Pâris, I., Petitjean, P., Ross, N. P., *et al.* 2017, *ApJ*, 597, A79
Veilleux, S. & Osterbrock, D. E. 1987, *ApJS*, 63, 295
Wright, E. L., Eisenhardt, P. R. M., Mainzer, A. K., *et al.* 2010, *AJ*, 140, 1868

Galaxy Evolution and Feedback across Different Environments
Proceedings IAU Symposium No. 359, 2020
T. Storchi-Bergmann, W. Forman, R. Overzier & R. Riffel, eds.
doi:10.1017/S1743921320002021

Feedback from central massive black holes in galaxies using cosmological simulations

Paramita Barai

Núcleo de Astrofísica (NAT) - Universidade Cruzeiro do Sul (UniCSul), Brazil
email: paramita.barai@iag.usp.br

Abstract. Gas accretion onto central supermassive black holes of active galaxies and resulting energy feedback, is an important component of galaxy evolution, whose details are still unknown especially at early cosmic epochs. We investigate BH growth and feedback in quasar-host galaxies at $z \geqslant 6$ by performing cosmological hydrodynamical simulations. We simulate the $2R_{200}$ region around a $2 \times 10^{12} M_{\odot}$ halo at $z = 6$, inside a (500 Mpc)^3 comoving volume, using the zoom-in technique. We find that BHs accrete gas at the Eddington rate over $z = 9$–6. At $z = 6$, our most-massive BH has grown to $M_{\rm BH} = 4 \times 10^9 M_{\odot}$. Star-formation is quenched over $z = 8$–6.

Keywords. galaxies: nuclei, quasars: supermassive black holes, cosmology: simulations

---

1. Introduction

Active galactic nuclei (AGN) emit enormous amounts of energy powered by the accretion of gas onto their central supermassive black holes (SMBHs) (Rees 1984). SMBHs of mass $\geqslant 10^9 M_{\odot}$ are observed to be in place in luminous quasars by $z \sim 6$, when the Universe was less than 1 Gyr old (Wu *et al.* 2015). It is difficult to understand how these early SMBHs formed over such short timescales (Matsumoto *et al.* 2015).

AGN feedback should operate mostly in the negative form quenching star formation (Schawinski *et al.* 2006). In the host galaxy of the quasar SDSS J1148+5251 at $z = 6.4$, Maiolino *et al.* (2012) detected broad wings of the [CII] line tracing a massive outflow with velocities up to ± 1300 km/s. The physical mechanisms by which quasar outflows affect their host galaxies remain as open questions.

We performed zoomed-in cosmological hydrodynamical simulations of quasar-host galaxies at $z \geqslant 6$ and study their outflows. Details of this work is in Barai *et al.* (2018).

2. Numerical method

The initial conditions are generated using the MUSIC† software (Hahn & Abel 2011). We use the code GADGET-3 (Springel 2005) to perform our zoom-in simulations. First, a dark-matter (DM) only low-resolution simulation is carried out of a (500 Mpc)^3 comoving volume, using 256^3 DM particles, from $z = 100$ up to $z = 6$. We select the most-massive halo at $z = 6$ to zoom-in, which has a total mass $M_{\rm halo} = 4.4 \times 10^{12} M_{\odot}$, and a virial radius $R_{200} \simeq 511$ kpc comoving. We execute a series of four simulations, all incorporating metal cooling, chemical enrichment, SF and SN feedback. The first run has no AGN included, while the latter three explore different AGN feedback models.

† MUSIC - Multi-scale Initial Conditions for Cosmological Simulations: https://bitbucket.org/ohahn/music

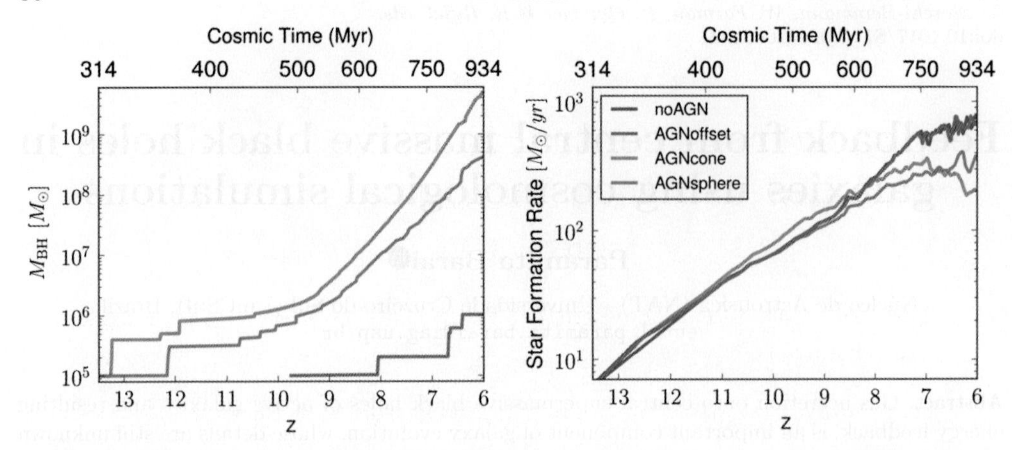

Figure 1. Left panel: BH mass growth with redshift of the most-massive SMBH in each run. Right panel: Sum total star formation rate (in $M_\odot yr^{-1}$) as a function of redshift. The different colours discriminate the various runs: *AGNoffset* - violet, *AGNcone* - red, *AGNsphere* - green.

3. Results and Discussion

3.1. *Black hole growth*

The redshift evolution of the most-massive SMBH mass in the three AGN runs is plotted in Fig. 1 - left panel. Each BH starts as a seed of $M_{\rm BH} = 10^5 M_\odot$, at $z \sim 14$ in the runs *AGNcone* and *AGNsphere* ($z \sim 10$ in *AGNoffset*). The dominant mode of BH growth occurs over $z = 9$–6 in runs *AGNcone* and *AGNsphere*, corresponding to Eddington-limited gas accretion where Eddington ratio $= 1$. The BH has grown to $M_{\rm BH} = 4 \times 10^9 M_\odot$ at $z = 6$ in run *AGNcone* (red curve). The BH grows 10 times more massive at $z = 6$ in the *AGNcone* case than in the *AGNsphere* run. This is because more gas can inflow along the perpendicular direction to the bi-cone, and accrete onto the BH.

3.2. *Star formation*

The star formation rate (total in the simulation box) versus redshift of the four simulations is displayed in Fig. 1 - right panel. The SFR rises with time in all the runs initially, and continues to increase in the *noAGN* case without a BH. The SFR in run *AGNoffset* is almost similar to that in the run *noAGN*, because the BHs are too small there to generate enough feedback. The models suppress SF substantially from $z \sim 8$ onwards, when the BHs have grown massive and generate larger feedback energy. Thus, we find that BHs need to grow to $M_{\rm BH} > 10^7 M_\odot$ in order to suppress star-formation. BH feedback causes a reduction of SFR up to 4 times at $z = 6$: from $800 M_\odot$/yr in the *noAGN* run, to $200 M_\odot$/yr in run *AGNcone*, and $350 M_\odot$/yr in run *AGNsphere*.

References

Barai, P., Gallerani, S., Pallottini, A., *et al.* 2018, *MNRAS*, 473, 4003
Hahn, O. & Abel, T. 2011, *MNRAS*, 415, 2101
Maiolino, R., *et al.* 2012, *MNRAS*, 425, L66
Matsumoto, T., Nakauchi, D., Ioka, K., *et al.* 2015, *ApJ*, 810, 64
Rees, M. J. 1984, *ARA&A*, 22, 471
Schawinski, K., *et al.* 2006, *Nature*, 442, 888
Springel, V. 2005, *MNRAS*, 364, 1105
Wu, X.-B., *et al.* 2015, *Nature*, 518, 512

Galaxy Evolution and Feedback across Different Environments
Proceedings IAU Symposium No. 359, 2020
T. Storchi-Bergmann, W. Forman, R. Overzier & R. Riffel, eds.
doi:10.1017/S1743921320001726

Substructure in black hole scaling diagrams and implications for the coevolution of black holes and galaxies

Benjamin L. Davis, Nandini Sahu and Alister W. Graham

Centre for Astrophysics and Supercomputing, Swinburne University of Technology, Hawthorn, VIC 3122, Australia
emails: `benjamindavis@swin.edu.au`, `nsahu@swin.edu.au`, `agraham@swin.edu.au`

Abstract. Our multi-component photometric decomposition of the largest galaxy sample to date with dynamically-measured black hole masses nearly doubles the number of such galaxies. We have discovered substantially modified scaling relations between the black hole mass and the host galaxy properties, including the spheroid (bulge) stellar mass, the total galaxy stellar mass, and the central stellar velocity dispersion. These refinements partly arose because we were able to explore the scaling relations for various sub-populations of galaxies built by different physical processes, as traced by the presence of a disk, early-type versus late-type galaxies, or a Sérsic versus core-Sérsic spheroid light profile. The new relations appear fundamentally linked with the evolutionary paths followed by galaxies, and they have ramifications for simulations and formation theories involving both quenching and accretion.

Keywords. black hole physics — galaxies: bulges — galaxies: elliptical and lenticular, cD — galaxies: evolution — galaxies: kinematics and dynamics — galaxies: spiral — galaxies: structure

---

We have identified 145 galaxies with directly-measured supermassive black hole (SMBH) masses obtained from stellar dynamics, gas dynamics, kinematics of mega-masers, proper motions (Sgr A*), or recent direct-imaging techniques (M87*). This sample comprises 96 early-type galaxies (ETGs) and 49 late-type galaxies (LTGs). 2D photometric models were generated for the galaxies using ISOFIT (Ciambur 2015) and their associated surface brightness profiles were modelled using PROFILER (Ciambur 2016). Their multi-component decompositions and stellar masses are presented in Davis *et al.* (2018, 2019) and Sahu *et al.* (2019a).

Sahu *et al.* (2019a) reported $M_{\rm BH} \propto M_{*,\rm sph}^{1.27\pm0.07}$ with a total scatter of $\Delta_{\rm rms|BH} = 0.52$ dex. Importantly, however, they discovered that the ES/S0-type galaxies with disks are offset from the E-type galaxies by more than a factor of ten (1.12 dex) in their $M_{\rm BH}/M_{*,\rm sph}$ ratios. Separately, each population follows a steeper relation with slopes of 1.86 ± 0.20 and 1.90 ± 0.20, respectively. The offset mass ratio is mainly due to the exclusion of each galaxy's disk mass, with the two populations offset by only a factor of two in their mean $M_{\rm BH}/M_{*,\rm gal}$ ratios and in the $M_{\rm BH}$–$M_{*,\rm gal}$ diagram (Figure 1).

By combining their data with LTGs from Davis *et al.* (2018, 2019); Sahu *et al.* (2019a) further showed a striking morphological distinction in the black hole scaling relations that separately govern LTGs and ETGs (Figure 2). Concerning the $M_{\rm BH}$–$M_{*,\rm gal}$ relation, LTGs follow scaling correlations with slopes approximately twice that of ETGs. In all cases, black holes and their host galaxies *do not* grow in lockstep; their coevolution is non-linear with scaling relation slopes greater than one. These varied growth mechanisms among different morphological types have consequences for galaxy/black hole formation

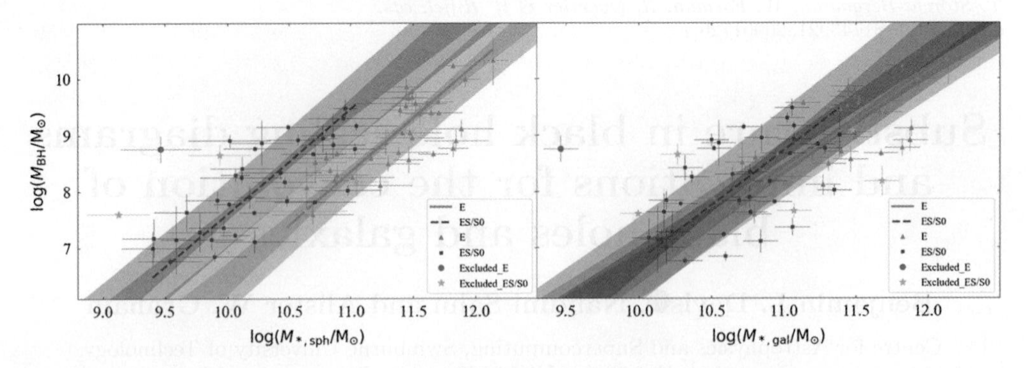

Figure 1. ETGs with (ES/S0) and without (E) disks. $M_{\rm BH}$–$M_{*,\rm sph}$ (left) – the relation for galaxies with disks is offset from the relation for galaxies without disks, revealing two different scaling relations for the two sub-morphological types (ES/S0 and E), with $\Delta_{\rm rms|BH}$ of 0.57 dex and 0.50 dex, respectively. $M_{\rm BH}$–$M_{*,\rm gal}$ (right) – both relations are consistent with each other, suggesting a single relation for galaxies with and without disks.

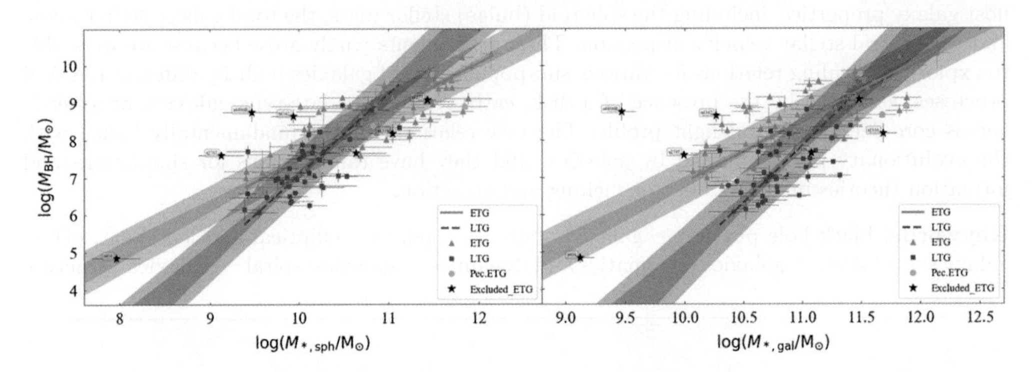

Figure 2. The $M_{\rm BH}$–$M_{*,\rm sph}$ and $M_{\rm BH}$–$M_{*,\rm gal}$ relations for ETGs and LTGs. $M_{\rm BH}$–$M_{*,\rm sph}$ (left) – $M_{\rm BH} \propto M_{*,\rm sph}^{1.27\pm0.07}$ for all ETGs combined and $M_{\rm BH} \propto M_{*,\rm sph}^{2.17\pm0.32}$ for LTGs. $M_{\rm BH}$–$M_{*,\rm gal}$ (right) – $M_{\rm BH} \propto M_{*,\rm gal}^{1.65\pm0.11}$ for ETGs and $M_{\rm BH} \propto M_{*,\rm gal}^{3.05\pm0.70}$ for LTGs.

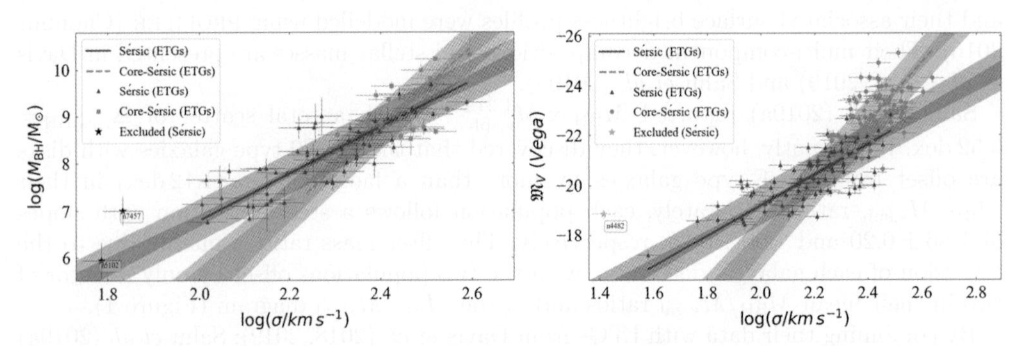

Figure 3. Left – $M_{\rm BH}$–σ diagram for Sérsic and core-Sérsic ETGs, with $M_{\rm BH} \propto \sigma^{5.75\pm0.34}$ and $M_{\rm BH} \propto \sigma^{8.64\pm1.10}$, respectively. Right – diagram of V-band absolute magnitude ($\mathfrak{M}_V$, Vega) vs. central velocity dispersion (σ) for Sérsic and core-Sérsic ETGs, with $L_V \propto \sigma^{2.44\pm0.18}$ and $L_V \propto \sigma^{4.86\pm0.54}$ for Sérsic and core-Sérsic ETGs, respectively.

theories, simulations, and predicting black hole masses. Sahu *et al.* (2019b) further found that Sérsic and core-Sérsic galaxies define two distinct relations in the $M_{\rm BH}$–σ diagram. They also reported how this yielded a consistency with the slopes and bends in the galaxy luminosity (L)–σ relation, due to Sérsic and core-Sérsic ETGs (Figure 3).

References

Ciambur, B. C. 2015, *ApJ*, 810, 120
Ciambur, B. C. 2016, *PASA*, 33, e062
Davis, B. L., Graham, A. W., & Cameron, E. 2018, *ApJ*, 869, 113
Davis, B. L., Graham, A. W., & Cameron, E. 2019, *ApJ*, 873, 85
Sahu, N., Graham, A. W., & Davis, B. L. 2019a, *ApJ*, 876, 155
Sahu, N., Graham, A. W., & Davis, B. L. 2019b, *ApJ*, 887, 10

Galaxy Evolution and Feedback across Different Environments
Proceedings IAU Symposium No. 359, 2020
T. Storchi-Bergmann, W. Forman, R. Overzier & R. Riffel, eds.
doi:10.1017/S1743921320001829

Classification and photometric redshift estimation of quasars in photometric surveys

L. M. Izuti Nakazono[1], C. Mendes de Oliveira[1], N. S. T. Hirata[2], S. Jeram[3], A. Gonzalez[3], S. Eikenberry[3], C. Queiroz[4], R. Abramo[4] and R. Overzier[5]

[1]Instituto de Astronomia, Geofísica e Ciências Atmosféricas da U. de São Paulo, Cidade Universitária, 05508-900, São Paulo, SP, Brazil

[2]Departamento de Ciência da Computação, Instituto de Matemática e Estatística da USP, Cidade Universitária, 05508-090, São Paulo, SP, Brazil

[3]Department of Astronomy, University of Florida, 211 Bryant Space Center, Gainesville, FL 32611, USA

[4]Departamento de Física Matemática, Instituto de Física, Universidade de São Paulo, SP, Rua do Matão 1371, São Paulo, Brazil

[5]Observatório Nacional / MCTIC, Rua General José Cristino 77, Rio de Janeiro, RJ, 20921-400, Brazil

Abstract. We present a machine learning methodology to separate quasars from galaxies and stars using data from S-PLUS in the Stripe-82 region. In terms of quasar classification, we achieved 95.49% for precision and 95.26% for recall using a Random Forest algorithm. For photometric redshift estimation, we obtained a precision of 6% using k-Nearest Neighbour.

Keywords. methods: statistical, catalogs, quasars: general

---

1. Introduction

Several techniques have been developed to classify extended and point-like sources on astronomical images. However, the classification of stars and quasars remains a challenge, as both are unresolved sources. Some techniques have been applied to perform star/quasar separation on photometric data, such as: colour-colour cuts (e.g. Pâris *et al.* 2018); Bayesian Statistics (e.g. Yang *et al.* 2017), and Machine Learning algorithms (e.g. Jin *et al.* 2019). In this work we use the data from 336 deg^2 of The Southern Photometric Local Universe Survey (S-PLUS) first data release (DR1) to create a new quasar catalog. This survey will cover 9300 deg^2 with the Javalambre optical filter system consisting of 7 narrow bands and 5 bands similar to the Sloan Digital Sky Survey (SDSS) broad-bands.

2. Methodology

The quasar sample from S-PLUS DR1 has 13 683 sources with near-infrared counterpart from Wide-field Infrared Survey (WISE) and their spectra were obtained from SDSS DR14 (Pâris *et al.* 2018). We retrieved from SDSS DR15 a total of 52 914 stars and 84 723 galaxies with infrared counterpart from AllWISE catalog. With these samples, we trained a Random Forest algorithm that classifies any S-PLUS observation in quasar, star, or galaxy. The performance of this classifier was evaluated through 4-Fold cross-validation in terms of precision/purity, recall/completeness and F-measure (harmonic mean of precision and recall). We tested the following feature spaces: (i) 12 S-PLUS bands;

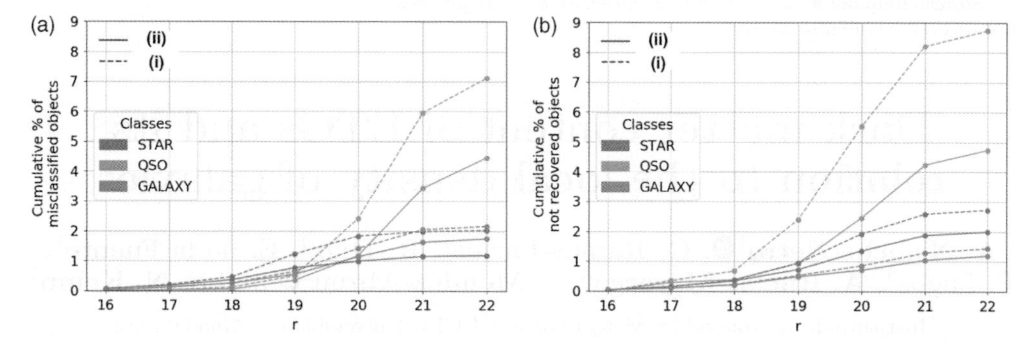

Figure 1. Cumulative percentages of (a) misclassified and (b) not recovered objects per magnitude and per class. The models referring to the feature spaces (ii) and (i) are represented by a solid and dashed curve, respectively.

(ii) 12 S-PLUS bands + 2 WISE bands; (iii) 5 S-PLUS broad-bands and (iv) 5 S-PLUS broad-bands + 2 WISE bands. We also included four morphological parameters: full width at half maximum (FWHM), major semi-axis, minor semi-axis and Kron radius. For the redshift estimation, we trained a k-Nearest Neighbours (kNN) algorithm. The performance of this model was evaluated through Holdout cross-validation in terms of normalized median absolute deviation (σ_{NMAD}). We considered the same feature spaces (i), (ii), (iii) and (iv) without the morphological parameters and with addition of 4 more spaces based on colors rather than magnitudes.

3. Results

For the classification problem, the best performances came from the feature spaces (ii) and (i). We achieved a macro-averaged F-measure of 97.44% (95.95%), with WISE (without WISE). In terms of quasar classification, we achieved 95.49% (92.83%) for precision and 95.26% (91.23%) for recall (Fig. 1). In terms of star classification, we achieved a precision of 98.82% (97.98%) and recall of 98% (97.26%). Finally for galaxies, a precision of 98.26% (97.83%) and a recall of 98.8% (98.56%) were obtained. The best precision for the photometric redshift estimation achieved is $\sigma_z = 6.56\%$ considering $k = 5$ when training kNN within the feature space 12 S-PLUS bands + 2 WISE bands with colors. The narrow-bands improved the star/quasar/galaxy classification, especially when no WISE information is available. We conclude that the S-PLUS optical system is a powerful tool for finding new quasars and for estimating their photometric redshifts.

References

Moore, J. A., Pimbblet, K. A., & Drinkwater, M. J. 2006, *PASA*, 23, 135
Pâris, I., Petitjean, P. Aubourg, É., *et al.* 2018, *A&A*, 613, A51
Heintz, K. E., Fynbo, J. P. U., Høg, E., *et al.* 2018, *A&A*, 615, L8
Yang, Q., Wu, X., Fan, X., *et al.* 2017, *AJ*, 154, 269
Jin, X., Zhang, Y., Zhang, J., *et al.* 2019, *MNRAS*, 485, 4539
Mendes de Oliveira, C., Ribeiro, T. , Schoenell, W., *et al.* 2019, *MNRAS*, 489, 241

Galaxy Evolution and Feedback across Different Environments
Proceedings IAU Symposium No. 359, 2020
T. Storchi-Bergmann, W. Forman, R. Overzier & R. Riffel, eds.
doi:10.1017/S1743921320004007

Dark matter content of ETGs and its relation to the local density of galaxies

A. Nigoche-Netro[1], G. Ramos-Larios[1], R. Díaz[4], E. de la Fuente[1], P. Lagos[5], A. Ruelas-Mayorga[3], J. Mendez-Abreu[2] and S. N. Kemp[1]

[1]Instituto de Astronomía y Meteorología, CUCEI, Universidad de Guadalajara, Guadalajara, Jal. 44130, México
email: anigoche@gmail.com

[2]Instituto de Astrofísica de Canarias, Calle Vía Láctea s/n, E-38205 La Laguna, Tenerife, Spain

[3]Instituto de Astronomía, Universidad Nacional Autónoma de México, Cd. Universitaria, México, D.F. 04510, México

[4]Gemini Observatory, 950 N Cherry Ave, Tucson AZ, USA

[5]Instituto de Astrofísica e Ciências do Espaço, Universidade do Porto, CAUP, Rua das Estrelas, 4150-762 Porto, Portugal

Abstract. We study the behaviour of the dynamical and stellar mass inside the effective radius as function of local density for early-type galaxies (ETGs). We use several samples of ETGs - ranging from 19000 to 98000 objects - from the ninth data release of the Sloan Digital Sky Survey. We consider Newtonian dynamics, different light profiles and different initial mass functions (IMF) to calculate the dynamical and stellar mass. We assume that any difference between these two masses is due to dark matter and/or a non-universal IMF. The main results are: (i) the amount of dark matter (DM) inside ETGs depends on the environment; (ii) ETGs in low-density environments span a wider DM range than ETGs in dense environments; (iii) the amount of DM inside ETGs in the most dense environments will be less than approximately 55-65 per cent of the dynamical mass; (iv) the accurate value of this upper limit depends on the impact of the IMF on the stellar mass estimation.

Keywords. Galaxies: fundamental parameters. Galaxies: photometry, distances and redshifts. Cosmology: dark matter

1. The sample of ETGs

We use a sample of approximately 98000 ETGs from SDSS-DR9 (York *et al.* 2000) in the g and r filters. The galaxies are distributed over a redshift interval $0.0024 < z < 0.3500$. This sample will be called hereafter, the "Total–SDSS–Sample". The selection criteria are the same we used in earlier papers (Hyde & Bernardi 2009a; Nigoche-Netro *et al.* 2010).

Selecting only ETGs from the morphological classification of the Zoospec catalogue (Lintott *et al.* 2008) and considering our selection criteria the sample is reduced to approximately 27000 ETGs. The galaxies with this added criterion have a higher probability of being ETGs. This sample shall be referred to as "Morphological–SDSS–Sample". In addition, if we want to control possible streaming motions, redshift bias, and evolutionary effects, we have to compile a relatively nearby and volume-limited sample. The redshift range $0.04 \leqslant z \leqslant 0.08$ corresponds to a volume that fits these characteristics (Nigoche-Netro *et al.* 2008, 2009). The resulting sample contains approximately

19 000 ETGs. This sample is approximately complete for $\log(\mathbf{M_{Virial}}/\mathbf{M_{Sun}}) > 10.5$ (Nigoche-Netro *et al.* 2010, 2011). We shall refer to it as the "Homogeneous–SDSS–Sample".

The photometry and spectroscopy of the galaxy samples drawn from the DR9 have been corrected for different biases and are the same that we have used in previous papers (Nigoche-Netro *et al.* 2015) for full details.

2. The stellar and virial mass of the ETGs

The stellar mass. The total stellar or luminous mass was obtained considering different stellar population synthesis models, using a universal IMF (Salpeter or Kroupa) and different brightness profiles (de Vaucouleurs or Sérsic). The combination of these ingredients results in three mass estimations, as follows (Nigoche-Netro *et al.* 2015):

- de Vaucouleurs Salpeter-IMF stellar mass,
- Sérsic Salpeter-IMF stellar mass,
- Kroupa-IMF stellar mass.

The virial mass. The total virial or dynamical mass was obtained using an equation from Poveda (1958). This method assumes Newtonian mechanics and virial equilibrium for the galaxies in question. The equation is as follows:

$$\mathbf{M_{Virial}} \sim K(n)\frac{r_e \sigma_e^2}{G}, \tag{2.1}$$

where $\mathbf{M_{Virial}}$, r_e and, σ_e represent the total virial mass, the effective radius and the velocity dispersion inside r_e, respectively. G is the gravitational constant and $K(n)$ is a scale factor that depends on the Sérsic *index* (n) as follows Cappellari *et al.* (2006):

$$K(n) = 8.87 - 0.831n + 0.0241n^2, \tag{2.2}$$

The amount of mass within an effective radius corresponds to 0.42 times the total mass previously calculated. This mass may or may not be luminous.

3. Density of galaxies

The projected density of galaxies (Σ_N) was computed following the method described in Aguerri *et al.* (2009). They used the projected co-moving distance to the Nth nearest neighbour (d_N) of the target galaxy as follows:

$$\Sigma_N \sim \frac{N}{\pi(d_N)^2}, \tag{3.1}$$

The local density was calculated using the third, fifth, eighth, and tenth nearest neighbours, for both the spectroscopic and photometric samples. In this work we use the local density considering the tenth nearest neighbours to avoid biases due to completeness of the SDSS survey.

4. Distribution of the difference between dynamical and stellar mass as function of local density of galaxies

From Fig. 1, we can see that the distribution of galaxies in the $\log(\mathbf{M_{Virial}}/\mathbf{M_{Sun}})$ – $\log(\mathbf{M_{Star}}/\mathbf{M_{Sun}})$ - density plane is similar for all samples, indicating that our results are not dependent on which galaxy sample is used or which galaxy is considered the nearest neighbour. We can see that the distribution is not random but rather has a bell-shape and that galaxies in the lowest density region cover the whole range of difference between virial and stellar mass. This range of the difference between masses decreases while the density increases. In all cases, we can see that the maximum of the density distribution is at about 0.35-0.45 in $\log(\mathbf{M_{Virial}}/\mathbf{M_{Sun}})$ – $\log(\mathbf{M_{Star}}/\mathbf{M_{Sun}})$. In linear values we find

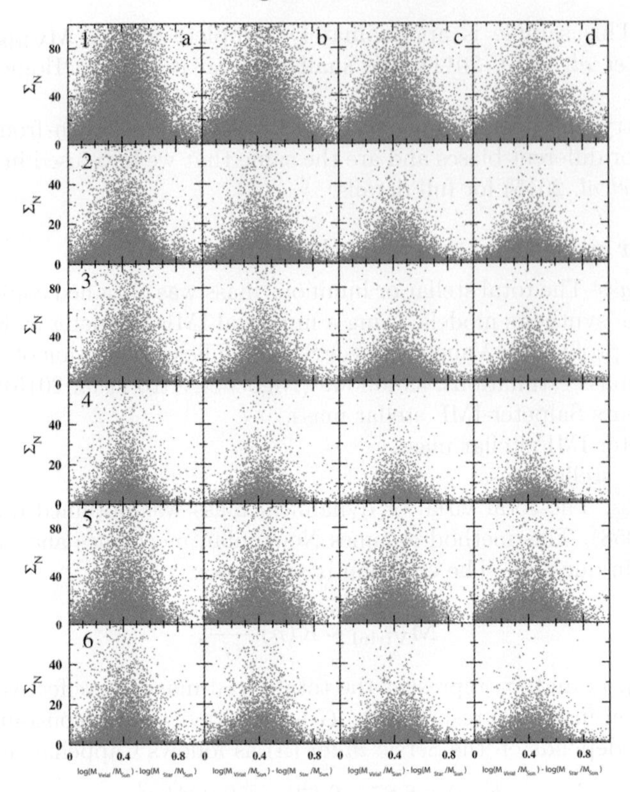

Figure 1. Distribution of the logarithmic difference between dynamical and stellar mass inside r_e as function of density of galaxies for different samples of ETGs considering a Kroupa-IMF stellar mass. Rows 1-2, 3-4 and 5-6 correspond to the total, morphological, and homogeneous SDSS samples, respectively. Rows 1, 3, and 5 are data from the photometric sample. Rows 2, 4 and 6 are data from the spectroscopic sample. Columns a, b, c, d are the data considering the third, fifth, eight, and tenth nearest neighbours respectively.

that the maximum of the density distribution has a difference between masses of about 55-65 per cent of the virial mass.

The previously discussed results were found using the Kroupa–IMF stellar mass, however we have found similar results for the de Vaucouleurs Salpeter–IMF and Sérsic Salpeter–IMF samples. There are two differences between the Kroupa and Salpeter IMF samples that do not change our general findings. The first one is that the average DM dispersion of the Sérsic-Salpeter-IMF samples ($\sigma_M \sim 0.12$) is less than the dispersion of the Kroupa IMF ($\sigma_M \sim 0.16$) and the de Vaucouleurs–Salpeter IMF ($\sigma_M \sim 0.16$) samples. This finding agrees with the one obtained by (Taylor *et al.* 2010) that find that the dynamical and stellar masses correlate best when the structure of the galaxy is taken into account. The second difference is that there is a small shift (approximately 10%) towards higher amounts of dark matter for the Salpeter IMF samples with respect to the Kroupa samples.

References

Aguerri, J. A. L., Méndez-Abreu, J., & Corsini, E. M., 2009, *A&A*, 495, 491
Cappellari, M., Bacon, R., Bureau, R., *et al.* 2006, *MNRAS*, 366, 1126
Hyde, J. B. & Bernardi, M. 2009, *MNRAS*, 394, 1978
Lintott, C. J., Schawinski, K., Slosar, A., *et al.* 2008, *MNRAS*, 389, 1179
Nigoche-Netro, A., Ruelas-Mayorga, A., & Franco-Balderas, A. 2008, *A&A*, 491, 731

Nigoche-Netro, A., Ruelas-Mayorga, A., & Franco-Balderas, A. 2009, *MNRAS*, 392, 1060
Nigoche-Netro, A., Aguerri, J. A. L., Lagos, P., *et al.* 2010, *A&A*, 516, 96
Nigoche-Netro, A., Aguerri, J. A. L., Lagos, P., *et al.* 2011, *A&A*, 534, 61
Nigoche-Netro, A., Ruelas-Mayorga, A., Lagos, P., *et al.* 2015, *MNRAS*, 446, 85
Poveda, A. 1958, *BOTT*, 2, 3 1
Taylor, E. N., Franx, M., Brinchmann, J., *et al.* 2010, *ApJ*, 722, 1
York, D. G., *et al.* 2000, *AJ*, 120, 1579

Galaxy Evolution and Feedback across Different Environments
Proceedings IAU Symposium No. 359, 2020
T. Storchi-Bergmann, W. Forman, R. Overzier & R. Riffel, eds.
doi:10.1017/S1743921320001878

Cosmological forecasts from photometric measurements of the angular correlation function for the Legacy Survey of Space and Time

Diogo H. F. de Souza[1,2] and **Basílio X. Santiago**[1,2]

[1]Instituto de Física, UFRGS, Caixa Postal 15051, Porto Alegre, RS - 91501-970, Brazil

[2]Laboratório Interinstitucional de e-Astronomia - LIneA, Rua Gal. José Cristino 77, Rio de Janeiro, RJ - 20921-400, Brazil

emails: diogo.henrique@ufrgs.br, basilio.santiago@ufrgs.br

Abstract. We aim to do forecasts for the Legacy Survey of Space and Time (LSST) with a theoretical modeling of the two point angular correlation function. The Fisher matrix is the starting point. This is a square matrix over the cosmological parameters, whose diagonal contains direct informations on the parameters expected uncertainties.

Keywords. Cosmology, large-scale structure of universe, cosmological parameters, LSST

1. Introduction

The LSST is located on the Cerro Pachón ridge in north-central Chile (Abell *et al.* (2009)). The LSST is expected to begin the operation in 2022. So one natural question is, can we extract information even before the release of the data? We aim to answer this question with a general formalism. We use the angular correlation function (ACF) to study forecasts for the errors of cosmological parameters obtained with LSST, the next future large-scale photometric survey. We intend to apply this study to the dark energy equation parameter w, cold dark matter density Ω_{cdm}, dark energy density Ω_w, baryonic matter density Ω_b, and the fluctuation amplitude at scale of $8h^{-1}$Mpc, σ_8.

2. Theoretical framework

The theoretical framework described here, in general, is the same for all photometric surveys, so the approach that we follows is very close to that described in Sobreira *et al.* (2011) that did forecasts for the Dark Energy Suvery (DES).

The starting point is the Fisher matrix $F_{\alpha\beta}$ described by

$$F_{\alpha\beta} = \frac{\partial w^i(\theta^n, p)}{\partial p_\alpha} \left[C^{-1}\right]^{ij}_{nm} \frac{\partial w^j(\theta^m, p)}{\partial p_\beta} + \frac{1}{2}\mathrm{Tr}\left[C^{-1}\frac{\partial C}{\partial p_\alpha}C^{-1}\frac{\partial C^{-1}}{\partial p_\beta}\right]. \tag{2.1}$$

In Eq. (2.1), $w^i(\theta^n, p)$ is the ACF evaluated at the i-th bin of redshift and in n-th angular bin, θ^n, and for a given particular set of cosmological parameters $\{p\}$. In general, the ACF is described by Eq. (2.2)

$$w(\theta) = \int_0^\infty dz_1 f(z_1) \int_0^\infty dz_2 f(z_2) \xi^s(r(z_1, z_2, \theta)), \tag{2.2}$$

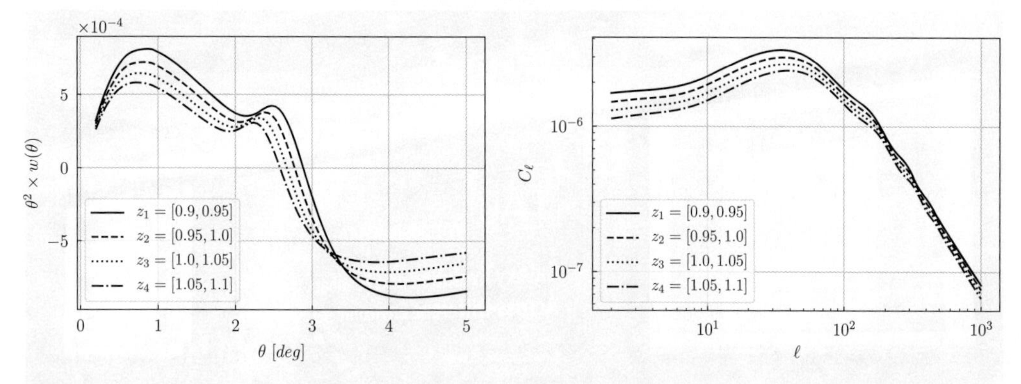

Figure 1. Analysis of ACF and APS, for the LSST, with redshift z ranging in the interval $[0.9, 1.1]$ with bin width $\Delta z = 0.05$, therefore $z = \cup_{i=1}^{4} z_i$.

where $f(z) = \phi(z)b(z)D(z)$ is a function of the photometric selection function, $\phi(z)$, the bias, $b(z)$, between galaxies and Dark Matter, and the growth function, $D(z)$. Finally, ξ^s, is the correlation function in redshift space.

The $C := \frac{2}{(4\pi)^2 f_{sky}} \sum_\ell [(2\ell+1)\mathcal{P}_\ell(cos\theta_n)\mathcal{P}_\ell(cos\theta_m)(C_\ell^{ij} + \frac{\delta_{ij}}{n_i})^2]$ is the covariance matrix, where C_ℓ^{ij} is the angular power spectrum (APS) given by

$$C_\ell^{ij} = \frac{2}{\pi} \int dk\, k^2 P(k) \Psi_\ell^i(k) \Psi_\ell^j(k). \tag{2.3}$$

$P(k)$ is the matter power spectrum and $\Psi_\ell^i(k)$ are functions, computed in the i-th redshift bin, that encode information about the survey and cosmology.

The information about the survey ($\Psi_\ell(k)$) includes the photometric selection function, the photometric redshift error $\mathrm{P}(z_p; z)$ and the galaxy redshift distributions $n(z)$. These two latter are given by (2.4)

$$\mathrm{P}(z_p; z) \propto \exp\left[-\frac{(z_p - z - \delta_z)^2}{2\sigma_z^2}\right], \; n(z) \propto z^\alpha \exp\left[-\left(\frac{z}{z^*}\right)^\beta\right]. \tag{2.4}$$

3. Modeling for LSST

To compute some of the equations presented here, we use specific existing code. The Code for Anisotropies in the Microwave Background (`CAMB`) is used to generate the $P(k)$ and other fundamental background functions, like Hubble parameter $H(z)$ and comoving distance χ. To generate ACF and APS, we use the Code Cosmology Library (`CCL`). Finally, we combine all these codes using the `Python` environment to automate the computations. Figure 1 shows the ACF and the APS for the LSST calculated for the $z = [0.9, 1.1]$. Currently we are applying this same method to obtain the complete covariance matrix, APS and ACF to the full redshift interval of interest that we are studying. This allows us to obtain the C^{-1} and the $\partial w / \partial p_\alpha$ to finally, by Eq (2.1), obtain the Fisher matrix.

References

Abell, P. A., Burke, D. L., Hamuy, M., *et al.* 2009, *Lsst science book*, version 2.0 (No. SLAC-R-1031)

Sobreira, F., de Simoni, F., Rosenfeld, R., *et al.* 2011, *Physical Review D*, 84, 103001

Karin Menendez-Delmestre

Karla Cutiva

Session 2: Cosmic noon

Galaxy Evolution and Feedback across Different Environments
Proceedings IAU Symposium No. 359, 2020
T. Storchi-Bergmann, W. Forman, R. Overzier & R. Riffel, eds.
doi:10.1017/S174392132000232X

INVITED LECTURE

The role of AGN feedback in the baryon cycle at z ∼ 2

Vincenzo Mainieri

European Southern Observatory, Karl-Schwarzschild-Strasse 2,
Garching bei München, Germany

Abstract. In this proceeding I will summarize our on-going observational campaign to characterize Active Galactic Nuclei (AGN) driven ionized gas outflows at z ∼ 2 and assess their impact on galaxy evolution. The results are mostly derived from a recently completed SINFONI/VLT Large Programme named SUPER, conducted with Adaptive Optics to reach a spatial resolution of ∼2 kpc at z ∼ 2.

Keywords. galaxies: active, galaxies: ISM, galaxies: kinematics and dynamics, galaxies: nuclei, (galaxies:) quasars: emission lines

1. Introduction

The study of outflows in the host galaxies of active galactic nuclei (AGN) have moved to the forefront of extra-galactic astronomy in recent years, since they are largely invoked on theoretical works to regulate the growth of galaxies (e.g. King & Pounds 2015 for a review). A crucial cosmic epoch is at $1 < z < 3$, corresponding to the peak of volume-averaged star formation and supermassive black hole (SMBH) accretion in the Universe (e.g., Madau & Dickinson 2014), when the energy injected by the central engine onto the host galaxy is expected to be at its maximum. Indeed, many observational studies have revealed that AGN-driven outflows are common at these redshifts (e.g. Brusa *et al.* 2015; Harrison *et al.* 2016; Kakkad *et al.* 2106; Leung *et al.* 2019; Davies *et al.* 2020). While the presence of these outflows out to galactic (kpc) scales is now undisputed, their impact on the gas content and kinematics of the host galaxy is highly debated. This is directly linked to the fact that having firm estimates for the mass of the outflowing gas (M_{out}) and of the energy associated with these outflows represent a major observational challenge. Determining the physical properties of the outflows requires, especially at high redshift, to make assumption on the outflow geometry, the filling factor of the outflowing gas, and the physical gas properties (e.g. density, metallicity, temperature).

Another highly debated topic in the literature is the physical mechanism responsible for generating and driving such outflows on kpc scales. Popular AGN-driven outflow models predict that the interaction of a nuclear wind with the Inter-Stellar Medium (ISM) will generate a reverse shock (Zubovas & King 2012; King & Pounds 2015). In the so-called momentum driven case, the shock cools efficiently and the velocity will decrease constantly with radius. In the energy-driven case, the energy injected by the nuclear wind is conserved and the outflow velocity remains high out to large radii.

It will be therefore upmost important to be able to have observations of outflows at z ∼ 2 with enough quality (e.g. spatial resolution, S/N) to minimize the assumptions and therefore the uncertainties on the physical quantities derived, as well as being able

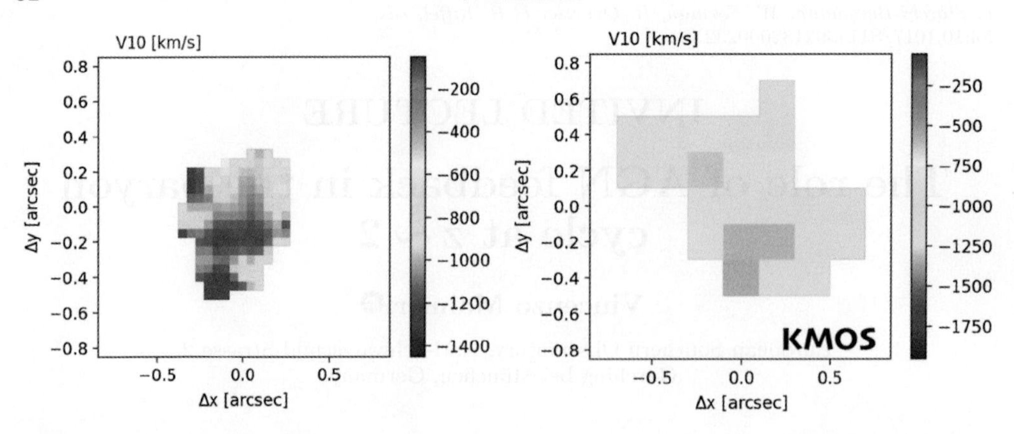

Figure 1. We show the velocity at the 10th percentile of the overall [OIII] line profile (v_{10}) maps for one of the objects in the SUPER survey obtained with SINFONI (*left*) with AO-assisted observations and KMOS (*right*) seeing limited observations in the H band. As it is clear the finer spatial resolution achievable with AO-assisted observations is crucial to have a more detailed characterization of the velocity structures and outflow geometry of the ionized gas, which allow to decrease the uncertainties in the derived quantities, as the mass outflow rate.

to trace the AGN-driven outflows from the proximity of the central SMBH out to the galaxy and possibly Circum-Galactic Medium (CGM) scales to constrain the models. In this proceeding, I will briefly summarize our observational effort to tackle this challenge.

2. The SUPER survey

SUPER† (PI: Mainieri - 196.A-0377) was a Large Programme at the ESO's Very Large Telescope (VLT). The survey has been allocated 280 hours of observing time in AO-assisted mode with the aim of providing high-resolution, spatially-resolved IFS observations of multiple emission lines for a carefully-selected sample of 39 X-ray AGN at z$\sim$2 (Circosta *et al.* 2018). The AO correction is performed in Laser Guide Star-Seeing Enhancer (LGS-SE) mode, which has demonstrated the capability to achieve a point spread function (PSF) full width at half maximum (FWHM) of ~0.2−0.3 arcsec under typical weather conditions in Paranal. This is a key feature of this survey that allows to resolve the kinematics of the ionized gas to a finer spatial scale than seeing-limited observations, and consequently decrease significantly the uncertainties on the derived physical properties of the detected outflows (see Fig. 1). The sample covers the redshift range of z = 2.1−2.5, which is the epoch of maximal activity of the volume averaged star formation in galaxies and the growth of black holes in the universe, making it ideal to study effects of radiative feedback from the black hole on the host galaxy. Thanks to the rich ancillary multi-wavelength data sets available, we were able to derive accurate measurements of the black hole and the host galaxy properties via spectral energy distribution fitting of UV-to-FIR photometry and X-ray spectral fitting (Circosta *et al.* 2018; Vietri *et al.* 2018). Of the 39 targets, 22 are classified as Type-1 (56%) and the remaining 17 as Type-2 (44%), based on the presence or absence of broad emission lines such as MgII or CIV in the rest-frame UV spectra. The overall SUPER sample span a wide range in AGN and host galaxy properties which allows us to identify any existing correlation between the outflow properties derived from the SINFONI data and those derived from the multi-wavelength ancillary data set (Circosta *et al.* 2018).

† http://www.super-survey.org

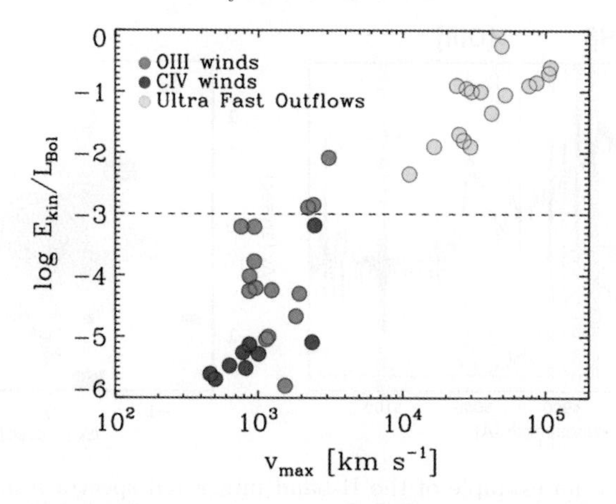

Figure 2. Kinetic coupling efficies obtained for the BLR winds (blue circles), NLR winds (red circles) for the SUPER targets (Vietri *et al.* 2020). We also show a collection of UFOs from Fiore *et al.* (2017). The horizontal dashed line shows the median kinetic coupling efficiency calculated in the simulations presented by Costa *et al.* (2018) (see also Harrison *et al.* 2018).

3. BLR winds

High ionization lines originating from the Broad Line Region (BLR) such as CIV, usually exhibit a shift of the peak to the blue, associated with gas in a non-virial motion (Gaskell 1982; Richards *et al.* 2011; Mejia-Restrepo *et al.* 2018; Vietri *et al.* 2018), which leads to biased estimation of the BH mass. Indeed, this line is known to be dominated by non virialized motions, making the profile asymmetric towards the blue-side of the line. We have therefore the remarkable opportunity to trace the motion of the ionized gas in the BLR (e.g. Vietri *et al.* 2018). In SUPER, we have used the broad Hα and Hβ lines from the SINFONI observations to estimate the radius of the BLR from the BLR radius-luminosity relation (Bentz *et al.* 2009). The inferred radius for the BLR for our AGN at z$\sim$2 is R$\sim$0.1-1 pc. We have measured the velocity shift of the CIV with respect to the laboratory wavelength, which is defined as v_{50}^{CIV}= (λ_{half}-1549.48)*c/1549.48, where λ_{half} is the wavelength that bisects the cumulative total line flux, and c the speed of light. We detected outflow velocities in the BLR up to 5000 km/s (Vietri *et al.* 2020). From those and the radius of the BLR we estimated the kinetic energy associated with these winds. In an attempt to compute the efficiency with which the AGN energy is transfered to the surrounding medium we derived the kinetic coupling efficiecy as the ratio between the kinetic energy and the AGN bolometric luminosity derived from the SED. We show in Fig. 2 the kinetic energy for the BLR and NLR winds from our survey, and a compilation of values for the Ultra-Fast Outflows (UFOs) from Fiore *et al.* (2017). We note that the lower coupling efficiency of BLR winds compared to the UFOs could be due to the combined effect of lower column densisty (N_H) and lower covering factor of the CIV outflowing gas. As discussed also in Harrison *et al.* (2018) the observed values of kinetic coupling efficiency for ionized outflows are lower than the energy injecition rate predicted by simulations, but this should be expected since a portion of the nuclear energy will be used to e.g. work against the gravitational potential.

4. NLR winds

We now move to trace the kinematics of the ionized gas on kpc scales using the [OIII]λ5007 line sampled with AO-assisted SINFONI observations in the H band (Kakkad *et al.* 2020). As it is detailed in the Kakkad *et al.* contribution to these proceedings,

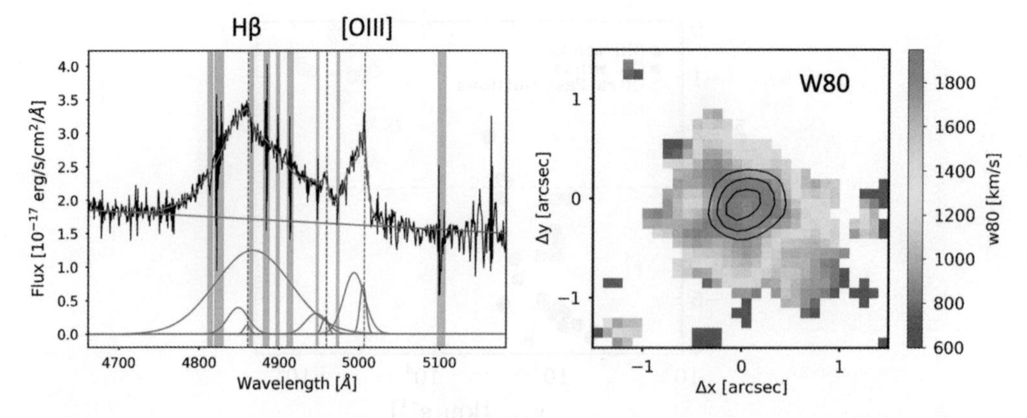

Figure 3. *Left panel:* an example of the H-band integrated spectra from the SUPER survey. The black curve shows the observed spectrum, the red curve shows the best model to reproduce the overall emission, the blue dashed curve shows the iron emission and the green curves show the continuum emission and the individual Gaussian components (Narrow, Broad and BLR) used to reproduce the profiles of various emission lines. The vertical grey regions mark the channels with strong sky lines which were masked during the fitting procedure.*Right panel:* an example of the velocity width (w_{80}) map.

we first verified that the [OIII] emission was actually extended taking into account beam smearing effects (Husemann *et al.* 2016; Villar-Martin *et al.* 2016). We used two techniques to asses if the emission is truly extended. The first consists in comparing the curve-of-growth (COG) of the total [OIII] emission of the AGN with that of an observed and modeled PSF. The second technique, which we called "PSF-subtraction" method, consists in producing a residual [OIII] map after subtracting the nuclear [OIII] spectrum spaxel-by-spaxel following a 2D PSF profile (see also Carniani *et al.* 2015). We found that ≈60% of the Type-1 AGN with S/N (OIII)>5 present kiloparsec-scale extended ionized gas emission. Flux and velocity maps of these resolved targets reveal outflows extended to kiloparsec scales, with indications of redshifted outflows in three objects. To study the [OIII] line profile properties spaxel-by-spaxel we adopted non-parametric measures, whose have the advantage that the parameter values do not depend on the fitting function adopted (e.g. the number of Gaussian components) that may strongly depend on the signal-to-noise of the spectrum under investigation (see e.g. Zakamska *et al.* 2014; Harrison *et al.* 2014 for more details). In particular, we measure the velocity width of the line that contains 80% of the line flux ($w_{80} = v_{90} - v_{10}$). Using a cut on $w_{80} > 600$ km/s, in the integrated spectra, we found that all the Type-1 AGN in the SUPER survey shows the presence of ionized outflows. The w_{80} maps (see Fig. 3 for an example) show a wide variety of structures and morphologies. A detailed kinematical modeling is on-going to properly characterize the different velocity components in these AGN hosts, and constrain the geometry of the outflow.

5. CO gas content

Finally, a promising way to assess the impact that such detected AGN outflows may have on the ability of the host galaxy to form new stars, may consist in assessing their impact on the moleacular gas reservoir. At the moment there is no consensus in the literature on this topic. Most local studies report no clear evidence for AGN to affect the ISM component of the host, by tracing the molecular phase (e.g. Husemann *et al.* 2017; Saintonge *et al.* 2017; Rosario *et al.* 2018), the atomic one (Ellison *et al.* 2019), and the dust mass as a proxy of the gas mass (Shangguan *et al.* 2019). AGN appear to follow the

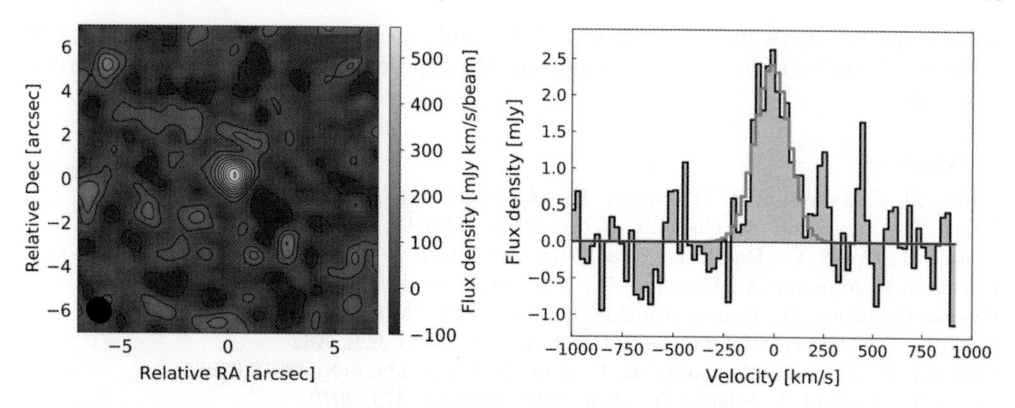

Figure 4. CO line emission maps (*left*) and spectra (*right*) extracted from the region above 2σ significance. Contours are in steps of one sigma for the left panels. On the right, the observed spectrum is plotted in black while the Gaussian model used to fit the spectrum is depicted in red.

same star-formation law of normal galaxies. On the other hand, studies at redshift $z>1$ present opposite results. In particular, some found reduced molecular gas fractions (i.e., the molecular gas mass per unit stellar mass, $f_{\rm gas}=M_{\rm mol}/M_*$) and depletion timescales of AGN compared with the parent population of normal galaxies (e.g. Kakkad *et al.* 2106; Fiore *et al.* 2017; Brusa *et al.* 2018; Perna *et al.* 2018). This has been interpreted as an evidence for highly efficient gas consumption possibly related to AGN feedback affecting the gas reservoir of the host galaxies. It is hard to directly compare all these studies, both at low and high redshift, given the different selection funcitions of the sample considered and the different techinique used to asses the molecular gas content in the host. As part of SUPER, we have started a systematic and uniform analysis of the molecular gas content of $z\sim2$ AGN to infer whether their activity affects the ISM of the host galaxy (Circosta *et al.* 2020; see Fig. 4). As a tracer, we use the CO(3-2) emission line, which is the lowest-J transition accessible with ALMA at $z\sim2$. We compare the CO emission properties of our AGN as traced by ALMA with those of star-forming galaxies. We have selected a stellar mass and SFR matched sample of comparison galaxies and limited to objects observed in the same CO transition to avoid further uncertainties introduced to the unknown spectral line energy distributions (SLEDs) for these objects.

By comparing the CO and FIR luminosities of our AGN and the control sample, as well as their stellar masses, we found that AGN are overall underluminous in CO. We quantified the CO deficit by: 1) performing a linear fit in the $L'_{\rm CO}$-$L_{\rm FIR}$ plane; 2) dividing our samples in bins of stellar mass and computing mean CO luminosities for each bin; 3) deriving the mean distribution of $L'_{\rm CO}/M_*$ (a proxy of gas fraction) for both samples. We interpreted this finding as an evidence for the effect of AGN activity, which may be able to excite, dissociate or deplete the gas reservoir of the host galaxies.

6. Conclusions

SUPER represents a major advancement in the systematic studies of AGN-driven outflows at a crucial cosmic epoch , $1<z<3$, corresponding to the peak of volume-averaged star formation and supermassive black holes accretion in the Universe. The ionized gas kinematics need to be complemented with a significant investment of ALMA time to trace the molecular phase of the outflows (e.g. Cicone *et al.* 2018). The coming years will see the development of facilites that will allow to extend such studies to higher redshift (e.g. the NIRSPec IFU on board of JWST) or to fainter magnitudes and higher spatial resolution (e.g HARMONI at the E-ELT). Finally, it will be very important to invest substantial resources in the modeling of the multi gas phases outflows, extending the theoretical studies to lower gas temperature than those of the tracers that are currently available.

References

Bentz, M. C., Peterson, B. M., Netzer, H., *et al.* 2009, *ApJ*, 697, 160
Brusa, M., Bongiorno, A., Cresci, G., *et al.* 2015, *MNRAS*, 446, 2394
Brusa, M., Cresci, G., Daddi, E., *et al.* 2018, *A&A*, 612, 29
Carniani, S., Marconi, A., Maiolino, R., *et al.* 2015, *A&A*, 580, A102
Cicone, C., Brusa, M., Ramos Almeida, C., *et al.* 2018, *NatAs*, 2, 176
Circosta, C., Mainieri, V., Padovani, P., *et al.* 2018, *A&A*, 620, A82
Circosta, C., Mainieri, V., Lamperti, I., *et al.* 2020, e-prints, arXiv:2012.07965
Costa, T., Rosdahl, J., Sijacki, D., *et al.* 2018, *MNRAS*, 479, 2079
Davies, R. L., Foerster Schreiber, N. M., Lutz, D., *et al.* 2020, *ApJ*, 894, 28
Ellison, S. L., Brown, T., Catinella, B., *et al.* 2019, *MNRAS*, 482, 5694
Fiore, F., Feruglio, C., Shankar, F., *et al.* 2017, *A&A*, 601, A143
Gaskell, C. M. 1982, *ApJ*, 263, 79
Harrison, C. M., Alexander, D. M., Mullaney, J. R., *et al.* 2014, *MNRAS*, 441,3306
Harrison, C. M., Alexander, D. M., Mullaney, J. R., *et al.* 2016, *MNRAS*, 456,1195
Harrison, C. M., Costa, T., Tadhunter, C. N., *et al.* 2018, *NatAs*, 2, 198
Husemann, B., Scharwaechter, J., Bennert, V. N., *et al.* 2016, *A&A*, 594, A44
Husemann, B., Davis, T. A., Jahnke, K., *et al.* 2017, *MNRAS*, 470, 1570
Kakkad, D., Mainieri, V., Padovani, P., *et al.* 2016, *A&A*, 592, A148
Kakkad, D., Mainieri, V., Vietri, G., *et al.* 2020, *A&A*, 642, A147
King, A. & Pounds, K. 2015, *ARA&A*, 53, 115
Leung, G. C. K., Coil, A. L., Aird, J., *et al.* 2019, arXiv e-prints, arXiv:1905.13338
Madau, P. & Dickinson, M. 2014, *ARA&A*, 52, 415
Mejía-Restrepo, J. E., Lira, P., Netzer, H., *et al.* 2018, *Nature Astronomy*, 2, 63
Perna, M., Sargent, M. T., Brusa, M., *et al.* 2018, *A&A*, 619, 90
Richards, G. T., Kruczek, N. E., Gallagher, S. C., *et al.* 2011, *AJ*, 141, 167
Rosario, D. J., Burtscher, L., Davies, R. I., *et al.* 2018, *MNRAS*, 473, 5658
Saintonge, A., Catinella, B., Tacconi, L. J., *et al.* 2017, *ApJS*, 233, 22
Shangguan, J. & Ho, L. C. 2019, *ApJ*, 873, 90
Vietri, G., Piconcelli, E., Bischetti, M., *et al.* 2018, *A&A*, 617, A81
Vietri, G., Mainieri, V., Kakkad, D., *et al.* 2020, *A&A*, 644, 175
Villar-Martin, M., Arribas, S., Emonts, B., *et al.* 2016, *MNRAS*, 460, 130
Zakamska, N. & Greene, J. E. 2014, *MNRAS*, 442, 784
Zubovas, K. & King, A. 2012, *ApJ*, 745, L34

Galaxy Evolution and Feedback across Different Environments
Proceedings IAU Symposium No. 359, 2020
T. Storchi-Bergmann, W. Forman, R. Overzier & R. Riffel, eds.
doi:10.1017/S1743921320002392

ORAL CONTRIBUTIONS

Quasar black hole masses and accretion rates across cosmic time

Michael Brotherton[1], Jaya Maithil[1], Adam Myers[1], Ohad Shemmer[2], Brandon Matthews[2], Cooper Dix[2], Pu Du[3] and Jian-Min Wang[3]

[1]Dept. of Physics & Astronomy, University of Wyoming, Laramie, WY 82071, USA
email: mbrother@uwyo.edu

[2]Department of Physics, University of North Texas, Denton, TX 76203, USA

[3]Key Laboratory for Particle Astrophysics, Institute of High Energy Physics, Chinese Academy of Sciences, 19B Yuquan Road, Beijing 100049, People's Republic of China

Abstract. Quasar black hole masses are most commonly estimated using broad emission lines in single epoch spectra based on scaling relationships determined from reverberation mapping of small samples of low-redshift objects. Several effects have been identified requiring modifications to these scaling relationships, resulting in significant reductions of the black hole mass determinations at high redshift. Correcting these systematic biases is critical to understanding the relationships among black hole and host galaxy properties. We are completing a program using the Gemini North telescope, called the Gemini North Infrared Spectrograph (GNIRS) Distant Quasar Survey (DQS), that has produced rest-frame optical spectra of about 200 high-redshift quasars ($z = 1.5-3.5$). The GNIRS-DQS will produce new and improved ultraviolet-based black hole mass and accretion rate prescriptions, as well as new redshift prescriptions for velocity zero points of high-z quasars, necessary to measure feedback.

Keywords. quasars: general, quasars: emission lines

---

1. Introduction

There exist a number of outstanding questions regarding supermassive black holes, where they come from, how they grow, and how they interact with their host galaxies. Two fundamental properties of populations of supermassive black holes that can be better determined with careful observations are their masses and accretion rates across cosmic time.

2. Quasar black hole mass estimation

While there are a number of methods of determining the mass of a supermassive black hole, reverberation mapping (RM) is the primary technique used to make direct mass measurements for luminous active galactic nuclei (AGNs) such as Seyfert galaxies and quasars (e.g., Peterson 1993). Spectroscopic monitoring can reveal the time delay between the variable continuum and the corresponding response from an extended broad-line region (BLR), which is governed by its spatial extent due to the finite speed of light. In combination with BLR velocities, obtained from the Doppler-broadened emission-line profiles, the time delays can be used to determine a virial mass for the central supermassive black hole:

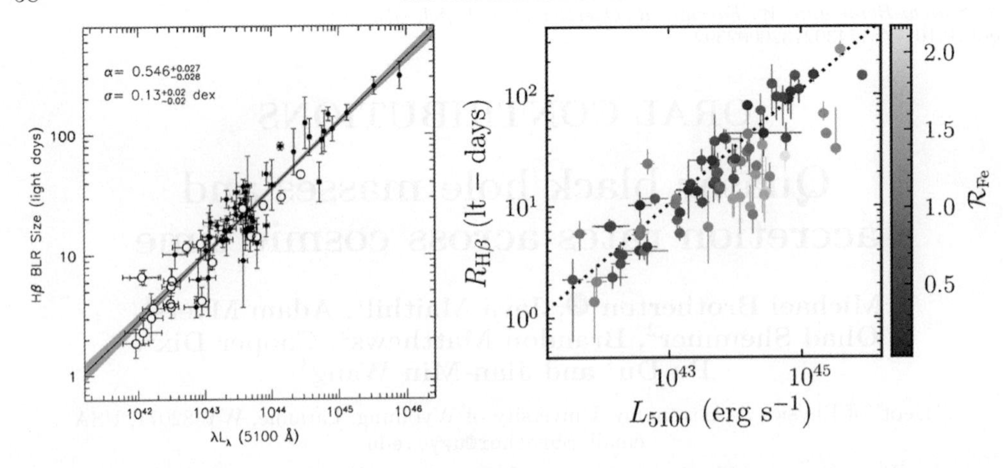

Figure 1. The radius-luminosity relationship for reverberation-mapped AGNs. Left. Bentz *et al.* (2013). Right. Du & Wang (2019), which includes a color scheme indicating the strength of the optical Fe II blend relative to the broad Hβ line. AGNs with strong Fe II have systematically shorter time lags.

$$M_{BH} = f_{BLR} \frac{R_{BLR} \Delta V^2}{G}, \tag{2.1}$$

where $R_{BLR} = c\tau_{BLR}$ is the emissivity-weighted radius of the BLR, c is the speed of light, G is the gravitational constant, and (ΔV) is measured from the FWHM or line dispersion (e.g., Peterson *et al.* 2004). The factor f_{BLR} is an empirically calibrated geometric correction factor of order unity (e.g., Woo *et al.* 2015). The value of f_{BLR} in individual objects likely differs due to effects such as inclination (e.g., Pancoast *et al.* 2014).

The recognition that there was a correlation between AGN luminosity and the Hβ time lag (e.g., Kaspi *et al.* 2000) led to the development of mass estimation based on single-epoch spectra (e.g., Vestergaard 2002; Vestergaard & Peterson 2006). Increasing reverberation mapping efforts led to a relatively tight relationship of: $\tau_{H\beta} \propto \lambda L_{\lambda 5100}^{0.54}$ (Bentz *et al.* 2013; Fig. 1, left).

Reverberation-mapped samples up to that time were biased, however, to have strong narrow [O III] λλ4959,5007 emission lines, which help with flux calibration due to their very slow variability. Reverberation mapping of AGNs with weak [O III] lines (which also have strong broad optical Fe II emission, see, e.g. Boroson & Green 1992), turn out to deviate from the scaling relationship of Bentz *et al.* (2013) and have much shorter than expected Hβ time lags by factors of 3-8 times in the most extreme objects (Fig. 1, right). They also have large accretion rates (Shen & Ho 2014; Du *et al.* 2018). This deviation in the time lag can be predicted by the strength of the ratio of optical Fe II to Hβ emission, R_{FeII} (Du & Wang 2019), leading to a revised scaling relationship unbiased by accretion rate: $\log\,(\tau_{H\beta}/\text{lt-days}) = 1.65 + 0.45\,\log\,(\lambda L_{\lambda_{5100}}/10^{44}\ \text{ergs/s}) - 0.35 R_{FeII}$.

Two parameters are used in describing accretion rates that depend on both the luminosity and the black hole mass. The Eddington fraction is L_{Bol}/L_{Edd}, which requires a bolometric correction, and scales simply as L/M_{BH}. Another normalized accretion rate can be estimated by the formula

$$\dot{M} = 20.1 \left(\frac{\ell_{44}}{\cos i} \right)^{3/2} m_7^{-2}, \tag{2.2}$$

where $m_7 = M_{BH}/10^7 M_\odot$, and i is inclination angle of the accretion disk (with $\cos i = 0.75$ as an average value for AGNs). Based on the standard α-disk model (Shakura & Sunyaev 1973), quasars with values of $\dot{M} > 3$ will be considered Super Eddington Accreting

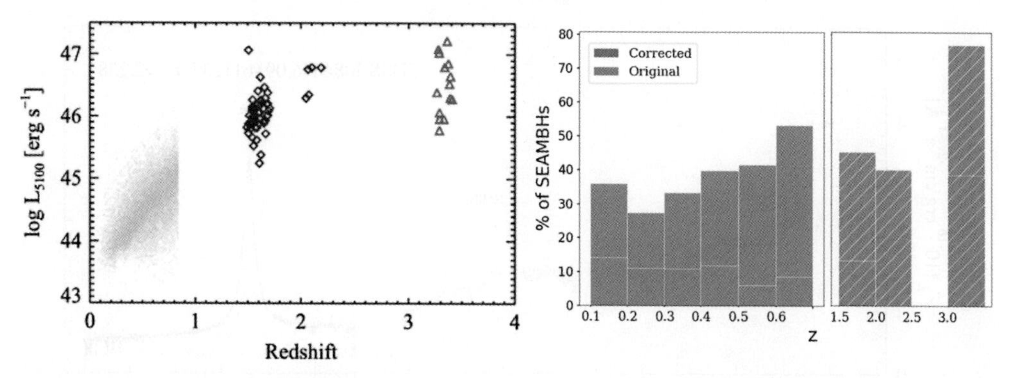

Figure 2. Left. The luminosity-redshift plane for our quasar samples. The gray points are from the SDSS sample using measurements from Shen *et al.* (2011). The points with $z > 1.5$ are from Shen (2016). Right. Histograms of the percentage of objects in each redshift bin with $\dot{M} > 3$ (Super-Eddington Accreting Massive Black Holes, or SEAMBHs) for the old scaling relation of Bentz *et al.* (2013) (red) and the new scaling relation of Du & Wang (2019) (green). Using the new scaling relation indicates a much higher fraction of highly accreting quasars that is likely higher at high-z.

Massive Black Holes (SEAMBHs) for the present discussion, and we note that this parameter scales as L/M_{BH}^2 (see Du & Wang 2019 and references therein for more discussion and details).

3. New black hole masses and accretion rates over cosmic time

Using the revised scaling relationship of Du & Wang (2019), we have calculated new black hole masses and accretion rates ($\dot{M}$) for two samples of quasars for which luminosities, Hβ line widths, and R_{FeII} have been measured. At lower redshifts we use the Shen *et al.* (2011) catalog for the Sloan Digital Sky Survey (SDSS) Data Release 7 Quasar Sample, only keeping objects with measurements having signal-to-noise ratios greater than 10 (totaling 1487 objects). At higher redshifts ($z = 1.5-3.5$) we use a compilation of 74 objects by Shen (2016) based on near-infrared spectra of the redshifted Hβ spectral region. Figure 2 shows the luminosity-redshift space and histograms of the fraction having super-Eddington accretion rates ($\dot{M} > 3$) using masses based on the Bentz *et al.* (2013) (red) and the Du & Wang (2019) (green) scaling relations. The masses for the latter are rather smaller on average and the accretion rates higher. We see that with previous calibrations, only a small fraction of quasars are SEAMBHs, but with the new calibration there is quite a large fraction that increases with redshift. This reflects the fact that high-z quasars have systematically different rest-frame optical properties with larger R_{FeII} than those at low-z, and that the luminous quasars at "cosmic noon" have larger accretion rates.

The SDSS provides many tens of thousands of observed-frame optical spectra for quasars with $z > 1.5$, but only a tiny minority of these have rest-frame optical spectra. Without an ultraviolet proxy for the accretion rate, which R_{FeII} acts as at optical wavelengths, the masses and accretion rates are systematically wrong by sometimes very significant factors.

4. The GNIRS Distant Quasar Survey (GNIRS GDQS)

In 2017, we launched a program to increase the quantity of high-quality near-infrared spectra of representative quasars between redshifts of 1.5 and 3.5 (Matthews *et al.* 2018, 2019, 2020). We selected bright, luminous objects from the SDSS that already possessed good quality rest-frame ultraviolet spectra, including radio-quiet, radio-loud, and broad

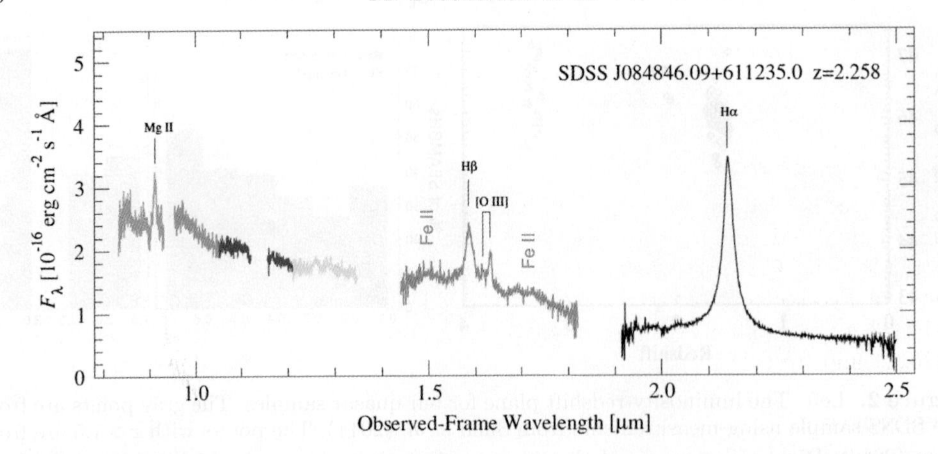

Figure 3. An example spectrum of a GNIRS DQS quasar selected from the SDSS. Prominent emission lines of interest are marked. The six GNIRS bands are distinguished using different colors. Regions of low signal-to-noise ratio where atmospheric absorption are large have been excluded.

absorption line (BAL) quasars. Over the last three years we have used the GNIRS instrument on Gemini-North to obtain approximately 200 high-quality near-infrared spectra of the rest-frame optical spectrum ensuring that the Hβ region, including the [O III] and Fe II lines, were generally centered in an observing band. Figure 3 shows an example GDQ spectrum.

A primary goal of this project is to assemble a data set that will enable us to calculate the most precise and accurate black hole masses, accretion rates, and redshifts of the luminous quasars at cosmic noon using the best methodologies, which rely on rest-frame optical spectral measurements. These in turn will be compared in a consistent manner with those of quasars at lower redshifts for a relatively unbiased look at quasars across cosmic time. Above and beyond simply establishing the black holes masses and accretion rates of the luminous quasars at cosmic noon, we will also be able to construct ultraviolet estimations for black hole mass and accretion rate that correct the systematic biases that exist in current formulations. We anticipate that a combination of measurements using the prominent Mg II and C IV ultraviolet emission lines can do the job. This work can then be used for the tens of thousands of high-z quasars with SDSS spectra to establish the demographics of actively accreting supermassive black holes.

The narrow [O III] λ5007 line is the preferred redshift reference for quasars at optical wavelengths (Boroson 2005). In the near-ultraviolet, the peak of the Mg II broad line provides a reasonably good measure of systematic redshift (Tytler & Fan 1992), but it is not always available in the spectra of distant quasars. We have also been pioneering a simple prescription using the prominent ultraviolet C IV line to measure accurate and precise redshifts despite the line's history as unsuitable due to its consistent blueshift (Mason, Brotherton & Myers 2017; Dix *et al.* 2020). The new GNIRS DQS data set will allow us to significantly refine our formula. These improved redshifts will better establish zero points for the absorption and emission features characteristic of outflows that demonstrate an extreme signature of quasar feedback. Better redshift estimates can also improve a range of studies at high redshift, for example, quasar clustering measurements.

5. Conclusions

We conclude that using the most up-to-date scaling relationships reveals that the black hole masses of the luminous quasars at cosmic noon have been significantly overestimated, and the accretion rates underestimated. With the GNIRS GDQS we will robustly quantify these statements and develop new and improved ultraviolet-based prescriptions for black hole mass and accretion rate, as well as redshift.

Acknowledgments

JMW acknowledges the support from the National Science Foundation of China (NSFC-11833008 and -11991054), from the National Key R&D Program of China (2016YFA0400701). PD acknowledges the support from NSFC-11873048, -11991051 and the Strategic Priority Research Program of the Chinese Academy of Sciences (XDB23010400). This work is supported by National Science Foundation grants AST-1815281 and AST-1815645.

References

Bentz, M. C., Denney, K. D., Grier, C. J., *et al.* 2013, *ApJ*, 767, 149
Boroson, T. 2005, *AJ*, 130, 381
Boroson, T. A. & Green, R. F. 1992, *ApJS*, 80, 109
Du, P. & Wang, J.-M. 2019, *ApJ*, 886, 42
Du, P., Zhang, Z.-X., Wang, K., *et al.* 2018, *ApJ*, 856, 6
Dix, C., Shemmer, O., Brotherton, M. S., *et al.* 2020, *ApJ*, 893, 14
Kaspi, S., Smith, P. S., Netzer, H., *et al.* 2000, *ApJ*, 533, 631
Matthews, B., Shemmer, O., Brotherton, M. S., *et al.* 2018, *American Astronomical Society Meeting Abstracts #232*, 318.09
Matthews, B., Shemmer, O., Brotherton, M. S., *et al.* 2019, *American Astronomical Society Meeting Abstracts #233*, 243.38
Matthews, B., Shemmer, O., Brotherton, M., *et al.* 2020, *American Astronomical Society Meeting Abstracts*, 381.06
Mason, M., Brotherton, M. S., & Myers, A. 2017, *MNRAS*, 469, 4675
Pancoast, A., Brewer, B. J., Treu, T., *et al.* 2014, *MNRAS*, 445, 3073
Peterson, B. M. 1993, *PASP*, 105, 247
Peterson, B. M., Ferrarese, L., Gilbert, K. M., *et al.* 2004, *ApJ*, 613, 682
Shakura, N. I. & Sunyaev, R. A. 1973, *A&A* 500, 33
Shen, Y. 2016, *ApJ*, 817, 55
Shen, Y. & Ho, L. C. 2014, *Nature*, 513, 210
Shen, Y., Richards, G. T., Strauss, M. A., *et al.* 2011, *ApJS*, 194, 45
Tytler, D. & Fan, X.-M. 1992, *ApJS*, 79, 1
Vestergaard, M. 2002, *ApJ*, 571, 733
Vestergaard, M. & Peterson, B. M. 2006, *ApJ*, 641, 689
Woo, J.-H., Yoon, Y., Park, S., *et al.* 2015, *ApJ*, 801, 38

Galaxy Evolution and Feedback across Different Environments
Proceedings IAU Symposium No. 359, 2020
T. Storchi-Bergmann, W. Forman, R. Overzier & R. Riffel, eds.
doi:10.1017/S1743921320002434

The merger-driven evolution of massive early-type galaxies

Carlo Cannarozzo[1,2], Carlo Nipoti[1], Alessandro Sonnenfeld[3], Alexie Leauthaud[4], Song Huang[5], Benedikt Diemer[6] and Grecco Oyarzún[4]

[1]Dipartimento di Fisica e Astronomia, Alma Mater Studiorum Università di Bologna,
Via Piero Gobetti 93/2, I-40129 Bologna, Italy
email: carlo.cannarozzo3@unibo.it

[2]INAF - Osservatorio di Astrofisica e Scienza dello Spazio di Bologna,
Via Piero Gobetti 93/3, I-40129 Bologna, Italy

[3]Leiden Observatory, Leiden University, Niels Bohrweg 2,
2333 CA Leiden, The Netherlands

[4]Department of Astronomy and Astrophysics, University of California, Santa Cruz,
1156 High Street, Santa Cruz, CA 95064 USA

[5]Department of Astrophysical Sciences, Princeton University,
4 Ivy Lane, Princeton, NJ 08544, USA

[6]NHFP Einstein Fellow, Department of Astronomy, University of Maryland,
College Park, MD 20742, USA

Abstract. The evolution of the structural and kinematic properties of early-type galaxies (ETGs), their scaling relations, as well as their stellar metallicity and age contain precious information on the assembly history of these systems. We present results on the evolution of the stellar mass-velocity dispersion relation of ETGs, focusing in particular on the effects of some selection criteria used to define ETGs. We also try to shed light on the role that in-situ and ex-situ stellar populations have in massive ETGs, providing a possible explanation of the observed metallicity distributions.

Keywords. Galaxies: elliptical and lenticular, cD; galaxies: evolution; galaxies: formation; galaxies: kinematics and dynamics.

1. Introduction

In the currently favoured model of galaxy formation, early-type galaxies (ETGs) are believed to assemble in two phases (Oser *et al.* 2010). The first phase ($z \gtrsim 2$) is dominated by the *in-situ* star formation. Afterwards, as a consequence of mostly dissipationless minor and major mergers, ETGs accrete stars formed *ex situ*. This scenario leads to intriguing questions including how the properties of ETGs evolve across cosmic time, whether and to what extent mergers modify the scaling relations of these massive galaxies and shape the distribution of the stellar populations within them.

In this proceeding we present some aspects of the evolution of massive ETGs based on two different works. In section 2 we report the results on the M_*-σ_e relation of ETGs obtained in Cannarozzo, Sonnenfeld & Nipoti (2020) (hereafter CSN), and study the effect of adopting two different selection criteria for ETGs. In section 3 we present preliminary results of a forthcoming paper (Cannarozzo *et al.* in preparation) aimed at studying the radial distributions of in-situ and ex-situ stellar components of massive ETGs.

2. The evolution of the stellar mass-velocity dispersion relation

The central stellar velocity dispersion σ_e of present-day ETGs is found to correlate with their stellar mass M_*. There are indications that this correlation evolves with redshift in the sense that, at given M_*, higher-z ETGs have, on average, higher σ_e (e.g., Tanaka *et al.* 2019). However, the detailed evolution of the M_*-σ_e relation is hard to determine, because of the difficulty of measuring σ_e in large samples of ETGs at high z.

CSN studied the evolution of the M_*-σ_e relation in massive ($\log(M_*/M_\odot) > 10.5$) ETGs in the redshift range $0 < z < 2.5$, using a Bayesian hierarchical approach. CSN considered a sample of galaxies composed by two main subsamples. The first subsample, named *fiducial sample*, consists of ETGs in the redshift range $0 \lesssim z \lesssim 1$ drawn from the Sloan Digital Sky Survey (SDSS, Eisenstein *et al.* 2011) and the Large Early Galaxy Astrophysics Census (LEGA-C, van der Wel *et al.* 2016; Straatman *et al.* 2018), homogeneously selected by performing a one-by-one visual inspection to include only objects with elliptical morphology and by applying a cut in the equivalent width (EW) of the emission line doublet [OII]$\lambda\lambda$3726, 3729, EW([OII]) $\geqslant -5$Å. The second subsample, named *high-redshift sample*, is a more heterogeneous collection of ETGs in the redshift range $0.8 \lesssim z \lesssim 2.5$ from previous works in literature (see CSN and references therein). CSN found that, for both the fiducial and the *extended* (fiducial + high-redshift) samples, the M_*-σ_e relation is well described by $\sigma_e \propto M_*^\beta (1+z)^\zeta$ with intrinsic scatter $\simeq 0.08$ dex in σ_e at given M_* and either $\beta \simeq 0.18$ independent of z or redshift-dependent β with $\mathrm{d}\beta/\mathrm{d}\log(1+z) \simeq 0.26$ for the fiducial sample and $\simeq 0.18$ for the extended sample; ζ, which measures the redshift dependence of σ_e at given M_*, is $\simeq 0.4$ for the fiducial sample ($0 \lesssim z \lesssim 1$) and $\simeq 0.5$ for the extended sample ($0 \lesssim z \lesssim 2.5$).

One of the properties of ETGs is to be passive and EW([OII]) is only one of the possible diagnostics for the star formation rate. Another indicator is the position of galaxies within the UVJ colour-colour diagram, in which the loci of passive and star-forming galaxies are separate (e.g., Cimatti, Fraternali & Nipoti 2019). For instance, Belli *et al.* (2014b), from which part of the galaxies of the high-redshift sample are taken, select using a UVJ-based criterion. In principle, this different selection criterion can induce spurious evolution when the extended sample is considered. Here we analyse the effect of adding a UVJ-based selection to the criteria used for the fiducial sample.

The model with the highest value of Bayesian evidence explored by CSN, named model $\mathcal{M}_{\mathrm{const,NES}}$, has six hyper-parameters: ζ, $\mu_{*,0}$, $\mu_{*,s}$, $\sigma_{*,0}$, $\sigma_{*,s}$ and α_* (see CSN for details). We repeated the analysis of CSN by considering a modified fiducial sample. In particular, we changed the selection criterion for the galaxies of the LEGA-C sample: in addition to the criteria used in CSN, we exclude galaxies that are star-forming on the basis of their position in the UVJ colour-colour diagram. In the top panel of Figure 1, the UVJ diagram for the LEGA-C sample of 178 ETGs used in CSN is shown (the UVJ colours are taken from the UltraVISTA catalogue of Muzzin *et al.* 2013). In this diagram the locus of passive galaxies is the area above and to the left of the broken line: about 90% of the galaxies of the LEGA-C sample of CSN are in this area. Excluding galaxies that are outside the locus of passive galaxies in the UVJ diagram of Figure 1, we end up with a modified fiducial sample, consisting of 161 instead of 178 LEGA-C galaxies, in addition to the SDSS galaxies. We applied model $\mathcal{M}_{\mathrm{const,NES}}$ to this modified fiducial sample (hereafter model $\mathcal{M}^{fid,UVJ}_{\mathrm{const,NES}}$) and compared the results with those obtained by CSN for the fiducial sample (hereafter model $\mathcal{M}^{fid}_{\mathrm{const,NES}}$). The posterior distributions of the hyper-parameters of models $\mathcal{M}^{fid,UVJ}_{\mathrm{const,NES}}$ and $\mathcal{M}^{fid}_{\mathrm{const,NES}}$, shown in the bottom panel of Figure 1, are in agreement within 1σ. In particular, for model $\mathcal{M}^{fid,UVJ}_{\mathrm{const,NES}}$, the

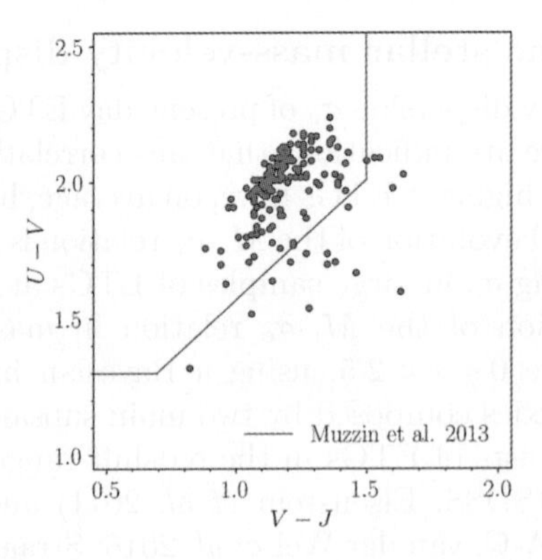

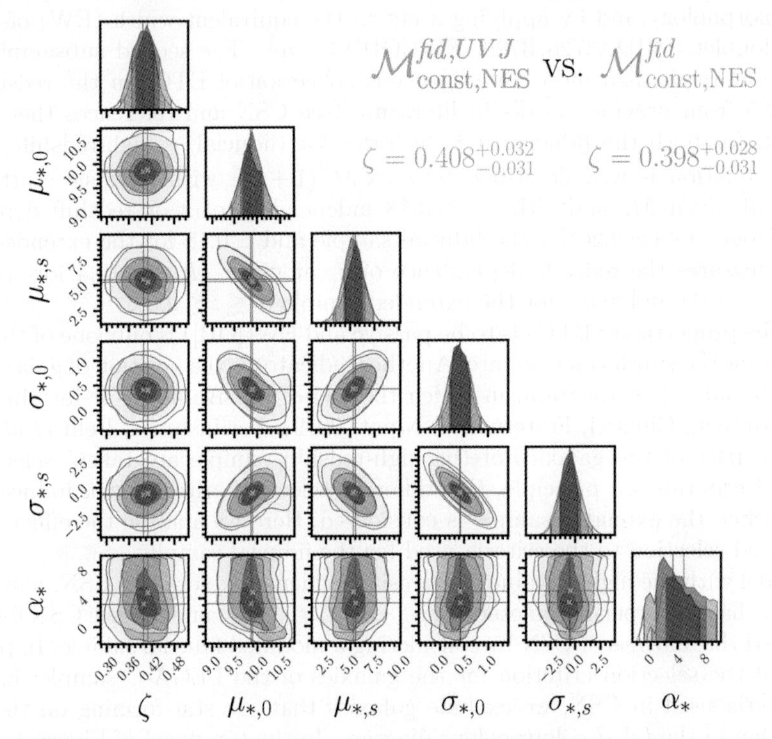

Figure 1. *Top panel: UVJ* colour-colour diagram for the LEGA-C sample of 178 ETGs (red dots). The broken line separates quiescent (upper-left region) and star-forming galaxies (lower-right region) as in Muzzin *et al.* (2013). *Bottom panel:* posterior probability distributions of the hyper-parameters for the M_*-σ_{e} models $\mathcal{M}^{fid}_{\mathrm{const,NES}}$ (purple contours) and $\mathcal{M}^{fid,UVJ}_{\mathrm{const,NES}}$ (blue contours). In the 1D distributions (top panel of each column) the vertical solid lines and colours delimit the 68, 95 and 99.7-th quantile based posterior credible interval. In the 2D distributions (all the other panels) the contours enclose the 68, 95 and 99.7 percent posterior credible regions. The lines indicate the median values of the hyper-parameters.

normalisation of the M_*-σ_{e} scaling relation evolves with $\zeta = 0.408^{+0.032}_{-0.031}$, consistent with $0.398^{+0.028}_{-0.031}$ obtained by model $\mathcal{M}^{fid}_{\mathrm{const,NES}}$. This analysis suggests that, at least as far as the *UVJ* selection is concerned, the results of the extended sample in CSN are not biased.

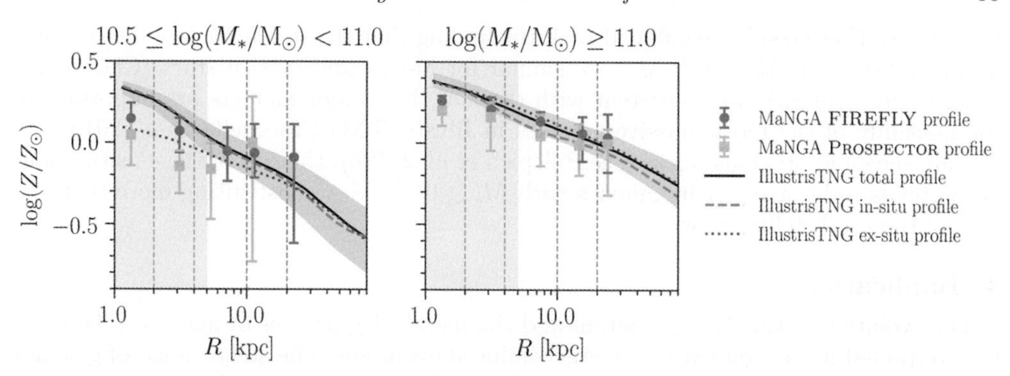

Figure 2. Mass-weighted metallicity radial profiles of massive ETGs with lower (left panel) and higher (right panel) stellar mass. The circles and squares represent the median estimates for MaNGA galaxies from FIREFLY and PROSPECTOR codes, respectively. The curves represent the median estimates for IllustrisTNG galaxies. The dashed and dotted curves represent the in-situ and the ex-situ stellar components, respectively, while the solid curve represents the total stellar components. The shaded area and the error bars indicate 1σ scatter.

3. In-situ and ex-situ stellar populations in ETGs

The evolution of metallicity, chemical abundances as well as the ages and other physical properties of stars in ETGs contain information on the evolutionary processes occurred across cosmic time. In this section we discuss the radial distribution of stellar metallicity in massive MaNGA (Bundy *et al.* 2015) ETGs in terms of in-situ and ex-situ stellar components, relying on simulated galaxies drawn from the magnetohydrodynamic cosmological simulation IllustrisTNG (Springel *et al.* 2018; Pillepich *et al.* 2018a; Nelson *et al.* 2018; Marinacci *et al.* 2018; Naiman *et al.* 2018).

We extracted from the MaNGA survey a sample of more than 700 ETGs, with $M_* \geqslant 10^{10.5}\,\mathrm{M}_\odot$, selected in $\mathrm{EW(H\alpha)} > -3\text{Å}$ and Sérsic index $n > 2.5$. In order to reduce the effects of systematic biases caused by different assumptions, priors and fitting methods (Conroy 2013), we rely on two estimates of metallicity derived by using the spectral fitting codes FIREFLY (e.g., Comparat *et al.* 2017) and PROSPECTOR (Leja *et al.* 2017). For a description of the derivation of stellar properties, we refer the reader to Ojarzún *et al.* (2019). To make a comparison with simulations, we extracted around 2800 MaNGA-like ETGs from the $z = 0.1$ snapshot of IllustrisTNG100. For each simulated galaxy, we choose randomly a line of sight and we build a 2D map of stellar properties by projecting the positions of stellar particles onto a 300×300 pixel area. The 1D profiles are derived from the 2D maps using concentric elliptic annuli with fixed ellipticity for each ETG. A detailed description of this fit procedure is provided in Huang *et al.* (2018).

In Figure 2, the median metallicity profiles for MaNGA and IllustrisTNG ETGs are shown for two stellar mass bins. Although the PROSPECTOR metallicity tends to be lower than the FIREFLY estimate (offset mainly due to different stellar models assumed), we find good agreement between the two measurements. Moreover, the observed profiles are well reproduced by the IllustrisTNG profiles, in particular at the high-mass end. In the case of IllustrisTNG galaxies, we can disentangle the in-situ and ex-situ stellar populations (see Rodriguez-Gomez *et al.* 2016) and measure for each component its metallicity profile. Looking at the behaviour of the in-situ and ex-situ metallicity distributions, we notice that for galaxies with $M_* \lesssim 10^{11}\,\mathrm{M}_\odot$ the inner regions are dominated by the in-situ component: the total and the in-situ metallicity profiles are indistinguishable out to 20 kpc. Instead, in ETGs with $M_* \gtrsim 10^{11}\,\mathrm{M}_\odot$, the ex-situ component is as relevant as (or even more relevant than) the in-situ component at all radii, and has, on average, higher metallicity. These results, combined with the finding that the stellar surface density profiles of ETGs with $M_* \gtrsim 10^{11}$ $M_\odot$ are similar for in-situ and ex-situ stars (Chowdhury *et al.* in preparation), are consistent with the fact that major mergers are important in the assembly of the most massive galaxies in IllustrisTNG (Tacchella *et al.* 2019). As already shown in previous works (see Pillepich *et al.* 2018b), the role of the ex-situ population tends to be stronger in galaxies with $M_* \gtrsim 10^{10.5}$ $M_\odot$, constituting more than the 50% of the total stellar mass.

4. Implications

The evolution of the M_*-σ_e relation and the metallicity profiles of massive ETGs can be interpreted in the context of a merger-induced evolution. The stellar mass of galaxies varies with cosmic time mainly as a consequence of star formation and accretion of stars. On the one hand, some internal mechanisms, such as stellar or active galactic nucleus feedback, can blow out part of the material, depriving the galaxy of the gas reservoir needed to form new stars and then limiting the growth of the stellar mass. On the other hand, phenomena like mergers can trigger star formation and bring in stars formed in other galaxies. In massive systems, like the ETGs considered in these works, the latter process is expected to be dominant. The results here presented underline the importance of having large and high-resolution observational surveys and cosmological simulations, both necessary to improve our understanding of the galaxy evolution. A self-consistent comparison between observations and simulations is crucial to draw robust conclusions.

References

Belli, S., Newman, A. B., Ellis, R. S., *et al.* 2014b, *ApJ*, 788, L29
Bundy, K., *et al.* 2015, *ApJ*, 798, 7
Cannarozzo, C., Sonnenfeld, A., & Nipoti, C., 2020, submitted to MNRAS, arXiv:1910.06987
Cannarozzo, C., Leauthaud, A., Huang, S., *et al.* in preparation
Chowdhury, R., Huang, S., Leauthaud, A., *et al.* in preparation
Cimatti, A., Fraternali, F., & Nipoti, C., 2019, *Introduction to Galaxy Formation and Evolution: From Primordial Gas to Present-Day Galaxies*, Cambridge University Press
Comparat, J., *et al.* 2017, *ArXiv e-prints*, arXiv:1711.06575
Conroy, C. 2013, *ARA&A*, 51, 393
Eisenstein, D. J., *et al.* 2011, *AJ*, 142, 72
Huang, S., *et al.* 2018, *MNRAS*, 475, 3348
Leja, J., Johnson, B. D., Conroy, C., *et al.* 2017, *ApJ*, 837, 170
Marinacci, F., *et al.* 2018, *MNRAS*, 480, 4
Muzzin, A., *et al.* 2013, *ApJ*, 777, 1
Naiman, J., *et al.* 2018, *MNRAS*, 477, 1
Nelson, D., *et al.* 2018, *MNRAS*, 475, 1
Ojarzún, G., *et al.* 2019, *MNRAS*, 475, 1
Oser, L., Ostriker, J. P., Naab, T., *et al.* 2010, *ApJ*, 725, 2
Pillepich, A., *et al.* 2018a, *MNRAS*, 475, 1
Pillepich, A., *et al.* 2018b, *MNRAS*, 475, 648
Rodriguez-Gomez, V., *et al.* 2016, *MNRAS*, 458, 3
Springel, V., *et al.* 2018, *MNRAS*, 475, 1
Straatman, C. M. S., *et al.* 2018, *ApJS*, 239, 27
Tacchella, S., *et al.* 2019, *MNRAS*, 487, 4
Tanaka, M., *et al.* 2019, *ApJL*, 885, 2
van der Wel, A., *et al.* 2016, *ApJS*, 223, 29

Galaxy Evolution and Feedback across Different Environments
Proceedings IAU Symposium No. 359, 2020
T. Storchi-Bergmann, W. Forman, R. Overzier & R. Riffel, eds.
doi:10.1017/S1743921320004214

GOODS-ALMA: AGNs and the slow downfall of massive star-forming galaxies at $z > 2$

Maximilien Franco[1,2]

[1]AIM, CEA, CNRS, Université Paris-Saclay, Université Paris Diderot, Sorbonne Paris Cité, F-91191 Gif-sur-Yvette, France

[2]Centre for Astrophysics Research, University of Hertfordshire, Hatfield, AL10 9AB, UK
email: m.franco@herts.ac.uk

Abstract. We present the results of a 69 arcmin^2 ALMA survey at 1.1 mm, GOODS-ALMA, matching the deepest HST-WFC3 H-band observed region of the GOODS-South field. The 35 galaxies detected by ALMA are among the most massive galaxies at $z = 2$–4 and are either starburst or located in the upper part of the galaxy star-forming main sequence. The analysis of the gas fraction, depletion time, X-ray luminosity and the size suggests that they are building compact bulges and are the ideal progenitors of compact passive galaxies at $z \sim 2$, and a slow downfall scenario is favoured in their future transition from star-forming to passive galaxies.

Keywords. galaxies: evolution, galaxies: high-redshift, galaxies: ISM, galaxies: nuclei, submillimeter

---

1. Introduction

The star-formation density at high-redshift remains relatively unknown. For galaxies located beyond the peak of the star formation rate (SFR) density ($z \sim 2$; Madau & Dickinson 2014), their SFRs are commonly estimated from UV measurements. Since the UV emission is highly sensitive to dust attenuation, these SFRs must be corrected to obtain the effective star formation rate. For this purpose an extinction law (e.g., Meurer *et al.* 1999; Calzetti *et al.* 2000) and a UV spectral slope (β) has been calibrated in the local universe. Several caveats persist at high redshift (e.g., Cowie *et al.* 1996; Pannella *et al.* 2009), especially for galaxies with high SFRs (e.g., Rodighiero *et al.* 2011), making the estimation of intrinsic SFR in high redshift galaxies highly uncertain.

It is for this reason that the observation of galaxies at infrared wavelengths through large surveys is essential to understand how galaxies build up their stellar mass across cosmic time. The millimetre and submillimetre wavelengths, benefiting from a strong negative K-correction across a wide redshift range $2 < z < 10$ (e.g., Blain *et al.* 2002), are particularly well-suited for probing high redshift dust-obscured star formation.

The advent of the Atacama Large Millimetre/submillimetre Array (ALMA) has led to a considerable improvement in the angular resolution and sensitivity of the detections. We present here the results of the largest survey obtained with ALMA, GOODS-ALMA, covering 69 arcmin^2 in the deepest part of Cosmic Assembly Near-infrared Deep Extragalactic Legacy Survey (CANDELS; Koekemoer *et al.* 2011; Grogin *et al.* 2011) field, within the Great Observatories Origins Deep Survey-South (GOODS-South) field.

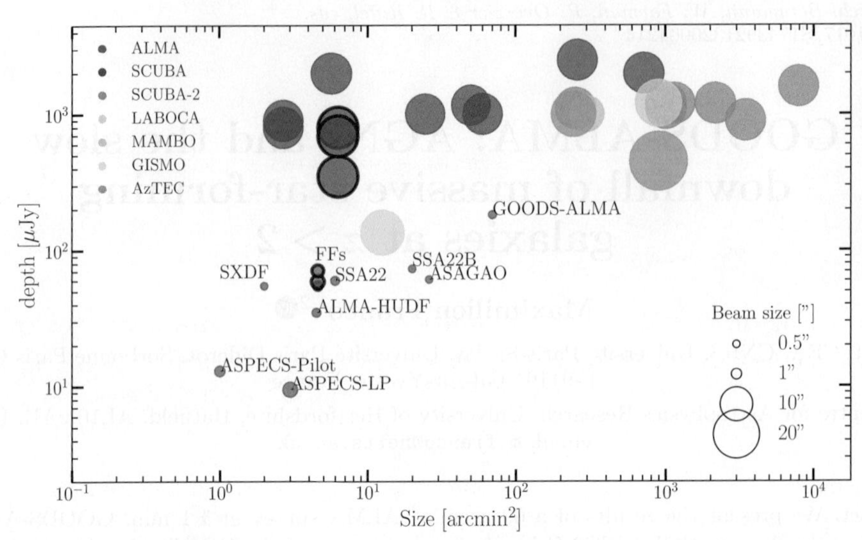

Figure 1. Size vs. depth for some of the main (sub)millimeter surveys at $850\mu\mathrm{m} < \lambda < 3\,\mathrm{mm}$. The circles are colour-coded according to the instrument used. The size of the circles corresponds to the size of the beam. The black outer line indicates that the survey covers a lensed area. This list is not exhaustive. We point out that to date, the GOODS-ALMA survey is the largest (sub)millimetre survey without confusion issues. The list of all the surveys shown, all the references, as well as the original figure, can be found in the PhD manuscript of M. Franco.

Since the commissioning of ALMA 8 years ago, several surveys have been obtained, in several regions of the sky The special feature of GOODS-ALMA can be seen clearly in Fig. 1; it is shallower but larger than other ALMA surveys. Thanks to this particularity, we are able to open a different parameter space than other surveys, notably towards massive high redshift star-forming galaxies.

However, it is challenging to obtain large observation areas with ALMA. A mosaic of 846 individual pointings was necessary to create our image. Several other instruments allow us to reach larger areas but at the price of a larger beam. The high angular resolution allows us to avoid blending, to be more confident in the identification of counterparts, to discover optically dark galaxies, and to constrain the sizes of galaxies. This last point, in addition to the computation of the gas mass, SFR, and the depletion time is key to understand galaxy evolution and, in particular, the transition between star-forming and quenched galaxies. We will explore the possibility that these galaxies detected by ALMA are the progenitors of passive elliptical galaxies at $z \sim 2$, which for the moment are largely unknown (e.g., Williams *et al.* 2014; Wang *et al.* 2019), and investigate the evidence we have to assess how this transition is taking place. We assume a Salpeter (1955) Initial Mass Function, and all magnitudes are quoted in the AB system (Oke & Gunn 1983).

2. GOODS-ALMA survey

We use ALMA observations (Project ID: 2015.1.00543.S; PI: D. Elbaz), covering 69 arcmin^2 within the GOODS-South field (Franco *et al.* 2018). We reach a median rms sensitivity $\sigma \simeq 0.18\,\mathrm{mJy\,beam^{-1}}$ in the mosaic tapered to 0.60". We chose to extract sources in this mosaic using two techniques. The first is a blind extraction down to a threshold of 4.8 σ, which assures us a purity (see Eq. 1 of Franco *et al.* 2018) of 80%. This blind extraction allows us to carry out systematic statistical analysis as well as number counts. This technique also allows us to exploit one of the great strengths of a blind survey - to detect galaxies that had not been detected by other surveys, in particular in deep fields with Hubble down to a 5σ limiting depth of H = 28.2 AB (HST/WFC3 F160W), also known as "HST-dark" or "optically dark" galaxies. These galaxies represent 20% (4/20) of our blind detections. We also extended the source detection in the GOODS-ALMA field down to a 3.5σ threshold using IRAC and the VLA. This allowed us to detect 16 additional galaxies in order to have a more complete view of the galaxies present in this field. The comparison between the number of detected galaxies and the number of galaxies expected by the number counts suggests that we detect more than 70–90% of the galaxies at 1.1 mm with fluxes greater than 0.65 mJy.

3. The slow downfall of star-formation in $z = 2-3$ massive galaxies

Thanks to these unique characteristics (large surface area and limited sensitivity compared to other ALMA surveys), GOODS-ALMA allowed us to detect distant and massive galaxies. As these galaxies are rare in terms of surface density, we need to cover large areas to detect them. These galaxies are more massive and more distant than those detected over a smaller surface during the same observation time. We have detected some of the most massive galaxies at $z = 2$–4 (see Fig. 12 in Franco *et al.* 2020a). The vast majority of the sample lie on or in the upper part of the main sequence (MS; e.g., Noeske *et al.* 2007; Elbaz *et al.* 2007; Schreiber *et al.* 2015, see Fig. 4 in Franco *et al.* 2020b). We also note that these galaxies are close to the limit between the power-law MS relation between SFR and stellar mass, and the bending of MS at high stellar mass (Abramson *et al.* 2014; Schreiber *et al.* 2016; Popesso *et al.* 2019).

These galaxies have high SFRs and cannot continue to form stars for long periods at this rate, otherwise we would see at $z \sim 1$, or in the local Universe, galaxies more massive than those we observe now. We are directly observing massive galaxies rapidly producing stars, and we know that in the "near" future (a few hundred million years at most), this rate of star formation may decrease. We therefore have an ideal laboratory to observe this critical period for galaxies, and investigate whether we have enough evidence using ALMA observations and rich multi-wavelength supporting data to determine if a clear scenario emerges for the decline of SFR in these galaxies.

3.1. *Low depletion time and low gas fraction*

We have investigated the molecular gas reservoirs of the galaxies detected with ALMA, as well as their depletion times. We extracted key parameters for galaxies with *Herschel* counterparts by constraining their spectral energy distributions using the SED fitting code `CIGALE` (Code Investigating Galaxies Emission; Boquien *et al.* 2019). We then converted the dust mass (Draine *et al.* 2014 models) into a gas mass using the Leroy *et al.* 2011 relations and the metallicity from Genzel *et al.* (2012). We consider only the galaxies that have *Herschel* counterparts (21 galaxies) in addition to a 1.1mm ALMA detection in order to avoid deriving two physical quantities (infrared luminosity and gas mass) with only one data point.

In order to understand whether our galaxy population has a molecular gas deficit or excess, we compared the gas masses of our galaxies with the relationship presented in Tacconi *et al.* (2018), derived from a large sample of 1444 star-forming galaxies between $z = 0 - 4$. We also compared the depletion times ($\tau_{dep} = \mathrm{M}_{gas}/\mathrm{SFR} = 1/\mathrm{SFE}$, where SFR is the total SFR = SFR_{IR} + SFR_{UV} and SFE is the star formation efficiency). This is the characteristic time a galaxy needs to empty its gas reserves, assuming a constant SFR and no gas replenishment, between our sample and Tacconi *et al.* (2018) (Fig. 2).

For the sake of clarity, we have displayed the Tacconi *et al.* (2018) relation corresponding to the median redshift ($z_{med} = 2.7$) and the median stellar mass ($\mathrm{M}_{\star,med} = 8.5 \times 10^{10}\mathrm{M}_{\odot}$) of our sample, and re-scaled the gas fraction and the depletion time of each

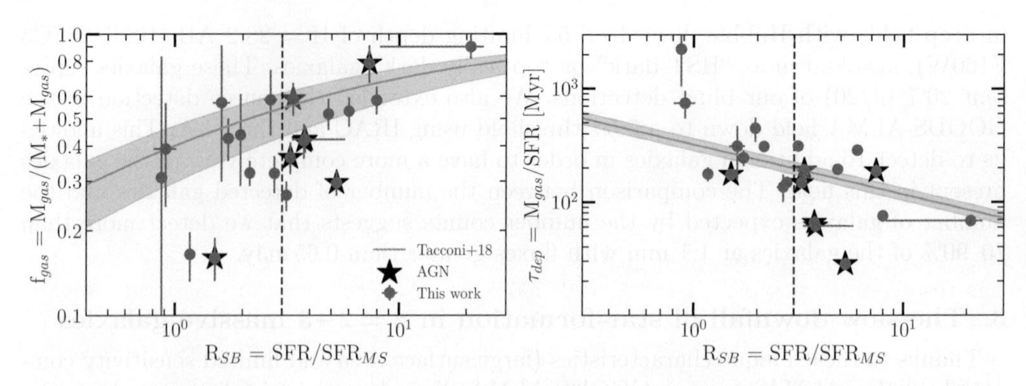

Figure 2. Evolution of the molecular gas fraction (f_{gas}) and the gas depletion timescale (τ_{dep}) as a function of the distance to the main sequence of star-forming galaxies ($R_{SB} = SFR/SFR_{MS}$) for galaxies detected in the GOODS-ALMA field. The solid blue line shows the relation obtained by Tacconi *et al.* (2018) for the median redshift and stellar mass of our sample. The uncertainty on the mean trend is obtained by Monte-Carlo simulations. In order to compare the gas fractions of all of the galaxies in our sample, we have rescaled our gas fractions according to the median redshift and stellar mass of our sample. Figure from Franco *et al.* (2020b).

galaxy individually to preserve the distance between the galaxy and the Tacconi *et al.* (2018) relation. We investigated whether galaxies hosting an AGN exhibited different characteristics to other galaxies. We used a criterion to discriminate the AGNs based on their X-ray luminosity using the 7Ms Chandra Deep Field-South Survey (Luo *et al.* 2017). The black stars on Fig. 2 represent galaxies with $L_{X,int} > 10^{43}\,\mathrm{erg\,s^{-1}}$.

We recover the global trend of an increase in gas fraction with distance to the main sequence ($R_{SB} = SFR/SFR_{MS}$). Our sample of galaxies exhibits a large scatter in the plane f_g- R_{SB}. Moreover, we find that a significant part of our sample, approximately (40%), lie below this relation. This is valid for both starburst and MS galaxies.

We also compared the gas depletion time with the Tacconi *et al.* (2018) relation, and find that a significant part of our sample has a low gas depletion time. For both depletion time and gas fraction, we see that galaxies hosting an AGN have, on average, a higher gas fraction and depletion time than galaxies with lower X-ray luminosities.

3.2. *Towards a reduction in the size of galaxies*

We determine whether our galaxies, that we see are compact at millimetre wavelengths, are also compact in H-band. Remarkably, we found that for the galaxies for which we have sizes measured in H-band (Van der Wel *et al.* 2014), these H band sizes are on the trend of star-forming galaxies with comparable redshifts and stellar masses (see Fig. 5 in Franco *et al.* 2020b). We note that in our sample, three galaxies are particularly compact in H-band. These three galaxies share a common characteristic - they host an AGN. Comparing galaxy ALMA sizes with H-band sizes, we find that ALMA sizes correlate well with the H-band trend of quenched galaxies with comparable redshifts and stellar masses. If we consider that ALMA traces the dust-obscured star formation, the fact that the ALMA sizes are more compact than the H-band sizes, and that the ALMA sizes match the trend of quenched galaxies, suggests that the dust-obscured star formation is taking place in the core of the galaxies. This process could morphologically transform a galaxy and therefore make it more compact.

4. Discussion and Conclusion

We had the opportunity to expand our knowledge of one of the most studied parts of the sky (the GOODS-South field) by adding a new layer - a new wavelength - over this region. The characteristics of the GOODS-ALMA survey allow us to detect a population of very massive star-forming galaxies at $z = 2$–4 ($z_{med} = 2.7$, $M_{\star,med} = 8.5 \times 10^{10} M_{\odot}$). We investigated their gas reservoirs, their depletion times and their sizes. We show that a significant part of our sample (~40%) exhibits abnormal low gas fractions. With their high star formation rates and without a gas refill mechanism, they will consume their gas reservoirs in a typical time of 100–200 Myrs. The compact submillimetre sizes of our sample are similar to the H-band sizes observed for $z \sim 2$ elliptical galaxies with comparable stellar masses, suggesting that they are building their compact bulges. All these different elements lead us to believe that the galaxies detected in the GOODS-ALMA survey are the ideal progenitors of passive compact galaxies at $z \sim 2$. The large fraction of galaxies with short depletion time, low gas fractions among those hosting an AGN, suggest that the AGN, by a starvation process, can prevent the gas refill of these galaxies. The transformation of the gas into the stars of these galaxies can induce a rapid transition between star-forming and passive galaxies, without needing to invoke an additional quenching mechanism.

References

Abramson, L. E., Kelson, D. D., Dressler, A., *et al.* 2014, *apjl*, 785, 36
Blain, A. W., Smail, I., Ivison, R. J., *et al.* 2002, *Phys. Rep.*, 369, 111
Boquien, M., Burgarella, D., Roehlly, Y., *et al.* 2019, *aap*, 622, 103
Calzetti, D., Armus, L., Bohlin, R. C., *et al.* 2000 *ApJ*, 533, 682
Cowie, L. L., Songaila, A., Hu, E. M., *et al.* 1996, *AJ*, 112, 839
Draine, B. T., Aniano, G., Krause, O., *et al.* 2019, *ApJ*, 780, 172
Elbaz, D., Daddi, E., Le Borgne, D., *et al.* 2007, *A&A*, 468, 33
Franco, M., Elbaz, D., Bethermin, M., *et al.* 2018, *A&A*, 620, 152
Franco, M., Elbaz, D., Zhou, L., *et al.* 2020, ArXiv:2005.03040
Franco, M., Elbaz, D., Zhou, L., *et al.* 2020, ArXiv:2005.03043
Genzel, R., Tacconi, L. J., Combes, F., *et al.* 2012, *ApJ*, 746, 69
Grogin, N. A., Kocevski, D. D., Faber, S. M., *et al.* 2011, *ApJS*, 197, 35
Koekemoer, A. M., Faber, S. M., Ferguson, H. C., *et al.* 2011, *ApJS*, 197, 36
Leroy, A. K., Bolatto, A., Gordon, K., *et al.* 2011, *ApJ*, 737, 12
Luo, B., Brandt, W. N., Xue, Y. Q., *et al.* 2017, *ApJS*, 228, 2
Madau, P. & Dickinson, M. 2014, *araa*, 52, 415
Meurer, G. R., Heckman, T. M., & Calzetti, D. 1999, *ApJ*, 521, 64
Noeske, K. G., Weiner, B. J., Faber, S. M., *et al.* 2007, *ApJ*, 660, 43
Oke, J. B. & Gunn, J. E. 1983, *ApJ*, 266, 713
Pannella, M., Carilli, C. L., Daddi, E., *et al.* 2009, *ApJ*, 698, 116
Popesso, P., Concas, A., Morselli, L., *et al.* 2019, *MNRAS*, 483, 3
Rodighiero, G., Daddi, E., Baronchelli, I., *et al.* 2011, *ApJ*, 739, 40
Salpeter, E. E. 1955, *ApJ*, 121, 161
Schreiber, C., Pannella, M., Elbaz, D., *et al.* 2015, *A&A*, 575, 74
Schreiber, C., Elbaz, D., Pannella, M., *et al.* 2016, *A&A*, 785, 36
Tacconi, L. J., Genzel, R., Saintonge, A., *et al.* 2018, *ApJ*, 853, 179
van der Wel, A., Franx, M., van Dokkum, P. G., *et al.* 2014, *ApJ*, 788, 28
Williams, C. C., Giavalisco, M., Cassata, P., *et al.* 2016, *ApJ*, 780, 1
Wang, T., Schreiber, C., Elbaz, D., *et al.* 2019, *Nature*, 572, 211

Galaxy Evolution and Feedback across Different Environments
Proceedings IAU Symposium No. 359, 2020
T. Storchi-Bergmann, W. Forman, R. Overzier & R. Riffel, eds.
doi:10.1017/S1743921320004238

The physics of galaxy evolution with SPICA observations

Luigi Spinoglio, Juan A. Fernández-Ontiveros and Sabrina Mordini

Istituto di Astrofisica e Planetologia Spaziali - INAF, Rome,
Via Fosso del Cavaliere 100, 00133, Roma, Italia
emails: luigi.spinoglio@iaps.inaf.it, j.a.fernandez.ontiveros@gmail.com,
sabrina.mordini@uniroma1.it

Abstract. The evolution of galaxies at Cosmic Noon ($1 < z < 3$) passed through a dust-obscured phase, during which most stars formed and black holes in galactic nuclei started to shine, which cannot be seen in the optical and UV, but it needs rest frame mid-to-far IR spectroscopy to be unveiled. At these frequencies, dust extinction is minimal and a variety of atomic and molecular transitions, tracing most astrophysical domains, occur. The Space Infrared telescope for Cosmology and Astrophysics (SPICA), currently under evaluation for the 5th Medium Size ESA Cosmic Vision Mission, fully redesigned with its 2.5-m mirror cooled down to $T < 8$ K will perform such observations. SPICA will provide for the first time a 3-dimensional spectroscopic view of the hidden side of star formation and black hole accretion in all environments, from voids to cluster cores over 90% of cosmic time. Here we outline what SPICA will do in galaxy evolution studies.

Keywords. telescopes, galaxies: evolution, galaxies: active, galaxies: starburst, quasars: emission lines, galaxies: ISM, galaxies: abundances, galaxies: high-redshift, infrared: galaxies

---

1. Introduction

Most of the activity in galaxy evolution, the formation of stars and supermassive black holes at the center of galaxies, took place more than six billion years ago, with a sharp drop to the present epoch (e.g., Madau & Dickinson 2014, Fig. 1). Since most of the energy emitted by stars and accreting SMBHs is absorbed and re-emitted by dust, understanding the physics of galaxy evolution requires infrared (IR) observations of large, unbiased galaxy samples spanning a range in luminosity, redshift, environment, and nuclear activity. From *Spitzer* and *Herschel* photometric surveys the global Star Formation Rate (SFR) and Black Hole Accretion Rate (BHAR) density functions have been *estimated* through measurements of the bolometric luminosities of galaxies (Le Floc'h *et al.* 2005; Gruppioni *et al.* 2013; Delvecchio *et al.* 2014). However, such integrated measurements could not separate the contribution due to star formation from that due to BH accretion (see, e.g., Mullaney *et al.* 2011). This crucial separation has been attempted so far through modelling of the spectral energy distributions and relied on model-dependent assumptions and local templates, with large uncertainty and degeneracy. On the other hand, determinations from UV (e.g. Bouwens *et al.* 2007) and optical spectroscopy (e.g., from the *Sloan* Digital Sky Survey, Eisenstein *et al.* 2011) track only marginally (∼10%) the total integrated light (Fig. 1). X-ray analyses of the BHAR, in turn, are based on large extrapolations and possibly miss a large fraction of obscured objects. Furthermore, the SFR density at $z > 2$–3 is very uncertain, since it is derived from UV surveys, highly affected

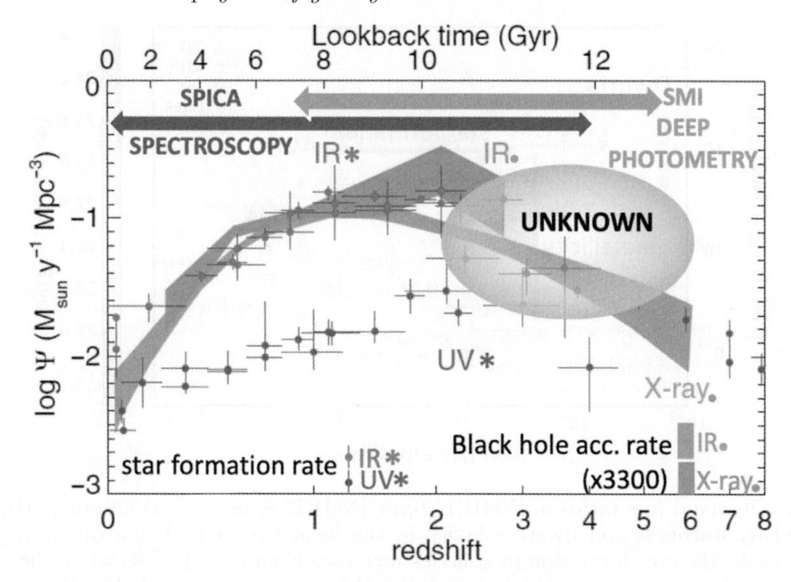

Figure 1. Estimated star-formation rate densities from the far-ultraviolet (blue points) and far-IR (red points) photometric surveys (figure adapted from Madau & Dickinson 2014). The estimated BHAR density, scaled up by a factor of 3300, is shown for comparison (in green shading from X-rays and light blue from the IR).

by dust extinction. As opposite, through IR emission lines, the contributions from stars and BH accretion can be separated. *SPICA* spectroscopy will allow us to directly measure redshifts, SFRs, BHARs, metallicities and dynamical properties of gas and dust in galaxies at lookback times down to about 12 Gyrs. *SPICA* spectroscopic observations will allow us for the first time to redraw the SFR rate and BHAR functions (Fig. 1) in terms of measurements directly linked to the physical properties of the galaxies.

The mid- to far-IR spectral range hosts a suite of atomic and ionic transitions, covering a wide range of excitation, density and metallicity, directly tracing the physical conditions in galaxies. Ionic fine structure lines (e.g. [NeII], [SIII], [OIII]) probe HII regions around hot young stars, providing a measure of the SFR and the gas density. Lines from highly ionized species (e.g. [OIV], [NeV]) trace the presence of AGN and can measure the BHAR (Spinoglio & Malkan 1992). Through line ratio diagrams, like the *new IR BPT diagram* (Baldwin *et al.* 1981; Fernández-Ontiveros *et al.* 2016, Fig. 2), IR spectroscopy can separate the galaxies in terms of both the source of ionization – either young stars or AGN excitation – and the gas metallicity, during the dust-obscured era of galaxy evolution ($0.5 < z < 4$).

2. SPICA observations of galaxy evolution

For a complete description of the SPace Infrared telescope for Cosmology and Astrophysics (*SPICA*), we refer to Roelfsema *et al.* (2018). We concentrate here on the work that *only* an IR observatory such as SPICA will be able to do to unveil galaxy evolution.

(*a*) SPICA will provide for the first time a 3-dimensional spectroscopic view of the hidden side of star formation and black hole accretion in all environments, from voids to cluster cores over 90% of cosmic time. This will lead to a complete census – as a function of cosmic time – of:

• the star formation in galaxies from low-mass dust-poor star-forming galaxies to high-mass heavily dust-obscured starbursts, setting a fundamental new benchmark for

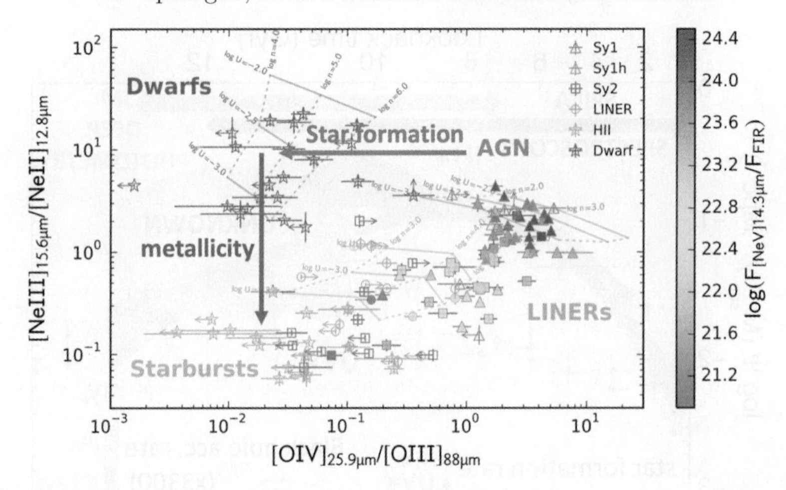

Figure 2. Observed line ratios of [NeIII]15.6μm/[NeII]12.8μm vs. [OIV]26μm/[OIII]88μm for AGN, LINER, starburst and dwarf galaxies in the local Universe. These ratios are powerful diagnostic tools: the star formation in galaxies increases from right to left, while the gas metallicity increases from the top to the bottom. For this reason we named this diagram the *new BPT diagram* in the IR (Fernández-Ontiveros *et al.* 2016). The active galaxies symbols have been color-coded from blue to red, according to the value of the ratio of the [NeV14.3]μm line to the FIR continuum as measured with *Herschel-PACS* spectra at $\sim 160\mu$m, corresponding the AGN dominance, taken from Fernández-Ontiveros *et al.* (2016). The solid and dashed lines show standard photoionization models for AGN, LINERS galaxies and dwarf *low-metallicity* galaxies.

cosmological models, allowing unambiguous tests of the physics driving star, galaxy and large-scale structure formation.

• the black hole accretion in the nuclei of galaxies, including also heavily obscured Compton thick nuclei.

From these observations, we will be able to derive both the average cosmic star-formation and accretion histories of the Universe as well as the physical conditions of each individual galaxy through their spectra in the large cosmic time interval since about 12 billion years (Spinoglio *et al.* 2017). SPICA will allow us to understand the physical origin of the dramatic change in the efficiency of star formation and accretion around cosmic noon, one of the major challenges for theoretical models in present-day cosmology.

(*b*) SPICA will measure if and how energetic (radiative and mechanical) feedback from starburst and AGN influences galaxy evolution. SPICA will determine the role of active galactic nuclei in quenching star formation in galaxies and address the origin of the low stellar to dark matter halo mass ratio in massive galaxies (González-Alfonso *et al.* 2017).

(*c*) SPICA will determine how galaxies build up their metals and dust during the last ten Gyrs of cosmic time and how are these metals recycled in the interstellar medium and injected into the circum-galactic medium (Fernández-Ontiveros *et al.* 2017). IR line based tracers will allow SPICA to peer through the heavily opaque medium of main-sequence galaxies at the cosmic noon and probe their true chemical ages. Galaxies at $z \sim 2-3$ show extremely low metallicities in optically-based traces, down to 0.8 dex below solar (Onodera *et al.* 2016), which is in conflict with the large amounts of dust inferred from IR photometry (Fig. 1), suggesting much higher metallicities. SPICA will also probe the first production of dust and metals, assess the presence of hot dust in the most distant galaxies ($z > 6$) and determine the physical properties of infrared bright galaxies and quasar hosts at the epoch of re-ionization.

2.1. *SPICA observational strategy to unveil galaxy evolution*

We briefly show here some of the results of the scientific work that has been done in preparation of the galaxy evolution studies that will be performed with the SPICA mission.

In the context of the development of a coherent observing strategy, we have considered a step by step observational sequence, aimed at optimising the measure of basic astrophysical quantities in galaxies through their evolution in the last 12 Gyrs. This sequence can be summarised as follows:

(i) Deep spectrophotometric and photometric surveys in the mid-IR, using the SPICA Mid-IR Instrument (SMI) of large enough fields (of order 1−10 deg^2) down to very faint flux limits (i.e. 3−12μJy in the continuum at 34μm and $2-5 \times 10^{-20}$ W/m^2 in the spectral lines). These observations will be complemented with the B-fields with BOlometers and Polarizers (B-BOP) camera at 70μm (at a 30μJy depth). *This survey will provide a 3-dimensional view of galaxy evolution.*

(ii) Identification of a sample of galaxies based on the outcome of the above surveys, in terms of redshift, mid-IR luminosity, classification (i.e. Starburst-dominated or AGN-dominated), characterisation of the stellar masses, dust masses and (where possible) bolometric luminosities using available multifrequency data.

(iii) Deep follow-up observations with the grating spectrometers, both the SPICA Far-Infrared Spectrometer (SAFARI) and SMI, at medium spectral resolution, of above defined sample. *This will provide a complete spectral atlas of galaxies as a function of redshift, luminosity and stellar mass.*

From the spectral atlas of galaxies we will determine their physical properties as a function of cosmic time: the SFR, the BHAR, the metal abundances, the outflow/infall occurrence, etc.

2.2. *Predictions of the SMI ultra-deep survey*

Following the work already presented in Kaneda *et al.* (2017) and Gruppioni *et al.* (2017), we are planning to perform an ultra-deep survey of 1 deg^2 with the SMI imaging low-resolution spectrometer (Kaneda *et al.* 2018) down to the photometric limit of 3μJy in the continuum at 34μm in a typical observing time of about ∼600 hours.

We show in Fig. 3 the predictions of how the SPICA SMI ultra-deep survey of 1 deg^2 will fill the luminosity-redshift plane for Star Forming Galaxies (left diagram) and for AGN (right diagram). The Star Forming galaxies are detected through the Polycyclic Aromatic Hydrocarbon (PAH) features at 6.2, 7.7, 8.6, 11.25 amd 17.0 μm, while the AGN can be detected through a set of high-ionization fine-structure lines, [NeVI]7.65μm, [SIV]10.5μm, [NeV]14.3 and 24.3μm, [NeIII]15.5μm, and [OIV]15.9μm.

We have used the far-IR luminosity functions as derived from *Herschel* photometric observations (Gruppioni *et al.* 2013) and the IR lines and PAH features calibrations to the total IR luminosities from observations in the local Universe. The details of these simulations can be found in Mordini *et al.* (2020, in preparation).

2.3. *Predictions of SAFARI follow-up pointed spectroscopic observations*

Following step *ii)* outlined in Section 2.1, i.e. the definition of a suitable sample of galaxies, of order 1,000 objects, we have made predictions of the observability of such sample with the SAFARI spectrometer (Roelfsema *et al.* 2018). We show these predictions in Fig. 4, where, on the left, the luminosity-redshift plane for Star Forming Galaxies through the detection of the Polycyclic Aromatic Hydrocarbon (PAH) feature at 17μm is shown. Analogously, on the left diagram of Fig. 4, the luminosity-redshift plane for

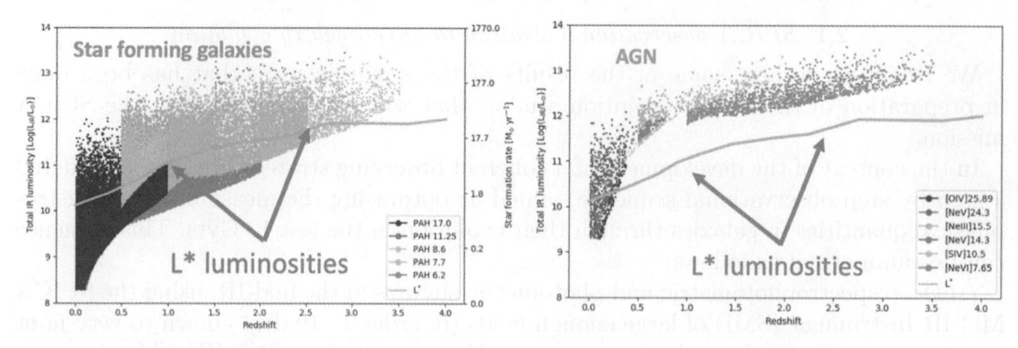

Figure 3. Left: Redshift-Luminosity diagram with the simulation of the SMI ultra-deep survey of 1 deg^2 showing the Star Forming Galaxies detections in the various PAH spectral features. The green solid line shows the knee of the Luminosity Functions, as a function of redshift. **Right:** Same, but simulated with the AGN detections in high-ionization fine-structure lines.

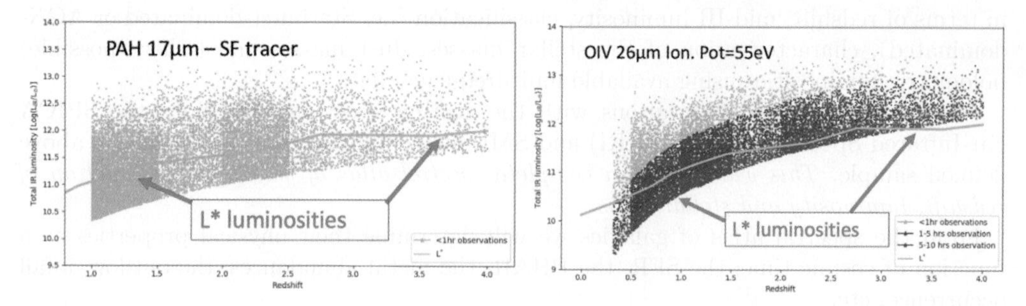

Figure 4. Left: Redshift-Luminosity diagram with the simulation of the SAFARI follow-up pointed observations of Star Forming Galaxies with the PAH 17μm spectral features. The green solid line shows the knee of the Luminosity Functions, as a function of redshift. **Right:** Same, but simulated with the AGN detections of the [OIV]26μm fine-structure line.

AGN is shown, where the detections by SPICA-SAFARI are made in the [OIV]25.9μm line. We can see from these predictions that SPICA will be able to measure the SFR and the BHAR of galaxies far below the knee of the luminosity functions at each redshift, in particular beyond $z \sim 4$ for Star Forming galaxies and up to $z \sim 3.5$ for AGN.

Acknowledgements

This paper is dedicated to the memory of Bruce Swinyard, who initiated the SPICA project in Europe, but unfortunately died on 22 May 2015 at the age of 52. He was ISO-LWS calibration scientist, Herschel-SPIRE instrument scientist, first European PI of SPICA and first design lead of SAFARI. We acknowledge the whole SPICA Collaboration Team, as without its multi-year efforts and work this paper could not have been possible. We also thank the SPICA Science Study Team appointed by ESA and the SPICA Galaxy Evolution Working Group. LS and JAFO acknowledge financial support by the Agenzia Spaziale Italiana (ASI) under the research contract 2018-31-HH.0.

References

André, Ph., *et al.* 2019, *PASA*, 36, e029
Baldwin, J. A., Phillips, M. M., Terlevich, R., *et al.* 1981, *PASP*, 93, 5
Bouwens, R. R., *et al.* 2007, *ApJ*, 670, 928
Delvecchio, I., *et al.* 2014, *MNRAS*, 439, 2736
Eisenstein, D. J., *et al.* 2011, *AJ*, 142, 72

Fernández-Ontiveros, J. A., *et al.* 2016, *ApJS*, 226, 19
Fernández-Ontiveros, J. A., *et al.* 2017, *PASA*, 34, e053
González-Alfonso, E., *et al.* 2017, *PASA*, 34, e054
Gruppioni, C., *et al.* 2013, *MNRAS*, 432, 23
Gruppioni, C., *et al.* 2017, *PASA*, 34, e055
Kaneda, H., *et al.* 2017, *PASA*, 34, e059
Kaneda, H., *et al.* 2018, Proceedings of the SPIE, Volume 10698, id. 106980C
Le Floc'h, E., *et al.* 2005, *ApJ*, 632, 169
Madau, P. & Dickinson, M. 2014, *ARA&A*, 52, 415
Mullaney, J. R., *et al.* 2011, *MNRAS*, 414, 1082
Onodera, M., *et al.* 2016, *ApJ*, 822, 42
Roelfsema, P. R., *et al.* 2018, *PASA*, 35, e030
Spinoglio, L. & Malkan, M. A. 1992, *ApJ*, 399, 504
Spinoglio, L., *et al.* 2017, *PASA*, 34, e057

Galaxy Evolution and Feedback across Different Environments
Proceedings IAU Symposium No. 359, 2020
T. Storchi-Bergmann, W. Forman, R. Overzier & R. Riffel, eds.
doi:10.1017/S1743921320004329

Distant quasar host galaxies and their environments with multi-wavelength 3D spectroscopy

Andrey Vayner

Department of Physics and Astronomy, Johns Hopkins University, Bloomberg Center, 3400 N. Charles St., Baltimore, MD 21218, USA

Abstract. We have conducted a multi-wavelength survey of distant ($1.3 < z < 2.6$) luminous quasars host galaxies using the Keck integral field spectrograph (IFS) OSIRIS and laser guide star adaptive optics (LGS-AO) system, ALMA, HST and VLA. Studying distant quasar host galaxies is essential for understanding the role of active galactic nuclei (AGN) feedback on the interstellar medium (ISM), and its capability of regulating the growth of massive galaxies and their supermassive black holes (SMBH). The combination of LGS-AO and OSIRIS affords the necessary spatial resolution and contrast to disentangle the bright quasar emission from that of its faint host galaxy. We resolve the nebular emission lines Hβ, [OIII], [NII], Hα, and [SII] at a sub-kiloparsec resolution to study the distribution, kinematics, and dynamics of the warm-ionized ISM in each quasar host galaxy. The goal of the survey was to search for ionized outflows and relate their spatial extent and energetics to the star-forming properties of the host galaxy. Combining ALMA and OSIRIS, we directly test whether outflows detected with OSIRIS are affecting the molecular ISM. We find that several mechanisms are responsible for driving the outflows within our systems, including radiation pressure in low and high column density environments as well as adiabatic and isothermal shocks driven by the quasar. From line ratio diagnostics, we obtain resolved measurements of the photoionization mechanisms and the gas-phase metallicity. We find that the quasars are responsible for photoionizing the majority of the ISM with metalicities lower than that of gas photoionized by AGN in the low redshift systems. We are now obtaining detailed observations of the circumgalactic medium (CGM) of these systems with the Keck Cosmic Web Imager (KCWI). The gas in the CGM may play an essential role in the evolution of these galaxies.

Keywords. galaxies: active, evolution, high-redshift, jets, kinematics and dynamics, nuclei

---

1. Introduction

Feedback from active galactic nuclei (AGN) has become an integral part of galaxy evolution. Feedback from AGN is often used to explain the correlation between the mass of the SMBH and the mass and velocity dispersion of the host galaxy (Ferrarese & Merritt 2000; Gebhardt *et al.* 2000; McConnell & Ma 2013). Theoretically, transferring 0.1-5% of the AGN bolometric luminosity into an outflow can have a significant impact on the star formation properties of the host galaxy, which can help establish some of the observed local scaling relations (Hopkins & Elvis 2010; Zubovas & King 2012).

We have surveyed 11 radio-loud quasars in the distant Universe (z$\sim$2) to study the impact of quasar driven outflows on their respective host galaxies during the peak epoch of galaxy and SMBH growth (Madau & Dickinson 2014; Delvecchio *et al.* 2014). The observations were undertaken with adaptive optics and an integral field spectrograph (IFS) at the W. M. Keck observatory. The primary goal of the survey was to search for

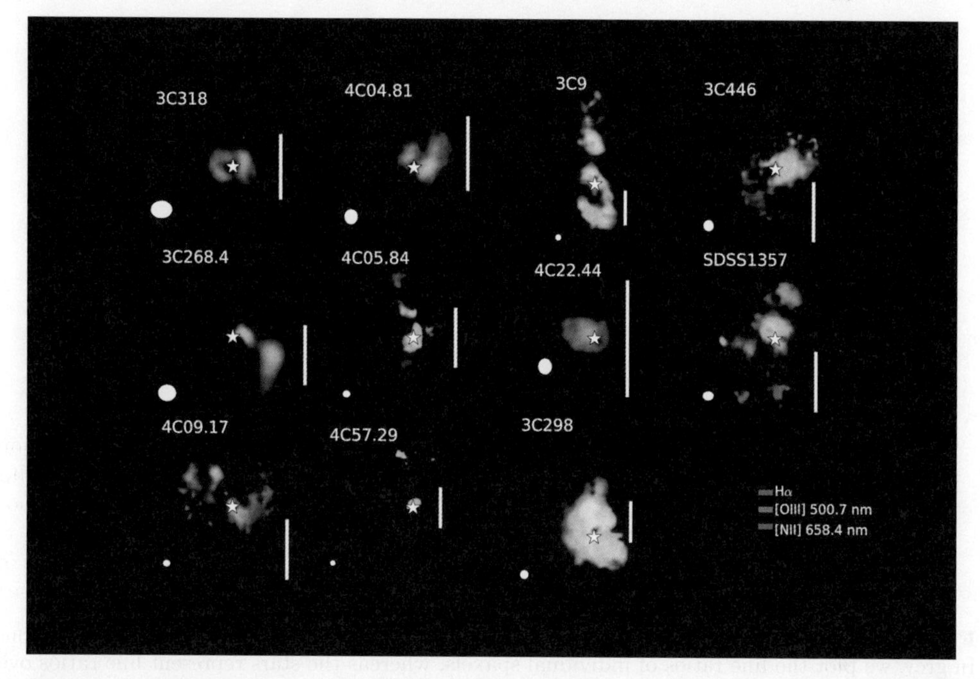

Figure 1. OSIRIS - LGS observations of 11 quasar host galaxies in our sample. Each image is a three-color composite where red is color-coded to Hα, green to [O III] and blue to [N II] . The location of the subtracted out quasar is represented by a white star, the ellipse in the lower-left corner shows the resolution of our observations (typically ~0.1″), and the scale bar to the right of each source represents 1″(8.4 kpc).

ionized outflows using nebular emission lines redshifted into the near-infrared, measure the galaxy masses, and explore the photoionization mechanism of the gas in these quasar host galaxies. In these proceedings, we present a snippet of the results from our survey. We present the details in two papers: Vayner *et al.* (2020a,b).

2. Discussion and Results

As part of our survey, we targeted optically luminous type-1 quasars with bolometric luminosities $>10^{46}$erg s^{-1}, with massive ($>10^9$ M$_\odot$) SMBHs. Such AGN typically outshine the light from their host galaxies; hence a careful removal of the unresolved quasar emission was necessary. We have devised a point spread function (PSF) subtraction routine using channels that contain emission from the broad-line region of the quasar to establish an image comprising of only the quasar emission. This image is then normalized and subtracted from the rest of the data channels leaving behind only the extended emission from the quasar host galaxy. Details on the PSF subtraction routine can be found in the following two papers from our survey: Vayner *et al.* (2016, 2020a).

Extended ionized emission is detected in 11 objects (Figure 1), comprising of photoionization from the quasar, massive young stars, and shocks. The quasar is responsible for producing the majority of the observed emission. We have placed line ratios from individual spaxels (spectral-pixels) on the BPT diagram as well as integrated values over the distinct regions to understand the spatially resolved photoionization mechanism and resolved gas-phase conditions (Figure 2). We find that the majority of our points lie outside the star formation and mixing sequence observed in the local Universe. Using photoionization models from Kewley *et al.* (2013a,b), we believe the main reason for the

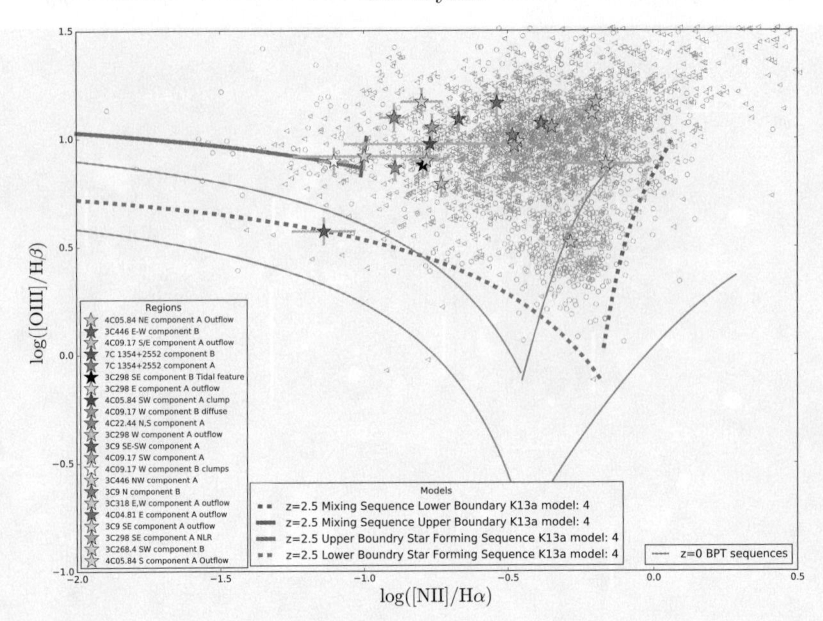

Figure 2. In this figure, we present line ratio diagnostics of individual resolved distinct regions. In grey, we plot the line ratios of individual spaxels, whereas the stars represent line ratios over integrated regions in each quasar host galaxy. The red and green curves show the best fit evolutionary models of the mixing and star-forming sequence from Kewley *et al.* (2013a), where on average, the gas photoionized by the quasar has a lower metallicity compared to that of galaxies with AGN in the local Universe. The boundary of the observed line ratios in the local Universe is shown with the teal curve.

offset is due to, on average lower metallicity in the gas photoionized by the quasar within our sample compared to that of local AGN.

We detect outflows in 10/11 sources on scales ranging from <1 kpc to 10 kpc. The outflow rates vary from 8-2400 $M_\odot$ yr^{-1}, with momentum flux ranging from 0.03-80 L_{AGN}/c, and energy rates of 0.01-1% L_{AGN}. In 5/11 sources, the momentum flux is above 2× L_{AGN}/c indicating that an adiabatic shock likely drives the outflows. The coupling efficiency in these objects is high enough for the outflows to have a significant impact on the star-forming properties of the host galaxy. In the rest of the objects, the outflows are likely driven by either radiation pressure on dust grains or an isothermal shock, and typically the coupling efficiencies in these sources are below the minimum value prescribed by theoretical work (Zubovas & King 2012). For the majority of the sources, the outflow rates at present are higher than those of star formation, and the paths of the outflow are inconsistent with regions of active star formation. The observed momentum flux ratios and coupling efficiencies between the kinetic luminosity of the outflow and bolometric luminosity of the quasar are within the range observed in other studies of ionized outflows in the distant universe (Figure 3).

For nine objects, we can measure both the galaxy velocity dispersion and the dynamical mass of the host galaxy. Combining with the mass of the SMBH, we can compare these systems to the local scaling relations between the mass of the SMBH and the velocity dispersion and mass of the galaxy. Our systems are both offset from the $M_\bullet$–σ and $M_\bullet$–M_* relationships. This offset indicates a substantial growth in stellar mass is necessary from z ∼ 2 to present-day if these systems are to evolve into the present-day elliptical galaxies. Combining with our previous results, we find that the galaxies experience feedback before assembling on the local scaling relations and before the ISM is enriched to a level observed in massive local galaxies.

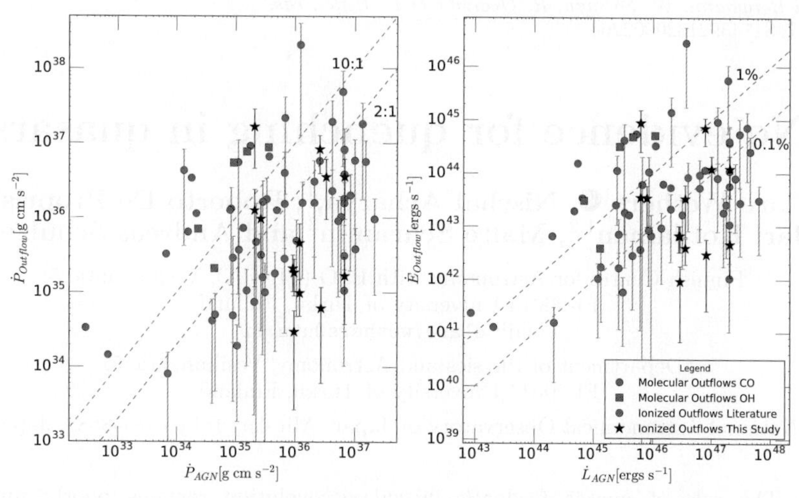

Figure 3. On the left, we plot the momentum flux of our outflows with a black star against the momentum flux of the quasar accretion disk. We plot a line of constant ratio at 2:1 between the momentum flux of the outflow and the accretion disk. Points above the 2:1 line represent outflows driven by an adiabatic shock, while points below the 2:1 are likely driven by an isothermal shock or radiation pressure on kpc scale. Red circles represent ionized outflows at z ∼ 2 computed in the same manner as our own. Blue points represent molecular outflows mainly at low redshift. On the right, we plot the kinetic luminosity of the outflow against the bolometric luminosity of the outflow. The green dashed curve represents the minimum coupling efficiency (0.1%) prescribed by theoretical work necessary for AGN driven outflows to affect the star-forming properties of their host galaxies.

We are currently undertaking an observing campaign to map the circum-galactic medium (CGM) of these systems to understand their subsequent evolution from z ∼ 2 to the present day. Likely, these systems host a massive gas reservoir in the CGM that can accrete material to fuel future star formation that will increase the stellar mass of the galaxies and enrich their ISM.

References

Delvecchio, I., Gruppioni, C., Pozzi, F., *et al.* 2014, *MNRAS*, 439, 2736–2754
Ferrarese, L. & Merritt, D. 2000, *ApJL*, 539, L9–L12
Gebhardt, K., Bender, R., Bower, G., *et al.* 2000, *ApJL*, 539, L13–L16
Hopkins, P. F. & Elvis, M. 2010, *MNRAS*, 401, 7–14
Kewley, L. J., Dopita, M. A., Leitherer, C. *et al.* 2013a, *ApJ*, 774, 100
Kewley, L. J., Maier, C., Yabe, K. *et al.* 2013b, *ApJL*, 774, L10
Madau, P. & Dickinson, M. 2014, *ARA&A*, 52, 415–486
McConnell, N. J. & Ma, C.-P. 2013, *ApJ*, 764, 184
Vayner, A., Wright, S. A., Do, T., *et al.* 2016, *ApJ*, 821, 64
Vayner, A., Wright, S. A., Murray, N., *et al.* 2020a, *ApJ*
Vayner, A., Wright, S. A., Murray, N., *et al.* 2020b, *ApJ*
Zubovas, K. & King, A. 2012, *ApJL*, 745, L34

Galaxy Evolution and Feedback across Different Environments
Proceedings IAU Symposium No. 359, 2020
T. Storchi-Bergmann, W. Forman, R. Overzier & R. Riffel, eds.
doi:10.1017/S1743921320002264

No evidence for quenching in quasars

Clare Wethers[1], Nischal Acharya,[1], Roberto De Propris[1], Jari Kotilainen[1,2], Malte Schramm[3] and Andreas Schulze[3]

[1]Finnish Centre for Astronomy with ESO (FINCA), Vesilinnantie 5, FI-20014 University of Turku, Finland
email: clare.wethers@utu.fi

[2]Department of Physics and Astronomy, Vesilinnantie 5, FI-20014 University of Turku, Finland

[3]National Astronomical Observatory of Japan, Mitaka, Tokyo 181-8588, Japan

Abstract. The role of quasar feedback in galaxy evolution remains poorly understood. Throughout this work, we explore the effects of negative feedback on star formation in quasar host galaxies, analysing two distinct populations of quasars. The first is a sample of high-redshift ($z > 2$) low-ionisation broad absorption line quasars (LoBALs) - a class of quasars hosting energetic mass outflows, in which we find evidence for prolific star formation ($>750 M_{\odot} yr^{-1}$) exceeding that of non-BAL quasars at the same redshift. The second is a population of 207 low-redshift ($z < 0.3$) quasars, in which we find an enhancement in the SFRs of quasar hosts compared to the underlying galaxy population, with no quasars residing in quiescent hosts over the last 2Gyr. Overall, we find no evidence for galaxy-wide quenching in either population, rather we suggest that the dominant effect of quasar activity is to enhance star formation in the galaxy.

Keywords. quasars: general, galaxies: general, galaxies: evolution, galaxies: active

1. Introduction

Understanding the interactions between quasars and their host galaxies is critical in building a coherent picture of galaxy evolution. In nearby galaxies tight correlations have been observed between the mass of the central super-massive black hole (M_{BH}) and that of the stellar bulge (e.g Kormendy & Ho 2013), leading to the idea that black holes and galaxies co-evolve. The mechanisms by which the black hole seemingly influences its host galaxy on scales beyond its sphere of influence remain poorly understood and as such, the origin of these tight correlations is still widely debated. Such observations are often explained by the presence of quasar feedback. Theoretical models predict quasar-driven outflows regulate black hole growth and star formation activity in the host by expelling gas from the galaxy which would otherwise fuel young stars and black hole accretion (e.g. Di Matteo *et al.* 2005; Fabian 2012; Carniani *et al.* 2016). In principle, this results in the quenching of star formation in the host. Perhaps the strongest evidence for this so-called *negative feedback* can be seen in the bright end of the galaxy luminosity function, where the number of bright galaxies is shown to decline more rapidly than models predict. Benson *et al.* (2003) for example demonstrate that basic cooling processes alone cannot account for this feature, finding instead that additional feedback processes are required. Indeed, observations have shown an anti-correlation between star formation in the host galaxy and the strength of quasar outflows (e.g. Farrah *et al.* 2010), indicating that these outflows act to suppress star formation in the host. On the other hand, semi-analytic models of galaxy assembly (Granato *et al.* 2004), invoke the same outflows to not only remove dense gas from the galaxy centre, but also to provide metal enrichment to the intergalactic medium (IGM). Such outflows may therefore also work to trigger regions of star formation in the galaxy by compressing cool, metal rich gas and allowing stars to form. Indeed, several studies have observed enhanced star formation in quasar hosts (e.g. Santini *et al.* 2012; Canalizo & Stockton 2001), implying that quasars may also act to enhance star formation in the galaxy via *positive feedback*. Understanding the interplay between negative and positive feedback mechanisms is important in building a comprehensive model of quasar-galaxy co-evolution. Here, we explore the impact of quasar feedback within two distinct quasar populations. The results presented are combined from a recent study on LoBALs at z$\sim$2 (Wethers *et al.* 2020) and an ongoing study of low-z quasars in GAMA (De Propris *et al.* 2020, *in prep*)., both of which focus the impact of quasar feedback on star formation in galaxies.

2. Quasars at z$\sim$2

Low-ionisation broad absorption line quasars (LoBALs) are an important, yet poorly understood subclass of quasars exhibiting direct evidence for energetic mass outflows. This makes them ideal laboratories in which to study the effects of quasar feedback. To this end, we make use of targeted *Herschel* SPIRE observations at 250, 350 and 500μm for a sample of 12 LoBALs at 2.0$<$z$<$2.5 - a peak epoch of both black hole accretion and star formation. Full details of the LoBAL sample, data reduction and methodology can be found in Wethers *et al.* (2020), along with a more thorough analysis of the results outlined in this section.

2.1. *Detection rates*

Using signal-to-noise maps of our LoBAL sample, we find three of the 12 LoBALs (25 per cent) are detected at $>5\sigma$ in all SPIRE bands. If quasar outflows are responsible for quenching star formation in the galaxy, we would expect a general decrease in the FIR detection rate of LoBALs compared to non-BAL quasar populations containing no such outflows according to an evolutionary BAL interpretation (e.g. Boroson *et al.* 1992), assuming the FIR emission is correlated with the star formation in the galaxy. To this end, we compare the detection rate of our LoBAL sample to that for a sample of 100 non-BAL quasars outlined by Netzer *et al.* (2016), which are broadly matched to our LoBAL sample in terms of both redshift and luminosity (L_{bol}). After adjusting for the different detection limit of each sample, we find the FIR detection rate of LoBALs to be higher than that of non-BALs by a factor of $\sim$1.6. We therefore conclude an enhancement in the FIR detection rate of LoBALs compared to their non-BAL counterparts, implying that quasar outflows work to enhance star formation (and thus the FIR flux) in its host on short timescales, rather than quenching the galaxy, although we cannot rule out quenching on timescales longer than the LoBAL lifetime.

2.2. *FIR SFRs*

To confirm whether the enhancement seen in the detection rate of LoBALs is indeed associated with higher star formation rates (SFRs) in these systems, we compare the inferred SFR at the detection limit of our sample to another sample of 20 non-BAL quasars from Schulze *et al.* (2017), for which SFRs have been derived from their 850μm fluxes. As such, we take the nominal 5σ flux threshold for our sample at 250μm (25.4mJy) and fit a modified blackbody (or *greybody*) curve to this single photometry point. The fitted greybody curve is integrated over the FIR wavelengths (8-1000μm) to calculate the FIR luminosity, L_{FIR}, which is then converted to a SFR following the methods outlined in

Table 1. The values and 1σ uncertainties of T_{DUST}, L_{FIR} and SFR inferred from the SED fitting for the sub-sample of detected LoBALs and for the stacked non-detections, where the 3σ upper limit on the SFR is instead given.

Name	$\mathbf{T_{DUST}}$ **[K]**	**log** $\mathbf{L_{FIR}}$ **[ergs^{-1}]**	$\mathbf{SFR_{FIR}}$ **[M$_\odot$yr^{-1}]**
SDSSJ0810+4806	$33.49^{+7.11}_{-5.70}$	$46.21^{+0.11}_{-0.12}$	740^{+220}_{-170}
SDSSJ0839+0454	$42.02^{+4.64}_{-4.32}$	$46.55^{+0.07}_{-0.08}$	1610^{+280}_{-260}
SDSSJ0943 -0100	$47.07^{+2.67}_{-2.50}$	$46.72^{+0.04}_{-0.04}$	2380^{+220}_{-210}
Stacked non-detections	$35.78\ ^{+13.98}_{-7.06}$	$45.77^{+0.09}_{-0.11}$	<440

Kennicutt & Evans (2012). This returns a crude lower limit on the SFRs of our detected targets of 640M$_\odot$yr^{-1}. Whilst three of the 12 LoBALs in our sample (25 per cent) are detected at $>5\sigma$ in all SPIRE bands, and thus lie above this lower SFR limit, just one target in Schulze *et al.* (2017) returns SFR$\,>640$M$_\odot$yr^{-1}, corresponding to 5 per cent of their sample. We therefore suggest that the enhancement we observe in the FIR detection rate of LoBALs, indicates an enhancement in the SFR of LoBALs compared to non-BAL quasars.

Having found evidence for enhanced star formation among LoBALs, we now seek to measure the individual SFRs for the three LoBALs in our sample detected at $>5\sigma$ in all *Herschel* SPIRE bands via SED fitting. To this end we combine the SPIRE photometry (250, 350 and 500μm) with additional photometry from *Herschel* PACS (70 and 100μm) and WISE (3.4, 4.6, 12.0 and 22.0μm). The full set of photometry (SPIRE + PACS + WISE) is fit with a two-component model, comprising a greybody template and a torus SED (Mor & Netzer 2012) to account for the potential contribution of quasar heating to the FIR emission. The fitting itself utilises a Markov-Chain Monte-Carlo (MCMC) method in order to obtain full posterior distributions on the best-fit model parameters and to marginalise over any nuisance parameters (Metropolis *et al.* 1953; Hastings 1970; Foreman-Mackey *et al.* 2013). Throughout the fitting we define three free parameters: the vertical scaling of both the torus SED (X_{TORUS}) and the greybody template (X_{SF}), and the dust temperature of the greybody template (T_{DUST}). Given the limited photometry tracing the cool dust emission, we adopt a fixed value of $\beta = 1.6$, consistent with the work of Priddey & McMahon (2001). Full details of the fitting routine can be found in Wethers *et al.* (2020). The SFR of each LoBAL is then estimated from the best-fit model by integrating over the FIR wavelengths (8-1000μm) of the greybody component. The resulting SFRs are presented in Tab. 1, along with the best-fit parameters derived from the fitting. The best-fit SEDs are also shown in Fig. 1. For each of the three detected LoBALs we derive high SFRs in the range 740-2380M$_\odot$yr^{-1}. These rates are consistent with the results of Pitchford *et al.* (2019), who find evidence for prolific star formation (SFR$\,\sim 2000$M$_\odot$yr^{-1}) in an FeLoBAL - a class of LoBAL with additional iron absorption features in their spectra.

2.3. *Stacking the non-detections*

Despite finding evidence for prolific star formation within our LoBAL sample, we note that the majority of our sample (75 per cent) remain undetected in at least one of the *Herschel* SPIRE bands. A mean weighted stack of these undetected targets returns a 3σ upper limit on the SFR of $\lesssim$440M$_\odot$yr^{-1}. Even among the non-detected targets we therefore cannot rule our prolific star formation. As such, we find no evidence to suggest that LoBAL outflows act to instantaneously suppress star formation in their hosts. Rather, star formation in LoBALs appears enhanced relative to non-BALs, suggesting that outflows may trigger an increase in star formation. However, due to the poor sensitivity of

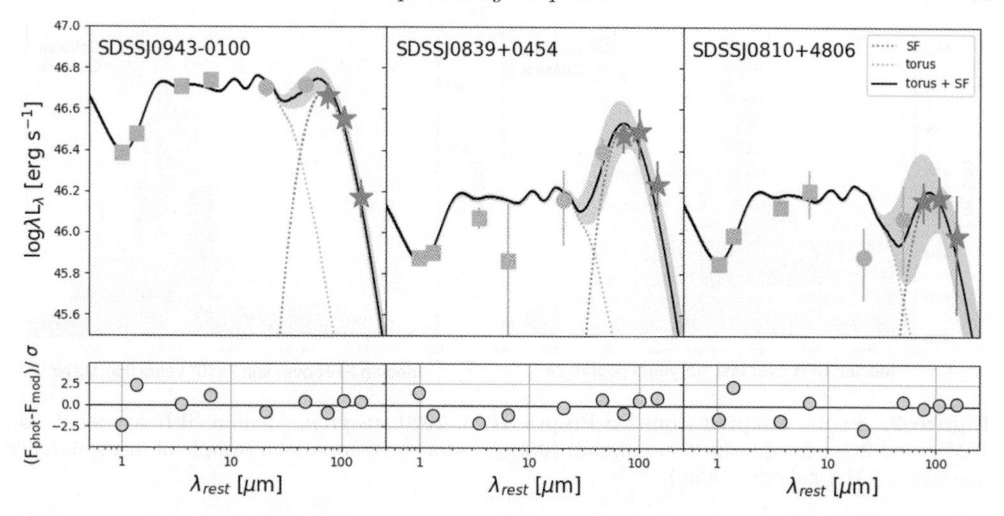

Figure 1. Figure taken from Wethers *et al.* (2020). **Upper:** Best-fit SED template based on the combined WISE (*blue squares*) + PACS (*orange circles*) + SPIRE (*pink stars*) photometry. The total model (*black*) is comprised of contributions from a hot torus (*dotted cyan*) and a star forming galaxy (*dotted pink*). Grey shaded regions denotes the 1σ uncertainty in the total model. Lower: Error weighted residuals of the best-fit model.

Herschel at z > 2, we cannot rule out the possibility of quenching in individual targets lying below our detection threshold. Furthermore, at these redshifts, we are unable to resolve the regions in which star formation is occurring, and so cannot rule out quenching within specific regions of the galaxy.

3. Quasars at z < 0.3

Having found no evidence that LoBAL quasars reside in quenched galaxies at z > 2, we now seek to test the impact of quasar feedback on star formation at low redshift. To this end, we make use of 207 confirmed quasars from the Large Quasar Astrometric Catalogue (LQAC-4) (Gattano *et al.* 2018) at redshifts $0.1 < z < 0.3$, overlapping the three equatorial fields (G09, G12 and G15) of the Galaxy and Mass Assembly survey (GAMA) - the redshifts and survey regions over which GAMA is most complete. Based on the mass and redshift distribution of these quasars, we select 100 realisations of N galaxies as a matched comparison sample, where $N = 207$: the number of quasars in our sample. One major advantage of using GAMA is the large amount of supplementary data available providing information on the SFRs, stellar masses and star formation histories (SFHs) of the catalogued galaxies (Liske *et al.* 2015). Based on this information, we compare the distribution of specific SFRs (sSFRs) within our quasar sample to that of the matched galaxy sample, finding higher sSFRs in quasars than in the underlying galaxy population. Furthermore, whilst a small fraction of the quasars in our sample reside in the so-called *green valley*, we find no evidence that any of our quasars reside in fully quiescent hosts. Furthermore, based on the SFH information provided by GAMA, we conclude that this apparent enhancement in SFR has occurred within the last 100Myr, whilst stellar populations older than ∼1Gyr appear largely indistinguishable from those of the matched inactive galaxies (Fig. 2). We therefore suggest that not only do the quasars in our sample reside in actively star-forming hosts, they also have not resided in quiescent galaxies at any point over the last 2Gyr.

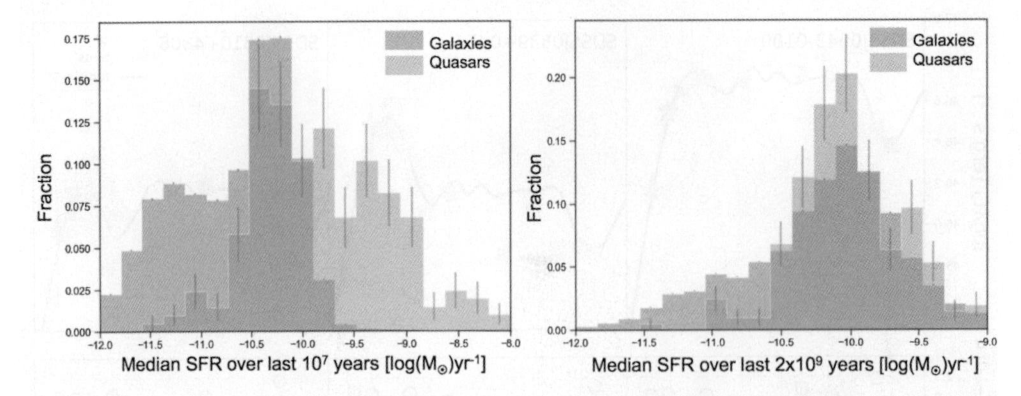

Figure 2. Figure adapted from De Propris *et al.* 2020, *in prep*. Median SFR averaged over 10Myr and 2Gyr for quasar host galaxies (*orange*) and the comparison sample of mass-matched inactive GAMA galaxies (*blue*).

4. Summary

Overall, we find no evidence for instantaneous galaxy-wide quenching either at $z > 2$ or at $z < 0.3$. Specifically, LoBALs at $z > 2$ exhibit higher SFRs than non-BALs, implying energetic mass outflows enhance star formation over short timescales. However, our results do not exclude quenching over longer timescales or galaxy-scale quenching in individual systems lying below our detection threshold. Similarly, we detect an enhancement in the SFRs of $z < 0.3$ quasars compared to the underlying galaxy population, finding no evidence for quasars existing in quiescent galaxies over the last 2Gyr. In future, spectra with high angular resolution will serve to resolve regions of quenching and active star formation in these quasars.

References

Benson, A. J. & Piero M. 2003, *MNRAS*, 344.3, 835–846
Boroson, T. A. 1992, *ApJ*, 399, L15–L17
Canalizo, G. & Stockton, A. 2001, *ApJ*, 555, 719
Carniani, S., *et al.* 2016, *A&A*, 591, A28
De Propris, R., Arachya, N., Kotilainen, J., *et al.* 2020, *in prep*
Di Matteo, T., Springel, V., & Hernquist, L. 2005, *Nature*, 433, 604
Fabian, A. C. 2012, *A&A*, 50, 455–489
Farrah, D., *et al.* 2010, *ApJ*, 717.2, 868
Foreman-Mackey, D., *et al.* 2013, *Publ. Astron. Soc. Pac.*, 125, 306
Gattano, C., *et al.* 2018 *A&A*, 614, A140
Granato, G. L., *et al.* 2004, *ApJ*, 600.2, 580
Hastings, W. K. 1970, 97–109
Kennicutt Jr, R. C. & Evans, N. J. 2012 *A&AR*, 50, 531–608
Kormendy, J. & Ho, L. C. 2013, *A&A*, 51, 511–653
Liske, J., *et al.* 2015, *MNRAS*, 452.2, 2087–2126
Metropolis, N., *et al.* 1953, *J. Chem. Phys.*, 21.6, 1087–1092
Mor, R. & Netzer, H. 2012, *MNRAS*, 420, 526–541
Netzer, H., Lani, C., Nordon, R., *et al.* 2016, *ApJ*, 819, 123
Pitchford, L. K., *et al.* 2019, *MNRAS*, 487.3, 3130–3139
Priddey, R. S. & McMahon, R. G. 2001, *MNRAS*, 324, L17
Santini, P., *et al.* 2012, *A&A*, 540, A109
Schulze, A., *et al.* 2017, *ApJ*, 848.2, 104
Wethers, C. F., Kotilainen, J., Schramm, M., *et al.* 2020, *MNRAS*, 498, 1469

Cristina Furlanetto

Maximilien Franco

Clare Wethers

Vincenzo Mainieri

Andrey Vayner

Session 3: Evolution of galaxies and AGN in high-density environments from high to low redshifts

Galaxy Evolution and Feedback across Different Environments
Proceedings IAU Symposium No. 359, 2020
T. Storchi-Bergmann, W. Forman, R. Overzier & R. Riffel, eds.
doi:10.1017/S1743921320002227

INVITED LECTURES

High density galaxy environments — the radio view

Martin J. Hardcastle

Centre for Astrophysics Research, University of Hertfordshire,
College Lane, Hatfield AL10 9AB, UK
email: m.j.hardcastle@herts.ac.uk

Abstract. Radio-loud active galaxies are widely believed to have a strong impact on their environments, and often lie in groups and clusters of galaxies. In this article I summarize what we can understand about the sources' effects on their surroundings from the perspective of radio galaxy physics, with special reference to the energetics of the impact on the external medium and its inference from large statistical studies of radio galaxies.

Keywords. galaxies: active, galaxies: jets, galaxies: clusters: general, radio continuum: galaxies

1. Introduction

In this article I aim to summarize the evidence for 'feedback' from the kpc-scale lobes of powerful radio galaxies on the gas-rich environments that typically surround them. I will discuss our current understanding of the properties of radio lobes, derived from observations and modelling of individual sources, and the statistical constraints that it is currently possible to place on the energetic impact of the population of radio galaxies as a whole. Due to space limitations, I do not consider the topic of how the active galactic nucleus (AGN) is connected to the large-scale ambient conditions, which would be necessary for a complete understanding of the feedback loop. For reviews on cluster feedback in general, the reader is referred to McNamara & Nulsen (2012). Hardcastle & Croston (2020) review radio lobe and AGN physics and some of the discussion in this article summarizes points of view that are presented in more detail there.

2. Physics of radio galaxies

Radio galaxies (also radio-loud AGN or RLAGN) are AGN-driven systems in which powerful relativistic jets are generated close to the central black hole and propagate to very large (up to Mpc) physical scales. Radio emission is the result of synchrotron radiation generated by relativistic electrons/positrons and magnetic fields in the jets and the lobes that surround them. A toy (and probably excessively simplistic) model of a radio galaxy is one in which jets switch on with some kinetic power Q at time $t = 0$ and maintain constant power until they switch off at some time T which may be of the order of 10^8 years later (Hardcastle 2018). From the point of view of feedback studies, we want to know what Q and T are and what fraction of the total released power QT is injected into the various components of the environment of the RLAGN. Of course, in reality, the jet power Q may vary substantially over the lifetime of the large-scale radio structures, perhaps giving rise to multiple outbursts of RLAGN activity, but the simple model above provides a starting point for discussion.

The fact that the kpc-scale components of powerful RLAGN must have *some* effect on their environments on these scales was realised very early on (e.g. Scheuer 1974). Since radio lobes are polarized at low frequencies, they cannot be filled with material at the same density and pressure as the hot external medium; instead they must push the external medium out of the way as they expand. In doing so, at the very least, they must do the work involved in compressing and lifting that material. To do this, the lobes must be at least in pressure balance with ambient material, but their internal pressure could be significantly higher (e.g. Begelman & Cioffi 1989) and so they could in principle drive strong shocks through the ambient medium, transferring significantly more energy to the medium than a naive $p\delta V$ estimate would suggest (Kaiser & Alexander 1997). All of this was known before the first detailed X-ray observations were made.

X-ray observations with the required sensitivity to study radio galaxy dynamics began to be available in large numbers with the advent of *Chandra*, and gave us three key pieces of information:

(*a*) X-ray cavities are observed associated with many radio lobes, confirming that the lobes do indeed remove ambient medium at X-ray temperatures from the space they occupy. This was already starting to become clear with *ROSAT* data, see e.g. Böhringer *et al.* 1993; Hardcastle *et al.* 1998, and is now very well known as a result of *Chandra* studies (e.g. Fabian *et al.* 2006; Bîrzan *et al.* 2012)

(*b*) Strong shocks are rare (though not unknown: Croston *et al.* 2007). However, weak shock-like enhancements of X-ray emission are seen around many radio lobes (Kraft *et al.* 2003; Forman *et al.* 2007; Croston *et al.* 2009; Randall *et al.* 2015).

(*c*) Crucially, measurements of inverse-Compton emission from lobes allow us to measure the internal energy density in the radiating particles and field and to show that in general, for powerful sources, the internal pressure of the lobes does not *greatly* exceed that of the external material (Hardcastle *et al.* 2002b).

For the most powerful sources (Fanaroff-Riley class II or FRII objects: Fanaroff & Riley 1974), a consistent picture emerges from these observations. Jets are light (probably electron-positron) bulk-neutral outflows, and the energy density on large scales is entirely explained by the relativistic leptons and magnetic field that give rise to the synchrotron and inverse-Compton emission (Hardcastle *et al.* 2002b). Initially these sources drive a strong shock into their environment, but the internal pressure in the lobes and the ram pressure due to the jet both drop with lobe size (i.e. with time) and by the time sources have grown to scales of hundreds of kpc they are in rough pressure balance transversely while continuing to expand supersonically longitudinally (Ineson *et al.* 2017). They are, however, surrounded by a shell of swept up, shocked and compressed gas and it is this that can be seen in X-ray observations. Numerical modelling shows that the fraction of the total energy in this shocked shell is about equal to the total energy stored in the lobes themselves (Hardcastle & Krause 2013; English *et al.* 2016) and all of this energy will eventually be available to heat the external medium, though it remains unclear whether it does so on the most useful scales (Omma & Binney 2004; Hardcastle & Krause 2013, and see further Section 4.3, below). One can gauge the energetic impact of a radio galaxy of this type, and/or its environment simply by making radio observations and inferring the internal energy density or pressure (Croston *et al.* 2017).

Low-power radio galaxies (Fanaroff-Riley class I, FRI) present a more complex picture. Firstly, inverse-Compton observations that would allow a direct measurement of magnetic field strengths do not exist in general (Hardcastle & Croston 2010). Secondly, it has been clear for some time that the radiating particles and field cannot on their own balance the pressure of the X-ray observed external environment for these objects (e.g. Morganti *et al.* 1988; Böhringer *et al.* 1993; Hardcastle *et al.* 1998; Worrall & Birkinshaw 2000; Croston *et al.* 2003; Dunn *et al.* 2005). In other words, there must be some other, dominant contribution to the internal pressure that we are not seeing, as discussed in detail by Croston *et al.* (2018). Various possibilities exist. In my view the most likely is that the entrained baryonic matter that is required to decelerate the large-scale jets from relativistic speeds is heated to high temperatures and comes to dominate the energetics of the source (Wykes *et al.* 2013; Croston & Hardcastle 2014). This material is not directly observable and so other methods, such as the 'cavity power' method of Bîrzan *et al.* (2012) must be applied to infer the radio source's energetic impact: these rely on expensive X-ray observations. Even when, as in the best-studied FRI objects, jet power can be inferred, it is not always clear that the jet is having any effect at all on the rapidly cooling central regions of the host environment (Hardcastle *et al.* 2002a). In systems such as Perseus A, where the radio source undergoes continuous, small-scale outbursts (Fabian *et al.* 2006) the AGN is certainly well coupled to the cooling material, but that does not appear always to be the case.

3. The impact of next-generation radio surveys

To make a quantitative estimate of the effect of radio galaxies (or more generally radio-loud AGN, which include the radio-loud quasars) it is first necessary to be able to find and characterize the RLAGN population. For many years radio astronomy has lagged behind other wavebands (particularly optical and infrared) in terms of its capability to survey the sky. Although wide-area sky surveys such as the 3C survey and its descendants (Laing *et al.* 1983) appeared very early on in the history of radio astronomy, these were both extremely shallow (thus selected only the most radio-luminous sources at any given redshift) and very low in angular resolution, so that radio sources that they detected could not easily be identified with their optical counterparts. The radio telescopes that succeeded these early survey instruments, with the capability to make high-resolution images (such as the 5-km telescope, now the Ryle Telescope, or the Very Large Array (VLA)) also tended to be based on large dishes and had correspondingly small fields of view, and so were of limited use in carrying out surveys, although extremely valuable in following up sources that were already known. The VLA survey with the best resolution to date, FIRST (Becker *et al.* 1995), unfortunately has little or no sensitivity to the extended structures that act as calorimeters for powerful RLAGN.

Next-generation radio surveys improve, or will improve, on the VLA's survey capabilities by having some or all of the following features:

- Few-arcsec resolution for source identification
- Much larger instantaneous field of view
- Much higher sensitivity
- Much better instantaneous *uv* plane coverage, in particular a combination of long and short baselines giving sensitivity to extended structure.

Here I focus on the LoTSS survey (Shimwell *et al.* 2017). This is at the time of writing the furthest advanced of the next-generation continuum surveys with arcsec resolution, and is carried out with the Low-Frequency Array (LOFAR: van Haarlem *et al.* 2013). LoTSS has a wide tier and a deep tier — only the first of these will be discussed in detail here. The wide tier aims to survey the whole sky above $\delta = 0°$ at 144 MHz, initially with 6-arcsec resolution, to a sensitivity better than 100 μJy beam^{-1} at favourable declinations. At the time of writing the second LoTSS data release (DR2), which will cover 5,700 deg^2 at a typical rms noise level of 70 μJy beam^{-1} and contains around 4.4 million radio sources, is in preparation. The scientific results presented here are taken from LoTSS DR1, covering 424 deg^2 in the HETDEX Spring field (Shimwell *et al.* 2019) with comparable quality to the DR2 area. Not only is the depth of the LoTSS survey much better than that of FIRST (for synchrotron emission with $\alpha = 0.7$ it goes ten times deeper while having essentially the same spatial resolution) but the images, though surface-brightness limited, are sensitive to structure up to degree scale. LoTSS thus allows us to carry out the best characterization yet of the low-z RLAGN population.

4. Feedback results from LoTSS

4.1. *Ubiquity of radio AGN activity*

Sabater *et al.* (2019) investigated the level of RLAGN activity in the overlap between the well-known MPAJHU SDSS-based spectroscopic sample of galaxies (Brinchmann *et al.* 2004), which also provides SDSS-based galaxy mass estimates, and the DR1 dataset. A crucial complication in extragalactic surveys is to separate radio emission originating in the AGN from that which is generated by star formation. In most cases, only the radio luminosity and the multiwavelength properties of the host galaxy exist to allow us to make this distinction. Where a star formation rate can be estimated, the radio/star-formation-rate relation (Gürkan *et al.* 2018) can be used to try to select objects whose radio emission is dominated by AGN activity, since these will have radio luminosity significantly greater than the level predicted from their star-formation rate. However, star-formation rates are best estimated using far-infrared data, and so Sabater *et al.* (2019) were forced to use various proxies of AGN activity that correlate with a radio excess. With this caveat, their result is still very striking. They reproduce the well-known result that RLAGN incidence depends on galaxy mass (Auriemma *et al.* 1977; Best *et al.* 2005) but, because of the sensitivity of LOFAR, are able to show for the first time that *all* galaxies above $\sim 10^{11} M_{\odot}$ have excess radio emission at the level of 10^{21} W Hz^{-1} or higher. The nature of this AGN activity, which in some cases is at very low levels, and its effect on the host environment remains to be investigated, but the fact that it is there at all is of great interest in feedback models.

4.2. *Energetic impact of the RLAGN population*

Hardcastle *et al.* (2019b) considered the larger RLAGN dataset obtained by cross-matching with the PanSTARRS/*WISE* optical/IR data in the DR1 field (Williams *et al.* 2019) and obtained a sample of 22,000 candidate RLAGN with usable spectroscopic or photometric redshift, using selections based on those of Sabater *et al.* (2019) where possible and *WISE* colour selection to exclude star-forming galaxies otherwise.

The objective of this study was to model the actual jet power of the population, and derive the jet kinetic luminosity function, the volume density of energy being generated by RLAGN as a function of jet power. Inference of jet power from the radio emission is a complex process, and has generally been done in the past by making use of simple scaling relations between radio luminosity and jet power, which cannot be correct in detail. Numerical modelling (Hardcastle & Krause 2014; English *et al.* 2016) shows that radio luminosity for a source with a given jet power Q evolves in a complex way with time and also depends on host environment and redshift, so inferring jet power from observations is not trivial. However, observed physical size is a proxy for source age, so jet power can be inferred by first using the observed size distribution to obtain the lifetime distribution, and then using that plus the radio luminosity distribution to infer jet power.

In detail, we used the analytic models of Hardcastle (2018) to represent the evolution of radio luminosity and physical size with lifetime. We constructed grids of models over the redshift range spanned by the RLAGN sample ($0 < z < 0.7$) and marginalized over the known distribution of angle to the line of sight and an assumed distribution of environmental richness (we assume that RLAGN environments are drawn from the group/cluster mass function). For each radio source in the LOFAR sample, we could then match the

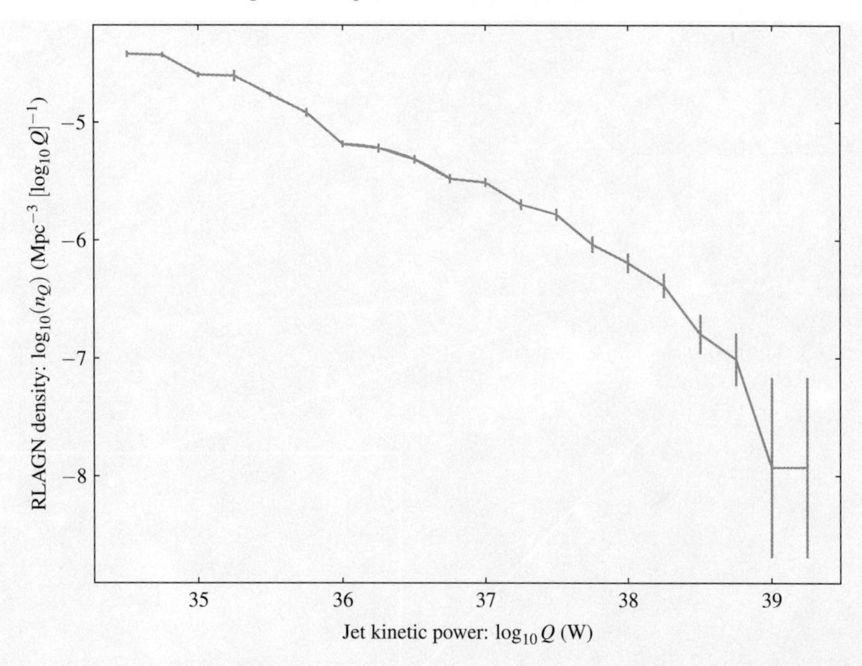

Figure 1. The kinetic luminosity function computed by Hardcastle *et al.* (2019b) for $z < 0.7$ LoTSS AGN. Note the low-power power-law behaviour and the steepening at the highest jet powers, which means that energy injection is dominated by sources with jet powers around $Q = 10^{38}$ W.

radio luminosity, observed size and redshift to points on this grid and read off an estimate of the jet power and its uncertainty. Constructing an ordinary luminosity function based on these jet powers (with the usual sample completeness corrections) leads to a jet kinetic luminosity function for jet powers in the range 10^{34} to 10^{40} W (Fig. 1).

The kinetic luminosity function can be directly integrated to obtain the power density injected by all radio-loud AGN in the local universe per unit volume: 7×10^{31} W Mpc^{-3}. Remarkably, this is comparable to the total X-ray cooling power of all groups and clusters in the local universe, estimated from the X-ray luminosity function of Böhringer *et al.* (2014) to be $\sim 2 \times 10^{31}$ W Mpc^{-3}. Thus the radio AGN population is capable of offsetting all the local radiative cooling of their environments. This is an essential test of feedback models in which RLAGN prevent the cooling of the hot phase of their environments. Our results are consistent with those of Best *et al.* (2006) and Smolčić *et al.* (2017), who used simple radio luminosity-based jet power inference rather than a sophisticated model. The latter authors also showed that the AGN heating density that we (and they) infer is consistent with the expectations from the SAGE model of Croton *et al.* (2016).

There remain questions about the detailed modelling of these sources, particularly the inference of jet power for low-luminosity FRI-type objects (see Section 2) but the overall picture is very encouraging and the next stage would be to investigate feedback effects e.g. as a function of host environment and redshift to see whether they are consistent with the predictions of galaxy formation models in more detail.

4.3. *The fate of the cosmic rays*

In active sources, as discussed in Section 2, about half the energy that is injected by the jet goes into the internal energy of the radio lobes, i.e. into the cosmic rays that dominate the energy density. These may be essentially pure electron/positron in the case of FRIIs or

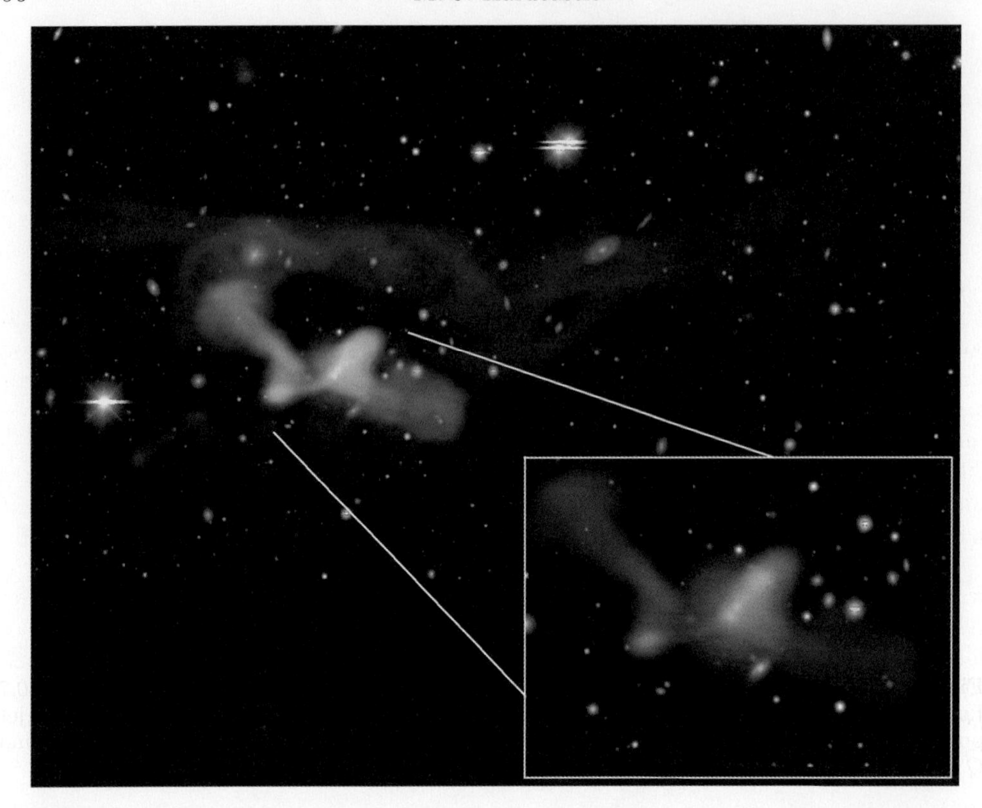

Figure 2. NGC 326 as studied with LOFAR (Hardcastle *et al.* 2019a); radio emission (orange/purple) is superposed on an RGB image from the Legacy survey. A trail of radio emission, 700 kpc long in projection and with complex internal structure, leaves the radio lobes and extends into the source's merging cluster environment. LOFAR images, taken from the unpublished LoTSS Data Release 2, allow both the large-scale structure and (inset, lower right) the details of the inner radio lobes and jets to be seen.

there may be high-energy hadrons in the mixture as argued above for FRIs. In either case, the long-term feedback effects of these particles is unclear. Numerical studies of sources in which the jet becomes disconnected in a simple environment show a significant loss of internal energy as the lobes 'coast' outwards, dragging external material behind them (English *et al.* 2019) on the way out towards neutral buoyancy at very large radii: any energy dissipated on these large (>Mpc) scales can have no effect in offsetting cooling. But there is a good deal of observational evidence in low-power sources (e.g. Fig. 2) that the process is more complex than that, and that substantial hydrodynamical mixing is taking place on hundred-kpc scales, presumably because the host galaxies of these systems are *not* at the centre of spherically symmetric hot-gas halos in pure hydrostatic equilibrium, but rather in a much more complex and unrelaxed hydrodynamical environment. New deep radio surveys are revealing the connection between radio galaxies of this type and the complex structures seen in radio emission in clusters of galaxies (van Weeren *et al.* 2019) and it seems clear that there is much more modelling work to do to understand the physics of feedback as mediated by processes like cosmic ray heating in such objects.

5. Prospects for future work

In the short term we plan a good deal of further work both on the LOFAR surveys and on modelling. In wide fields, we already have an order of magnitude more sky area in hand than covered by Hardcastle *et al.* (2019b): the combination of this with better optical data (from the Legacy survey), with better photometric redshifts, and with spectroscopic data for WEAVE-LOFAR will make the ideal parent survey for a detailed study of radio feedback in the local universe e.g. as a function of environment, with many thousands of LOFAR AGN even at low z. Do radio galaxies actually provide feedback energy at a higher rate in systems where radiative cooling is more effective? The new data will allow us to find out.

Crucial to improving the inference of jet powers is the currently little exploited long baseline capability of LOFAR, which in principle can provide a resolution ten times higher than we currently achieve, giving us a sensitive proxy for lobe physical size and so source age. Polarization at low frequencies also provides environmental information that has so far not been exploited. In the LoTSS deep fields, we are now working towards a study of the evolution of the kinetic luminosity function out to $z \sim 2$; here there are important synergies with deep surveys being carried out with MeerKAT (Jarvis *et al.* 2016) and with the JVLA. On the modelling front, our main concern is to obtain a better understanding of the low-power source population — if we greatly underestimate their jet powers, that will complicate any attempt to infer their feedback effects.

In the longer term, the Square Kilometer Array (both low-frequency and high-frequency components) should be able to carry out still more sensitive surveys, though its resolution will not approach that of LOFAR at low frequencies for some time — its use for deriving an understanding of radio source life cycles is discussed by Kapinska *et al.* (2015). The development of models for accurate bulk inference of feedback effects incorporating all the information provided by the SKA (and the deep optical data available from e.g. the LSST) is our long-term goal.

Acknowledgements

I acknowledge support from the UK Science and Technology Facilities Council under grant ST/R000905/1, and contributions from the many colleagues who have been involved in the work described here.

References

Auriemma, C., Perola, G. C., Ekers, R. D., *et al.* 1977, *A&A*, 57, 41
Becker, R. H., White, R. L., & Helfand, D. J. 1995, *ApJ*, 450, 559
Begelman, M. C. & Cioffi, D. F. 1989, *ApJ*, 345, L21
Best, P. N., Kauffmann, G., Heckman, T. M., *et al.* 2005, *MNRAS*, 362, 25
Best, P. N., Kaiser, C. R., Heckman, T. M., *et al.* 2006, *MNRAS*, 368, L67
Bîrzan, L., Rafferty, D. A., Nulsen, P. E. J., *et al.* 2012, *MNRAS*, 427, 3468
Böhringer, H., Voges, W., Fabian, A. C., *et al.* 1993, *MNRAS*, 264, L25
Böhringer, H., Chon, G., Collins, C. A., *et al.* 2014, A&A, 570, A31
Brinchmann, J., Charlot, S., White, S. D. M., *et al.* 2004, *MNRAS*, 351, 1151
Croston, J. H. & Hardcastle, M. J. 2014, *MNRAS*, 438, 3310
Croston, J. H., Hardcastle, M. J., Birkinshaw, M., *et al.* 2003, *MNRAS*, 346, 1041
Croston, J. H., Kraft, R. P., Hardcastle, M. J., *et al.* 2007, *ApJ*, 660, 191
Croston, J. H., *et al.* 2009, *MNRAS*, 395, 1999
Croston, J. H., Ineson, J., Hardcastle, M. J., *et al.* 2017, *MNRAS*,
Croston, J. H., Ineson, J., Hardcastle, M. J., *et al.* 2018, *MNRAS*, 476, 1614
Croton, D. J., *et al.* 2016, *ApJS* 222, 22
Dunn, R. J. H., Fabian, A. C., & Taylor, G. B. 2005, *MNRAS*, 364, 1343
English, W., Hardcastle, M. J., Krause, M. G. H., *et al.* 2016, *MNRAS*, 461, 2025
English, W., Hardcastle, M. J., Krause, M. G. H., *et al.* 2019, *MNRAS*, 490, 5807
Fabian, A. C., Sanders, J. S., Taylor, G. B., *et al.* 2006, *MNRAS*, 366, 417
Fanaroff, B. L. & Riley, J. M. 1974, *MNRAS*, 167, 31P

Forman, W., *et al.* 2007, *ApJ*, 665, 1057
Gürkan, G., *et al.* 2018, *MNRAS*, 475, 3010
Hardcastle, M. J. 2018, *MNRAS*, 475, 2768
Hardcastle, M. J. & Croston, J. H. 2010, *MNRAS*, 404, 2018
Hardcastle, M. J. & Croston, J. H. 2020, arXiv e-prints, p. arXiv:2003.06137
Hardcastle, M. J. & Krause, M. G. H. 2013, *MNRAS*, 430, 174
Hardcastle, M. J. & Krause, M. G. H. 2014, *MNRAS*, 443, 1482
Hardcastle, M. J., Worrall, D. M., Birkinshaw, M., *et al.* 1998, *MNRAS*, 296, 1098
Hardcastle M. J., Worrall D. M., Birkinshaw M., *et al.* 2002a, *MNRAS*, 334, 182
Hardcastle, M. J., Birkinshaw, M., Cameron, R., *et al.* 2002b, *ApJ*, 581, 948
Hardcastle, M. J., *et al.* 2019a, *MNRAS*, 488, 3416
Hardcastle, M. J., *et al.* 2019b, *MNRAS*, 622, A12
Ineson, J., Croston, J. H., Hardcastle, M. J., *et al.* 2017, *MNRAS*, 467, 1586
Jarvis, M. *et al.* 2016, in Proceedings of MeerKAT Science: On the Pathway to the SKA. 25-27 May, 2016 Stellenbosch, South Africa (MeerKAT2016). Online at https://pos.sissa.it/cgi-bin/reader/conf.cgi?confid=277. p. 6 (arXiv:1709.01901)
Kaiser, C. R. & Alexander, P. 1997, *MNRAS*, 286, 215
Kapinska, A. D., Hardcastle, M., Jackson, C., *et al.* 2015, in Advancing Astrophysics with the Square Kilometre Array (AASKA14). p. 173 (arXiv:1412.5884)
Kraft, R. P., Vázquez, S., Forman, W. R., *et al.* 2003, *ApJ*, 592, 129
Laing, R. A., Riley, J. M., Longair, M. S., *et al.* 1983, *MNRAS*, 204, 151
McNamara, B. R. & Nulsen, P. E. J. 2012, *New Journal of Physics*, 14, 055023
Morganti, R., Fanti, R., Gioia, I. M., *et al.* 1988, *A&A*, 189, 11
Omma, H. & Binney, J. 2004, *MNRAS*, 350, L13
Randall, S. W., *et al.* 2015, *ApJ*, 805, 112
Sabater, J. *et al.* 2019, *A&A*, 622, A17
Scheuer, P. A. G. 1974, *MNRAS*, 166, 513
Shimwell, T. W., *et al.* 2017, *A&A*, 598, A104
Shimwell, T. W., *et al.* 2019, *A&A*, 622, A1
Smolčić, V., *et al.* 2017, *A&A*, 602, A2
Williams, W. L., *et al.* 2019, *A&A*, 622, A2
Worrall, D. M. & Birkinshaw, M. 2000, *ApJ*, 530, 719
Wykes, S., *et al.* 2013, *A&A*, 558, A19
van Haarlem, M. P., *et al.* 2013, *A&A*, 556, A2
van Weeren, R. J., de Gasperin, F., Akamatsu, H., *et al.* 2019, *Sp. Science Reviews*, 215, 16

Galaxy Evolution and Feedback across Different Environments
Proceedings IAU Symposium No. 359, 2020
T. Storchi-Bergmann, W. Forman, R. Overzier & R. Riffel, eds.
doi:10.1017/S174392132000407X

The effects of outbursts from Supermassive Black Holes: A close look at M87

C. Jones and W. Forman

Harvard-Smithsonian Center for Astrophysics
60 Garden Street, Cambridge, MA, USA
email: cjones@cfa.harvard.edu

Abstract. Supermassive black holes (SMBHs) play fundamental roles in the evolution of galaxies, groups, and clusters. The fossil record of supermassive black hole outbursts is seen through the cavities and shocks that are imprinted on these gas-rich systems. For M87, the central galaxy in the Virgo cluster, deep Chandra observations illustrate the physics of AGN feedback in hot, gas-rich atmospheres and allow measurements of the age, duration, and power of the outburst from the supermassive black hole in M87 that produced the observed cavities and shocks in the hot X-ray atmosphere.

Keywords. galaxies: elliptical and lenticular, cD, jets, evolution; X-rays: galaxies

1. Introduction

With masses that can exceed 10^{15} solar masses, clusters of galaxies are the largest gravitationally bound objects in the Universe. Clusters form at the nodes of the cosmic web, through the infall of groups and small clusters and occasionally through the merger of massive clusters. The gas is clusters is heated primarily by the energy released during their initial gravitational collapse. While some of this gas has cooled to form galaxies, most of the baryons, especially in massive clusters, remain in the form of a hot intracluster gas.

Early X-ray imaging observations from Einstein and ROSAT allowed astronomers to map the density of the hot gas in clusters of galaxies (see review by Forman & Jones 1982). In many clusters, the gas density increases toward the cluster center. Often the gas density in the cluster core is so high that the gas in these regions is cooling rapidly and should accrete onto the cluster centers (see Fabian & Nulsen 1977 and Fabian 1994). However since the predicted large amounts of cool gas in cluster cores were not detected, there was, at that time, a perceived "cooling flow" problem.

Two major X-ray observatories, Chandra and XMM-Newton, both launched 20 years ago, allow astronomers to map the density and temperature of the gas in early type galaxies, groups and clusters. These measurements of the gas density and temperature are then used to map the distribution of the total mass in groups and clusters, which is primarily dark matter. Chandra's arcsecond spatial resolution also has allowed the detection and study of very energetic outbursts from supermassive black holes in the cores of galaxies, groups and clusters (e.g. Churazov *et al.* 2000, 2005; Fabian *et al.* 2003; McNamara *et al.* 2005; Fabian 2012).

A large early-type galaxy lies at the center of nearly all cool-core clusters and groups. At the center of this galaxy is a supermassive black hole (SMBH). At very early epochs, these massive black holes grew rapidly through mass accretion and are observed as very luminous quasars. At the present epoch, the SMBHs in the centers of clusters accrete matter as the hot gas in the cluster cores cools. Unlike quasars, these supermassive black

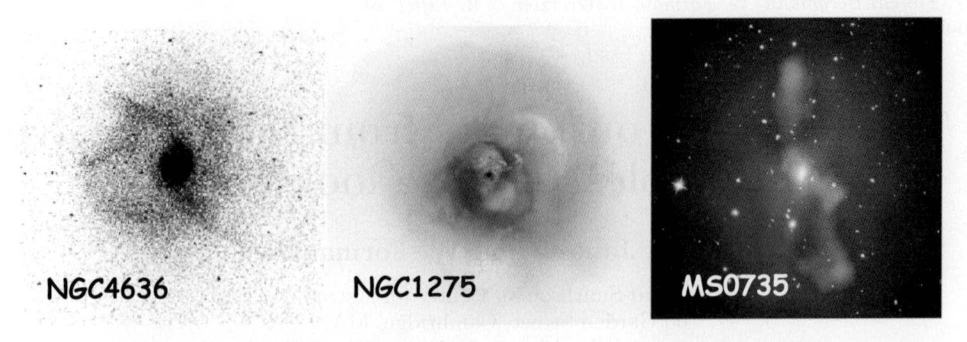

Figure 1. Chandra observations show cavities produced by outbursts of supermassive black holes in the hot gas atmospheres of (left) NGC4636 (adapted from Baldi *et al.* 2009), (center) the Perseus cluster (Fabian *et al.* 2000), and (right) MSO735+74 (McNamara *et al.* 2009).

holes are generally radiatively faint and are in a state of "maintenance feedback," in which the SMBHs, although accreting at levels well below the Eddington mass accretion rate, can undergo AGN outbursts which create cavities in the hot gas halos of early type galaxies and clusters (see Figure 1). By measuring the volumes of these cavities, observers have calculated the energy required to displace the hot gas and have determined that the kinetic energies of the local SMBHs are far larger than their radiative energies.

For M87, with a SMBH mass of 3-6 $\times 10^9 M_{sun}$ (Harms *et al.* 1994; Ford *et al.* 1994; Gebhardt *et al.* 2011; Walsh *et al.* 2013), the Eddington luminosity of M87's SMBH is 4-8 $\times\, 10^{47}$ ergs s^{-1}. However the current observed bolometric luminosity is only 3×10^{42} ergs s^{-1}, five orders of magnitude lower than the expected Eddington luminosity. In addition the estimated jet mechanical power is significantly higher, $\approx 10^{44}$ ergs s^{-1}(e.g. Owen, Eilek & Kassim 2000; Stawarz *et al.* 2006). This conclusion is supported M87's spectral energy distribution (e.g. Reynolds *et al.* 1996; Yuan *et al.* 2009; Mościbrodzka *et al.* 2016). Together, these properties of M87's SMBH suggest M87 has a hot, radiatively inefficient accretion flow (e.g. Yuan & Narayan 2014).

Outbursts from a central SMBH can reheat much of the radiatively cooling gas in the cores of elliptical galaxies, groups, and clusters and thus prevent new stars from forming in the cores. This process leads to the separation of the red elliptical galaxies, which are massive and reside in dense, gas rich environments, from the generally lower mass spiral galaxies, which lie in the field or in poor galaxy groups that have little or no hot intracluster gas. However, there are a small number of clusters that have relatively cool gas in their cores and intense star formation in their central galaxy. A primary example is the Phoenix Cluster where the high level of star formation in the core is produced by a "runaway" cooling flow (McDonald *et al.* 2012, 2015).

The family of systems with hot diffuse gas and cavities produced by outbursts from their central supermassive black hole is illustrated in Figure 1. These systems range from relatively isolated elliptical and S0 galaxies (e.g. NGC4636; Jones *et al.* 2002; Baldi *et al.* 2009) with relatively low X-ray luminosities, cool gas temperatures ($\sim 10^7$ K) and a low fraction of gas mass to stellar mass, to galaxy groups and finally to the massive clusters. Examples of massive clusters include Perseus (Fabian *et al.* 2000) and MSO735+74 (McNamara *et al.* 2005), that have X-ray luminosities up to several 10^{45} ergs s^{-1}, high gas temperatures (10^8 K) and seven to ten times the mass in hot gas compared to the mass in all the cluster galaxies. Although in clusters, the hot X-ray emitting gas dominates the stellar mass, the hot gas is only about 15% of the total cluster mass. Most of the matter in clusters is Dark Matter.

Figure 2 shows the ROSAT map of the X-ray emission from the Virgo cluster (Böhringer *et al.* 1994). The ROSAT X-ray image clearly shows that M87 is the central

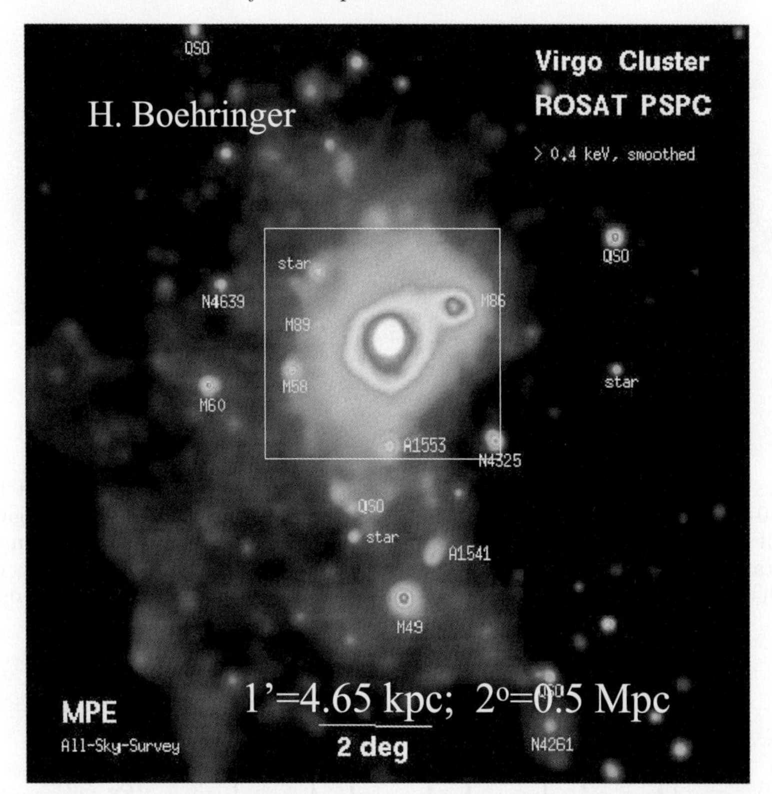

Figure 2. The X-ray emission of the Virgo Cluster as mapped by ROSAT (Böhringer *et al.* 1994) shows the brightest X-ray emission centered on M87, with other bright regions of extended X-ray emission associated with the elliptical galaxies M86, M49 and M60.

dominant galaxy in the Virgo Cluster, even though M49 (NGC4472) is optically slightly more luminous than M87. M87 also hosts a supermassive black hole in its core, which was observed in April 2017 with the Event Horizon Telescope (Event Horizon Telescope Collaboration *et al.* 2019), as well as a classic cooling flow. These characteristics, as well as the proximity of the Virgo cluster, make M87 an ideal system to study the interaction of the SMBH and the hot gas. A detailed analysis of the M87 jet was carried out by Marshall *et al.* (2002), while the comprehensive analysis of the AGN outburst published by Forman *et al.* (2017) forms the basis of this review.

2. SMBH feedback in M87

While we now know that feedback from supermassive black holes is key to understanding how galaxies evolve (see review by McNamara & Nulsen 2007), we still do not know exactly how this feedback works. In Figure 3, Chandra images of M87 in the "soft" (0.2-2.0 keV) energy band (left panel) and in the "hard" (2.0-3.5 keV) energy band (middle panel), along with the optical image of M87 (right panel) are all shown on the same physical scale. The "soft" band image shows the X-ray bright core and two long X-ray filaments of cool gas. The "hard" band image shows the bright central core of the cluster and the hot gas that was shock heated by the AGN outburst that occurred about 13 million years ago.

Figure 4 shows many possible feedback paths for a massive galaxy in a hot gaseous atmosphere with a supermassive black hole in its core. Although the correct evolutionary path is not yet known, what is currently agreed, is that the mechanical input, as seen

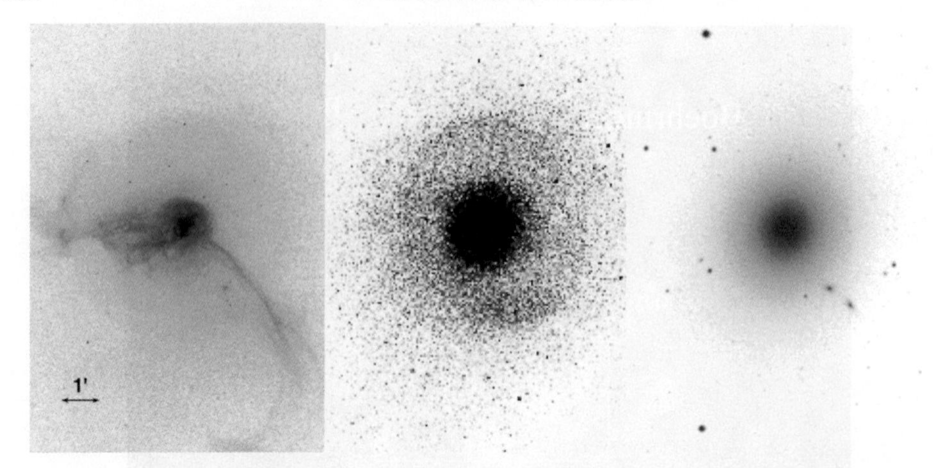

Figure 3. Chandra images of M87 in the "soft" (0.2-2.0 keV) energy band (left image) and "hard" (2.0-3.5 keV) energy band (center image), and the optical image of the galaxy (right image). All three images are on the same physical scale and for the same region of the sky. At the distance of M87, 1′ is ∼5 kpc. The soft band shows long extended filaments of cool gas, while the hard band image clearly shows the location of the shock from the AGN outburst.

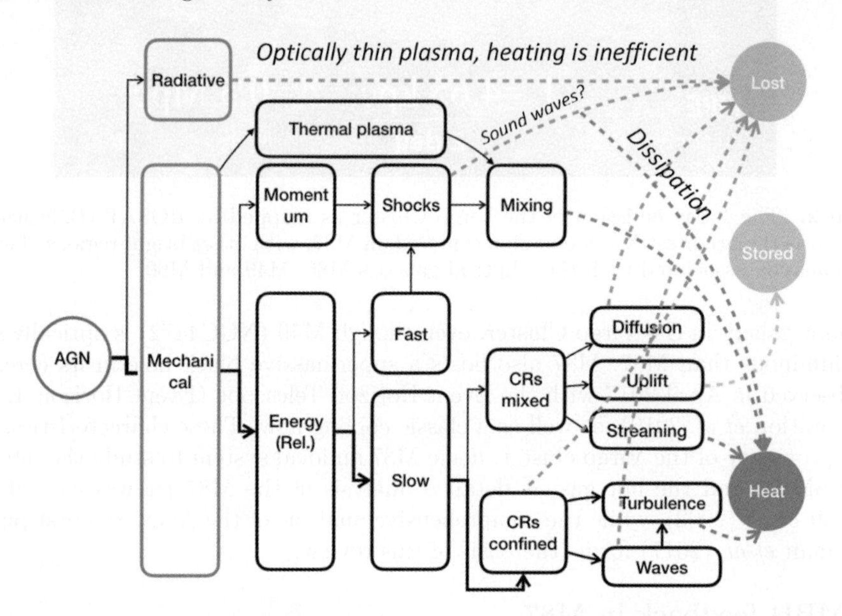

Figure 4. The process of gas cooling onto a supermassive black hole and generating outbursts can follow many possible paths, as illustrated by this figure that was adapted from an earlier version made by Eugene Churazov.

through the inflation of bubbles and occasionally the presence of shocks in the gas, in crucial to reheating the cooling gas, while nearly all of the radiative energy from the AGN is lost from the system. One possible "path" to reheating relies on turbulence generated by the rising buoyant bubbles (see Churazov *et al.* 2001 and Zhuravleva *et al.* 2014). While the exact path that the mechanical energy takes is still unclear, the deep Chandra observations of M87 provide important constraints on both the duration of the AGN outburst and the heating of the intracluster gas (Forman *et al.* 2017).

At the present epoch, the mechanical power of supermassive black holes dominates the radiated power in clusters, groups and gas-rich early type galaxies. As described below,

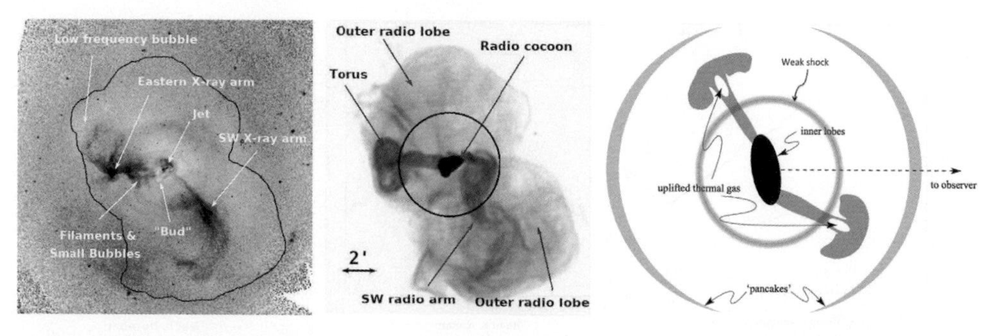

Figure 5. The left panel shows the Chandra X-ray image of M87, divided by the average X-ray radial profile to better highlight the faint X-ray structures, particularly the "arms", jet, and small bubbles. The black outline of the radio emission derived from the 90 cm VLA observation shown in the center panel (adapted from Owen, Eilek & Kassim 2000), is superposed on the X-ray image in the left panel. The right panel shows a schematic of the M87 system with the inner radio lobes, the "arms" of uplifted thermal gas, the shock shown in red, and the outer radio lobes, which are labeled as "pancakes".

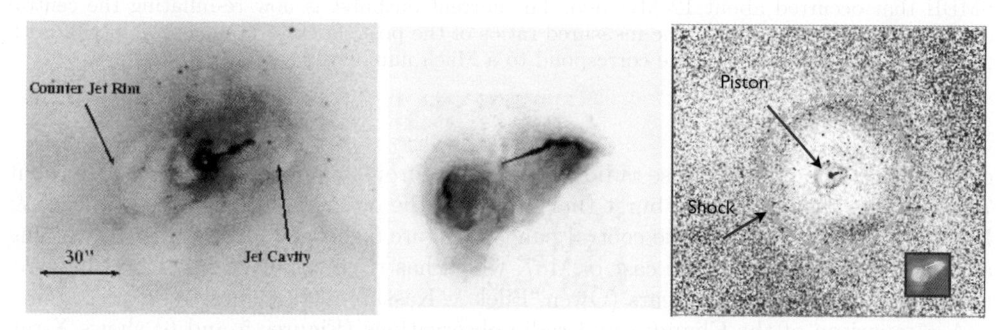

Figure 6. The left panel shows the Chandra image of the core of M87 with cavities in the hot gas and surrounding filaments created by outbursts from the supermassive black hole. The central panel shows the radio jet and radio lobes in the core of M87. The right panel shows the result of dividing the counts in the X-ray image by the average radial surface brightness profile, which "flattens" the field and enhances the features in the core, in particular the emission from the X-ray jet, as well as the hot X-ray emitting inner rim, labeled here as the "piston". The X-ray enhancement at the shock, produced by the AGN outburst, is marked by the blue dotted circle and is clearly visible.

the X-ray and radio observations of M87 together can chronicle the history of AGN outbursts from the supermassive blackhole over the past 150 Myr. The left and center images in Figure 5 show the Chandra X-ray and radio images of M87.

In the left image in Figure 5, the faint X-ray structures in M87 have been highlighted, compared to what can be seen in the original deep Chandra observation, by dividing the background subtracted image by the average X-ray radial surface brightness. This image clearly shows the X-ray "arms" and small bubbles, as well as the X-ray jet which extends $20''$ to the northwest.

The M87 jet has filled the central cavity with relativistic plasma that is seen in the JVLA radio image (Owen, Eilek & Kassim 2000), in the central panels of Figures 5 and 6. The outline of the JVLA extended radio emission is shown superposed on the Chandra image in the left panel of Figure 5. The two large, outer radio lobes northeast and southwest of the M87 nucleus have ages of $\approx$100$-$150 million years. Thus these observations provide evidence for an outburst from the SMBH about 150 million years ago. A filamentary radio arm, extending southwest of the nucleus, also can be seen in the

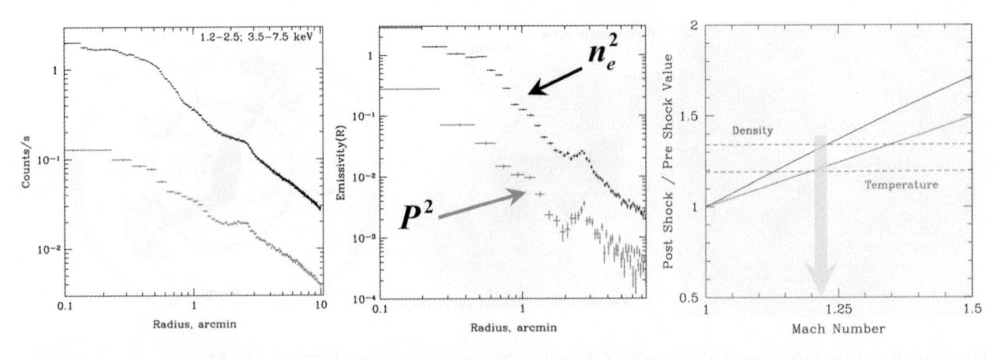

Figure 7. The left panel shows the radial surface brightness profile in the energy band 1.2-2.5 keV (in black) and in the hard energy band 3.5-7.5 keV (in red), extracted from a 90 degree azimuth centered on north, with point sources removed and corrected for telescope vignetting and exposure time. The "bumps" at a radius of 2.7′ (13 kpc, at the distance of M87), in both the surface brightness profiles (left panel) and in the plots of the square of the density and pressure (central panel), are the strongest features in the X-ray surface brightness profiles. These bumps correspond to the shock at a radius of 13 kpc produced by an outburst from the SMBH that occurred about 12 Myr ago. The current outburst is now re-inflating the central cavity. The right panel shows the measured ratios of the post shock to pre-shock values of both the gas density and temperature correspond to a Mach number of 1.2.

central panel of Figure 5. These radio features result from an outburst of the SMBH about 70 million years after the outburst that produced the outer radio lobes (Owen, Eilek & Kassim 2000). In addition, the central panel of Figure 5 shows a radio torus (resembling a "mushroom cloud"), to the east of M87, which has risen about 20 kpc from the core, in the last 40 to 70 million years (Owen, Eilek & Kassim 2000; Churazov *et al.* 2001).

A comparison of the Chandra and radio observations (Figures 5 and 6) shows X-ray filaments that are coincident with the radio structures. While Figure 5 shows the structures on the larger scales, Figure 6 shows the the detailed structure in the core seen in the deep Chandra observations (left panel) and in the deep radio observations (middle panel). Also a bubble (labeled as "bud" in Figure 5 and shown on a larger scale in Figure 6) that is now separating from the central cocoon as seen in both the X-ray and radio images, while filamentary X-ray structures, extending to the east of the nucleus, are likely the remnants of small bubbles produced by less energetic nuclear outbursts. The X-ray emitting gas in the X-ray arms is cooler than the gas in the outer regions, thus supporting the idea that the X-ray arms are uplifted from the core by rising bubbles created in an outburst by the SMBH. While the shock from the AGN outburst is clearly visible in the Chandra "hard" band image (3.5 to 7 keV), as shown in the central image of Figure 5, the largest bubbles seen in the radio observation are not apparent in the X-ray images of the hot gas.

The left panel in Figure 7 shows the radial surface brightness profile in the soft energy band (1.2-2.5 keV) in black and in the hard energy band (3.5-7.5 keV) in red (for additional details, see the caption for Figure 7). The center panel shows the radial profiles of the square of the gas density and the square of the gas pressure. "Bumps" in both profiles, at a radius of 2.7′ (13 kpc, at the distance of M87) are the strongest features seen in the four profiles and show the existence of a shock at a radius of 13 kpc that was produced by the current outburst from the M87 SMBH that is now re-inflating the central X-ray cavity. The right panel in Figure 7 shows that the measured ratios of the post-shock to pre-shock values for both the gas density and temperature correspond to a Mach number of 1.2 for the shock that produced the features at a radius of 13 kpc.

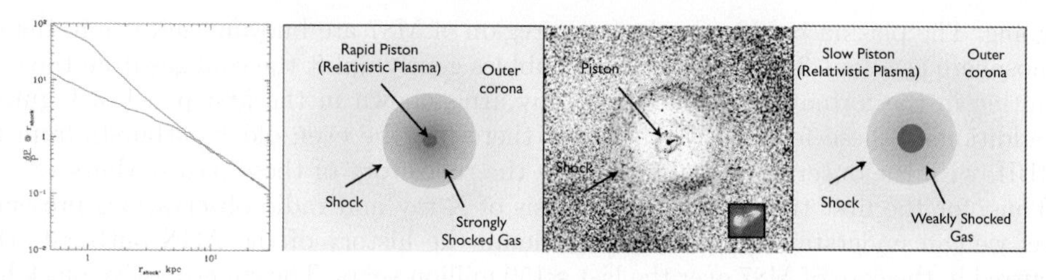

Figure 8. The left panel shows the evolution of the shock strength, parameterized as the pressure jump for a short (0.6 million year) duration outburst in blue, compared to a long duration outburst (2.2 million years) in red, extracted from a 90 degree azimuth centered on north, with point sources removed and corrected for telescope vignetting and exposure. In the second panel, the skematic drawing illustrates the strongly shocked gas in the galaxy halo that would result from a short intense SMBH outburst, compared the weakly shocked gas in the galaxy halo that would result from a "slow piston," illustrated in the fourth panel.

A total of four outbursts from the supermassive black hole in M87 have been identified, with ages up to ≈150 million years, with the most recent outburst still ongoing. The plasma bubbles in the outer region of M87 are buoyant and rise in the hot atmosphere at about 300 km s^{-1}. These bubbles uplift the cool gas from the core, resulting in the formation of the cool X-ray arms shown in the first panel of Figure 5. In addition to these four AGN outbursts, there may be even older outbursts from the SMBH, with the outer radio bubbles being the repository of the energy from these past outbursts.

The fate of the energy released from a SMBH outburst is gas motion. The rising bubble created by the outburst forms a torus. The plot in Figure 8 shows how much bubble enthalpy remains as a function of distance from the SMBH. By the time the bubble is 10 kpc from the SMBH, it has lost about half its enthalpy, which has gone into gas mostions and ultimately into heating the gas. This "solves" what was called "the cluster cooling flow problem." In Virgo and most cool core clusters, the cooling gas is reheated by outbursts from the central SMBH (see Zhuravleva *et al.* 2014).

The left panel in Figure 8 shows the evolution of the shock strength, parameterized as the pressure jump for a short (0.6 million year) duration outburst from the SMBH in blue, compared to a long duration outburst (2.2 million years) in red. Although the two curves differ dramatically in the core of M87, they both match the observed shock at a radius of 13 kpc.

The two senarios for the SMBH outburst in M87, in particular a short (0.6 million year) outburst verses a long (2.2 million year) outburst, are shown in the second and fourth panels of Figure 8. The second panel illustrates the case of a short duration (10^5 years), powerful outburst, with a strong shock that would be driven into the surrounding hot atmosphere. In this scenario, at the present time, the region interior to the shock would be hot. The fourth panel of Figure 8 shows the case of a longer duration (2.2×10^6 years) and thus more "gentle" outburst that would still generate the same magnitude shock at 13 kpc as that of the short duration outburst. However in this case, the gas interior to the shock would be only weakly shocked and the central plasma-filled piston would be larger, than in the case of a short duration outburst. The third panel shows the observed X-ray emission from the core of M87. In this image, the X-ray emission has been "flattened" by dividing the observed emission by the smoothed radial profile. This analysis of the image highlights the central ring of X-ray emission (labeled as "piston") as well as the increased X-ray emission at the shock.

In summary, a total of four outbursts from the supermassive black hole in M87 have been identified, with ages up to ≈150 million years, with the most recent outburst still ongoing. The plasma bubbles in the outer region of M87 are buoyant and rise in the hot atmosphere at about 300 km s^{-1}. These bubbles gently uplift the cool gas from the core, resulting in the formation of the cool X-ray arms shown in the first panel of Figure 5. In addition to these four AGN outbursts, there may be even older outbursts from the SMBH, with the outer radio bubbles being the repository of these past outbursts.

Thus, for the first time, from the analysis of X-ray and radio observations presented here, we can understand in considerable detail the history of the AGN outbursts that occurred in the core of M87 over the last ≈150 million years. The supermassive black hole in M87 has had at least four outbursts. M87 hosts old radio bubbles, produced 100-150 million years ago, as well as the radio torus and arms and the X-ray arms, which were produced by an AGN outburst about 40 Myrs ago. In addition the hard X-ray image shows a shock in the hot gas, which resulted from the AGN outburst that occurred 12 million years ago. This relatively recent AGN outburst also created the central small X-ray cavity in the core. Finally, there is evidence that the jet is now reinflating the central cavity that drove the main shock that is observed at a radius of 13 kpc. Thus, in total, the supermassive black hole in M87 has had at least four outbursts in the past ≈150 million years, and maybe many more older outbursts, with the outer bubbles being the repository of the energy from these past outbursts. In addition to understanding the history of outbursts from the SMBH in M87, from the Chandra observations of M87, we also understand that the absence of a strongly shock-heated region exterior to the central cavity implies a "gentle", relatively long outburst (see Forman *et al.* 2017 for details). Quantitatively, the size of the cavity and the shock strength constrain the outburst duration to be about 2 Myr.

Acknowledgements

We acknowledge support from the Smithsonian Institution, the Smithsonian Astrophysical Observatory, and the Chandra High Resolution Camera project, supported by NAS8-03060.

References

Baldi, A., Forman, W., Jones, C., *et al.* 2009, *ApJ*, 707, 1034
Böhringer, H., Briel, U. G., Schwarz, R. A., *et al.* 1994, *Nature*, 368, 828
Churazov, E., Forman, W., Jones, C., *et al.* 2000, *A&A*, 356, 788
Churazov, E., Brüggen, M., Kaiser, C. R., *et al.* 2001, *ApJ*, 554, 261
Churazov, E., Sazonov, S., Sunyaev, R., *et al.* 2005, *MNRAS*, 363, L91
Event Horizon Telescope Collaboration, Akiyama, K., Alberdi, A., *et al.* 2019, *ApJL*, 875, L1
Fabian, A. C. & Nulsen, P. E. J. 1977, *MNRAS*, 180, 479
Fabian, A. 1994, *ARAA*, 32, 277
Fabian, A. C., Sanders, J. S., Allen, S. W., *et al.* 2003, *MNRAS*, 344, L43
Fabian, A. C., Sanders, J. S., Taylor, G. B., *et al.* 2006, *MNRAS*, 366, 417
Fabian, A. 2012, *ARAA*, 50, 455
Ford, H. C., Harms, R. J., Tsvetanov, Z. I., *et al.* 1994, *ApJL*, 435, L27
Forman, W. & Jones, C. 1982, *ARAA*, 20, 547
Forman, W., Churazov, E., Jones, C., *et al.* 2017 *ApJ*, 844, 122
Gebhardt, K., Adams, J., Richstone, D., *et al.* 2011, *ApJ*, 729, 119
Harms, R. J., Ford, H. C., Tsvetanov, Z. I., *et al.* 1994, *ApJL*, 435, L35
Jones, C., Forman, W., Vikhlinin, A., *et al.* 2002, *ApJL*, 567, L115
Marshall, H. L., Miller, B. P., Davis, D. S., *et al.* 2002, *ApJ*, 564, 683
McDonald, M., Bayliss, M., Benson, B. A., *et al.* 2012, *Nature*, 488, 349
McDonald, M., McNamara, B. R., van Weeren, R. J., *et al.* 2015, *ApJ*, 811, 111
McNamara, B. R., Nulsen, R. E .J., Wise, M. W., *et al.* 2005, *Nature*, 433, 7021, 45
McNamara, B. R. & Nulsen, P. E. J. 2007, *ARA&A*, 45, 117

McNamara, B. R., Kazemzadeh, F., Rafferty, D. A., *et al.* 2009, *ApJ*, 698, 594
Mościbrodzka, M., Falcke, H., & Shiokawa, H. 2016, *A&A*, 586, A38
Owen, F. N., Eilek, J. A., & Kassim, N. E. 2000, *ApJ*, 543, 611
Reynolds, C. S., Di Matteo, T., Fabian, A. C., *et al.* 1996, *MNRAS*, 283, L111
Stawarz, Ł., Aharonian, F., Kataoka, J., *et al.* 2006, *MNRAS*, 370, 981
Walsh, J. L., Barth, A. J., Ho, L. C., *et al.* 2013, *ApJ*, 770, 86
Yuan, F., Yu, Z., & Ho, L. C. 2009, *ApJ*, 703, 1034
Yuan, F. & Narayan, R. 2014, *ARAA*, 52, 529
Zhuravleva, I., Churazov, E., Schekochihin, A. A., *et al.* 2014, *Nature*, 515, 85

Galaxy Evolution and Feedback across Different Environments
Proceedings IAU Symposium No. 359, 2020
T. Storchi-Bergmann, W. Forman, R. Overzier & R. Riffel, eds.
doi:10.1017/S1743921320002070

The role of environment on quenching, star formation and AGN activity

Bianca M. Poggianti[1], Callum Bellhouse[1], Tirna Deb[2], Andrea Franchetto[1,3], Jacopo Fritz[4], Koshy George[5], Marco Gullieuszik[1], Yara Jaffé[6], Alessia Moretti[1], Ancla Mueller[7], Mario Radovich[1], Mpati Ramatsoku[8], Benedetta Vulcani[1] and the rest of the GASP team†

[1]INAF-Osservatorio Astronomico di Padova, vicolo dell'Osservatorio 5, 35122, Padova, Italy
email: bianca.poggianti@inaf.it

[2]Kapteyn Astronomical Institute, University of Groningen, Postbus 800, NL-97009 AV, Groningen, The Netherlands

[3] Dipartimento di Fisica e Astronomia, Universitá di Padova, vicolo dell'Osservatorio 3, 35122 Padova, Italy

[4]Instituto de radioastronomia y Astrofisica, UNAM, Campus Morelia, A.P. 3-72, 58089, Mexico

[5]Faculty of Physics, Ludwig-Maximilians-Universitat, Scheinerstr. 1, 81679, Munich, Germany

[6]Instituto de Fisica y Astronomia, Universidad de Valparaiso, Avda. Gran Bretana 1111, Valparaiso, Chile

[7]Ruhr University Bochum, Faculty of Physics and Astronomy, Universitatsstr. 150, 44801 Bochum, Germany

[8]Department of Physics and Electronics, Rhodes University, PO Box 94, Makhanda, 6140, South Africa‡

Abstract. Galaxies undergoing ram pressure stripping in clusters are an excellent opportunity to study the effects of environment on both the AGN and the star formation activity. We report here on the most recent results from the GASP survey. We discuss the AGN-ram pressure stripping connection and some evidence for AGN feedback in stripped galaxies. We then focus on the star formation activity, both in the disks and the tails of these galaxies, and conclude drawing a picture of the relation between multi-phase gas and star formation.

Keywords. Galaxies: active, galaxies: evolution, galaxies: clusters: general

1. Introduction

Spiral galaxies in clusters and groups lose their gas due to the ram pressure exerted by the hot intergalactic medium on the galaxy interstellar and circumgalactic medium. The effects of ram pressure stripping (RPS) on the disk gas have been observed at several different wavelengths (e.g. Gavazzi 1989; Kenney *et al.* 2004; Yagi *et al.* 2010; Sun *et al.* 2010; Smith *et al.* 2010; Ebeling *et al.* 2014; Gavazzi *et al.* 2018; Boselli *et al.* 2020) and have been predicted by both analytical approaches and hydrodynamical simulations (Gunn & Gott 1972; Tonnesen & Bryan 2009; Roediger & Brüggen 2008; Roediger *et al.* 2014). Stripped galaxies offer a great opportunity to study several fundamental physical processes in astrophysics, especially thanks to recent integral-field spectroscopic studies.

† http://web.oapd.inaf.it/gasp/index.html
‡ INAF-Osservatorio Astronomico di Cagliari, via della Scienza 5, 09047 Selargius (CA), Italy

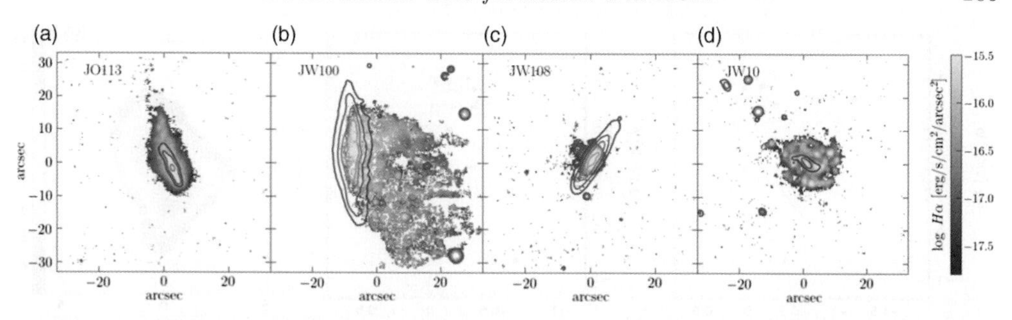

Figure 1. MUSE Hα surface brightness maps for four GASP galaxies in different conditions of RPS: a galaxy undergoing moderate stripping (JO113); a jellyfish galaxy (JW100); an advanced stage of stripping, with gas left only in the central region of the disk (JW108) and a galaxy that is disturbed but not stripped (JW10): the latter is a merger, as testified by the stellar velocity map (not shown). Red contours delimit the galaxy stellar disk. From Jaffé *et al.* 2018.

Hereafter, we will discuss the latest results of ram pressure studies concerning three fields of research: the triggering of AGN activity; the star formation process within and outside of galaxy disks; and the baryonic cycle between multi-phase gas and star formation. As we discuss below, unexpected findings were uncovered for each of these fields.

Our summary is mostly based on results from the survey GASP (GAs Stripping Phenomena in galaxies, Poggianti *et al.* 2017a, http://web.oapd.inaf.it/gasp/index.html), which includes a MUSE integral-field ESO Large Program and follow-up multiwavelength programs investigating the molecular gas (APEX, ALMA), the neutral gas (JVLA, MeerKAT) and the young stellar content (UVIT@ASTROSAT). The GASP sample includes cluster galaxies at different stages and different strengths of the stripping process (Fig. 1), from initial to peak stripping to the late phases with little gas left, and even fully stripped post-starburst galaxies and an undisturbed control sample. GASP also includes a group and field subsample of galaxies, which are not discussed here (Vulcani *et al.* 2017, 2018a,c, 2019a,b).

2. AGN

An unexpected result was the high incidence of AGN among the so called "jellyfish galaxies", defined as galaxies with one-sided tails of ionized gas (longer than the stellar disk diameter). MUSE data demonstrate that the tails are due to RPS. Six out of the seven GASP jellyfish galaxies studied hosted an AGN (one of them is an optical LINER). This AGN incidence is much higher than in general cluster and field samples, suggesting that ram pressure can cause gas to flow towards the center and trigger the AGN activity (Poggianti *et al.* 2017b, Fig. 2).

The exact physical mechanism responsible for the gas inflow still needs to be pinpointed. It may be due to a loss of angular momentum of the galactic gas when it interacts with the non-rotating intracluster-medium (Tonnesen & Bryan 2012), or it can be generated by oblique shocks in a disk flared by the magnetic field (Ramos-Martínez, Gómez & Pérez-Villegas 2018). Very recent high resolution simulations of a galaxy cluster also find that ram pressure triggers enhanced accretion onto the central black hole (Ricarte *et al.* 2020).

In this context, it is relevant to ask: a) how sure is the presence of the AGN, and could the gas ionization be due to shocks or other mechanisms? Based on the comparison with AGN, shocks and HII-region photoionization models and using different line ratios, Radovich *et al.* (2019) confirmed the univocal interpretation of the presence of AGN. The same work found iron coronal lines (Fig. 2) and extended (>10kpc)

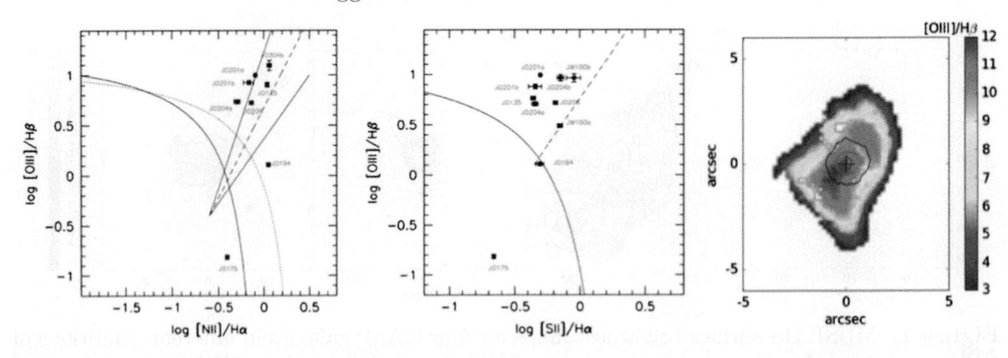

Figure 2. Left and center. BPT line-ratio diagnostic diagrams for the jellyfish sample from Poggianti *et al.* (2017b). Most galaxies lie in the AGN region of this diagram. Right: The high [OIII]/Hβ ratio of the central region of the JO201 jellyfish galaxy indicates the presence of the AGN. The black contour shows the region with emission of the coronal [Fe VII]λ6087 line, also indicative of an AGN. From Radovich *et al.* (2019).

AGN-powered ionization cones in some of these galaxies, as well as AGN outflows extending out to 1.5-2.5 kpc from the center, with outflow velocities in the range 250-550km s^{-1}. b) The sample published in Poggianti *et al.* (2017b) is small, and consists of quite massive galaxies ($\geqslant 4 \times 10^{10} M_{\odot}$). How significant is the enhancement of the AGN fraction, and is that confirmed by further studies? Is the RPS-enhanced AGN activity present only under certain circumstances, e.g. in a certain stage of stripping (when it is strongest), or for certain orbits within the cluster, etc? Or does it occur only in galaxy clusters with certain intracluster medium properties? For example, Roman-Oliveira *et al.* (2019), in their study of the A901/2 supercluster, find only 5 AGN host galaxies in their sample of 58 jellyfish candidates with an assigned classification (see also Roman-Oliveira in these proceedings). The analysis of the whole GASP sample is underway, and studies of other samples/redshifts will help clarify this point, keeping in mind that the detection of a Seyfert2/LINER AGN depends crucially on the data quality and sensitivity. Furthermore, for several galaxies there is evidence for a large amount of dust in their nuclear regions: in this case, the optical line diagnostic ratios provided even by deep MUSE data sometimes may not reveal the dust obscured AGN (e.g. Fritz *et al.* 2017), and X-ray data would be required for its detection.

Moreover, the combination of MUSE and multiwavelength data has provided strong evidence for the effects of AGN feedback in the jellyfish galaxy JO201 (George *et al.* 2019, see also Bellhouse *et al.* 2017, 2019). The central 8kpc region of JO201 is depleted of both molecular gas (as traced by a CO ALMA observation) and of recent and ongoing star formation (as traced by NUV and FUV imaging with UVIT@ASTROSAT) (Fig. 3). This region is filled with gas ionized by the AGN (as seen by MUSE). Evidence for a similar effect in other GASP jellyfish galaxies is present and is currently under investigation.

3. Star formation

The effects of RPS on the star formation activity are variegated and in a sense counterintuitive, since RPS removes gas which is the fuel for the formation of new stars.

On a galaxy-wide scale, generally the star formation rate (SFR) in the disks of galaxies undergoing stripping is slightly but significantly enhanced with respect to undisturbed galaxies of similar mass, i.e. galaxies undergoing stripping tend to lie above the SFR-stellar mass relation (Vulcani *et al.* 2018b, Fig. 4). Moreover, jellyfish galaxies follow the mass-metallicity relation of non-stripped cluster galaxies, with metallicities higher than

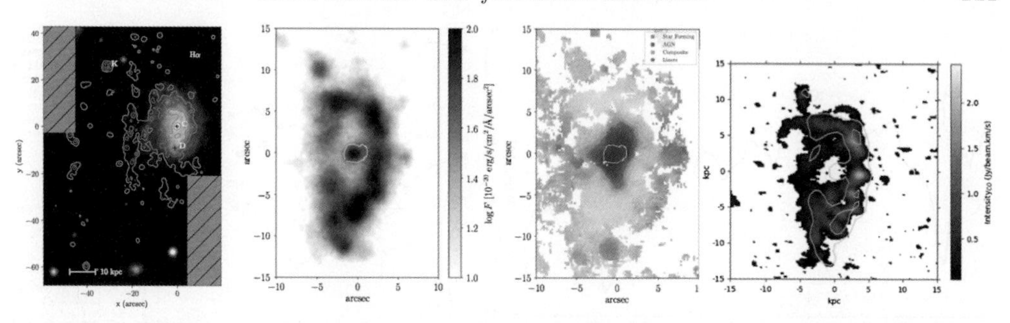

Figure 3. Left. The jellyfish galaxy JO201 Hα contours superimposed on the stellar image. The long extraplanar tails of ionized gas are visible (Bellhouse *et al.* 2017, 2019). Other three panels: a zoom on the disk of (from left to right) NUV emission, ionization source map and CO map (George *et al.* 2019). The 8 kpc central hole in UV and CO emission corresponds to the AGN-powered Hα emission.

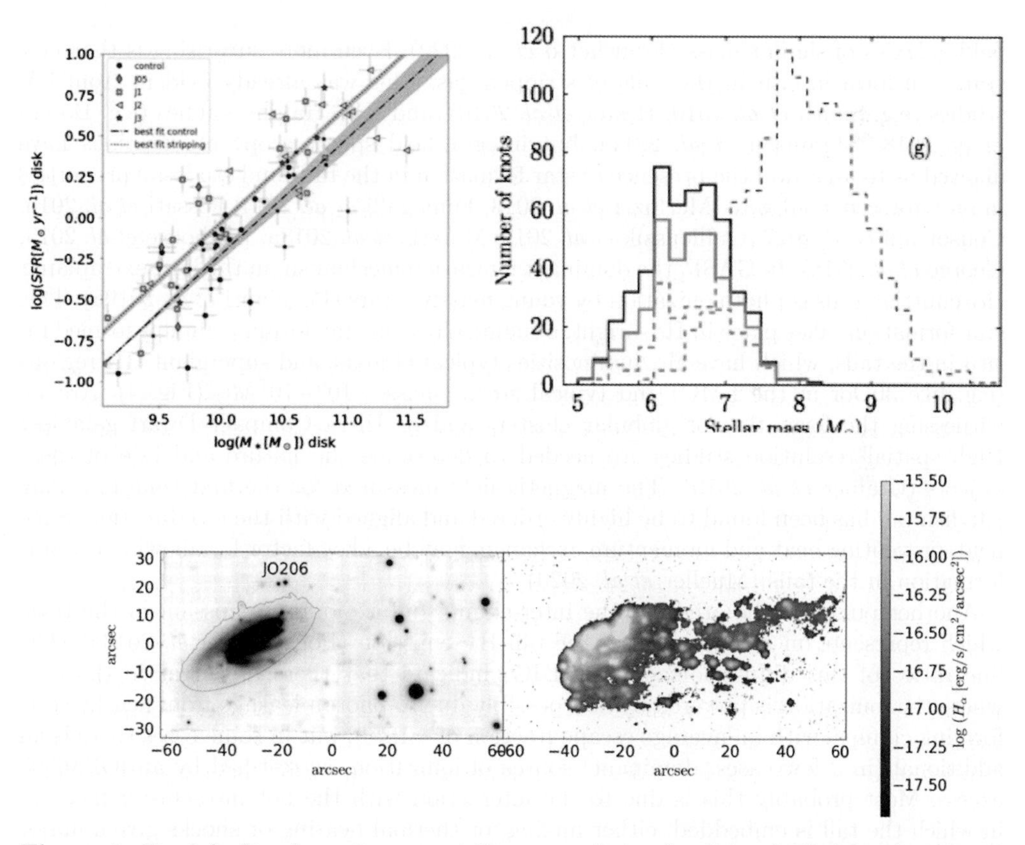

Figure 4. Top left: Star formation rate-stellar mass relation for disks of jellyfish galaxies compared with undisturbed galaxies, from Vulcani *et al.* 2018b. Top right: Stellar mass distribution of star-forming clumps in the tails of jellyfish galaxies as solid histograms (red: only clumps that are star-forming according to the BPT diagrams; black: all clumps). For comparison, the dashed histogram is for clumps in the disks. Bottom: the jellyfish galaxy JO206 with its 90kpc-long tail of Hα emitting gas (right), and its optical image dominated by the stellar disk (left). The star-forming clumps stand out in the Hα image, where also the diffuse emission is visible.

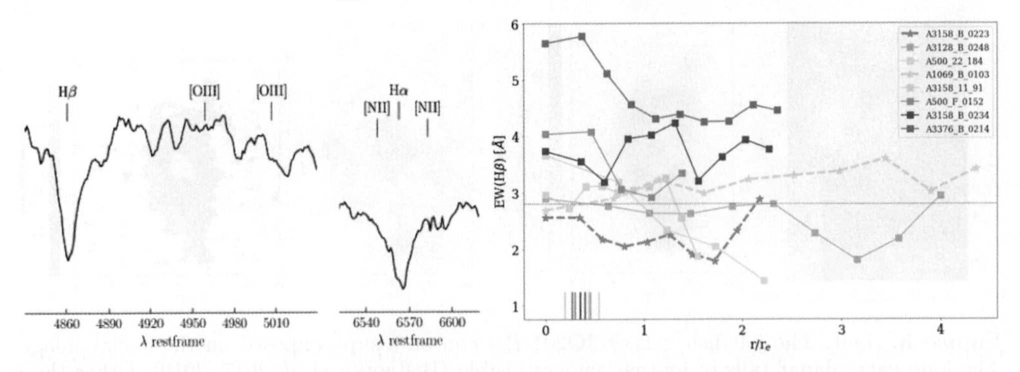

Figure 5. Left: Strong Balmer absorption lines in the outskirts of jellyfish disks, where gas has already been stripped (from Gullieuszik *et al.* 2017). Right: The galactocentric radial distribution of the Hβ equivalent width in absorption in GASP post-starburst/post-starforming galaxies. The vertical bars at the bottom of the right panel indicate the size of 1 kpc in units of r_e for each galaxy. From Vulcani *et al.* (2020).

field galaxies of similar mass (Franchetto *et al.* 2020). Even more surprising is that new stars can form in situ in the tails of stripped gas. This was already evident from UV studies (e.g. Smith *et al.* 2010; Hester *et al.* 2010), and UV+Halpha studies (e.g. Boselli *et al.* 2018; Abramson *et al.* 2011), but integral-field spectroscopy observations have allowed us to ascertain the presence of star formation in the tails and study its properties in an unprecedented way (Merluzzi *et al.* 2013; Fumagalli *et al.* 2014; Fossati *et al.* 2016; Consolandi *et al.* 2017; Gullieuszik *et al.* 2017; Moretti *et al.* 2018a; Bellhouse *et al.* 2019; George *et al.* 2018). In GASP, the dominant ionization mechanism in the long extraplanar Hα-emitting tails is photoionization by young massive stars (Poggianti *et al.* 2019a). This star formation takes place in Hα-bright, dynamically cold star-forming clumps formed in-situ in the tails, which have Hα luminosities typical of giant and supergiant HII regions (e.g. like 30Dor in the LMC) and typical stellar masses $10^6-10^7 M_\odot$ (Fig. 4). Are we witnessing the formation of globular clusters and/or Ultra Compact Dwarf galaxies? High spatial resolution studies are needed to determine the nature and fate of these objects (Cramer *et al.* 2019). The magnetic field measured for the first time in a long jellyfish tail has been found to be highly ordered and aligned with the tail direction. Such field, preventing heat and momentum exchange, may be a key factor for allowing the star formation in the tails (Mueller *et al.* 2020).

Another puzzle is the origin of the inter-clump, diffuse ionized emission in the tails, which represents on average 50% of the tail Hα emission (Poggianti *et al.* 2019a). The line ratios of this diffuse ionized gas (DIG) indicate that there are areas in the tails where the ionization is powered by SF (possibly due to photon leakage from nearby star-forming clumps, with an average escape fraction of $\sim$18%), but in some cases there is an additional (in a few cases, dominant) source of ionization, as testified by an [OI]λ6300 excess. Most probably this is due to the interaction with the hot intracluster medium in which the tail is embedded: either mixing, or thermal heating or shocks give a major contribution to the tail ionization in the jellyfish galaxy JW100 (Poggianti *et al.* 2019b), and this might be the case also for other jellyfish examples for which line ratio data is missing (Boselli *et al.* 2016).

After gas is removed by ram pressure, star formation comes to an end. A clear signature for a recent truncation of the star formation activity are the strong Balmer lines in absorption typical of post-starburst/post-starforming spectra. Such a signature is present in the outer regions of the disk of several jellyfish galaxies (e.g. Gullieuszik *et al.* 2017; Poggianti *et al.* 2019b) and is observed throughout the disk of those non-starforming

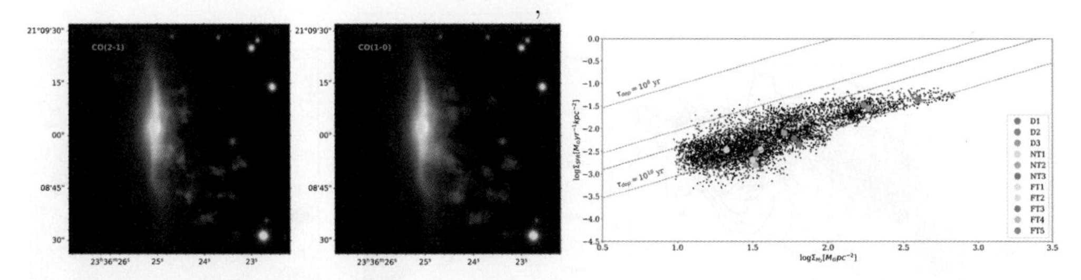

Figure 6. Left and center: CO(2-1) and CO(1-0) emission on top of the I-band image of the galaxy JW100. Right: SFR surface density versus H_2 surface density (1kpc scale) for spaxels of JW100, with a few regions of interest in the disk and tails highlighted. In the right panel, the red dashed lines are fixed depletion times (10^8, 10^9, 10^{10}yr from top to bottom), while the blue dashed line is the average relation for normal nearby disk galaxies at 1 kpc resolution. From Moretti *et al.* (2020).

galaxies that have recently finished to be stripped (Vulcani *et al.* 2020) (Fig. 5). These are totally devoid of emission lines, are typically located between 0.5 and 1 cluster virial radii (Owers *et al.* 2019; Vulcani *et al.* 2020) and have been quenched outside-in (the disk outskirts first) as expected in the ram pressure stripping scenario (Gavazzi *et al.* 2013).

4. Multi-phase gas

The number of ram pressure stripped galaxies with CO data is still rather small, but a picture is emerging: large masses of molecular gas have been detected both in disks and tails (Jáchym *et al.* 2014, 2017, 2019; Verdugo *et al.* 2015; Lee & Chung 2018; Moretti *et al.* 2018b, 2020). Following the old debates about whether the molecular gas can be stripped by ram pressure (Kenney & Young 1989; Boselli *et al.* 1997, 2014), the ALMA resolution has recently allowed to study large individual CO clumps/complexes of $10^6-10^9 M_\odot$ masses of H_2 in the tails (Jáchym *et al.* 2019; Moretti *et al.* 2020). These studies suggest that while the cold gas observed close to the disk may be stripped, that observed further out in the tail forms there (see also Verdugo *et al.* 2015).

The amount of molecular gas in some jellyfishes is impressive, ($>10^9-10^{10} M_\odot$). The GASP galaxy JW100 contains $2.5 \times 10^{10} M_\odot$ of molecular gas (8% of the galaxy stellar mass), of which 30% is in the tail (Fig. 6, Moretti *et al.* 2020). Interestingly, the CO-star formation efficiency, defined as the ratio between the SFR and the molecular gas mass, is low, both on the galaxy scale and on a 1kpc spatially resolved scale, yielding depletion timescales up to 10^{10} yr (e.g. Vollmer *et al.* 2008; Jáchym *et al.* 2014; Verdugo *et al.* 2015; Moretti *et al.* 2018b, 2020, see Fig. 6).

In the tails there is a general correspondance between the spatially resolved distribution of the various tracers related to star formation (UV light, Hα emission and CO emission), but it is also possible to observe directly the "star formation sequence", with CO-only clumps, CO+Hα+UV clumps, Hα+UV and UV-only clumps, representing the different stages of the star formation process (Poggianti *et al.* 2019b).

As far as the neutral gas is concerned, HI observations paved the way to ram pressure studies, with milestones results showing the deficiency of HI in cluster galaxies (Haynes *et al.* 1984; Cayatte *et al.* 1990; Vollmer *et al.* 2001; Chung *et al.* 2009, to name a few). However, the number of jellyfish galaxies with multiwavelength data, probing neutral, molecular and ionized gas in the same system, is very limited, thus the origin and the conditions allowing the presence of multi-phase tails are still to be clarified. Generally, when an Hα tail has been observed, sufficiently deep HI data has also revealed a neutral gas tail. However, the morphologies of the Hα and the HI tail can be very different (see Fig. 7 for three example galaxies), and the kinematical decoupling of HI and Hα can be

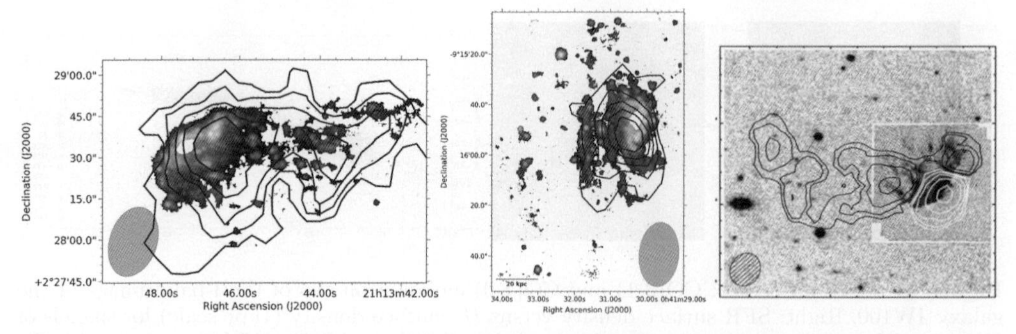

Figure 7. HI contours on top of the MUSE Hα image (plus the optical broad band image outside of the inset in the right panel) of three GASP jellyfish galaxies (from left to right, Ramatsoku *et al.* 2019, 2020 and Deb *et al.* 2020). In the right panel, the black contour is HI in emission and the white contour in absorption. The HI absorption is due to neutral gas along the line of sight between us and the central AGN. In some cases the HI and Hα tails are roughly co-spatial (left panel, tails are 90-kpc long), sometimes the Hα is much more extended than the HI (middle), and sometimes it is the opposite (right).

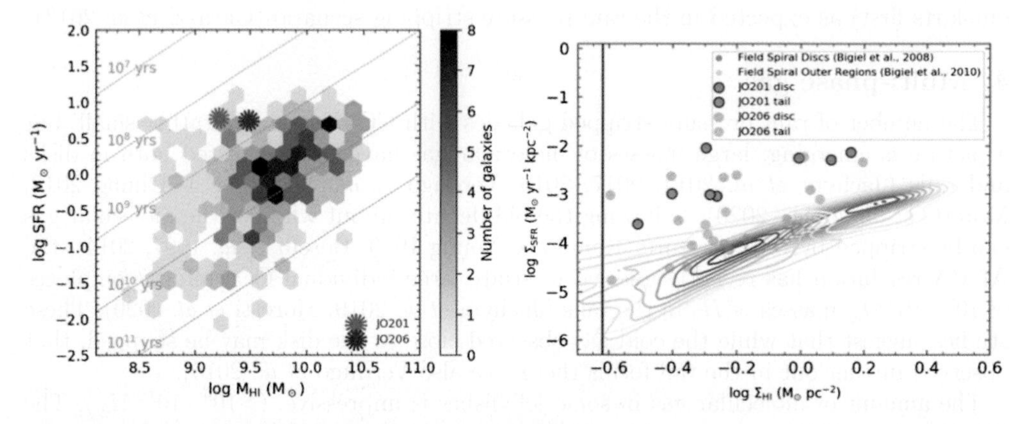

Figure 8. Left: SFR vs HI mass for two GASP jellyfish galaxies (green and blue stars), compared with a control sample of spirals. Right: SFR and HI surface densities of the disks and tails (separately) of the same galaxies, compared with field spirals disks and outskirts. The contours in the right panel are for the inner regions (orange) and outer regions (blue) of field spiral galaxies, both convolved with the HI beam. From Ramatsoku *et al.* 2020.

significant (Deb *et al.* 2020). The GASP jellyfish galaxies for which HI data is available suggest that a) during the jellyfish phase these galaxies still possess large amounts of HI gas (they are only slightly HI deficient, Ramatsoku *et al.* 2020), but the HI is clearly displaced from the disk, spatially and/or kinematically, and b) there is an excess of SFR for the HI content, compared to normal spirals, both globally and on a 1kpc scale (Fig. 8). In other words, the HI-star formation efficiency (ratio of SFR over HI mass) is higher than in normal spirals. Thus, to recap, the SFR is in excess with respect to both the HI content and the stellar mass, but the CO emission is in excess with respect to the SFR. As a consequence, in jellyfish galaxies the star formation efficiency is unusually low for molecular gas, but unusually high for neutral gas, suggesting a very efficient transformation of neutral into molecular gas in these systems (Moretti *et al.* submitted).

We have no space here to deal with tails at yet other wavelengths (X-ray; radio continuum), but we note that multi-λ studies of jellyfish galaxies including these components are growing, after the pioneering studies of e.g. Sun *et al.* (2010); Gavazzi & Jaffé (1985).

To conclude, the study of ram pressure stripped galaxies is informing us on several physical processes which are fundamental for astrophysics in general. We have mostly focused on the GASP results, but there is a broad, high quality (and growing) literature on these fascinating systems, which we hope the reader will be encouraged to explore by this contribution of ours. We apologize for not being able to report on the questions received after BP's talk, due to the difficulty in hearing them remotely. BP sincerely thanks the organizers for their kind invitation and for allowing her to give her presentation from the other side of the world.

Acknowledgements

This project has received funding from the European Reseach Council (ERC) under the Horizon 2020 research and innovation programme (grant agreement N. 833824). Based on observations collected at the European Organisation for Astronomical Research in the Southern Hemisphere under ESO programme 196.B-0578.

References

Abramson, A., Kenney, J. D. P., Crowl, H. H., *et al.* 2011, *AJ*, 141, 164
Bellhouse, C., Jaffé, Y. L., Hau, G. K. T., *et al.* 2017, *ApJ*, 844, 49
Bellhouse, C., Jaffé, Y. L., McGee, S. L., *et al.* 2019, *MNRAS*, 485, 1157
Boselli, A., Gavazzi, G., Lequeux, J., *et al.* 1997, *A&A*, 327, 522
Boselli, A., Cortese, L., Boquien, M., *et al.* 2014, *A&A*, 564, A66
Boselli, A., Cuillandre, J. C., Fossati, M., *et al.* 2016, *A&A*, 587, A68
Boselli, A., Fossati, M., Cuillandre, J. C., *et al.* 2018, *A&A*, 615, A114
Boselli, A., Fossati, M., Longobardi, A., *et al.* 2020, *A&A*, 634, L1
Cayatte, V., van Gorkom, J. H., Balkowski, C., *et al.* 1990, AJ, 100, 604
Chung, A., van Gorkom, J. H., Kenney, J. D. P., *et al.* 2009, AJ, 138, 1741
Consolandi, G., Gavazzi, G., Fossati, M., *et al.* 2017, *A&A*, 606, A83
Cramer, W. J., Kenney, J. D. P., Sun, M., *et al.* 2019, *ApJ*, 870, 63
Deb, T., Verheijen, M. A. W., Gullieuszik, M., *et al.* 2020, *MNRAS*, doi:10.1093/mnras/staa968
Ebeling, H., Stephenson, L. N., & Edge, A. C. 2014, *ApJL*, 781, L40
Fossati, M., Fumagalli, M., Boselli, A., *et al.* 2016, *MNRAS*, 455, 2028
Franchetto, A., Vulcani, B., Poggianti, B. M., *et al.* 2020, arXiv e-prints, arXiv:2004.11917
Fritz, J., Moretti, A., Gullieuszik, M., *et al.* 2017, *ApJ*, 848, 132
Fumagalli, M., Fossati, M., Hau, G. K. T., *et al.* 2014, *MNRAS*, 445, 4335
Gavazzi, G. & Jaffé, W. 1985, *ApJL*, 294, L89
Gavazzi, G. 1989, *ApJ*, 346, 59
Gavazzi, G., Fumagalli, M., Fossati, M., *et al.* 2013, *A&A*, 553, A89
Gavazzi, G., Consolandi, G., Gutierrez, M. L., *et al.* 2018, *A&A*, 618, A130
George, K., Poggianti, B. M., Gullieuszik, M., *et al.* 2018, *MNRAS*, 479, 4126
George, K., Poggianti, B. M., Bellhouse, C., *et al.* 2019, *MNRAS*, 487, 3102
Gullieuszik, M., Poggianti, B. M., Moretti, A., *et al.* 2017, *ApJ*, 846, 27
Gunn, J. E. & Gott, J. R. 1972, *ApJ*, 176, 1
Haynes, M. P., Giovanelli, R., & Chincarini, G. L. 1984, *ARA&A*, 22, 445
Hester, J. A., Seibert, M., Neill, J. D., *et al.* 2010, *ApJL*, 716, L14
Jáchym, P., Combes, F., Cortese, L., *et al.* 2014, *ApJ*, 792, 11
Jáchym, P., Sun, M., Kenney, J. D. P., *et al.* 2017, *ApJ*, 839, 114
Jáchym, P., Kenney, J. D. P., Sun, M., *et al.* 2019, *ApJ*, 883, 145
Jaffé, Y. L., Poggianti, B. M., Moretti, A., *et al.* 2018, *MNRAS*, 476, 4753
Joshi, G. D., Pillepich, A., Nelson, D., *et al.* 2020, arXiv e-prints, arXiv:2004.01191
Lee, B. & Chung, A. 2018, *ApJL*, 866, L10
Kenney, J. D. P. & Young, J. S. 1989, *ApJ*, 344, 171
Kenney, J. D. P., van Gorkom, J. H., & Vollmer, B. 2004, *AJ*, 127, 3361
Merluzzi, P., Busarello, G., Dopita, M. A., *et al.* 2013, *MNRAS*, 429, 1747

Moretti, A., Poggianti, B. M., Gullieuszik, M., *et al.* 2018a, *MNRAS*, 475, 4055
Moretti, A., Paladino, R., Poggianti, B. M., *et al.* 2018b, *MNRAS*, 480, 2508
Moretti, A., Paladino, R., Poggianti, B. M., *et al.* 2020, *ApJ*, 889, 9
Mueller, A., Poggianti, B. M., Pfrommer, C., *et al.* 2020, submitted
Owers, M. S., Hudson, M. J., Oman, K. A., *et al.* 2019, *ApJ*, 873, 52
Poggianti, B. M., Moretti, A., Gullieuszik, M., *et al.* 2017a, *ApJ*, 844, 48
Poggianti, B. M., Jaffé, Y. L., Moretti, A., *et al.* 2017b, *Nature*, 548, 304
Poggianti, B. M., Gullieuszik, M., Tonnesen, S., *et al.* 2019a, *MNRAS*, 482, 4466
Poggianti, B. M., Ignesti, A., Gitti, M., *et al.* 2019b, *ApJ*, 887, 155
Radovich, M., Poggianti, B., Jaffé, Y. L., *et al.* 2019, *MNRAS*, 486, 486
Ramatsoku, M., Serra, P., Poggianti, B. M., *et al.* 2019, *MNRAS*, 487, 4580
Ramatsoku, M., Serra, P., Poggianti, B. M., *et al.* 2020, *A&A*, submitted
Ramos-Martínez, M., Gómez, G. C., & Pérez-Villegas, Á. 2018, *MNRAS*, 476, 3781
Ricarte, A., Tremmel, M., Natarajan, P., *et al.* 2020, arXiv e-prints, arXiv:2003.05950
Roediger, E. & Brüggen, M. 2008, *MNRAS*, 388, 465
Roediger, E., Bruggen, M., Owers, M. S., *et al.* 2014, *MNRAS*, 443, L114
Roman-Oliveira, F. V., Chies-Santos, A. L., Rodríguez del Pino, B., *et al.* 2019, *MNRAS*, 484, 892
Smith, R. J., Lucey, J. R., Hammer, D., *et al.* 2010, *MNRAS*, 408, 1417
Sun, M., Donahue, M., Roediger, E., *et al.* 2010, *ApJ*, 708, 946
Tonnesen, S. & Bryan, G. L. 2009, *ApJ*, 694, 789
Tonnesen, S. & Bryan, G. L. 2012, *MNRAS*, 422, 1609
Verdugo, C., Combes, F., Dasyra, K., *et al.* 2015, *A&A*, 582, A6
Vollmer, B., Cayatte, V., van Driel, W., *et al.* 2001, *A&A*, 369, 432
Vollmer, B., Braine, J., Pappalardo, C., *et al.* 2008, *A&A*, 491, 455
Vulcani, B., Moretti, A., Poggianti, B. M., *et al.* 2017, *ApJ*, 850, 163
Vulcani, B., Poggianti, B. M., Moretti, A., *et al.* 2018a, *ApJ*, 852, 94
Vulcani, B., Poggianti, B. M., Gullieuszik, M., *et al.* 2018b, *ApJL*, 866, L25
Vulcani, B., Poggianti, B. M., Moretti, A., *et al.* 2019a, *MNRAS*, 487, 2278
Vulcani, B., Poggianti, B. M., Moretti, A., *et al.* 2019b, *MNRAS*, 488, 1597
Vulcani, B., Poggianti, B. M., Jaffé, Y. L., *et al.* 2018c, *MNRAS*, 480, 3152
Vulcani, B., Fritz, J., Poggianti, B. M., *et al.* 2020, *ApJ*, 892, 146
Yagi, M., Yoshida, M., Komiyama, Y., *et al.* 2010, *AJ*, 140, 1814

Galaxy Evolution and Feedback across Different Environments
Proceedings IAU Symposium No. 359, 2020
T. Storchi-Bergmann, W. Forman, R. Overzier & R. Riffel, eds.
doi:10.1017/S1743921320002161

ORAL CONTRIBUTIONS

The gas-loss evolution in dwarf spheroidal galaxies: Supernova feedback and environment effects in the case of the local group galaxy Ursa Minor

Anderson Caproni and Gustavo Amaral Lanfranchi

Núcleo de Astrofísica, Universidade Cidade de São Paulo, R. Galvão Bueno 868, Liberdade, São Paulo, SP, 01506-000, Brazil
email: anderson.caproni@cruzeirodosul.edu.br

Abstract. In this work, we performed two distinct non-cosmological, three-dimensional hydrodynamic simulations that evolved the gas component of a galaxy similar to the classical dwarf spheroidal galaxy Ursa Minor. Both simulations take into account types II and Ia supernovae feedback constrained by chemical evolution models, while ram-pressure stripping mechanism is added into one of them considering an intergalactic medium and a galactic velocity that resemble what is observed nowadays for the Ursa Minor galaxy. Our results show no difference in the amount of gas left inside the galaxy until 400 Myr of evolution. Moreover, the ram-pressure wind was stalled and inverted by thermal pressure of the interstellar medium and supernovae feedback during the same interval.

Keywords. galaxies: dwarf — galaxies: evolution — galaxies: individual(Ursa Minor) — galaxies: ISM — hydrodynamics — methods: numerical

---

1. Methodology and Results

Similarly to other classical dwarf spheroidal (dSph) galaxies, Ursa Minor is deficient in neutral gas (e.g., Grcevich & Putman 2009). The physical mechanism responsible for gas depletion in dwarf galaxies is still a matter of debate in the literature. Supernovae feedback and ram-pressure stripping are examples of mechanisms that are able of removing gas from those objects (e.g., Wada & Venkatesan 2003; Emerick *et al.* 2016).

In this work, we explored types II and Ia supernovae feedback and ram-pressure stripping as the main drivers of the gas loss in a dSph galaxy similar to Ursa Minor. We used the numerical code PLUTO (Mignone *et al.* 2007) to solve the three-dimensional hydrodynamic (HD) differential equations that rules the temporal evolution of the gas content. We simulated a cubic region of 3 kpc $\times$ 3 kpc $\times$ 3 kpc, using a non-uniform grid with 330 points in each Cartesian direction (spatial resolution of 8 pc per cell between -0.6 and 0.6 kpc and 10 pc per cell elsewhere). We also adopted a cored, static dark matter gravitational potential due to an isothermal, spherically symmetric mass density profile in our simulations.

Types II and Ia SNe rates used in our two simulations were derived by a chemical evolution model that reproduces several observational constraints (abundances of several chemical elements, the present day gas mass, stellar metallicity distribution) of the dSph galaxy Ursa Minor (Lanfranchi & Matteucci 2004, 2007). In addition, our code follows

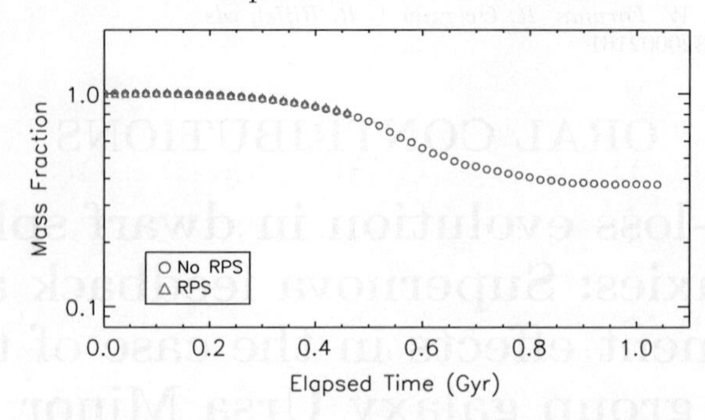

Figure 1. Comparison between the gas mass fraction inside the tidal radius of the galaxy for the HD simulations with SN feedback and with SN feedback plus RP stripping.

independently types II and Ia SNe rates, in contrast to Caproni *et al.* (2017). In the HD simulation including ram-pressure stripping (RPS), a wind is injected with constant number density of 10^{-4} cm^{-3} (in agreement with Grcevich & Putman 2009) and speed of 162 km s^{-1} (in agreement with Piatek *et al.* 2005) from one of the computational boundaries. This mimics roughly the motion of Ursa Minor through the hot halo of the Milky Way.

2. Final remarks

The two three-dimensional HD simulations discussed in this work were performed using one of the highest numerical resolution in the context of dSph galaxies up to now. SN feedback injects turbulence into ISM, producing complex patterns in the gas motions, as reported in previous works (e.g., Ruiz *et al.* 2013; Caproni *et al.* 2015; Caproni *et al.* 2017). No difference in the gas losses was detected during the first 400 Myr of evolution when RPS is included in the calculations, as it can be seen in Fig. 1. Moreover, RP wind is stalled and inverted by thermal pressure of the interstellar medium and SN feedback during the same first 400 Myr of evolution. Additional HD simulations have been conducted to verify whether this behavior is sustained for a longer period of time.

A.C. thanks FAPESP grants 2017/25651-5 and 2014/11156-4. The authors acknowledge the National Laboratory for Scientific Computing (LNCC/MCTI, Brazil) for providing HPC resources of the SDumont supercomputer, which have contributed to the research results reported within this paper. URL: http://sdumont.lncc.br. This work has made use of the computing facilities of the Laboratory of Astroinformatics (IAG/USP, NAT/UCS), whose purchase was made possible by the Brazilian agency FAPESP (grant 2009/54006-4) and the INCT-A.

References

Caproni, A., Lanfranchi, G. A., Luiz da Silva, A., *et al.* 2015, *ApJ*, 805, 109
Caproni, A., Lanfranchi, G. A., Campos Baio, G. H., *et al.* 2017, *ApJ*, 838, 99
Emerick, A., Mac Low, M., Grcevich, J., *et al.* 2016, *ApJ*, 826, 148
Grcevich J. & Putman M. E., 2009, *ApJ*, 696, 385
Lanfranchi, G. A. & Matteucci, F. 2004, *MNRAS*, 351, 1338
Lanfranchi, G. A. & Matteucci, F. 2007, *A&A*, 468, 927
Mignone, A., Bodo, G., Massaglia, S., *et al.* 2007, *ApJS*, 170, 228
Piatek, S., Pryor, C., Bristow, P., *et al.* 2005, *AJ*, 130, 95
Ruiz, L. O., Falceta-Gonçalves, D., Lanfranchi, G.A., *et al.* 2013, *MNRAS*, 429, 1437
Wada, K. & Venkatesan, A. 2003, *ApJ*, 591, 38

Galaxy Evolution and Feedback across Different Environments
Proceedings IAU Symposium No. 359, 2020
T. Storchi-Bergmann, W. Forman, R. Overzier & R. Riffel, eds.
doi:10.1017/S1743921320004081

Supermassive Black Hole feedback in early type galaxies

W. Forman[1], C. Jones[2], A. Bogdan[2], R. Kraft[2], E. Churazov[3], S. Randall[2], M. Sun[4], E. O'Sullivan[2], J. Vrtilek[2] and P. Nulsen[2]

[1]SAO-CfA, 60 Garden St. Cambridge, MA, USA
email: wforman@cfa.harvard.edu

[2]SAO-CfA, 60 Garden St. Cambridge, MA, USA

[3]Space Research Institute, Profsoyuznaya 84/32, Moscow, Russia Max Planck Institute for Astrophysics, Karl Schwarzschild Strasse 1, Garching, Germany

[4]University of Alabama in Huntsville, 301 Sparkman Drive, Huntsville, Alabama

Abstract. Optically luminous early type galaxies host X-ray luminous, hot atmospheres. These hot atmospheres, which we refer to as *coronae*, undergo the same cooling and feedback processes as are commonly found in their more massive cousins, the gas rich atmospheres of galaxy groups and galaxy clusters. In particular, the hot coronae around galaxies radiatively cool and show cavities in X-ray images that are filled with relativistic plasma originating from jets powered by supermassive black holes (SMBH) at the galaxy centers. We discuss the SMBH feedback using an X-ray survey of early type galaxies carried out using Chandra X-ray Observatory observations. Early type galaxies with coronae very commonly have weak X-ray active nuclei and have associated radio sources. Based on the enthalpy of observed cavities in the coronae, there is sufficient energy to "balance" the observed radiative cooling. There are a very few remarkable examples of optically faint galaxies that are 1) unusually X-ray luminous, 2) have large dark matter halo masses, and 3) have large SMBHs (e.g., NGC4342 and NGC4291). These properties suggest that, in some galaxies, star formation may have been truncated at early times, breaking the simple scaling relations.

Keywords. galaxies: elliptical and lenticular, cD X-rays: galaxies, galaxies: jets, galaxies: evolution

---

1. Early type galaxies in the family of dark matter halos

Across the range of dark matter halo masses, massive early type galaxies often lie at the centers of gas rich atmospheres (see Fig. 1 for two prominent examples of gas rich atmospheres around individual early type galaxies). From brightest cluster galaxies (BCGs) in massive clusters to groups and individual galaxies, central galaxies with supermassive black holes (SMBHs) play an important role (see contribution from C. Jones in this volume for a discussion of more massive systems). With the discovery of the X-ray bright atmospheres, the problem of preventing over cooling and massive amounts of star formation became an important focus of study. Early studies argued that many clusters, with high central gas densities, should have gas cooling and settling on the central galaxies (e.g., Fabian & Nulsen 1977). These were the so-called "cooling flows". Despite careful searches for cool gas or newly formed stars, the predicted amounts of material remained undetected (Fabian 1994).

Only at the turn of the millennium did the first hints of a robust solution to the problem begin to appear with the comparison of high angular resolution X-ray and radio

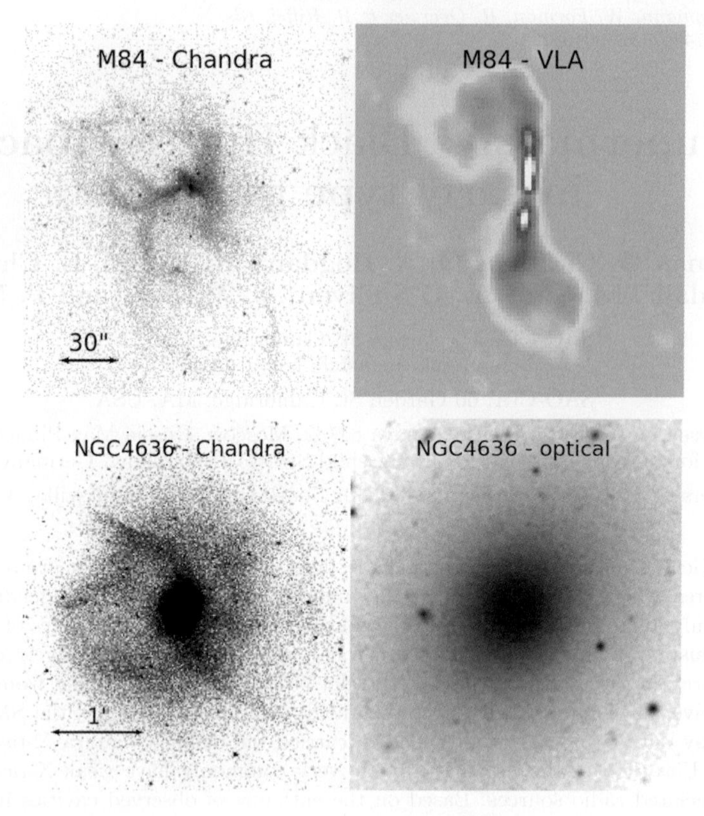

Figure 1. Examples of feedback in early type galaxies from active galactic nuclei (AGN). (top) The Chandra and VLA images (at matched scales) of M84 = NGC4374 showing the very disturbed X-ray atmosphere surrounding the galaxy and the corresponding radio image. The radio plasma fills the large northern and southern cavities in the X-ray surface brightness (see Finoguenov *et al.* 2008 for more details). (bottom) Chandra and digital sky survey images (at matched scales) of NGC4636. The Chandra image shows the very disturbed X-ray atmosphere surrounding this otherwise optically "normal" galaxy (see Baldi *et al.* 2009 for more details).

observations (Böhringer *et al.* 1993; Churazov *et al.* 2000, 2001; McNamara *et al.* 2000; Fabian *et al.* 2000). These analyses showed that supermassive black holes (SMBH) at the centers of X-ray bright atmospheres were radiatively faint, but mechanically powerful. In this new paradigm, SMBHs drive jets that inflate buoyant bubbles of relativistic plasma which appear as cavities in cluster, group, and galaxy X-ray images (see reviews Fabian 2012; Bykov *et al.* 2015; Werner *et al.* 2019). In fact, the power required to inflate the bubbles is sufficient to compensate for the radiative cooling (e.g., Bîrzan *et al.* 2004). The detailed mechanism for transferring the bubble energy to the hot gas remains a topic of intense discussion, but one very strong candidate is turbulent motion of the hot gas as the buoyant bubbles rise in the hot atmospheres (see Zhuravleva *et al.* 2014).

In this contribution, we focus on the lower end of the mass scale of the family of dark matter halos. Hot galactic coronae were first detected from Einstein Observatory observations (Forman *et al.* 1979, 1985; Nulsen *et al.* 1984). As in more X-ray luminous systems, early type galaxies have short cooling times and, in the absence of a heating mechanism, are expected to be the sites of significant star formation (Thomas *et al.* 1986). Although the feedback process in single galaxies is not as powerful as in rich galaxy clusters, the signatures of AGN feedback - cavities in the X-ray gas distribution, filled with relativistic plasma from the SMBH - are commonly seen (see Fig. 1 for examples of

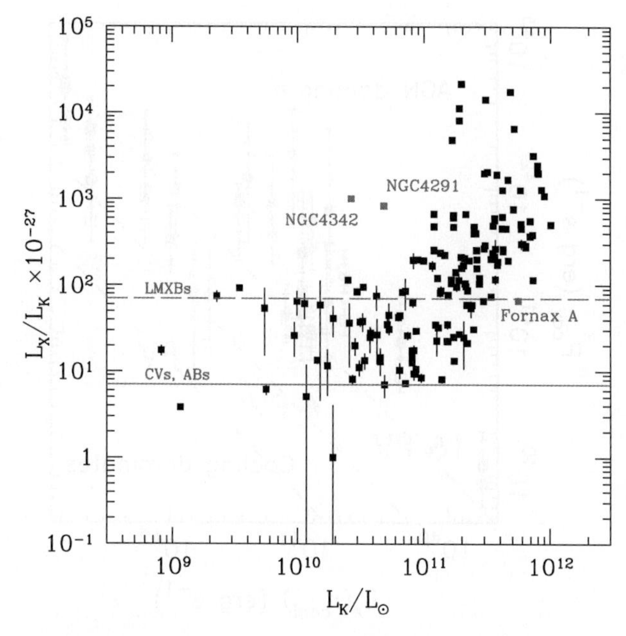

Figure 2. The X-ray luminosity (ergs sec^{-1}) of galaxies per unit K-band luminosity (in solar luminosities) plotted vs. K-band luminosity from a large sample of galaxies, observed with the Chandra X-ray Observatory. The figure shows that luminous systems (in the K-band, a good proxy for mass; Bell *et al.* 2003), which typically have massive dark matter halos, are able to contain the hot gas, although they do exhibit a wide range of luminosity. At fainter optical luminosities, the galaxies are X-ray faint and likely drive galactic winds. Three labeled galaxies are discussed in the text. Although the sample excludes brightest cluster galaxies (BCGs), galaxies in groups are included. The two horizontal lines show the predicted contribution from low mass X-ray binaries (LMXBs) and cataclysmic variables and active stellar systems (ABs). In the survey, when exposures are sufficiently deep, any detected LMXBs are excluded.

bubbles and cavities in galaxy scale dark matter halos; see Jones & Forman, this volume, for a discussion of feedback in M87).

2. Chandra survey of early type galaxies

Fig. 2 shows the X-ray luminosity of galaxies per unit K-band luminosity plotted vs. K-band luminosity from a large sample, observed with the Chandra X-ray Observatory, that excludes galaxy clusters but does include galaxies at the centers of poor galaxy groups (e.g., NGC4636; see Fig. 1). This sample was taken from observations of optically bright galaxies that were targeted by Chandra. As a class, luminous early type galaxies (brighter than $L_K{\sim}10^{11}$ L$_\odot$) are also luminous in X-rays. At a given L_K, the figure shows that the galaxies display a wide range of X-ray luminosity. Prior to the discovery of hot coronae, early type galaxies were believed to be gas poor as a result of galactic winds (e.g., Mathews & Baker 1971 and Faber & Gallagher 1976). However, X-ray observations have shown that these galaxies are *not gas poor* systems but have gaseous atmospheres as massive as the interstellar mediums of their spiral galaxy cousins (e.g., Forman *et al.* 1985). In this sample of more than 100 early type galaxies, 70% of the galaxies have radio sources associated with their nuclei. In addition, about 80% of the galaxies have X-ray bright sources located within a few arcseconds of the galaxy center. These nuclear sources have luminosities ranging from ${\sim}10^{38}-10^{42}$ ergs s^{-1}. Using a simple scaling relation of K-band luminosity to stellar mass and stellar mass to SMBH mass, we can derive the Eddington ratios for the active nuclei. We find $L_x/L_{edd}{\sim}10^{-5}-10^{-9}$. Hence, these are

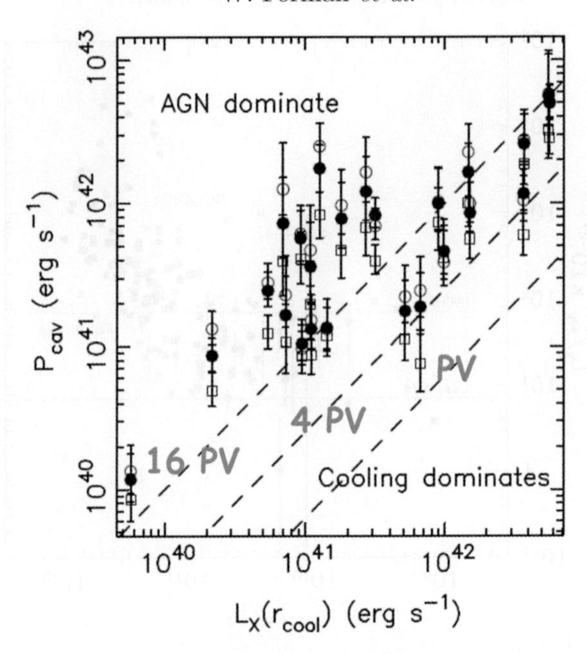

Figure 3. Cavity power vs cooling power for outbursts detected in the Chandra sample (adapted from Nulsen *et al.* 2009). Filled markers, open circles and open squares show three different estimates of the power using different timescales: P_{sonic}, $P_{buoyancy}$ and P_{refill}, the sonic, buoyancy, and refill timescales, respectively, with 90% confidence ranges. Dashed lines (as labeled) indicate different conversion factors that account for the ratio of the cavity power to the total power injected (see Fig. 8 in Werner *et al.* 2019 for an extension to galaxy clusters).

consistent with radiatively inefficient, but mechanically powerful, SMBH outbursts (for models of this class see, for example: Ichimaru 1977; Rees *et al.* 1982; Yuan & Narayan 2014).

Of the X-ray bright systems, about 30% exhibit X-ray cavities similar to those seen in Fig. 1. The discovery of cavities provides a method for measuring the energy, injected by the jets from SMBHs, into the hot, X-ray emitting gas, as the relativistic plasma inflates cavities. By measuring the pressure, P, and volume, V of the cavities, the PV work could be computed (Churazov *et al.* 2000). First for galaxy clusters and then for groups, studies of cavities provided convincing evidence that the SMBH feedback was sufficient to dramatically reduce the cooling of the X-ray emitting atmospheres (see for example, Bîrzan *et al.* 2004 and Fig. 5 of Fabian 2012).

On galaxy scales too, the cavities were shown to provide sufficient energy to "balance" cooling. Fig. 3 (adapted from Nulsen *et al.* 2009) shows that for many optically and X-ray luminous galaxies, the cavity power is sufficient to balance radiative cooling. The analysis uses three estimates to derive the cavity timescale. These three timescales are 1) the sonic timescale, the sound crossing time, 2) the buoyancy time, the time for the buoyant bubble to rise to its present position at the buoyant terminal speed, and 3) the refill time, the time for the observed volume to be refilled as the bubble rises. Fig. 3 shows all three estimates for 24 galaxies observed with Chandra. These estimates of the power are plotted vs. the cooling luminosity (for details see Nulsen *et al.* 2009). Three diagonal, dashed lines are drawn where the cavity power and the cooling luminosity are equal, with the cavity power assumed to be PV, $4PV$, or $16PV$. Note that jets could, and sometimes do, inflate bubbles supersonically which drives shocks into the surrounding gas, providing additional energy beyond just the work required to inflate the bubbles (for example, see

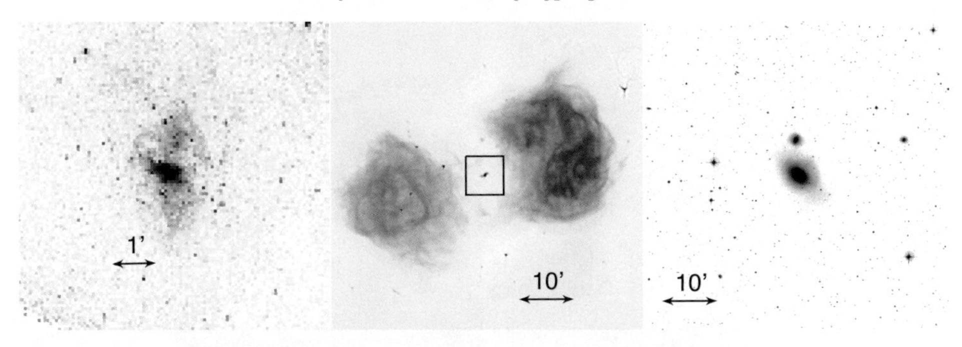

Figure 4. Chandra, VLA, and optical images of Fornax A (NGC1316). Note the very different scale of the Chandra image compared to radio and optical. (left) Chandra image of Fornax A shows a very disturbed gas distribution. Lanz *et al.* (2010) estimate an outburst energy of $\sim 10^{58}$ ergs. The X-ray luminosity is reduced by the outburst that disperses the hot gas from the core, reducing the density and hence the luminosity. (center) 20 cm VLA image (Fomalont *et al.* 1989) showing the large scale radio emission and the central core. The square box at the center delineates the field of view of the Chandra image at left. (right) optical (red) image of NGC1316 at the same scale as the radio image. At a distance of 23 Mpc, $1' = 6.5$ kpc).

Baldi *et al.* 2009; Fabian *et al.* 2000; Forman *et al.* 2017, and Randall *et al.* 2015). For those galaxies with cavities, there is more than sufficient energy to offset radiative cooling. While cavities are not detected in all hot galaxy atmospheres, the cycle of cooling and heating outbursts is irregular, but on average, the SMBHs appear to be balancing radiative cooling.

While the feedback process is very often "balanced", there are notable exceptions on scales from clusters to galaxies. On cluster scales, the Phoenix cluster (McDonald *et al.* 2012), with 800 $M_{\odot}$ yr^{-1} of star formation, shows a SMBH that has failed to offset cooling. Among the galaxies, Fornax A (labeled in blue in Fig. 2) has one of the lowest X-ray luminosities for its K-band luminosity (a proxy for stellar mass). Fig. 4 shows the X-ray and radio images of Fornax A. The SMBH at the center of the galaxy NGC1316, that hosts Fornax A, has undergone a remarkable outburst and dramatically disturbed the X-ray atmosphere which serves to decrease the X-ray luminosity (since the luminosity depends on the square of the gas density). Detailed multi-wavelength studies (Mackie & Fabbiano 1998 and Lanz *et al.* 2010) have discussed the outburst history and the merger event that occurred in NGC1316 within the last 2 Gyr. Lanz *et al.* (2010) suggest that observed cavities require an outburst energy of $\sim 10^{58}$ ergs. With a massive SMBH, possibly fueled by mergers, SMBHs are capable of dramatically impacting a hot atmosphere.

3. NGC4342 and NGC4291 - optically faint, X-ray bright galaxies

Two remarkable galaxies, NGC4342 and NGC4291, both members of poor groups, are shown in Fig. 2 (see red identifiers). These two galaxies are unusually X-ray luminous for their K-band luminosities (see Bogdán *et al.* 2012 for a detailed discussion of these two galaxies). Both galaxies have dynamically measured SMBH masses (Schulze & Gebhardt 2011; Cretton & van den Bosch 1999) that are 60 and 13 times larger than expected from the mean SMBH mass vs. optical bulge mass relation (given by Häring & Rix 2004). The uniqueness of these galaxies is visually emphasized in Fig. 5 that compares the X-ray and optical properties of several comparable galaxies around NGC4342. Although optical properties are similar, NGC4342 has a remarkable X-ray corona. For both NGC4342 and NGC4291, the X-ray data can be used to derive the total gravititating mass, assuming

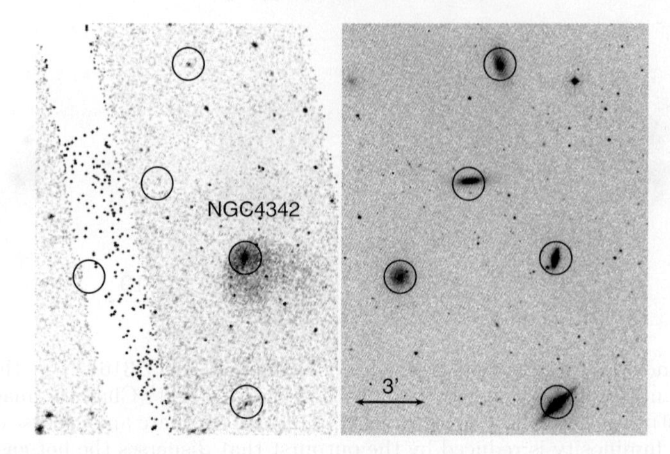

Figure 5. Chandra and optical images of NGC4342 at matched scales. (left) Chandra image of NGC4342 and its surroundings with other optically comparable galaxies that are all members of the W' cloud (Mei *et al.* 2007 and Bogdán *et al.* 2012). (right) Optical image matching the Chandra image. Optical galaxies are circled in both the X-ray and optical images. Clearly, NGC4342 is "special" - of these comparable galaxies, several of which also lie in the Chandra field, only NGC4342 hosts a hot X-ray corona and, therefore, a significant dark matter halo. As described in the text, its SMBH is also unusually massive for the mass of its measured stellar bulge.

hydrostatic equilibrium, which shows they both lie in substantial dark matter halos (10 and 5 times the stellar mass for NGC4342 and NGC4291, respectively; Bogdán *et al.* 2012).

As shown above for Fornax A, SMBHs are capable of powerful outbursts when properly "fed". One possible scenario to create systems like NGC4342 and NGC4291, is for an aggressively growing SMBH to have had an unusually large outburst at an early epoch, before star formation was completed, which prematurely terminated star formation. Some simulations (e.g., Bonoli *et al.* 2014 and Katz *et al.* 2015) suggest that SMBH growth precedes the growth of the stellar mass. If this scenario is correct, then we should expect to see more systems like NGC4342 and NGC4291.

Another class of interesting galaxies that host X-ray bright coronae and massive SMBHs are "relics", the present epoch counterparts of high redshift "red nuggets" (e.g., Buote & Barth 2018, 2019). The variety of hot coronae will provide new insights on galaxy evolution and yield powerful constraints for the growth of SMBHs.

4. Summary and Conclusions

At the low end range of the mass distribution of dark matter halos, early type galaxies host gas rich atmospheres. However, unlike their spiral counterparts, these atmospheres are hot and easily recognizable as extended in X-ray images. Furthermore, like their more massive cousins, groups and clusters, early type galaxies (those brighter than $L_K{\sim}10^{11}$ $L_\odot$) undergo cycles of feedback with outbursts from SMBH's inflating bubbles that provide energy to the radiatively cooling gas. A few rare objects, e.g., NGC4342 and NGC4291, appear to have become overly "enthusiastic", at early epochs, and may have terminated star formation before the stellar mass grew to the characteristic size needed to lie on the stellar mass - SMBH mass relation. New scaling relations (e.g., Bogdán & Goulding 2015; Bogdán *et al.* 2018; Phipps *et al.* 2019, and Gaspari *et al.* 2019) suggest that the stellar mass may not be the fundamental driver of SMBH mass. X-ray all-sky surveys from *eROSITA* will provide large samples of X-ray bright galaxies to fully explore the present epoch population of dark matter halos and their hot gas content.

Acknowledgments

We acknowledge support from the Smithsonian Institution, the Smithsonian Astrophysical Observatory, and the Chandra High Resolution Camera project, supported by NAS8-03060.

References

Baldi, A., Forman, W., Jones, C., *et al.* 2009, *ApJ*, 707, 1034
Bell, E. F., McIntosh, D. H., Katz, N., *et al.* 2003, *ApJS*, 149, 289
Bîrzan, L., Rafferty, D. A., McNamara, B. R., *et al.* 2004, *ApJ*, 607, 800
Bogdán, Á., Forman, W. R., Zhuravleva, I., *et al.* 2012, *ApJ*, 753, 140
Bogdán, Á., Forman, W. R., Kraft, R. P., *et al.* 2012, *ApJ*, 755, 25
Bogdán, Á. & Goulding, A. D. 2015, *ApJ*, 800, 124
Bogdán, Á., Lovisari, L., Volonteri, M., *et al.* 2018, *ApJ*, 852, 131
Böhringer, H., Voges, W., Fabian, A., *et al.* 1993, *MNRAS*, 264, L25
Bonoli, S., Mayer, L., & Callegari, S. 2014, *MNRAS*, 437, 1576
Buote, D. A. & Barth, A. J. 2018, *ApJ*, 854, 143
Buote, D. A. & Barth, A. J. 2019, *ApJ*, 877, 91
Bykov, A. M., Churazov, E. M., Ferrari, C., *et al.* 2015, *Sp. Sci. Rev.*, 188, 141
Churazov, E., Forman, W., Jones, C., *et al.* 2000, *A&A*, 356, 788
Churazov, E., Brüggen, M., Kaiser, C. R., *et al.* 2001, *ApJ*, 554, 261
Cretton, N. & van den Bosch, F. C. 1999, *ApJ*, 514, 704
Faber, S. M. & Gallagher, J. S. 1976, *ApJ*, 204, 365
Fabian, A. C. & Nulsen, P. E. J. 1977, *MNRAS*, 180, 479
Fabian, A. C. 1994, *ARA&A*, 32, 277
Fabian, A. C., Sanders, J. S., Ettori, S., *et al.* 2000, *MNRAS*, 318, L65
Fabian, A. C. 2012, *ARA&A*, 50, 455
Finoguenov, A., Ruszkowski, M., Jones, C., *et al.* 2008, *ApJ*, 686, 911
Fomalont, E. B., Ebneter, K. A., van Breugel, W. J. M., *et al.* 1989, *ApJL*, 346, L17
Forman, W., Schwarz, J., Jones, C., *et al.* 1979, *ApJL*, 234, L27
Forman, W., Jones, C., & Tucker, W. 1985, *ApJ*, 293, 102
Forman, W., Churazov, E., Jones, C., *et al.* 2017, *ApJ*, 844, 122
Gaspari, M., Eckert, D., Ettori, S., *et al.* 2019, *ApJ*, 884, 169
Häring, N. & Rix, H.-W. 2004, *ApJL*, 604, L89
Ichimaru, S. 1977, *ApJ*, 214, 840
Katz, H., Sijacki, D., & Haehnelt, M. G. 2015, *MNRAS*, 451, 2352
Lanz, L., Jones, C., Forman, W. R., *et al.* 2010, *ApJ*, 721, 1702
Mackie, G. & Fabbiano, G. 1998, *AJ*, 115, 514
Mathews, W. G. & Baker, J. C. 1971, *ApJ*, 170, 241
McDonald, M., Bayliss, M., Benson, B. A., *et al.* 2012, *Nature*, 488, 349
McNamara, B. R., Wise, M., Nulsen, P. E. J., *et al.* 2000, *ApJL*, 534, L135
Mei, S., Blakeslee, J. P., Côté, P., *et al.* 2007, *ApJ*, 655, 144
Nulsen, P. E. J., Stewart, G. C., & Fabian, A. C. 1984, *MNRAS*, 208, 185
Nulsen, P., Jones, C., Forman, W., *et al.* 2009, *American Institute of Physics Conference Series*, 198
Phipps, F., Bogdán, Á., Lovisari, L., *et al.* 2019, *ApJ*, 875, 141
Randall, S. W., Nulsen, P. E. J., Jones, C., *et al.* 2015, *ApJ*, 805, 112
Rees, M. J., Begelman, M. C., Blandford, R. D., *et al.* 1982, *Nature*, 295, 17
Schulze, A. & Gebhardt, K. 2011, *ApJ*, 729, 21
Thomas, P. A., Fabian, A. C., Arnaud, K. A., *et al.* 1986, *MNRAS*, 222, 655
Werner, N., McNamara, B. R., Churazov, E., *et al.* 2019, *Sp. Sci. Rev.*, 215, 5
Yuan, F. & Narayan, R. 2014, *ARA&A*, 52, 529
Zhuravleva, I., Churazov, E., Schekochihin, A. A., *et al.* 2014, *Nature*, 515, 85

Galaxy Evolution and Feedback across Different Environments
Proceedings IAU Symposium No. 359, 2020
T. Storchi-Bergmann, W. Forman, R. Overzier & R. Riffel, eds.
doi:10.1017/S174392132000174X

The light side of proto-cluster galaxies at $z \sim 4$

Kei Ito

Department of Astronomical Science, The Graduate University for Advanced Studies, SOKENDAI, Mitaka, Tokyo, 181-8588, Japan
email: kei.ito@grad.nao.ac.jp

Abstract. Overdense regions at high redshift, which are often called "protoclusters", are thought to be a place where the early active structure formations are in progress. Thanks to the wide and deep-sky survey of Hyper Suprime-Cam Subaru Strategic Program, we have selected 179 protocluster candidates at $z \sim 4$, enabling us to statistically discuss high-z overdense regions. I report results of the HSC-SSP protocluster project, focusing on a couple of results on the bright-end of protocluster galaxies. We identify the UV-brightest galaxies, which are likely progenitors of Brightest Cluster Galaxies. We find that these are dustier and larger than field galaxies. This suggests that galaxies in protoclusters have experienced different star formation histories at $z \sim 4$. Also, the UV luminosity function of galaxies in protoclusters (PC UVLF) has a significant excess on the bright-end from field UVLF. The PC UVLF suggests that protoclusters contribute $\sim$5−16% of the total cosmic SFRD at $z \sim 4$. The result implies that early galaxy formation occurs in protoclusters.

Keywords. galaxies: evolution, galaxies: clusters: general, galaxies: high-redshift

---

1. Introduction

Properties of galaxies in galaxy clusters are known to be different from those of galaxies in the field (e.g., Dressler 1980). Searching progenitors of these clusters at high redshift and investigating properties of their galaxies can give some clues for unveiling origins of such an effect of overdense regions. Therefore, overdense regions at high redshift, so-called protoclusters, which are defined as structures that will collapse into virialized objects with $\geqslant 10^{14} M_{\odot}$ at $z \geqslant 0$, are unique targets. These have been found through a large variety of selection techniques and various tracers at $z \sim 2$–7.

Observational studies show that protocluster galaxies at $z \sim 2$ have different properties compared to field galaxies at the same epoch. They tend to have enhancements of star formation rates (SFRs) (e.g., Shimakawa *et al.* 2018; Koyama *et al.* 2013), with larger stellar mass (above references and Cooke *et al.* 2014; Hatch *et al.* 2011; Steidel *et al.* 2005). Several theoretical simulations support these trends and suggest that galaxies in protoclusters experience earlier formation and are a significant contribution to the cosmic star formation rate density (SFRD) at $z \geqslant 2$ (e.g., Chiang *et al.* 2017; Lovell *et al.* 2018; Muldrew *et al.* 2015). Extending the study of protocluster galaxies properties is essential to understand the origin of the environmental effect. However, due to the low number density of protocluters and the various techniques to detect them, they had not yet been systematically investigated at $z > 3$.

Recently, we have conducted a new protocluster survey from the photometric data of the Hyper Suprime-Cam (HSC) Subaru Strategic Program (HSC-SSP) (Aihara *et al.* 2018). Our protocluster survey is based on Lyman break galaxies (LBGs), and we have already selected 179 protocluster candidates from g-dropout galaxies over an area of 121 deg^2 (Toshikawa *et al.* 2018). Based on this sample, we have conducted several follow-up studies, investigating the relationship between overdensity and bright quasars in Uchiyama *et al.* (2018), and quasar pairs in Onoue *et al.* (2018), and using the stacked infrared (IR) properties of protoclusters to probe obscured star formation and active galactic nuclei in Kubo *et al.* (2019). Here, we report two results of rest-ultraviolet (rest-UV) bright galaxies in these protoclusters; the rest-UV properties of UV-brightest galaxies Ito *et al.* (2019) and the rest-UV luminosity function (UVLF) of protocluster galaxies (PC UVLF) (Ito *et al.* submitted). According to the star-formation main sequence, the UV-brightest galaxies are expected to be the most massive among other g-dropout members, which means that they can be progenitors of Brightest Cluster Galaxies in the local universe (hereafter proto-BCGs). Various papers have studied the UVLF of field galaxies, and they are the dominant diagnostics of the cosmic SFRD at $z \sim 3$–8. The PC UVLF enables us to estimate the contribution of protoclusters to the cosmic SFRD and examine whether protocluster galaxies have different properties, even at $z \sim 4$.

2. The UV-brightest Protocluster Galaxies

Since we are focusing on the significantly brightest galaxies compared to other protocluster members, we select the UV-brightest galaxies, which are 1 mag brighter than the fifth brightest galaxies in each protocluster in the rest-UV. As a result, we select 63 brightest protocluster galaxies from 179 protoclusters. For the detailed selection method, including the protocluster selection, see Ito *et al.* (2019).

We compare two properties; rest-UV color ($i - z$ for $z \sim 4$) and rest-UV size. When comparing the rest-UV color, we match the UV luminosity of reference field galaxies to that of proto-BCGs in order to exclude the effect of the relation of the brightness and color (e.g., Bouwens *et al.* (2009)) The left panel of Figure 1 shows the distribution of $i - z$ colors of proto-BCGs and field galaxies. We can see that the proto-BCGs are redder than field galaxies. The rest-UV color of galaxies is believed to be primarily related to their dust extinction. This suggests that proto-BCGs are dustier than field galaxies with the same luminosity.

Next, we estimate the average rest-UV size for proto-BCGs. We use i band images taken from HSC-SSP since it has the best image quality in terms of seeing and depth. For maximizing the signal to noise ratio, the images of all proto-BCGs were average-stacked. The average radial profile of the magnitude matched field galaxies sample is also derived in the same manner. The radial profile of proto-BCGs looks more extended than that of field galaxies. To quantitatively discuss this feature, we fit the 2D galaxy surface profile to stacked images by using `GALFIT`. We employ the Sérsic profile (Sérsic 1963) assuming the Sérsic index to be -1.5, following Shibuya *et al.* (2015). We plot our proto-BCGs value compared to the size-luminosity relation of field galaxies from Shibuya *et al.* (2015) in the right panel of Figure 1. The value of field galaxies estimate in this study is also overplotted, and they are consistent with Shibuya *et al.* (2015), suggesting that our estimation is in good agreement with previous studies. Proto-BCGs are 28% larger than field galaxies.

In this UV-brightest protocluster galaxy study, we find that UV-brightest galaxies in HSC-SSP protoclusters at $z \sim 4$ are dustier and have larger sizes in rest-UV. The result suggests that the environmental effect has occurred at $z \sim 4$, at least for the UV-brightest galaxies. The proper study for all protocluster members is crucial for understanding the whole picture of the environmental effect.

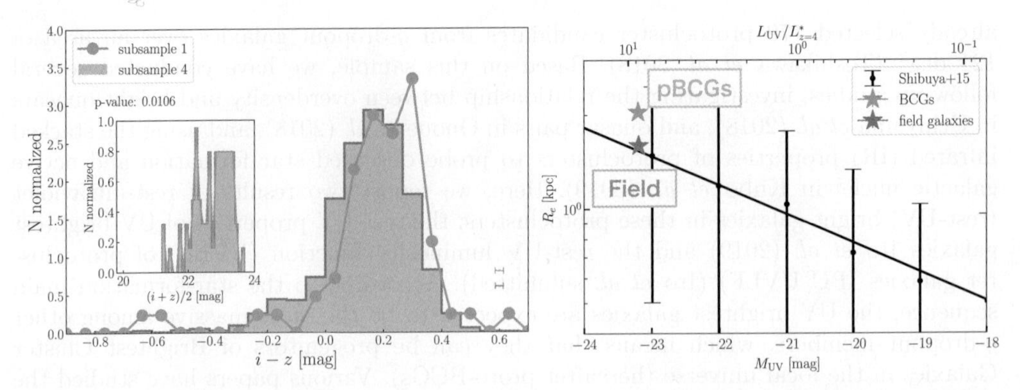

Figure 1. Left Panel: The $i-z$ color distribution of proto-BCGs (red line) and field galaxies (blue histogram) at the same brightness, edited from Ito *et al.* (2019). The typical UV magnitude ($(i+z)/2$) is shown in the inset. The result of the Anderson-Darling test suggests that the p-value $p = 1.1 \times 10^{-2}$, so we reject the null hypothesis that these two color distributions are the same. Right Panel: The average size-luminosity relation of proto-BCGs (red star). A blue star is that of magnitude-matched field galaxies. Black dots correspond to a size-luminosity relation of LBGs at $z \sim 4$ from Shibuya *et al.* (2015).

3. The rest-UV luminosity function of protocluster galaxies

We derive a UVLF of g-dropout galaxies in HSC-SSP protocluster at $z \sim 4$. Here, we define protocluster galaxies as objects that are located within $1'.8$ from overdensity peak of each of the 179 protocluster. This size corresponds to the typical size of protoclusters predicted in theoretical simulations (Chiang *et al.* 2013).

Here, we describe the procedure for deriving the PC UVLF. We first derive the completeness function of our g-dropout galaxy sample and the UVLF of field galaxies. Then from the number count of protocluster members and these functions, we derive the PC UVLF $\Phi_{\rm PC}(M_{\rm UV})$ following this equation:

$$\Phi_{\rm PC}(M_{\rm UV}) = \frac{1}{F(M_{\rm UV})}\left(\frac{n_{\rm obs,PC}(M_{\rm UV})}{V_{\rm eff}(M_{\rm UV})} - \Phi_{\rm field}(M_{\rm UV})\right) \tag{3.1}$$

Here, $n_{\rm obs,PC}(M_{\rm UV})$ is the observed number of g-dropout galaxies in protocluster regions, $\Phi_{\rm field}(M_{\rm UV})$ is the luminosity function of field galaxies without the contamination correction, and $V_{\rm eff}(M_{\rm UV})$ is the effective volume of g-dropout galaxies derived from the completeness function. $F(M_{\rm UV})$ is the correction factor of the effective volume of protoclusters from the entire g-dropout galaxies since they are located in a significantly smaller volume than field galaxies. HSC-SSP protoclusters are located in 5 fields (GAMA15H, HECTOMAP, VVDS, WIDE12H, and XMM), and since the depth of each field can be different, we estimate PC UVLF for each field separately.

Figure 2 shows our PC UVLF. Two differences found for the PC UVLF can be seen from this figure. First, the amplitude of the PC UVLF is higher than that of field galaxies. This is because we focus on protoclusters, which are galaxy overdense regions. Secondly, also the shape of the PC UVLF is different from that of the field UVLF. We can see the PC UVLF has flatter shape than that of field galaxies. This implies that protocluster galaxies are brighter in the rest-UV than field galaxies. It is known that rest-UV light can be converted to the star formation rate (SFR) (Kennicut 1998), so brighter galaxies in protoclusters means higher SFRs for protocluster galaxies, which supports the existence of the environment effect on galaxy properties even at $z \sim 4$.

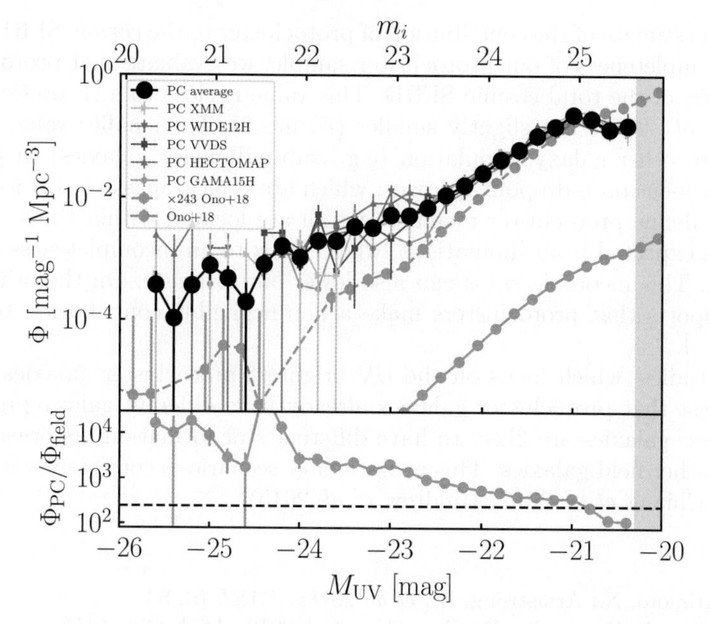

Figure 2. The UV luminosity function of galaxies in protocluster candidates at $z \sim 4$ from Ito et al. submitted. The color-coded lines represent the PC UVLF for each survey field. The black circles show the average of all fields. For reference, we show the field UVLF of Ono *et al.* (2018) (solid gray line) and shifted upward to match the PC UVLF (gray dotted line with circles). The bottom panel shows the ratio of the PC UVLF and scaled field UVLF (red circles). The black dashed line shows the value of the ratio of the sum of each UVLF. For both panels, the magnitude range that is fainter than the depth is shaded in gray.

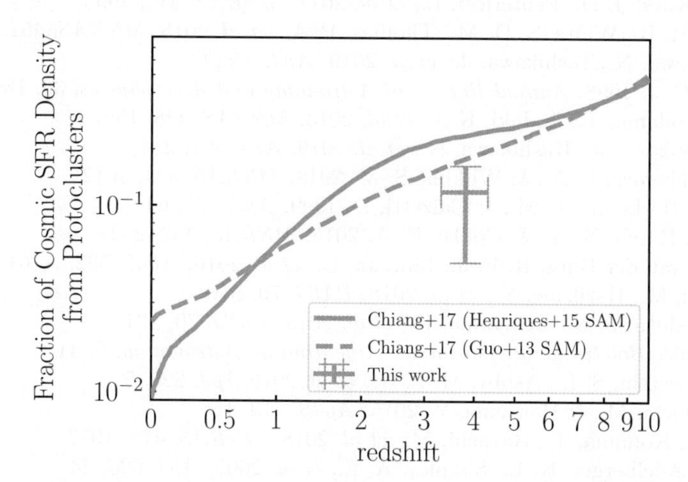

Figure 3. The fraction of the cosmic SFRD in protoclusters, edited from Ito et al. submitted. A red cross represents our estimated value for HSC-SSP protoclusters at $z \sim 4$. The gray solid and dashed lines are its predicted evolution in Chiang *et al.* (2017) with the use of the semi-analytical model of Henriques *et al.* (2015) and Guo *et al.* (2013), respectively.

The SFRD in protocluster regions can be obtained from the PC UVLF. From the PC UVLF and β–$M_{\rm UV}$ relation of protocluster galaxies, we estimate the average UV luminosity density and the average FIR luminosity density via the IRX-β-M_* relation of $z \sim 3$ LBGs (Álvarez-Márquez *et al.* 2019). The UV/FIR luminosity density is converted to SFRD from the relation in Kennicut (1998). The estimated SFRD in protocluster regions leads us to an estimate of the contribution of protocluster to the cosmic SFRD. Correcting purity and completeness of our protocluster sample, we evaluate that protoclusters contribute 5–16% of the total cosmic SFRD. This value is close to a theoretical prediction in Chiang *et al.* (2017) but slightly smaller (Figure 3). The smaller value of our result can be due to other galaxy population (e.g., sub-millimeter galaxies) in protoclusters since we only focus on g-dropout galaxies, which are typical massive star forming galaxies. Also, we define protocluster members which are located within the smallest size of protoclusters expected from simulations, which leads to an incompleteness of protocluster members. This incompleteness can also lower our estimate. On the other hand, our estimate supports that protoclusters make a non-negligible contribution to the cosmic SFRD at $z \sim 4$.

Our two studies, which focus on the UV-brightest protocluster galaxies and the PC UVLF, suggest that protocluster galaxies already have different galaxy properties even at $z \sim 4$. These galaxies are likely to have different star formation histories; they evolve earlier than other field galaxies. This evolutionary scenario is consistent with theoretical predictions (Chiang *et al.* 2017; Muldrew *et al.* 2015).

References

Aihara, H., Arimoto, N., Armstrong, R., *et al.* 2018a, *PASJ*, 70, S4
Álvarez-Márquez, J., Burgarella, D., Buat, V., *et al.* 2019, *A&A*, 630, A153
Bouwens, R. J., Illingworth, G. D., Franx, M., *et al.* 2009, *ApJ*, 705, 936
Chiang, Y.-K., Overzier, R. A., Gebhardt, K., *et al.* 2017, *ApJL*, 844, L23
Chiang, Y.-K., Overzier, R. A., & Gebhardt, K. 2013, *ApJ*, 779, 127
Cooke, E. A., Hatch, N. A., Muldrew, S. I., *et al.* 2014, *MNRAS*, 440, 3262
Dressler, A. 1980, *ApJ*, 236, 351
Guo, Q., White, S., Angulo, R. E., *et al.* 2013, *MNRAS*, 428, 1351
Hatch, N. A., Kurk, J. D., Pentericci, L., *et al.* 2011, *MNRAS*, 415, 2993
Henriques, B. M. B., White, S. D. M., Thomas, P. A., *et al.* 2015, *MNRAS*, 451, 2663
Ito, K., Kashikawa, N., Toshikawa, J., *et al.* 2019, *ApJ*, 878, 68
Kennicutt, R. C. J. 1998, *Annual Review of Astronomy and Astrophysics*, 36, 189
Koyama, Y., Kodama, T., Tadaki, K.-i., *et al.* 2013, *MNRAS*, 428, 1551
Kubo, M., Toshikawa, J., Kashikawa, N., *et al.* 2019, *ApJ*, 887, 214
Lovell, C. C., Thomas, P. A., & Wilkins, S. M. 2018, *MNRAS*, 474, 4612
Meurer, G. R., Heckman, T. M., & Calzetti, D. 1999, *ApJ*, 521, 64
Muldrew, S. I., Hatch, N. A., & Cooke, E. A. 2015, *MNRAS*, 452, 2528
Nantais, J. B., van der Burg, R. F. J., Lidman, C., *et al.* 2016, *A&A*, 592, A161
Ono, Y., Ouchi, M., Harikane, Y., *et al.* 2018, *PASJ*, 70, S10
Onoue, M., Kashikawa, N., Uchiyama, H., *et al.* 2018, *PASJ*, 70, S31
Sérsic, J. L. 1963, *Boletin de la Asociacion Argentina de Astronomia*, 6, 41
Song, M., Finkelstein, S. L., Ashby, M. L. N., *et al.* 2016,*ApJ*, 825, 5
Shibuya, T., Ouchi, M., & Harikane, Y. 2015, *ApJS*, 219, 15
Shimakawa, R., Kodama, T., Hayashi, M., *et al.* 2018, *MNRAS*, 473, 1977
Steidel, C. C., Adelberger, K. L., Shapley, A. E., *et al.* 2005, *ApJ*, 626, 44
Toshikawa, J., Uchiyama, H., Kashikawa, N., *et al.* 2018, *PASJ*, 70, S12
Uchiyama, H., Toshikawa, J., Kashikawa, N., *et al.* 2018, *PASJ*, 70, S32

Galaxy Evolution and Feedback across Different Environments
Proceedings IAU Symposium No. 359, 2020
T. Storchi-Bergmann, W. Forman, R. Overzier & R. Riffel, eds.
doi:10.1017/S1743921320001660

Mass outflow of the X-ray emission line gas in NGC 4151

S. B. Kraemer[1], T. J. Turner[2], D. M. Crenshaw[3], H. R. Schmitt[4], M. Revalski[5] and T. C. Fischer[5]

[1]Department of Physics, Institute for Astrophysics and Computational Sciences, The Catholic University of America, Washington, DC 20064, USA
email: kraemer@cua.edu

[2]Department of Physics, University of Maryland Baltimore County, Baltimore, MD 21250 U.S.A

[3]Department of Physics and Astronomy, Georgia State University, 25 Park Place, Room 631, Atlanta, GA 30303, USA

[4]Naval Research Laboratory, Washington, DC 20375, USA

[5]Space Telescope Science Institute, Baltimore, MD 21218, USA

Abstract. We have analyzed *Chandra*/High Energy Transmission Grating spectra of the X-ray emission line gas in the Seyfert galaxy NGC 4151. The zeroth-order spectral images show extended H- and He-like O and Ne, up to a distance $r \sim 200$pc from the nucleus. Using the 1st-order spectra, we measure an average line velocity $\sim$230 km s^{-1}, suggesting significant outflow of X-ray gas. We generated Cloudy photoionization models to fit the 1st-order spectra; the fit required three distinct emission-line components. To estimate the total mass of ionized gas (M) and the mass outflow rates, we applied the model parameters to fit the zeroth-order emission-line profiles of Ne IX and Ne X. We determined an $M \approx 5.4 \times 10^5 M_{\odot}$. Assuming the same kinematic profile as that for the [O III] gas, derived from our analysis of *Hubble Space Telescope*/ Space Telescope Imaging Spectrograph spectra, the peak X-ray mass outflow rate is approximately 1.8 $M_{\odot}$ yr^{-1}, at $r \sim 150$pc. The total mass and mass outflow rates are similar to those determined using [O III], implying that the X-ray gas is a major outflow component. However, unlike the optical outflows, the X-ray emitting mass outflow rate does not drop off at $r > 100$pc, which suggests that it may have a greater impact on the host galaxy.

Keywords. galaxies:active – galaxies: individual: NGC 4151 – galaxies: Seyfert – X-rays: galaxies

---

1. Introduction

Active galactic nuclei (AGN) are powered by the accretion of matter onto a supermassive black hole (SMBH), which generates huge amounts of radiation from a very small volume. AGN are capable of generating powerful outflows, or winds, which are believed to be critical to the structure, energetics, and evolution of AGN and their connection to their host galaxies. Specifically, the relation between bulge mass and black hole mass, the so-called M_{BH}–σ_* relation (Gebhardt *et al.* 2000), is thought to be regulated by AGN outflows, i.e., "AGN feedback" (Begelman 2004).

The narrow line regions (NLRs) in AGN can extend from a few parsecs to several kpc, depending on luminosity, and hence offer unique insight into the interaction of the AGN with its host galaxy. NLR-scale outflows have been studied via *Hubble Space Telescope (HST)*/Space Telescope Imaging Spectrograph (STIS) observations (e.g., Fischer *et al.* 2013, 2018) and ground-based spectra, such as those obtained with the *Gemini*/Near Infrared Integral Field Spectrograph (e.g., Storchi-Bergmann *et al.* 2010; Riffel *et al.* 2013; Fischer *et al.* 2017). The NLR outflows extend to kpc scales (Fischer *et al.* 2018), where much of the nuclear star-formation occurs. Furthermore, most of the mass in these outflows is accelerated in situ, due to the radiation pressure from the AGN accelerating material in the disk of the host galaxy. Hence, the NLR-scale outflows directly reveal the interaction of the AGN with the host galaxy.

Although the NLR outflows are often massive (e.g., mass of $\sim 10^6$ M$_\odot$, Revalski *et al.* 2018b), and may inject large amounts of kinetic energy and momentum into the interstellar medium (ISM) of the host galaxy (e.g., Revalski *et al.* 2018a), they do not appear to be able to escape the inner kpc of the host galaxy bulge (Fischer *et al.* 2018). This calls into question the effectiveness of AGN-driven outflows in feedback. However, these results are generally derived from optical or near-IR emission line kinematics, they do not take into account the role of higher-ionization gas. For example, X-ray winds, in the form of Ultra Fast Outflows (e.g. Tombesi *et al.* 2010), appear to be able to drive large-scale molecular outflows in ULIRGs (Tombesi *et al.* 2015). Therefore, it is plausible that the X-ray gas can entrain and accelerate the optical emission-line gas.

Chandra/ACIS images of the nearby Seyfert galaxies NGC 4151 (Ogle *et al.* 2000) and NGC 1068 (Young *et al.* 2001) revealed extended soft X-ray emission, co-located with the [O III] emission-line gas. Bianchi *et al.* (2006) determined that this was the case for most Seyfert galaxies. *Chandra* imaging has been used to map and model the NLR X-ray emission in several Seyferts, e.g. NGC 4151 - Wang *et al.* (2011a,b,c), NGC 3393 - Maksym *et al.* (2019) and Mrk 573 - Gonzalez-Martin *et al.* (2010). By isolating bands dominated by emission-lines, e.g., Ne IX 13.7 Å, these authors were able to derive constraints on the ionization structure of the X-ray emission-line regions. In particular, Maksym et al. and Wang et al. suggest that there is evidence for shocks, indicative of the interaction of the X-ray with the ISM of the host galaxy. However, the image data, as opposed to spectra obtained with the *Chandra*/High Energy Transmission Grating (HETG), lack the spectral resolution to constrain the kinematics of the X-ray emitting gas. In contrast, Kallman *et al.* (2014) performed a detailed photoionization analysis of HETG spectra of NGC 1068 and were able to derive estimates of the total mass and mass outflow rates of the emission-line gas $M \approx 3.7 \times 10^5$ M$_\odot$ and $\dot{M}_{out} \sim 0.3$ M$_\odot$ yr^{-1}, respectively. Here we summarize the results of our analysis of the 240 ksec HETG observation of NGC 4151 (Kraemer *et al.* 2020).

2. Observations and Analysis

We obtained a 240 ksec *Chandra*/HETG observation of NGC 4151 (OSBID 16089/ 19060; 2014 February). These data were split into two epochs due to constraints in the roll angle alignment to meet the observation goals, i.e., to allow us to have the cross-dispersion direction correctly oriented to observe the extended emission of NGC 4151 (see Ogle *et al.* 2000). The full, unfolded spectrum is shown in Figure 1 (left hand panel). The intrinsic continuum was heavily absorbed and in a low-flux state (Couto *et al.* 2016), which enabled the isolation of numerous emission features, such as H- and He-like lines from O, Ne, Mg, and Si, Fe Kα, Si Kα and various radiative recombination continua.

The emission-line morphology is revealed via the zero-th order spectra, as shown in Figure 1 (right hand panel). The continuum, both at 2.05-2.25 keV and 5.60-6.00 keV, is roughly symmetric and centered on the nucleus. The Si Kα and Fe Kα lines are also strongly centrally peaked, again with a slight extension to the NE. The location of the peak of the emission encompasses the innermost region of the AGN, consistent an origin in the putative torus. The lines with energies >1.3 keV, e.g., Si XIV, Si XIII, and Mg XII, are also quite compact. On the other hand, there is evidence that Mg XI is extended

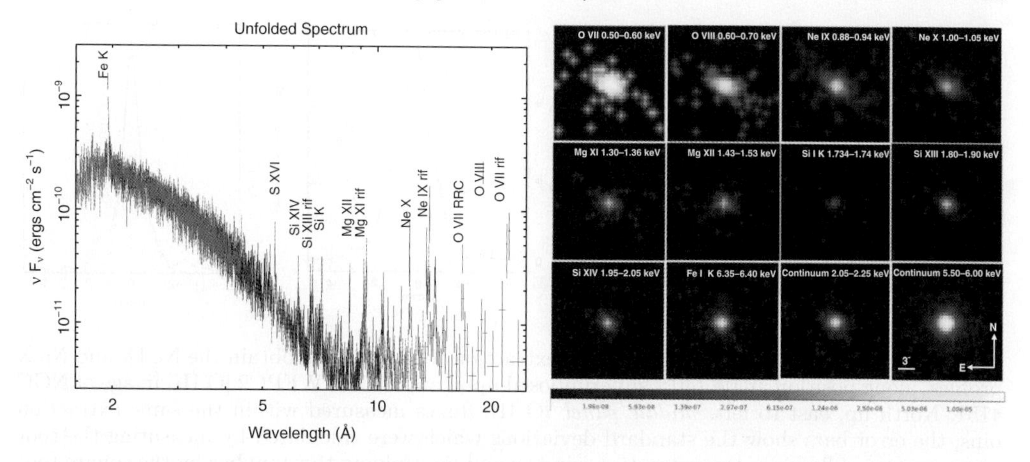

Figure 1. *left hand panel*: Full, $\pm 1^{st}$ order, HETG spectrum from 1.59 Å–24.8 Å, from the summed OBSIDs. The red and black points are from the HEG and MEG, respectively. The data were binned to a minimum of 10 counts (per bin). The turn-over towards lower energies (greater wavelengths) is due to intrinsic absorption (see Couto *et al.* 2016). Emission at wavelengths >6Å is emission-line dominated. *right hand panel*; Emission profiles for hydrogen and helium-like O, Ne, Mg, and Si lines, and Si Kα and Fe Kα lines, from the zeroth-order HETG spectra. Line-free regions of continuum in the ranges 2.05-2.25 keV and 5.50-6.00 keV are also shown. The images have been smoothed with a Gaussian function with kernel radius of $1''$, and are equally scaled for comparison in the colorbar in terms of total number of counts. The spatial scale and orientation are shown in the lower right panel.

along the NE-SW direction. Lines with energies $\leqslant$1.05 keV, i.e., Ne X, Ne IX, O VIII, and O VII, are all clearly extended along a NE-SW direction, out to a projected distance of $>$200 pc.

To measure line fluxes, we used the combined 1st-order spectra. We fitted a Gaussian model and local continuum (using a power-law, with an un-constrained slope) for a restricted bandpass close to the line, using χ^2 statistics. The width of the bandpass used depended on the particular region, i.e. whether it was crowded or the line was isolated. The line flux and observed energy were then fitted for most lines. The average radial velocity, with upper and lower limits, is $v_r = -230^{-90}_{-370}$ km s^{-1}. Although there is evidence for a relationship between velocities and the ionization potential for optical and UV lines in some Seyfert galaxies (Kraemer & Crenshaw 2000; Kraemer *et al.* 2009), which suggests a multi-component NLR, we found no strong evidence of such an relationship in NGC 4151 (Kraemer *et al.* 2000). Therefore, while the X-ray emission line gas consists of components of different ionization, there is no evidence for kinematic differences among them.

The details of the spectral fitting are given in Kraemer *et al.* (2020). In summary, we generated grids of Cloudy photoionization models (Ferland *et al.* 2017) over a range of values of ionization parameter, U, and column density, N_H, using the default energy resolution. We converted the Cloudy output to fittable grids using the CLOUDY-to-XSPEC interface (Porter *et al.* 2006). The emission features were fit via the additive emission components (ATABLES). The final ATABLE parameters, log$U$, logN_H, were: 1.0, 22.5; 0.19, 22.5; $-$0.50, 23.0.

We then applied the 1st-order results to fit the extended emission, using the zeroth order HETG image. We measured the Ne X and Ne IX fluxes in a region oriented along the major axis of the optical NLR, as shown in Figure 2 (left hand panel). The emission line profiles show structure similar to that of [O III] (Figure 2, middle and right hand panels), indicating that the X-ray and optical emission-line gas have similar morphologies. Using the density law from Crenshaw *et al.* (2015), we generated photoionization models

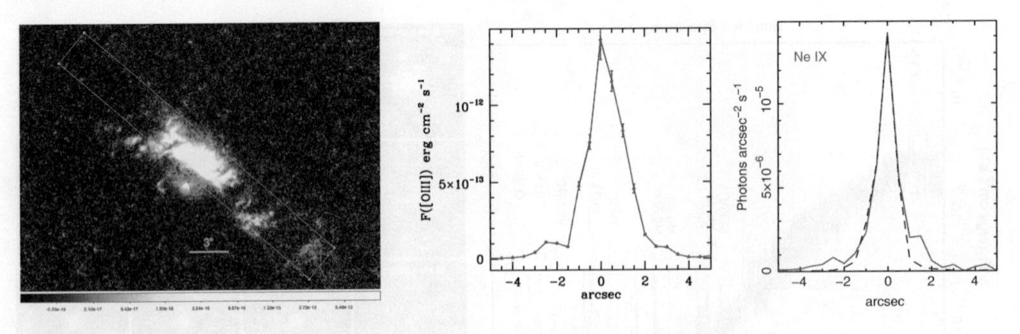

Figure 2. *left hand panel*: The 0″.5 X 3″.0 extraction region used to obtain the Ne IX and Ne X profiles, along position angle 140°, superimposed upon an archival WFPC2 [O III] image of NGC 4151. North up, east to left. *middle panel*: [O III] fluxes measured within the same extraction bins; the error bars show the standard deviations which were calculated by measuring the root mean squares of fluxes in the extraction window and multiplying this number by the square root of the number of pixels in this extraction region. Positive positions are those southwest (SW) of the nucleus. *right hand panel* Ne IX emission-line profile (in red) as a function of position. The fluxes were measured from the zeroth order HETG spectrum. The blue dashed line is the continuum; positive positions are SW of the nucleus

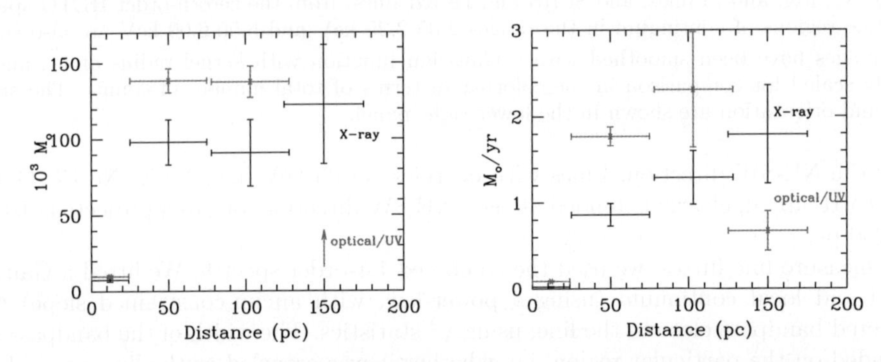

Figure 3. *left hand panel*: Computed M values at the same radial distances (SW and NE values summed) for the HETG analysis (black crosses), compared to those from the STIS optical/UV analysis (red asterisks), from Crenshaw *et al.* (2015) and Revalski *et al.* (2018b). The optical/UV points are summed to correspond with the HETG extraction bin sizes. Uncertainties in the optical/UV points are those from Crenshaw et al., and have been added in quadrature to account for the binning. *right hand panel*: Computed values of $\dot{M}_{out}$ as a function of distance, from the HETG analysis (black crosses), compared to those from the STIS optical/UV analysis (red asterisks). For both datasets, the $\dot{M}_{out}$ values were computed using flux-weighted velocities within each bin from the STIS analysis. Uncertainties include both those of the mass values and the flux-weighted velocities.

for each extraction bin in order to determine M as a function of de-projected radial distance, r. In Figure 3, M and $\dot{M}_{out}$ of the X-ray and optical gas are shown for $r < 200$ pc. The peak X-ray mass outflow rate is $\approx$1.8 M yr^{-1}, at $r \sim 150$ pc. While the M and $\dot{M}_{out}$ values are similar for both components, the X-ray emitting mass outflow does not appear to diminish for $r > 100$ pc.

3. Summary and Conclusions

The similarity in the mass profile, total mass, and mass outflow rates between the X-ray and optical gas suggest that the former is dynamically important in the inner nucleus of NGC 4151. Nevertheless, several open issues remain.

1) The origin of the X-ray gas is uncertain. One possibility is that it is formed via thermal expansion of the [O III] emission-line gas. Crenshaw *et al.* (2015) and Fischer *et al.* (2017) have suggested that the optical NLR gas in Seyferts originates in the disk of the host galaxy, and is ionized and accelerated, by the AGN, in situ. Also, the [O III] knots do not appear to travel far from their point of origin (Fischer *et al.* 2017). An intriguing possibility is that the optical outflows evolve into X-ray emitting winds (Trindade Falcão et al., in preparation).

2) A major limit to this analysis (see also Gonzalez-Martin *et al.* 2010) is the lack of spatially resolved X-ray kinematics. However, highly ionized optical emission lines, e.g, [Fe X] λ6374, [Fe IX] λ7892, and [Fe XIV] λ5303, are formed in the same range of ionization as the X-ray emission-line gas. Hence, they are potential "footprints" of X-ray winds. These lines can be mapped with *HST*/STIS (e.g., Kraemer *et al.* 2000) and offer the possibility of tracking the X-ray wind kinematics.

References

Begelman, M. C. 2004, in Carnegie Observatories Astrophysics Series, Vol. 1, Coevolution of Black Holes and Galaxies, from the Carnegie Observatories Centennial Symposia, ed. L. Ho, 374–390

Bianchi, S., Guainazzi, M., & Chiaberge, M. 2006, *A&A*, 448, 499

Couto, J., Kraemer, S., Turner, T., *et al.* 2016, *ApJ*, 833, 191

Crenshaw, D. M., Fischer, T. C., Kraemer, S. B., *et al.* 2015, *ApJ*, 799, 83

Ferland, G. J., Chatzikos, M., Guzman, F., *et al.* 2017, *Rev.Mexicana AyA*, 53, 385

Fischer, T. C., Crenshaw, D. M., Kraemer, S. B., *et al.* 2013, *ApJS*, 209, 1

Fischer, T. C., Machuca, C., Diniz, M. R., *et al.* 2017, *ApJ*, 834, 30

Fischer, T. C., Kraemer, S. B., Schmitt, H. R., *et al.* 2018, *ApJ*, 856, 102

Gebhardt, K., Bender, R., Bower, G., *et al.* 2000, *ApJ*, 539, L13

Gonzalez-Martin, O., Acosta-Pulido, J. A., Perez Garcia, A. M., *et al.* 2010, *ApJ*, 723, 1748

Kallman, T. R., Evans, D. A., Marshall, H., *et al.* 2014, *ApJ*, 780, 121

Kraemer, S. B. & Crenshaw, D. M. 2000, *ApJ*, 532, 256

Kraemer, S. B., Crenshaw, D. M., Hutchings, J. B., *et al.* 2000, *ApJ*, 531, 278

Kraemer, S. B., Trippe, M. L., Crenshaw, D. M., *et al.* 2009, *ApJ*, 698, 106

Kraemer, S. B., Turner, T. J., Couto, J. D., *et al.* 2020, *MNRAS*, 493, 3893

Maksym, W. P., Fabbiano, G., Elvis, M., *et al.* 2019, *ApJ*, 872, 94

Ogle, P. M., Marshall, H. L., Lee, J. C., *et al.* 2000, *ApJ*, 545, L81

Porter, R. L., Ferland, G. J., Kraemer, S. B., *et al.* 2006, *PASP*, 118, 920

Revalski, M., Crenshaw, D. M., Kraemer, S. B., *et al.* 2018a, *ApJ*, 856, 46

Revalski, M., Dashtamirova, D., Crenshaw, D. M., *et al.* 2018b, *ApJ*, 867, 88

Riffel, R. A., Storchi-Bergmann, T., & Winge, C. 2013, *MNRAS*, 430, 2249

Storchi-Bergmann, T., Lopes, R. D. S., McGregor, P. J., *et al.* 2010, *MNRAS*, 402, 819

Tombesi, F., Cappi, M., Reeves, J. N., *et al.* 2010, *A&A*, 521, A57

Tombesi, F., Melédez, M., Veilleux, S., *et al.* 2015, *Nature*, 519, 436

Wang, J., Fabbiano, G., Elvis, M., *et al.* 2011a, *ApJ*, 736, 62

Wang, J., Fabbiano, G., Risaliti, G., *et al.* 2011b, *ApJ*, 729, 75

Wang, J., Fabbiano, G., Elvis, M., *et al.* 2011c, *ApJ*, 742, 23

Young, A. J., Wilson, A. S., & Shopbell, P. L. 2001, *ApJ*, 556, 6

Galaxy Evolution and Feedback across Different Environments
Proceedings IAU Symposium No. 359, 2020
T. Storchi-Bergmann, W. Forman, R. Overzier & R. Riffel, eds.
doi:10.1017/S1743921320004226

Cold gas studies of a z = 2.5 protocluster

Minju M. Lee[1], Ichi Tanaka[2] and Rohei Kawabe[3,4,5]

[1]Max-Planck-Institut für Extraterrestrische Physik (MPE), Giessenbachstr. 1, D-85748 Garching, Germany

[2]Subaru Telescope, National Astronomical Observatory of Japan, 650 North Aohoku Place, Hilo, HI 96720, USA

[3]National Astronomical Observatory of Japan, 2-21-1 Osawa, Mitaka, Tokyo 181-8588, Japan

[4]The Graduate University for Advanced Studies (SOKENDAI), 2-21-1 Osawa, Mitaka, Tokyo 181-8588, Japan

[5]Department of Astronomy, The University of Tokyo, 7-3-1 Hongo, Bunkyo, Tokyo 113-0033, Japan

Abstract. We present studies of a protocluster at $z = 2.5$, an overdense region found close to a radio galaxy, 4C 23.56, using ALMA. We observed 1.1 mm continuum, two CO lines (CO (4–3) and CO (3–2)) and the lower atomic carbon line transition ([C I]($^3P_1$-$^3P_0$)) at a few kpc (0″.3-0″.9) resolution. The primary targets are 25 star-forming galaxies selected as Hα emitters (HAEs) that are identified with a narrow band filter. These are massive galaxies with stellar masses of $>10^{10}\ M_\odot$ that are mostly on the galaxy main sequence at $z = 2.5$. We measure the molecular gas mass from the independent gas tracers of 1.1 mm, CO (3–2) and [C I], and investigate the gas kinematics of galaxies from CO (4–3). Molecular gas masses from the different measurements are consistent with each other for detection, with a gas fraction ($f_{\rm gas} = M_{\rm gas}/(M_{\rm gas} + M_{\rm star})$) of $\simeq$0.5 on average but with a caveat. On the other hand, the CO line widths of the protocluster galaxies are typically broader by ~50% compared to field galaxies, which can be attributed to more frequent, unresolved gas-rich mergers and/or smaller sizes than field galaxies, supported by our high-resolution images and a kinematic model fit of one of the galaxies. We discuss the expected scenario of galaxy evolution in protoclusters at high redshift but future large surveys are needed to get a more general view.

Keywords. galaxies: clusters: general, galaxies: evolution, galaxies: high-redshift, galaxies: ISM, galaxies: kinematics and dynamics, large-scale structure of universe

---

1. Introduction

Over the past two decades, multi-wavelength studies have provided a detailed picture of 'baryonic cycle'. Dust and CO observations revealed that the global (i.e., a few kpc scale) gas content, which fuels the star-forming activity, increases toward higher-z. The cosmic star-formation rate density peaked at $z \approx 1.5-2.5$, and has declined since then by a factor of 10-15 (e.g., Madau & Dickinson 2014). Together with a moderate or little evolution of star-forming efficiency, the observed gas content explains the observed cosmic star-formation rate density (e.g., Scoville *et al.* 2013; Genzel *et al.* 2015; Tacconi *et al.* 2018). A simple bath tub model (e.g., Bouché *et al.* 2010) or gas regulator model (e.g., Lilly *et al.* 2013) works remarkably well to describe such trends. However, such measurements are obtained for field galaxies and observations of members of dense, massive clusters and their progenitors are needed for a complete picture of 'baryonic cycle' across the Universe.

The kinematical properties are additional parameters that one needs to take into account to fully understand galaxy evolution. Typically, massive quiescent galaxies are short of star-forming activities with higher bulge-to-total mass ratio, meaning that star-forming galaxies need to not only stop their star formation activities but also change their appearance to become quiescent galaxies. The morphology-density relation has been known for four decades which shows an observational trend that the fraction of early-type galaxies, which are massive quiescent galaxies, increases in denser environment of clusters (Dressler 1980). However, it is still an open question whether environment is an additional parameter regulating the galaxy evolution which is more elusive for higher redshift. To aim for answering the question, we investigate both the gas content and gas kinematics using ALMA targeting a protocluster at $z = 2.5$ as a pilot study to fully understand galaxy evolution.

2. 4C 23.56 protocluster and parent sample

Protocluster 4C 23.56 was identified as an overdense region of the narrow-band(NB) selected Hα emitters in the vicinity of radio galaxy, 4C 23.56. It was targeted as a part of the MAHALO-Subaru (MApping HAlpha and Lines of Oxygen with Subaru) survey (Kodama *et al.* 2015).

There are rich ancillary data sets from X-ray to radio, which makes the protocluster an ideal target for a pilot study. Currently, the protocluster is known to have (projected) overdensities of differently selected galaxy populations, for example, mass-selected distant red galaxies (DRGs) (Kajisawa *et al.* 2006), extremely red objects (EROs; Knopp & Chambers 1997), IRAC (Mayo *et al.* 2012), MIPS (Galametz *et al.* 2012) sources, and SMGs observed at 1.1 mm (K. Suzuki 2013 PhD thesis; Zeballos *et al.* 2018).

We focus on HAEs as a primary target because they have relatively secure redshift information compared to other galaxy populations. We use the broadband emissions in J and Ks bands and NB Hα emissions to derive stellar masses and star-formation rate (SFR), respectively. The massive ($>10^{10}\,M_\odot$) galaxies are on the galaxy main sequence (MS) at $z = 2.5$ except for two. One of these two galaxies is the radio galaxy whose SFR may be overestimated by the AGN contamination. More details are described in Lee *et al.* (2017).

3. Gas content

We observed 1.1 mm dust continuum, CO (3–2) and [C I]($^3P_1$-$^3P_0$) (hereafter, [C I] (1–0)) using ALMA to measure the global gas content of the galaxies. Seven HAEs are detected in CO (3–2) line, four out of which are also detected in 1.1 mm continuum. Provided that the HAEs are mostly on the main-sequence, the typical gas recipes of dust (Scoville *et al.* 2013) and CO (Genzel *et al.* 2015) are applied to measure the gas content. We confirm that the methods are still valid for the members of the protocluster, and two different tracers provide consistent values within a factor 3 (Lee *et al.* 2017).

These measurements are revisited using [C I] (1–0) line, which is another independent gas tracer (Lee *et al.* 2021 in preparation). We detected the line in three galaxies. These include galaxies previously detected in CO (3–2) and two out of them also have dust continuum detection that allow us to self-consistently weigh the gas content from three different tracers. Figure 1 shows how the CO (3–2) luminosities are scaled with the [C I] luminosities that are both proxies for gas mass. For [C I] detections, they are aligned with field galaxies. Assuming a nominal abundance ratio that is applied to milky-way like galaxies (i.e., $\log([\mathrm{C\,I}]/[\mathrm{H_2}]) = -4.8$, see e.g., Valentino *et al.* 2018), we measure the molecular gas masses that are consistent within a factor of 3 compared to the previous

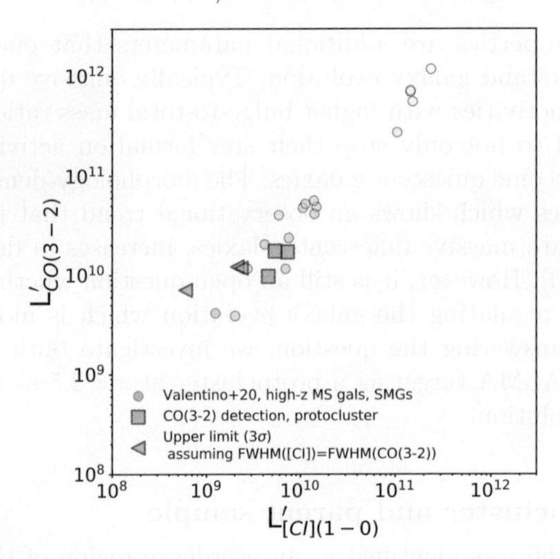

Figure 1. CO (3–2) line luminosity versus [C I] (1–0) line luminosity. Square symbol is for the [C I] (1–0) detection. The upper limit of [C I] (1–0) luminosity is obtained assuming the line width (FWHM) is the same as CO (3–2). For comparison, we also plot literature values compiled in Valentino *et al.* (2020) (Lee *et al.* 2020, in preparation).

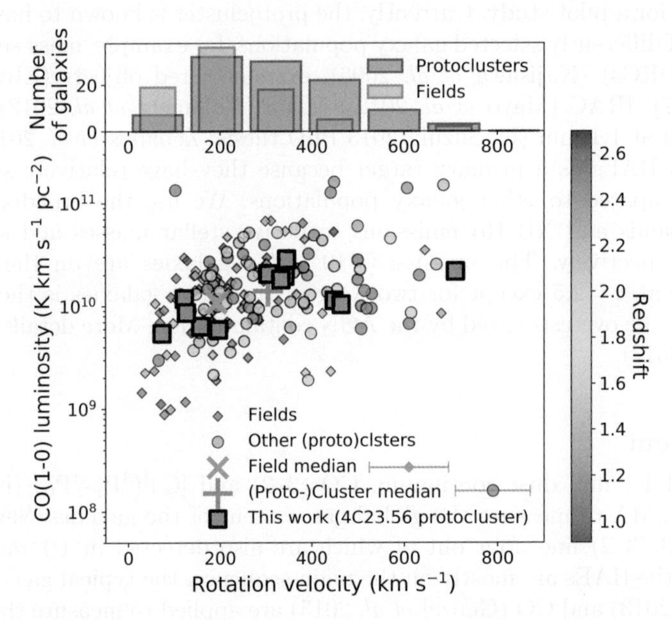

Figure 2. The distribution of the CO (1–0) luminosity and rotation velocity in the comparison with field galaxies. The rotation velocity is converted from the FWHM by taking the isotropic virial estimate of the circular velocity (FWHM$\times\sqrt{3/8\ln 2}$) to match with galaxies in Tacconi *et al.* (2013). The CO (1–0) luminosities are converted by assuming the line luminosity ratio of $R1J = 1.2$, 1.8, and 2.4 for $J = 2$, 3, and 4, respectively. Adapted from Lee *et al.* (2019).

measurements. We note that these are relatively gas-rich galaxies and less massive compared to the other [C I] non-detection galaxies. For [C I] non-detection, however, we find larger discrepancies, suggestive of different gas conditions or excitation toward more massive, gas-poorer systems, which needs deeper, other CO/[C I] observations to confirm. The details of the [C I] results are presented in Lee *et al.* (2020, in preparation). Overall, the (cold) gas fraction of the protocluster galaxies is $f_{\rm gas} = 0.5$ on average for the simultaneous detections (Lee *et al.* 2017) but there is a signature of different gas conditions for more massive, gas-poorer galaxies.

4. Gas kinematics

To probe the gas kinematics, we observed the protocluster galaxies with CO (4–3) at higher angular resolution of $0''.4$. Eleven HAEs are detected in CO (4–3), including six HAEs that were previously detected in CO (3–2) at a coarser angular resolution. The detections in both CO lines are broadly consistent in the line widths and the redshifts, confirming both detections. The CO (4–3) line detections confirm the redshifts of 11 HAEs, giving a constraint on the protocluster halo mass, which is a few $\times 10^{13}\, M_\odot$, supporting that the protocluster is a progenitor of Virgo-like clusters (Lee *et al.* 2019).

The CO line widths are on average broader by $\approx$50% compared to field galaxies, while the median CO luminosities are similar (Figure 2). Based on the resolved source structures and spectral analysis, we conclude that the broader line widths can be ascribed to unresolved gas-rich mergers and/or compact gas distribution. The compact gas distribution may be the result of gas-rich mergers but also of gas-stripping, which is difficult to confirm with the current available data sets and needs deeper observations to verify. Meanwhile, the best-fit kinematic parameters of one of the galaxies indicate that the specific angular momentum is similar to that of field populations during the cluster assembly and in the existence of gas-rich mergers, and hence “dissipational” processes are needed if it is a progenitor of early-type galaxies that are abundant in clusters (Lee *et al.* 2019).

5. Conclusions

We studied gas properties of galaxies using dust, CO and [C I] as a pilot project of studying protocluster galaxies. While the number of galaxies are limited, the cold gas studies provided rich information on the cold gas content and kinematical properties in detail. The different gas tracers of dust/CO/[C I] give consistent values of gas content with each other when [C I] line is detected. For protocluster 4C 23.56 at $z = 2.5$, the measured gas fraction is $f_{\rm gas} = 0.5$ on average comparable to field main-sequence galaxies at similar stellar mass range, but there is a hint of deviation when the stellar mass becomes larger ($\gtrsim 10^{11}\, M_\odot$), supported by the non-detection of [C I] and a tentative trend of a stronger stellar mass dependency of the gas depletion time scale. This also indicates the change of gas properties accordingly that one needs to confirm with other CO/[C I] transitions. We note that cold gas studies on different (proto)clusters at slightly lower redshift ($z = 1-2$) argued different gas fraction (larger/smaller than fields) that might be related to different evolutionary stages of cluster assembly (e.g., Noble *et al.* 2017; Coogan *et al.* 2018). Such differences may be an indicative of the change of the relative role of environment in different cluster phases at early times. For 4C 23.56, galaxies experience gas-rich mergers and some have compact sizes than field populations that may lead to fast consumption of gas and morpho-kinematic transformation, if most of the galaxies are progenitors of local early-type galaxies in clusters. Systematic surveys with a larger number of galaxies are needed to test the general picture of galaxy evolution in overdense regions.

References

Bouché, N., Dekel, A., Genzel, R., *et al.* 2010, *ApJ*, 718, 1001
Coogan, R. T., Daddi, E., Sargent, M. T., *et al.* 2018, *MNRAS*, 479, 703
Dressler, A. 1980, *ApJ*, 236, 351
Galametz, A., Stern, D., De Breuck, C., *et al.* 2012, *ApJ*, 749, 169
Genzel, R., Tacconi, L. J., Lutz, D., *et al.* 2015, *ApJ*, 800, 20
Kajisawa, M., Kodama, T., Tanaka, I., Yamada, T., & Bower, R. 2006, *MNRAS*, 371, 577

Knopp, G. P. & Chambers, K. C. 1997, *ApJS*, 109, 367
Kodama, T., Hayashi, M., Koyama, Y., *et al.* 2015, in IAU Symposium, Vol. 309, Galaxies in 3D across the Universe, ed. B. L. Ziegler, F. Combes, H. Dannerbauer, & M. Verdugo, 255–258
Lee, M. M., Tanaka, I., Kawabe, R., *et al.* 2017, *ApJ*, 842, 55
—. 2019, *ApJ*, 883, 92
Lee, M. M., Tanaka, I., Iono, D., *et al.* 2021, e-prints, arXiv:2101.04691
Lilly, S. J., Carollo, C. M., Pipino, A., Renzini, A., & Peng, Y. 2013, *ApJ*, 772, 119
Madau, P. & Dickinson, M. 2014, *ARA&A*, 52, 415
Mayo, J. H., Vernet, J., De Breuck, C., *et al.* 2012, *A&A*, 539, A33
Noble, A. G., McDonald, M., Muzzin, A., *et al.* 2017, *ApJL*, 842, L21
Scoville, N., Arnouts, S., Aussel, H., *et al.* 2013, *ApJS*, 206, 3
Tacconi, L. J., Neri, R., Genzel, R., *et al.* 2013, *ApJ*, 768, 74
Tacconi, L. J., Genzel, R., Saintonge, A., *et al.* 2018, *ApJ*, 853, 179
Valentino, F., Magdis, G. E., Daddi, E., *et al.* 2018, *ApJ*, 869, 27
—. 2020, *ApJ*, 890, 24
Zeballos, M., Aretxaga, I., Hughes, D. H., *et al.* 2018, *MNRAS*, 479, 4577

Galaxy Evolution and Feedback across Different Environments
Proceedings IAU Symposium No. 359, 2020
T. Storchi-Bergmann, W. Forman, R. Overzier & R. Riffel, eds.
doi:10.1017/S1743921320004287

The recurrent nuclear activity of Fornax A and its interaction with the cold gas

F. M. Maccagni[1], P. Serra[1], M. Murgia[1], F. Govoni[1], K. Morokuma-Matsui[2] and D. Kleiner[1]†

[1]INAF – Osservatorio Astronomico di Cagliari, via della Scienza 5, 09047, Selargius (CA), Italy
email: filippo.maccagni@inaf.it

[2]Institute of Astronomy, Graduate School of Science, The University of Tokyo, 2-21-1 Osawa, Mitaka, Tokyo 181-0015, Japan

Abstract. Sensitive (noise ∼16 μJy beam^{-1}), high-resolution (∼10″) MeerKAT observations of Fornax A show that its giant lobes have a double-shell morphology, where dense filaments are embedded in a diffuse and extended cocoon, while the central radio jets are confined within the host galaxy. The spectral radio properties of the lobes and jets of Fornax A reveal that its nuclear activity is rapidly flickering. Multiple episodes of nuclear activity must have formed the radio lobes, for which the last stopped 12 Myr ago. More recently (∼3 Myr ago), a less powerful and short ($\lesssim$1 Myr) phase of nuclear activity generated the central jets. The distribution and kinematics of the neutral and molecular gas in the centre give insights on the interaction between the recurrent nuclear activity and the surrounding interstellar medium.

Keywords. galaxies: individual: (Fornax A, NGC 1316), galaxies: active, radio continuum: galaxies, galaxies: jets, radiation mechanisms: non-thermal

---

1. Introduction

The energy released by Active Galactic Nuclei (AGNs) into the surrounding interstellar medium (ISM) through radiation and/or relativistic jets of radio plasma can drastically change the fate of its host galaxy by removing or displacing the gas in the galaxy and preventing it from cooling to form new stars (e.g. Fabian 2012). This mechanism is commonly referred to as 'AGN feedback'. Numerical simulations of galaxy evolution indicate that only multiple phases of nuclear activity, and therefore recurring episodes of AGN feedback, may prevent the hot circumgalactic gas from cooling back onto the galaxy, and explain the rapid quenching of star formation in early-type galaxies (Werner *et al.* 2019).

The radio emission of AGNs allows us to measure the duty cycle of the nuclear activity. In particular, the steepening of the radio spectrum is often interpreted as radiative ageing of the electron population in the relativistic plasma (e.g., Murgia *et al.* 1999; Harwood *et al.* 2013; Kolokythas *et al.* 2015).

In this proceeding, we summarize the study of the radio spectrum of the nearby ($D_{\rm L} \sim 20$ Mpc) radio galaxy Fornax A to determine the timescale and the duty cycle of its nuclear activity. This analysis is shown in detail in Maccagni *et al.* (2020).

† This project has received funding from the European Research Council (ERC) under the European Union's Horizon 2020 research and innovation programme (grant agreement no. 679627).

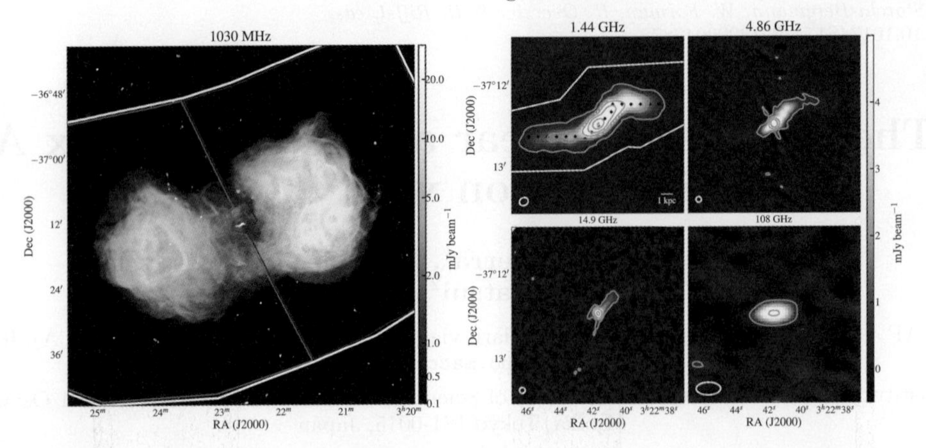

Figure 1. *Left panel*: Fornax A seen by MeerKAT at 1.03 GHz. The red and blue contours mark the region where we measure flux density of the east and west lobes, respectively. The synthesised beam of the image is $11.2'' \times 9.1''$. *Right panel*: Central emission of Fornax A seen at 1.44 GHz by MeerKAT (*top left*), at 4.86 GHz (*top right*), 14.9 GHz (*bottom left*) by the VLA and at 108 GHz (*bottom right*) by ALMA. The PSF of the images is shown in white. (Maccagni *et al.* 2020).

Fornax A is one of the most fascinating radio sources in the local Universe because of its filamentary extended radio lobes ($\sim$1.1°, Fomalont *et al.* 1989). The MeerKAT (Jonas *et al.* 2016) observation (Fig. 1) at 1.03 GHz shows that the lobes are embedded in a diffuse cocoon, with a 'bridge' of synchrotron emission connecting them. In the centre, two radio jets are confined within the host galaxy ($r \lesssim 6$ kpc) and exhibit an s-shaped morphology. Most of the radio emission is produced in the extended lobes. At 1.4 GHz, their total flux density is 121 Jy while that of the jets is $\sim$300 mJy.

Fornax A is hosted by the giant early-type galaxy NGC 1316, which is the brightest member of a galaxy group at the outskirts of the Fornax cluster. NGC 1316 underwent through a major merger that likely brought large amounts of dust, cold molecular gas (Horellou *et al.* 2001; Galametz *et al.* 2014; Morokuma-Matsui *et al.* 2019), and neutral hydrogen (Horellou *et al.* 2001; Serra *et al.* 2019) into the centre and around the galaxy. This merger occurred $\sim$1–3 Gyr ago (e.g., Sesto *et al.* 2018), and it may have triggered the nuclear activity of Fornax A (e.g. McKinley *et al.* 2015). Nevertheless, large uncertainties remain on the timescale of formation of the radio lobes. Moreover, this past merger event does not properly explain the properties of the central emission, nor the soft X-ray cavities between the lobes and the host galaxy (Lanz *et al.* 2010).

The goal of this study is to measure, over a wide range of frequencies, the flux density distribution of the radio lobes and the central emission to characterise the AGN activity history that created them. For the lobes we need wide-field-of-view observations sensitive to their diffuse emission (i.e. good *uv*-coverage on the short baselines), while arc-second resolution is not needed. Hence, between 84 and 200 MHz we chose observations Murchison Widefield Array survey (Hurley-Walker *et al.* 2017). We use the MeerKAT observation to generate images of Fornax A at 1.03 (Fig. 1, left panel) and 1.44 GHz. At 1.5 GHz we chose archival Very Large Array observations (Fomalont *et al.* 1989). Between 5.7 and 6.9 GHz we use observations from the Sardinia Radio Telescope (Prandoni *et al.* 2017). Between 70 GHz and 217 GHz, we selected images from the *Planck* foreground maps (Planck Collaboration IV, 2018).

To study the central emission we selected observations with arc-second resolution (Fig. 1, right panel): the MeerKAT images at 1.03 and 1.44 GHz, archival VLA observations at 4.8 and 15 GHz (Geldzahler & Fomalont 1984), and an observation at 108 GHz

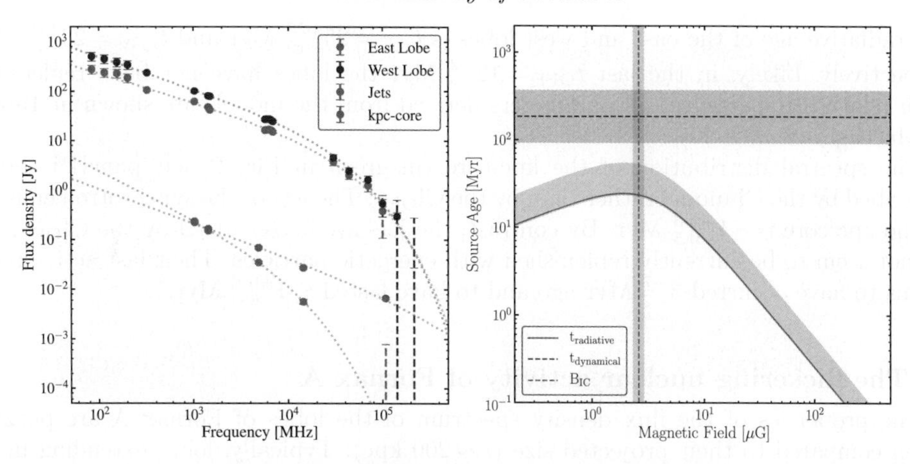

Figure 2. *Left panel*: Radio spectrum of the different components of Fornax A. The east lobe is shown in red, the west lobe in blue. The jets and kpc-core are in magenta and green, respectively. The dashed lines show the $\mathrm{CI_{OFF}}$ model of injection that best fits the flux distributions. The spectral shape of the lobes is very different from that of the inner components. *Right panel*: Comparison of radiative and dynamical age of Fornax A (red and blue shaded regions) as a function of the magnetic field. The radiative age is derived from the spectral break of the two lobes. Their magnetic field ($B_{IC} \sim 2.6\mu$G) is shown in green. (Maccagni *et al.* 2020).

(Morokuma-Matsui *et al.* 2019) taken with the Atacama Large Millimeter and submillimeter Array. The left panel of Fig. 2 shows the spectral flux densities of the lobes and of the central emission we measured using these two samples. Given the morphology of the central emission, we measure its flux density by dividing it into two parts, the central unresolved component (hereafter, the *kpc-core*) and the extended component forming the emission (the *jets*).

2. Spectral analysis of the main components of Fornax A

In the simplest scenario of AGN activity, the lobes (or jets) are continuously injected with particles (continuous injection model, CI). Assuming that radiative energy losses from synchrotron and inverse Compton radiation dominate over expansion losses, the radio spectrum shows a sharp cut-off whose frequency (ν_{break}) depends on the age of the radiation (t_s, Kardashev 1962).

A more complicated scenario can be the continuous injection plus turn off model ($\mathrm{CI_{OFF}}$). The injection of high-energy particles from the nucleus starts at $t=0$ and at the time t_{CI} it is switched off. After that, the *off phase* of the AGN begins, and the total age of the radiation is: $t_{\mathrm{s}} = t_{\mathrm{CI}} + t_{\mathrm{OFF}}$ (Murgia *et al.* 2011). Compared to the CI model, the spectral shape is characterised by a second break-frequency ($\nu_{\mathrm{break,\,high}}$), beyond which the radiation spectrum drops exponentially. This frequency depends on the ratio between the dying phase (t_{OFF}) and the total age of the source ($\nu_{\mathrm{break,\,high}} = \nu_{\mathrm{break}}(t_s/t_{\mathrm{OFF}})^2$). To determine the best-fit models of the flux density spectrum of the lobes and central emission of Fornax A, we use the software package SYNAGE++ (Murgia *et al.* 2011).

For both lobes, the spectral flux density shows a sharp cut-off at high frequencies (see Fig. 2, left panel). This, along with the $\tilde{\chi}^2$ value of the $\mathrm{CI_{OFF}}$ models closer to 1 than the $\tilde{\chi}^2$ of the CI models, suggests that the radio spectrum of both radio lobes is best described by the $\mathrm{CI_{OFF}}$ model and that currently the radio lobes are not being injected with relativistic particles.

According to the $\mathrm{CI_{OFF}}$ model, the dying-to-total-age ratio is $t_{\mathrm{OFF}}/t_s = 0.49^{+0.08}_{-0.42}$. Assuming that the magnetic field of the lobes is $2.6 \pm 0.3\mu$G (Tashiro *et al.* 2009), the radiative age of the east and west lobes is $t_{s,\,\mathrm{E}} = 25^{+23}_{-19}$ Myr and $t_{s,\,\mathrm{W}} = 23^{+20}_{-17}$ Myr, respectively. Likely, in the last $t_{\mathrm{OFF}} = 12^{+2}_{-9}$ Myr the lobes have not been replenished with relativistic particles (all parameters derived from the models are shown in Table 6 of Maccagni *et al.* 2020).

The spectral distribution of the kpc-core (in green in Fig. 2, left panel) is better described by the CI model rather than by the $\mathrm{CI_{OFF}}$. The age of the synchrotron emission of the kpc-core is $\sim 1^{+0.3}_{-0.5}$ Myr. By contrast, the jets are better fitted by the $\mathrm{CI_{OFF}}$, and do not seem to be currently replenished with energetic particles. Their last active phase seems to have occurred 3^{+7}_{-2} Myr ago and to have lasted $\lesssim 1^{+6}_{-0.5}$ Myr.

3. The flickering nuclear activity of Fornax A

The properties of the flux density spectrum of the lobes of Fornax A are puzzling when compared to their projected size ($r \sim 200$ kpc). Typically, lobes extending in the IGM for hundreds of kiloparsecs are either the remnant of an old nuclear activity, and show a steep spectrum with low break frequency, or they are currently being injected with relativistic particles, and show a jet or stream of particles connecting the AGN with the lobes. The most remarkable properties of the lobes of Fornax A are the flat spectral shape and high break frequency ($\gtrsim$20 GHz) of their radio emission, and that the nuclear activity that was replenishing the lobes with high-energy particles was short ($\sim$24 Myr) and has recently stopped ($\sim$12 Myr ago). The main open question therefore pertains to how these large lobes have formed in such a short time.

The right panel of Figure 2 indicates that the radiative age of the lobes of Fornax A and the dynamical age (assuming transonic expansion) are incompatible. This, along with an axial ratio of the lobes close to 1, disfavours the hypothesis that only a recent episode of activity formed the lobes. If, instead, the lobes, filled with low-energy particles and under-pressured with respect to the surrounding IGM, were already present because of previous activities, a new nuclear phase may rapidly fill them with new high-energy particles which now dominate the radio emission of the source. This scenario would explain the overall flat radio spectrum of Fornax A and its high break-frequency. Two separate AGN outbursts have also been proposed by Lanz *et al.* (2010) to explain the location of the X-ray cavities relative to the radio lobes of Fornax A.

Besides the multiple activities that may have formed them, the lobes have been off for $\sim$9 Myr. More recently ($\sim$3 Myr ago) the AGN turned on again for a very short phase ($\lesssim$1 Myr) that formed the central jets. Given the short timescales of the different nuclear activities, Fornax A is likely rapidly flickering between an active phase and a non-active one. The recurrent activity of Fornax A may fit well in the theoretical scenario of AGN evolution whereby the central engine is active for short periods of time (10^{4-5} years), and that these phases repeteadly occur over the total lifetime of the AGN (10^8 years; e.g. Schawinski *et al.* 2015; Morganti 2017).

After the major merger, NGC 1316 went through several accretion events and minor mergers of smaller companions (Iodice *et al.* 2017). These numerous interactions may have regulated the switching on and off of the multiple episodes of activity that formed the lobes as we see them now. Merger and interaction events are often invoked to explain the triggering of powerful AGNs (e. g., Ramos-Almeida *et al.* 2012; Sabater *et al.* 2013).

4. Neutral and molecular gas in the centre of Fornax A

The kinematics and distribution of the cold gas in the innermost 6 kpc (Fig. 3) provide further information on the last episode of the recurrent activity of Fornax A, which generated the central jets. In the centre, ALMA observations detect 5.6×10^8 $\mathrm{M_\odot}$ of molecular

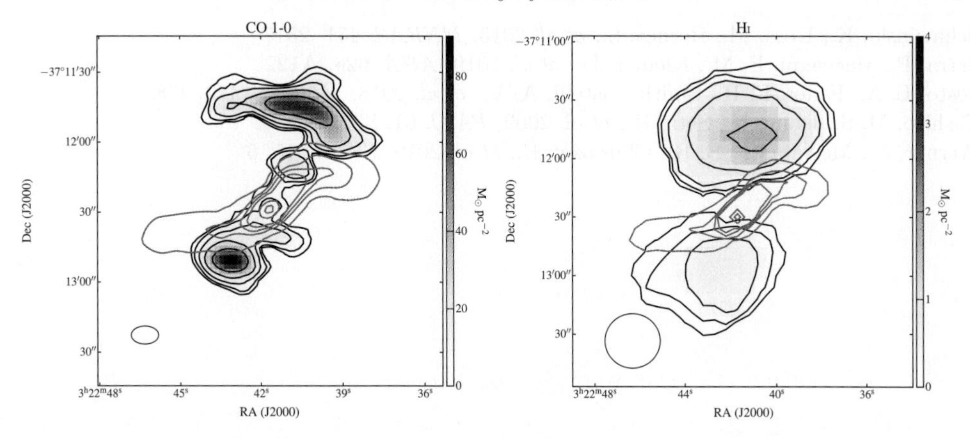

Figure 3. *Left panel*: Surface brightness map of the CO 1-0 line detected by ALMA (Morokuma-Matsui *et al.* 2019), overlaid with the radio jets. Contour levels are 3×2^n M$_\odot$ pc^{-2}, (n = 0, 1, 2, ...). *Right panel*: Surface brightness map of the Hɪ detected by MeerKAT (Serra *et al.* 2019), overlaid with the radio jets. Contour levels are 0.1×2^n M$_\odot$ pc^{-2}, (n = 0, 1, 2, ...).

hydrogen (H_2) distributed in a clumpy shell around the jets (Morokuma-Matsui *et al.* 2019). Neutral hydrogen (Hɪ) clouds (4×10^7 M$_\odot$) are closely associated with the molecular gas (Serra *et al.* 2019). The Hɪ seems to be more extended than the H_2, forming a halo where the molecular clouds are embedded. This is confirmed by new, yet unpublished, MeerKAT higher resolution observations which reveal a diffuse Hɪ component surrounding the radio jets. The jets bend in the denser regions of the gas distribution (in proximity of H_2 clouds with irregular kinematics) toward sparser regions (where no molecular clouds are detected but only diffuse Hɪ, Fig. 3 right panel) suggesting a tight interplay between the nuclear activity and the surrounding cold interstellar medium.

References

Fabian, A. C. 2012, *ARAA*, 50, 455
Fomalont, E. B., Ebneter, K. A., van Breugel, *et al.* 1989, *APJ* (Letters), 346, L17
Galametz, M., Albrecht, M., *et al.* 2014, *MNRAS*, 439, 2542
Geldzahler, B. J. & Fomalont, E. B. 1984,*AJ*, 89, 1650
Harwood, J. J., Hardcastle, M. J., Croston, J. H., *et al.* 2013, *MNRAS*, 435, 3353
Horellou, C., Black, J. H., van Gorkom, J. H., *et al.* 2001, *A&A*, 376, 837
Hurley-Walker, N., Callingham, J. R., Hancock, P. J., *et al.* 2017, *MNRAS*, 464, 1146
Iodice, E., Spavone, M., Capaccioli, M., *et al.* 2017, *ApJ*, 839, 21
Jonas, J. & MeerKAT Team 2016, *Proceedings of MeerKAT Science*, 1
Kardashev, N. S. 1962, *Soviet Astron.*, 6, 317
Kolokythas, K., O'Sullivan, E., Giacintucci S., *et al.* 2015,*MNRAS*, 450, 1732
Lanz, L., Jones, C., Forman, W. R., *et al.* 2010, *ApJ*, 721, 1702
Maccagni, F. M., Murgia, M., Serra, P., *et al.* 2020, *A&A*, 634, A9
McKinley, B., Yang, R., López-Caniego, M., *et al.* 2015, *MNRAS*, 446, 3478
Morganti, R. 2017, *Nature Astronomy*, 1, 596
Morokuma-Matsui, K., Serra, P., Maccagni, F. M., *et al.* 2019, *PASJ*, 71, 85
Murgia, M., Fanti, C., Fanti, R., *et al.* 1999, *A&A*, 345, 769
Murgia, M., Parma, P. Mack, H. K., *et al.* 2011, *A&A*, 526, A148
Planck Collaboration, Akrami, Y., Ashdown, M., *et al.* 2018, *arXiv e-prints*, arXiv:1807.06208
Prandoni, I., Murgia, M., Tarchi, A., *et al.* 2017, *A&A*, 608, A40
Ramos-Almeida, C., Bessiere, P. S., Tadhunter, C. N., *et al.* 2012, *MNRAS*, 419, 687
Sabater, J., Best, P. N., & Argudo-Fernández, M. 2013, *MNRAS*, 430, 638

Schawinski, K., Koss, M., Berney, S., *et al.* 2015, *MNRAS*, 451, 2517
Serra, P., Maccagni, F. M., Kleiner, D., *et al.* 2019, *A&A*, 628, A122
Sesto, L. A., Faifer, F. R., Smith Castelli, A. V., *et al.* 2018, *MNRAS*, 479, 478
Tashiro, M. S., Isobe, N., Seta, H., *et al.* 2009, *PASJ*, 61, S327
Werner, N., McNamara, B. R., Churazov, E., *et al.* 2019, *SSRv*, 215, 5

Galaxy Evolution and Feedback across Different Environments
Proceedings IAU Symposium No. 359, 2020
T. Storchi-Bergmann, W. Forman, R. Overzier & R. Riffel, eds.
doi:10.1017/S1743921320002252

Diving deeper into jellyfish: The rich population of jellyfish galaxies in Abell 901/2

Fernanda Roman de Oliveira[1], Ana L. Chies-Santos[1], Fabrício Ferrari[2] and Geferson Lucatelli[2]

[1]Departamento de Astronomia, Universidade Federal do Rio Grande do Sul,
Av. Bento Gonçalves 9500, 91501-970, Porto Alegre, RS, Brazil
email: fernanda.oliveira@ufrgs.br

[2]Instituto de Matemática, Estatística e Física, Universidade Federal do Rio Grande,
Rio Grande, RS, Brazil

Abstract. Jellyfish galaxies are the most striking examples of galaxies undergoing ram pressure stripping – the removal of gas as a result of a hydrodynamic friction in dense environments. As part of the OMEGA (OSIRIS Mapping of Emission-line Galaxies in Abell 901/2) survey, we have identified the largest sample of jellyfish galaxies in a single system to this date, located in the Abell 901/2 multi-cluster system at $z \sim 0.165$. We present our results with a detailed description of this sample regarding their very high star formation rates and their unique spatial distribution pattern that can be explained as a result of the merging system triggering ram pressure stripping events. Furthermore, we also show the results of our most recent morphometric studies where we use Morfometryka as a tool to characterise the morphologies and structural evolution of jellyfish galaxies. Our morphometric analysis shows that jellyfish galaxy candidates have peculiar concave regions in their surface brightness profiles. Therefore, these profiles are less concentrated (lower Sérsic indices) than other star forming galaxies that are not experiencing such extreme ram pressure effects.

Keywords. galaxies: evolution, galaxies: structure, galaxies: clusters: general, galaxies: intergalactic medium

1. Introduction

The evolution of a galaxy is significantly shaped by the environment. The absence of late-type galaxies in the densest environments suggests that environmental mechanisms are an important factor when it comes to galaxy quenching and morphological evolution (Dressler 1980). By interacting with its surroundings, a galaxy can undergo gravitational and hydrodynamic effects, such as tidal interactions, mergers and harassment (Barnes 1992; Moore *et al.* 1996) or ram pressure stripping (Gunn & Gott 1972). The most extreme cases of galaxies undergoing ram pressure stripping are known as jellyfish galaxies. They received this name as they can have extensive tails of material being stripped that also host intense star formation (Ebeling *et al.* 2014; Owers *et al.* 2012). These galaxies are excellent objects to understand the role of ram pressure stripping in the scenario of galaxy evolution within dense galaxy clusters.

2. Overview

We present the results of a systematic search and analysis of 73 ram pressure stripping candidates in the Abell 901/2 multi-cluster system (Roman-Oliveira *et al.* 2019). The sample was selected through visual inspection in the Hubble Space Telescope F606W

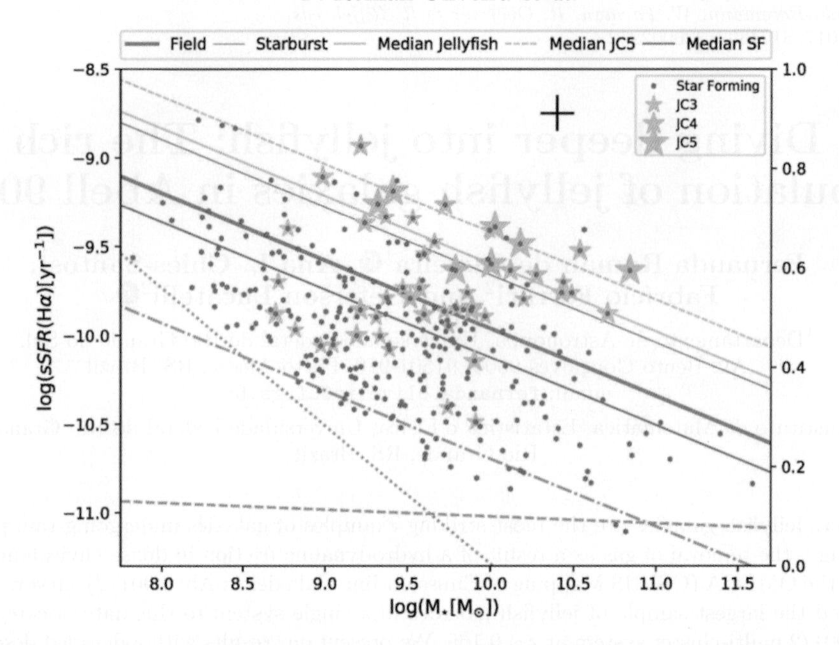

Figure 1. Specific star formation rate versus mass plot (adapted from Roman-Oliveira *et al.* 2019). We represent different JClasses by green star markers according to the legend. The blue dots represent the star forming galaxies without ram pressure stripping morphological features. The star formation main sequence for the SDSS field galaxies is shown as a thick red line, the thin red line is used to identify starbursts. The green and blue lines represent the median SSFR for the ram pressure stripping and star forming populations. The black dash-dot, dotted line, and dashed lines are the observation limits of OMEGA and the cross represents the uncertainty of the data points.

band images following identification methods used previously in Ebeling *et al.* (2014) and Poggianti *et al.* (2016). This selection also accounts for categories of intensity of the morphological evidence of ram pressure stripping, known as JClasses. These categories range from 1 to 5, being that 5 is the strongest case possible. Our sample of 73 candidates only comprises JClasses 3 to 5, the strongest cases. We also visually assigned trail vectors to the galaxies, which is a vector that infers the direction of motion for a jellyfish galaxy on the plane of the sky.

2.1. *Star Formation Rates and AGN activity*

To probe the effects of ram pressure stripping on the star formation activity in the galaxies, we show in Figure 1 a plot of the Specific Star Formation Rate (SSFR) versus mass. We compare the sample of ram pressure stripping candidates against other star forming galaxies that do not show jellyfish morphological features. We see that the jellyfish galaxy candidates show a systematic enhancement in the SSFR, which is in agreement with other recent studies (Vulcani *et al.* 2018; Poggianti *et al.* 2016). Another interesting result, is that the scatter is correlated with the JClasses of the candidates: JClass 5 galaxies show much higher specific star formation rates than the JClass 3 galaxies.

As for the Active Galactic Nuclei (AGN), we did not find any evidence for a correlation between ram pressure stripping and AGN. In fact, only 5 of our candidates are hosts to an AGN. This is in conflict with what was found in Poggianti *et al.* (2017b), in which 5 of the 7 strongest candidates in the GAs Stripping Phenomena in galaxies with MUSE survey (GASP) (Poggianti *et al.* 2017a) hosted an AGN and another one was a Low-Ionization Nuclear Emission-line Region (LINER). To fully investigate these results, we compare in

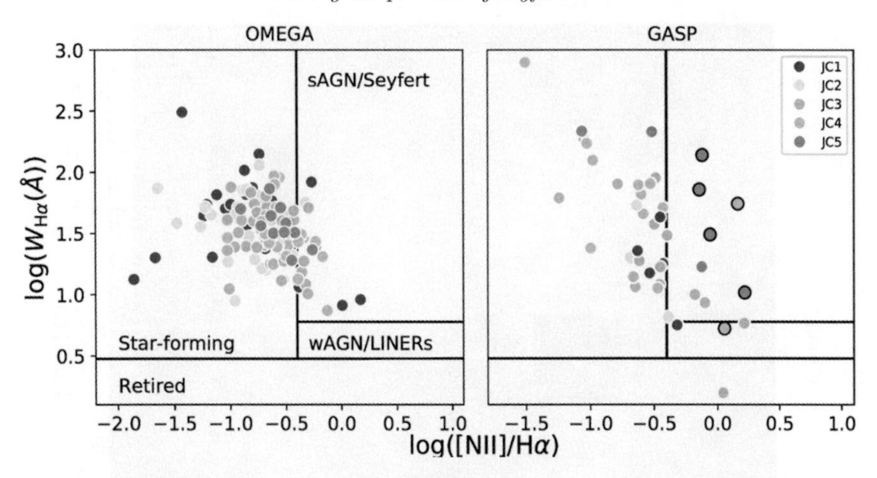

Figure 2. The WHAN diagram for the Abell 901/2 ram pressure stripping candidates in the left panel and for the public GASP sample in the right panel (adapted from Roman-Oliveira *et al.* 2019). The markers with a black ring are the galaxies present in Poggianti *et al.* (2017b).

Figure 2 the Abell 901/2 ram pressure stripping candidates to the GASP public sample in a WHAN diagram (equivalent Width of Hα versus [NII]/Hα). We find that the scatter and AGN fraction of both samples are similar and that the majority of the galaxies are star forming and do not host an AGN.

We suggest that the strong presence of AGN in the 7 strongest GASP jellyfish galaxy cases might be due to a bias in stellar mass or environment rather than the ram pressure stripping phenomenon. However, it is important to note that these 7 galaxies were selected due to their very extended tails, a selection we cannot reproduce in our sample.

2.2. *Environment*

When it comes to the environment, we could not find any evidence for a spatial distribution pattern nor on the infalling direction of the galaxies provided by the trail vectors. This differs significantly from the pattern in the jellyfish galaxies found in Smith *et al.* (2010), in which the majority of the sample was directed to the centre of the Coma cluster.

We have simulated the Abell 901/2 multi-cluster system assuming that the four main substructures will be merging in the future. Our results reveal the existence of a region in which the ram pressure stripping phenomenon is highly enhanced, as illustrated in Fig. 3 (Ruggiero *et al.* 2019). This happens because of the high relative velocity between the galaxies and the merging subclusters. These narrow regions can enhance the efficiency of the ram pressure stripping by a factor of a thousand in a few kiloparsecs, being an optimal trigger to ram pressure stripping events. When we compare the spatial distribution of the observed ram pressure stripping candidates, we find that they are systematically closer to these boundaries when compared to the other cluster members. We then propose that the multi-cluster system, or merging systems in general, can act as triggers to the creation of new jellyfish galaxies around these boundaries. This reinforces tentative findings from previous works (McPartland *et al.* 2016; Owers *et al.* 2012).

2.3. *Morphology*

In order to explore the morphological transformation of jellyfish galaxies, we have performed a morphometric analysis with MORFOMETRYKA (Ferrari *et al.* 2015). Our first result is the proposition of a new way of characterising trail vectors in a robust and

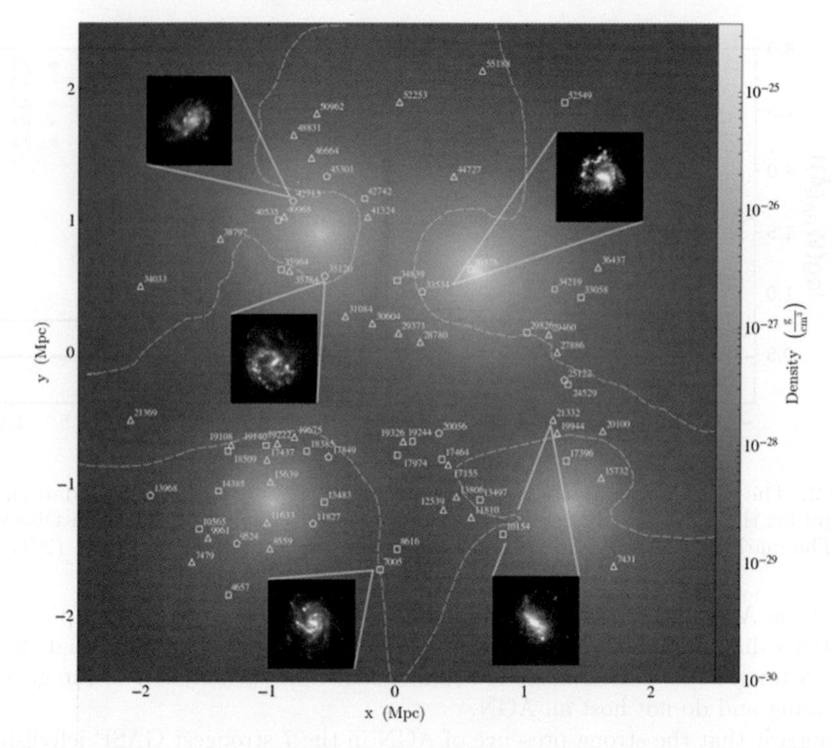

Figure 3. Spatial distribution of the Abell 901/2 ram pressure stripping candidates over the simulated system (adapted from Ruggiero *et al.* 2019). The dashed lines are the locations of the ram pressure boundaries. The small inset images are HST thumbnails of jellyfish galaxies and the triangles, squares, pentagons are the JClass 3, 4 and 5 galaxies, respectively (for more details see Ruggiero *et al.* 2019).

automatic manner, completely independent of visual inspection. This can simply be done by tracing a vector from the centre of light to the peak of light in an image of a galaxy, as can be seen in the illustration shown in Figure 4 left panel.

By investigating the surface brightness profiles of the candidates we find low Sérsic indices for many of the galaxies. This translates into a concave feature in their surface brightness profiles, like the one shown in Figure 4 right panel. One way of quantifying this occurrence is by using the curvature tool designed in Lucatelli *et al.* (2019). This tool analyses the concavity of a surface brightness profile curve revealing high and low light concentration features in galaxies, that can be related to structural components such as bulges or discs. By measuring the curvature of our ram pressure stripping candidates, we find that they have systematically more concave regions that are related to a low concentration in the surface brightness, or a broader profile. Our preliminary and tentative results suggest that ram pressure stripping alters the galaxy morphology by broadening the surface brightness profiles effectively creating galaxies that have the stellar component less concentrated than a pure disc.

3. Implications

The main findings of this study are:

• We find a systematic enhancement of the specific star formation rates in the jellyfish galaxy candidates. This suggests that the ram pressure stripping phenomenon can be an efficient trigger of star formation before it depletes the interstellar gas and can even lead to a starburst period.

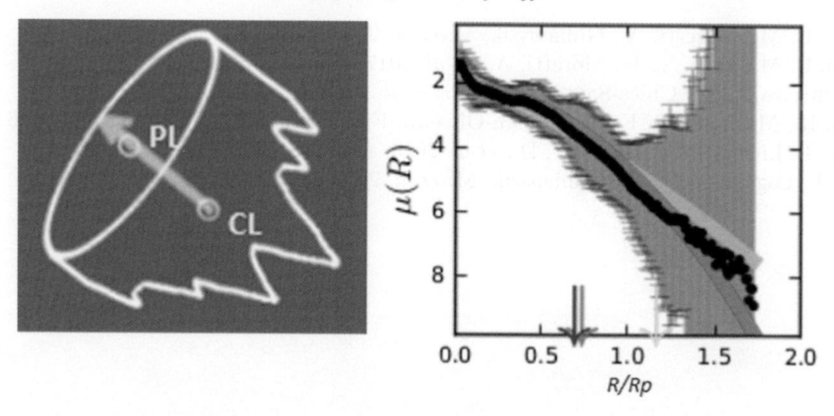

Figure 4. Left panel: Illustration of how to automatically define a trail vector. PL stands for peak of light, which is the brightest pixel in an image. CL stands for centre of light, which is the average point, weighted according to the intensity of light. Right panel: Surface brightness profile for one of the Abell 901/2 ram pressure stripping candidates. The black dots are the measurements from MORFOMETRYKA and the pale yellow and red lines are the 1D and 2D Sérsic law fits, respectively. Error bars on the background represent the error propagation for the surface brightness. The arrows represent different measured radii, for more details refer to Ferrari *et al.* (2015).

- We also could not find a strong correlation between ram pressure stripping and AGN (Roman-Oliveira *et al.* 2019).
- We propose a scenario in which merging clusters can efficiently trigger new ram pressure stripping episodes. These are the best laboratories to find and study more candidates (Ruggiero *et al.* 2019).
- We propose a robust and automatic way of defining trail vectors that is independent of visual inspection (Roman-Oliveira *et al.* submitted).
- Our surface brightness results suggest that the extreme ram pressure that produces jellyfish features also serves to broaden the surface brightness profiles, sometimes creating concave surface brightness profiles (Roman-Oliveira *et al.* submitted).

To validate our tentative findings, we plan on further investigating the morphology of jellyfish galaxies by performing the morphometric analysis (with MORFOMETRYKA and ELLIPSE) on the OMEGA Hα maps. With this, we could identify the Hα morphologies (Koopmann *et al.* 2004), discovering the extent and concentration of the star formation spatially, whether it is being enhanced or suppressed in different regions and maybe retrieving more information on how the morphology is being affected by looking into different regions of the spectrum.

References

Barnes, J. E. 1992, *ApJ*, 393, 484
Dressler, A. 1980, *ApJ*, 236, 351
Ebeling, H., Stephenson, L. N., & Edge, A. C. 2014, *ApJ*, 781, 40
Ferrari, F., de Carvalho, R. R., & Trevisan, M. 2015, *ApJ*, 814, 55
Gunn, J. E. & Gott, J. R. 1972, *ApJ*, 176, 1
Koopmann, R. A. & Kenney, J. D. P. 2004, *ApJ*, 613, 866
Lucatelli, G. & Ferrari, F. 2019, *MNRAS*, 489, 1161
McPartland, C., Ebeling, H., Roediger, E., *et al.* 2016, *MNRAS*, 455, 2994
Moore, B., Katz, N., Lake, G., *et al.* 1996, *Nature*, 379, 613
Owers, M. S., Couch, W. J., Nulsen, P. E. J., *et al.* 2012, *ApJ*, 750, 230
Poggianti, B. M., Fasano, G., Omizzolo, A., *et al.* 2016, *AJ*, 151, 78

Poggianti, B. M., Moretti, A., Gullieuszik, M.,*et al.* 2017a, *ApJ*, 844, 48
Poggianti, B. M., Jaffé, Y. L., Moretti, A., *et al.* 2017b, *Nature*, 548, 304
Roman-Oliveira, F. V., Chies-Santos, A. L. P., *et al.* 2019, *MNRAS*, 484, 892
Ruggiero, R., Machado, R, E. G., Roman-Oliveira, F. V., *et al.* 2019, *MNRAS*, 484, 906
Smith, R. J., Lucey, J. R., Hammer, D., *et al.* 2010, *MNRAS*, 408, 1417
Vulcani, B., Poggianti, B. M., Gullieuszik, M. *et al.* 2018, *ApJ*, 866, 25

Galaxy Evolution and Feedback across Different Environments
Proceedings IAU Symposium No. 359, 2020
T. Storchi-Bergmann, W. Forman, R. Overzier & R. Riffel, eds.
doi:10.1017/S1743921320002409

The intriguing case of Was 49b

Henrique R. Schmitt[1], Nathan J. Secrest[2], Laura Blecha[3], Barry Rothberg[4,5] and Jacqueline Fischer[5]

[1]Remote Sensing Division, Naval Reseearch Laboratory, 4555 Overlook Ave. NW, Washington DC-20375, USA
email: henrique.schmitt@nrl.navy.mil

[2]U.S. Naval Observatory, 3450 Massachusetts Ave. NW, Washington, DC20392, USA

[3]Dept. of Physics, University of Florida, P.O. Box 118440, Gainesville, FL32611, USA

[4]LBT Observatory, University of Arizona, 933 N. Cherry Ave., Tucson AZ85721, USA

[5]Dept. of Physcis and Astronomy, George Mason University, MS3F3, 4400 University Drive, Fairfax, VA22030, USA

Abstract. We present results of a multiwavelength study of the isolated dual AGN system Was 49. Observations show that the dominant component in this interacting system, Was 49a, is a spiral galaxy, while Was 49b is hosted in a dwarf galaxy located at 8 kpc from the nucleus of Was 49a, at the edge of its disk. The intriguing fact about this system is the luminosity of their corresponding AGNs. While Was 49a hosts a low luminosity Seyfert 2 with $L_{bol} \sim 10^{43} erg\ s^{-1}$, Was 49b has a Seyfert 2 with $L_{bol} \sim 10^{45} erg\ s^{-1}$, in the luminosity range of Quasars. Furthermore, estimates of the black hole and host galaxy masses of Was 49b indicate a black hole significantly more massive than one would expect from scaling relations. This result is in contrast with findings that the most luminous merger-triggered AGNs are found in major mergers and that minor mergers predominantly enhance AGN activity in the primary galaxy.

Keywords. galaxies: active – galaxies: bulges – galaxies: dwarf – galaxies: interactions – galaxies: nuclei – galaxies: Seyfert

1. Introduction

The dual-AGN system Was 49, first described by Bothun *et al.* (1989), is composed of a disk galaxy, Was 49a, and a dwarf galaxy, Was 49b, co-rotating in the disk of Was49a, at a projected distance of ~8 kpc from its nucleus (Moran *et al.* 1992). While Was 49a shows signs of nuclear activity consistent with its size, a low-luminosity Seyfert 2, Was 49b hosts the most luminous AGN in this binary system, with luminosity levels typical of Quasars. Spectropolarimetric observations of Was 49b (Tran 1995) show polarized broad emission lines (Hα and Hβ) with FWHM$\sim$6000km s^{-1}, as well as a strong featureless continuum, where the stellar component is estimated to contribute $\lesssim$15% of the light. This system is also known to be relatively isolated, with Secrest *et al.* (2017) reporting the non-detection of other galaxies of similar size ($M_r \lesssim -20$mag) within a projected distance of 1Mpc.

These peculiarities of the Was 49 system make it an excellent candidate for a case study. The high luminosity of the AGN on Was 49b, combined with the low luminosity of its host, suggests that this system may not follow the usuall black hole *vs* bulge mass relations. Furthermore, the fact that the smallest component of this minor merger hosts the most luminous AGN, is contrary to empirical results, which find that minor mergers

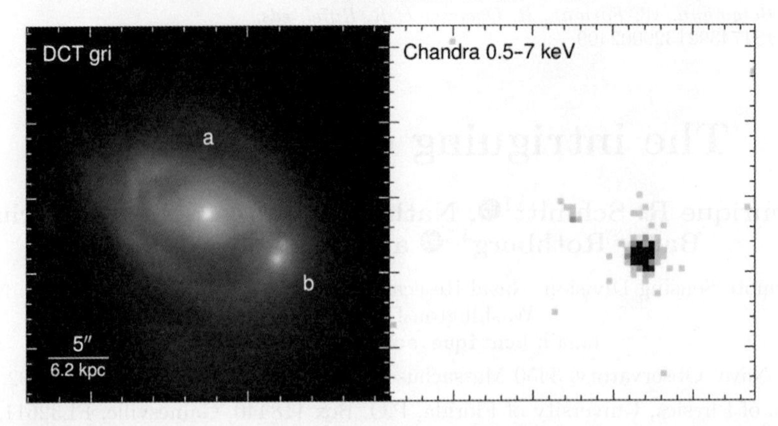

Figure 1. Optical (left) and X-ray (right) images of the Was 49 system from Secrest *et al.* (2017). The optical image is an RGB composite using images obtained with the 4.3 m Discovery Channel Telescope, using the color representation scheme from Lupton *et al.* (2004). At the redshift of Was 49 (z = 0.06328), [OIII] and Hβ fall within the g (blue) filter, the r (green) filter is predominantly continuum, and the i (red) filter contains Hα+[NII]. Note the extensive ionization region around Was 49b, which results in a pink/magenta color, indicating that the optical light is dominated by line emission. Images oriented north up, east left.

predominantly enhance the AGN activity the most massive component of the system Ellison *et al.* 2011. Here we summarize some of our previous results Secrest *et al.* 2017, and discuss some of our ongoing efforts.

2. Ground-based observations

In order to better constrain the mass of the host galaxy of Was 49b and compare this value to the mass of the black hole, we obtained deep ground-based images of this system using the 4.3 m Lowell Discovery Channel Telescope on the night of 2016 April 03. These observations were done under excellent seeing conditions ($\sim$0.5″), which allowed us to obtain images of this target in the g',r', i', z' bands, significantly deeper and much sharper than the one available in the SDSS archive ($\sim$1.4″). A color composite image is presented in Fig. 1, where we can see the structure of the system in better detail. The main galaxy, Was 49a, is a spiral galaxy with a pseudobulge and bar, as well as some morphological disturbances that can be associated with the interaction with the small companion, Was 49b. In this higher resolution image we see that the companion galaxy is split into 2 components. Given the redshift of this system, the i' image is dominated by the Hα+[NII] emission, while the r' band is dominated by continuum emission. This suggests that most of the NW component of Was 49b is due to line emission, while the SE component is a mixture of both line and continuum emission.

Using the r' image and GALFIT (Peng *et al.* 2002), we were able to decompose the surface brightness profile of this system using seven components, four for Was 49a (nucleus, bulge, disk and tidal component) and three for Was 49b (nucleus, bulge and ionized component). The results of the best fitting model can be seen in Secrest *et al.* (2017). The component that can be identified with the bulge of Was 49b has a Sérsic index $n = 1.07$, consistent with a pseudobulge, an effective radius of 1.62 kpc (1.31″) and $M_r = -19.1$ mag, consistent with a dwarf galaxy classification.

3. X-ray observations

Further evidence of the peculiarity of the Was 49 system can be seen in the right panel of Fig. 1. Here we present the *Chandra* 0.5-7 keV image, where one can see that Was 49a has very weak emission, only 7 counts in this energy band, while Was 49b is the

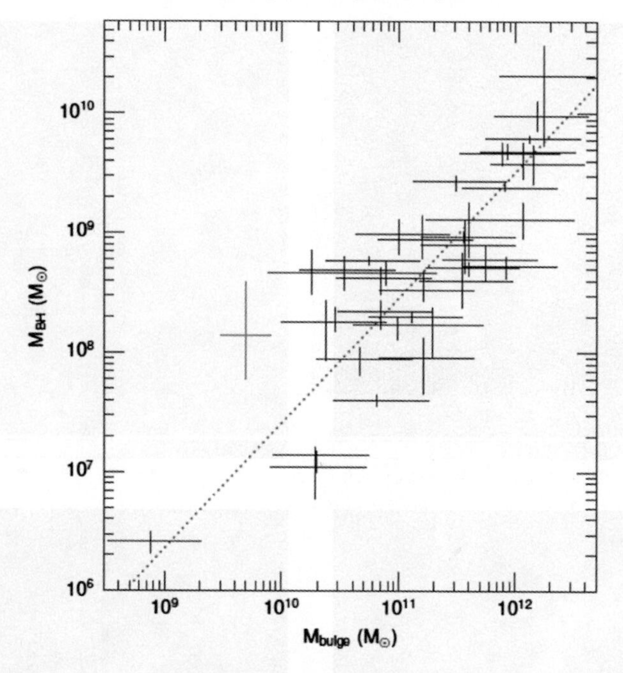

Figure 2. Black hole mass as a function of bulge mass. Data from McConnell & Ma (2013). Note that the black hole of Was 49 is overmassive relative to its bulge.

brightest component in this binary system, with 215 counts. Combining the information from *Chandra* with X-ray spectra from *NuSTAR*, *Swift BAT* and *ASCA*, we were able to fit the 0.5-195 kev spectrum of Was 49b with a power law with $\Gamma = 1.6 \pm 0.1$ and $N_H = 2.3 \times 10^{23}$, resulting in $L_{0.5-195keV} = 2.4 \times 10^{44}$ erg s^{-1}, and $L_{14-195keV} = 1.7 \times 10^{44}$erg s^{-1} (Secrest *et al.* 2017). Using the relation between L_{bol} and $L_{14-195keV}$ from Winter *et al.* (2012), we can calculate that Was 49b has $L_{bol} = 1.3 \times 10^{45}$erg s^{-1}, which puts it in the luminosity range of quasars. This high X-ray luminosity is consistent with the [OIII]λ5007Å luminosity for this source $L_{[OIII]} = 1.4 \times 10^{42}$ erg s^{-1} (Meléndez *et al.* 2008).

4. Bulge and black hole mass

The stellar mass of the bulge of Was 49b was calculated using the mass-to-light relations from Bell *et al.* (2003) and the correlation between *g-r* and Sérsic index n (Blanton & Moustakas 2009), given that the g band image is strongly contaminated by line emission. Based on these relations we adopted $M/L = 1.82$ and determined that the mass of the bulge of Was 49b is $5.6 \times 10^9 M_\odot$.

Determining the black hole mass of Was 49b was complicated by the fact that this is a Seyfert 2 object. As was previously pointed out by Tran (1995) and Moran *et al.* (1992), the spectrum of this galaxy is dominated by emission lines and a mostly featureless continuum. We were able to detect a weak Ca II K line absorption on the SDSS spectrum, however, attempts to constrain the stellar velocity dispersion were unsuccessfull. In order to solve this problem we measured the FWHM of the broad Hα component on the SDSS spectrum, which originates from reflected nuclear radiation (Tran 1995), and estimated how much of the reflected continuum is due to the nucleus. Since the line-of-sight emission is heavily-obscured, we used the absorption-corrected X-ray luminosity to calculate the size of the BLR using the relation from Kaspi *et al.* (2005). This allowed us to determine a black hole mass $M_{BH} = 1.3 \times 10^8\ M_\odot$. In Fig. 2 we show the $M_{BH} \times M_{bulge}$ diagram,

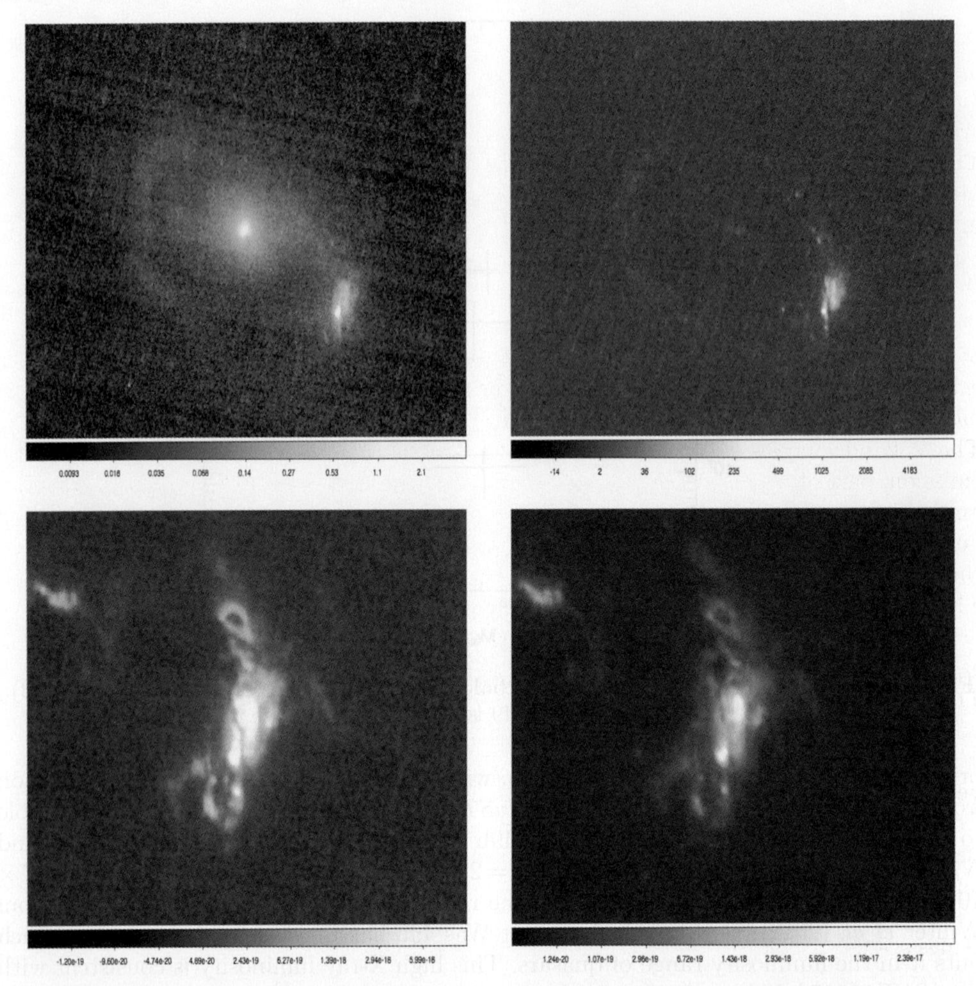

Figure 3. The top panels show the HST continuum images of the Was 49 system, rest frame V band on the left and ultraviolet λ2100Å on the right. The bottom panels show the HST emission line images. [OIII]λ5007 on the left and Hα+[NII] on the right. The images are oriented N up and East to the left. Note that the emission line images are zommed relative to the continuum ones, to better show details of the emission around the nucleus of the two galaxies.

where we can see that Was 49b has an overmassive black hole relative to its bulge mass by a factor of $\sim$20. It should be noted that our bulge mass estimate should be considered an upper limit, due to line and recombination continuum contamination to the r' filter.

5. Discussion

Was 49b is a high luminosity AGN, hosted in a dwarf galaxy. One possibility to be explored is whether this AGN is a candidate for a recoiling black hole, kicked out of the nucleus of Was 49a during coalescence (Blecha *et al.* 2011). However, such an event would also kick the black hole on Was 49a away from its nucleus, indicating that this possibility may not be applicable in this case. Another intriguing property of this system is the fact that it is a minor merger where the lowest mass component hosts the highest luminosity AGN. Simulations of coplanar gas-rich minor mergers (e.g. Callegari *et al.* 2011; Capelo *et al.* 2015) find that the secondary AGN rarely exceed a luminosity of 10^{43}erg s^{-1}. However, these simulations assume that the black hole masses of the two components follow the usual scaling relations. If one were to scale the mass of the secondary galaxy black holes to that of Was 49b, these simulations could reach the luminosity range of 10^{44}–10^{45} erg s^{-1}, in line with the observed values. Another interesting property of this system is its relative isolation, suggesting that Was 49b may have evolved differently. This could be either related to the way the black hole evolves relative to the bulge at high redshifts (Volonteri *et al.* 2016), or how the black hole/bulge scaling relation evolve in systems with low bulge mass (Jahnke & Macciò 2011).

6. Future work

We are currently analysing new *Hubble Space Telescope* line and continuum images of this system, as well as *Gemini* long-slit spectra, *VLA* and *VLBA* radio images. An example of some of these observations is shown in Fig. 3, where we present line free V band and ultraviolet continuum images, as well as [OIII] and Hα+[NII] images from *HST*. The V band image shows that Was 49b has a bright component, surrounded by diffuse emission, part of which can be associated with recombination continuum. The ultraviolet image shows a similar structure, as well as a ring of emission around Was 49a, which can be identified with older star forming regions, possible ignitied by the interaction between the two galaxies. The continuum images are being used to refine the bulge mass estimate. The emission line images show a large range of structure, associated with the nucleus of both Was 49a and Was 49b. The ionized gas associated with Was 49b extends for $\sim 3''$ on each side of the host galaxy, being ~ 2 times more extended than the host galaxy itself. These observations are being combined with long-slit spectroscopic and radio continuum information to determine the origin of this gas, if it is a wind driven by the nuclues, or if it is gas in the disk of the two galaxies being ionized by Was 49b. The radio images will be used to determine the importance of the radio jets on ionizing and driving this gas. Considering the characteristics of Was 49b, overmassive black hole relative to the host, this galaxy may represent a good low-redshift surrogate to study the effects of feedback on high-redshift targets.

References

Bell, E. F., *et al.* 2003, *ApJS*, 149, 289
Blanton, M. R. & Moustakas, J. 2009, *ARA&A*, 47, 159
Blecha, L., *et al.* 2011, *MNRAS*, 412, 2154
Bothun, G. D., Schmitz, M., Halpern, J. P., *et al.* 1989, *ApJS*, 70, 271
Callegari, S., *et al.* 2011, *ApJ*, 729, 85
Capelo, P. R., *et al.* 2015, *MNRAS*, 477, 2123
Ellison, S. L., *et al.* 2011, *MNRAS*,418, 2043
Gebhardt, K., *et al.* 2000, *ApJ*, 539, L13
Jahnke, K. & Macciò, A. V. 2011 *ApJ*, 734, 92
Kaspi, S., *et al.* 2005, *ApJ*, 629, 61
Lupton, R., *et al.* 2004, *PASP*, 116, 133
Magorrian, J., *et al.* 1998, *AJ*, 115, 2285
McConnell, N. J. & Ma, C.-P. 2013, *ApJ*, 764, 184
Meléndez, M., *et al.* 2008, *ApJ*, 682, 94
Moran, E. C., Halpern, J. P., Bothun, G. D., *et al.* 1992, *AJ*, 104, 990
Peng, E. Y., Ho, L. C., Impey, C. D., *et al.* 2002, *AJ*, 124, 266
Secrest, N. J., Schmitt, H. R., Blecha, L., *et al.* 2017, *ApJ*, 836, 183
Tran, H. D. 1995, *ApJ*, 440, 565
Tran, H. D. 1995, *ApJ*, 440, 578
Volonteri, M., *et al.* 2016, *MNRAS*, 460, 2979
Winter, L. M., *et al.* 2012 2012, *ApJ*, 745, 107

Galaxy Evolution and Feedback across Different Environments
Proceedings IAU Symposium No. 359, 2020
T. Storchi-Bergmann, W. Forman, R. Overzier & R. Riffel, eds.
doi:10.1017/S1743921320001738

H_2 content of galaxies inside and around intermediate redshift clusters

Damien Spérone-Longin

Laboratoire d'Astrophysique, Ecole Polytechnique Fédérale de Lausanne (EPFL),
1290 Sauverny, Switzerland
email: damien.sperone-longin@epfl.ch

Abstract. Dense environments have an impact on the star formation rate of galaxies. As stars form from molecular gas, looking at the cold molecular gas content of a galaxy gives useful insights on its efficiency in forming stars. However, most galaxies observed in CO (a proxy for the cold molecular gas content) at intermediate redshifts, are field galaxies. Only a handful of studies focused on cluster galaxies. I present new results on the environment of one medium mass cluster from the EDisCS survey at $z \sim 0.5$. 27 star-forming galaxies were selected to evenly sample the range of densities encountered inside and around the cluster. We cover a region extending as far as 8 virial radii from the cluster center. Indeed there is ample evidence that star formation quenching starts already beyond 3 cluster virial radii. I discuss our CO(3-2) ALMA observations, which unveil a large fraction of galaxies with low gas-to-stellar mass ratios.

Keywords. galaxies: evolution – galaxies: clusters: general – submillimeter: galaxies

1. SEEDisCS

We initiated the first systematic study of galaxy properties along the large scale structures (LSS) feeding galaxy clusters, so called Spatially Extended EDisCS (SEEDisCS). It focuses on 2 clusters from the ESO Distant Cluster Survey (EDisCS, White *et al.* 2005): CL1301.7−1139 and CL1411.1−1148. They are located at redshifts $z_{\rm cl} = 0.4828$ and 0.5195 and have velocity dispersions of $\sigma_{\rm cl} = 681$ and 710 km/s, respectively. Deep u, g, r, i, z and $K_{\rm s}$ images were gathered with CFHT/MEGACAM and WIRCam. They cover a region that extends up to around 10 times the cluster virial radius, 10 R_{200}. This survey is divided into 3 main steps. The first one consists of the identification of the large scale structures around the clusters thanks to accurate photometric redshifts. The second is a spectroscopic follow-up of these large scale structures to study the properties of the galaxy stellar populations. The last step is made up of ALMA programs to study the cold molecular gas reservoir status of galaxies. Here, we focus on one cluster : CL1411.1−1148, since analysis is still underway for the other one.

2. ALMA sample

We selected 27 star-forming galaxies from the spectroscopic sample within $5 \times R_{200}$ of CL1411.1−1148 to be observed with ALMA. This is the largest sample of galaxies with direct cold gas measurements at intermediate redshift and the only one of galaxies in interconnected cosmic structures around a galaxy cluster. Those galaxies were selected to have similar colours as the PHIBSS2 *field* main-sequence star-forming galaxies sample (Freundlich *et al.* 2019) with redshift between $z = 0.5$ and 0.6. Figure 1 shows the position of our targets in the $g - i$ vs i diagram with respect to the other cluster galaxies and the PHIBSS2/COSMOS sample. Those 27 targets also sample the range of densities

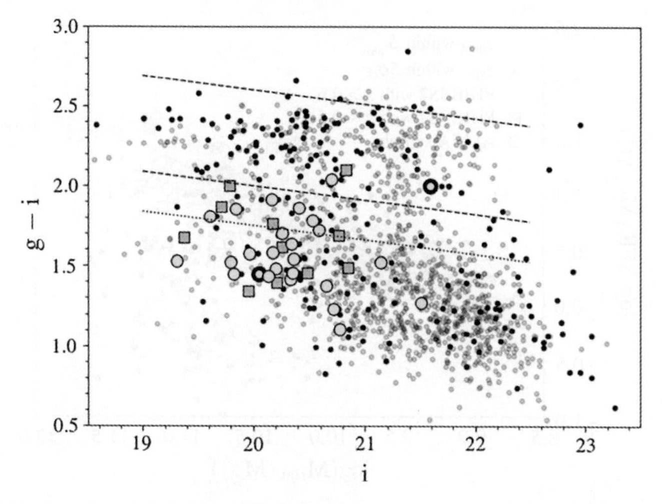

Figure 1. Observed colour-magnitude diagram, $g-i$ as a function of i for the CL1411.1−1148 galaxies. Our ALMA sample is represented by the circles. The markers with the larger edges are for our *Spitzer*-observed galaxies. The squares are for parts of the PHIBSS2 sample. The small gray dots are galaxies with photometric redshifts and the black ones are for galaxies with spectroscopic redshifts, within the cluster limit. The dashed lines delimit the red sequence (± 0.3 mag above De Lucia *et al.* (2007) relation) and the dotted line delimits the transition zone between the blue clump and the red sequence (0.3 mag below the red sequence).

encountered inside and around the cluster. They were observed at 226 GHz to detect the $CO_{J=3\to2}$ transition. We benefited from 23h of observations during Cycle 3 and 5, and reported 27 detections.

The star formation rates (SFR) and stellar masses (M_{star}) were derived using `MAGPHYS` (da Cunha *et al.* 2008) for our targets and all the galaxies within the cluster redshift limits. Figure 2 shows the SFR-M_{star} diagram with the position of our ALMA sample, circles, compared to the position of the PHIBSS2 sample, squares and diamonds. The limits in M_{star} and SFR are similar for both samples, as expected from our colour-based selection. Hence, our galaxies are located on the main-sequence or slightly below. Our sample is also evenly sampling the main-sequence between $10.3 \leqslant \log(M_{star}/M_{\odot}) \leqslant 11.3$.

From the ALMA observations, the CO luminosities were derived following the Solomon & Vanden Bout (2005) recipe. To derive the cold molecular gas masses, M_{H_2}, we used

$$M_{H_2} = \alpha_{CO} \frac{L'_{CO(3\to2)}}{r_{31}}, \tag{2.1}$$

where $r_{31} = 0.5$ is the conversion factor from the third rotational transition of CO to the first (Genzel *et al.* 2015, Chapman *et al.* 2015, Carleton *et al.* 2017 and Tacconi *et al.* 2018), and $\alpha_{CO} = \alpha_{MW} = 4.36\,M_{\odot}(\mathrm{K\,km/s\,pc^2})^{-1}$ is the CO(1-0) luminosity-to-molecular-gas-mass conversion factor, considering a 36% correction to account for interstellar helium (Leroy *et al.* 2011; Bolatto *et al.* 2013; Carleton *et al.* 2017).

3. Results

In order to compare our results to the literature and the PHIBSS2 sample, we used the cold molecular gas-to-stellar mass ratio: $\mu_{H_2} = M_{H_2}/M_{star}$. Figure 3 shows the cold molecular gas-to-stellar mass ratio as a function of the stellar mass for our galaxies and comparison samples. One can see that while 67% of our targeted galaxies have comparable gas fractions as their field counterparts at the same redshift, 33% are populating a new area of low μ_{H_2}. They are represented by pentagons in Fig. 3. They fall below the 1σ

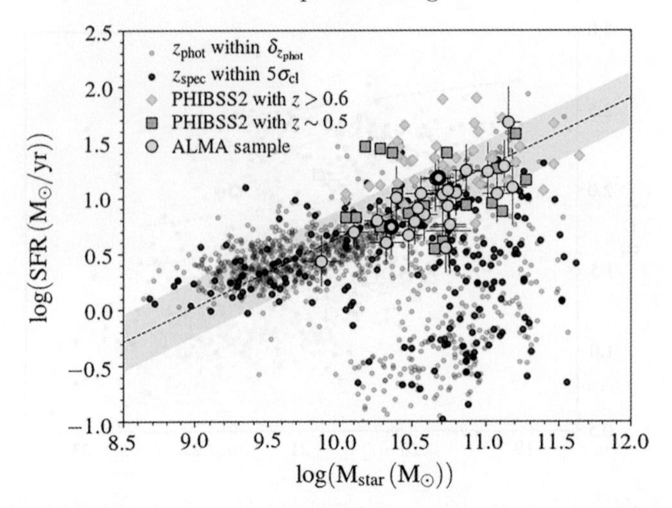

Figure 2. Location of the CL1411.1−1148 (dots) and ALMA (circles) galaxies in the stellar mass-SFR plane. The markers with thicker black edges are for our *Spitzer*-observed galaxies. The squares are the PHIBSS2 galaxies at our cluster's redshift and the diamonds are the rest of the low-z PHIBSS2 sample ($0.6 < z \leqslant 0.8$). The middle line is the Speagle *et al.* (2014) main-sequence at our cluster redshift with the corresponding ±0.3dex scatter as the shaded area.

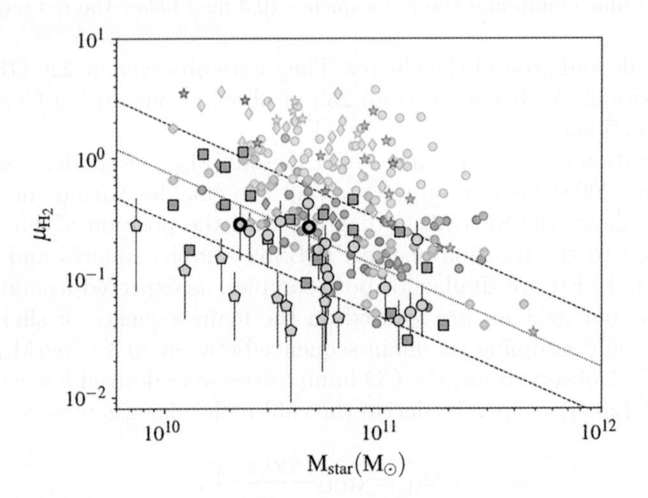

Figure 3. Fraction of cold molecular gas with respect to stellar masses. The circles and pentagons are our ALMA sample. The squares and diamonds are the same as in Fig. 2. The gray markers in the background are other galaxies from the literature. The dotted line is the fit of the M_{star}-μ_{H_2} relation. The dashed line is its dispersion. The pentagons are our galaxies with cold gas ratios below the dispersion of the previous relation.

dispersion of the relation between μ_{H_2} and M_{star}, which was derived using the PHIBSS2 sample at $z \sim 0.5$. Their large number indicates they are not simply part of the tail of the distribution of the field galaxies but rather a new, different population, which implies a different relation between μ_{H_2} and M_{star}, specifically in the $10.2 \leqslant \log(M_{star}/M_\odot) \leqslant 10.85$ mass range.

Figure 4 shows how the cold gas ratio scales with the specific SFR (sSFR), normalized to the one of the main-sequence, of galaxies at different redshifts. At $z \sim 0.5$, a trend can be noticed. Most of our targets follow it, but here again, some of them are populating a

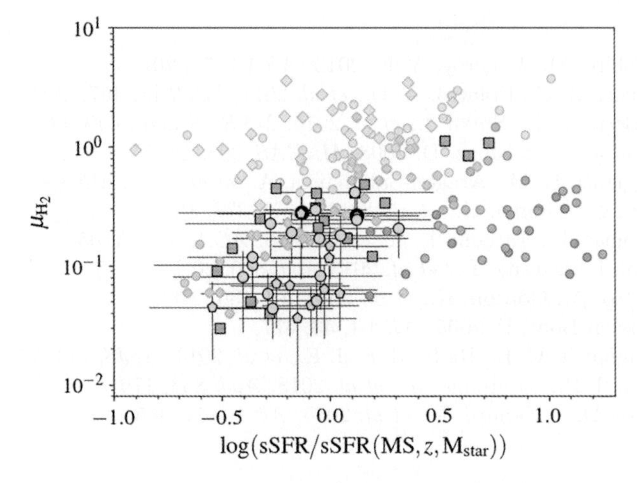

Figure 4. Fraction of cold molecular gas with respect to specific SFRs. Markers are the same as on Fig. 3

new area, which corresponds to galaxies lying within or very close to the main-sequence but depleted in cold molecular gas.

A question we can ask is whether the values chosen for the different factors used to derive M_{H_2} can have an influence on our μ_{H_2}.

$\underline{\alpha_{CO}}$: we chose to use the same coefficient as for the Milky-Way, whereas, in PHIBSS2, Freundlich *et al.* (2019) are using a coefficient based on the mass-metallicity relation derived by Genzel *et al.* (2015). If we were to use this recipe, we would have $3.8 \leqslant \alpha_{CO} \leqslant 4.9$, which would lower our already low μ_{H_2}.

$\underline{r_{31}}$: We took the same value as Genzel *et al.* (2015). But Dumke *et al.* (2001), by looking at local spirals, showed that r_{31} could vary within a spiral galaxy. Indeed, it would be closer to 0.8 in the center of those galaxies, which would lower our ratios by 38%. This coefficient can be closer to 0.4 in the outer parts of the disk of spiral galaxies, which would lead to increase our ratios by 25%, but this is not enough to put our low μ_{H_2} back into the "normal" sequence.

Moreover, those low gas fraction galaxies have narrower CO line widths compared to their PHIBSS2 counterparts. Therefore, the former potentially have more extended morphologies ($\Delta V \propto$ Mass/Radius) than the normal μ_{H_2} and field galaxies. This could be a signature of environmental effects at play in the interconnected structures surrounding the cluster.

4. Summary

We report the first large sample of galaxies with direct cold molecular gas measurements within the same cluster environment at intermediate redshift. Those 27 galaxies were selected to have the same colours as the PHIBSS2 galaxies. No prior selection on SFR was applied as it is usually the case for such surveys and larger. A large portion, 67%, of our targets have similar gas content as other field galaxies (PHIBSS2) at $z \sim 0.5$. The rest of them, 33%, have low cold gas to stellar mass ratio despite being main-sequence star-forming galaxies. The latter have lower CO line widths than their field counterparts, which is an indication of more extended morphologies. To summarize, we unveiled a new population of low gas content star-forming galaxies, using a different selection criterion, and located in the different environments encountered around a cluster.

References

Bolatto, A. D., Wolfire, M., & Leroy, A. K. 2013, *ARAA*, 51, 207
Carleton, T., Cooper, M. C., Bolatto, A. D., *et al.* 2017, *MNRAS*, 467, 4886
Chapman, S. C., Bertoldi, F., Smail, I., *et al.* 2015, *MNRAS Lett.*, 449, L68
da Cunha, E., Charlot, S., & Elbaz, D. 2008, *MNRAS*, 388, 1595
De Lucia, G., Poggianti, B. M., Aragòn-Salamanca, A., *et al.* 2007, *MNRAS*, 374, 809
Dumke, M., Nieten, C., Thuma, G., *et al.* 2001, *A&A*, 373, 853
Freundlich, J., Combes, F., Tacconi, L. J., *et al.* 2019, *A&A*, 622, A105
Genzel, R., Tacconi, L. J., Lutz, D., *et al.* 2015,*ApJ*, 800, 20
Leroy, A. K., Bolatto, A., Gordon, K., *et al.* 2011 *AJ*, 136, 2782
Solomon, P. & Vanden Bout, P. 2005, *ARAA*, 43, 677
Speagle, J. S., Steinhardt, C. L., Radford, S. J. E., *et al.* 2014, *ApJS*, 214, 15
Tacconi, L. J., Genzel, R., Saintonge, A., *et al.* 2018, *ApJ*, 853, 179
White, S. D., Clowe, D. I., Simard, L., *et al.* 2005, *A&A*, 444, 365

Galaxy Evolution and Feedback across Different Environments
Proceedings IAU Symposium No. 359, 2020
T. Storchi-Bergmann, W. Forman, R. Overzier & R. Riffel, eds.
doi:10.1017/S1743921320001891

POSTERS

Effects of AGN feedback on galaxy downsizing in different environments

Amirnezam Amiri[1,2,3,4], Kastytis Zubovas[5,6], Alessandro Marconi[1,2], Saeed Tavasoli[4] and Habib G. Khosroshahi[3]

[1]Dipartimento di Fisica e Astronomia, Universita di Firenze, via G. Sansone 1, I-50019, Sesto Fiorentino (Firenze), Italy

[2]INAF-Osservatorio Astrosico di Arcetri, Largo E. Fermi 2, I-50125, Firenze, Italy

[3]School of Astronomy, Institute for Research in Fundamental Sciences (IPM), PO Box 19395-5746 Tehran, Iran

[4]Physics Dept., Kharazmi University, Tehran, Iran

[5]Center for Physical Sciences and Technology, Vilnius LT-10257, Lithuania

[6]Vilnius University Observatory, Vilnius LT-10257, Lithuania

emails: amirnezam.amiri@unifi.it, amirnezamamiri@gmail.com

Abstract. We have investigated the role of AGN feedback on galaxy downsizing in cluster and void environments, using the sample from Amiri *et al.* (2019). Our results indicate that, at least in the local universe, the correlation between black hole mass and (specific) star formation rate is statistically indistinguishable in the two environments. Therefore, the role of the environment in modulating AGN feedback effects on the host galaxy star formation is negligible.

Keywords. AGN, black holes, environments, stellar parameters

---

1. Introduction

Galaxies in cluster and void environments differ in their morphology, luminosity, star formation rates and other related parameters (Kreckel *et al.* 2012; Ricciardelli *et al.* 2014). Based on observations, the evidence of the relation between AGN and star formation in the host galaxy has been mixed. On long timescales, AGN are connected with quenching of star formation in their host galaxies (Siriri chawinski 2009), although they might induce brief starbursts as well (Silk 2005). On shorter timescales, the correlation between AGN luminosity and the host galaxy star formation rate (SFR) is generally weak, except in the highest-luminosity cases, where there is a positive correlation although it is probably just a consequence of the brightest AGN residing in the largest galaxies. The reason for this lack of correlation might be that AGN vary on timescales much shorter than the host galaxy can respond, therefore meaningful connections might only be possible when both AGN luminosity and SFR are averaged on several Myr timescales (Zubovas 2018). An AGN outflow regulates the host galaxy's SFR by removing gas out to large distances. If the galaxy resides in a denser environment, the outflow should stall at a smaller distance and gas can fall back into the galaxy on a shorter timescale. Therefore, we may expect the overall suppressing effect of AGN on the host galaxy's SFR to be smaller in cluster galaxies than in field/void galaxies. Moreover, The differential evolution of bright and faint AGNs with redshift has been described as downsizing (Barger *et al.* 2005). This implies that AGN activity in the low-z Universe is dominated by either high-mass BHs

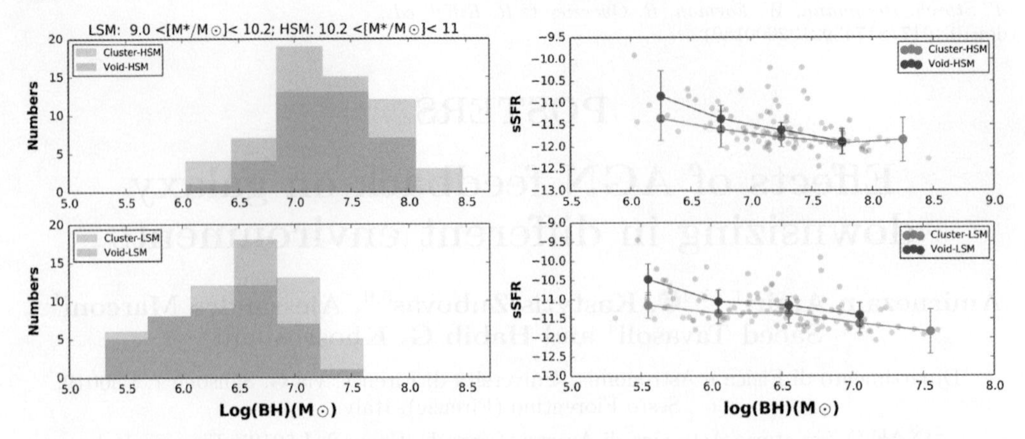

Figure 1. Left panels: Distribution of black hole masses for AGN samples in clusters and voids. The upper and lower panels consider high and low mass AGN hosts, respectively. The red histogram shows cluster galaxies and the blue one shows void galaxies. A KS test shows that there is no significant difference between the BH mass distributions of AGN in the HSM and LSM regimes, in both environments. Right panels: The relation between sSFR ans BH mass. The upper and lower panels are for high and low mass respectively. Transparent circles are the actual AGN, while solid-colour circles are bin averages. The only clear indication is that the correlation between sSFR versus BH mass is negative in all cases. When considering both a linear fit and a KS test, we can conclude that there is a negligible role of the environment for the sSFR of AGN hosts galaxies.

accreting at low rates or low-mass BHs growing rapidly. To analyse these issues, we look into the possible connections between the galactic environment and the co-evolution of its SMBH and stellar components.

2. Sample selection

Amiri *et al.* (2019) selected a volume limited sample of AGN host galaxies in both void and cluster environments, limited to objects brighter than $M_r = -18$ and covering a redshift range 0.01–0.04, which have been divided into low stellar mass (LSM, $9.0 \leqslant \log(M_\star/M_\odot) < 10.2$) and high stellar mass (HSM, $10.2 \leqslant \log(M_\star/M_\odot) < 11.0$) subsamples. Samples include 60 and 40 galaxies in LSM and HSM samples, respectively. For determining velocity dispersion, sSFR and BH mass, we have used the Max-Planck-Institute for Astrophysics (MPA)-Johns Hopkins University (JHU) SDSS DR7 catalogue Brinchmann *et al.* (2004). Also, we have used the Saulder *et al.* (2013) correction for velocity dispersion. Finally, We have estimated black hole masses by using the $M_{BH} - \sigma$ relation from Gultekin *et al.* (2011) (Fig. 1).

3. Conclusion

We have addressed the unsettled issue of the effects of AGN feedback on galaxy downsizing in different environments by comparing the galaxies hosting AGNs in the local universe, as a function of the stellar mass of the host galaxies. Based on Amiri *et al.* (2019), as figure 1 shows, we could conclude that there is a negligible role of the environment for the sSFR of AGN hosts galaxies.

References

Amiri, A., *et al.* 2019, *ApJ*, 874 140
Barger, A. J., *et al.* 2005, *Aj*, 129, 578
Brinchmann, *et al.* 2004, *MNRAS*, 351, 1151

Gultekin, K., *et al.* 2011, *ApJ*, 741, 38
Kreckel, K., Platen, E., *et al.* 2012, *AJ*, 144, 16
Ricciardelli, E., *et al.* 2014, *MNRAS*, 445, 4045
Saulder, C., *et al.* 2013, *A&A*, 557, 21
Schawinski, K. 2009, *American Institute of Physics Conference Series*, 1201
Silk, J. 2005, *MNRAS*, 364, 4
Zubovas, K. 2018, *MNRAS*, 473, 3525

Galaxy Evolution and Feedback across Different Environments
Proceedings IAU Symposium No. 359, 2020
T. Storchi-Bergmann, W. Forman, R. Overzier & R. Riffel, eds.
doi:10.1017/S1743921320001933

A systematic search for galaxy proto-cluster cores at $z \sim 2$

Makoto Ando[1], Kazuhiro Shimasaku[1,2] and Rieko Momose[1]

[1]Department of Astronomy, Graduate School of Science, The University of Tokyo, 7-3-1 Hongo, Bunkyo-ku, Tokyo 113-0033, Japan

[2]Research Center for the Early Universe, The University of Tokyo, 7-3-1 Hongo, Bunkyo-ku, Tokyo 113-0033, Japan

Abstract. A proto-cluster core is the most massive dark matter halo (DMH) in a given proto-cluster. To reveal the galaxy formation in core regions, we search for proto-cluster cores at $z \sim 2$ in $\sim$1.5 deg^2 of the COSMOS field. Using pairs of massive galaxies ($\log(M_*/M_\odot) \geqslant 11$) as tracers of cores, we find 75 candidate cores. A clustering analysis and the extended Press-Schechter model show that their descendant mass at $z = 0$ is consistent with Fornax-like or Virgo-like clusters. Moreover, using the IllustrisTNG simulation, we confirm that pairs of massive galaxies are good tracers of DMHs massive enough to be regarded as proto-cluster cores. We then derive the stellar mass function and the quiescent fraction for member galaxies of the 75 candidate cores. We find that stellar mass assembly and quenching are accelerated as early as $z \sim 2$ in proto-cluster cores.

Keywords. galaxies: clusters: general – galaxies: high-redshift – galaxies: formation

---

1. Proto-cluster cores

The most massive and largest dark matter haloes (DMHs) in today's universe are called galaxy clusters. The mass of galaxy clusters is typically $\gtrsim 10^{14}\ M_\odot$ and a mature cluster hosts hundreds to thousands of galaxies with large fraction of quiescent galaxies and/or elliptical galaxies. To reveal the galaxy formation of cluster galaxies, progenitors of local clusters at $z \gtrsim 2$, proto-clusters, should be investigated. They are defined as a whole structure that will collapse into a cluster by $z = 0$. A proto-cluster typically extends to more than 20 comoving Mpc at $z \sim 2$, being split into a number of DMHs and unbound regions. Among those substructures, we define the "core" of the proto-cluster as the most massive DMH. A theoretical study has studied galaxy evolution in proto-clusters (Muldrew *et al.* 2018). They have found that galaxies in core regions have different properties from those in fields and the rest of the proto-cluster regions: a more top-heavy stellar mass function, a higher fraction of quiescent galaxies.

In this study, we propose a new method to find proto-cluster cores at $z \sim 2$, the epoch when massive cores appear (Chiang *et al.* 2017). The extended Press-Schechter model predicts that a DMH whose mass is $\gtrsim 2\text{–}3 \times 10^{13}\ M_\odot$ at $z \sim 2$ typically evolves into the cluster mass regime, $\gtrsim 10^{14}\ M_\odot$, by $z = 0$. We regard DMHs with $\gtrsim 2\text{–}3 \times 10^{13}\ M_\odot$ at $z \sim 2$ as proto-cluster cores and search for such massive systems. We use a "pair" of massive galaxies ($M_* \geqslant 10^{11}\ M_\odot$) as a tracer of cores.

2. Proto-cluster core candidates

We use data from the COSMOS2015 galaxy catalogue (Laigle *et al.* 2016). We only use galaxies with mag(K_s) $\leqslant 24.0$ and $1.5 \leqslant z \leqslant 3.0$. First, we explore pairs of massive galaxies ($M_* \geqslant 10^{11}\, M_\odot$) whose angular and redshift separations are less than 30″ and 0.12, respectively. We find 75 pairs as core candidates, among which 54% are estimated to be real. Then, we estimate their halo mass by clustering analysis. Using cross-correlation technique, we derive $2.6^{+0.9}_{-0.8} \times 10^{13}\, M_\odot$, or $4.0^{+1.8}_{-1.5} \times 10^{13} M_\odot$ after contamination correction. Therefore, pairs of massive galaxies are good tracers of massive haloes at least at $z \sim 2$. We also calculate the descendant halo mass of the pairs. Using the extended Press-Schechter model, we show they grow into haloes with mass of $\gtrsim 10^{14}\, M_\odot$ at $z = 0$. Therefore, we regard the pairs of massive galaxies as proto-cluster cores.

3. Implications from simulation

We further investigate the effectiveness of pairs of massive galaxies as tracers of proto-cluster cores. For this purpose, we employ a mock galaxy catalogue of the IllustrisTNG project (Nelson *et al.* 2019). We extract mock galaxies with $M_* \geqslant 10^{11}\, M_\odot$ at $z = 2$, and search for massive galaxy pairs whose separations are less than 0.3 pMpc. What we find is the following:

- M_* of central galaxy versus host halo mass: we check the relationship between stellar mass of the central galaxy and host halo mass. At fixed stellar mass, DMHs which host pairs are more massive than those which do not by ~0.2 dex, suggesting that pairs are good tracers of massive haloes.
- Pair-host fraction: we derive the fraction of DMHs which host pairs of massive galaxies as a function of halo mass. We find that more than 50% of massive haloes with $\gtrsim 3 \times 10^{13}\, M_\odot$ host pairs of massive galaxies.
- Mass growth of pair-host haloes: we find that 62% of pair-host haloes grow into clusters ($M_{\rm DMH} \geqslant 10^{14}\, M_\odot$) at $z = 0$.

4. Properties of member galaxies of core regions

We examine the stellar mass function (SMF) for galaxies in the detected cores. To calculate the SMF, we assume that DMHs hosting a pair are spheres with a radius of 0.3 pMpc. It is found that the SMFs of total and star-forming galaxies in the cores have a flat shape below $\log(M_*/M_\odot) < 11$. We also calculate the ratio between the SMFs of core member galaxies and those of field galaxies. We normalise this ratio by total mass. We find that galaxies in the cores have a more top-heavy SMF than the field. This suggests that the formation of high- (low-) mass galaxies can be enhanced (suppressed) in the core regions given the fact that the ratio of the SMFs is above (marginally below) unity for higher- (lower-) mass galaxies than $\log(M_*/M_\odot) = 10$.

In addition to the SMF, we also examine the fraction of quiescent galaxies. We find clear excess at $\log(M_*/M_\odot) \lesssim 10.6$ compared to field galaxies although no excess at $\log(M_*/M_\odot) \gtrsim 10.6$. This suggest that core environment quenches low-mass galaxies more efficiently than high-mass ones. Considering the mass range $9 \leqslant \log(M_*/M_\odot) \leqslant 11$, the fraction of quiescent galaxies in the cores is three times higher than that of the field. We conclude that mass assembly and quenching are accelerated in a proto-cluster core.

References

Chiang, Y.-K., Overzier, R. A., Gebhardt, K., *et al.* 2017, *ApJ*, 844, L23
Laigle, C., *et al.* 2016, *ApJS*, 224, 24
Muldrew, S. I., Hatch, N. A., Cooke, E. A., *et al.* 2018, *MNRAS*, 473, 45, 2335
Nelson, D., *et al.* 2019, *Computational Astrophysics and Cosmology*, 6, 2

Galaxy Evolution and Feedback across Different Environments
Proceedings IAU Symposium No. 359, 2020
T. Storchi-Bergmann, W. Forman, R. Overzier & R. Riffel, eds.
doi:10.1017/S1743921320001830

Isolated groups of extremely blue dwarf galaxies

Vitor Bootz[1], Marina Trevisan[1], Trinh Thuan[2], Yuri Izotov[3], Angela Krabbe[4] and Oli Dors Jr.[4]

[1]Federal University of Rio Grande do Sul, Porto Alegre, RS, Brazil
emails: vitor.bootz@ufrgs.br, marina.trevisan@ufrgs.br
[2]University of Virginia, Charlottesville, VA, United States
email: txt@virginia.edu
[3]Bogolyubov Institute for Theoretical Physics, Kyiv, Ukraine
email: yizotov@bitp.kiev.ua
[4]Vale do Paraiba University, São José dos Campos, SP, Brazil
emails: angela.krabbe@gmail.com, olidors@univap.br

Abstract. Interactions and mergers between dwarf galaxies are mostly gas-rich and should be marked by an intense star formation activity. But these processes, which are expected to be common at earlier times, are very difficult to observe at low redshifts. To investigate that, we look in the *Sloan Digital Sky Survey* (SDSS) for compact groups that contain one luminous compact galaxy (LCG) with very high specific star formation rate (sSFR) and at least two other blue galaxies. We found 24 groups that satisfy these criteria, among which 12 groups have SDSS spectroscopic data for at least 2 member galaxies. Here we want to investigate, using the tidal strength estimator Q, how interactions between neighbouring galaxies affect the sSFR and concentration of each LCG. Statistical tests reveal a correlation between Q and their sSFR, indicating that tidal forces between neighbouring galaxies might be inducing bursts of star formation in the LCGs.

Keywords. Galaxies: dwarf, Galaxies: evolution, Galaxies: interactions

1. Introduction

Pairs and groups of interacting dwarf galaxies provide a unique window to address the hierarchical, gas-dominated assembly and the buildup of stellar mass in low-mass galaxies. However, fewer than 5% of dwarf galaxies are observed to have close companions, and most galaxy surveys are not deep enough to detect the companions even when they are present. Besides that, the existence of these dwarf groups at low redshifts is not expected and, to understand their nature and possible fate, a more detailed study of their properties is needed.

To investigate that, we look in the *Sloan Digital Sky Survey* (SDSS) for compact groups ($R_{\rm group} < 80$ kpc) that contain only galaxies with $g-i$ colours far below the red sequence ($> 4\sigma$), one luminous compact galaxy (LCG) with very high specific star formation rate (sSFR, $-9.5 < \log(\rm sSFR/yr) < -7.6$) and at least two other blue galaxies. We found 24 groups that satisfy these criteria. In this paper, however, we present the properties of 12 groups that have SDSS spectroscopic data for at least 2 member galaxies. Group candidates that are not complete were or will be observed with GMOS@Gemini: 6 groups were observed during 2018A and 2019B (the data are currently being reduced), and 5 will be observed during 2020A (time already granted).

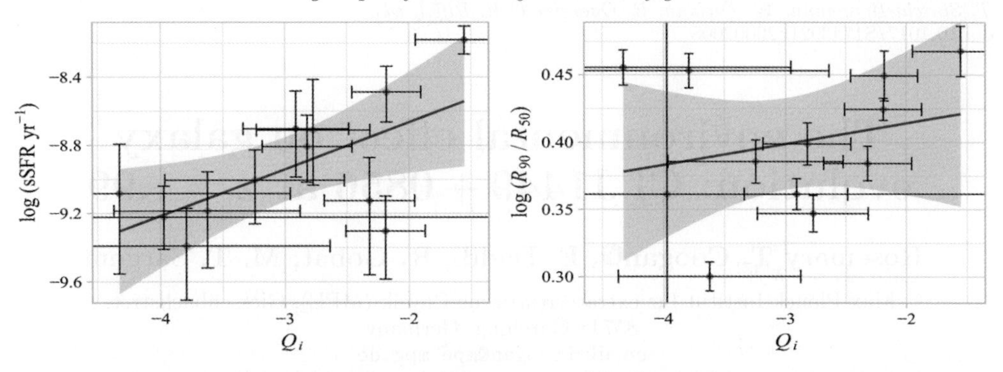

Figure 1. Specific star formation rate (left) and concentration (right) as a function of the tidal strength estimator Q. The plot shows only the groups for which we have at least two galaxies with SDSS spectroscopic observations. The black line represents a linear regression of the data and the shaded region represents the 95% confidence interval. The Kendall and Spearman correlation coefficients are $\tau = 0.42$ (p-value $= 0.06$) and $\rho = 0.54$ (p-value $= 0.07$) for the left image and $\tau = 0.18$ (p-value $= 0.46$) and $\rho = 0.19$ (p-value $= 0.56$) for the right image.

2. How the interactions affect the LCG sSFR and morphology?

To investigate how tidal interactions affect the properties of the LCGs residing in each of the 12 groups, we computed the tidal strength estimator, Q, from the relation (Goddard 2016):

$$Q_i = \frac{F_{tidal}}{F_{binding}} \propto \frac{M_i}{M_{LCG}} \left(\frac{D_{LCG}}{R_{i,LCG}}\right)^3 \tag{2.1}$$

where M_i is the stellar mass of the neighbour galaxy, $M_{\rm LCG}$ is the stellar mass of the LCG, $D_{\rm LCG}$ is the LCG diameter of the region containing 90% of the Petrosian flux in the r-band and $R_{i,\rm LCG}$ is the projected distance between the neighbour galaxy and the LCG. The galaxy stellar masses and radii were retrieved from the SDSS database.

In Fig. 1 left, we show how the LCG sSFR correlates with Q. The sSFRs are corrected for the age of the burst ($\rm sSFR_0$, Izotov 2011). A Kendall and Spearman correlation tests indicate a strong correlation between these quantities, with coefficients $\tau = 0.42$ (p-value $= 0.06$) and $\rho = 0.54$ (p-value $= 0.07$). On the other hand, the LCG concentration is weakly correlated to Q. This is shown in Fig. 1 right, where we plot the ratio $\log R_{90}/R_{50}$ (radius containing 90% and 50% of the Petrosian flux of the LCG) versus the parameter Q. The weak correlation is evidenced by the Kendall and Spearman correlation tests, with coefficients $\tau = 0.18$ (p-value $= 0.46$) and $\rho = 0.19$ (p-value $= 0.56$), respectively.

The correlation between Q and sSFR indicates that tidal forces between neighbouring galaxies might be inducing bursts of star formation in the LCGs. The fact that we are not computing the total tidal forces acting on the LCG (i.e., we include only the spectroscopically confirmed group members) might be washing out the correlation between Q parameter and the concentration of the LCGs. Therefore, the observations that are being carried out using GMOS@Gemini are important to correctly estimate both relations.

References

Goddard, *et al.* 2016, *MNRAS*, 465, 688
Izotov, *et al.* 2011, *ApJ*, 728, 161

Galaxy Evolution and Feedback across Different Environments
Proceedings IAU Symposium No. 359, 2020
T. Storchi-Bergmann, W. Forman, R. Overzier & R. Riffel, eds.
doi:10.1017/S1743921320002008

The environmental effect on galaxy evolution: Cl J1449 + 0856 at z = 1.99

Rosemary T. Coogan, E. Daddi, R. Gobat, M. T. Sargent

Max-Planck-Institut für extraterrestrische Physik (MPE), Giessenbachstr.1, 85748 Garching, Germany
email: rcoogan@mpe.mpg.de

Abstract. This work focuses on understanding the formation of the first massive, passive galaxies in clusters, as a first step to the development of environmental trends seen at low redshift. Cl J1449 + 0856 is an excellent case to study this - a galaxy cluster at redshift z = 1.99 that already shows evidence of a virialised atmosphere. Here we highlight two recent results: the discovery of merger-driven star formation and highly-excited molecular gas in galaxies at the core of Cl J1449, along with the lowest-mass Sunyaev-Zel'dovich detection to date.

Keywords. galaxies: evolution, galaxies: clusters: individual, galaxies: ISM, galaxies: starburst

1. Cl J1449 + 0856: A mature cluster in the early universe

In order to understand the origins of the most massive, early-type galaxies that dominate the cores of local galaxy clusters, we must trace these structures back in time, to the epoch of peak galaxy and cluster assembly at $z \gtrsim 1.5$ (e.g. De Breuck *et al.* 2004; Wang *et al.* 2016; Castignani *et al.* 2018; Lee *et al.* 2019). In Coogan *et al.* (2018); Coogan *et al.* (2019), Strazzullo *et al.* (2018) and Gobat *et al.* (2019) we use Atacama Large Millimeter/submillimeter Array (ALMA) and Karl G. Jansky Very Large Array (VLA) observations to study dust-obscured star-formation, interstellar medium (ISM) content and the intra-cluster medium (ICM) in Cl J1449+0856. Cl J1449, an X-ray detected galaxy cluster at z = 1.99, is one of the highest redshift mature clusters discovered to date, with the mass of a typical Coma-like progenitor (Gobat *et al.* 2011). Unlike unrelaxed proto-clusters more commonly found at this redshift, Cl J1449 already contains a large fraction of passive galaxies at its core, in addition to a diverse population of highly star-forming galaxies.

1.1. *Merger-driven star formation in the core of Cl J1449 + 0856*

In Coogan *et al.* (2018) we measure the star-formation activity and ISM content of 11 galaxies in the core of Cl J1449 at z = 1.99 - a crucial epoch for mass assembly, where no consensus on the environmental effect on galaxy evolution has yet been reached. Molecular gas masses are calculated from both dust and CO[1-0] emission, and we discover that the fuel for star-formation in these galaxies will be depleted within ~100-400 Myrs, as observed for starburst galaxies at z ~ 2. A key result of Coogan *et al.* (2018) is the discovery of a large fraction of galaxies in the core of Cl J1449 with highly excited molecular gas (the ratio of denser, star-forming gas at high quantum state J > 1, to the total molecular gas reservoir), revealed through CO Spectral Line Energy Distribution modelling. When compared with expectations for co-eval field galaxies, we conclude that this cluster contains a strongly enhanced fraction of excited, starburst-like

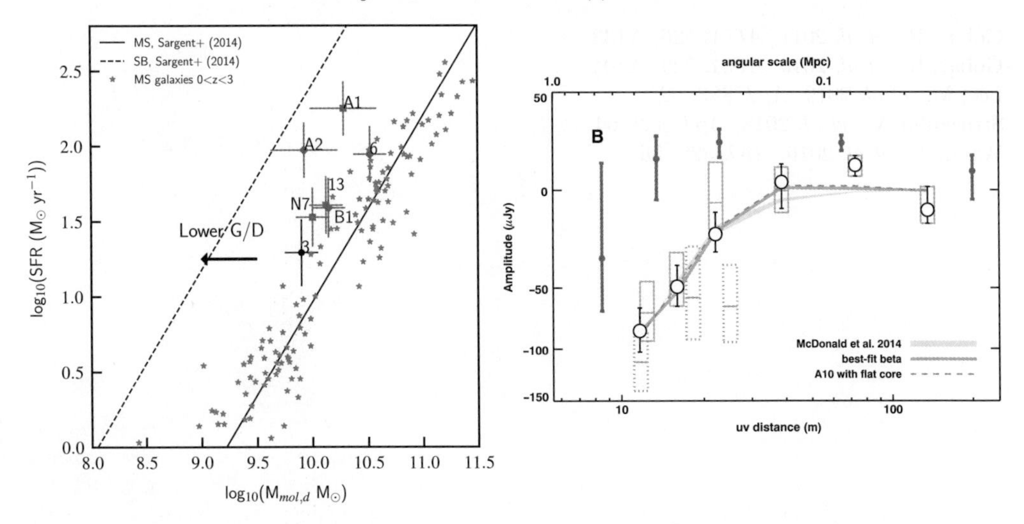

Figure 1. Left: SFR vs. molecular gas mass for Cl J1449 (red/blue/black, Coogan *et al.* 2018). We see a high fraction of cluster galaxies with excited CO (red), and the cluster galaxies lie at enhanced SFE compared to Main Sequence galaxies (green), regardless of their CO excitation. **Right**: The SZ pressure profile in uv-space of Cl J1449 (white circles, Gobat *et al.* 2019). Red points show the pressure profile if positive galaxy emission is not removed before analysis.

galaxies compared to the field, driven by the high fraction of mergers, interactions and Active Galactic Nuclei (AGN) in the core. This also leads to increased star-formation efficiencies (SFEs, Fig. 1 left, Coogan *et al.* 2018), and depletion of molecular gas on short timescales.

1.2. *Sunyaev-Zel'dovich detection of Cl J1449: The pressure profile in uv-space*

The next step was to investigate the large-scale cluster environment, using 92 GHz observations of the Sunyaev-Zel'dovich (SZ) decrement arising from the ICM, presented in Gobat *et al.* (2019). As there is a significant amount of positive 92 GHz continuum present in the core of Cl J1449 (originating from galaxy dust emission), observations were made using both ALMA and the ALMA Compact Array (ACA), in order to first subtract the positive emission from known galaxies using the smaller ALMA beam. Having done this - without disrupting the SZ signature itself - the ALMA and ACA data were combined, uncovering a 5σ extended SZ decrement at $z = 1.99$, shown in the right panel of Fig. 1 (Gobat *et al.* 2019). The pressure profile of the SZ decrement as a function of spatial scale can give constraints on the physics of the ICM and the cluster halo mass, and the total mass recovered for Cl J1449 is found to be consistent with that derived from X-ray observations ($\sim 6 \times 10^{13} M_\odot$, Gobat *et al.* 2011). This is the lowest mass single SZ detection to date. We compare the pressure profile at $z = 1.99$ with models from local galaxy clusters, and do not find strong evidence for an evolution with redshift. However, a slight tension at small-to-intermediate spatial scales suggests a flattened central profile, which could potentially be related to the effects of energy injection by AGN.

References

Castignani, G., *et al.* 2018, *A&A*, 617, A103

Coogan, R. T., *et al.* 2018, *MNRAS*, 479, 703

Coogan, R. T., *et al.* 2019, *MNRAS*, 485, 2092

De Breuck, C., *et al.* 2004, *A&A*, 424, 1

Gobat, R., *et al.* 2011, *A&A*, 526, A133
Gobat, R., *et al.* 2019, *A&A*, 629, A104
Lee, M., *et al.* 2019, *ApJ*, 883, 92
Strazzullo, V., *et al.* 2018, *ApJ*, 862, 64
Wang, T., *et al.* 2016, *ApJ*, 828, 56

Galaxy Evolution and Feedback across Different Environments
Proceedings IAU Symposium No. 359, 2020
T. Storchi-Bergmann, W. Forman, R. Overzier & R. Riffel, eds.
doi:10.1017/S1743921320001908

The co-responsibility of mass and environment in the formation of lenticular galaxies

A. Cortesi[1], L. Coccato[2], M. L. Buzzo[3], K. Menéndez-Delmestre[1], T. Goncalves[1], C. Mendes de Oliveira[3], M. Merrifield[4] and M. Arnaboldi[2]

[1]Observatório do Valongo, Ladeira do Pedro Antônio 43, Rio de Janeiro, RJ, Brazil
email: aricorte@astro.ufrj.br

[2]European Southern Observatory, Karl-Schwarzchild-str., 2, D-85748 Garching b. Muenchen, Germany

[3]Universidade de São Paulo, IAG, Rua do Matão 1226, São Paulo, Brazil

[4]School of Physics and Astronomy, University of Nottingham, University Park, Nottingham NG7 2RD, UK

Abstract. We present the latest data release of the Planetary Nebulae Spectrograph Survey (PNS) of ten lenticular galaxies and two spiral galaxies. With this data set we are able to recover the galaxies' kinematics out to several effective radii. We use a maximum likelihood method to decompose the disk and spheroid kinematics and we compare it with the kinematics of spiral and elliptical galaxies. We build the Tully- Fisher (TF) relation for these galaxies and we compare with data from the literature and simulations. We find that the disks of lenticular galaxies are hotter than the disks of spiral galaxies at low redshifts, but still dominated by rotation velocity. The mechanism responsible for the formation of these lenticular galaxies is neither major mergers, nor a gentle quenching driven by stripping or Active Galactic Nuclei (AGN) feedback.

Keywords. galaxies: elliptical and lenticular, cD, galaxies: evolution, galaxies: kinematics and dynamics

1. Introduction

Lenticular or S0 galaxies form a composite class of objects, whose evolutionary path appears to be driven by their mass. They are composed of a disk and a spheroid, and might present several subcomponents. The best way to recover lenticular galaxies' formation histories is to study the evolution of their disks and spheroids in different environments. If the progenitors of S0 galaxies are spiral galaxies whose gas was gently stripped or consumed, their disk kinematics would be similar to the kinematics of the disks of spiral galaxies. Their spheroid, though, might have been enhanced by a last sparkle of star formation, triggered by the infall of the stripped material onto the spheroid. On the other hand, S0 galaxies could be the result of mergers or interactions; in this case the disk kinematics would be hotter than in spiral galaxies. Higher random motion than in spiral galaxies disks might also be explained by clumpy disk formation. It is particularly promising to combine photometric and kinematic information to break the degeneracies in disentangling the spheroid and disk components, but for early type galaxies it is rather difficult to get kinematics information at large radii due to the steep decrease of the galaxy luminosity and the lack of a stable gaseous disk. Only discrete

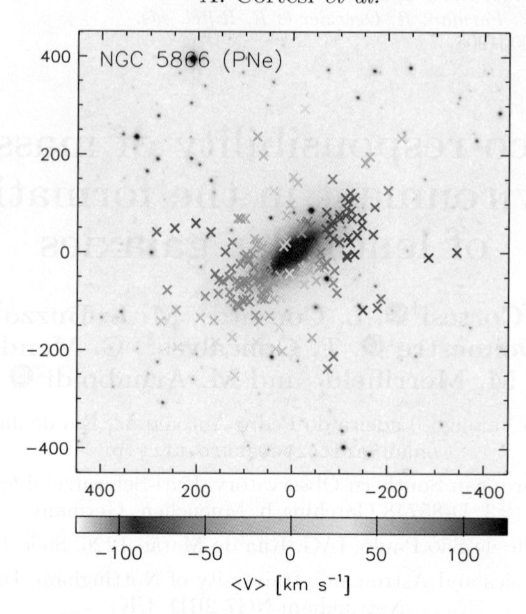

Figure 1. Digitised Sky Survey (DSS) image of NGC 5866, as an example of the galaxies in the sample, with the PN positions marked with crosses. The colours of the crosses represent the measured radial velocity of the PNe, after applying a kernel smoothing, according to the velocity scale below the panel.

tracers, such as planetary nebulae (PNe) and globular clusters (GCs), allow us to recover the kinematics of early type galaxies at large radii (Coccato *et al.* 2009, among others).

2. Analysis & Conclusions

We use PNe as tracers of the stellar kinematics of 10 lenticular and 2 spiral galaxies (see Fig. 1). To separate the kinematics of the disk and spheroid components, we use a maximum-likelihood method that combines the discrete kinematic data with a photometric component decomposition. The results of this analysis reveal that: the discs of S0 galaxies are rotationally supported; however, the amount of random motion in these discs is systematically higher than in comparable spiral galaxies. We compare the recovered TF relation with simulations that explain lenticular galaxies as major merger remnants (Tapia *et al.* 2017), or the results of merger of clumps at $z \sim 2$ (Saha & Cortesi 2018). All of these findings are consistent with a scenario in which spirals are converted into S0s through a process that heats up the disk, ruling out ram pressure stripping and AGN feedback. Yet, the high value of rotation over dispersion velocity can hardly be reproduced by a recent major merger event. We conclude that mild harassment or minor mergers can lead to the formation of S0 galaxies. Today' s low-mass ($log(M/M) < 9.5$) isolated S0s could be primordial galaxies formed at $z = 2$ by clumpy disk formation (Saha & Cortesi 2018), where clump migration builds a prominent bulge with no spiral structure formation as an intermediate step. We here confirm Morgan's suggestion that the SO classification type is a repository of physically quite distinct sorts of objects (Van den Bergh 1990).

References

Coccato, L., *et al.* 2009, *MNRAS*, 394, 1249
Tapia, T., Eliche-Moral, M. C., Aceves, H., *et al.* 2017, *A&A*, 604, A105
Saha, K. & Cortesi, A. 2018, *ApJL*, 862, L12
van den Bergh, S. 1990, *ApJ*, 348, 57

Galaxy Evolution and Feedback across Different Environments
Proceedings IAU Symposium No. 359, 2020
T. Storchi-Bergmann, W. Forman, R. Overzier & R. Riffel, eds.
doi:10.1017/S1743921320004019

Cosmic magnetism evolution using cosmological simulations

Stela Adduci Faria[1], Elisabete M. de Gouveia Dal Pino[1] and Paramita Barai[2]

[1]Instituto de Astronomia, Geofísica e Ciências Atmosféricas, University of São Paulo (USP), São Paulo, Brazil
e-mail: stela.faria@usp.br

[2]Núcleo de Astrofísica - Universidade Cruzeiro do Sul (NAT - UNICSUL), São Paulo, Brazil

Abstract. The Intergalactic Medium (IGM) is the region comprising the environment between the galaxies. Gamma-ray observations have provided lower limits to IGM magnetic fields of the order of $\gtrsim 10^{-16}$ G. Magnetic fields are continuously ejected from galaxies by jets and galactic winds. However, the origin and evolution of cosmic magnetic fields in the more diffuse regions, like voids, is still debated. The difficulties in directly measuring magnetic fields and their coherent scales, make hydrodynamic and magnetohydrodynamic (MHD) cosmological simulations useful tools to shed light on this debate. As a first approach, we have performed hydrodynamic cosmological simulations assuming energy equipartition as an initial condition between the baryonic gas and the magnetic field, starting at $z = 8$, to track the evolution of magnetic fields, and compare with results of MHD simulations. We have found that for halos and cores, our results are comparable to the MHD description. For the less dense regions, the equipartition condition clearly overestimates the observed limits. In forthcoming work, we will investigate MHD simulations of cosmological evolution and amplification of seed magnetic fields, considering all relevant feedback processes and exploring turbulent dynamo amplification versus primordial mechanisms across cosmological timescales.

Keywords. Magneto-hydrodynamics, Cosmic Ray Propagation, Cosmological hydrodynamical simulations, Origin of magnetic fields

---

1. Numerical method and Results

The IGM is observed to contain diffuse magnetic fields that may have been seeded by starburst (SBs) galaxies, jets from radio galaxies, dynamo action, mergers and tidal interactions between galaxies, or they may have had a primordial origin (e.g. de Gouveia Dal Pino 2011; Barai & de Gouveia Dal Pino 2019). However, it is still not clear if such processes are able to reach the more diffuse regions of the IGM, i.e., the large scale voids where magnetic fields of $\gtrsim 10^{-16}$ G have been inferred from gamma-ray observations on scales of the order of Mpc (Fermi-LAT Collaboration 2018). Our aim is to investigate the origin of cosmic magnetic fields in the diffuse intergalactic medium and clusters of galaxies. In a first approach here we consider pure hydrodynamical simulations with passive magnetic fields in equipartition with the gas thermal pressure in order to compare with existing MHD cosmological simulations and check whether this hypothesis is valid.

We use the hydrodynamic version of the Lagrangian SPH code GADGET-3 (Dolag & Stasyszyn 2009) to simulate cosmological boxes with constant volume (2 Mpc)^3. The equipartition assumption is given by $u = \frac{u_E}{2} = \frac{B^2}{8\pi}$, where u is the magnetic energy density,

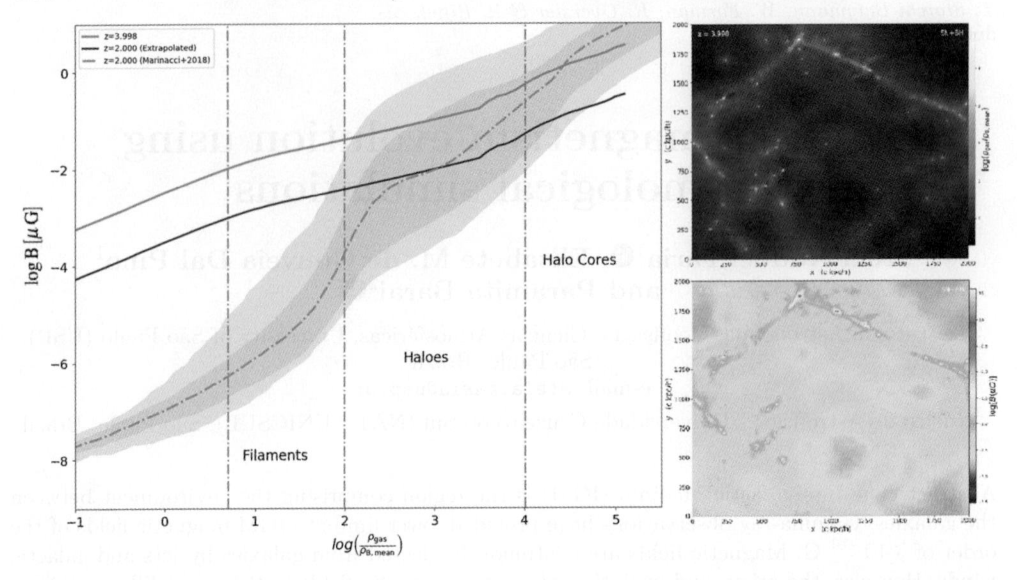

Figure 1. (Left) MHD simulations from Marinacci *et al.* (2018) at z = 2, shown as the pink dot-dashed curve (shaded areas are the 1σ (gray) and 2σ (cyan) scatter). We show our simulated mean magnetic field at z = 4 (red solid line) which we extrapolate to z = 2 (blue solid curve) for comparison to Marinacci *et al.* (2018). (Right) Projected 2D maps of the gas overdensity and the magnetic field, top and bottom respectively, are shown. We observe that the equipartition assumption is valid for the denser regions, i.e., the MFs follow a dependence with gas density that is compatible with cosmological MHD simulations, thus indicating that in such regions the gas and the MFs are in equipartition.

u_E the gas energy density, and B the magnetic field. To test our assumption, we compare our results with the MHD simulations of Marinacci *et al.* (2018).

In Fig. 1, we show our simulated results at $z = 4$ (red curve) for a range of densities from galaxy halos to voids. Marinacci *et al.* (2018) performed their simulations between $z = 2$ and 0. We take their results for $z = 2$ (represented in Fig. 1 by the dashed-pink curve) to compare with our results. Since our simulation stopped at $z = 4$ (Fig. 1, red solid curve), we extrapolated our results to $z = 2$ (Fig. 1, blue solid curve), assuming magnetic flux conservation and cosmological expansion of the IGM, using $\log B_1 = \log B_2 + \log \frac{(1+z_1)^2}{(1+z_2)^2}$, so that the magnetic field drops according to $\propto \frac{1}{a^2} = (1+z)^2$. The equipartition assumption yields magnetic field intensities which are highly overestimated for the less dense, large scale structures, while it gives consistent values for the denser regions, the halos and cores. Further studies performing MHD simulations are still needed in order to confirm these results. They also evidence that the dynamical effects of MFs *cannot* be neglected at halo cores scales. Therefore, at these scales HD simulations are not an optimal approach to describe the evolution of these systems. The astrophysical processes that enrich the IGM are local and could feed the voids, however, the potential MF in these regions are far from the equipartition condition with the thermal pressure of matter, as one should expect. Our calculations cannot eliminate the hypothesis that the field present in the voids is of primordial origin. We will continue to investigate these magnetic fields in the low density regions through cosmological MHD numerical studies of the formation and evolution of large scale structures. For this purpose, we will use the MHD version of the GADGET SPH code, along with its connection with turbulence at smaller scales (Santos-Lima *et al.* 2014), cosmic ray (CR) propagation, and all the important feedback ingredients (Alves Batista, Saveliev & de Gouveia Dal Pino 2019).

Acknowledgements

We thank the referee Bill Forman for his useful comments.

References

Alves Batista, R., Saveliev, A., & de Gouveia Dal Pino, E. M. 2019, *MNRAS*, 489, 3836
Barai, P. & de Gouveia Dal Pino, E. M. 2019, *MNRAS*, 487, p.5549–5563
de Gouveia Dal Pino, E. M. 2011, *ISBN 978-1-908106-12 Hardback*, pp.37–52
Dolag, K. & Stasyszyn, F. 2009, *MNRAS*, 398, 1678
Fermi-LAT Collaboration, 2018, *ApJS*, 237, 36
Grasso, D. & Rubinstein, H. R. 2001, *Phys. Rep.*, 348, 163–266
Marinacci, F. *et al.* 2018, *MNRAS*, 480, 5113M
Santos-Lima, R. *et al.* 2014, *ApJ*, 781, 84

Galaxy Evolution and Feedback across Different Environments
Proceedings IAU Symposium No. 359, 2020
T. Storchi-Bergmann, W. Forman, R. Overzier & R. Riffel, eds.
doi:10.1017/S1743921320001805

Propagation of cosmic rays and their secondaries in the intracluster medium

Saqib Hussain[1], Rafael Alves Batista[2], Elisabete Maria de Gouveia Dal Pino[1] and Klaus Dolag[3]

[1]Institute of Astronomy, Geophysics and Atmospheric Sciences (IAG), University of São Paulo (USP), São Paulo, Brazil
emails: s.hussain2907@gmail.com, dalpino@iag.usp.br

[2]Institute for Mathematics, Astrophysics, and Particle Physics, Radboud University Nijmegen, Netherlands
email: r.batista@astro.ru.nl

[3]Max Planck Institute for Astrophysics, Karl-Schwarzschild-Str 1, 85741 Garching, Germany

Abstract. We present results of the propagation of high-energy cosmic rays (CRs) and their secondaries in the intracluster medium (ICM). To this end, we employ three-dimensional cosmological magnetohydrodynamical simulations of the turbulent intergalactic medium to explore the propagation of CRs with energies between 10^{14} and 10^{19} eV. We study the interaction of test particles with this environment considering all relevant electromagnetic, photohadronic, photonuclear, and hadronuclear processes. Finally, we discuss the consequences of the confinement of high-energy CRs in clusters for the production of gamma rays and neutrinos.

Keywords. cosmic rays, gamma rays, neutrinos, cluster of galaxies

1. Simulation of CRs propagation

In this work we study the propagation of CRs in the turbulent environments built out of three-dimensional magnetohydrodynamical (MHD) cosmological simulations of the distribution of filaments and clusters of galaxies in the local universe within a 260 Mpc scale, performed with the GADGET code (see Dolag *et al.* 2005). Then, we employ the code CRPropa 3 (Alves Batista *et al.* 2016) to study the particles propagation in this environment. We consider all relevant CR interactions, namely, photopion production, photodisintegration, and Bethe-Heitler pair production with the background photon fields that include the cosmic microwave background (CMB), the extragalactic background light (EBL), and the thermal Bremsstrahlung field at X-rays from clusters of galaxies primarily due to the emission of hot plasma with temperatures 10^6-10^8 K. We also consider CR proton interactions with the background gas of the ICM.

2. Results and Discussion

The spectra of CRs and neutrinos are given in Fig. 1. We considered different clusters of distinct masses and magnetic field distributions in the local universe ($z < 0.1$). We find a significant suppression in the flux of CRs (Fig. 1) at an energy around 10^{17} eV, which indicates a trapping of CRs within the clusters at energies of this order and below. This upper limit is more constraining than the one we obtained previously when neglecting the CR interactions with the background protons and thermal photon field (Alves Batista *et al.* 2019). It is found that the CR interactions with thermal Bremsstrahlung radiation are

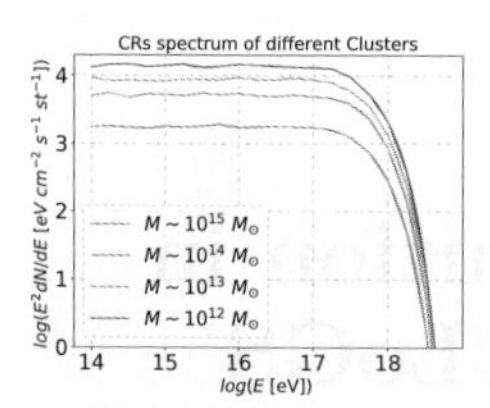

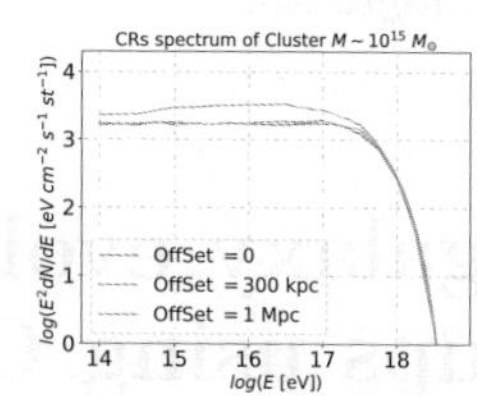

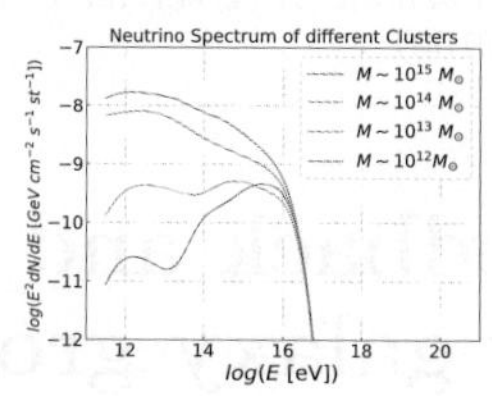

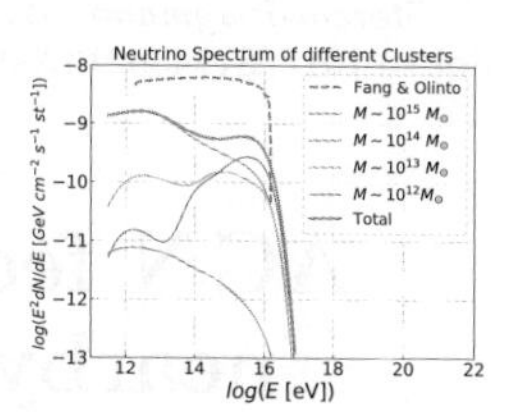

Figure 1. The first panelv(from left) shows the CR spectrum for clusters of different masses. The second shows the spectrum of CRs for one cluster of mass $M = 2 \times 10^{15}$, computed at its edge, assuming sources at: the centre of the cluster, and at 300 kpc and 1 Mpc away from the centre. The third and fourth panels show, respectively, the neutrino spectra for individual clusters, and the total spectrum. The label 'Total' in the last panel corresponds to the sum of the fluxes of neutrinos from all clusters of the background simulation. Our results are compared to Fang & Olinto (2016).

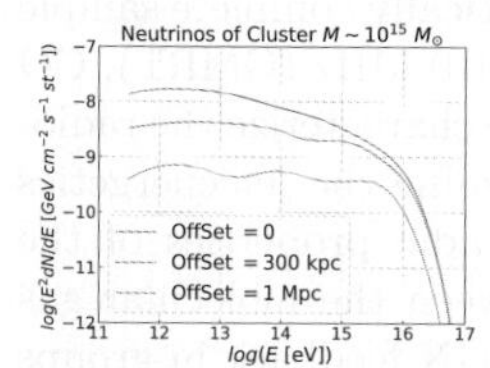

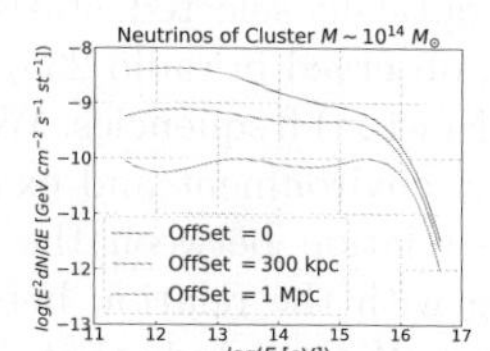

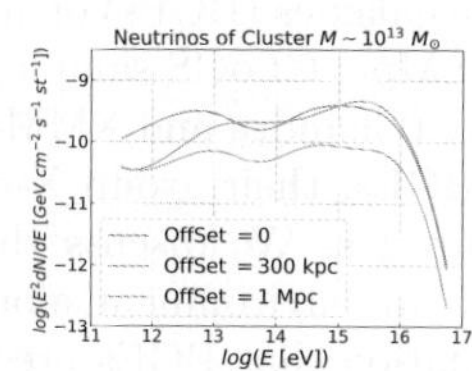

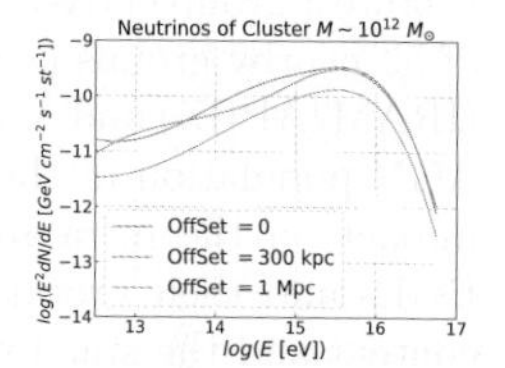

Figure 2. Spectrum of neutrinos for individual clusters of distinct masses, computed at their edge, assuming sources: in the centre of the cluster, at 300 kpc and 1 Mpc away from the centre.

negligible compared to the proton-proton interactions. Clusters are unique environments. Due to their magnetic field, the confinement of CRs for long periods of time enhances the interaction rates, increasing the production of secondary particles including neutrinos and gamma rays (e. g., Brunetti & Jones 2014, Blasi 2013 and Amato & Blasi 2018). These high-energy gamma rays and neutrino fluxes can be measured with the Cherenkov Telescope Array (CTA), the IceCube Neutrino Observatory, and the Giant Radio Array for Neutrino Detection (GRAND). We also present the flux of neutrinos generated by the interactions of CR protons with the background protons of the ICM (Figs. 2 & 1). We plotted both, the flux of neutrinos for individual clusters in Fig. 2 and the total flux from all clusters in Fig. 1, and compared with Fang & Olinto 2016 (therein see Fig. 3, $0.01 < z < 0.3$). In the latter, the authors employed a simplified model of the baryon distribution and the turbulent magnetic field, and then extrapolated their results for the entire distribution of galaxy clusters. Besides, our total flux has been integrated up to $z < 0.1$, while in their work, this integration stops at $z = 0.3$. This may explain the minor discrepancy between their and our results. Finally, we note that these results are still preliminary and we are still limited by our poor statistics, which is a consequence of the high computational load required by the particle-by-particle approach employed. Any potential statistical fluctuations will be mitigated in our future works.

References

Alves Batista, R., *et al.* 2016, *J. Cosmol. Astropart. Phys.*, 05, 038

Alves Batista, R., de Gouveia Dal Pino, E. M., Dolag, K., *et al.* 2019, *Proceedings IAU*: New Insights in Extragalactic Magnetic Fields, arXiv:1811.03062

Amato, E. & Blasi, P. 2018, *Adv. Space Res.*, 62 2731

Blasi, P. 2013, *Astron. Astrophys. Rev.*, 21, 70

Brunetti, G. & Jones, T. W. 2014, *Int. J. Mod. Phys. D*, 23, 1430007

Dolag, K., *et al.* 2005, *J. Cosmol. Astropart. Phys.*, 01, 009

Fang, K. & Olinto, A. V. 2016, *Astrophys. J.*, 828, 37

Galaxy Evolution and Feedback across Different Environments
Proceedings IAU Symposium No. 359, 2020
T. Storchi-Bergmann, W. Forman, R. Overzier & R. Riffel, eds.
doi:10.1017/S1743921320001507

AGN feedback and galaxy evolution in nearby galaxy groups using CLoGS

Konstantinos Kolokythas† and **CLoGS team**

[1]Centre for Space Research, North-West University, Potchefstroom 2520, South Africa
emails: k.kolok@nwu.ac.za (KK), CLoGS: http://www.sr.bham.ac.uk/ ejos/CLoGS.html

Abstract. Much of the evolution of galaxies takes place in groups where feedback has the greatest impact on galaxy formation and evolution. We summarize results from studies of the central brightest group early-type galaxies (BGEs) of an optically selected, statistically complete sample of 53 nearby groups (<80 Mpc; CLoGS sample), observed in radio 235/610 MHz (GMRT), CO (IRAM/APEX) and X-ray (Chandra and XMM-Newton) frequencies. We characterize the radio-AGN population of the BGEs, their group X-ray environment and examine the jet energetics impact on the intra-group gas. We discuss the relation between the radio properties of the BGEs and their group X-ray environment along with the relation between the molecular gas content and the star formation that BGEs present. We conclude that AGN feedback in groups can appear as relatively gentle near-continuous thermal regulation, but also as extreme AGN activity which could potentially shut down cooling for longer periods.

Keywords. AGN feedback, galaxy groups, galaxy evolution, jets

---

1. Scientific background

Galaxy groups are gravitationally bound systems in which the majority of galaxies and stars in the Universe reside (Eke *et al.* 2006). They contain >50% of all galaxies and are not simply scaled-down galaxy clusters (e.g., Ponman, Cannon & Navarro 1999), but possess shallow gravitational potentials and low velocity dispersions, which are conducive to the galaxy mergers and tidal interactions that drive rapid galaxy evolution (e.g., Alonso *et al.* 2012). As such, galaxy groups are the most important laboratories for the study of galaxy formation and evolution, where the crucial effects of baryon physics, such as cooling, galactic winds, and AGN feedback, are most evident.

The nature of the feedback is one of the most important unresolved questions in extragalactic astronomy (see McNamara & Nulsen 2007). While the required mechanical power from AGN activity needed to produce X-ray cavities has been argued as sufficient to balance gas cooling, in some systems star formation continues in the central galaxy. The mechanism of transfer of energy between the AGN and the intra-group medium (IGM) is poorly understood, and requires a combination of multi-wavelength data, including radio, CO and high-quality X-rays to provide insight into the processes involved. Towards this end, we defined the Complete Local-Volume Groups Sample (CLoGS), a statistically complete, optically-selected set of 53 groups within 80 Mpc creating the first truly representative survey of groups in the local Universe (see O'Sullivan *et al.* 2017 for more details on the sample). CLoGS is ideal for understanding the balance between hot and cold gas, AGN activity and star formation in groups, and therefore the evolution of galaxies as a whole, as nearby groups provide the best angular resolution for a detailed study.

† KK acknowledges support from the Grants for IAU Symposia and Regional IAU Meetings

2. A multi-wavelength view of galaxy evolution with CLoGS

X-ray observations from *XMM* and *Chandra* reveal that almost half (49%) of CLoGS groups present a full scale X-ray halo; hot gas extending >65 kpc (26/53; O'Sullivan *et al.* 2017, 2020 in prep). More than half of these X-ray bright groups (14/26) are unknown or misidentifed as single galaxies prior to our observations, which implies that 30% of the X-ray bright groups in the local universe might still be unknown. Examining in detail the radio-AGN population of the central BGEs using GMRT observations at 235/610 MHz and archival VLA data, a high radio detection rate of 87% (46/53) is found, with the BGEs presenting a wide range in radio power (10^{20}–10^{25} W/Hz) and projected size (~3 kpc to 2 Mpc). Of the radio detected BGEs, 53% present point-like radio emission, followed by 19% having jets with non-detections at 13%. We find that radio morphology correlates with the dynamical youth of the groups as radio point sources are more common in the dominant galaxies of spiral-rich systems whereas jet sources show no preference of their close environment (Kolokythas *et al.* 2018, 2019).

Combining radio and X-rays reveals that 11/26 (~42%) of X-ray bright groups host jet systems with the radio non-detections appearing in X-ray faint groups. The jet occurence in X-ray bright groups implies an AGN duty cycle >1/3 with these central jet sources seen in systems that possess cool cores with short central cooling times (t_{cool} <7.7 Gyr; O'Sullivan *et al.* 2017) and low entropies in their central region (jet activity hasn't increased dramatically the entropy in their cores). Examining the balance between heating from AGN and cooling from X-rays we conclude that AGN feedback can manifest in groups as smooth near-continuous thermal regulation, but also as extreme outbursts which could potentially shut down cooling for long periods of time (Kolokythas *et al.* 2018).

Examinination of the cold gas content of the CLoGS groups' dominant galaxies using the IRAM 30m and APEX telescopes shows a high detection rate for CO (~40%), but a short depletion time indicating that group-central galaxies must replenish their molecular gas reservoirs on timescales ~100 Myr. The majority of the BGEs are found to be AGN (instead of star formation) dominated with at least half of them containing HI as well as molecular gas (O'Sullivan *et al.* 2015, 2018a).

Lastly, due to its proximity, CLoGS provides also the opportunity to study in detail the AGN outburst properties and energetics of individual strong radio jet sources (e.g., NGC 4261, Kolokythas *et al.* 2015) as well as the examination of galaxy interactions and their important role in the development of a galaxy group and its formation history (see e.g., NGC 5903, O'Sullivan *et al.* 2018b and NGC 1550, Kolokythas *et al.* 2020).

References

Alonso, S., Mesa, V., Padilla, N., *et al.* 2012, *A&A*, 539, 46
Eke, V. R., Baugh, C. M., Cole, S., *et al.* 2006, *MNRAS*, 370, 1147
Kolokythas, K., O'Sullivan, E., Giacintucci, S., *et al.* 2015, *MNRAS*, 450, 1732
Kolokythas, K., O'Sullivan, E., Raychaudhury, S., *et al.* 2018, *MNRAS*, 481, 1550
Kolokythas, K., O'Sullivan, E., Intema, H., *et al.* 2019, *MNRAS*, 489, 2488
Kolokythas, K., O'Sullivan, E., Giacintucci, S., *et al.* 2020, *MNRAS*, 496, 1471
McNamara, B. & Nulsen, P. 2007, *ARA&A*, 45, 117
O'Sullivan, E., Combes, F., Hamer, S., *et al.* 2015, *A&A*, 573, 111
O'Sullivan, E., Ponman, T. J., Kolokythas, K., *et al.* 2017, *MNRAS*, 472, 1482
O'Sullivan, E., Combes, F., Salome, P., *et al.* 2018a, *A&A*, 618, 126
O'Sullivan, E., Kolokythas, K., Kantharia, N. G., *et al.* 2018b, *MNRAS*, 473, 5248
Ponman, T. J., Cannon, D. B., & Navarro, J. F. 1999, *Nature*, 397, 135, 33

Galaxy Evolution and Feedback across Different Environments
Proceedings IAU Symposium No. 359, 2020
T. Storchi-Bergmann, W. Forman, R. Overzier & R. Riffel, eds.
doi:10.1017/S1743921320002355

Ubiquitous cold and massive filaments in brightest cluster galaxies

Valeria Olivares and Philippe Salomé

Observatoire de Paris, PSL University, CNRS, Sorbonne University, UPMC

Abstract. The origin of the mysterious multiphase filamentary structures surrounding Brightest Cluster Galaxies (BCGs) remains unknown. We present Atacama Large Millimeter/submillimeter Array (ALMA) and Multi Unit Spectroscopic Explorer (MUSE) observations for a sample of 15 BCGs to investigate the origin and life-cycle of the gas. Those observations show clumpy and massive molecular filaments, preferentially located around the radio bubbles inflated by the active galactic nuclei (AGN). We investigate where the cold gas condenses from the intra-cluster medium, by comparing the radial extent of the filaments with predictions from numerical simulations.

Keywords. galaxies: active, galaxies: jets, galaxies: clusters: intracluster medium

1. Motivation

The classical cooling-flow model fails in the absence of an external non-gravitational heating mechanism needed to compensate for catastrophic radiative cooling in the intra-cluster medium (ICM) of galaxy clusters. Feedback from an AGN contributes to offset the cooling through bubbles inflated by radio jets launched from massive black holes. However, it cannot completely offset the cooling as the BCGs harbors a complex multi-phase medium of extended warm and cold reservoirs of gas, whose physical origin remains unknown. Recent theoretical analyses and simulations have suggested that the hot atmospheres can become thermally unstable locally when the ratio of the cooling to free-fall timescales, t_{cool}/t_{ff} falls below ∼10−20 (e.g., Voit *et al.* 2017), or when the ratio of the cooling time over the eddy turn-over timescale, t_{cool}/t_{eddy}, is close to unity (Gaspari *et al.* 2018). The cold gas has likely cooled in-situ from (i) either low-entropy gas that has been uplifted by the bubbles at an altitude where it becomes thermally unstable (ii) or by direct thermal instability of small perturbation in the hot halo (see the introduction of Tremblay *et al.* (2018) for a detailed description of the different scenarios). To investigate the nature and life-cycle of those enigmatic filaments we used ALMA observations that map the kinematics and morphology of the cold molecular gas phase of 15 BCGs (3 from new ALMA observations and 11 gathered from the ALMA archive). We compare ALMA observations with new MUSE data that map the kinematics and morphology of the warm ionised gas.

2. Results and Discussion

Molecular filaments: From the sample studied with ALMA, we found that most of the sources show extended unrelaxed structures with disturbed motions along the filaments. In those systems, the molecular distribution usually consists of a nuclear emission component closely related to the BCG's core and a set of extended massive clumpy filaments, with cold molecular masses of a few $\sim 10^{8}-10^{10}$ $M_{\odot}$, which are preferentially

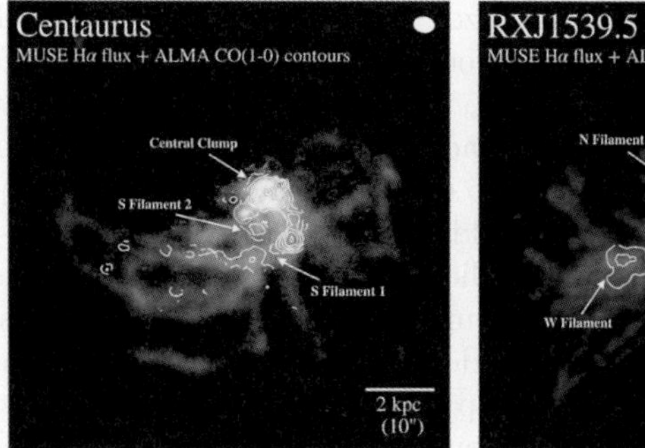

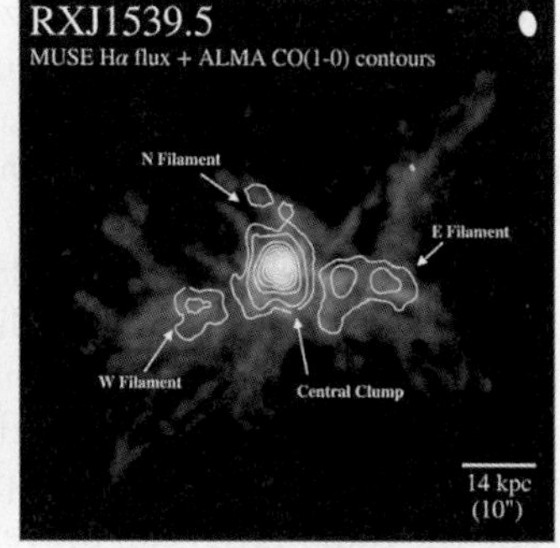

Figure 1. Hα flux maps from MUSE observations overlaid with contours from the CO(1-0) integrated intensity maps for two of the new ALMA sources: Centaurus (left panel) and RXCJ1539.5-8335 (right panel). The co-spatial and morphological correlation between the warm ionised and cold molecular nebulae is clear in these maps. From Olivares *et al.* (2019)

located around the radio bubbles inflated by the AGN or beneath the X-ray cavities. While two systems, Abell 262 and Hydra-A, are well described by relaxed structures showing ordered motions within a compact ($\sim$2$-$5 kpc length) thin ($\sim$kpc) rotating disc located at the very centre of the BCG. This cold gas disc may be an essential step in driving the gas into the vicinity of the AGN. In an AGN-regulated scenario, one indeed expects that a fraction of the cooled gas will eventually fuel the SMBH to maintain the powerful jets which are injecting mechanical energy into the ICM. The cold molecular gas in some of the galaxy cluster cores shows disturbed velocities, while others show smooth velocity gradients in substructures indicating either inflow or outflow of gas. The velocities of the cold clouds are slow through the molecular gas, lying in a range of 100$-$400 km s^{-1}. Those molecular velocities are inconsistent with the scenario of simple freely in-falling gas, where higher velocities are expected. These small velocity being below the escape velocity of the galaxy, the outflowing cold clouds should eventually fall back and fuel the central AGN. The superposition of inflowing and outflowing filaments along the line of sight can mix and cancel velocities structures.

Evidence supporting that cold and ionised emission arises from the same ensemble of clouds: The molecular gas is generally spatially distributed along the brightest emission from the warm ionised nebula (see Fig. 1). It also appears that the cold molecular and warm ionised gas share the same overall velocity structure. The comparison between Hα and CO(1-0) emissions indicates that the Hα velocity dispersion is broader than the CO one by a factor of $\sim$2. This can occur because the lines-of-sight are likely to intersect more warm gas than cold clouds, and also that the warm gas is more likely to be turbulent with higher velocities. The Hα-to-CO flux ratios are close to unity all along the nebula with a lack of significant radial gradients, which indicate a local excitation mechanism. It is also possible that the molecular filaments are as long as those seen in Hα. We derived expected total molecular masses, based on the Hα-to-CO flux ratios, which are higher than those one observed from ALMA by a factor of 1.2 or up to 6. Future high sensitivity ALMA observations are needed to confirm this. Such correlations can also be interpreted as the manifestation of a common origin, as the condensation of low-entropy gas via the top-down multiphase condensation cascade through thermal instabilities.

Origin of the cold gas: Using the ACCEPT (Archive of Chandra Cluster Entropy Profile Tables) sample X-ray ICM properties (Cavagnolo *et al.* 2008), we found that filaments always lie within the low-entropy and short cooling-time gas. As described in the

Chaotic Cold Accretion (Gaspari & Churazov 2013) or precipitation models (Voit *et al.* 2017), an important radius is when the cooling time exceeds the free-fall time by no more than a factor of 10, or when the t_{cool} is below 1 Gyr. We find that the extent of the filaments roughly corresponds to the radius where t_{cool}/t_{ff} is close to its minimum, which is always between 10 and 20. We also compared the ratio, t_{cool}/t_{eddy} as a function of the extent of the filaments, where the eddy turnover time is related to the turbulence injection scale. We find that the filaments lie in regions where this ratio is less than ~1. So the AGN bubbles may be powering the turbulent energy, which triggers gas cooling by compression and thermal instabilities. In addition, we showed that the energy contained in the AGN-cavities is enough to drag up some low-entropy gas in a region distant from the center. So the radio-AGN can prevent an overcooling on large scales, and it is also the engine that may trigger the cold accretion along filamentary structures and provide the material to feed the regulated feedback cycle. The full condensation process includes bubbles inflation, uplift and cocoon shocks. As a result, one or several mechanisms may dominate, i.e., precipitation, stimulated feedback, or even sloshing. In our sample, all of these processes seem to be responsible (to some degree) for the condensation of the cold gas. Further deep observations and velocity structure information of the hot phase will be crucial to clear up the dominant process.

References

Cavagnolo, K. W., Donahue, M., Voit, G. M., *et al.* 2008, *ApJ*, 683, L107
Gaspari, M. & Churazov, E. 2013, *A&A*, 559, A78
Gaspari, M., McDonald, M., Hamer, S. L., *et al.* 2018, *ApJ*, 854, 167
Olivares, V., Salome, P., Combes, F., *et al.* 2019, *A&A*, 631, A22
Tremblay, G. R., Combes, F., Oonk, J. B. R., *et al.* 2018, *Apj*, 865, 13
Voit, G. M., Meece, G., Li, Y., *et al.* 2017, *ApJ*, 845, 80

Galaxy Evolution and Feedback across Different Environments
Proceedings IAU Symposium No. 359, 2020
T. Storchi-Bergmann, W. Forman, R. Overzier & R. Riffel, eds.
doi:10.1017/S1743921320002343

Excitation mechanism in the intracluster filaments surrounding the Brightest Cluster Galaxies

Fiorella L. Polles

Observatoire de Paris, LERMA, Collège de France, CNRS, PSL University,
Sorbonne University, UPMC, Paris
email: fpolles@usra.edu

Abstract. Multi-phase filamentary structures surrounding giant elliptical galaxies at the center of cool-core clusters, the Brightest Cluster Galaxies (BCGs), have been detected from optical to submillimeter wavelengths. The source of the ionisation in the filaments is still debated. Studying the excitation of these structures is key to our understanding of Active Galactic Nuclei (AGN) feedback in general, and more precisely of the impact of environmental and local effects on star formation. One possible contributor to the excitation of the filaments is the thermal radiation from the cooling of the hot plasma surrounding the BCGs, the so-called cooling flow.

Keywords. (galaxies:) cooling flows, intergalactic medium

---

1. Introduction

In the centers of cool-core clusters lie giant elliptical galaxies, the Brightest Cluster Galaxies (BCGs). X-ray observations revealed huge Intra-Cluster Medium cavities, which are produced by the jet of the central black hole (e.g. Boehringer *et al.* 1993). Optical observations show that the BCGs are often surrounded by a system of filaments (e.g. Heckman *et al.* 1989; Olivares *et al.* 2019), whose shapes suggest a relation between these structures and the cavities (e.g. Fabian *et al.* 2008). The filaments have been observed in a wide range of wavelengths, illustrating their multi-phase nature. Many of these filaments do not have strong on-going star formation and the photoionization by stellar emission as well as Active Galactic Nucleus (AGN) cannot reproduce their emission (e.g. Conselice *et al.* 2001). Ferland *et al.* (2009) proposed, as the main heating mechanism, collisions of the cold gas in the filaments with ionizing particles. This mechanism can reproduce the observations (optical-to-infrared), but the effect of the penetration of the radiation inside the clouds has not been explored. Combining state-of-the-art models with recent multi-wavelength observations, we investigate the contribution of the thermal radiation from the cooling of the hot plasma surrounding the BCGs, to the excitation of the filaments, and explore the outcomes due to photoionisation at different depths in the cloud.

2. Modeling

Using the spectral synthesis code CLOUDY (Ferland *et al.* 2017), we model a slab of gas of extinction $A_V \leqslant 30$ mag at constant pressure, to reproduce self-consistently all of the gas phases. The ionizing source is the soft X-ray and EUV radiation emitted by the cooling gas (Polles *et al.* 2020a in prep.). Our grid of models samples different metallicities (from $0.3\,Z_\odot$ to $1\,Z_\odot$), turbulent heating rates (v_{tur} from 0 km s^{-1} to

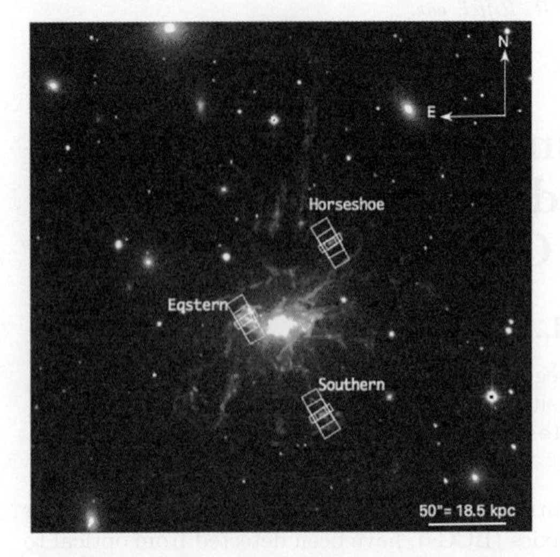

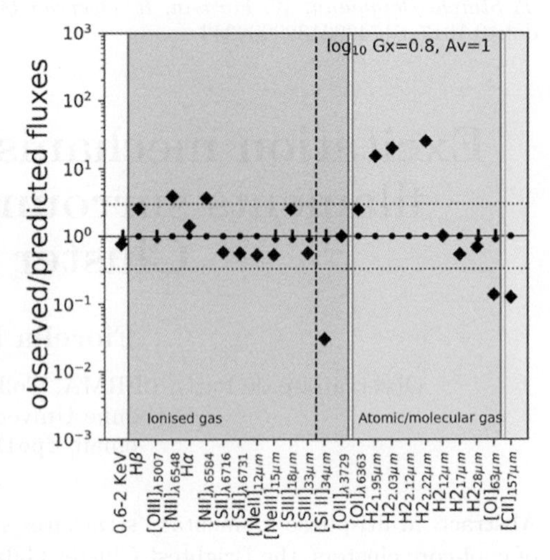

Figure 1. Left: Hα map of NGC 1275 (SITELLE/SN3). The *Spitzer*/IRS high-resolution slits are overlaid. Right: modeling results of the Horseshoe region. Comparison of the observed fluxes and the predicted fluxes from the best model: $G_X = 10^{0.8}$, $A_V = 1$ mag, $Z = Z_\odot$ and $v_{tur} = 2$ km s^{-1}. The points are the observed values normalised to one and the arrows indicate the upper limits. The diamonds are the observed/predicted ratios. The dashed vertical line separates the lines used as constraints, on the left, from those predicted, on the right. The dashed horizontal lines highlight an agreement within a factor of 3 between fluxes from the model and observations.

100 km s^{-1}) and X-ray intensities (G_X between 10^{-2} to 10^3; $G_X = 1$ corresponds to the integrated intensity of 1.6×10^{-3} erg cm^{-2} s^{-1} in the 0.6-2 keV band).

Constraining the model with line emission that traces different gas phases, is of fundamental importance to identify the physical properties of the filaments and to investigate the different excitation mechanisms. The multi-wavelength data set available for NGC 1275, the BCG at the center of the Perseus cluster, makes its nebula the perfect object for our modeling. We focus on three regions: Horseshoe, Southern, and Eastern (Figure 1; Polles *et al.* 2020b in prep.). The *Spitzer* observations of these regions gives us access to mid-infrared line emission. We collect all the available data from optical to far-infrared, and we combine them with our grid of models. We start using only the line emission tracing the ionised gas to constrain the models. The best model results for the Horseshoe region, as an example, is shown in Figure 1. The best model reproduces well all of the ionised gas tracers, except [Si II]λ 34 μm, which is a line that can arise from ionised gas as well as from a neutral medium. The H_2 pure rotational lines are also well reproduced, while the H_2 ro-vibrational lines as well as [C II]λ 157 μm and [O I]λ 63 μm lines are not. The overestimation of [CII] and [OI] could indicate that the filling factor of the neutral phase, which is traced by these lines, is <1 (default assumption). The underestimation of the H_2 ro-vib lines suggests that an additional heating source, e.g. shocks or cosmic rays, could be necessary to reproduce all of the line emission.

3. Conclusion

Combining the multi-phase models produced by the code Cloudy with the most recent multi-wavelength observations, we have shown that the thermal radiation from the cooling plasma can reproduce most of the observables of the filaments surrounding NGC 1275. This study is part of the LYRICS project (gas Life cYcle around galaxies : oRIgin and state of Cold accretion Streams).

References

Boehringer, H., Voges, W., Fabian, A. C., *et al.* 1993, *MNRAS*, 264, L25
Conselice, C. J., Gallagher, J. S., Wyse, R. F. G., *et al.* 2001, *ApJ*, 122, 2281
Fabian, A. C., Johnstone, R. M., Sanders, J. S., *et al.* 2008, *Nature*, 454, 968
Ferland, G. J., Fabian, A. C., Hatch, N. A., *et al.* 2009, *MNRAS*, 392, 1475
Ferland, G. J., Chatzikos, M., Guzmán, F., *et al.* 2017, *Rev. Mexicana AyA*, 53, 385
Heckman, T. M., Baum, S. A., van Breugel, W. J. M., *et al.* 1989, *ApJ*, 338, 48
Olivares, V., Salome, P., Combes, F., *et al.* 2019, *A&A*, 631, 22

Galaxy Evolution and Feedback across Different Environments
Proceedings IAU Symposium No. 359, 2020
T. Storchi-Bergmann, W. Forman, R. Overzier & R. Riffel, eds.
doi:10.1017/S1743921320001945

Investigating the properties of a galaxy group at z = 0.6

Daniela Hiromi Okido, Cristina Furlanetto, Marina Trevisan and Mônica Tergolina

Universidade Federal do Rio Grande do Sul, Instituto de Física
Av. Bento Gonçalves, 9500, 91501-970, Porto Alegre, Brazil
email: hiromi.okido@ufrgs.br

Abstract. Galaxy groups offer an important perspective on how the large-scale structure of the Universe has formed and evolved, being great laboratories to study the impact of the environment on the evolution of galaxies. We aim to investigate the properties of a galaxy group that is gravitationally lensing HELMS18, a submillimeter galaxy at $z = 2.39$. We obtained multi-object spectroscopy data using Gemini-GMOS to investigate the stellar kinematics of the central galaxies, determine its members and obtain the mass, radius and the numerical density profile of this group. Our final goal is to build a complete description of this galaxy group. In this work we present an analysis of its two central galaxies: one is an active galaxy with $z = 0.59852 \pm 0.00007$, while the other is a passive galaxy with $z = 0.6027 \pm 0.0002$. Furthermore, the difference between the redshifts obtained using emission and absorption lines indicates an outflow of gas with velocity $v = (278.0 \pm 34.3)$km/s relative to the galaxy.

Keywords. galaxies: groups, galaxies: high-redshift, galaxies: general

---

1. Introduction

Characterizing the mass distribution of structures across the full mass range, from single galaxies to galaxy clusters, can be used to test and constrain the model of structure formation and evolution (Bartelmann *et al.* 2013). Galaxy groups bridge the gap between individual galaxies and galaxy clusters, so they offer a new window of investigation in the mass spectrum. So far, few studies have been done to measure the mass distribution of groups, despite being the most common structures in the Universe, hosting at least half of all galaxies in the local Universe. Furthermore, galaxy groups are great laboratories to study the influence of the environment in the galaxy evolution. The main goal of this project is to investigate the properties of a galaxy group that is gravitationally lensing HELMS18, a submillimeter galaxy at $z = 2.39$ from Herschel's HerMES Large Mode Survey (HELMS; Nayyeri *et al.* 2016). We aim to have a complete description of this galaxy group, including its members, mass, redshift, radius and the numerical density profile. Hence we hope to shed some light on how the mass distribution varies across the mass spectrum, from individual galaxies to galaxy clusters.

2. Data and Methods

Multi-object spectroscopy data were obtained for this galaxy group with Gemini-GMOS (Gemini Multi-Object Spectrographs) to investigate its properties. We used the R400-G5325 diffraction grating, whose resolution is R$\sim$1000. The grating covers the rest-frame spectral interval from $\sim$3500$-$6000Å, allowing the observation of absorption features and emission lines. The size of the slit used was $1''$. Two masks were made in

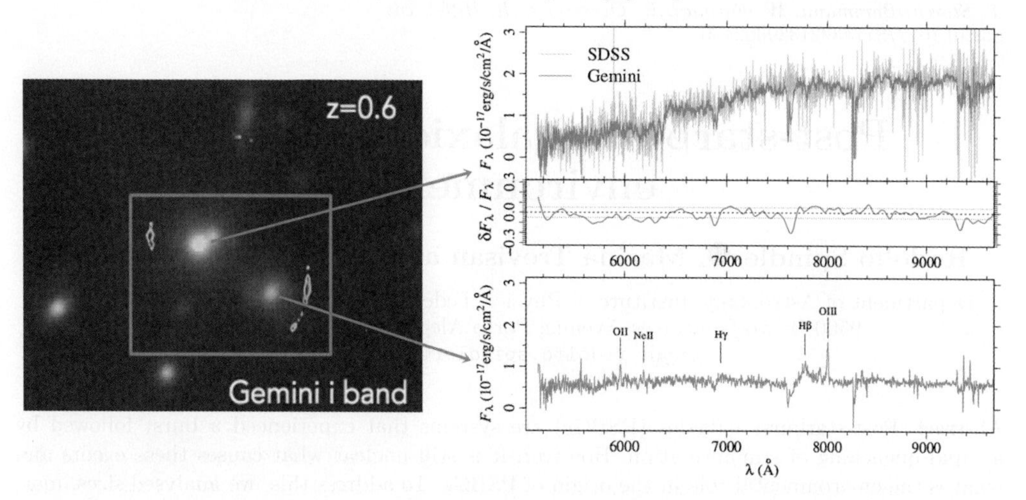

Figure 1. *Left:* Gemini i-band image, showing the innermost region of the group and its two central galaxies. The overlapping white contours correspond to the images of the HELMS18 submillimeter lensed galaxy, obtained with ALMA. *Right:* spectra of the two central galaxies. The upper panel shows the comparison between the SDSS and Gemini spectra of the passive galaxy. The bottom panel shows the spectrum of the active galaxy and the emission lines detected.

order to maximize the number of objects observed. Thus, we observed 55 galaxies. The total exposure time for the two masks was 8 hours.

The data reduction was done in IRAF (Image Reduction and Analysis Facility) following the standard Gemini data reduction pipeline, using the GMOS tasks. We present the results of the two central galaxies (a quasar and a passive galaxy), while data reduction is ongoing for the remaining galaxies. The redshifts were determined using the Radial Velocity Package developed at the Smithsonian Astrophysical Observatory (`RVSAO`), which is a package to measure radial velocities from spectra using the emission lines (EM) or the cross-correlation (XC) technique. The redshift of the quasar was determined using `EMSAO` and `XCSAO` tasks, while we used only `XCSAO` task for the passive galaxy. We used the spectra of 10 galaxy as templates for the cross-correlation procedure.

3. Results

We obtained the spectra of the two central galaxies, being one of them a quasar, as can be seen in the spectrum in the bottom right panel of Figure 1. We determined that the redshift of the quasar is $z = 0.59852 \pm 0.00007$. The redshift of the elliptical galaxy is $z = 0.60246 \pm 0.00004$, which is in accordance with the value of the Sloan Digital Sky Survey (SDSS), which is $z = 0.6027 \pm 0.0002$. The difference between the redshifts obtained using the emission lines ($z = 0.59852 \pm 0.00007$) and absorption lines ($z = 0.59945 \pm 0.00009$) indicates an outflow of gas with velocity 278.0 ± 34.3 km/s relative to the galaxy.

We intend to finalize the data reduction and measure the redshifts of remaining galaxies to identify the members of the group. Then, we plan to obtain a complete description of the properties of this galaxy group.

References

Bartelmann, M., Limousin, M., Meneghetti, M., *et al.* 2013, *Space Science Reviews*, 177, 3
Nayyeri, H., Keele, M., Cooray, A., *et al.* 2016, *ApJ* 823, 17

Galaxy Evolution and Feedback across Different Environments
Proceedings IAU Symposium No. 359, 2020
T. Storchi-Bergmann, W. Forman, R. Overzier & R. Riffel, eds.
doi:10.1017/S1743921320002136

Post-starburst galaxies in different environments

Rodolfo Spindler, Marina Trevisan and Allan Schnorr-Müller

Department of Astronomy, Institute of Physics, Federal University of Rio Grande do Sul, 9500 Bento Gonçalves Avenue, Porto Alegre, RS, 91501-970, Brazil
email: rodolfo.spindler@ufrgs.br

Abstract. Post-starburst galaxies (PSBGs) are systems that experienced a burst followed by a rapid quenching of star formation. However, it is still unclear what causes these events and what is the environmental role in the origin of PSBGs. To address this, we analysed sizes, morphologies, ages, and metallicities of PSBGs at $0.05 \leqslant z \leqslant 0.1$ in groups and clusters of galaxies. We find a statistically significant excess of compact PSBGs in groups compared to a control sample of passive galaxies. Satellite PSBGs in groups tend to be more compact compared to their counterparts in clusters. Additionally, the PSBGs in groups have smaller T-type values and are likely to be found in inner group regions compared to PSBGs in clusters. Our results are compatible with dissipative wet merger events being an important mechanism responsible for the origin of PSBGs in groups, but other – less dissipative – processes may be producing PSBGs in cluster environments.

Keywords. galaxies: post-starburst, galaxies: quenching, galaxies: environment

1. Introduction

Post-starbust galaxies (PSBGs) are a rare class of objects with atypical spectral properties, such as strong Balmer absorption lines – a signature of A-type stars (Dressler & Gunn 1983). Studies have shown that these spectral features can only be reproduced by models of a recent burst followed by a rapid quenching of the star formation (e.g. Wild *et al.* 2007; von der Linden *et al.* 2010). However, it is still unclear what mechanisms are responsible for these events.

There is evidence pointing to a major-merger origin of PSBGs, such as disturbed morphologies and the high PSBG frequency in poor galaxy groups (Zabludoff *et al.* 1996; Blake *et al.* 2004). Alternatively, the redshift evolution of the PSBG number density (Wild *et al.* 2009) is not compatible with that of the major-merger rate (de Ravel *et al.* 2018). Additionally, PSBGs are found in rich clusters, where merger events are rare due to high velocity dispersions (Dressler *et al.* 2013). In this work, we investigate the physical processes responsible for the origin of PSBGs by analysing how the properties of PSBGs depend on the environment.

2. Data and Sample selection

To define the sample of PSBGs, we used the data from the *Sloan Digital Sky Survey - Data Release 12* (SDSS-DR12). We selected galaxies: *i*) at $0.05 \leqslant z \leqslant 0.1$; *ii*) brighter than $M_r \leqslant -20.4$, where M_r is the *k*-corrected absolute magnitude in the *r* band; *iii*) with low Hα equivalent widths (EW[Hα] $\leqslant 1$ Å); and *iv*) high Hδ_A index values (H$\delta_{\rm A} \geqslant 1.5$ Å). The galaxy ages and metallicities were inferred from the SDSS spectra using the STARLIGHT code (Cid Fernandes *et al.* 2005) with spectral models by (Vazdekis *et al.* 2015).

We retrieved the galaxy effective radii and morphologies from the catalogues by Simard *et al.* (2011) and by Dominguéz Sanchéz *et al.* (2018), respectively. We used an updated version† of the catalogue of groups and clusters by Yang *et al.* (2007) to identify the galaxies that are centrals and satellites in groups ($M_{\rm halo} \leqslant 10^{14}\,{\rm M}_\odot$), and satellites in clusters ($M_{\rm halo} > 10^{14}\,{\rm M}_\odot$). Finally, the control sample galaxies (CSGs) were selected to have similar distributions of stellar masses and specific star formation rates, and to reside in similar environments as the PSBGs.

3. Results and Perspectives

Our results can be summarized as follows:

• Using a mass-size relation of passive galaxies defined by van der Wel *et al.* (2014), we find that the central PSBGs are more compact than the central passive CSGs.

• The PSBGs that are satellites in groups also tend to be more compact than the satellite PSBGs in clusters. In addition, the PSBGs in groups have smaller T-type values (galaxy morphology index related to the Hubble sequence) and are more likely to be found in the inner group regions compared to satellite PSBGs in clusters.

• The PSBGs are young and metal-rich systems, regardless of the environment where they reside. The ages of the PSBGs are similar to those of the star-forming CSGs, but the PSBG metallicities are more compatible with those of the passive CSGs.

The small sizes of the PSBGs and their position within the host group are compatible with dissipative wet-merger events being an important mechanism responsible for the origin of the PSBG population in groups of galaxies. However, the differences that we find between the PSBGs in groups and in clusters suggest that other physical mechanisms produce PSBGs in these environments, as already suggested by Dressler *et al.* (2013). We will continue our investigation by performing a detailed morphological and structural analysis of group and cluster PSBGs.

It has been proposed that PSBGs are a transitioning population between star-forming and passive galaxies (Wild *et al.* 2009), and the ages and metallicities of our PSBGs compared to those of the CSGs lead us to two possible scenarios: *i*) the progenitors of PSBGs were star-forming systems that were enriched very efficiently; or *ii*) they were passive galaxies that got rejuvenated. To continue this investigation, we will trace the chemical enrichment histories of PSBGs following the approach by Trevisan *et al.* (2012).

References

Blake, C., Pracy, M. B., Couch, W. J., *et al.* 2004, *MNRAS*, 355, 713
Cid Fernandes, R., Mateus, A., Sodré, L. Jr., *et al.* 2005, *MNRAS*, 358, 363
de Ravel, L., Le Fèvre, O., Tresse, L., *et al.* 2009, *A&A*, 498, 379
Dominguéz Sanchéz, H., Huertas-Company, M., Bernardi, M., *et al.* 2018, *MNRAS*, 476, 3661
Dressler, A. & Gunn, J. E. 1983, *ApJ*, 270, 7
Dressler, A., Oemler, Jr., A., Poggianti, B. M., *et al.* 2013, *ApJ*, 770, 62
Simard, L., Mendel, J. T., Patton, D. R., *et al.* 2011, *ApJS*, 196, 11
Trevisan, M., Ferreras, I., de La Rosa, I. G., *et al.* 2012, *ApJL*, 752, L27
van der Wel, A., Franx, M., van Dokkum, P. G., *et al.* 2014, *ApJ*, 788, 28
Vazdekis, A., Coelho, P., Cassisi, S., *et al.* 2015, *MNRAS*, 449, 1177
von der Linden, A., Wild, V., Kauffmann, G., *et al.* 2010, *MNRAS*, 404, 1231
Wild, V., Kauffmann, G., Heckman, T., *et al.* 2007, *MNRAS*, 381, 543
Wild, V., Walcher, C. J., Johansson, P. H., *et al.* 2009, *MNRAS*, 395, 144
Yang, X., Mo, H. J., van den Bosch, F. C., *et al.* 2007, *ApJ*, 671, 153
Zabludoff, A. I., Zaritsky, D., Lin, H., *et al.* 1996, *ApJ*, 466, 104

† Available at www.astro.umass.edu/ xhyang/Group.html.

Galaxy Evolution and Feedback across Different Environments
Proceedings IAU Symposium No. 359, 2020
T. Storchi-Bergmann, W. Forman, R. Overzier & R. Riffel, eds.
doi:10.1017/S1743921320001842

Testing the link between mergers and AGN in the Arp 245 system

Elismar Lösch and Daniel Ruschel-Dutra

Universidade Federal de Santa Catarina, UFSC, Brazil
email: elismar@astro.ufsc.br

Abstract. Galaxy mergers are known to drive an inflow of gas towards galactic centers, potentially leading to both star formation and nuclear activity. In this work we aim to study how a major merger event in the ARP 245 system is linked with the triggering of an active galactic nucleus (AGN) in the NGC galaxy 2992. We employed three galaxy collision numerical simulations and calculated the inflow of gas through four different concentric spherical surfaces around the galactic centers, estimating an upper limit for the luminosity of an AGN being fed the amount of gas crossing the innermost spherical surface. We found that these simulations predict reasonable gas inflow rates when compared with the observed AGN luminosity in NGC 2992.

Keywords. galaxies: active, galaxies: evolution, galaxies: interactions

1. Introduction

Galactic interactions are among the mechanisms which can cause an inflow of gas towards the galactic center, potentially triggering phenomena such as starbursts, formation of bulges and spheroids and fuel active galactic nuclei (AGN).

Using the library of galaxy collision numerical simulations GalMer (Chilingarian *et al.* 2010), we studied the inflow of gas to the galactic centers in Arp 245, an interacting system consisting of two spiral galaxies: NGC 2992, an active galaxy recently classified as Seyfert 1.8 by Schnorr-Müller *et al.* (2016), and NGC 2993. The system is observed at an early stage of the interaction, about 100 Myr after perigalacticon (Duc *et al.* 2000).

Considering the gas inflow rate at 10 pc around the supermassive black hole (SMBH), we estimated an upper limit for the luminosity of that AGN, comparing it with the bolometric luminosity of the AGN present in NGC 2992.

2. Methods and Results

We applied three different galaxy collision numerical simulations with varying values of orbital angular momentum available at GalMer library and relied on the work of Duc *et al.* (2000) to choose zero orbital energy ($E=0$) prograde encounters and with galaxies' spin vectors making a 75° angle with respect to each other. Then, we calculated the gas inflow through four different concentric spherical surfaces of radii 1 kpc, 0.5 kpc, 0.1 kpc and 0.01 kpc around galactic centers during the collision. The result is shown in figure 1.

Applying $\dot{E}=\epsilon\dot{m}c^2$, where $\epsilon=0.1$ is the matter-energy conversion efficiency of a black hole, we estimated the luminosity of an AGN with a positive inflow rate through the spherical surface of radius 10 pc, and then compared it with the bolometric luminosity of NGC 2992' AGN. Results are presented in figure 2.

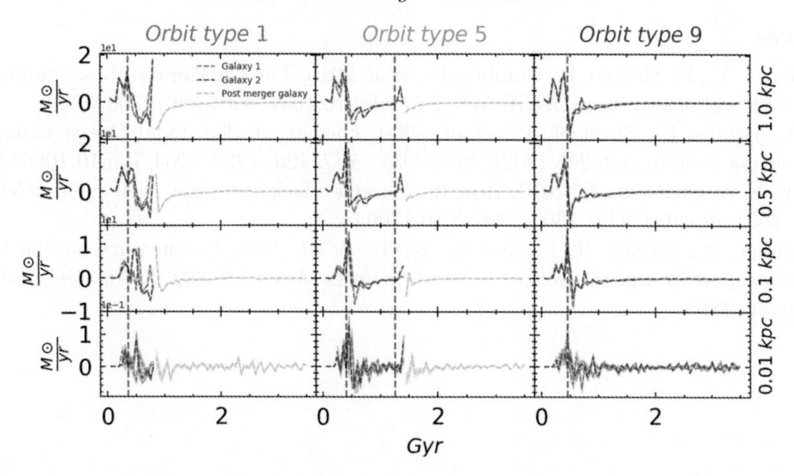

Figure 1. Gas flux through spherical surfaces of radii $r = \{1\ kpc, 0.5\ kpc, 0.1\ kpc, 0.01\ kpc\}$ around galactic centers for each simulation. Orbital angular momentum of the simulations increase from left to right. Vertical dashed lines mark perigalacticons.

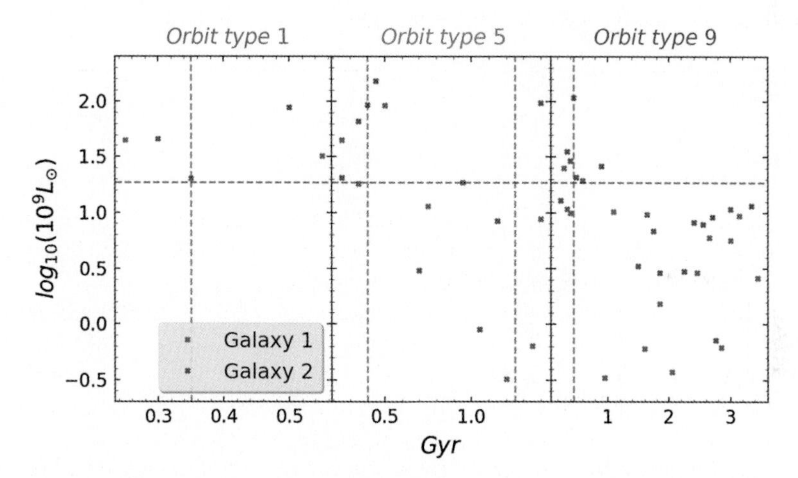

Figure 2. Luminosities of an AGN, considering 100% gas transportation efficiency from $10pc$ down to the SMBH at the centre of the simulated galaxies, compared to the bolometric luminosity of NGC 2992' AGN (horizontal dashed line). Vertical lines mark perigalacticons.

3. Conclusions

Considering the estimated luminosities at 100 Myr after perigalacticon, the simulation with intermediate value of orbital angular momentum ($L = 80.0\,km\,kpc\,s^{-1}$) is the one which better describes the observed NGC 2992 AGN luminosity. However, this would imply a scenario where 20% of the gas at 10 pc is accreted by the black hole. This transport efficiency from 10 pc to the SMBH is slightly larger than expected by current numerical simulations which are able to resolve scales of 0.1 pc (Hopkins & Quataert (2010)). This can probably be attributed both to the lack of AGN feedback and the uncertainty of the parameters used in our model.

References

Chilingarian, I. V., Di Matteo, P., Combes, F., *et al.* 2010, The GalMer database: galaxy mergers in the virtual observatory, *A&A*, 518, A61, doi:10.1051/0004-6361/200912938

Duc, P. A., Brinks, E., Springel, V., *et al.* 2000, Formation of a Tidal Dwarf Galaxy in the Interacting System Arp 245 (NGC 2992/93), *ApJ*, 120, 1238–1264, doi:10.1086/301516

Hopkins, P. F. & Quataert, E. 2010, How do massive black holes get their gas? *MNRAS*, 407, 1529–1564, doi:10.1111/j.1365-2966.2010.17064.x

Schnorr-Müller, A., Davies, R. I., Korista, K. T., *et al.* 2016, Constraints on the broad-line region properties and extinction in local Seyferts, *MNRAS*, 462, 3570–3590, doi:10.1093/mnras/stw1865

Galaxy Evolution and Feedback across Different Environments
Proceedings IAU Symposium No. 359, 2020
T. Storchi-Bergmann, W. Forman, R. Overzier & R. Riffel, eds.
doi:10.1017/S1743921320004263

Nuclear Star Clusters in Coma confirmation of an unusually high nucleation fraction

Emilio Zanatta[1], **Ruben Sanchez-Janssen**[2] **and Ana L. Chies-Santos**[1]

[1]Departamento de Astronomia, Instituto de Física, UFRGS, Porto Alegre, R.S., Brazil
email: emiliojbzanatta@ufrgs.br

[2]STFC UK Astronomy Technology Centre, Royal Observatory, Edinburgh, UK

Abstract. Nuclear star clusters (NSCs) are stellar systems similar in size to globular clusters (GCs) but extremely dense, comparable only to some GCs and ultra-compact dwarfs. They are present in galaxies with a wide range of masses, morphologies and gas content. There are several formation scenarios proposed for the formation of such objects, such as the merger of GCs or extreme star formation caused by the inflow of gas. Recent studies show that the presence of an NSC is related to galaxy stellar mass. Moreover, it has been suggested that NSCs are more often found in high density environments. In our work, we use deep imaging of the core regions of the Coma cluster down to an absolute magnitude of −8.2 and found that in this environment the nucleation fraction is higher than in the Virgo and Fornax clusters. We find nucleated galaxies in Coma as faint as −11.2 mag.

Keywords. galaxies:star clusters, galaxies:evolution, galaxies:nuclei

1. Introduction and Objectives

In the central regions of galaxies with a wide range of masses, luminosities and morphological types, there exists a class of compact stellar systems known as nuclear star clusters (NSCs). Such objects have half-light radii in the range of 1-50 pc and extreme stellar densities Drinkwater *et al.* (2000). The formation of NSCs has been suggested to derive from two non-exclusive scenarios: The dry merging of star clusters in the early stages of galaxy formation, and the inflow of gas to the central region of galaxies (Antonini (2013)). Sanchez-Janssen *et al.* (2019) showed that the fraction of galaxies that harbour NSCs is a strong function of galaxy stellar mass (see left panel of Fig. 1). A secondary dependence on environment has been suggested, since the nucleation fraction in the Coma cluster was higher at least in the range of $log(M/M\odot) = 10^7$ to 10^9. This work investigates the nucleation fraction in Coma with deep Hubble Space Telescope (HST) images, to detect NSCs even in very faint galaxies.

2. HST data and Photometry

The data used in this work were obtained with the WFC/ACS instrument on HST in both the F814W and F475W filters, centered on NGC 4874 and NGC 4889. To detect the faintest galaxies possible, we subtracted the brightest galaxies in the images and performed two SExtractor (Bertin *et al.* 1996) runs: a first one focusing on extracting point-sources and a second one on the *background* image generated from the first run.

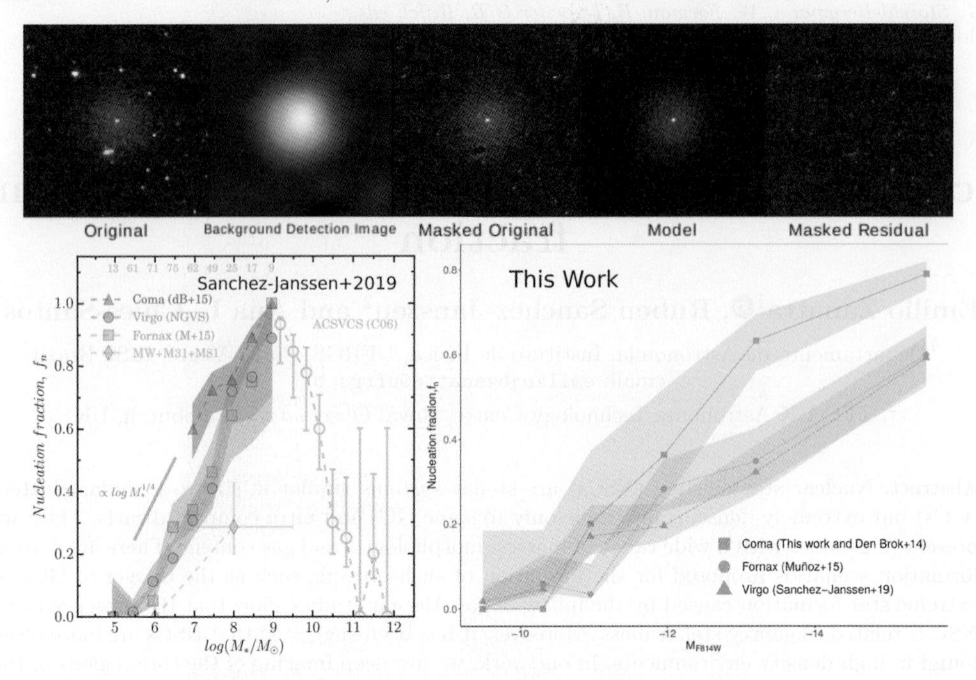

Figure 1. *Top:* Example of the detection and photometry method used in this work. *Bottom left:* Nucleation fraction for Virgo, Coma and Fornax clusters from Sanchez-Janssen *et al.* (2019). *Bottom right:* Results of this work, with the complete nucleation fraction in Coma down to the fainter galaxies, compared to Virgo and Fornax.

Then, we proceed to visual inspection. We were left with 57 galaxies that were modelled with GALFIT (Peng *et al.* (2002)). An illustration of the method is shown in the top panel of Fig. 1.

3. Results

We joined our data with those from Den Brok *et al.* (2014) and compared the nucleation fraction for galaxies in Coma with Virgo and Fornax (bottom right panel of Fig. 1). We found that the nucleation fraction in Coma is higher for galaxies with $M_{F814W} > -11$ mag. For fainter galaxies, the nucleation fraction between the clusters is equal within uncertainties. We consider this as a confirmation of an environmental influence on the formation and evolution of NSCs that might also be related to the host galaxy mass, such that the least massive galaxies are not affected. Further work is needed to better constrain the origins of this result.

References

Antonini, F. 2013, *ApJ*, 763, 62
Bertin, E. & Arnouts, S. 1996, *AAS*, 117, 393
den Brok, M., Peletier, R. F., Seth, A., *et al.* 2014, *MNRAS*, 445, 2385
Drinkwater, M. J., Jones, J. B., Gregg, M. D., *et al.* 2000, *PASA*, 17, 227
Peng, C. Y., Ho, L. C., Impey, C. D., *et al.* 2002, *AJ*, 124, 266
Sánchez-Janssen, R., Coté, P., Ferrarese, L., *et al.* 2019, *ApJ*, 878, 18

Key Ito

Minju Lee

Damien Spérone-Longin

Christine Jones

Fernanda Roman de Oliveira

Henrique Schmitt

William Forman

Session 4: Secular evolution and internal processes: Mass quenching, stellar and AGN feedback over different z's

Galaxy Evolution and Feedback across Different Environments
Proceedings IAU Symposium No. 359, 2020
T. Storchi-Bergmann, W. Forman, R. Overzier & R. Riffel, eds.
doi:10.1017/S1743921320001696

INVITED LECTURES

Establishing the impact of powerful AGN on their host galaxies

C. M. Harrison[1], S. J. Molyneux[2], J. Scholtz[3] and M. E. Jarvis[4,5,6]

[1]School of Mathematics, Statistics and Physics, Newcastle University, Newcastle Upon Tyne, NE1 7RU, United Kindgom
email: christopher.harrison@newcastle.ac.uk

[2]Astrophysics Research Institute, Liverpool John Moores University, 146 Brownlow Hill, Liverpool L3 5RF, UK

[3]Department of Space, Earth and Environment, Chalmers University of Technology, Onsala Space Observatory, SE-43992 Onsala, Sweden

[4]Max-Planck Institut für Astrophysik, Karl-Schwarzschild-Str. 1, 85748 Garching, Germany

[5]European Southern Observatory, Karl-Schwarzschild-Str. 2, 85748 Garching, Germany

[6]Ludwig Maximilian Universität, Professor-Huber-Platz 2, 80539 Munich, Germany

Abstract. Establishing the role of active galactic nuclei (AGN) during the formation of galaxies remains one of the greatest challenges of galaxy formation theory. Towards addressing this, we summarise our recent work investigating: (1) the physical drivers of ionised outflows and (2) observational signatures of the impact by jets/outflows on star formation and molecular gas content in AGN host galaxies. We confirm a connection between radio emission and extreme ionised gas kinematics in AGN hosts. Emission-line selected AGN are significantly more likely to exhibit ionised outflows (as traced by the [O III] emission line) if the projected linear extent of the radio emission is confined within the spectroscopic aperture. Follow-up high resolution radio observations and integral field spectroscopy of 10 luminous Type 2 AGN reveal moderate power, young (or frustrated) jets interacting with the interstellar medium. We find that these sources live in highly star forming and gas rich galaxies. Additionally, by combining ALMA-derived dust maps with integral field spectroscopy for eight host galaxies of $z\approx2$ X-ray AGN, we show that Hα emission is an unreliable tracer of star formation. For the five targets with ionised outflows we find no dramatic in-situ shut down of the star formation. Across both of these studies we find that if these AGN do have a negative impact upon their host galaxies, it must be happening on small (unresolved) spatial scales and/or an observable galaxy-wide impact has yet to occur.

Keywords. galaxies: jets, ISM: jets and outflows, galaxies: active, galaxies: evolution

---

1. Introduction

Supermassive black holes (with masses $\gtrsim 10^6$ M$_\odot$) reside at the centre of galaxies and they release extraordinary amounts of energy when they grow through mass accretion events. During these growth periods they are identified as active galactic nuclei (AGN) using observations across the electromagnetic spectrum (see review in Alexander & Hickox 2012). To reproduce realistic galaxy populations, successful cosmological models of galaxy evolution require that the energy released from AGN heats and/or removes the gas in the host galaxy and/or surrounding halos in the process called "AGN feedback" (e.g., Bower *et al.* 2006; Schaye *et al.* 2015; Pillepich *et al.* 2019).

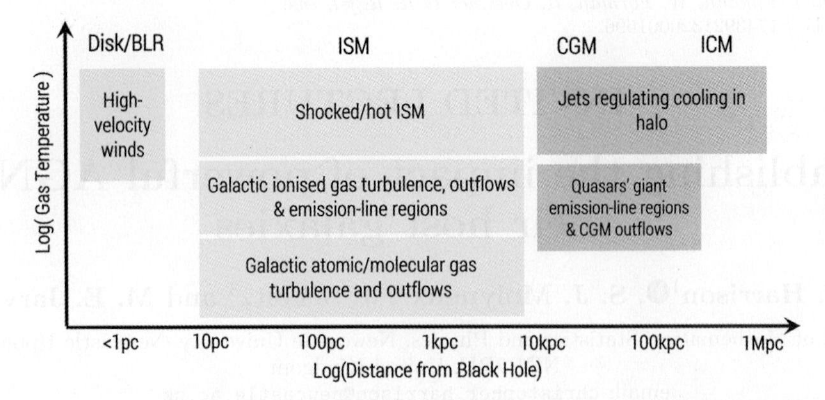

Figure 1. A schematic diagram to summarise the broad observations, that cover a wide range of gas temperatures, that have shown that AGN can inject energy into the gas in and around their host galaxies over a huge range of spatial scales.

From an observational perspective, AGN have been seen to interact with gas over a wide range of spatial scales and with gas at a large range of temperatures (see schematic in Fig. 1). For example: spectroscopic observations of extremely high velocity winds (at a significant fraction of the speed of light) in the regions around the accretion disk or broad line region (BLR; e.g., King & Pounds 2015); in the interstellar medium (ISM) AGN-driven outflows have been observed across a huge range of temperatures from cold molecular gas to hot X-ray gas (e.g., Cicone *et al.* 2018); in the circumgalactic medium (CGM) outflows are also observed and powerful AGN (i.e., quasars) are able to ionise gas out to ≈100 kpc (e.g., Arrigoni Battaia *et al.* 2019) and; on the scales of the intracluster medium (ICM) powerful radio jets are seen to be regulating gas cooling, at least in the densest environments (e.g., McNamara & Nulsen 2012). Despite these efforts there remains little consensus on many key aspects of the physical processes behind, and impact of, AGN feedback. In this article we focus on two key questions: (1) what physical process is most important for AGN to inject energy into the ISM (e.g., radiation, collimated jets of particles, or wide-angle disk winds; see Wylezalek & Morganti 2018) and; (2) is there observational evidence that AGN directly impact upon the molecular gas or star formation in their host galaxies (e.g., Harrison 2017; Cresci & Maiolino 2018)?

2. The radio–ionised outflow connection

Previous studies have shown a link between the prevalence and/or velocities of ionised outflows (predominantly traced using the [O III] emission line) and the radio luminosity of AGN (e.g., Mullaney *et al.* 2013; Villar Martín *et al.* 2014; Zakamska & Greene 2014). As the basis for our work, Mullaney *et al.* (2013) used a sample of 24264 optical emission line selected $z < 0.4$ AGN from the Sloan Digital Sky Survey (SDSS). They used a combination of stacking analyses and multi-component fits to the emission-line profiles and found that there is a higher prevalence of the most powerful outflows when the radio luminosity is higher. Specifically, AGN of moderate radio luminosity ($L_{1.4\mathrm{GHz}} = 10^{23}$–$10^{25}$ W Hz^{-1}) were found to have the broadest [O III] profiles and AGN with $L_{1.4\mathrm{GHz}} > 10^{23}$ W Hz^{-1} are ≈5 times more likely to have extremely broad [O III] lines (i.e., FWHM$_{\mathrm{Avg}} > 1000$ kms^{-1})† compared to lower radio luminous AGN. However, some other similar statistical works have claimed little-to-no correspondence between the radio emission and ionised outflow properties in low redshift AGN (e.g., Woo *et al.* 2016; Kauffmann & Maraston 2019).

† FWHM$_{\mathrm{Avg}}$ is the flux-weighted average full-width-half maximum of the two individual Gaussian components that were fitted to the [O III] emission-line profiles.

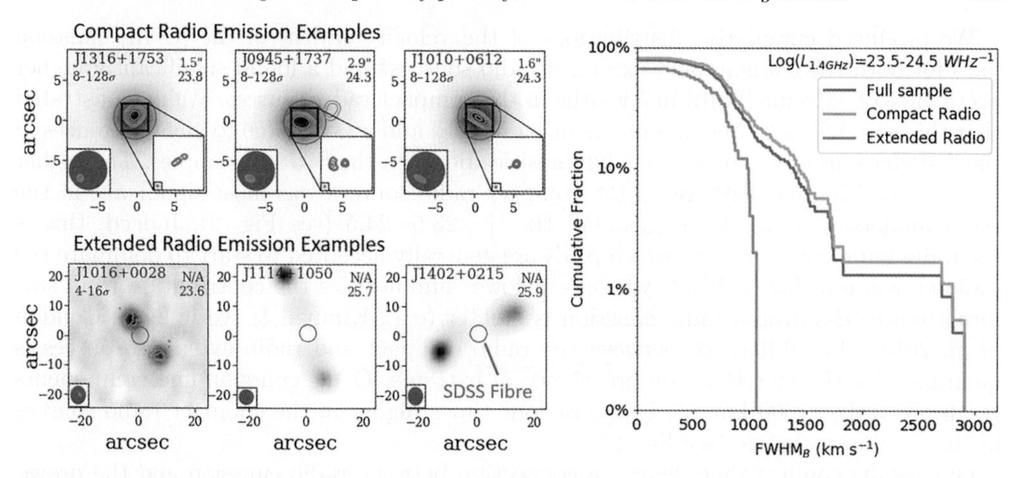

Figure 2. *Left:* Example 1.4 GHz images ($\approx$5 arcsec resolution) and, where available, overlaid green contours from $\approx$1 arcsec resolution 1.4 GHz images. The insets show $\approx$0.3 arcsec resolution 6 GHz images (with blue contours). Synthesised beams are represented by appropriately coloured ellipses. Compact radio sources (top row) have their radio emission dominated within the SDSS fibre extent (magenta circles), whilst extended sources (bottom row) show significant emission outside of this. *Right:* Cumulative fraction of AGN with [O III] emission-line components with FWHM$_B$ greater than a given value, for: all sources (blue curve), compact radio sources (orange curve) and extended radio sources (green curve). Extreme ionised gas velocities are more prevalent for compact sources. Figure adapted from Molyneux *et al.* (2019).

In the following subsections, we describe how we investigated the radio–outflow connection in greater detail by including spatial information.

2.1. *Extreme ionised outflows are more prevalent when the radio emission is compact*

In Molyneux *et al.* (2019) we conducted a study into the relationship between the size of the radio emission (i.e., projected linear spatial extent) and the prevalence of extreme ionised outflows, as traced by [O III] emission-lines from SDSS spectra. We identified the $\approx$3000 $z < 0.2$ AGN from Mullaney *et al.* (2013), introduced above, with a radio detection in all-sky 1.4 GHz radio surveys (for more details see Molyneux *et al.* 2019).

To characterise the extent of the radio emission we combined two different approaches. We used: (1) sizes from simple Gaussian models and; (2) an automated morphological classification scheme (THE FIRST CLASSIFIER; Alhassan *et al.* 2018). It was necessary to combine both of these approaches because whilst the former method has the advantage of providing a quantitative measure of the radio sizes, and corresponding uncertainties, it has the disadvantage of missing structures that are not well characterised by a single elliptical Gaussian model (such as spatially separated radio lobes). Using these two measures we defined *compact radio* sources as those with radio emission concentrated within 3 arcsec or they were classified as *extended radio* sources, otherwise (see Molyneux *et al.* 2019 for full details). This size is equivalent to $\lesssim$5 kpc, at the median redshift of the sample, and is therefore of the order of the galaxy sizes. Importantly, this cut-off size corresponds to the aperture size of the SDSS fibres, and allows for the most direct comparison to be made between the radio emission and the emission-line properties observed in the SDSS spectra (see Fig. 2). Reassuringly, using our follow-up radio data, with $\approx$0.3$-$1 arcsec spatial resolution, of a subset of targets, from the Very Large Array, we were able to confirm our size classification method was successful and identified jet-like structures within $\approx$5 kpc for the compact radio sources (see Fig. 2).

We produced cumulative distributions of the velocity widths of the [O III] emission lines for both the compact and extended radio sources† and a found significantly higher prevalence of extreme [O III] line widths in the compact radio sources. We demonstrated that this was driven by the differences in radio sizes and was not due to any differences in the distributions of radio or [O III] luminosities between the two sub-samples. The higher prevalence of ionised outflows in the compact radio sources was most significant in the radio luminosity range $\log[L_{1.4\mathrm{GHz}}/\mathrm{W\ Hz}^{-1}] = 23.5-24.5$ (see Fig. 2). Indeed, this is the radio luminosity range in which AGN are generally accepted to start to dominate the radio emission at low redshift, whereas at lower luminosities the contribution from star formation to the overall radio emission is greater (e.g., Kimball & Ivezić 2008; Condon *et al.* 2013). The difference between the radio compact and radio extended sources is quantified by the fact that the prevalence of extreme [O III] emission-line components (with $\mathrm{FWHM}_B > 1000\ \mathrm{kms}^{-1}$) is almost four times higher in the compact radio sources in this luminosity range (see Fig. 2).

Our results confirm that there is a connection between radio emission and the prevalence of extreme ionised gas kinematics (outflows) in AGN host galaxies. Importantly, we find that this difference is most extreme in the luminosity range $\log[L_{1.4\mathrm{GHz}}/\mathrm{W\ Hz}^{-1}] = 23.5-24.5$ and we suggest that statistical samples which are not complete in this luminosity range may not find such a strong outflow–radio connection (see discussion in Molyneux *et al.* 2019). One explanation for this result is that small scale (young or frustrated) radio jets are strongly interacting with the interstellar medium (also see Holt *et al.* 2008), whilst larger scale radio jets are depositing their energy on larger scales (at least outside of the area covered by the spectroscopic fibre). Indeed, low-to-moderate radio sources with bright optical emission (i.e., radiatively efficient AGN), may represent a key galaxy evolution phase (e.g., Pierce *et al.* 2020). To investigate this further requires a combination of radio imaging and spatially-resolved spectroscopy.

2.2. *Prevalent jet-ISM interactions in radiatively efficient AGN host galaxies*

For a more direct investigation into the relationship between ionised outflows and radio emission, in Jarvis *et al.* (2019), we combined ≈0.3–1 arcsec resolution radio observations with integral field spectroscopy on a sample of 10 Type 2 AGN, selected from the sample in Molyneux *et al.* (2019) described above. These sources are in the key radio luminosity range $\log[L_{1.4\mathrm{GHz}}/\mathrm{W\ Hz}^{-1}] = 23.5-24.5$, are radiatively efficient (with quasar-level bolometric luminosities) and have signatures of outflows in their SDSS spectra (i.e., [O III] $\mathrm{FWHM}_B > 700\,\mathrm{km\,s}^{-1}$). Data for three example targets are shown in Figure 3. In 9/10 of the targets, we confirmed that $\gtrsim$90% of the radio emission could not be attributed to star formation (see Jarvis *et al.* 2019 for details). Furthermore, we found that 80–90% of these targets show spatially-extended radio structures on 1–25 kpc scales and these structures are aligned with the gas distribution and/or gas kinematics (e.g., Fig. 3).

In Figure 3 we give specific examples of the data. The target J1316+1753 exhibits a strong velocity gradient in [O III] emission, where the blue and red shifted gas is brightest at the termination of each jet. Similarly J0958+1439 shows a kinematic splitting and co-spatial jets/lobes. The potential jet we observe in J0945+1737, terminates at an [O III] brightened and blue-shifted gas cloud (indicated by the grey box in the figure). This is evidence of jets hitting a cloud of gas, both pushing the gas away and deflecting the jet. Such observations are signatures of jet-driven outflows (e.g., Rosario *et al.* 2010).

† We found the same results using both the FWHM of secondary broad Gaussian components (FWHM_B) and the flux-weighted average line widths of two components ($\mathrm{FWHM}_{\mathrm{Avg}}$)

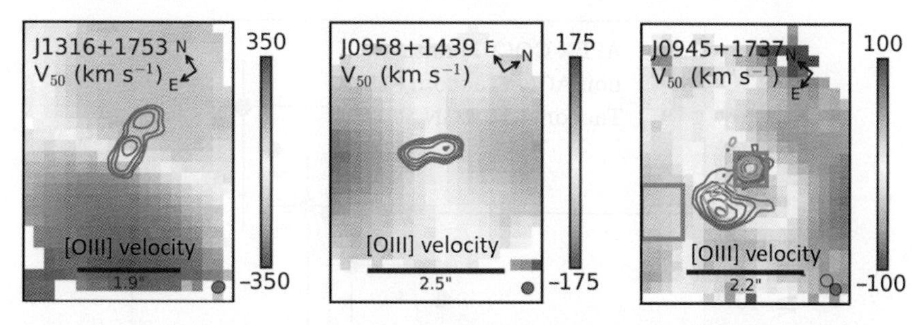

Figure 3. Example radio data and ionised gas velocity maps from integral field spectroscopy for $z \sim 0.1$ Type 2 quasars, adapted from Jarvis *et al.* (2019). In each case, the radio contours are overlaid (blue and orange) and the corresponding beam(s) are shown in the lower right corner. The scale bar in each image represents 5 kpc. The ionised gas shows distinct kinematics at the location of the spatially-extended jet/lobe structures indicating jet-ISM interactions.

Overall, our results are consistent with jet-ISM interactions as the cause of the radio–outflow connection seen in statistical samples (e.g., Mullaney *et al.* 2013; Molyneux *et al.* 2019; Fig. 2). The combined observations suggest that compact, low-power radio jets, young or frustrated by interactions with the host galaxy ISM, may be responsible for the high-velocity ionised gas, in line with some recent model predictions (e.g., Mukherjee *et al.* 2018a). However, we can not rule out other possible processes, such as nuclear wide-angle winds, that may contribute to producing the radio emission and outflows in the wider sample (e.g., see Zakamska *et al.* 2016), or in the very nuclear regions of the targets we investigated (see discussion in Jarvis *et al.* 2019). A spatially-resolved investigation of the radio–outflow connection, with larger samples, is required to determine how prevalent jet-ISM interactions are across the entire AGN population.

3. Investigating the impact of AGN on their host galaxies

The simple *identification* of radio jets and AGN-driven ionised outflows does not immediately establish that they have any appreciable *impact* on the evolution of their host galaxies (see e.g., Harrison 2017). Towards understanding the possible impact of jets and outflows, we investigated the molecular gas content of the $z \approx 0.1$ Type 2 AGN from Jarvis *et al.* (2019) (summarised in Section 3.1) and the spatial correlation between outflows and star formation for $z \approx 2$ X-ray identified AGN (summarised in Section 3.2).

3.1. *Outflows and jets associated with high molecular gas fractions at* $z \approx 0.1$

For the 9 out of the 10 $z \approx 0.1$ AGN in Jarvis *et al.* (2019) that showed AGN-dominated radio emission (see Section 2), we used APEX to observe the CO(2–1) emission line (as a tracer of the global molecular gas content; Jarvis *et al.* submitted). We found that these targets have CO luminosities consistent with the overall galaxy population for their observed infrared luminosities and once you account for their high specific star formation rates ($\mathrm{SFR}/M_\star = 0.1$–$5.9\,\mathrm{Gyr}^{-1}$) and corresponding position relative to the 'main sequence' of star forming galaxies (Δ_{MS}; e.g., Sargent *et al.* 2014).† Converting these CO luminosities to molecular gas masses implies that the host galaxies have high molecular gas fractions ($M_{gas}/M_\star = 0.1$–1.2; Fig. 4; Jarvis *et al.* submitted). These gas fractions are consistent with, or moderately higher than, the non-active galaxies with the same specific star formation rates and position relative to the star-forming main sequence (and after carefully applying consistent assumptions) from Tacconi *et al.* (2018) (Fig. 4).

† We note that the infrared luminosities we use have been corrected for contamination from the AGN.

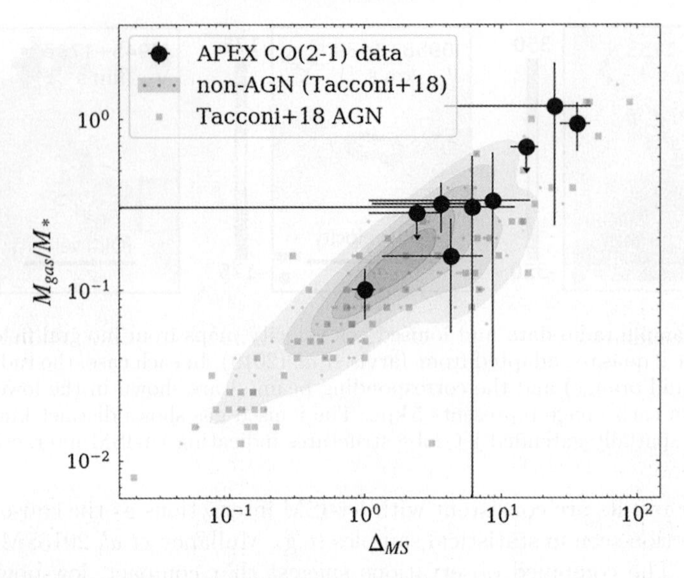

Figure 4. Molecular gas fraction as a function of distance to the main sequence, for our sample (black circles) compared to a redshift-matched comparison sample of galaxies from Tacconi *et al.* (2018) (small green circles and contours, with AGN host galaxies marked with pink squares). Our powerful AGN, containing both outflows and jets, follow the overall trends seen in the comparison sample, with moderately higher (~0.1 dex) average gas fractions compared to the general star-forming population. Figure adapted from Jarvis *et al.* (submitted).

Our sources may represent a key phase in galaxy evolution where there are rapid levels of star formation and black hole growth, but the feedback processes we observe (i.e., ionised outflows and associated jets) have not yet been able to have an observable impact upon the global gas content or star formation in their host galaxies (full discussion in Jarvis *et al.* submitted).

3.2. *No dramatic in-situ impact on star formation by ionised outflows at $z\sim 2$*

One approach to determine the impact of AGN feedback on star formation, is to use spatially-resolved observations to map both the AGN driven outflows and the star formation in or around the outflows. For example, Cresci *et al.* (2015) suggest both *suppressed* star formation at the location of an ionised outflow and *enhanced* star formation around the edges of the outflow for a $z = 1.6$ X-ray identified AGN. Similar findings were presented for three $z = 2.5$ extremely powerful (and consequently rare) quasars by Cano-Diaz *et al.* (2012) and Carniani *et al.* (2016). These works used high-velocity [O III] emission-line components to map the ionised outflows and narrow (i.e., not associated with the BLR) Hα emission to map the spatial distribution of the star formation.

However, these studies have only mapped the *un-obscured* star formation in these AGN host galaxies which is sensitive to potentially missing any dust-obscured star formation. Furthermore, these studies mostly focus on the extremely luminous systems with the most extreme ionised gas outflows. In our pilot study, presented in Scholtz *et al.* (2020), we combined integral field spectroscopy (to trace the emissions lines) with ALMA sub-mm interferometry (to trace the dust). In this study we focus on more moderate luminosity, representative, AGN without any pre-selection on the presence of an outflow.

For Scholtz *et al.* (2020) we selected z = 1.5−2.6 AGN from the KASHz (KMOS AGN at High-z) survey Harrison *et al.* (2016), which is an integral field spectroscopic survey of X-ray AGN. The targets were selected to be detected in both Hα and [O III] emission

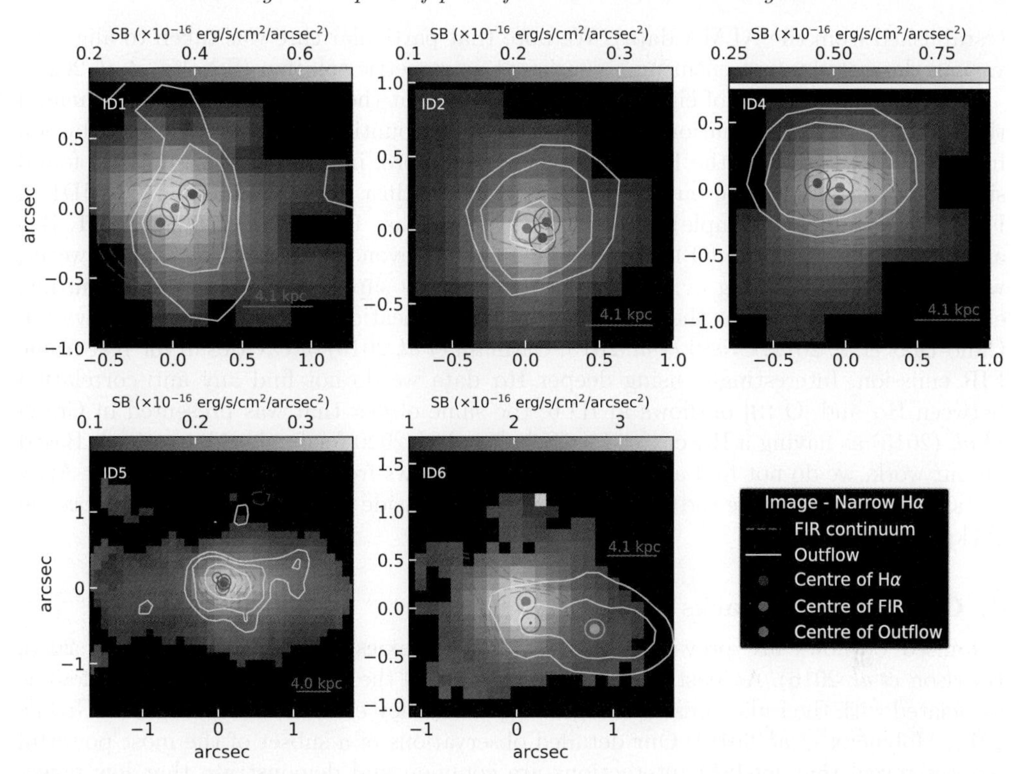

Figure 5. For the five X-ray AGN, where we identified outflows, we show maps of surface brightness of the narrow line Hα, with the red-dashed contours showing the distribution of FIR emission (from Scholtz *et al.* 2020). White contours show the distribution of ionised outflows (3,4,5 σ levels), as defined by the high-velocity wings of the [O III] emission lines. The blue, red and green points show the peak of the Hα, FIR and the outflow, respectively. We do not see any significant suppression of the Hα emission or the dust emission at the location of the outflows. Figure taken from Scholtz *et al.* (2020).

lines and an archival detection in ALMA Band 6 or 7 continuum (corresponding to the rest-frame far-infrared [FIR] emission; $\approx$300$-$400μm; see Scholtz *et al.* 2020). Our final sample of eight targets has X-ray luminosities of 10^{43}-$10^{45.5}$ erg s^{-1} and [O III] emission-line widths of $W_{80} = 300$–850 kms^{-1}, representative of the parent X-ray AGN population at this redshift (Scholtz *et al.* 2020).† We established that the rest-frame FIR (ALMA) emission in our targets was tracing star formation heated dust using spectral energy distribution decomposition (i.e., by assessing the possible contribution from AGN emission; Scholtz *et al.* 2020).

We fitted the datacubes containing Hα emission with spectral line models for narrow Hα, Hα originating from the BLR and for the [N II] doublet. This way, we were able to construct a map of the narrow Hα emission, without contamination. Furthermore, we detected [O III] emission-line outflows (i.e., asymmetric emission-line profiles) in five of the eight targets (which is consistent with the $\approx$60% outflow detection fraction in this luminosity range; Harrison *et al.* 2016). For these five targets 5 we display the resulting maps of the narrow Hα and the spatial distributions of the ionised outflows (defined as the blue wings of the [O III] emission line; see Scholtz *et al.* 2020) in Figure 5. The red contours in Figure 5 represent the spatial distribution of the dust continuum emission

† W_{80} is the velocity width containing 80% of the total emission-line flux.

(established from the ALMA data). We note that particular care was taken to align the various datasets by first ensuring a consistent astrometric solution (Scholtz *et al.* 2020).

Across the full sample of eight targets we found that the Hα emission underestimated the total global star formation rates, even before accounting for the strong contribution from AGN ionisation to the Hα emission. Furthermore, in half of the sample we found significant offsets (1.4 kpc on average) between the dust and Hα emission (see ID1 in Figure 5 for a clear example). In Figure 5 we find that three of the targets (ID1, ID5 and ID6) show significant [O III] outflows elongated beyond the central regions. However, we do not see any strong evidence that the outflows suppress the star formation; i.e., either through cavities in the Hα emission at the location of the ionised outflows (cf. Cano-Diaz *et al.* 2012; Cresci *et al.* 2015, Carniani *et al.* 2016) or cavities in the rest-frame FIR emission. Interestingly, using deeper Hα data we do not find any anti-correlation between Hα and [O III] outflows in ID 6, the same object that was presented in Cresci *et al.* (2015) as having a Hα cavity (see Scholtz *et al.* 2020 for further discussion). Based on our work, we do not find any evidence that outflows from moderate luminosity AGN instantaneously influence the in-situ star formation inside their host galaxies, at least on $\gtrsim$4 kpc scales.

4. Concluding remarks

Ionised outflows are prevalent in AGN host galaxies (e.g., Mullaney *et al.* 2013; Harrison *et al.* 2016). At least at low redshift ($z < 0.2$) the prevalence of these outflows is associated with the radio emission (Fig. 2; e.g., Mullaney *et al.* 2013; Zakamska & Greene 2014; Molyneux *et al.* 2019). Our detailed observations of a subset of the most powerful sources reveal that jet-ISM interactions are common and demonstrate that low power radio jets are an important feedback mechanism, even in radiatively efficient AGN (e.g., quasars; Fig. 3; Jarvis *et al.* 2019). However, the relative role of jets over other processes, such as wind-angle winds, is still to be determined. We find the AGN-driven outflows and jets do not have a dramatic galaxy-wide impact on the the host galaxy molecular gas content (at least for our $z \approx 0.1$ sample; Fig. 4; Jarvis *et al.* submitted) or in-situ star formation (at least for our $z \approx 2$ sample; Fig. 5; Scholtz *et al.* 2020). However, some impact could be occurring on spatial scales below those to which we are sensitive. Alternatively, the cumulative impact of AGN outflows or jets may regulate future galaxy properties over longer timescales even if they have little impact on the *in-situ* star formation (e.g., see discussion in Harrison 2017; Scholtz *et al.* 2018, 2020).

References

Alexander, D. M. & Hickox, R. C. 2012, *New Astron. Reviews*, 56, 93
Alhassan, W., Taylor, A. R., Vaccari, M., *et al.* 2018, *MNRAS*, 480, 2085
Arrigoni Battaia, F., Obreja, A., Prochaska, J. X., *et al.* 2019, *A&A*, 631, 18
Bower, R. G., Benson, A. J., Malbon, R., *et al.* 2006, *MNRAS*, 370, 645
Carniani, S., Marconi, A., Maiolino, R., *et al.* 2016, *A&A*, 591, A28
Cano-Díaz, M., Maiolino, R., Marconi, A., *et al.* 2012, *A&A*, 537, L8
Cicone, C., Brusa, M., Ramos Almeida, C., *et al.* 2018, *Nature Astronomy*, 2, 176
Condon, J. J., Kellermann, K. I., Kimball, A. E., *et al.* 2013, *ApJ*, 768, 37
Cresci, G., Mainieri, V., Brusa, M., *et al.* 2015, *ApJ*, 799, 1
Cresci, G. & Maiolino, R. 2018, *Nature Astronomy*, 2, 179
Harrison, C. M., Alexander, D. M., Mullaney, J. R., *et al.* 2016, *MNRAS*, 1, 165
Harrison, C. M. 2017, *Nature Astronomy*, 1, 165
Holt, J., Tadhunter, C. N., & Morganti, R. 2008, *MNRAS*, 387, 639
Jarvis, M. E., Harrison, C. M., Thomson, A. P. *et al.* 2019, *MNRAS*, 485, 2710
Kauffmann, G. & Maraston, C. 2019, *MNRAS*, 489, 1973

Kimball, A. E. & Ivezić, Ž. 2008, *AJ*, 136, 684
King, A. & Pounds, K. 2015, *ARA&A*, 53, 115
McNamara, B. R. & Nulsen, P. E. J. 2012, *New Journal of Physics*, 14, 055023
Molyneux, S. J., Harrison, C. M., Jarvis M. E., *et al.* 2019, *A&A*, 631, A132
Mukherjee, D., Bicknell, G. V., Wagner, A. Y., *et al.* 2018, *MNRAS*, 479, 5544
Mullaney, J. R., Alexander, D. M., Fine, S., *et al.* 2013, *MNRAS*, 433, 622
Pierce, J. C. S., Tadhunter, C. N., Morganti, R., *et al.* 2020, *MNRAS*, 492 2053
Pillepich, A., Nelson, D., Springer, V., *et al.* 2019, *MNRAS*, 490, 3196
Rosario, D. J., Shields, G. A., Taylor, G. B., *et al.* 2010, *ApJ*, 716, 131
Sargent, M. T., Daddi, E., Béthermin, M., *et al.* 2014, *ApJ*, 793, 19
Schaye, J., Crain, R., Bower, R. G., *et al.* 2015, *MNRAS*, 446, 521
Scholtz, J., Alexander, D. M., Harrison, C. M., *et al.* 2018, *MNRAS*, 475, 1288
Scholtz, J., Harrison, C. M., Rosario, D. J., *et al.* 2020, *MNRAS*, 492, 3194
Tacconi, L. J., Genzel, R., Saintonge, A., *et al.* 2018, *ApJ*, 853, 179
Villar Martín, M., Emonts, B., Humphrey, A., *et al.* 2014, *MNRAS*, 440, 3202
Woo, J.-H., Bae, H.-J., Son, D., & Karouzos, M. 2016, *ApJ*, 817, 108
Wylezalek, D. & Morganti, R. 2018, *Nature Astronomy*, 2, 181
Zakamska, N. L. & Greene, J. E. 2014, *MNRAS*, 442, 784
Zakamska, N. L., Lampayan, K., Petric, A., *et al.* 2016, *MNRAS*, 455, 4191

Galaxy Evolution and Feedback across Different Environments
Proceedings IAU Symposium No. 359, 2020
T. Storchi-Bergmann, W. Forman, R. Overzier & R. Riffel, eds.
doi:10.1017/S1743921320002203

The physical properties and impact of AGN outflows from high to low redshift†

Giacomo Venturi[1,2] **and Alessandro Marconi**[3,2]

[1]Instituto de Astrofísica, Pontificia Universidad Católica de Chile,
Avda. Vicuña Mackenna 4860, 8970117, Macul, Santiago, Chile
email: gventuri@astro.puc.cl

[2]INAF - Osservatorio Astrofisico di Arcetri, Largo E. Fermi 5, I-50125, Firenze, Italy

[3]Dipartimento di Fisica e Astronomia, Università degli Studi di Firenze,
Via G. Sansone 1, I-50019, Sesto Fiorentino, Firenze, Italy

Abstract. Feedback from active galactic nuclei (AGN) on their host galaxies, in the form of gas outflows capable of quenching star formation, is considered a major player in galaxy evolution. However, clear observational evidence of such major impact is still missing; uncertainties in measuring outflow properties might be partly responsible because of their critical role in comparisons with models and in constraining the impact of outflows on galaxies. Here we briefly review the challenges in measuring outflow physical properties and present an overview of outflow studies from high to low redshift. Finally, we present highlights from our MAGNUM survey of nearby AGN with VLT/MUSE, where the high intrinsic spatial resolution (down to ∼10 pc) allows us to accurately measure the physical and kinematic properties of ionised gas outflows.

Keywords. galaxies: evolution, galaxies: Seyfert, galaxies: individual (NGC 1365, Circinus), galaxies: ISM, galaxies: kinematics and dynamics, techniques: spectroscopic

---

1. Introduction

Feedback from active galactic nuclei (AGN) on their host galaxies is considered a critical element to explain many key observed properties of galaxies during their evolution over cosmic time. These are: 1) the observed discrepancy in the luminosity function of galaxies at high masses between observations and predictions from models without AGN feedback (e.g. Kormendy & Ho 2013 and references therein); 2) the scaling relations observed between the mass of supermassive black holes (BHs) and the properties of their host galaxies (e.g. Behroozi *et al.* 2019); 3) the similarity between the BH accretion history and the star formation (SF) history throughout cosmic time (e.g. Aird *et al.* 2015); 4) the bimodality of galaxies, divided in two distinct populations in the colour versus stellar mass diagram, i.e. the so-called "blue cloud" and the "red sequence" (e.g. Schawinski *et al.* 2014). Feedback from AGN is thus routinely included in models and simulations of galaxy formation and evolution, as it is able to explain the above properties (e.g. Ciotti *et al.* 2010; Schaye *et al.* 2015). AGN feedback is commonly divided in two different modes (see e.g. Fabian 2012 and references therein): a "radiative" (or "quasar") mode, operating during a luminous AGN phase through powerful outflows which sweep away the gas reservoir from the host galaxy, thus quenching SF; a "kinetic" (or "radio") mode, acting steadily on longer timescales through radio jets which heat the gas halo surrounding massive galaxies, thereby preventing its cooling and reaccretion on the host

† Based on observations made with ESO Telescopes at the La Silla Paranal Observatory under program ID 094.B-0321(A).

and consequently further SF. While clear evidence for the kinetic mode has been found in massive central cluster galaxies, where cavities in the X-ray emitting hot ionised gas, filled by radio jets propagating from the AGN, are observed (e.g. McNamara *et al.* 2000; Bîrzan *et al.* 2012), the radiative mode is more elusive and convincing evidence for the impact of outflows on host galaxies is still missing.

Here we provide a brief review on AGN outflows and their role in the context of AGN feedback from an observational point of view, focusing on outflows on galactic scales, where their effects on host galaxies are expected to be observable.

2. Measuring outflow physical properties

Measuring the physical properties of outflows with high accuracy and well established methods is of primary importance for constraining their impact on host galaxies. Different assumptions in calculating outflow properties lead in fact to a large spread in their inferred kinematics and energetics. The spread remains quite large even when using the same method, since some outflow properties (such as density, extension, inclination) are often unknown and must be assumed (e.g. Harrison *et al.* 2018). Such uncertainties and different adopted methods make then difficult the comparison between different works and with the predictions from models and simulations of outflows and feedback.

The mass outflow rate through a spherical (or conical or multi-conical) surface can be calculated from the fluid continuity equation, as:

$$\dot{M} = \Omega r^2 \rho v \simeq f \frac{Mv}{r}, \tag{2.1}$$

where Ω is the angle subtended by the outflow as seen by the AGN, r its radius, ρ the outflowing gas density, v its velocity, M its mass and f a factor depending on the outflow geometry, equal to 3 in the case of a volume filled with outflowing clouds or 1 for a simple shell geometry (see e.g. Maiolino *et al.* 2012). Velocity, mass and size of the outflow are thus what needs to be measured to determine the mass outflow rate.

Velocities can be obtained through a multi-Gaussian fitting of the emission-line profiles to isolate the outflowing component. However, in some cases this approach may fail in isolating outflows; parametric velocities from the total line profile can be used such as percentile velocities (e.g. v_{10}, the velocity containing 10% of the total emission line flux) and the corresponding velocity widths (W80 = $v_{90}-v_{10}$; e.g. Harrison *et al.* 2014). Such measurements are based on the observed line-of-sight velocity distribution of the gas and thus, in order to try accounting for projection effects, the far wing of the line profiles (e.g. v_{05}) or a combination of this with the line width are often used. However, to properly recover the intrinsic velocities across the outflow, which also likely vary with distance and angle, a full kinematic modelling would be required (e.g. Crenshaw *et al.* 2015).

Measuring the outflow mass firstly deals with the challenge of isolating the fraction of spectral line profiles associated with the outflow, which may not be a trivial task, being affected by degeneracies. The mass, in the case of molecular outflows, is mainly obtained from CO millimetre/sub-millimetre emission lines as $M_{\rm mol} = \alpha_{\rm CO} L_{\rm CO(1-0)}$, where the CO-to-$H_2$ conversion factor $\alpha_{\rm CO}$, usually assumed, is uncertain [$\sim$0.8–4 $M_\odot$ (K km s^{-1} $pc^2)^{-1}$, based on values for local spiral galaxies or mergers; e.g. Bolatto *et al.* 2013]; in the case of higher J rotational transitions, uncertainties up to a factor of $\sim$3 are introduced by the conversion of their luminosity to $L_{\rm CO(1-0)}$ (e.g. Brusa *et al.* 2018). H_2 near-IR emission lines only trace a minor warm transitory phase and are not representative of the total molecular mass budget (e.g. Tadhunter *et al.* 2014). The warm ionised outflow mass is usually obtained from the luminosity of optical emission lines, such as [O III]λ5007 or Balmer lines, and depends on electron temperature, element abundance and ionisation state (except for Balmer lines) and, above all, electron density (e.g. Carniani *et al.* 2015), the latter constituting the largest source of uncertainty in the ionised mass determination (e.g. Harrison *et al.* 2018).

Finally, the ability to measure the size of the outflow is strictly dependent on the spatial resolution of observations: in many cases outflows are often only barely resolved or completely unresolved and consequently the size must be assumed.

3. Outflows from high to low redshift

Pieces of evidence for the presence of outflows in AGN are found all the way from high to low z. Here we highlight some significant examples from the many studies available in the literature.

High redshift. At $z \sim 6$ very few indications of outflows exist so far. Direct detection of outflows at high z comes in the form of faint broad [C II]λ 157 μm wings in quasar (QSO) sub-mm spectra, both in few individual objects (e.g. Maiolino *et al.* 2012) or stacked spectra (Bischetti *et al.* 2019) or from OH in absorption in the far-IR (Herrera-Camus *et al.* 2020). The cold gas kinematics of QSOs, traced by CO or [C II]λ 157 μm, indicates turbulent thick gas discs, possibly due to dynamically hot discs and/or outflows (Pensabene *et al.* 2020). High-resolution (~400 pc) kinematics of a $z \sim 6.5$ QSO revealed [C II]λ 157 μm cavities around the galaxy centre, a potential signature of QSO feedback in action (Venemans *et al.* 2019).

"Intermediate" redshift. Outflows at $z \sim 1$–3, around the peak of AGN and SF activity ("cosmic noon"), are mainly observed in the near-IR band, where the optical emission lines are redshifted. Evidence for outflows suppressing SF in the host was also found in a few objects, from the spatial anti-correlation between the distribution of [O III] and CO high velocity emission tracing outflows and narrow Hα emission tracing SF (Cresci *et al.* 2015; Carniani *et al.* 2016; Brusa *et al.* 2018), which has recently been questioned in one case (Scholtz *et al.* 2020). A number of observational campaigns such as the WISSH of hyper-luminous QSOs (Bischetti *et al.* 2017), the SUPER survey (Circosta *et al.* 2018), the KASH-z survey of X-ray selected AGN (Harrison *et al.* 2016), found ubiquitous ionised outflows at these redshifts. Moreover, the SINS zC-SINF plus KMOS3D (Förster Schreiber *et al.* 2018, 2019) and the MOSDEF (Leung *et al.* 2019) surveys found that the incidence of fast ionised outflows increases with stellar mass jointly with AGN fraction, and so their incidence in AGN is independent on stellar mass.

Low redshift. A plethora of outflow studies exists at $z < 1$. These span from QSOs around $z \sim 0.5$ (e.g. Villar-Martín *et al.* 2011) or $z \sim 0.2$ (e.g. Harrison *et al.* 2014) down to the very local Universe. In particular, recent spatially resolved studies of local AGN, mostly making use of integral field spectroscopic observations, proved themselves crucial in characterising and constraining the properties of ionised outflows, by providing the extension and geometry of outflows and allowing to resolve their kinematics. This has been possible both on small scales, through few-arcsec scale integral field spectrographs (such as Gemini GMOS and NIFS; e.g. Riffel *et al.* 2015; Freitas *et al.* 2018) or *HST* imaging + spectroscopy (e.g. Revalski *et al.* 2018), and on large scales where, in particular, the optical and near-IR integral field spectrograph MUSE at VLT has been a breakthrough. In fact, its combination of wide field of view ($1' \times 1'$) and spectral coverage provides a wealth of gas diagnostic lines which, together with the 8m-telescope high sensitivity, allows to study in details the outflow kinematics and the gas physical properties from tens of pc up to few kpc, as shown for instance by our MAGNUM survey (e.g. Venturi *et al.* 2017, 2018, Mingozzi *et al.* 2019) - which we will delve into in Section 4 - and by the AMUSING++ compilation (López-Cobá *et al.* 2020).

Neutral outflows in local objects are either studied in their neutral atomic phase through the Na I D optical absorption line doublet (e.g. Perna *et al.* 2019) and H I (Morganti *et al.* 2016), or, mostly, in their molecular phase through broad wings of species like CO or OH, tracing the cold gas reservoir (e.g. Cicone *et al.* 2014; García-Burillo *et al.* 2014; Lutz *et al.* 2020), and H_2 for what concerns the warm molecular phase (e.g. Emonts *et al.* 2014). The origin of molecular gas in outflows, with supersonic velocities of $\sim$1000 km/s and mass outflow rates of a few 1000s $M_\odot$/yr, and whether it is accelerated directly at the source or formed by gas cooling within the outflow, is debated. Some models (e.g. Zubovas & King 2014) predict that outflowing hot gas is unstable and would eventually lead to a two-phase medium, formed by cold dense molecular clumps surrounded by hot tenuous gas. Moreover, efficient gas cooling would result in stars forming within the outflow itself, with potential important implications for galaxy formation and evolution. Such phenomenon was recently confirmed from observations of gas ionised by young stars within outflows, or through direct kinematic signatures of outflowing young stars (e.g. Maiolino *et al.* 2017; Gallagher *et al.* 2019).

Multi-phase studies. Having information from all the gas phases is of primary importance for determining the total mass and energetic budget of the outflows, comparing with predictions from feedback models and determining their impact on host galaxies. However, very few multi-phase outflow studies exist up to now (Cicone *et al.* 2018). The prototypical example of a multi-phase outflow is Mrk 231, the closest QSO known, studied in the ionised phase (Rupke & Veilleux 2011), in the neutral atomic phase from both Na I D (Rupke & Veilleux 2011) and H I (Morganti *et al.* 2016) and in the molecular phase from both CO (Feruglio *et al.* 2015) and OH (Fischer *et al.* 2010). When the total mass budget of the galactic-scale outflow is available, it is indeed possible, by comparing with the nuclear unresolved X-ray outflow, to infer whether the outflow is, for instance, momentum- or energy-conserving (e.g. Feruglio *et al.* 2015).

Summary of local outflow properties. Studies of $z < 1$ AGN from SDSS (e.g. Mullaney *et al.* 2013; Woo *et al.* 2016) and MaNGA (Wylezalek *et al.* 2020) have shown that ionised outflows are ubiquitous, that their fraction is higher in AGN than in inactive galaxies and that it increases with AGN luminosity and/or Eddington ratio. Despite their widespread presence, their global effect on SFR in the host does not seem to be significant (e.g. Balmaverde *et al.* 2016), except maybe in high-SFR objects (Wylezalek & Zakamska 2016).

Studies of samples of AGN outflows in multiple gas phases (Carniani *et al.* 2015; Fiore *et al.* 2017; Fluetsch *et al.* 2019) found relations for outflow velocity, mass, momentum and kinetic energy rates with AGN luminosity across several orders of magnitude and in all phases analysed. Moreover, they found that ionised outflow masses are usually $\sim$10 times smaller than those of molecular outflows in AGN, while they are similar in SF galaxies. The mass in the neutral atomic phase is generally comparable with that in the molecular phase. They also concluded that, while gas depletion times increases with AGN luminosity, the outflow escape fraction is small ($\lesssim$5%) and so most of the outflowing gas is not able to leave the galaxy. Finally, short flow times ($\sim 10^6$ yr) suggest intermittent AGN activity and, indeed, evidence for "fossil" outflows (i.e. from a faded AGN stronger in the past) is found (e.g. Fluetsch *et al.* 2019; Audibert *et al.* 2019).

4. MAGNUM survey

Here we present some highlights from our MAGNUM survey (Measuring Active Galactic Nuclei Under MUSE Microscope; e.g. Venturi *et al.* 2017; Mingozzi *et al.* 2019) of nearby Seyfert galaxies observed with the wide-field optical and near-IR integral field spectrograph Multi Unit Spectroscopic Explorer (MUSE; Bacon *et al.* 2010) at the Very Large Telescope (VLT). The vicinity of the targets, combined with the unique capabilities of MUSE, allowed us to map the ionised gas down to $\sim$10 pc in several nebular emission lines, revealing ubiquitous kpc-scale outflows, and to study their physical properties.

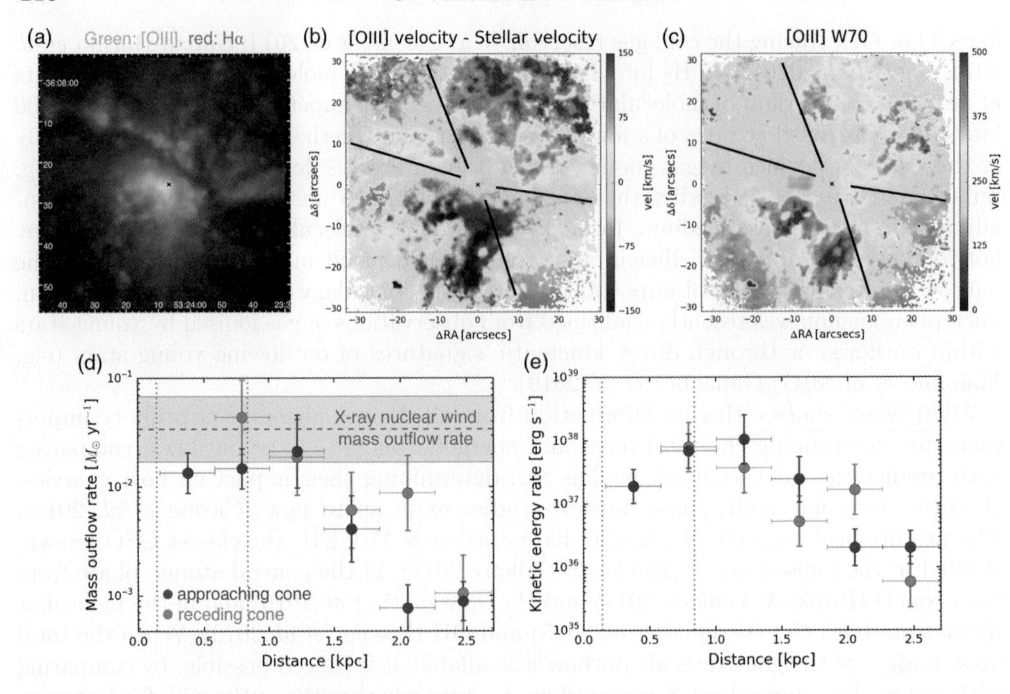

Figure 1. VLT/MUSE maps of NGC 1365. a) Emission-line maps of ionised gas, [O III]λ5007 (green) and Hα (red). b) [O III] velocity, after subtraction, spaxel-by-spaxel, of the stellar velocity. c) [O III] W70 (line width). The black lines are meant to guide the eye. d) Radial profile of the ionised mass outflow rate and e) kinetic energy rate as a function of distance from the AGN. The green dashed line indicates the mass outflow rate of the nuclear unresolved X-ray wind.

Dissecting the properties of the ionised outflow in NGC 1365. In Venturi *et al.* (2018) we focused on the detailed study of a specific source, NGC 1365, a barred Seyfert galaxy located at 17.3 Mpc from Earth hosting a low-luminosity AGN (~2×10^{43} erg/s). MUSE observations, covering its ~5.3×5.3 kpc^2 central region, are shown in Fig. 1. Panel a) reports the [O III]λ5007 (green) and Hα (red) emission, the former mostly tracing the gas ionised by the AGN, the latter stemming from SF regions. The ionisation cones host a bi-conical ionised outflow, as shown in panels b) and c), where the maps of [O III] velocity and of [O III] W70, measure of the line velocity width (difference between the 85% and 15% percentile velocities of the total line profile), are displayed, respectively. In order to isolate the gas motions in excess to rotation, the stellar velocity has been subtracted spaxel-by-spaxel from the [O III] velocity. We then extracted radial profiles of the outflow kinematics and energetics as a function of distance from the AGN. We show the mass outflow rate and the kinetic energy rate in panels d) and e). The mass outflow rate was calculated by radially slicing the two outflowing cones and using Eq. 2.1, where f = 1, $r = \Delta$r is the shell radial width and M and v the outflow mass and velocity in each radial slice, respectively, obtained from the centroid and Hα flux of the fitted Gaussian component associated to the outflow (blueshifted to the south-east, redshifted to the north-west), respectively. The comparison of the galactic-scale outflow with the mass outflow rate of the highly-ionised nuclear unresolved outflow, measured from Fe XXV and Fe XXVI X-ray absorption lines and having velocities of ~3000 km s^{-1}, provides insights on the outflow driving mechanisms. We could thereby exclude a momentum- or an energy-driven mechanism as the origin of the outflow acceleration. The former is in fact predicted on scales $\lesssim$1 kpc (e.g. King & Pounds 2015), smaller than those observed for the outflow in NGC 1365. The latter is ruled out since energy conservation between the large-scale

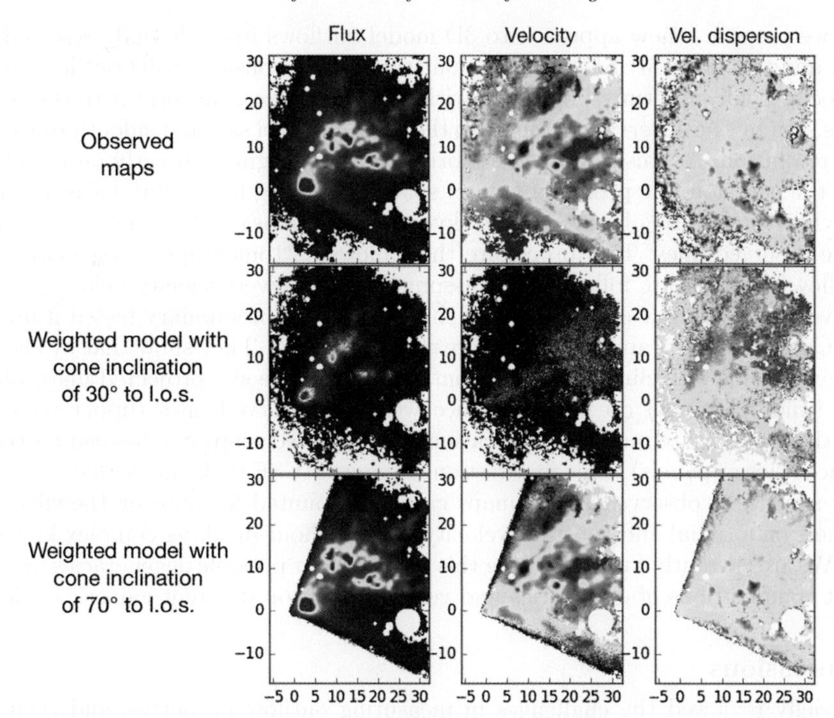

Figure 2. Test of our new 3D outflow "tomographic" reconstruction on Circinus MUSE data. Maps of [O III] flux (left panels), velocity (middle; first-order moment of velocity of line profile) and velocity dispersion (right), respectively, from observations (top panels) and reconstructed from two models with constant velocity and hollow conical outflow geometry with inner and outer apertures of 40° and 60°, respectively, one model having an inclination to line of sight (l.o.s.) of 30° (middle panels), the other of 70° (lower panels).

kinetic rate in all gas phases and the nuclear one would require a fraction of neutral atomic and molecular gas a factor $\gtrsim 10^3$ larger than that in the ionised phase. On the other hand, direct acceleration by the AGN radiation pressure on dusty clouds could, in principle, be a possible driver of the outflow. The AGN photon momentum ($L_{\rm bol}/c$) is in fact $\sim$20 times larger than the peak of the optical galactic outflow momentum rate ($\dot{p}$), and models predict $\dot{p} \sim 1{-}5 L_{\rm bol}/c$ (e.g. Ishibashi *et al.* 2018), implying that a fraction of neutral atomic plus molecular gas a factor 20−100 larger than the ionised one would be necessary, which is feasible based on literature studies (e.g. Fluetsch *et al.* 2019).

New 3D kinematic model. The extensive spatially-resolved information provided by MUSE on outflow kinematics allows us to obtain information on their geometry and structure. For instance, in Venturi *et al.* (2017) the comparison between the observed MUSE maps and a very simple kinematic toy model suggested a hollow conical geometry for an outflow. Only a few attempts of 3D-modelling of outflows from spatially resolved data, to take into account projection effects and infer the outflow kinematic structure, exist in the literature (e.g. Crenshaw *et al.* 2015). These usually adopt a velocity field defined as a function of distance from the nucleus and a smoothly distributed medium, convolve the model with instrument properties (PSF, bin on slit or spaxel size) and fit it to spatially resolved data. However, the actual observed distribution of gas in outflows is not smooth but really clumpy and the velocity fields may be quite complex (e.g. Fig. 2, top panels). Such complex velocity fields may either arise from a real defined 3D velocity structure of the outflow or be an effect of the clumpy line-emission along the line of sight.

Here we propose a new approach to 3D-model outflows from spatially resolved observations, and test it on MUSE data from our survey. We consider a 3D outflow geometry and velocity field, uniformly populated with gas clouds, we transform it to the reference frame as seen by the observer, we bin it to the MUSE spaxel size and smooth to its seeing-limited resolution. At this point the uniform model is weighted directly on the observed cube of the emission-line profile in (x,y,v) space, according to the flux measured in each spaxel where a cloud is "observed", and sky-projected maps can be generated to compare with the observed ones. This procedure then allows a "tomographic" reconstruction of the outflow 3D structure following the assumption of a given velocity field.

To investigate possible degeneracies of this process, we preliminary tested it on MUSE data of Circinus galaxy, and the results are shown in Fig. 2. They show that by choosing a wrong geometry and inclination for the outflow model, the sky-projected maps obtained from it (middle panels) do not reproduce well the observed ones (upper panels). By finding the appropriate model configuration, the observed maps can instead be very well reproduced (lower panels). This first test on Circinus MUSE data shows that the complex velocity structures observed in the maps can be accounted for only by the effect of gas clumpiness on a radial and constant velocity field, without invoking complex kinematical effects. We are currently further testing this approach for possible degeneracies by probing different combinations of geometries and velocity fields on different galactic outflows.

5. Conclusions

We briefly reviewed the challenges in measuring outflow properties and their observational evidence from high to low redshift. Outflows are found to be ubiquitous in the ionised gas and multi-phase. The molecular and neutral atomic phases, when detected, dominate the mass of outflowing gas on galactic scales. While few indications of outflows damping star formation are found, clear observational evidence that they are capable of globally affecting star formation processes in the galaxy population are still missing. On one hand, reducing the uncertainties on measured physical properties and obtaining the total outflow mass budget from multi-phase studies is pivotal for a comparison with models and to better constrain the impact of outflows on host galaxies. On the other hand, the effect of outflows on star formation might be more subtle, for instance by having an effect on different timescales than those when outflows and star formation coexist ("delayed" feedback) or by acting on longer timescales through a cumulative effect made up of multiple outflow and AGN activity episodes.

References

Aird, J., Coil, A. L., Georgakakis, A., *et al.* 2015, *MNRAS*, 451, 1892
Audibert, A., Combes, F., García-Burillo, S., *et al.* 2019, *A&A*, 632, A33
Bacon, R., Accardo, M., Adjali, L., *et al.* 2010, in Ground-based and Airborne Instrumentation for Astronomy III, Vol. 7735 *Proc. SPIE*, 131–139
Balmaverde, B., Marconi, A., Brusa, M., *et al.* 2016, *A&A*, 585, A148
Behroozi, P., Wechsler, R. H., Hearin, A. P., *et al.* 2019, *MNRAS*, 488, 3143
Bîrzan, L., Rafferty, D. A., Nulsen, P. E. J., *et al.* 2012, *MNRAS*, 427, 3468
Bischetti, M., Maiolino, R., Carniani, S., *et al.* 2019, *A&A*, 630, A59
Bischetti, M., Piconcelli, E., Vietri, G., *et al.* 2017, *A&A*, 598, A122
Bolatto, A. D., Wolfire, M., & Leroy, A. K. 2013, *ARA&A*, 51, 207
Brusa, M., Cresci, G., Daddi, E., *et al.* 2018, *A&A*, 612, A29
Carniani, S., Marconi, A., Maiolino, R., *et al.* 2015, *A&A*, 580, A102
Carniani, S., Marconi, A., Maiolino, R., *et al.* 2016, *A&A*, 591, A28
Cicone, C., Brusa, M., Ramos Almeida, C., *et al.* 2018, *Nature Astronomy*, 2, 176
Cicone, C., Maiolino, R., Sturm, E., *et al.* 2014, *A&A*, 562, A21

Ciotti, L., Ostriker, J. P., & Proga, D. 2010, *ApJ*, 717, 708
Circosta, C., Mainieri, V., Padovani, P., *et al.* 2018, *A&A*, 620, A82
Crenshaw, D. M., Fischer, T. C., Kraemer, S. B., *et al.* 2015, *ApJ*, 799, 83
Cresci, G., Mainieri, V., Brusa, M., *et al.* 2015, *ApJ*, 799, 82
Emonts, B. H. C., Piqueras-López, J., Colina, L., *et al.* 2014, *A&A*, 572, A40
Fabian, A. C. 2012, *ARA&A*, 50, 455
Feruglio, C., Fiore, F., Carniani, S., *et al.* 2015, *A&A*, 583, A99
Fiore, F., Feruglio, C., Shankar, F., *et al.* 2017, *A&A*, 601, A143
Fischer, J., Sturm, E., González-Alfonso, E., *et al.* 2010, *A&A*, 518, L41
Fluetsch, A., Maiolino, R., Carniani, S., *et al.* 2019, *MNRAS*, 483, 4586
Förster Schreiber, N. M., Renzini, A., Mancini, C., *et al.* 2018, *ApJS*, 238, 21
Förster Schreiber, N. M., Übler, H., Davies, R. L., *et al.* 2019, *ApJ*, 875, 21
Freitas, I. C., Riffel, R. A., Storchi-Bergmann, T., *et al.* 2018, *MNRAS*, 476, 2760
Gallagher, R., Maiolino, R., Belfiore, F., *et al.* 2019, *MNRAS*, 485, 3409
García-Burillo, S., Combes, F., Usero, A., *et al.* 2014, *A&A*, 567, A125
Harrison, C. M., Alexander, D. M., Mullaney, J. R., *et al.* 2016, *MNRAS*, 456, 1195
Harrison, C. M., Alexander, D. M., Mullaney, J. R., *et al.* 2014, *MNRAS*, 441, 3306
Harrison, C. M., Costa, T., Tadhunter, C. N., *et al.* 2018, *Nature Astronomy*, 2, 198
Herrera-Camus, R., Sturm, E., Graciá-Carpio, J., *et al.* 2020, *A&A*, 633, L4
Ishibashi, W., Fabian, A. C., & Maiolino, R. 2018, *MNRAS*, 476, 512
King, A. & Pounds, K. 2015, *ARA&A*, 53, 115
Kormendy, J. & Ho, L. C. 2013, *ARA&A*, 51, 511
Leung, G. C. K., Coil, A. L., Aird, J., *et al.* 2019, *ApJ*, 886, 11
López-Cobá, C., Sánchez, S. F., Anderson, J. P., *et al.* 2020, *AJ*, 159, 167
Lutz, D., Sturm, E., Janssen, A., *et al.* 2020, *A&A*, 633, A134
Maiolino, R., Gallerani, S., Neri, R., *et al.* 2012, *MNRAS*, 425, L66
Maiolino, R., Russell, H. R., Fabian, A. C., *et al.* 2017, *Nature*, 544, 202
McNamara, B. R., Wise, M., Nulsen, P. E. J., *et al.* 2000, *ApJL*, 534, L135
Mingozzi, M., Cresci, G., Venturi, G., *et al.* 2019, *A&A*, 622, A146
Morganti, R., Veilleux, S., Oosterloo, T., *et al.* 2016, *A&A*, 593, A30
Mullaney, J. R., Alexander, D. M., Fine, S., *et al.* 2013, *MNRAS*, 433, 622
Pensabene, A., Carniani, S., Perna, M., *et al.* 2020, arXiv e-prints, arXiv:2002.00958
Perna, M., Cresci, G., Brusa, M., *et al.* 2019, *A&A*, 623, A171
Revalski, M., Crenshaw, D. M., Kraemer, S. B., *et al.* 2018, *ApJ*, 856, 46
Riffel, R. A., Storchi-Bergmann, T., & Riffel, R. 2015, *MNRAS*, 451, 3587
Rupke, D. S. N. & Veilleux, S. 2011, *ApJL*, 729, L27
Schawinski, K., Urry, C. M., Simmons, B. D., *et al.* 2014, *MNRAS*, 440, 889
Schaye, J., Crain, R. A., Bower, R. G., *et al.* 2015, *MNRAS*, 446, 521
Scholtz, J., Harrison, C. M., Rosario, D. J., *et al.* 2020, *MNRAS*, 492, 3194
Tadhunter, C., Morganti, R., Rose, M., *et al.* 2014, *Nature*, 511, 440
Venemans, B. P., Neeleman, M., Walter, F., *et al.* 2019, *ApJL*, 874, L30
Venturi, G., Marconi, A., Mingozzi, M., *et al.* 2017, *Front. Astron. Space Sci.*, 4, 46
Venturi, G., Nardini, E., Marconi, A., *et al.* 2018, *A&A*, 619, A74
Villar-Martín, M., Humphrey, A., Delgado, R. G., *et al.* 2011, *MNRAS*, 418, 2032
Woo, J.-H., Bae, H.-J., Son, D., *et al.* 2016, *ApJ*, 817, 108
Wylezalek, D., Flores, A. M., Zakamska, N. L., *et al.* 2020, *MNRAS*, 492, 4680
Wylezalek, D. & Zakamska, N. L. 2016, *MNRAS*, 461, 3724
Zubovas, K. & King, A. R. 2014, *MNRAS*, 439, 400

Raffaella Morganti

Christopher Harrison

Galaxy Evolution and Feedback across Different Environments
Proceedings IAU Symposium No. 359, 2020
T. Storchi-Bergmann, W. Forman, R. Overzier & R. Riffel, eds.
doi:10.1017/S1743921320002215

ORAL CONTRIBUTIONS

Kiloparsec-scale jet-driven feedback in AGN probed by highly ionized gas: A MUSE/VLT perspective

A. Rodríguez-Ardila[1,2] and M. A. Fonseca-Faria[2]

[1]Laboratório Nacional de Astrofísica, R. dos Estados Unidos,
CEP 37504-364, Itajubá - MG, Brazil
email: aardila@lna.br

[2]Instituto Nacional de Pesquisas Espaciais, Av. dos Astronautas, CEP 12227-010,
São José dos Campos - SP, Brazil

Abstract. We employ optical spectroscopy from the Multi Unit Spectroscopic Explorer (MUSE) combined with X-ray and radio data to study the highly-ionized gas (HIG) phase of the feedback in a sample of five local nearby Active Galactic Nuclei (AGN). Thanks to the superb field of view and sensitivity of MUSE, we found that the HIG, traced by the coronal line [Fe VII] λ6089, extends to scales not seen before, from 700 pc in Circinus and up to ∼2 kpc in NGC 5728 and NGC 3393. The gas morphology is complex, following closely the radio jet and the X-ray emission. Emission line ratios suggest gas excitation by shocks produced by the passage of the radio jet. This scenario is further supported by the physical conditions derived for the HIG, stressing the importance of the mechanical feedback in AGN with low-power radio jets.

Keywords. active galaxy nuclei, Seyfert, emission lines, spectroscopic

---

1. Introduction

Jet-driven outflows are now recognized as an important ingredient in the active galactic nuclei (AGN) feedback scenario. The effects of such a mechanism in low-luminosity radio-quiet AGN are not yet clear. Recent evidences gathered from NIR AO observations (Rodríguez-Ardila *et al.* 2017; May *et al.* 2018; Jarvis *et al.* 2019) indicate that their impact to the ISM cannot be underestimated. Indeed, studies made on low-power radio sources ($\leqslant 10^{24}$ W Hz^{-1}), which dominate the radio sky, show that jets can significantly impact their surrounding gaseous medium (Sridhar *et al.* 2020). This is because low power sources tend to have low velocity jets. As a result, their jets are characterised by a larger component of turbulence compared to jets in powerful radio sources. Therefore, this kinetic channel can be more relevant for galaxy evolution than previously thought (Wylezalek & Morganti 2018).

Traditionally, the identification of outflows associated to radio jets in the warm, ionized phase of AGN has been done by means of the [O III] λ5007 line (Greene *et al.* 2011). However, this line may also carry the contribution from a starburst component, the galaxy disk and the part of the NLR that is not participating in the outflow (Rodríguez-Ardila & Fonseca-Faria 2020). Isolating the different contributions to the line profile is tricky and sometimes subject to large uncertainties due to the lack of enough spectral and angular resolution.

In this respect, Rodríguez-Ardila *et al.* (2006) showed that coronal lines (CLs) such as [Fe VII] λ6087 in the optical or [Si VI] 1.963μm in the near-infrared (NIR) are excellent tracers of the ionized component of the outflows. The energy required for their production ($\chi \geqslant 100$ eV, where χ is the ionization potential required to produce the ion) rules out stellar or galactic origin. Moreover, as shown by Ferguson *et al.* (1987), when powered solely by photoionization by the central source in local, low-luminosity AGN ($L_{\rm bol} < 10^{43}$ erg s^{-1}, where $L_{\rm bol}$ is the bolometric luminosity), the CL emission cannot extend to distances larger than a few hundred of parsecs from the central source. Thus, their detection outside the circumnuclear region is a signature of outflowing gas associated to jet-driven shocks, capable of ionizing gas at large distances and possibly pushing outwards the ionized gas.

With the above in mind, we started a program aimed at: (*i*) studying the role of outflow/jet-induced mode feedback in a local sample of low-luminosity AGN known to have low-power jets and strong coronal line emission; (*ii*) characterize the physical properties of the highest ionized component of the outflow in these sources.

2. The sample, observations and data analysis

The sample employed in this work is composed of five local Seyfert 2 AGN ($z < 0.02$, where z is the redshift), widely known for displaying a low-luminosity radio-jet and a narrow line region (NLR) characterized by a bi-conical structure, previously mapped through the [O III] λ5007 emission. This is the case of the Circinus Galaxy, NGC 5728, IC 5063, NGC 3393 and NGC 5643. They all display [Fe VII] λ6087 in the nuclear spectrum and a NLR that extends to at least 2 kpc from the AGN. Integral Field Unit (IFU) data for the above 5 targets were obtained using MUSE/VLT and retrieved from the European Southern Observatory science portal. The IFU cube for each source is fully reduced, including calibration in flux (in absolute units) and wavelength. Details of the observations and data reduction are provided elsewhere (i.e Mingozzi *et al.* 2019).

Each data cube was analyzed making use of a set of custom PYTHON scripts developed by us as well as software publicly available in the literature. First, we rebinned the cube, reducing the total number of spaxels to ~10000. We then removed the stellar continuum across the whole spectral range of MUSE (4700−9100 Å) using STARLIGHT (Cid-Fernandes *et al.* 2005) and the Bruzual & Charlot (2003) stellar libraries. This procedure left us with spectra dominated by the nebular emission, allowing us to focus only on the gas emission properties, free of any continuum contamination.

Thereafter, we measured the emission line fluxes of Hα and Hβ at every spaxel on each galaxy to determine the extinction (Galactic and intrinsic) affecting the gas. This was done assuming an intrinsic line ratio H$\alpha/H\beta = 3.1$ and the Cardelli *et al.* (1987) extinction law. All integrated line fluxes measured at each spaxel were corrected by the extinction measured accordingly.

We then constructed maps of the flux distribution for the most important lines such as [O III] λ5007, Hβ, Hα and [Fe VII] λ6087. Their inspection revealed that the mid- and high-ionized gas is usually arranged in bi-conical structures, with apex at the AGN.

In the following sections, we will describe the main results, first for the Circinus galaxy and then for the remainder of the sample.

3. The case of the Circinus Galaxy

At the adopted distance of 4.2 Mpc (1" ~20.4 pc), Circinus is the closest Seyfert 2 galaxy to us. Because of its proximity, angular resolution on scales of a few tens of parsecs can be reached even at seeing-limited conditions.

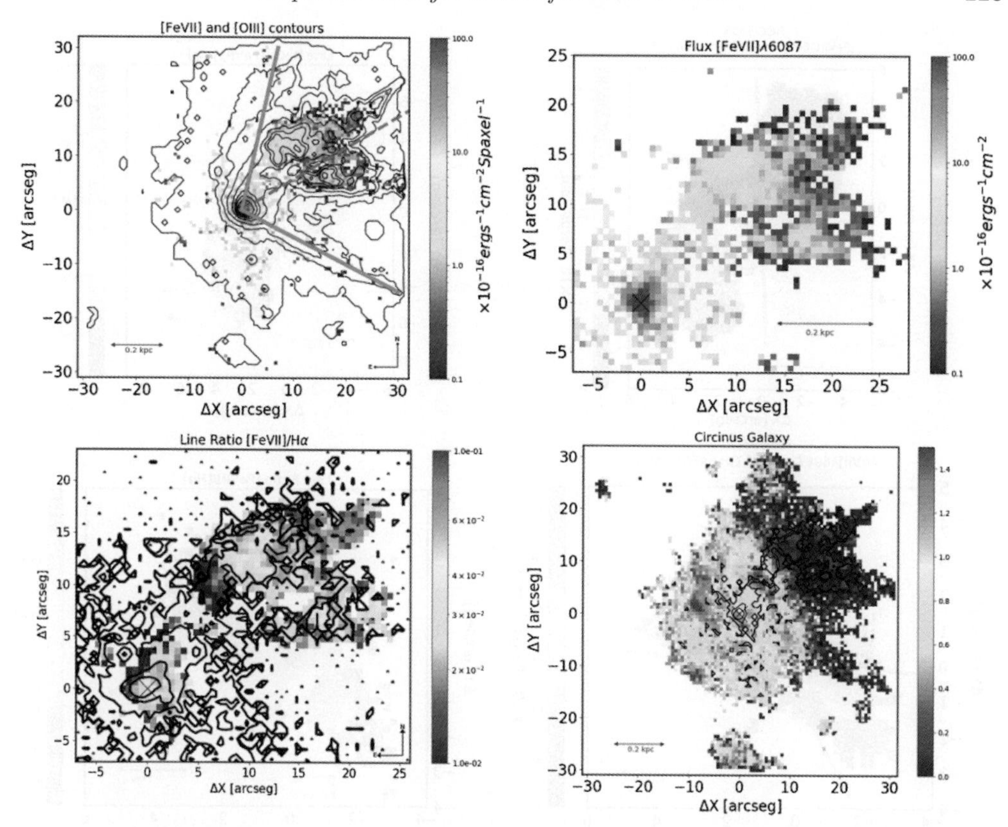

Figure 1. Upper left: Emission map of the [Fe VII] $\lambda 6087$, overlaid to [O III] $\lambda 5007$ contours. The red lines mark the edges of the ionization cone (Mingozzi *et al.* 2019). The dashed magenta line indicates the PA $= 295 \pm 5^{\circ}$ of the radio continuum reported by Elmouttie *et al.* (1998). Upper right: zoom to the NW ionization cone, emphasizing the extended [Fe VII] emission. Bottom left: Extinction corrected emission line flux ratio [Fe VII]/Hα for the highest ionized portion of the cone. The region in white corresponds to values with signal to noise ratio (S/N) < 3 or where the [Fe VII] line is not detected. The cross marks the position of the AGN. The contours represent *Chandra* ACIS image in the 0.5–8 keV band from Smith & Wilson (2001). Bottom right: extinction map (A_V, in units of mag) of Circinus overlaid to the extended [Fe VII] emission (black contours).

Circinus is widely known for its prominent CL spectrum (Moorwood *et al.* 1996; Rodríguez-Ardila *et al.* 2006). Adaptic optics (AO) observations have revealed that the coronal line region (CLR) in this object extends from the nucleus up to 30 pc (Prieto *et al.* 2005; Müller-Sánchez *et al.* 2006). Moreover, Oliva *et al.* (1999), using optical spectroscopy, reported extended [Fe VII] emission up to 22" from the center at a PA $= 318^{\circ}$. This result was recently confirmed by Mingozzi *et al.* (2019), who found extended [Fe VII] emission associated to the ionization cone, but no information about its full extension, morphology and gas physical properties were presented.

The [Fe VII] $\lambda 6087$ flux distribution constructed for Circinus from MUSE reveals the most extended high-ionization emission ever observed in that AGN. Figure 1 shows that the coronal gas extends up to 700 pc from the central engine. The gas emission appears clumpy, with several knots of emission, concentrated in the innermost region of the ionization cone, following the radio jet axis. The gas displays a complex kinematics (see Rodríguez-Ardila & Fonseca-Faria 2020), implying that it is out of the galaxy plane.

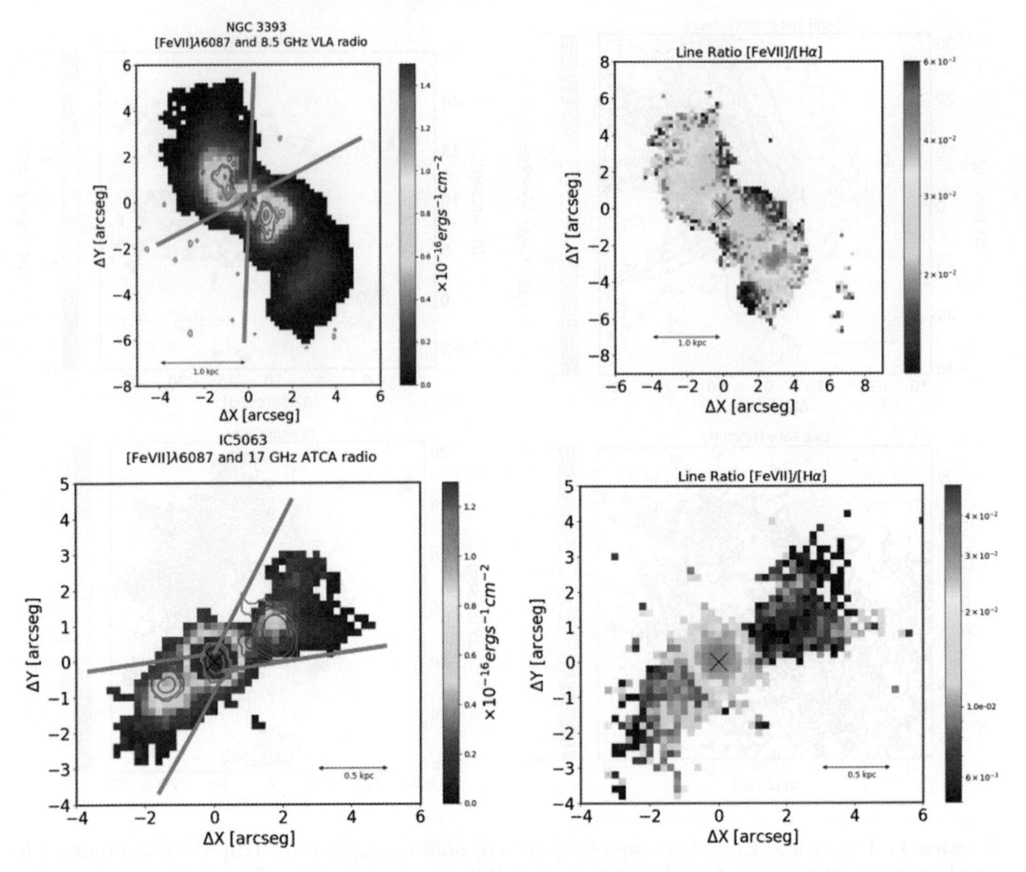

Figure 2. [Fe VII] λ6087 emission line maps for NGC 3393 (upper left) and IC 5063 (bottom left). The full red lines indicate the edges and orientation of the ionization cone. The contours correspond to the radio emission (R. Morganti, *private communication*, 2020). The upper and bottom right panels show the line ratio maps (excitation) [Fe VII]/Hα for NGC 3393 and IC 5063, respectively. White areas correspond to masked values with S/N < 3 or where [Fe VII] is not detected. In all panels, North is up and East to the left.

Figure 1 also shows that the coronal gas is little affected by dust extinction. Moreover, the spatial coincidence with extended thermal X-ray emission, and its orientation along the radio jet axis, indicates that it is likely the remnant of shells inflated by the passage of the radio jet. The gas velocity dispersion of $\sim$300 km s^{-1} supports this hypothesis.

The emission line flux ratio [Fe VII]/Hα, which directly reflects the ionization state of the gas, is shown in the bottom left panel of Figure 1. It can be seen that the ratio increases outwards, varying from $\sim$0.01 at $\sim$77 pc from the AGN to $\sim$0.1 in the parcel of gas located at 700 pc from the nucleus. This result cannot be explained by photoionization from the AGN but it is consistent with a scenario where the jet is inducing shocks capable of ionizing gas at large distances and possibly pushing outwards the ionized gas.

4. First detection of kiloparsec-scale [Fe VII] emission in local AGN

A similar analysis as the one described in the preceding section, was applied to NGC 5728, IC 5063, NGC 3393 and NGC 5643.

Figure 2 displays the [Fe VII] flux distribution for NGC 3393 and IC 5063, evidencing the extension of the coronal gas. It can be seen that this energetic emission is arranged within the bi-cones already mapped by means of the [O III] line. In NGC 3393, the [Fe VII] emission is detected up to 1.3 kpc NE and 2.2 kpc SW from the AGN. Similarly, in IC 5063, the coronal gas is found up to 1 kpc NW and 0.9 kpc SE. In NGC 5728, the CL extends up to 2 kpc SE and 1.6 kpc NW (not shown here). Finally, in NGC 5643, [Fe VII] is observed up to 0.8 kpc E and 0.9 kpc W to the AGN. To the best of our knowledge, these four Seyfert 2s display the largest CLR already detected in samples of local AGN.

Figure 2 also evidences that the gas distribution is clumpy. Although the CL emission peaks at the AGN position, secondary peaks are also detected along the bi-cones. This is better seen in the upper and bottom right panels of the same figure. They show the excitation map [Fe VII]/Hα. In NGC 3393, in addition to the nucleus, two strong peaks of emission are observed at $\sim$1 kpc NE and SW of the AGN. In IC 5063, an increase of the gas excitation is evident outside the nuclear region, at $\sim$500 pc NW and SE of the AGN.

The strong relationship with the radio jet is evident in Figure 2. Radio emission from the jet traces cavities that enhances [Fe VII]/Hα, suggesting a role in the excitation of these hot spots. The values of that ratio, between 10^{-2} to 10^{-1}, cannot be easily explained by photoionization from the central source. Instead, shock models of Contini & Viegas (2001) are able to reproduce the observed gas excitation, along the region where the coronal gas is detected.

5. Final remarks

• Extended [Fe VII] emission, at kpc scales, highlights the relevance of the kinetic channel as an important way of depositing energy to the ISM of low-luminosity AGN, even when driven by radiatively poor radio jets.

• The five sources examined in this work are a showcase of this scenario. We found a conspicuous [Fe VII] emission filling the inner cavities of the ionization cone. It extends up to 2 kpc from the AGN in NGC 3393, NGC 5728 and IC 5063.

• We show that extended coronal emission is associated to the presence of radio jets, lack of dust and extended X-ray emission. The central source plays a fundamental role for gas at distances $R <$ 100-200 pc from the AGN, but at larger R, shocks driven by the jet is fundamental to enhance the coronal emission.

References

Bruzual, G. & Charlot, S. 2003, *MNRAS*, 344, 1000
Cardelli, J. A., Clayton, G. C., Mathis, J. S., *et al.* 1989, *ApJ*, 345, 245
Cid-Fernandes, R., Mateus, A., Sodré, L., *et al.* 2005, *MNRAS*, 358, 363
Contini, M. & Viegas, S. M. 2001, *Apj*, 132, 211
Elmouttie, M., Koribalski, B., Gordon, S., *et al.* 1998, *MNRAS*, 297, 49
Ferguson, K., Korista, T., Ferland, G. 1997, *ApJS*, 110, 287
Jarvis, M. E., Harrison, C. M., Thomson, A. P., *et al.* 2019, *MNRAS*, 485, 2710
Greene, J. E., Zakamska, N. L., Ho, L. C., *et al.* 2011, *ApJ*, 732, 9
May, D., Rodríguez-Ardila, A., Prieto, M. A., *et al.* 2018, *MNRAS*, 481, L105
Mingozzi, M., Cresci, G., Venturi, G., *et al.* 2019, *A&A*, 622, A146
Moorwood, A. F. M., Lutz, D., Oliva, E., *et al.* 1996, *A&A*, 315, L109
Müller-Sánchez, F., Davies, R. I., Eisenhauer, F., *et al.* 2006, *A&A*, 454, 481
Oliva, E., Marconi, A., Moorwood, A. F. M. 1999, *A&A*, 342, 87
Prieto, M. A., Marco, O., Gallimore, J. 2005, *MNRAS*, 346, L28
Rodríguez-Ardila A., Prieto M. A., Viegas S., *et al.* 2006, *ApJ*, 6
Rodríguez-Ardila, A., Prieto, M. A., Mazzalay, X., *et al.* 2017, *MNRAS*, 470, 2845
Rodríguez-Ardila, A. & Fonseca-Faria, M. A. 2020, *ApJ*, 895, L9
Sridhar, S. S., Morganti, R., Nyland, K., *et al.* 2020, *A&A*, 634A, 108
Smith, D. A. & Wilson, A. S. 2001, *ApJ*, 557, 180
Wylezalek, D. & Morganti, R. 2018, *Nature Astronomy*, 2, 181

Galaxy Evolution and Feedback across Different Environments
Proceedings IAU Symposium No. 359, 2020
T. Storchi-Bergmann, W. Forman, R. Overzier & R. Riffel, eds.
doi:10.1017/S1743921320002276

Ionized outflows in local luminous AGN: Density and outflow rate

R. Davies[1], D. Baron[2], T. Shimizu[1] and H. Netzer[2]

[1]Max-Planck-Institut für extraterrestrische Physik, Postfach 1312, 85741, Garching, Germany
[2]School of Physics and Astronomy, Tel-Aviv University, Tel Aviv 69978, Israel

Abstract. We use the LLAMA survey to study the density and outflow rate of ionized gas in a complete volume limited sample of local (<40 Mpc) luminous ($43.0 < \log L_{\rm AGN}(erg/s) < 44.5$) AGN selected by very hard 14-195 keV X-rays. The detailed data available for this survey enable us to measure the density of the outflowing ionized gas in the central 300 pc of these AGN using three different and independent methods (the standard [SII] doublet ratio; a method comparing [OII] and [SII] ratios that include auroral and transauroral lines; and a recently proposed method based on the ionization parameter). For each method there is, as expected, a modest spread of densities among the AGN in the sample. But remarkably, the median densities for each method differ hugely, by an order of magnitude from below 400 cm^{-3} to almost 5000 cm^{-3}. We discuss how the derived densities can be reconciled, and what the impact is on the implied outflow rate.

Keywords. Galaxies: active, Galaxies: ISM, Galaxies: nuclei, Galaxies: Seyfert

1. Introduction

That outflows driven by star formation and AGN play a fundamental role in the evolution of galaxies is undisputed. This is highlighted by cosmological models of galaxay evolution, which demonstrate the role that outflows are expected to have in order for such models to reproduce observed galaxy scaling relations (Somerville & Davé 2015). For these models to halt accretion onto the central supermassive black hole and quench star formation, the kinetic power of the gas outflow should be typically 5% of the AGN bolometric luminosity (Springel *et al.* 2005). Whether this is achieved through ionized outflows or cool outflows of neutral and molecular gas is not yet established (Veilleux *et al.* 2020). As pointed out by Harrison *et al.* (2018), a large part of this uncertainty is linked to the density of the gas in ionized outflows. The reason is simply that the mass $M_{\rm out}$ of ionized gas derived from a measurement of line luminosity $L_{\rm line}$ is inversely proportional to the density $n_{\rm e}$ adopted: $M_{\rm out} \propto L_{\rm line} / (\gamma_{\rm line}\, n_{\rm e})$, where $\gamma_{\rm line}$ is the volume emissivity. In the literature, a very wide range of densities has been assumed, adopted, or measured. Often, the density is estimated using the [SII] doublet ratio. This is applicable up to $\sim 10^4$ cm^{-3} for stellar HII regions; but may be biased in AGN photoionized gas because the high energy photons penetrate deep into the clouds and create a partially ionized zone, which is responsible for the enhanced emisson from low excitation lines around AGN. The impact of this on the density one derives has recently been assessed by Davies *et al.* (2020).

The LLAMA (Local Luminous AGN with Matched Analogues) survey (Davies *et al.* 2015) survey of active and inactive galaxies is an ideal sample in which to assess the true density of gas in ionized outflows because it enables the measurement to be made in several different ways for the same galaxies. The sample was taken from the all-sky flux

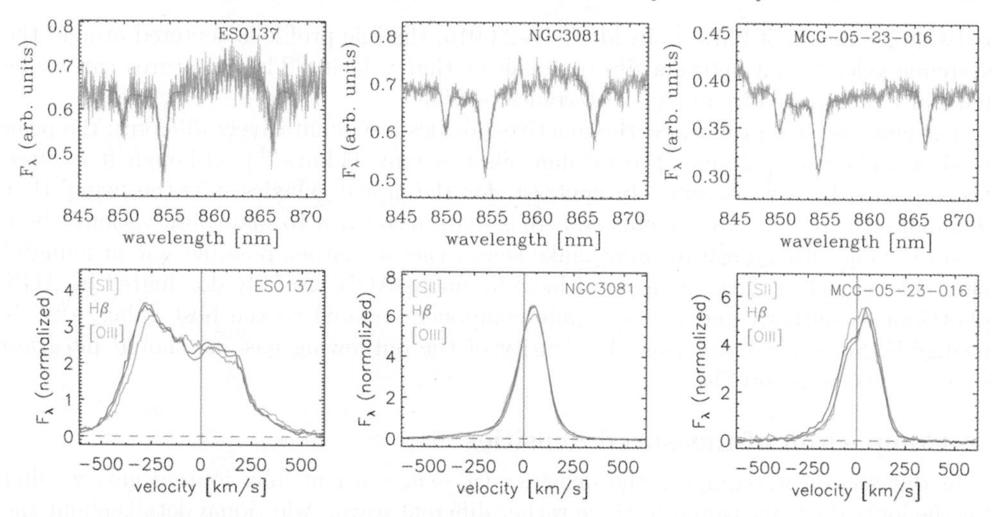

Figure 1. Examples of the Ca II triplet stellar absorption (top row) used as the velocity reference for the more complex emission line profiles (bottom row). These are ordered in terms of luminosity, from $L_{\rm AGN} = 10^{43.4}$ erg s^{-1} for ESO 137-G034 (left), to $10^{44.1}$ erg s^{-1} for NGC 3081, and $10^{44.3}$ erg s^{-1} for MCG-05-23-016 (right).

limited 14-195 keV 58-month Swift BAT survey in such a way as to create a volume limited sample of active galaxies that is as unbiased as possible, for detailed study using optical spectroscopy and adaptive optics integral field near-infrared spectroscopy. The sole selection criteria were $z < 0.01$ (corresponding to a distance of ∼40 Mpc), $\log L_{14-195\rm keV}$ [erg s^{-1}] > 42.5 (using redshift distance), and $\delta < 15°$ so that they are observable from the VLT. This yielded 20 AGN. A set of inactive galaxies were selected to match them in terms of host galaxy type, mass (using H-band luminosity as a proxy), inclination, presence of a bar, and distance. Although small, this volume limited sample is sufficient for detailed studies of emission line ratios, the molecular and ionized gas kinematics and distributions, as well as the stellar kinematics and populations, in the nuclear and circumnuclear regions. And the ability to compare the results to a matched sample of inactive galaxies has been essential in many of the studies so far, including the analysis presented here. Here we focus on the optical spectroscopy from Xshooter (Vernet *et al.* 2011) of a subset comprising Seyfert 2 and inactive galaxies, using spectra extracted from a $1.8'' \times 1.8''$ aperture that corresponds to sizes of 200–350 pc at the distances of the targets.

2. Emission line characteristics

This section focusses on whether the AGN emission line profiles are dominated by the interstellar medium or by the outflow. An important ingredient to this is an accurate velocity reference for the emission lines, for which we make use of the Ca II triplet lines around 860 nm. As can be seen from the examples in Fig. 1, these stellar absorption features provide a robust measurement of the systemic velocity because they are strong, well defined, and unbiased by interstellar absorption. The emission lines are rather complex, exhibiting a range of characteristics illustrated by the examples given in the figure. The [SII], Hβ, and [OIII] lines shown have remarkably similar profiles, which do not typically peak at the systemic velocity. In ESO 137-G034, the blue side of the profile is stronger, peaking more than 250 km s^{-1} from systemic, and there is a suggestion of a dip around the systemic velocity. In NGC 3081, without a velocity reference one might conclude that the line peaks at systemic and has a blue wing; but the peak is in fact offset to the red side by almost $50\,\mathrm{km\,s^{-1}}$. In MCG-05-23-016, the line profile is centered around the systemic velocity but shows a dip or break at that velocity. These features can all be naturally reproduced by models of biconical outflow.

The emission line profiles for the inactive galaxies are qualitatively different: the peak is always close to systemic (the median offset is only $11\,\mathrm{km\,s^{-1}}$), although it is often accompanied by a weak wing. In contrast, for the active galaxies it is the 'wing' that dominates the profile. Our conclusion here is that, in contrast to some observations where a 'core+wing' decomposition may make sense, this is neither possible nor meaningful here. These nuclear measurements of local luminous AGN are fully dominated by AGN photoionized outflow, and any systemic component related to the host galaxy disk is negligible. So, when measuring the density of the outflowing gas, we should integrate over the whole line profile.

3. Comparison of measured densities

In this section we compare the densities ($n_\mathrm{e} \sim n_\mathrm{p} \sim n_\mathrm{H}$ in fully ionized gas) we find for the ionized gas measured in three rather different ways. Additional details about the methods, and of the line fitting and modelling can be found in Davies *et al.* (2020).

The most commonly used tracer of n_e uses the [SII] λ6716,6731 Å doublet, which only requires a measurement of the ratio of two strong emission lines in a convenient and clean part of the optical spectrum. The physics of the excitation and de-excitation means that density – covering a range commonly found in H II regions – dominates the emitted line ratio; and because the lines are necessarily close in wavelength, the result is unaffected by extinction. These lines are well fitted in our data, yielding a median value of $n_\mathrm{e} \sim 350\,\mathrm{cm^{-3}}$ for the AGN and $190\,\mathrm{cm^{-3}}$ for the inactive galaxies.

An alternative method was proposed by Holt *et al.* (2011) as a way to avoid the limitations of the [SII] doublet ratio. In particular it is sensitive to high densities because it uses the transauroral lines [SII] λ4069,4076 and the auroral lines [OII] λ7320,7331 which have higher critical densities. These are used together with the stronger lines to give the ratios [SII] λ(4069+4076) / [SII] λ(6716+6731) and [OII] λ(3726+3729) / [OII] λ(7320+7331). How these lines are used differs fundamentally from the standard doublet ratio above: advantageously, the lines within each doublet are summed, and the ratios of the total flux in the doublets are compared to photoionization models. Such models take into account internally they way in which n_e (on which the emitted lines depend) varies through the cloud, and hence how the resulting cumulative line ratios are related to n_H. In this way the method traces n_H, but typically one equates that with n_e as for fully ionized gas. This means the method is dependent on the photoionization model; and also, because of the wide wavelength range covered by the lines, on extinction and the choice of extinction model. It is notable that a grid of models for n_e and A_V has almost orthogonal axes, and comparable dependency on both line ratios. The main limitation of this method is the weakness of the auroral and transauroral lines, which means it can only be applied to high signal-to-noise data. Using this method, the the median value for our data is $n_\mathrm{e} \sim 1900\,\mathrm{cm^{-3}}$.

A new method developed by Baron & Netzer (2019) is based on the definition of the ionization parameter as the number of ionising photons per atom. Re-arranging the definition gives $n_\mathrm{H} \propto L_\mathrm{AGN}\, r^{-2}\, U^{-1}$, so that the density can be derived from the AGN luminosity, the distance of the ionized gas from the AGN, and the ionization parameter. At the same time, Baron & Netzer (2019) showed that, observationally, U can be derived from the strong line ratios N[II]/Hα and [OIII]/Hβ. Alternatively, for luminous AGN and/or clouds that are close to the AGN, one can adopt a limiting value of $U = 0.01$ (Dopita *et al.* 2002). This method is widely applicable, and well suited to spatially resolved data. However, one needs to be aware of the impact of uncertainties in

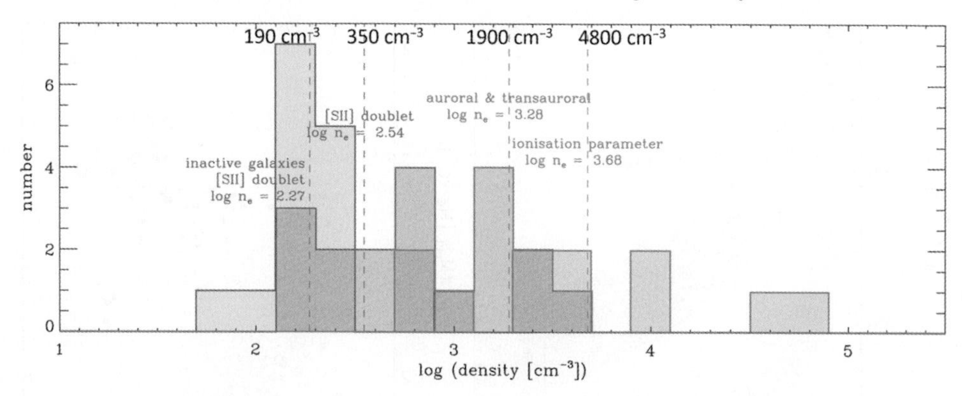

Figure 2. Derived ionized gas densities. Grey: inactive galaxies using the [SII] doublet ratio. Blue: AGN using the doublet ratio. Red: AGN using the auroral/transauroral ratios. Green: AGn using the ionization parameter method.

the AGN luminosity, and of projection effects on the apparent distance from the AGN to the gas. The use of aperture measurements, as is done in this work, will also add to the uncertainty. Our data yield a median density of $n_e \sim 4800\,\mathrm{cm}^{-3}$.

The outcome of this comparison is apparent in Fig. 2, which shows that for these AGN, the [SII] doublet underestimates the true ionized gas density by a factor 5–10. The reason for this stems from the high energy photons produced by an AGN, which can penetrate deep into clouds, creating a partially ionized zone. As explained by Davies *et al.* (2020), the ionic fraction of S^+ remains high even at depths where the fraction of H^+ is already greatly reduced. As an illustration, in a cloud with $N_H = 10^3\,\mathrm{cm}^{-3}$, only about 25% of the [SII] emission arises from the fully ionized gas; about half comes from the narrow region around the ionization front where the electron density descreases by a factor 10; and the rest arises from mostly neutral gas deeper in the cloud, where the electron density is only 1–10% that of the fully ionized gas. Photoionization models suggest that for such a cloud, n_e in the fully ionized gas can be underestimated by as much as a factor 3 when using the [SII] doublet. In denser clouds, the divergence increases rapidly, and can become very extreme. This is reasonably consistent with our empirical finding that for a true density of $2000\,\mathrm{cm}^{-3}$ or more, using the [SII] doublet leads to an underestimate of at least a factor 5.

As such, one should be very cautious when using the [SII] doublet to estimate the density of gas that is photoionized by an AGN. If the data have high enough signal-to-noise one can use the auroral/transauroral method to obtain a more reliable estimate. Otherwise one should assess the density also using the ionization parameter method.

4. Implication for outflow rate

The outflow rate $\dot{M}_{\mathrm{out}}$ depends on the ionized gas mass M_{out}, defined previously, together with the outflow speed v_{out} and size r_{out}. It is given as $\dot{M}_{\mathrm{out}} \propto M_{\mathrm{out}}\, v_{\mathrm{out}} \,/\, r_{\mathrm{out}}$, where the constant of proportionality is typically taken to be 1 or 3 as described in Lutz *et al.* (2020). We derive the outflow rate using the [OIII] line luminosity and outflow velocity, setting the outflow size to be the radius of the aperture with a modest correction for projection (since these are Seyfert 2s and hence oriented more edge-on), and adopting the density from the ionization parameter method. The resulting rates are 0.001–$0.5\,\mathrm{M}_\odot\,\mathrm{yr}^{-1}$. However, comparing these to published relations such as that of Fiore *et al.* (2017) is problematic. The reason is that, when compiling their data from the literature, in order to deal with the range of values adopted for n_e, these authors re-scaled

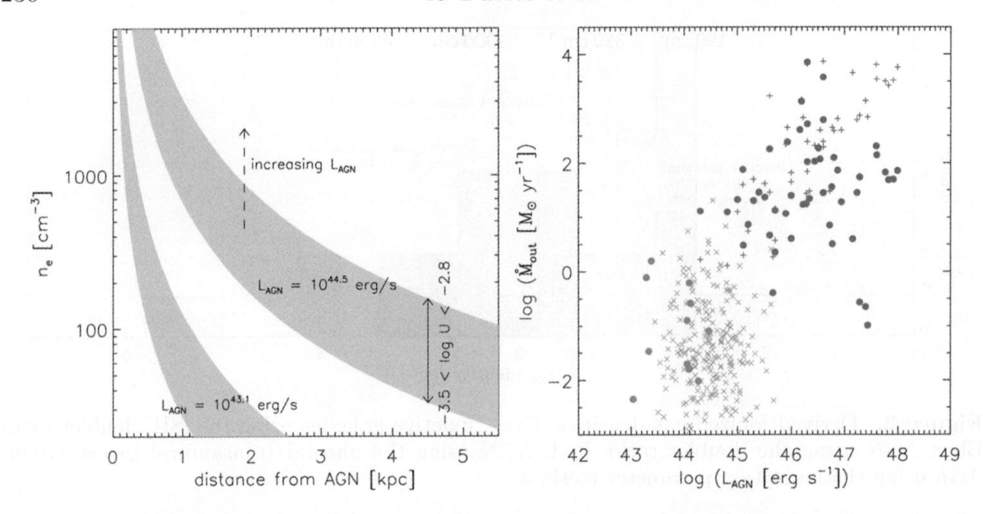

Figure 3. left: Illustration of how inverting the definition of the ionization parameter to give $n_H \propto L_{AGN}\, r^{-2}\, U^{-1}$ means that the gas density at any location in an outflow can be calculated from the AGN luminosity, the distance to the AGN, and the ionization parameter. Examples for 2 luminosities are shown, with the grey region indicative of the uncertainty if the ionization parameter is not known. Right: A modification to the well known $L_{AGN} - \dot{M}_{out}$ relation. Black plusses denote the original relation from Fiore *et al.* (2017). Filled blue circles are our modfication based on our best estimate of the density. Green crosses are from Baron & Netzer (2019); and the filled red circles are for the AGN in this study and Davies *et al.* (2020).

all the rates to a constant value of $n_e = 200\,\mathrm{cm}^{-3}$. This is an order of magnitude different from our typical value.

We therefore explore an alternative representation of the relation: instead of fixing a constant density for all objects, we adopt an estimate of the density based on the ionization parameter method. We use the AGN luminosites L_{AGN} and outflow sizes r_{out} given in Fiore *et al.* (2017). And we adopt a value of $\log U = -2.7$, which is the median for outflows in which both [OIII] and Hβ luminosities are available. We then make an order-of-magnitude estimate of n_e as illustrated in the left panel of Fig. 3. The key point of this figure is that it emphasizes the enormous variation that can be expected for n_e in different outflows. It highlights how, in the central few hundred parsecs one can find densities in the range 10^3–$10^4\,\mathrm{cm}^{-3}$; and that while in some AGN this may fall to $\sim$100 cm^{-3} at kiloparsec scales, for luminous quasars it may still be in the $\sim$1000 cm^{-3} range.

The impact on the relation between $\dot{M}_{out}$ and L_{AGN} is shown in the right panel of Fig. 3. The black plusses represent data for the original relation published by Fiore *et al.* (2017). The filled blue circles denote our adjustment to those points, in which we have used our best estimate of the density. The green crosses are from Baron & Netzer (2019); and the filled red circles are the data for our local luminous AGN in Davies *et al.* (2020). It can be seen that the scatter of this proposed revision to the relation is rather larger, and the outflow rates are lower by about a factor three.

5. Conclusions

We have analysed high signal-to-noise spectra from Xshooter at a resolution of $R\sim$ 10000 covering the central $\sim$300 pc (1.8″) of 11 local luminous AGN ($43.0 \leqslant L_{AGN} \leqslant 44.5$). And we have measured the density of the ionized gas in three independent ways. In this data the [SII] doublet underestimates the density by a factor 5–10. It has been known for a long time that in AGN photoionized gas, the low excitation lines ([SII],

[NII], [OI]) are enhanced in a partially ionized zone where the electron density drops rapidly to $< 0.1 n_\mathrm{H}$ and the gas becomes mostly neutral. Photoionization models confirm that because of this, the [SII] doublet is unreliable as a tracer of n_e in the fully ionized gas, and so one should be very cautious of using it to estimate density. Instead, if the signal-to-noise is sufficiently high, one can consider the method that uses auroral and transauroral lines. Alternatively, the ionization parameter method can yield an estimate based on $n_\mathrm{H} \propto L_\mathrm{AGN}/(r^2 U)$. The higher densities, $n_\mathrm{e} \sim 3000\,\mathrm{cm}^{-3}$, we have found indicate the outflowing ionized gas mass is lower than often reported; but it is also associated with a significant mass of neutral gas. Updating the well known $L_\mathrm{AGN} - \dot{M}_\mathrm{out}$ relation based on our best estimate of the density, suggests that the scatter is larger than previously thought, and the outflow rates lower.

References

Baron, D. & Netzer, H. 2019, *MNRAS*, 486, 4290
Davies, R., Burtscher, L., Rosario, D., *et al.* 2015, *ApJ*, 806, 127
Davies, R., Baron, D., Shimizu, T., *et al.* 2020, *MNRAS*, 498, 4150
Dopita, M., Groves, B., Sutherland, R., *et al.* 2015, *ApJ*, 806, 127
Fiore, F., Feruglio, C., Shankar, F., *et al.* 2017, *A&A*, 601, A143
Harrison, C., Costa, T., Tadhunter, C., *et al.* 2018, *NatAs*, 2, 198
Lutz, D., Sturm, E., Janssen, A., *et al.* 2020, *A&A*, 633, A134
Holt, J., Tadhunter, C., Morganti, R., *et al.* 2011, *MNRAS*, 410, 1527
Somerville, R. & Davé, R. 2015, *ARA&A*, 53, 51
Springel, V., Di Matteo, T., & Hernquist, L. 2005, *MNRAS*, 361, 776
Veilleux, S., Maiolino, R., Bolatto, A., *et al.* 2020, *A&ARv*, 28, 2
Vernet, J., Dekker, H., D'Odorico, S., *et al.* 2011, *A&A*, 536A, 105

Galaxy Evolution and Feedback across Different Environments
Proceedings IAU Symposium No. 359, 2020
T. Storchi-Bergmann, W. Forman, R. Overzier & R. Riffel, eds.
doi:10.1017/S1743921320004330

Outflows & Feedback from Extremely Red Quasars

Fred Hamann[1], Serena Perrotta[2] and Nadia Zakamska[3]

[1]Department of Physics & Astronomy, University of California, Riverside,
900 University Ave., Riverside, CA 92521, USA
email: fhamann@ucr.edu

[2]Center for Astrophysics & Space Sciences, University of California, San Deigo,
9500 Gilman Drive, La Jolla, CA 92093, USA
email: s2perrotta@ucsd.edu

[3]Department of Physics & Astronomy, Johns Hopkins University,
3400 N. Charles Street Baltimore, MD 21218, USA
email: zakamska@jhu.edu

Abstract. Feedback from accreting supermassive black holes is often invoked in galaxy evolution models to inhibit star formation, truncate galaxy growth, and establish the observed black-hole/bulge mass correlation. We are studying outflows and feedback in a unique sample of extremely red quasars (ERQs) during the peak epoch of galaxy formation (at redshifts $2.0 < z < 3.4$). We identified ERQs in the Sloan Digital Sky Survey III (SDSS-III) Baryon Oscillation Spectroscopic Survey (BOSS) quasar catalog based on their extremely red i–W3 colors, but we find that ERQs typically have a suite of other extreme properties including 1) a high incidence of blueshifted broad absorption lines, 2) broad emission lines with unusually large rest equivalent widths (REWs), peculiar "wingless" profiles, and frequent large blueshifts (reaching $\sim$8740 km s^{-1}), and 3) characteristically very broad and blueshifted [OIII] 4959,5007Å lines that trace ionized outflows at speeds up to $\sim$6700 km s^{-1}. We propose that these ERQs represent a young quasar population with powerful outflows on the precipice of causing important disruptive feedback effects in their host galaxies.

Keywords. galaxies: evolution, quasars: emission lines, quasars: general

---

1. Introduction

Luminous quasars have powerful outflows that might regulate star formation and mass assembly in their host galaxies (Weinberger *et al.* 2017; Hamann *et al.* 2019). Evolution models predict that early SMBH growth and accompanying quasar/AGN activity occurs mostly in obscurity, deep inside dusty starburst host galaxies. Outflows driven by star formation and the central quasar/AGN activity then combine with consumption of the gas and dust to quench star formation and reveal visibly luminous quasars in galactic nuclei (Sanders *et al.* 1988; Veilluex *et al.* 2009).

Quasars that are obscured and reddened by dust can provide important tests of this galaxy/quasar evolution picture because they are expected to appear preferentially during the brief transition/blowout phase when quasars are still partially embedded in a dusty starburst and, perhaps, the effects of quasar-driven feedback to the host galaxies are most prominent (Glikman *et al.* 2015; Assef *et al.* 2015; Banerji *et al.* 2015). It is helpful to exclude Type 2 quasars from this discussion because they are generally expected to be normal quasars with obscuration caused by orientation effects (Netzer 2015). In contrast,

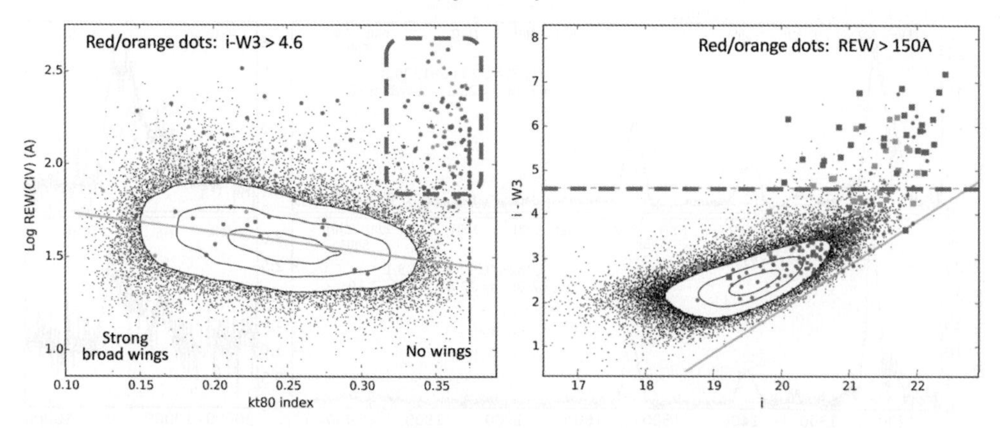

Figure 1. *Left panel:* Log REW(CIV) versus the CIV profile kurtosis, kt80, for BOSS quasars at redshifts $2.0 < z < 3.4$. The red and orange symbols mark Type 1 and 2 ERQs, respectively defined by i–W3 > 4.6. They are strongly clustered to the upper right, e.g., with large REWs and wingless line profiles. *Right panel:* i–W3 color versus i magnitude for these same BOSS quasars. The solid green line shows approximate sensitivity limit of WISE in the W3 filter. The red and orange dots again indicate Type 1 and 2 quasars, respectively, here with REW(C IV) > 150 Å. The red and orange squares additionally indicate "wingless" profiles with kt80 > 0.33. The dashed green line marks the ERQ threshold at i–W3 > 4.6. See Hamann *et al.* (2017).

red Type 1 quasars provide direct views of quasars whose enhanced obscuration might be connected to a young dusty stage of host galaxy evolution.

2. Extremely Red Quasars (ERQs)

Our team discovered a unique new population of extremely red quasars (ERQs) at redshifts $z \sim 2.0$–3.4 in the SDSS-III BOSS and WISE mid-IR surveys (Ross *et al.* 2015; Hamann *et al.* 2017). We identify ERQs based on extremely red colors in i–W3 > 4.6 (AB), as shown in the right panel of Figure 1. The i and W3 filters sample rest-frame ∼0.2μm and ∼3.4 μm, respectively, at the median redshift of our sample. Our ability to find even redder quasars is limited by the i $\lesssim$ 22 limit of the BOSS survey. ERQs selected this way have large bolometric luminosities, $L \gtrsim 10^{47}$ ergs/s, normal radio emission (roughly 8% radio loud), and sky densities that are a few percent of similarly-luminous blue (unobscured) quasars (Hamann *et al.* 2017). These basic characteristics are similar to other red quasar samples such as hot dust-obscured galaxies (HotDOGs, Assef *et al.* 2015; Tsai *et al.* 2015) and highly-reddened Type 1 quasars (Banerji *et al.* 2013, 2015), and they are consistent with ERQs representing a brief obscured phase of quasar activity.

However, many ERQs have a suite of other extreme properties unlike any known quasar population. Most obvious in the BOSS spectra are UV broad emission lines with 1) very large rest equivalent widths (REWs), 2) unusual "wingless" profiles, and 3) peculiar line strength ratios. Figure 1 shows, specifically, how ERQs with i–W3 > 4.6 have a strong preference for large REWs and wingless profiles (large kurtosis, kt80) in the CIV 1549Å broad emission line. Roughly 50% of the reddest ERQs, with i–W3 > 5.6, have REW(CIV) > 150 Å compared to only 0.4% of similarly-luminous blue quasars (Hamann *et al.* 2017). The left panels in Figure 2 show examples of this behavior in two ERQs. Notice, in particular, the dramatically stronger broad emission lines and unusual line profiles in the ERQs compared the median SDSS quasar (from vanden Berk *et al.* 2001). Also notice the unusually weak broad Lyα emission lines and the great strength of NV 1240Å relative to CIV 1549Å.

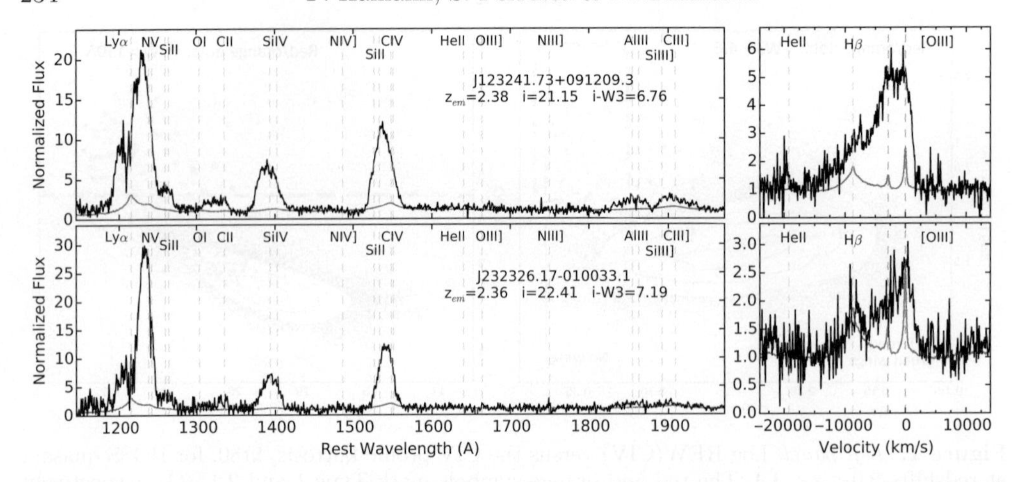

Figure 2. Normalized spectra of two ERQs in the rest UV (left panels) and across the [OIII] 4959,5007A and Hβ lines in the visible (right). The magenta curves show the composite spectrum of normal blue quasars in SDSS (vanden Berk *et al.* 2001). Compared to this normal quasar spectrum, ERQs have very strong and blueshifted NV and CIV broad emission lines, peculiar line profiles, and spectacularly broad and blueshifted [OIII] lines at speeds reaching >6000 km s^{-1}. The [OIII] lines identify powerful outflows that could be driving important feedback effects in the host galaxies (Zakamska *et al.* 2016; Hamann *et al.* 2017; Perrotta *et al.* 2019.

Another common property of ERQs is large blueshifts in CIV 1549Å and other broad emission lines. These blueshifts identify outflows in quasar broad emission-line regions (BELRs, Richards *et al.* 2011). The two ERQs shown in Figure 2 have CIV blueshifts >2500 km s^{-1} relative to a rest frame (blue dashed vertical lines) defined by the low-ionization emission lines of SiII and MgII (not shown). Figure 3 shows more examples of large CIV blueshifts measured relative to narrow emission-line spikes in Lyα, that appear (from Keck KCWI imaging spectroscopy) to be part of extended Lyα emission halos. These results provide additional confirmation of large CIV blueshifts in ERQs, including the largest blueshift ever reported in a quasar spectrum at v$\approx$8740 km s^{-1} (bottom panel in Fig. 3). Overall, we estimate that CIV blueshifts >2500 km s^{-1} are roughly 50 times more common in ERQs than normal blue quasars at the same redshifts and luminosities (Hamann *et al.* 2017).

We also find that ERQs have a high incidence of blueshifted broad *absorption* lines (BALs) and other BAL-like features in their spectra. We estimate that the incidence of these outflow absorption lines is roughly 3 times greater in ERQs compared to the normal blue quasar population (Hamann *et al.* 2017).

Perhaps the most remarkable property of ERQs is their strong tendency to have broader and more blueshifted [OIII] 4959,5007Å emission lines than blue quasar samples. This includes the broadest and most blueshifted [OIII] lines ever reported in quasars, with FWHMs and blueshifted wings reaching ~6700 km s^{-1} (Zakamska *et al.* 2016; Perrotta *et al.* 2019). The right-hand panels in Figure 2 show examples of this extreme [OIII] behavior compared to the median spectrum of SDSS quasars. The left panel in Figure 4 shows more generally how the large [OIII] outflow speeds (e.g., v_{98} measured from the [OIII] line wings) are strongly correlated with red i–W3 colors, e.g., much more so than a previously-known weaker relationship to quasar luminosities. The right panel in Figure 4 shows further that, with reasonable standard assumptions about the outflow densities and sizes, the [OIII] outflows in ERQs have kinetic powers at least several times larger than similarly-luminous blue quasars. Importantly, the derived [OIII] outflow kinetic luminosities, roughly in the range 1–10% of the quasar bolometric luminosities,

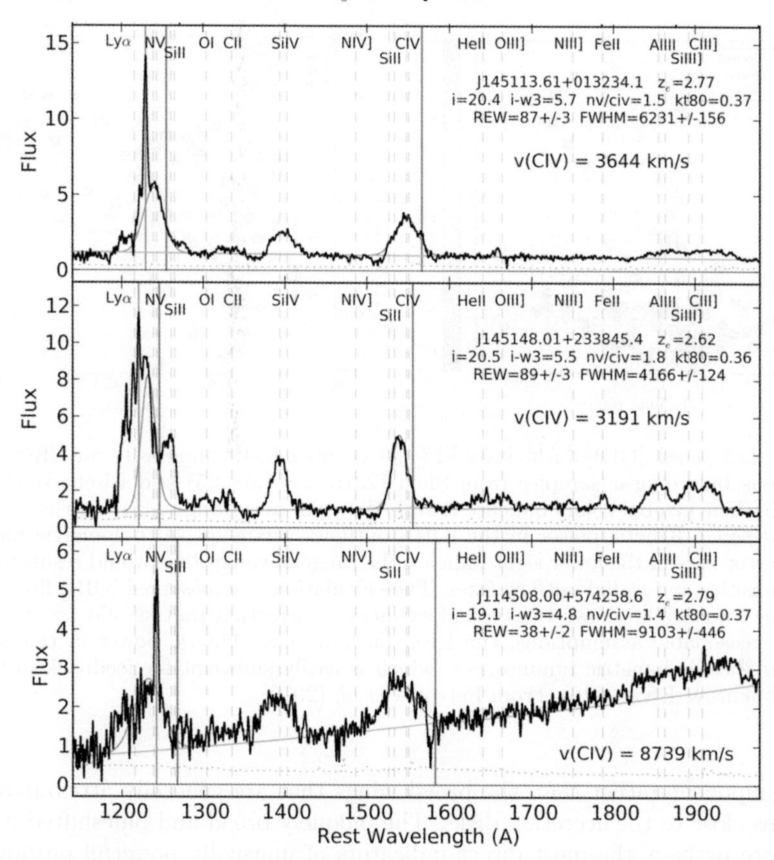

Figure 3. Rest UV spectra of several ERQ with large CIV blueshifts measured relative to narrow Lyα emission-line "spikes" (FWHM$<$1000 km s^{-1}) that form in spatially extended regions around the quasars in the host galaxies and circumgalactic halos. The vertical red lines mark the expected wavelengths of Lyα, NV, and CIV in the this halo frame. The CIV blueshifts relative to this frame, indicated by v(CIV), including the largest emission-line blueshift ever reported at 8739 km s^{-1} (bottom panel). Other emission-line data for CIV are included in the upper right of each panel. These large emission-line blueshifts combined with a high incidence of blueshifted BALs and consistently large [OIII] outflow speeds indicate that ERQs have unusually powerful outflows across spatial scales from $\lesssim$1 pc to $\gtrsim$0.5 kpc.

are larger than recent estimates of the minimum needed for important feedback to the host galaxies (Hopkins & Elvis 2010, see Perrotta *et al.* 2019).

3. Implications

The exotic spectral properties of ERQs might be manifestations of exceptionally powerful accretion-disk outflows, perhaps during an early dusty stage of galaxy/quasar evolution. The higher incidence of BALs and BAL-like outflow lines is consistent with enhanced outflow activity, although orientation effects cannot be ruled out for these absorption-line features. The broad emission lines with unusually large REWs, wingless profiles, and frequent large blueshifts are similarly consistent with enhanced outflow activity in the BELRs (Hamann *et al.* 2017). In particular, the large blueshifts are direct outflow signatures, while the large REWs and wingless profiles might result from unusually extended BELR outflows that intercept and reprocess more of the continuum flux into line radiation (leading to large REWs) with narrower profiles that emphasize the

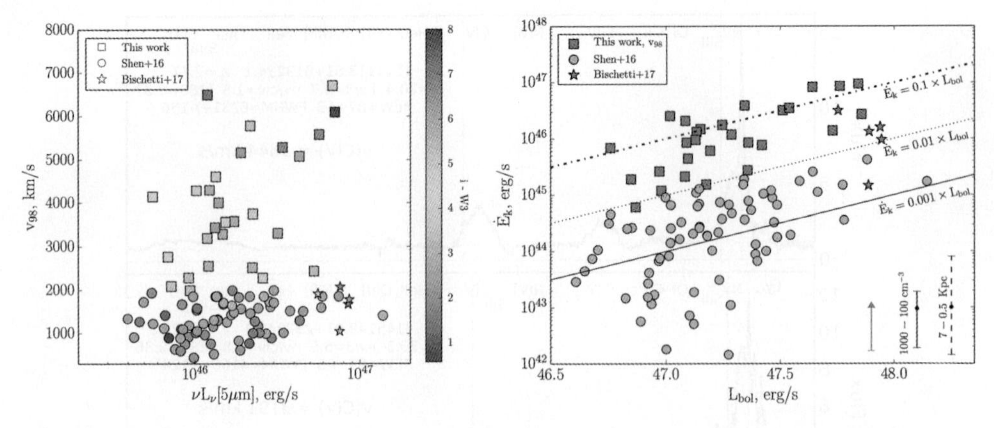

Figure 4. *Left panel:* [OIII] outflow speed (v_{98}) versus mid-IR luminosity for ERQs (squares), and luminous blue quasar samples from Shen (2016, $z \sim 1.5$–3.5) and Bischetti *et al.* (2017, $z \sim 2.3$–3.5). The symbol colors indicate their i–W3 color according to the blue–red color to the right. *Right panel:* Kinetic power in the [OIII] outflows versus quasar bolometric luminosities. The black error bars in the lower right indicate the range of values that would result for different assumed densities and radial outflow sizes. The calculations use *observed* [OIII] fluxes. The red arrow in the lower right shows the typical extinction correction that should be applied to the ERQs. For reasonable assumptions, the ERQ outflows have kinetic power in the range ~1% to ~10% of their bolometric luminosities, which is easily sufficient for feedback in theoretical models (Hopkins & Elvis 2010). From Perrotta *et al.* (2019).

outflow component rather than the broad wings that are generally attributed to emission regions close to the accretion disks. The uniquely broad and blueshifted [OIII] lines in ERQs are perhaps the most direct indicators of unusually powerful outflows. These lines require relatively low densities and therefore extended emitting regions $\gtrsim$0.5 kpc from the central quasars, where they might directly cause feedback to the host galaxies (Zakamska *et al.* 2016; Perrotta *et al.* 2019). However, a tentative result from our ongoing IFU observations with Keck OSIRIS is that the [OIII] outflows in ERQs are compact on scales $\lesssim$1.5 kpc. This is consistent with ERQs being young objects, where the outflows have not had time to expand into the host galaxies. It might also explain the unusually high [OIII] outflow speeds if this gas has not yet impacted and swept up substantial material in the host galaxies. Our working hypothesis is that ERQs are indeed young objects with powerful outflows whose feedback effects have not yet reached their extended host galaxies.

References

Assef, R. J., *et al.* 2015, *ApJ*, 804, 27
Banerji, M., McMahon, R. G., Hewett, P. C., *et al.* 2013, *MNRAS*, 429, L55
Banerji, M., Alaghband-Zadeh, S., Hewett, P. C., *et al.* 2015, *MNRAS*, 447, 3368
Bischetti, M., *et al.* 2017, *AA*, 598, A122
Glikman, E., Simmons, B., Mailly, M., *et al.* 2015, *ApJ*, 806, 218
Hamann, F., *et al.* 2017, *MNRAS*, 464, 3431
Hamann, F., Herbst, H., Paris, I., *et al.* 2019, *MNRAS*, 483, 1808
Hopkins, P. F. & Elvis, M. 2010, *MNRAS*, 401, 7
Netzer, H. 2015, *ARAA*, 53, 365
Perrotta, S., *et al.* 2019, *MNRAS*, 488, 4126
Richards, G. T., *et al.* 2011, *AJ*, 141, 167
Ross, N. P., *et al.* 2015, *MNRAS*, 453, 3932

Sanders, D. B., Soifer, B. T., Elias, J. H., *et al.* 1988, *ApJ*, 325, 74
Shen, Y. 2016, *ApJ*, 817, 55
Tsai, C.-W., *et al.* 2015, *ApJ*, 805, 90
vanden Berk, *et al.* 2001, *ApJ*, 122, 549
Veilleux, S., *et al.* 2009, *ApJS*, 182, 628
Weinberger, R., *et al.* 2017, *MNRAS*, 465, 3291
Zakamska, N. L., *et al.* 2016, *MNRAS*, 459, 3144

Galaxy Evolution and Feedback across Different Environments
Proceedings IAU Symposium No. 359, 2020
T. Storchi-Bergmann, W. Forman, R. Overzier & R. Riffel, eds.
doi:10.1017/S1743921320002240

Feeding and feedback from little monsters: AGN in dwarf galaxies

Mar Mezcua[1,2]

[1]Institute of Space Sciences (ICE, CSIC), Campus UAB, Carrer de Magrans, 08193 Barcelona, Spain
email: marmezcua.astro@gmail.com

[2]Institut d'Estudis Espacials de Catalunya (IEEC), Carrer Gran Capità, 08034 Barcelona, Spain

Abstract. Detecting the seed black holes from which quasars formed is extremely challenging; however, those seeds that did not grow into supermassive should be found as intermediate-mass black holes (IMBHs) of $100-10^5$ $M_\odot$ in local dwarf galaxies. The use of deep multiwavelength surveys has revealed that a population of actively accreting IMBHs (low-mass AGN) exists in dwarf galaxies at least out to $z \sim 3$. The black hole occupation fraction of these galaxies suggests that the early Universe seed black holes formed from direct collapse of gas, which is reinforced by the possible flattening of the black hole-galaxy scaling relations at the low-mass end. This scenario is however challenged by the finding that AGN feedback can have a strong impact on dwarf galaxies, which implies that low-mass AGN in dwarf galaxies might not be the untouched relics of the early seed black holes. This has important implications for seed black hole formation models.

Keywords. galaxies: dwarf, active, accretion, nuclei, evolution

---

1. Introduction

The discovery more than 20 years ago of two low-mass (black hole mass $M_{\rm BH} \lesssim 10^6$ $M_\odot$) active galactic nuclei (AGN; NGC 4395 and POX 52; Filippenko & Sargent 1989; Kunth *et al.* 1987) triggered a quest that has yielded today more than 500 sources. Most of these low-mass AGN are hosted either by disky (Greene *et al.* 2008; Jiang *et al.* 2011; Chilingarian *et al.* 2018) or dwarf galaxies (with stellar mass $M_* \leqslant 3 \times 10^9$ $M_\odot$; Reines *et al.* 2013; Moran *et al.* 2014). They have been identified based either on narrow emission-line diagnostic diagrams accompanied by the detection of broad emission lines (from which a black hole mass measurement has been obtained; e.g. Greene & Ho 2004, 2007; Reines *et al.* 2013; Chilingarian *et al.* 2018), on high-ionization optical/infrared emission lines (e.g. Satyapal *et al.* 2008; Marleau *et al.* 2017), on the detection of X-ray or radio emission indicative of AGN accretion (e.g. Schramm *et al.* 2013; Lemons *et al.* 2015; Mezcua *et al.* 2016, 2018a, 2019; Reines *et al.* 2020), or a combination of all (e.g. see review by Mezcua 2017; Greene *et al.* 2019). A few more tens have been recently identified based on optical variability (e.g. Baldassare *et al.* 2018; Martínez-Palomera *et al.* 2020).

The finding of such a number of (low-mass) AGN in dwarf galaxies poses a challenge to galaxy/black hole formation models. How have dwarf galaxies, with their shallow potential well, been able to form a $\sim 10^4-10^6$ $M_\odot$ black hole at their center? Actually recent studies show that AGN in dwarf galaxies are wandering in their host (e.g. Reines *et al.* 2020), so how have dwarf galaxies been able to assemble an off-nuclear AGN? The answer seems to come from high redshifts ($z\sim10-20$). Low-mass black holes in local dwarf galaxies are thought to be the ungrown relics of the seed black holes formed in the early Universe, with $M_{\rm BH}$ ranging from 100 to $\lesssim10^6$ M$_\odot$ (e.g. Volonteri 2010, 2012; Greene 2012). Such seed black holes have been invoked to explain the finding of quasars hosting supermassive black holes of 10^9-10^{10} M$_\odot$ by $z\sim7$ (Bañados *et al.* 2018; Matsuoka *et al.* 2019) and the presence of overmassive black holes in local brightest cluster galaxies (McConnell *et al.* 2011; Mezcua *et al.* 2018b). They could have formed from the first Population III stars (e.g. Bromm & Larson 2004) or from the collapse of metal-free halos and subsequent formation and death of a supermassive star (direct collapse black holes; e.g. Loeb & Rasio 1994; Hosokawa *et al.* 2013) among other possible scenarios (see reviews by Mezcua 2017; Woods *et al.* 2019).

Stellar and supernova (SN) feedback is assumed to be responsible for hampering the growth of the high-z seed black holes via winds that deplete gas from the center (e.g. Volonteri *et al.* 2008; van Wassenhove *et al.* 2010; Habouzit *et al.* 2017), so that neither the seed black hole nor its host dwarf galaxy grow much through cosmic time and we can observe them today as relics of the first galaxies and first black holes. Recent studies are however starting to challenge this scenario. Both simulations (Smethurst *et al.* 2016; Dashyan *et al.* 2018; Barai & de Gouveia Dal Pino 2019; Koudmani *et al.* 2019; Regan *et al.* 2019) and observations (Bradford *et al.* 2018; Penny *et al.* 2018; Dickey *et al.* 2019; Mezcua *et al.* 2019) are starting to find that AGN feedback can be equally, or even more, significant than SN feedback in dwarf galaxies.

2. AGN vs SN feedback in dwarf galaxies

In cosmological simulations AGN feedback is a crucial ingredient in order to reproduce the observed properties of massive galaxies and the galaxy luminosity function, found to break at L_* (e.g. Croton *et al.* 2006; Bower *et al.* 2017; Choi *et al.* 2017). AGN feedback is also required to explain the baryon cooling efficiency of massive galaxies, while in the low-mass regime SN feedback is sufficient (Behroozi *et al.* 2013). The change of slope from SN- to AGN-regulated regimes is found to occur at L_*, or at $M_*\sim3\times10^{10}$ M$_\odot$, close to the transitional mass typically used to distinguish between massive and dwarf galaxies. Observational evidence for AGN feedback regulation of massive galaxies comes from the spatial coincidence between the large-scale X-ray cavities of galaxy clusters and the radio jets of their central supermassive black holes (e.g. Fabian *et al.* 2000; McNamara *et al.* 2000; Hlavacek-Larrondo *et al.* 2012) and from the finding that the star-formation rate of local massive galaxies depends on supermassive black hole mass (Martín-Navarro *et al.* 2018). The no dependence of star-formation rate with black hole mass for local dwarf galaxies hosting low-mass AGN was instead taken as evidence for SNe being the dominant source of feedback governing such galaxies (Martín-Navarro & Mezcua 2018) as so far assumed in most numerical simulations. This was reinforced by the finding that the $M_{\rm BH}-\sigma$ correlation changed its slope when moving to the low-mass end (i.e. below stellar velocity dispersion $\sigma\sim100$ km s^{-1}) at a transitional stellar mass of $M_*\sim5\times10^{10}$ M$\odot$ that was fully consistent with that of the break in the galaxy luminosity function and change of regimes in baryon cooling efficiency (Martín-Navarro & Mezcua 2018).

However, independent studies performed at the same time indicated opposite results: simulations show that the stellar debris from tidal disruption events could fuel and grow seed black holes (Alexander & Bar-Or 2017; Zubovas 2019; Pfister *et al.* 2020) whose feedback could become relevant and have significant effects on the host galaxy (Zubovas 2019). Observationally, long-slit and integral-field spectroscopy studies of two different samples of quiescent dwarf galaxies revealed that they possibly host AGN, which could be preventing the formation of stars in such galaxies (Dickey *et al.* 2019; Penny *et al.* 2018). AGN feedback could also explain the finding, based on HI observations, of a

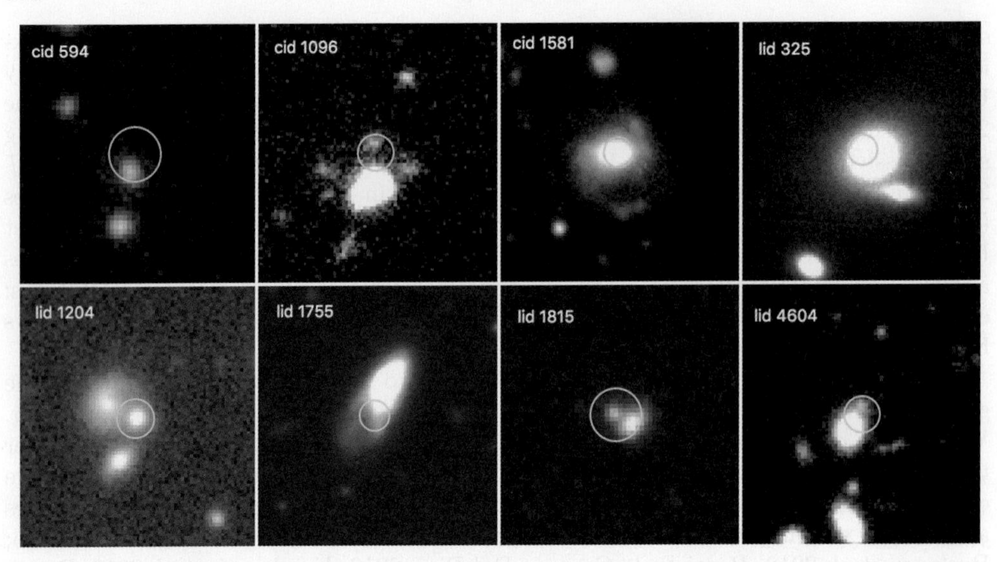

Figure 1. Subaru Hypercam images of eight of the 23 X-ray AGN dwarf galaxies at $z < 0.3$ of Mezcua *et al.* (2018a) showing possible companions or possibly undergoing a merger. The *Chandra* X-ray position is marked with a green circle of radius 1 arcsec.

sample of gas-depleted isolated dwarf galaxies possibly hosting AGN (Bradford *et al.* 2018). X-ray or radio observations are however required to confirm the presence of AGN in these quiescent dwarf galaxies. Based on deep radio observations of the COSMOS field (Smolčić *et al.* 2017), Mezcua *et al.* (2019) found a sample of radio AGN dwarf galaxies whose radio jets have powers and efficiencies as high as those of massive galaxies. This indicates that AGN feedback could be as significant in dwarf galaxies as in more massive ones. In massive galaxies AGN feedback can both prevent and trigger star formation on pc or kpc scales around the black hole (e.g. Silk 2013; Querejeta *et al.* 2016; Maiolino *et al.* 2017), which can affect the amount of material available for the black hole to grow. If AGN feedback is also significant in dwarf galaxies, it could be that seed black hole growth is not hampered by SN feedback but enhanced by AGN feedback (Mezcua 2019).

3. Dwarf galaxy mergers

Dwarf galaxy mergers are another factor to be taken into account. Some AGN are found in dwarf galaxies undergoing a merger (e.g. Bianchi *et al.* 2013; Secrest *et al.* 2017) and several IMBH candidates are located in the outskirts of large galaxies, which suggests they are the nucleus of a dwarf galaxy stripped in the course of a merger (e.g. Farrell *et al.* 2009; Mezcua *et al.* 2013a,b, 2015, 2018c). Such minor mergers are expected to be very common and to trigger up to 50 % of the local star formation activity (Kaviraj 2014). Cosmological simulations show that mergers of dwarf galaxies with similar mass (i.e. dwarf-dwarf mergers) can nonetheless also be very frequent (Fakhouri *et al.* 2010; Deason *et al.* 2014). Studies of individual systems (e.g. Paudel *et al.* 2015, 2020) and large surveys (e.g. Stierwalt *et al.* 2015; Paudel *et al.* 2018) indeed show that dwarf galaxies can be commonly found as interacting or merging pairs. If in the course of such dwarf-dwarf galaxy mergers the two IMBHs coalesce and high accretion rates are triggered, the resulting black hole could have a mass significantly enhanced with respect to that of the initial seeds (Deason *et al.* 2014; Mezcua 2019).

The finding of low-mass AGN in such dwarf-dwarf galaxy mergers is however scarce. Reines *et al.* (2014) find an AGN with $M_{\rm BH}{\sim}10^5-10^7$ M$_\odot$ in the southern member of the dwarf galaxy pair Mrk709, but no black hole is detected for the northern galaxy.

Statistically, 9% of the low-mass AGN in Jiang *et al.* (2011) are found in dwarf galaxies with a possible companion and eight out of the 23 (i.e. 35%) AGN dwarf galaxies at $z < 0.3$ of Mezcua *et al.* (2018a) seem also to have a companion or to be undergoing a merger (see Fig. 1). However, in all these systems the stellar mass of the companion galaxy and whether it hosts a low-mass AGN is unknown. Whether dual AGN are formed or AGN activity is triggered during the merger of two dwarf galaxies is thus far from clear. Even if a dual low-mass AGN was formed, it is yet unclear whether the merger of the two IMBHs would occur as in dwarf galaxies dynamical friction might not be efficient enough to remove the necessary angular momentum to form a close black hole binary so that the black holes might stall and not merge (Tamfal *et al.* 2018). Further studies are thus required to probe the role of dwarf galaxy mergers in seed black hole growth.

4. Conclusions

A myriad of low-mass AGN are being found in dwarf galaxies both in the local Universe and at the peak of cosmic star formation history. Such low-mass AGN could host the ungrown leftover of the first seed black holes formed in the early Universe and invoked to explain the rapid growth of supermassive black holes by $z \sim 7$. AGN feedback, tidal disruption events, and dwarf galaxy mergers can nonetheless yield significant growth of these primordial seeds, in which case local low-mass AGN in dwarf galaxies should not be considered the untouched relics of the high-z seed black holes. This has crucial implications not only for seed black hole formation models, but also for understanding the mechanisms governing black hole-galaxy evolution in the realm of dwarf galaxies.

References

Alexander, T. & Bar-Or, B. 2017, *Nature Astronomy*, 1, 0147
Baldassare, V. F., Geha, M., & Greene, J. 2018, *ApJ*, 868, 152
Bañados, E., Venemans, B. P., Mazzucchelli, C., *et al.* 2018, *Nature*, 553, 473
Barai, P. & de Gouveia Dal Pino, E. M. 2019, *MNRAS*, 487, 5549
Behroozi, P. S. Wechsler, R. H., & Conroy, C. 2013, *ApJ*, 770, 57
Bianchi, S., Piconcelli, E., Pérez-Torres, M. Á., *et al.* 2013, *MNRAS*, 435, 2335
Bower, R. G., Schaye, J., Frenk, C. S., *et al.* 2017, *MNRAS*, 465, 32
Bradford, J. D., Geha, M. C., Greene, J. E., *et al.* 2018, *ApJ*, 861, 50
Bromm, V. & Larson, R. B. 2004, *ARA&A*, 42, 79
Chilingarian, I. V., Katkov, I. Y., Zolotukhin, I. Y., *et al.* 2018, *ApJ*, 863, 1
Choi, E., Ostriker, J. P., Naab, T., *et al.* 2017, *ApJ*, 844, 31
Croton, D. J., Springel, V., White, S. D. M., *et al.* 2006, *MNRAS*, 365, 11
Dashyan, G., Silk, J., Mamon, G. A., *et al.* 2018, *MNRAS*, 473, 5698
Deason, A., Wetzel, A., & Garrison-Kimmel, S. 2014, *ApJ*, 794, 115
Dickey, C. M., Geha, M., Wetzel, A., *et al.* 2019, *ApJ*, 884, 180
Fabian, A. C., Sanders, J. S., Ettori, S., *et al.* 2000, *MNRAS*, 318, L65
Fakhouri, O., Ma, C.-P., & Boylan-Kolchin, M. 2010, *MNRAS*, 406, 2267
Farrell, S. A., Webb, N. A., Barret, D., *et al.* 2009, *Nature*, 460, 73
Filippenko, A. V. & Sargent, W. L. W. 1989, *ApJL*, 342, L11
Greene, J. E. 2012, *Nature Communications*, 3, 1304
Greene, J. E. & Ho, L. C. 2004, *ApJ*, 610, 722
—. 2007, *ApJ*, 670, 92
Greene, J. E., Ho, L. C., & Barth, A. J. 2008, *ApJ*, 688, 159
Greene, J. E., Strader, J., & Ho, L. C. 2019, arXiv:1911.09678
Habouzit, M., Volonteri, M., & Dubois, Y. 2017, *MNRAS*, 468, 3935
Hlavacek-Larrondo, J., Fabian, A. C., Edge, A. C., *et al.* 2012, *MNRAS*, 421, 1360
Hosokawa, T., Yorke, H. W., Inayoshi, K., *et al.* 2013, *ApJ*, 778, 178
Jiang, Y.-F., Greene, J. E., Ho, L. C., *et al.* 2011, *ApJ*, 742, 68

Kaviraj, S. 2014, *MNRAS*, 437, L41
Koudmani, S., Sijacki, D., Bourne, M. A., *et al.* 2019, *MNRAS*, 484, 2047
Kunth, D., Sargent, W. L. W., & Bothun, G. D. 1987, *AJ*, 93, 29
Lemons, S. M., Reines, A. E., Plotkin, R. M. *et al.* 2015, *ApJ*, 805, 12
Loeb, A. & Rasio, F. A. 1994, *ApJ*, 432, 52
Maiolino, R., Russell, H. R., Fabian, A. C., *et al.* 2017, *Nature*, 544, 202
Marleau, F. R., Clancy, D., Habas, R., *et al.* 2017, *A&A*, 602, A28
Martín-Navarro, I., Brodie, J. P., Romanowsky, A. J., *et al.* 2018, *Nature*, 553, 307
Martín-Navarro, I. & Mezcua, M. 2018, *ApJL*, 855, L20
Martínez-Palomera, J., Lira, P., Bhalla-Ladd, I., *et al.* 2020, *ApJ*, 889, 113
Matsuoka, Y., Onoue, M., Kashikawa, N., *et al.* 2019, *ApJL*, 872, L2
McConnell, N. J., Ma, C.-P., Gebhardt, K., *et al.* 2011, *Nature*, 480, 215
McNamara, B. R., Wise, M., Nulsen, P. E. J., *et al.* 2000, *ApJL*, 534, L135
Mezcua, M. 2017, *International Journal of Modern Physics D*, 26, 1730021
—. 2019, *Nature Astronomy*, 3, 6
Mezcua, M., Civano, F., Fabbiano, G., *et al.* 2016, *ApJ*, 817, 20
Mezcua, M., Civano, F., Marchesi, S., *et al.* 2018a, *MNRAS*, 478, 2576
Mezcua, M., Farrell, S. A., Gladstone, J. C., *et al.* 2013a, *MNRAS*, 436, 1546
Mezcua, M., Hlavacek-Larrondo, J., Lucey, J. R., *et al.* 2018b, *MNRAS*, 474, 1342
Mezcua, M., Kim, M., Ho, L. C., *et al.* 2018c, *MNRAS*, 480, L74
Mezcua, M., Roberts, T. P., Lobanov, A. P., *et al.* 2015, *MNRAS*, 448, 1893
Mezcua, M., Roberts, T. P., Sutton, A. D., *et al.* 2013b, *MNRAS*, 436, 3128
Mezcua, M., Suh, H., & Civano, F. 2019, *MNRAS*, 488, 685
Moran, E. C., Shahinyan, K., Sugarman, H. R., *et al.* 2014, *AJ*, 148, 136
Paudel, S., Duc, P. A., & Ree, C. H. 2015, *AJ*, 149, 114
Paudel, S., Sengupta, C., Yoon, S.-J., *et al.* 2020, *AJ*, 159, 141
Paudel, S., Smith, R., Yoon, S. J., *et al.* 2018, *ApJs*, 237, 36
Penny, S. J., Masters, K. L., Smethurst, R., *et al.* 2018, *MNRAS*, 476, 979
Pfister, H., Volonteri, M., Lixin Dai, J., & Colpi, M. 2020, arXiv:2003.08133
Querejeta, M., Schinnerer, E., García-Burillo, S., *et al.* 2016, *A&A*, 593, A118
Regan, J. A., Downes, T. P., Volonteri, M., *et al.* 2019, *MNRAS*, 486, 3892
Reines, A. E., Condon, J. J., Darling, J., *et al.* 2020, *ApJ*, 888, 36
Reines, A. E., Greene, J. E., & Geha, M. 2013, *ApJ*, 775, 116
Reines, A. E., Plotkin, R. M., Russell, T. D., *et al.* 2014, *ApJL*, 787, L30
Satyapal, S., Vega, D., Dudik, R. P., *et al.* 2008, *ApJ*, 677, 926
Schramm, M., Silverman, J. D., Greene, J. E., *et al.* 2013, *ApJ*, 773, 150
Secrest, N. J., Schmitt, H. R., Blecha, L., *et al.* 2017, *ApJ*, 836, 183
Silk, J. 2013, *ApJ*, 772, 112
Smethurst, R. J., Lintott, C. J., Simmons, B. D., *et al.* 2016, *MNRAS*, 463, 2986
Smolčić, V., Novak, M., Bondi, M., *et al.* 2017, *A&A*, 602, A1
Stierwalt, S., Besla, G., Patton, D., *et al.* 2015, *ApJ*, 805, 2
Tamfal, T., Capelo, P. R., Kazantzidis, S., *et al.* 2018, *ApJL*, 864, L19
van Wassenhove, S., Volonteri, M., Walker, M. G., *et al.* 2010, *MNRAS*, 408, 1139
Volonteri, M. 2010, *A&Ar*, 18, 279
—. 2012, *Sci*, 337, 544
Volonteri, M., Lodato, G., & Natarajan, P. 2008, *MNRAS*, 383, 1079
Woods, T. E., Agarwal, B., Bromm, V., *et al.* 2019, *PASA*, 36, e027
Zubovas, K. 2019, *MNRAS*, 483, 1957

Galaxy Evolution and Feedback across Different Environments
Proceedings IAU Symposium No. 359, 2020
T. Storchi-Bergmann, W. Forman, R. Overzier & R. Riffel, eds.
doi:10.1017/S1743921320001775

Taking snapshots of the jet-ISM interplay with ALMA

Raffaella Morganti[1,2], Tom Oosterloo[1,2] and Clive N. Tadhunter[3]

[1]ASTRON, the Netherlands Institute for Radio Astronomy, Oude Hoogeveensedijk 4, 7991PD Dwingeloo, The Netherlands
email: morganti@astron.nl

[2]Kapteyn Astronomical Institute, University of Groningen, Postbus 800, 9700 AV Groningen, The Netherlands

[3]Department of Physics and Astronomy, University of Sheffield, Sheffield, S7 3RH, UK

Abstract. We present an update of our ongoing project to characterise the impact of radio jets on the interstellar medium (ISM). This is done by tracing the distribution, kinematics and excitation of the molecular gas at high spatial resolution using ALMA. The radio active galactic nuclei (AGN) studied are in the interesting phase of having a recently born radio jet. In this stage, the plasma jets can have the largest impact on the ISM, as also predicted by state-of-the-art simulations. The two targets we present have quite different ages, allowing us to get snapshots of the effects of radio jets as they grow and evolve. Interestingly, both also host powerful quasar emission, making them ideal for studying the full impact of AGN. The largest mass outflow rate of molecular gas is found in a radio galaxy (PKS 1549−79) hosting a newly born radio jet still in the early phase of emerging from an obscuring cocoon of gas and dust. Although the molecular mass outflow rate is high (few hundred $M_\odot$ yr^{-1}), the outflow is limited to the inner few hundred pc region. In a second object (PKS 0023−26), the jet is larger (a few kpc) and is in a more advanced evolutionary phase. In this object, the distribution of the molecular gas is reminiscent of what is seen, on larger scales, in cool-core clusters hosting radio galaxies. Interestingly, gas deviating from quiescent kinematics (possibly indicating an outflow) is not very prominent, limited only to the very inner region, and has a low mass outflow rate. Instead, on kpc scales, the radio lobes appear associated with depressions in the distribution of the molecular gas. This suggests that the lobes have broken out from the dense nuclear region. However, the AGN does not appear to be able, at present, to stop the star formation observed in this galaxy. These results support the idea that the effects of the radio source start in the very first phases by producing outflows which, however, tend to be limited to the kpc region. After that, the effects turn into producing large-scale bubbles which could, in the long term, prevent the surrounding gas from cooling. Thus, our results provide a way to characterise the effect of radio jets in different phases of their evolution and in different environments, bridging the studies done for radio galaxies in clusters.

Keywords. galaxies: active, galaxies: jets, radio continuum: galaxies, ISM: jets and outflows

---

1. Introduction

The evolution of massive galaxies appears to be strongly influenced by the energy released during the active phase of their super massive black hole (SMBH). This process, known as feedback, is considered the one regulating, and quenching, their star formation (e.g. Harrison 2017). Feedback is believed to work in two main modes, both aimed at reducing the amount of cold gas and the related star formation: "quasar" mode, with gas outflows driven by the active galactic nucleus (AGN), clearing the gas from the host galaxy, and the "maintenance" mode, where the energy released by the AGN prevents the cooling of the gas on larger scales, from the hot halo or from the intergalactic medium. In the commonly assumed picture of AGN feedback, the role of radio jets is considered to be mostly connected to the latter mode (see e.g. McNamara & Nulsen 2012). This mode complements the effect of outflows/winds considered to dominate in radiatively efficient AGN. However, the picture we are getting from the growing number of detailed observations tracing multiple phases of the gas appears to be more complex and these two modes appear intertwined. An example of such complexity is the fact that radio jets can also produce gaseous outflows, thus having an impact on (sub) kpc-scales. Furthermore, their relative importance may even change during the life of the AGN. *Thus, radio AGN with jets represent ideal objects to trace the feedback while they evolve from sub-kpc to many tens of kpc scales.*

The impact of jets in producing outflows has been observed in an increasing number of both high- and low-power (including radio quiet) radio sources (Morganti 2020 and refs therein). Particularly interesting is that jet-driven outflows are more prominent when the jets are in their initial phase (see e.g. Holt *et al.* 2009; Shih *et al.* 2013; Morganti & Oosterloo 2018; Molyneux *et al.* 2019 and many others). Newly born or young radio jets can be identified by the characteristics of their radio emission (e.g. size, and peaked radio spectrum, for more details see e.g. O'Dea 1998; Orienti 2016). The impact of jets in their starting phase is also predicted by numerical simulations. The work of Wagner *et al.* (2012); Bicknell *et al.* (2018); Mukherjee *et al.* (2016, 2018a) has shown the strong coupling of the newly born jet with the surrounding clumpy interstellar medium (ISM). Furthermore, this interaction is predicted to produce a cocoon of shocked gas expanding perpendicular to the jet, thus impacting a much larger volume of the host galaxy than just the region of the jets themselves.

Finally, the *impact of jets can be dominant even when a radiatively efficient AGN is present.* The best example of this is IC 5063, a Seyfert 2 galaxy, with strong emission lines. This galaxy hosts low radio power jets ($P_{1.4\ \mathrm{GHz}} \sim 10^{23.4}$ W Hz^{-1}). Despite the low radio power, they provide one of the clearest examples of jet-induced outflows, where the radio plasma is disturbing the kinematics of *all the phases of the gas* (see Tadhunter *et al.* 2014; Morganti *et al.* 2015, for an overview). The region co-spatial with the radio emission is where the most kinematically disturbed ionised, molecular and HI gas is located (with velocities deviating up to 600 km s^{-1} from regular rotation). This jet-ISM interaction also affects the physical conditions of the gas (Oosterloo *et al.* 2017). The properties observed have been well reproduced by hydrodynamic simulations and the details of the comparison between the data and the simulation is presented in Mukherjee *et al.* (2018a).

All this strongly indicates the relevance of radio jets, particularly in the initial phases of their evolution. However, after 1-2 kpc (typically after a few Myr) the jet breaks out from the dense central core and the way it interacts with the surrounding medium changes. *Thus, should we expect that the impact will change with jet evolution?* This question is particularly important, also because there appears to be a general consensus that most observed outflows (regardless their origin) are largely limited to the central kpc region while only a small fraction of the outflowing gas is actually leaving the galaxy. Thus, gas outflows may not be enough to supply the required feedback from AGN. Here we focus on two powerful radio sources ($P \sim 10^{26}$–10^{27} W Hz^{-1}) with jets in different phases of their *initial* evolution. Interestingly, the sources also host radiativly efficient AGN. Thus, in these objects there is no shortage of energy released by their active SMBH. We use molecular gas as tracer of the impact of the AGN because it is typically found to carry most of the outflowing mass. This is part of a larger project to understand the impact of radio jet as function of their properties (i.e. power, age, environment etc.; see also Maccagni *et al.* 2018; Oosterloo *et al.* 2017 for other objects studied).

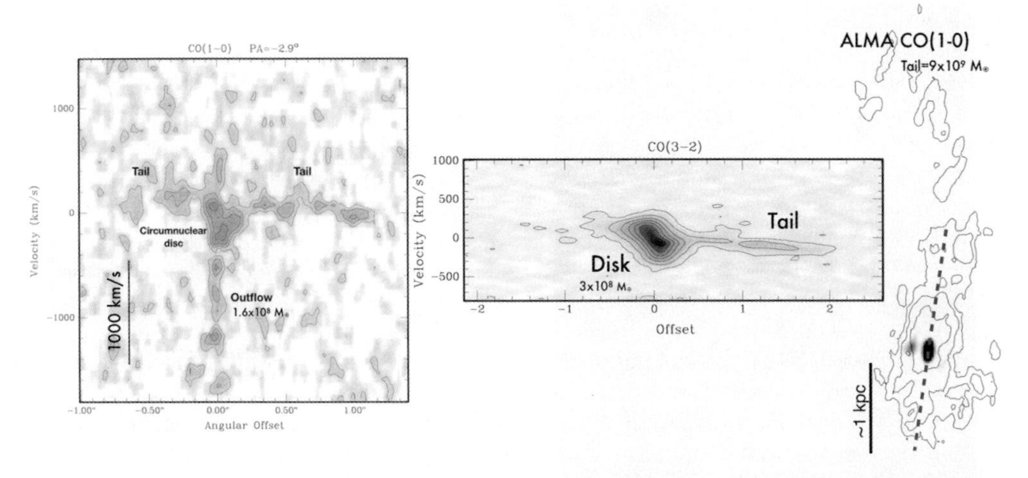

Figure 1. The distribution and kinematics of the molecular gas in the central few hundred pc of PKS 1549−79 as obtained with ALMA, see Oosterloo *et al.* (2019) for details. ALMA CO(1-0) and CO(3-2) detected in emission with spatial resolution ranging from 0.05 arcsec (∼100 pc) to 0.2 arcsec. The position-velocity plots are made along the dashed line, Morganti (2020).

2. Accretion and feedback in an obscured, young radio quasar

The first object shown here is PKS 1549−79 ($z = 0.150$), having a young radio jet (∼300 pc in size) hosted by an obscured, far-IR bright quasar and, therefore, in a particularly crucial early stage in the evolution (Holt *et al.* 2006). As expected, a fast outflow of warm ionised gas as well as a Ultra Fast Outflow in X-rays (Tombesi *et al.* 2014) are present, but the kinetic power of the warm outflow $\sim 4 \times 10^{-4}\, L_{\rm edd}$.

The ^{12}CO(1–0) and CO(3-2) ALMA high resolution observations (Oosterloo *et al.* 2019) show the presence of three gas structures, which can be seen in Fig. 1. Kiloparsec-scale tails are observed, resulting from an on-going merger which provides gas that is accreting onto the centre of PKS 1549–79. At the same time, a circum-nuclear disc has formed in the inner few hundred parsec, and a very broad (>2300 km s^{-1}) component associated with fast outflowing molecular gas is detected at the position of the AGN. As expected, the outflow is massive (∼600 M$_\odot$ yr^{-1}) but, despite the fact that PKS 1549−79 should represent an ideal case of feedback in action, it is limited to the inner 200 pc. These results illustrate that the impact on the surrounding medium of the energy released by the AGN is not always as expected from the feedback scenario. The outflow of warm, ionised gas is slightly more extended (see Oosterloo *et al.* 2019), but modest in terms of mass outflow rate (∼2 M$_\odot$ yr^{-1}; Holt *et al.* 2006, Santoro *et al.* in prep).

Both the jet and the radiation could drive the outflow. Circumstantial evidence suggests that the jet may play the prominent role. We observed a strongly bent component of the jet, characterised by a very steep radio spectrum. This suggests that this part of the jet is a remnant structure, possibly resulting from a strong interaction that has temporarily destroyed the jet. This interaction could have produced the massive outflow. This interaction is also impacting the conditions of the gas, as seen from the high ratio CO(3-2)/CO(1-0) found in the central regions. The depletion time is relatively short (10^5−10^6 yr), suggesting that the outflow will last only for a relatively short time.

3. PKS 0023−26: a few kpc-scale young radio AGN

A second object, PKS 0023−26 ($z = 0.32188$), was selected because, although still a young radio source, it is in a more evolved phase having reached already a few kpc in

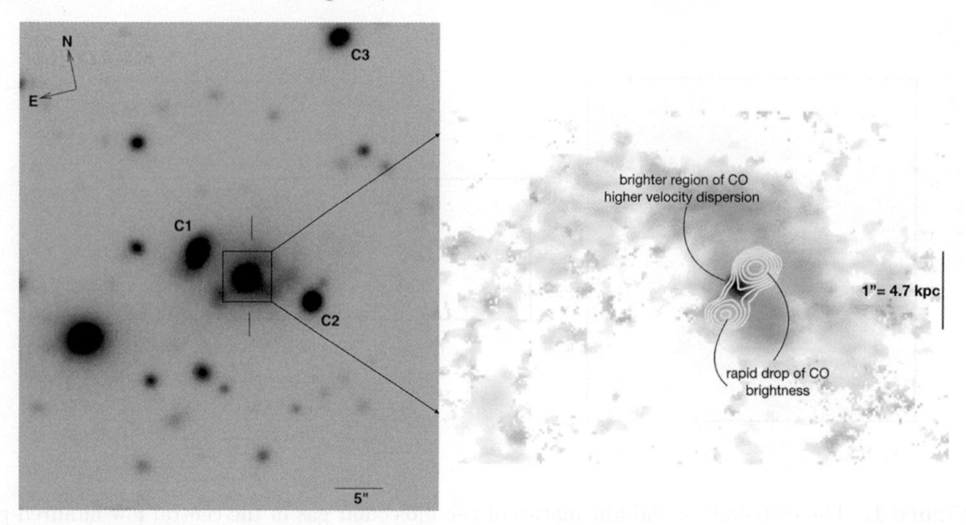

Figure 2. Left Optical image from Gemini GMOS-S. Marked as C1, C2, C3 galaxies confirmed to have redshifts similar to PKS 0023−26. More objects and tails are seen even closer (∼10 kpc) to the target galaxy, Ramos Almeida *et al.* (2013); **Right** Total intensity of the molecular gas (orange scale) with superimposed contours (cyan) of the continuum emission. The center(core) of the radio emission is coincident with the peak of the molecular gas.

size. Perhaps unusual for powerful radio galaxies, it is located in a dense environment (see Fig. 2, left; Ramos Almeida *et al.* 2013). Also interesting is the presence in the host galaxy of an extended region with a very young stellar population (∼30 Myr, Holt *et al.* 2006; Tadhunter *et al.* 2011). The corresponding star formation rate (∼30 $M_\odot$ yr^{-1}) is consistent with what is expected for main-sequence star forming galaxies of similar stellar mass as the host galaxy. The deep, high resolution (0.2 arcsec) ALMA CO(2-1) observations reveal that PKS 0023−26 is embedded in 5×10^{10} $M_\odot$ of molecular gas, distributed over about 20 kpc (see Fig. 2, right). Interestingly, the distribution is reminiscent of those seen in cool-core clusters (e.g. Russell *et al.* 2019), because it appears offset from the centre of the galaxy (and radio source). Part of the gas distributed in filaments with relatively smooth velocity gradients reaching out some of the companion galaxies. However, the large amount of molecular gas detected, and the high velocity dispersions observed, suggest that, at least part of the gas is coming from the cooling of the hot X-ray halo (tentatively detected with XMM; Mingo *et al.* 2014).

The central region is brightest in CO (Fig. 2, right), either because the molecular gas has piled up there, or because it has higher excitation due to the stronger impact of the AGN (as seen in other objects; Oosterloo *et al.* 2017, 2019). Indeed, in this object the gas with velocities deviating from the quiescent kinematics is also located in the central sub-kpc region. However, the velocities are low (not more than ∼300 km s^{-1}). If associated with an outflow, the mass outflow rate is much more modest than in PKS 1549−79. Interestingly, this appears to follow the trend found by Holt *et al.* (2008) for the ionised gas: the amplitude of the outflows decreases as the radio jets expand. Outside this central region, the brightness of the molecular gas drops rapidly in the regions of the radio lobes. A possible explanation is that the jets have already broken out from the dense, central region and are now starting to create radio plasma bubbles which, at a certain point time, will prevent the cooling of the hot ISM. Thus, the AGN (optical and radio) does not have *at present* any substantial impact on the gas on galaxy scales of tens of kpc where substantial star formation is ongoing from the large reservoir of molecular gas.

Possibly, the radio source is still too young and is in an early phase of interaction with the rich gaseous medium and *only starting* to affect it.

4. Connecting the two objects: evolution of the impact of the jets?

Based on the results on these two objects (and other cases studied in detail in the literature, see Morganti 2020 for an overview), we suggest that in the first phases (i.e. in the sub-kpc region and for ages $<10^6$yr) the radio jets are expanding in the inner dense, clumpy ISM where the coupling between the jet and the ISM is very strong. In this phase, the jets can drive fast and massive outflows. The meandering of the jet through the ISM also creates a cocoon of shocked gas expanding in the direction perpendicular to the jet (see Mukherjee *et al.* 2018a and refs therein). Although the mass outflow rate of the molecular gas can be large in very young jets (as found in PKS 1549−79), the size of the region affected can be limited to a few hundred pc. Furthermore, the speed of the outflows appears to decrease as the jet expands as seen in PKS 0023−26 (and found for the ionised gas; Holt *et al.* 2009). However, the impact appears to change as the jet evolves. When the radio jets expand further, i.e. outside the 1-2 kpc region, they break out from the dense central gas and the type of interaction changes, becoming more similar to the one observed e.g. in X-rays with the formation of cavities and jet-driven expanding bubbles in the host galaxy and in the IGM.

In the case of PKS 0023−26, the gas at kpc scales is still forming stars, unaffected by the AGN. This will likely continue until the available molecular gas is depleted by this process ($\sim10^9$ yr) while in the meantime the effect of the growing jets will likely increase. When they reach larger scales (tens of kpc on time scales of a few $\times10^7-10^8$ yr), they can become more efficient in preventing more of the hot gas in the halo from cooling and, therefore, quenching future star formation. It interesting that studies focusing on the star formation rate and AGN luminosity are reaching similar conclusions (see Harrison *et al.* 2019). In order to confirm the trends found so far and to test this scenario, we need to expand the number of radio galaxies studied, while covering a large parameter space in terms of age and radio power, as well as exploring the properties of the hot gas.

References

Bicknell, G. V., Mukherjee, D., Wagner, A. Y., *et al.* 2018, *MNRAS*, 475, 3493
Harrison, C. M. 2017, *Nature Astronomy*, 1, 0165
Harrison, C. M., Alexander, D. M., Rosario, D. J., *et al.* 2019, arXiv e-prints, arXiv:1912.01020
Holt, J., Tadhunter, C., Morganti, R., *et al.* 2006, *MNRAS*, 370, 1633
Holt, J., Tadhunter, C. N., & Morganti, R. 2008, *MNRAS*, 387, 639;
Holt, J., Tadhunter, C. N., & Morganti, R. 2009, *MNRAS*, 400, 589
Maccagni, F. M., Morganti, R., Oosterloo, T. A., *et al.* 2018, *A&A*, 614, A42
McNamara, B. R. & Nulsen, P. E. J. 2012, *New Journal of Physics*, 14, 055023
Mingo, B., Hardcastle, M. J., Croston, J. H., *et al.* 2014, *MNRAS*, 440, 269
Molyneux, S. J., Harrison, C. M., & Jarvis, M. E. 2019, *A&A*, 631, A132
Morganti, R., Oosterloo, T., Oonk, J. B. R., *et al.* 2015, *A&A*, 580, A1
Morganti, R. & Oosterloo, T. 2018, *A&ARew*, 26, 4
Morganti, R. 2020, *proceedings of IAU Symp 356*, arXiv e-prints, arXiv:2001.02675
Mukherjee, D., Bicknell, G. V., Wagner, A. Y. *et al.* 2018a, *MNRAS*, 479, 5544
Mukherjee, D., Bicknell, G. V., Sutherland, R., *et al.* 2016, *MNRAS*, 461, 967;
O'Dea, C. P. 1998, *PASP*, 110, 493
Oosterloo, T., Raymond Oonk, J. B., Morganti, R., *et al.* 2017, *A&A*, 608, A38
Oosterloo, T., Morganti, R., Tadhunter, C., *et al.* 2019, *A&A*, 632, A66
Orienti, M. 2016, *Astronomische Nachrichten*, 337, 9
Ramos Almeida, C., Bessiere, P. S., Tadhunter, C. N., *et al.* 2013, *MNRAS*, 436, 997
Russell, H. R., McNamara, B. R., Fabian, A. C., *et al.* 2019, *MNRAS*, 490, 3025

Shih, H.-Y., Stockton, A., & Kewley, L. 2013, *ApJ*, 772, 138
Tadhunter, C., Holt, J., González Delgado, R., *et al.* 2011, *MNRAS*, 412, 960
Tadhunter, C., Morganti, R., Rose, M., *et al.* 2014, *Nature*, 511, 440
Tombesi, F., Tazaki, F., Mushotzky, R. F., *et al.* 2014, *MNRAS*, 443, 2154
Wagner, A. Y., Bicknell, G. V., Umemura, M., *et al.* 2012, *ApJ*, 757, 136

Galaxy Evolution and Feedback across Different Environments
Proceedings IAU Symposium No. 359, 2020
T. Storchi-Bergmann, W. Forman, R. Overzier & R. Riffel, eds.
doi:10.1017/S1743921320002446

Nuclear ionised outflows in a sample of 30 local galaxies

D. Ruschel-Dutra[1], T. Storchi-Bergmann[2] and A. Schnorr-Müller[2]

[1]Departamento de Física, Universidade Federal de Santa Catarina, P.O. Box 476,
88040-900, Florianópolis, SC, Brazil
e-mail:daniel.ruschel@ufsc.br

[2]Instituto de Física, Universidade Federal do Rio Grande do Sul, Av. Bento Goncalves 9500,
91501-970 Porto Alegre, RS, Brazil

Abstract. Understanding active galactic nuclei (AGN) feedback is essential for building a coherent picture of the evolution of the super massive black hole and its host galaxy. To that end we have analysed the inner kiloparsec of a sample of 30 local AGN with spatially resolved optical spectroscopy. In this talk I will review the analysis of the ionised gas for the galaxies in our sample, including kinematical maps, emission line ratios and fluxes. The W_{80} kinematical index is used to trace outflows, and also to provide an estimate for the outflowing velocity. Electron densities, derived from the [S II] $\lambda\lambda$6716, 6731Å lines, along with Hα luminosities and the sizes of the outflowing regions are employed in estimates of the outflowing gas mass. We find a median mass outflow rate of $\dot{M} = 0.3\ \mathrm{M_\odot\ yr^{-1}}$ and median outflow power of $\log[P/(\mathrm{erg\,s^{-1}})] = 40.4$.

Keywords. active galactic nuclei, feedback, ionised gas

---

1. Introduction

The discovery of correlations between the mass of the central supermassive black hole (SMBH) and properties of the host galaxy, such as the host spheroid mass and velocity dispersion (Ferrarese & Merritt 2000; Gültekin *et al.* 2009; Kormendy & Ho 2013; van den Bosch 2016), and the similar evolution of the cosmic star formation rate (SFR) density and the black hole accretion rate density (Madau & Dickinson 2014) points to the growth of the SMBH being closely linked to the assembly of stellar mass of the host galaxy. It is believed that this link emerges due to both the mass transfer to the inner region of the galaxy that also feeds the SMBH (Storchi-Bergmann & Schnorr-Müller 2019) and the regulating effect of feedback from the triggered active galactic nuclei (AGN) on the star formation in the host galaxy. The effect of AGN feedback is supported by cosmological simulations and models of galaxy evolution (Springel *et al.* 2005; Vogelsberger *et al.* 2014; Schaye *et al.* 2015), as they require such feedback to reproduce observables such as the shape of the galaxy luminosity function, the colour bimodality of the galaxy population in the local universe, and the low star formation efficiency in the most massive galaxies (Alexander & Hickox 2012; Fabian 2012; Harrison 2017).

The main goal here is to probe the gas excitation and kinematics within the inner kiloparsec of the host galaxies at spatial resolution down to a few tens of parsecs in order to resolve the relevant processes of feeding and feedback of the AGN at the nucleus. In this contribution, mass-outflow rates and outflow kinetic power are based on the W_{80} index, defined as the width in velocity space that encompasses 80 per cent of the emission line

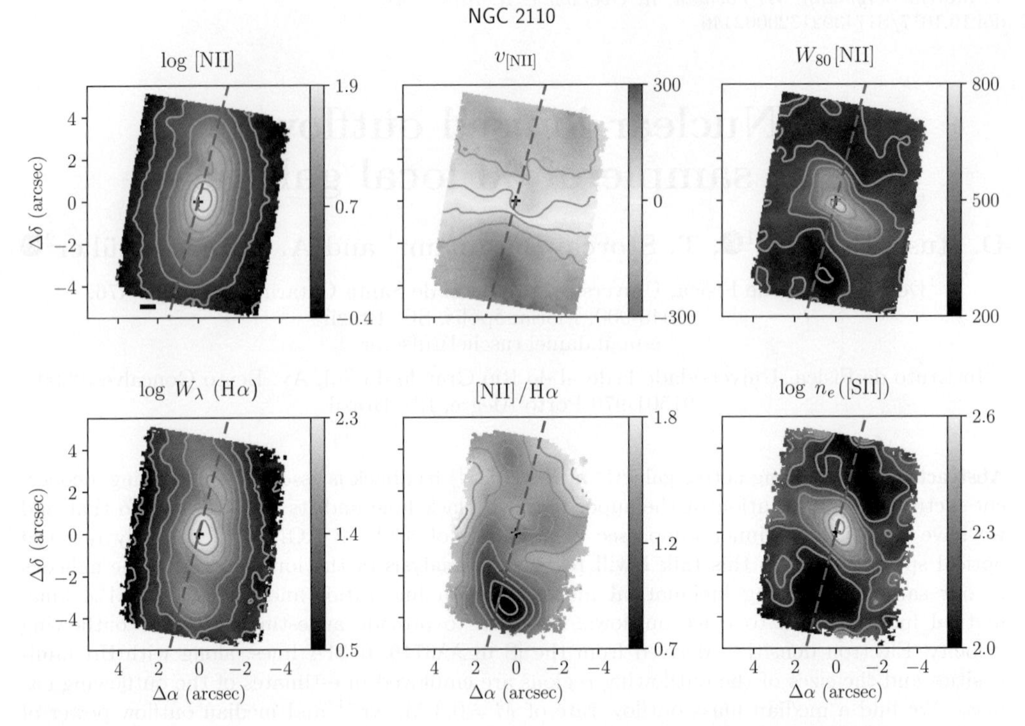

Figure 1. Example of quantities derived from the emission line fitting for NGC 2110. From left to right and top to bottom the panels show: i) the logarithm of the flux in units of 10^{-15} erg s^{-1} cm^{-2} arcsec^{-2}; ii) the radial velocity in units of km s^{-1}; iii) the W_{80} index also in units of km s^{-1}; iv) the logarithm of the equivalent width of the Hα line in Å; v) the flux ratio between the [N II] and the Hα lines; vi) and the electron density derived from the [S II] lines. The green contour in the W_{80} panel highlights the region where $W_{80} \geqslant 600$ km s^{-1}, and thus spectra within this contour are classified as outflow dominated.

flux. We also compare our results with those of previous studies relating these quantities to the AGN luminosity (e.g. Fiore *et al.* 2017).

2. Data

Our sample comprises 30 AGN observed with the Gemini instruments GMOS-IFUs (North and South), up until the observing semester 2017A, and limited to a redshift of $z \leqslant 0.03$. Out of the 30 galaxies in the sample, 17 have a counterpart in the 70 month Swift/BAT (hereafter SB70) catalogue (Baumgartner et al. 2013). The remaining 7 less luminous sources are either LINERS or Sy2's, including at least one known galaxy (NGC 4180) harbouring a Compton-thick AGN.

The data used in this study comes from many observing runs, although with similar setups, obtained with the Gemini Multi-Object Spectrographs (GMOS) integral field units (IFUs) (Allington-Smith et al. 2002) both at the northern and southern Gemini telescopes. The gratings used in these observations, namely B600 and R400, have a resolving power of $R \sim 1800$, which translates to emission-line widths FWHM of ~50 km/s. [O III]and Hβ lines are available for 19 galaxies of the sample observed in single slit mode, covering the wavelength range ≈ 4800–7000Å, while the other 11 targets have spectra in the range ≈5600–7000Å. Angular resolutions vary between 0.6 and 1.0 arcsec, depending on the seeing. Fig. 1 shows an example of the quantities derived from the emission line fitting of the data cubes.

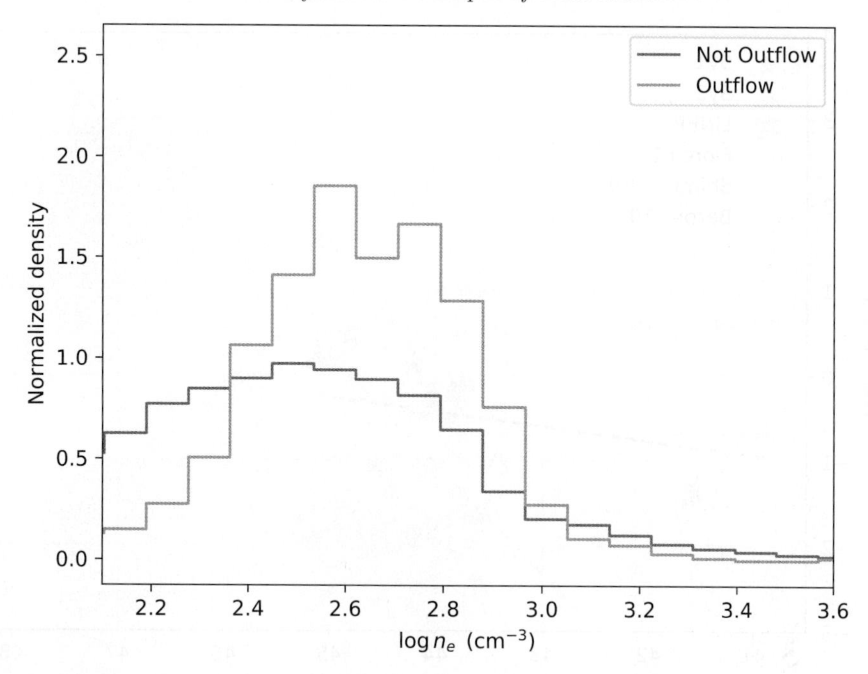

Figure 2. Histogram of electron densities measured in spaxels identified as outflow dominated (*orange*), and not outflow dominated (*blue*), in units of cm^{-3}. Spectra for which the NLR has an outflow component are usually denser and have a narrower range of densities.

3. Results and Discussion

In order to quantify the outflows, we identified the spaxels in which the ionised gas kinematics is not compatible with disk rotation. A spaxel is defined as being part of an outflow, or having its nebular emission dominated by outflowing gas, based on the following criteria: i) $W_\lambda(H\alpha) > 6$Å to ensure an accurate value for $L_{\mathrm{H}\alpha}$; ii) the velocity dispersion, measured by W_{80} of the [N II] λ6583Å or [O III] 5007Å line, must be above 600 km s^{-1}, a limit which excludes reasonable expectations for bound orbits even in the most massive galaxies; iii) it has at least four contiguous neighbouring spaxels also matching the first two criteria, which warrants against spurious detection, since the FWHM of the point spread function is at best 3 spaxels. Out of 28 galaxies with well constrained emission line fits, 12 are classified as having outflows if we apply the above mentioned criteria to the [N II] λ6583Å line. The two excluded galaxies are NGC 1068 and Mrk 6 which have some caveats that are beyond the scope of this communication.

Using the [S II] lines λλ6716, 6731Å we have estimated the electron densities for all the spaxels in the sample, adopting a fixed standard electron temperature of 10^4 K, and applying the relation described in Proxauf *et al.* (2014). Figure 2 shows the histogram of electron densities for spectra identified as outflows and for those we do not classify as outflows. The histogram is given in units of probability density, which means that the integral of the histogram equals unity. Median values are 257^{+351}_{-157} cm^{-3} and 414^{+282}_{-170} cm^{-3} for the non-outflow and outflow samples respectively, where the upper and lower limits are the distances from the median to the 16% and 84% percentiles. Our main conclusion in this regard is that there is a clear tendency towards higher density for regions where the ionised gas emission is characteristic of outflows. This could be interpreted as the result of shocks between the outflowing gas and the interstellar medium. However, it is also possible that in some cases the high values of W_{80} are caused by two gas systems which are superimposed on our line of sight, and not necessarily interacting.

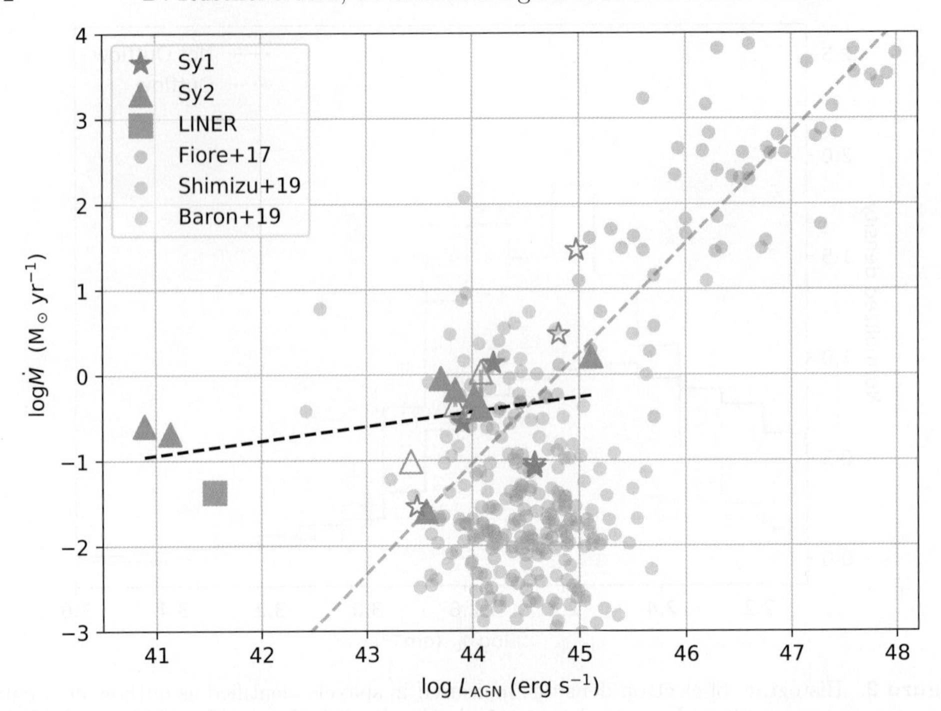

Figure 3. Relation between AGN bolometric luminosities and mass outflow rates based on [O III](*open symbols*) and [N II] (*filled symbols*) emission. The blue line represents the best fit from Fiore *et al.* (2017), and the circles are points from Fiore *et al.* (2017) (*blue*), Shimizu *et al.* (2019) (*orange*) and Baron & Netzer (2019) (*green*).

We begin our discussion of mass outflow rates by first considering the ionised gas mass, which was evaluated for each individual spaxel of each galaxy as given by

$$M = \frac{m_p L_{\mathrm{H}\alpha}}{n_e j_{\mathrm{H}\alpha}(T)} \tag{3.1}$$

where m_p is the proton mass, n_e is the number density of electrons from the [S II] lines flux ratio, $L_{\mathrm{H}\alpha}$ is the Hα luminosity and $j_{\mathrm{H}\alpha}$ is the Hα emissivity in erg cm^3 s^{-1} for a given temperature. Since we did not measure the temperature based on the nebular emission, we assume a standard value of 10^4 K. Mass outflow rates are given by equation 3.2, which follows from the basic assumption that the outflow velocity $v = W_{80}/2$, is approximately the average velocity of the gas since it left the vicinity of the AGN. Therefore, the time it took for the gas to reach its current distance from the central engine is R/v.

$$\dot{M} = M\frac{v}{R} \tag{3.2}$$

The distance R is assumed to be the distance, projected onto the plane of the sky, of the farthest spectrum classified as being outflow dominated. Underlying this assumption is a model which considers the outflow to be spherically symmetrical. This rather crude simplification ignores any differences in the geometry of the outflows, however it does allow for a homogenous treatment of the sample.

Our main results are shown in figures 3 and 4, which show the relation between mass outflow rates and outflow kinetic power to the bolometric luminosity of the AGN. Most of the galaxies in our sample, which were identified as having outflows, lie in the same region of the estimates from previous works on the subject (Fiore *et al.* 2017; Baron & Netzer 2019; Shimizu *et al.* 2019). However, we find that three objects with luminosities

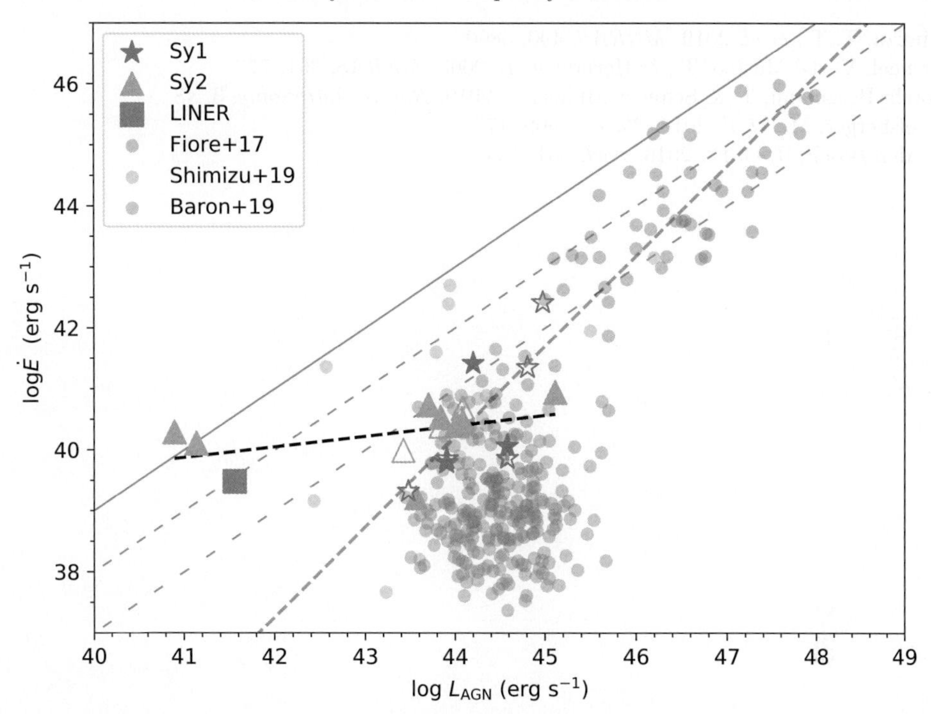

Figure 4. Outflow kinetic power vs. AGN bolometric luminosity. Solid and dashed black lines represent the ratios 1/10, 1/100 and 1/1000. The symbols are the same as in Fig. 3.

below 10^{42} erg s^{-1} show mass outflow rates that are unexpectedly high. We we look at the kinetic power carried out by these outflows (Fig. 4), we see that they are between having 1 and 10 per cent of the bolometric luminosity of the AGN. In the present study, estimates based on [O III] and [N II] emission yield similar results, for both mass outflow rates and outflow kinetic power, however, since only half of our sample has spectral coverage which includes the λ5007Å [O III] line, we will focus on the more representative [N II] emission. Based on the 12 galaxies for which the [N II] emission shows signs of outflowing gas, we find a median mass outflow rate of $\dot{\rm M} = 0.3$ M$_\odot$ yr^{-1}, and a median outflow power of $\log P = 40.4$ erg s^{-1}. Considering the bolometric luminosities we see that, for the majority of the sources considered, the energy carried out in the form of outflows is less than one per cent of the total energy irradiated by the AGN.

References

Alexander, D. M. & Hickox, R. C. 2012, *New A Rev.*, 56, 93
Allington-Smith, J., *et al.* 2002, *PASA*, 114, 892
Baron, D. & Netzer, H. 2019, *MNRAS*, 482, 3915
Baumgartner, W. H., Tueller, J., Markwardt, C. B., *et al.* 2013, *ApJS series*, 207, 19
Fabian, A. 2012, *ARA&A*, 50, 455
Ferrarese, L. & Merritt, D. 2000, *ApJ*, 539, L9
Fiore, F., *et al.* 2017, *A&A*, 601, A143
Gültekin, K., *et al.* 2009, *ApJ*, 698, 198
Harrison, C. M. 2017, *Nature Astronomy*, 1, 0165
Kormendy, J. & Ho, L. C. 2013, *ARA&A*, 51, 511
Madau, P. & Dickinson, M. 2014, *ARA&A*, 52, 415
Proxauf, B., Öttl, S., & Kimeswenger, S. 2014, *A&A*, 561, A10
Schaye, J., *et al.* 2015, *MNRAS*, 446, 521

Shimizu, T. T., *et al.* 2019, *MNRAS*, 490, 5860
Springel, V., Di Matteo, T., & Hernquist, L. 2005, *MNRAS*, 361, 776
Storchi-Bergmann, T. & Schnorr-Müller, A. 2019, *Nature Astronomy*, 3, 48
Vogelsberger, M., *et al.* 2014, *Nature*, 509, 177
van den Bosch, R. C. E., 2016, *ApJ*, 831, 134

Galaxy Evolution and Feedback across Different Environments
Proceedings IAU Symposium No. 359, 2020
T. Storchi-Bergmann, W. Forman, R. Overzier & R. Riffel, eds.
doi:10.1017/S1743921320004056

POSTERS

Stellar population synthesis of jellyfish galaxies

Gabriel M. Azevedo[1], Ana L. Chies Santos[1], Rogério Riffel[1], Augusto Lassen[1], Marina Trevisan[1], Nícolas Mallmann[1], Fernanda Oliveira[1] and Jean Gomes[2]

[1]Instituto de Física, Universidade Federal do Rio Grande do Sul, Av. Bento Gonçalves 9500, 91501-970, Porto Alegre, RS, Brazil

[2]Instituto de Física e Astronomia, Universidade do Porto, Porto, Portugal
email: gabriel.maciel.azevedo@gmail.com

Abstract. Jellyfish are the most extreme cases of galaxies undergoing ram-pressure stripping. In order to analyse the stellar populations distribution along these galaxies, we have performed stellar population synthesis in data cubes of jellyfish from the GASP programme, using both Starlight and FADO codes.

Keywords. galaxies: jellyfish, ram-pressure stripping, galaxies: stellar population galaxies:individual:JW108

1. Introduction

It is well known that galaxies are strongly affected by their environment. The morphology-density relation (Dressler 1980) shows that in dense groups and clusters there are more quiescent galaxies than "gas rich" ones. In those regions occur some phenomena that can change the physical properties of the galaxies. One of them is the ram-pressure stripping (RPS) (Gunn & Gott 1972): the loss of interstellar material (ISM) of a galaxy falling in a cluster or group, caused by the ram pressure with the intracluster material (ICM). The galaxies that are suffering the stripping are called jellyfish galaxies and have an unilateral asymmetry, forming "tails" with the escaping gas.

The RPS can enhance the star formation along all the galaxy (Vulcani *et al.* 2018; Roman Oliveira *et al.* 2019), even in the tails, that present HII regions (Poggianti *et al.* 2019), but it may end quenching the galaxy by removing most of its gas.

In order to study the stellar populations in galaxies we have performed spatially resolved stellar population synthesis (SPS) on public data cubes from MUSE. We are using public data from GASP (Poggianti *et al.* 2017), a large programme that observed jellyfish between redshifts 0.04 and 0.07 with the spectrograph MUSE in the Very Large Telescope (VLT). The field of view covers an area of 1' × 1', with spatial resolution 0.4" and spectral range from 4650 Å to 9300 Å.

2. Analysis and Results

To perform the SPS we used the Megacube code (Mallmann *et al.* 2017), that implements the Starlight code (Cid Fernandes *et al.* 2005) in datacubes. We have also used the FADO code (Gomes & Papaderos 2017) that performs the SPS using a differential evolutionary algorithm and takes into account the nebular emission, which is a greatly

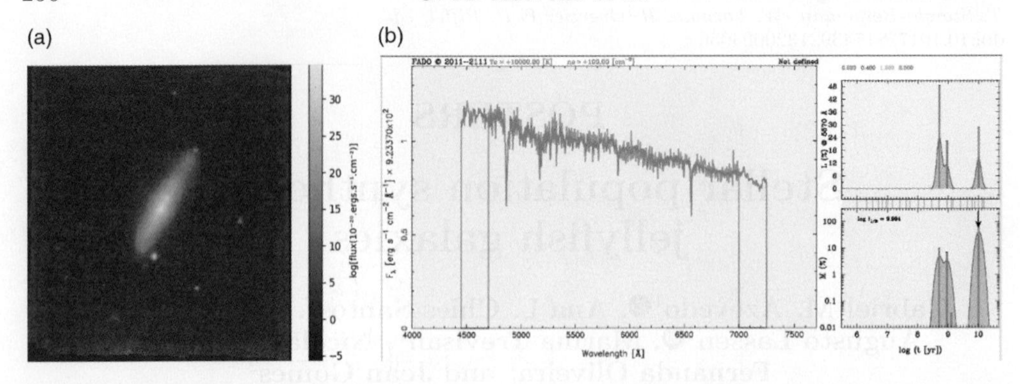

Figure 1. (a) Flux of Hα+NII emission lines in JW108. (b) In orange the spectrum of a region in the disk of JW108, in blue the spectrum fitted with FADO, in gray the stellar continuum and in red the nebular continuum. On the wright the bins of age weighted by mass and luminosity. It has an intermediate/old population (0.5−10 Gyr), and we see no nebular contribution, along with the lack of emission lines.

important constraint to synthesizing the spectra of star-forming regions, such as the tails of the jellyfish. In both codes we used a base from Bruzual & Charlot (2003) simple stellar populations.

During our analysis we found a galaxy, JW108, that seems to present star formation in the center, but not in the disk, and the cause of this is not fully understood. We speculate it lost part of the gas of its outskirts due to gravitational interaction with some nearby galaxy, but the second is not well understood yet. We intend to examine the causes of this problem in JW108, in addition to studying the profiles of the stellar populations along the different regions of the jellyfish.

References

Bruzual, G. & Charlot, S. 2003, *MNRAS*, 344, 1000
Dressler, A. 1980, *ApJ*, 236, 351
Fernandes, R. C., Mateus, A., Sodré, L., *et al.* 2005, *MNRAS*, 358, 363
Gomes, J. M. & Papaderos, P. 2017, *AAP*, 618, C3
Gunn, J. E. & Gott, J. R., III 1972, *ApJ*, 176, 1
Mallmann, N. D., Riffel, R., Storchi-Bergmann, T., *et al.* 2017, *MNRAS*, 478, 5491
Poggianti, B. M., Moretti, A., Gullieuszik, M.,*et al.* 2017, *ApJ*, 844, 48
Poggianti, B. M., Gullieuszik, M., Tonnesen, S., *et al.* 2019, *MNRAS*, 482, 4466
Roman-Oliveira, F. V., Chies-Santos, A. L., Rodríguez del Pino, B., *et al.* 2019, 484, 892
Vulcani, B., Moretti, A., Poggianti, B. M., *et al.* 2018, *ApJ*, 866, L25

Galaxy Evolution and Feedback across Different Environments
Proceedings IAU Symposium No. 359, 2020
T. Storchi-Bergmann, W. Forman, R. Overzier & R. Riffel, eds.
doi:10.1017/S1743921320001957

Searching for Ultra-diffuse galaxies in the low-density environment around NGC 3115

Marco Canossa-Gosteinski[1], Ana L. Chies-Santos[1], Cristina Furlanetto[1], Rodrigo F. Freitas[1] and William Schoenell[2]

[1]Universidade Federal do Rio Grande do Sul, Av. Bento Gonçalves, 9500, Porto Alegre - RS, Brazil
email: canossa.marco@gmail.com

[2]GMTO Corporation, 465 N. Halstead Street, Suite 250, Pasadena - CA, USA

Abstract. Ultra-diffuse galaxies (UDGs) are extremely low luminosity galaxies and some of them seem to have a lack of dark matter. Therefore, they can offer important clues to better understand galaxy formation and evolution. Little is known about UDGs in less dense environments, as most of the known UDGs have been found in very dense regions, in the outskirts of massive galaxies in galaxy clusters. In this work, we present the properties of UDGs candidates identified through visual inspection around the low-density environment of NGC 3115, the closest S0 galaxy from the Milky Way. We have measured the structural parameters of 41 UDGs candidates using images obtained with the Dark Energy Camera at the Blanco Telescope. Such structural parameters will be used to characterise and select the best UDG candidates, that will have their properties traced for future follow-up campaigns.

Keywords. Ultra-diffuse galaxies, Galaxy evolution

1. Introduction

Ultra-diffuse galaxies (UDGs) are extremely low luminosity galaxies but large in their size. They seem to be an extreme type of Low Surface Brightness dwarf galaxies (LSBd). Since most of the known UDGs have been found in very dense environments, we know more about their properties than about the ones in less dense environments UDGs. Therefore, to understand and determine the properties of these galaxies might offer important clues about galaxy formation and evolution. Previous works show that field UDGs share some properties with the ones found in groups (Jiang *et al.* 2019), namely, that the field galaxies seem to form through secular mechanisms, and others works indicate that UDGs in the field are predominantly blue and star-forming (Prole *et al.* 2019). In this work, we will present the structural parameters of UDGs identified through visual inspection around the low-density environment of NGC 3115, the closest S0 galaxy from Milky-Way. These parameters will be used to trace their properties in future work.

2. Method

Structural parameters of UDG candidates were measured using images obtained with the Dark Energy Camera at the Blanco Telescope. The images were inspected visually using the Ultra-Diffuse Object Candidates Search Tool developed by W. Schoenell. The tool generates four image files: a FITS file of the stamp for each band; an image of linear render with scales as the background of the stamp with a Gaussian kernel applied; two images that have SExtractor detection masked out; and an image with the colour combination of $g-r$ filter. SExtractor was used to mask other objects in the field. To measure

Figure 1. Example of models obtained using IMFIT, the original stamp on the left panel, the model in the middle panel and the residual on the right panel.

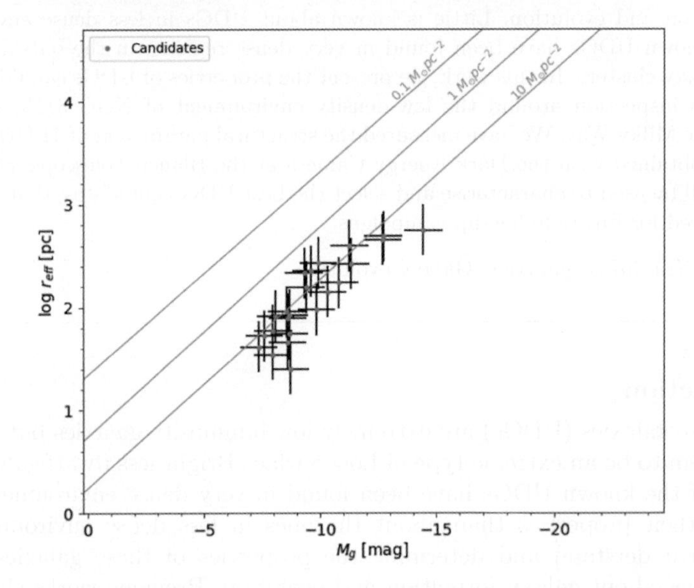

Figure 2. The effective radius versus absolute magnitude in g band of the candidates.

the structural parameters, we used IMFIT to fit an exponential function and a Sérsic function with an uniform background sky (Fig. 1). After the structural parameters of each candidate were obtained, the absolute magnitude in the g band of the candidates and their effective radius were measured.

3. Results and Discussion

To determine which of the candidates were more likely to be an UDG, we compared the absolute g band magnitude and their effective radius of our candidates with the ones of the UDGs presented in (Eigenthaler *et al.* 2019). The plot of effective radius versus absolute g band magnitude of our candidates are presented in Fig. 2. Based on this figure, we found around 10 plausible UDGs/LSBd candidates. In the future, we will use the structural parameters obtained in this work to select the best candidates for follow-up campaigns to confirm their UDG nature and characterise their properties.

References

Eigenthaler, P., *et al.* 2019, *AJ*, 885, 2
Jiang, F., *et al.* 2019, *MNRAS*, 487, 4
Prole, D. J., *et al.* 2019, *MNRAS*, 488, 2

Galaxy Evolution and Feedback across Different Environments
Proceedings IAU Symposium No. 359, 2020
T. Storchi-Bergmann, W. Forman, R. Overzier & R. Riffel, eds.
doi:10.1017/S174392132000191X

Constraining general relativity at z ~ 0.299 MUSE Kinematics of SDP.81

Carlos R. M. Carneiro, Cristina Furlanetto and Ana L. Chies-Santos

Departamento de Astronomia, Instituo de Física, UFRGS Porto Alegre, RS, Brasil
email: carlos.melo@ufrgs.br

Abstract. General Relativity has been successfully tested on small scales. However, precise tests on galactic and larger scales have only recently begun. Moreover, the majority of these tests on large scales are based on the measurements of Hubble constant (H_0), which is currently under discussion. Collett *et al.* (2018) implemented a novel test combining lensing and dynamical mass measurements of a galaxy, which are connected by a γ parameter, and found $\gamma = 0.97 \pm 0.09$, which is consistent with unity, as predicted by GR. We are carrying out this same technique with a second galaxy, SDP.81 at $z = 0.299$, and present here our preliminary results.

Keywords. gravitational lensing, stellar dynamics, gravitation

1. Introduction

General relativity (GR) has been tested in many ways. Precise tests in the Solar System and the Milky Way show that GR is currently the most successful theory of gravity (Will 2014; Ferreira 2019). However, on larger scales, there are few tests of this theory. Although GR is the most successful theory of gravity, some unresolved tensions remain, for example, inconsistent values for the Hubble constant (H_0) (Riess 2020) and the discovery of dark energy. For these reasons, modified theories of gravity have been proposed to address these tensions (Ishak 2019). One such approach is the Parameterized Post-Newtonian (PPN) formalism (Bertschinger 2011).

The PPN formalism has two important potentials: the classical Newtonian potential Φ, which acts on massive and non-relativistic particles; and Ψ, that can be interpreted as a curvature potential, more important for the motion of relativistic and massless particles. In the weak gravitational field approximation, GR predicts that these two potentials must be the same, $\Phi = \Psi$. Usually, the PPN approach is characterized by the parameter $\gamma = \Psi/\Phi$, that clearly is equal to one for GR.

One way of constraining this parameter is to measure the mass of a galaxy by different methods, such as gravitational lensing and dynamical modeling. The measurement of mass performed by gravitational lensing (M_{lens}) is sensitive to both potentials, while the dynamic mass measurement (M_{dyn}) is affected only by the classical Newtonian potential. In the PPN approach, these measurements are related by the simple relation $M_{dyn} = \frac{1+\gamma}{2} M_{lens}$.

2. SDP.81 data and MUSE Kinematics

To perform the test mentioned above, we use observations of SDP.81 obtained with three different instruments: Multi-Unit Spectroscopic Explorer (MUSE), Hubble Space Telescope (HST) and Atacama Large Millimeter/submillimeter Array (ALMA). SDP.81 (H-ATLAS J090311.6 + 003906) consists of an elliptical lens galaxy at $z_l = 0.299$, which

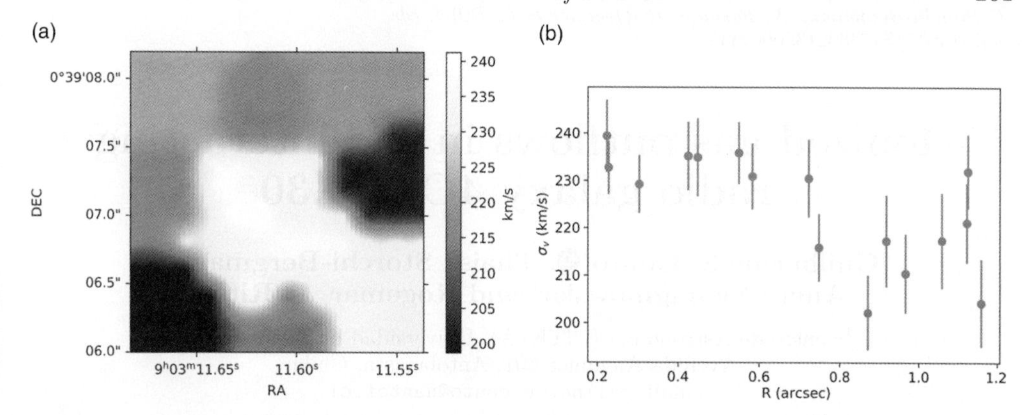

Figure 1. (a) velocity dispersion map of SDP.81. (b) radial profile of SDP.81 assuming the highest dispersion as the center of the galaxy. The error bar is the 1-σ deviation calculated through PPXF output. Note that from $R \approx 0.9''$ the velocity dispersion increases again.

is gravitationally lensing a submillimeter galaxy at $z_s = 3.04$. This system was observed with MUSE, covering the spectral range of 460−935 nm with 0.2″ spatial pixels. For the extraction of stellar kinematics, we use the VORONOI BINNING method (Cappellari & Copin 2003) in the central 2.5″. We use the PPXF method (Cappellari 2017) to determine the velocity dispersion in each bin in SDP.81 and the Medium-resolution Isaac Newton Telescope library of empirical spectra (MILES) (Sanchez-Blazquez *et al.* 2006) for modeling each binned spectrum in a wavelength range of 480−800nm, masking regions of emission and telluric lines. Finally, for each spectrum we derive a velocity V, velocity dispersion σ_V and 2 other moments (h_3,h_4). For the velocity dispersion, we construct a bi-dimensional map and a radial profile (Figure 1).

3. Discussion

As expected, the dispersion decreases from the galaxy center to the outskirts, but an unexpected result is the higher dispersion in the bottom right side of Figure 1a. This becomes clearer when we look at the radial profile in Figure 1b. Usually, these higher dispersions at the borders are caused by perturbations. However, in the SDP.81, there is no evidence for such perturbations. On the other hand, tests using higher S/N and different regions of the spectrum reproduce this same result, indicating that a problem with the fitting is unlikely. The next step is to construct a dynamical model based on HST images using the JAM CODE (Cappellari 2008), that could confirm or refute these previous results. Hereafter, we aim to combine these measurements obtained from kinematics with measurements obtained from the modeling of the gravitational lens to infer γ and test GR.

References

Bertschinger, E. 2011, *Phil. Trans. R. Soc. A.* (1957), 369, 4947–4961
Cappellari, M. 2008, *MNRAS*, 390, 71–86
Cappellari, M. 2017, *MNRAS*, 466, 798–811
Cappellari, M. & Copin, Y. 2003, *MNRAS*, 342, 345–354
Collett, T. E., Oldham, L. J., Smith, R. J., *et al.* 2018, *Science*, 360, 1342–1346
Ferreira, P. G. 2019, *ARAA*, 57 335–374
Ishak, M. 2019, *Living Rev. Relativity*, 22, 1
Riess, A. G. 2020, *Nat. Rev. Phys.*, 2, 10–12
Sanchez-Blazquez, P., Peletier, R. F., Jimenez-Vicente, J., *et al.* 2006, *MNRAS*, 371, 703–718
Will, C. M. 2014, *Living Rev. Relativity*, 17, 4

Galaxy Evolution and Feedback across Different Environments
Proceedings IAU Symposium No. 359, 2020
T. Storchi-Bergmann, W. Forman, R. Overzier & R. Riffel, eds.
doi:10.1017/S1743921320002112

Ionized gas outflows in the interacting radio galaxy 4C +29.30

Guilherme S. Couto[1], Thaisa Storchi-Bergmann[2], Aneta Siemiginowska[3] and Rogemar A. Riffel[4]

[1]Centro de Astronomía (CITEVA), Universidad de Antofagasta, Avenida Angamos 601, Antofagasta, Chile
email: guilherme.couto@uantof.cl

[2]Universidade Federal do Rio Grande do Sul, IF, CP 15051, Porto Alegre 91501-970, RS, Brazil

[3]Harvard Smithsonian Center for Astrophysics, 60 Garden St, Cambridge, MA 02138, USA

[4]Departamento de Física, Universidade Federal de Santa Maria, Centro de Ciencias Naturais e Exatas, 97105-900, Santa Maria, RS, Brazil

Abstract. We investigate the ionized gas excitation and kinematics in the inner $4.3 \times 6.2\,\mathrm{kpc}^2$ of the merger radio galaxy 4C +29.30. Using optical integral field spectroscopy with the Gemini North Telescope, we find signatures of gas outflows, including high blueshifts of up to $\sim$−650 km s^{-1} observed in a region $\sim 1''$ south of the nucleus, which also presents high velocity dispersion ($\sim$250 km s^{-1}). A possible redshifted counterpart is observed north from the nucleus. We propose that these regions correspond to a bipolar outflow possibly due to the interaction of the radio jet with the ambient gas. We estimate a total ionized gas mass outflow rate of $\dot{M}_{out} = 18.1^{+8.2}_{-5.3}$ M$_\odot$ yr^{-1} with a kinetic power of $\dot{E} = 5.8^{+7.6}_{-2.9} \times 10^{42}$ erg s^{-1}, which represents $3.9^{+5.1}_{-1.5}$% of the AGN bolometric luminosity. These values are higher than usually observed in nearby active galaxies and could imply a significant impact of the outflows on the evolution of the host galaxy.

Keywords. galaxies: individual: 4C +29.30 – galaxies: active – galaxies: nuclei – galaxies: kinematics and dynamics – galaxies: jets

1. Introduction

Active Galactic Nuclei (AGN) feedback is now thought to play a major role in galaxy evolution. In order to explain the observed scaling relationships between the mass-accreting supermassive black holes (SMBHs) and galaxy bulge properties (McConnell & Ma 2013; Kormendy & Ho 2013), AGN feedback is usually summoned (Fabian 2012). Although AGN feedback processes have been identified in recent works (Couto *et al.* 2017; Revalski *et al.* 2018), their impact must be quantified in order to determine whether or not they deliver effective power, capable of altering star formation rates and evacuating gas reservoirs (Harrison 2017; Zubovas & Bourne 2017).

In this work, we present results obtained from integral field spectroscopy observations (Gemini MultiObject Spectrograph instrument mounted on the Gemini North Telescope) of 4C +29.30, a radio galaxy with elliptical morphology at redshift z = 0.06, presenting an extended jet (up to $\sim$30 kpc from the nucleus) and moderate radio luminosity ($\sim 10^{42}$ erg s^{-1}). 4C +29.30 is possibly a merger system, displaying a characteristic dust lane passing in front of the central region in similar fashion to Centaurus A (Siemiginowska *et al.* 2012).

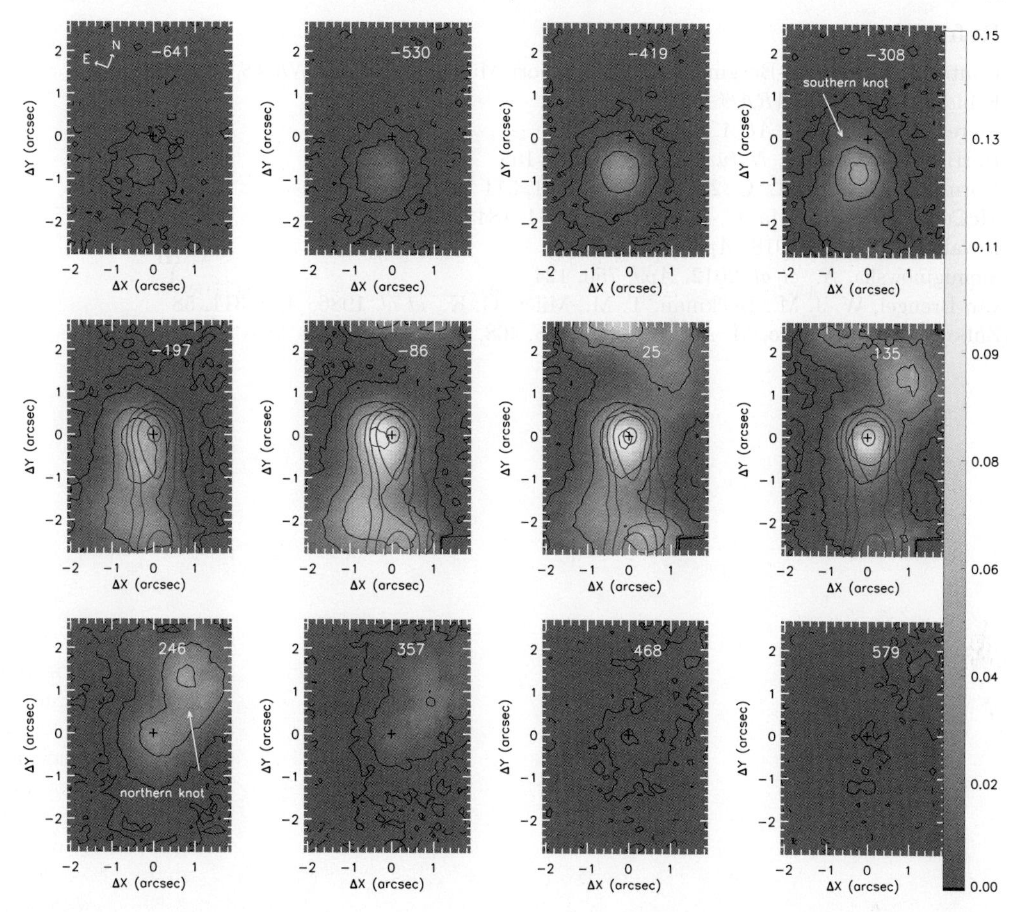

Figure 1. Channel maps along the [O III]λ5007 emission-line profile, in order of increasing velocities shown at the top of each panel in units of km s^{-1}. Flux units are 10^{-16} erg s^{-1} cm^{-2} spaxel^{-1}. The blue contours display the VLA 4.8 GHz emission from the radio jet Breugel*et al.* (1986).

2. Results and discussion

Fig. 1 shows channel maps along the [O III]λ5007 emission-line profile. High blueshifts (∼650 km s^{-1}) are observed in a region ∼1″ south of the nucleus, while a northern counterpart is observed with velocities of up to ∼580 km s^{-1}. These components also present high velocity dispersion (not shown in this proceedings, Couto et al. in preparation), with up to $\sigma \sim 250$ km s^{-1} in the southern region.

We interpret that these components are a bipolar outflow structure, that could be related to the interaction of the radio jet with non-symmetrical density environment, given the orientation of the outflows in relation to the jet. This is supported by the presence of the highest electron densities (∼400 cm^{-3}, also not shown here) in a region spatially correlated with the radio jet 1″ south-west from the nucleus. We estimate that the outflow mass rate is $\dot{M}_{out} = 18.1^{+8.2}_{-5.3}$ M$_\odot$ yr^{-1} and the outflow kinetic power corresponds to $3.9^{+5.1}_{-1.5}$% of 4C +29.30 AGN bolometric luminosity ($L_{\rm bol} = 1.4 \pm 0.8 \times 10^{44}$ erg s^{-1}). These values are above the trend observed by Fiore *et al.* (2017), and indicate that the outflows in 4C +29.30 are powerful and could affect the galaxy evolution.

References

Couto, G. S., Storchi-Bergmann, T., & Schnorr-Müller A. 2017, *MNRAS*, 469, 1573
Fabian, A. C. 2012, *ARA&A*, 50, 455
Fiore, F., *et al.* 2017, *A&A*, 601, A143
Harrison, C. M. 2017, *Nature Astronomy*, 1, 165
Kormendy, J. & Ho, L. C. 2013, *ARA&A*, 51, 511
McConnell, N. J. & Ma, C.-P. 2013, *ApJ*, 764, 184
Revalski, M., *et al.* 2018, *ApJ*, 867, 88
Siemiginowska, A., *et al.* 2012, *ApJ*, 750, 124
van Breugel, W. J. M., Heckman, T. M., Miley, G. K., *et al.* 1986, *ApJ*, 311, 58
Zubovas, K. & Bourne, M. A. 2017, *MNRAS*, 468, 4956

Galaxy Evolution and Feedback across Different Environments
Proceedings IAU Symposium No. 359, 2020
T. Storchi-Bergmann, W. Forman, R. Overzier & R. Riffel, eds.
doi:10.1017/S1743921320004275

Gauging the effect of feedback from QSOs on their host galaxies

Bruno Dall'Agnol de Oliveira and Thaisa Storchi-Bergmann

Departamento de Astronomia, Universidade Federal do Rio Grande do Sul, IF,
CP 15051, 91501-970 Porto Alegre, RS, Brazil
email: bruno.ddeo@gmail.com

Abstract. Often associated with the regulation of star formation in galaxies, active galactic nuclei (AGN) play a fundamental role in the evolution of galaxies through their feedback effects. To investigate the impact of these effects, we analysed the optical emission-line properties of 8 type II AGNs with bolometric luminosities $L_{Bol} > 10^{45}$ erg s^{-1}, using integral field spectroscopy (IFS) observations with Gemini Multi-Object Spectrograph (GMOS). The gas kinematics was obtained by fitting Gaussian components to the profiles of the emission lines of the ionized gas. Using only the broadest component – that we associate with the gas in outflow – we calculated the mass outflow rate ($\dot{M}_{out}$), finding values of up to 10 M$_\odot$ yr^{-1}. The outflow kinetic power ($\dot{E}_{out}$) reaches maximum values between 10^{41} and 10^{43} erg s^{-1}, which correspond to feedback efficiencies of ~0.001–0.1 % of L_{Bol}. These values are below that required to quench the star formation during the evolution of galaxies in simulations and analytical models. We also investigated the effect of uncertainties on the values of the physical quantities used in the calculations – such as the electron density – on the final values of $\dot{M}_{out}$ and $\dot{E}_{out}$.

Keywords. galaxies: active, galaxies, kinematics and dynamics; quasars: emission lines

1. Introduction

AGNs are a major player in the evolution of galaxies, given the feedback energy released during the accretion of matter onto the central supermassive black hole of their host galaxies (Fabian 2012), where part of this energy couples with the local gas, lowering the star formation rate. The implementation of AGN feedback in cosmological simulations (e.g. Nelson *et al.* 2019) and semi-analytical models (e.g. Croton *et al.* 2016) is needed to reproduce some properties observed in the local Universe – such as the number of massive galaxies (Silk & Mamon *et al.* 2012). However, since simulations with different implementations can reproduce these results (Somerville & Davé 2015), it is important to use observations to constrain the recipes used in these studies.

Sample. To address this question, we studied the extended gas kinematics of a sample of 8 type II QSOs, with luminosities $10^{45.5} < L_{Bol} < 10^{46.5}$ erg s^{-1} and $0.1 < z < 0.5$, using Gemini GMOS integral field spectroscopy, with the purpose of gauging the feedback power of the AGN via ionized gas outflows. These objects have extended narrow line regions (ENLRs) reaching distances beyond the limits of their host galaxies (Storchi-Bergmann *et al.* 2018; Fischer *et al.* 2018).

Emission-line fitting. We used multiple Gaussians to model the emission-line profiles of the ionized gas in the ENLR. In general, the broadest component – that we assumed is tracing an outflow – has high velocity dispersion (up to 750 km s^{-1}) and negative velocities relative to the systemic velocity. The additional narrower components show complex kinematics, consistent with the presence of interactions with nearby galaxies, previously observed in the Hubble Space Telescope images.

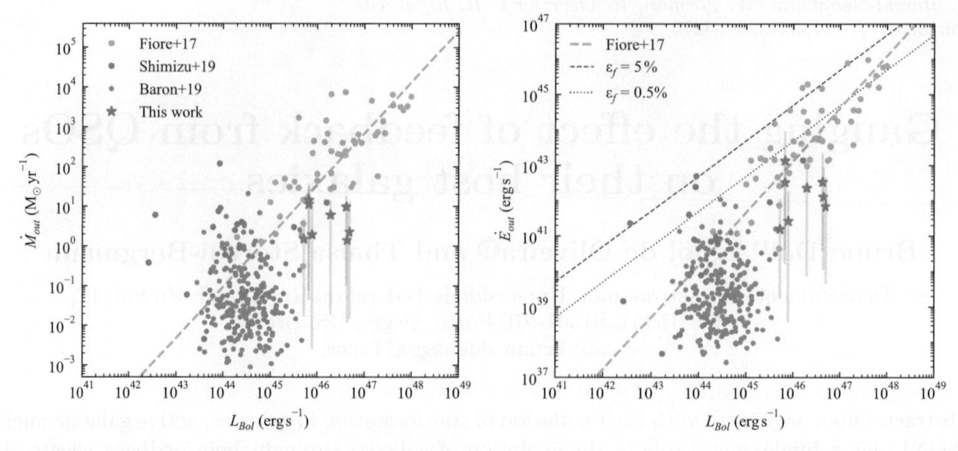

Figure 1. Mass outflow rate ($\dot{M}_{out}$, left) and outflow kinetic power ($\dot{E}_{out}$, right) as a function of the AGN bolometric luminosity. Blue stars: our data (with uncertainties due to different methods and assumptions). Orange circles and best-fit dashed lines: Fiore *et al.* (2017, ionized outflows). Brown circles: Shimizu *et al.* (2019). Green circles: Baron & Netzer (2019). Dotted and dashed black lines: outflow feedback efficiencies of 0.5 and 5 %.

2. Results

Outflow properties. Using only the broadest component, we calculated the mass outflow rates ($\dot{M}_{out}$) finding values of 1–10 $M_\odot$ yr^{-1}, with corresponding outflow kinetic powers ($\dot{E}_{out}$) reaching values between 10^{41} and 10^{43} erg s^{-1}. On average, these values are below previous ones from the literature for the same AGN luminosity (L_{Bol}) range (see Fig. 1).

Uncertainties. Testing the effect of different methods/assumptions used to obtain the outflow properties, we found that they can lead to a variation of 2–3 orders of magnitude in the final values of $\dot{M}_{out}$ and $\dot{E}_{out}$ (errorbars in Fig. 1). In particular, using higher densities (Baron & Netzer 2019) these quantities decrease by a factor of up to 100.

Feedback efficiency ($\varepsilon_f = \dot{E}_{out}/L_{Bol}$). These are in the range 0.001–0.1 %, below those found to be able to quench star formation in simulations (e.g. Hopkins & Elvis 2010). However, our calculations consider only the contribution from the ionized gas to the outflow power, which may represent only a fraction of the total feedback power (Harrison *et al.* 2018).

References

Baron, D. & Netzer, H. 2019, *MNRAS*, 486, 4290
Croton, D. J., Stevens, A. R. H., Tonini, C., *et al.* 2016, *ApJS*, 222, 22
Fabian, A. C. 2012, *ARA&A*, 50, 455
Fiore, F., Feruglio, C., Shankar, F., *et al.* 2017, *A&A*, 601, A143
Fischer, T. C., Kraemer, S. B., Schmitt, H. R., *et al.* 2018, *ApJ*, 856, 102
Harrison, C. M., Costa, T., Tadhunter, C. N., *et al.* 2018, *Nature Astronomy*, 2, 198
Hopkins, P. F. & Elvis, M. 2010, *MNRAS*, 401, 7
Nelson, D., Pillepich, A., Springel, V. *et al.* 2019, *MNRAS*, 490, 3234
Shimizu, T. T., Davies, R. I., Lutz, D., *et al.* 2019, *MNRAS*, 490, 5860
Silk, J. & Mamon, G. A. 2012, *RAA*, 12, 917
Somerville, R. S. & Davé R. 2015, *ARA&A*, 53, 51
Storchi-Bergmann, T., Dall'Agnol de Oliveira, B., Longo Micchi, L. F., *et al.* 2018, *ApJ*, 868, 14

Galaxy Evolution and Feedback across Different Environments
Proceedings IAU Symposium No. 359, 2020
T. Storchi-Bergmann, W. Forman, R. Overzier & R. Riffel, eds.
doi:10.1017/S1743921320004305

Feedback from ionised gas outflows in the central kpc of nearby active galaxies

Edwin David and Thaisa Storchi-Bergmann

Universidade Federal do Rio Grande do Sul, Postbus 91501-970, Porto Alegre - RS - Brasil
email: `edwin.godavid@gmail.com`

Abstract. We use integral-field spectroscopy obtained with the Gemini instrument GMOS-IFU (Gemini Multi-Object Spectrograph Integral Field Unit) to map the gas distribution, excitation and kinematics in the central kpc of 11 nearby active galaxies. We use channel maps to quantify the ionised gas masses, mass outflow rates and powers of the outflows in order to gauge the feedback effect of these outflows on the host galaxies. We compare this method with others previously used to calculate the feedback power of such outflows.

Keywords. galaxies: active; active galactic nuclei; AGN: feedback, AGN: outflows

1. Introduction and Goals

One of the most important phenomena occurring in the center of galaxies are the feeding and feedback processes associated with Active Galactic Nuclei (AGN) that may explain how nuclear activity is connected with the evolution of the host galaxy (Madau & Dickinson 2014). In this work we quantify AGN feedback via ionised gas outflows, and show a comparison between two different methods to calculate the mass outflow rate $\dot{M}$ and outflow power P. We use data from the Gemini instrument GMOS-IFU to derive these quantities from emission-line channel maps (Fig. 1), under the assumption that ionised gas that has velocities $|v| \geqslant 300\,\mathrm{km\,s^{-1}}$ is outflowing. We use the equations below:

$$M = \frac{m_p L_{H\alpha}}{n_e j_{H\alpha}(T)}, \quad \dot{M} = M\frac{v}{R} \quad \text{and} \quad P = \frac{1}{2}\dot{M}v^2 \tag{1.1}$$

where M is the ionised gas mass, m_p is the proton mass, $L_{H\alpha}$ the $H\alpha$ luminosity, n_e the electronic density obtained from the [SII] lines ratio, and $j_{H\alpha}$ is the $H\alpha$ emissivity in $\mathrm{erg\,cm^3}\,s^{-1}$ for a typical temperature of 10^4K. $\dot{M}$ is the mass outflow rate, calculated in units of solar masses per year, v is the velocity of the channel and R is the distance of each pixel to the galaxy center. P is the outflow power, calculated in units of $\mathrm{erg}\,s^{-1}$.

2. Results

We use the hypothesis that the outflows are mapped only via the channels with velocities $|v| \geqslant 300\,\mathrm{kms^{-1}}$, with lower velocities assumed to correspond to orbits in the galaxy disk. The calculated $\dot{M}$ and P values are shown in Fig. 2, compared to previous values from the literature, showing that the points approximately follow the relation with the AGN luminosity L_{Bol} obtained by Fiore *et al.* (2017) for active galaxies with similar L_{Bol}.

Hopkins & Elvis (2010) argue that $P/L_{Bol} \geqslant 0.005$ is necessary to impact the galaxy evolution. We did not find any galaxy in our sample with $P/L_{Bol} \geqslant 0.005$, and concluded that these outflows do not have significant impact on the host galaxies.

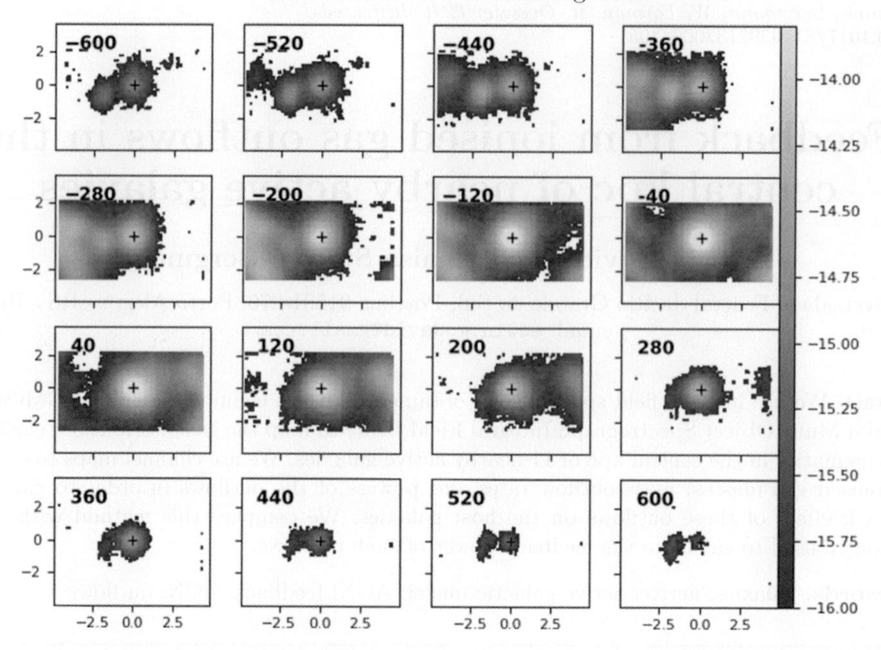

Figure 1. [OIII]λ5007Å channel maps of the inner kpc of the nearby Seyfert 1 galaxy NGC 3516. Angular units are arcseconds and the color bar gives the fluxes in logarithmic units of erg cm^{-2} s^{-1}.

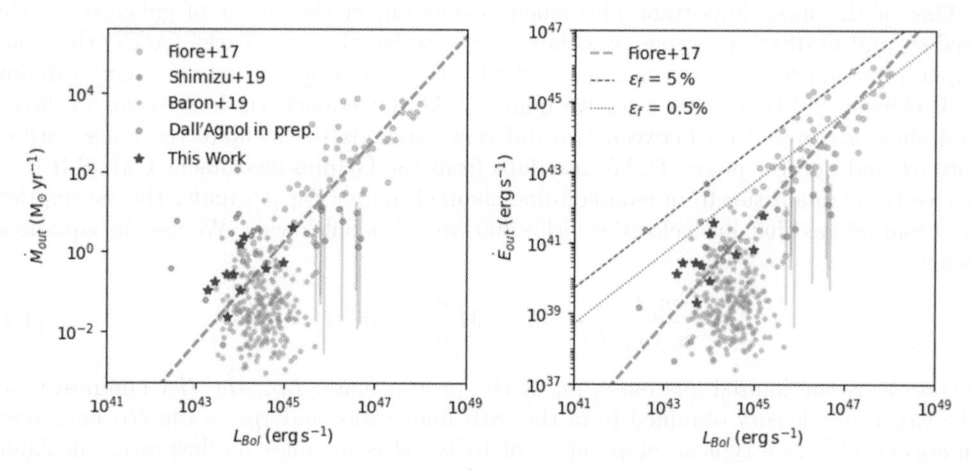

Figure 2. Our results (magenta stars) for $\dot{M}$ and P as a function of the AGN luminosity as compared with others from the literature. Adapted from Dall'Agnol *et al.* (2020).

References

Dall'Agnol de Oliveira, B. 2020, in prep
Fiore, F. 2017, *A&A*, 601, A143
Hopkins, P. F. & Elvis, M. 2010, *MNRAS*, 401, 7
Madau, P. & Dickinson, M. 2014, *ARA&A*, 52, 145

Galaxy Evolution and Feedback across Different Environments
Proceedings IAU Symposium No. 359, 2020
T. Storchi-Bergmann, W. Forman, R. Overzier & R. Riffel, eds.
doi:10.1017/S1743921320004196

HST observations of [O III] emission in nearby QSO2s: Physical properties of the outflows

Anna Trindade Falcao[1], S. B. Kraemer[1], T. C. Fischer[2], D. M. Crenshaw[3], M. Revalski[2] and H. R. Schmitt[4]

[1]Institute for Astrophysics and Computational Sciences, Department of Physics, The Catholic University of America, Washington, DC, USA

[2]Space Telescope Science Institute, Baltimore, MD, USA

[3]Department of Physics and Astronomy, Georgia State University, Atlanta, GA, USA

[4]Naval Research Laboratory, Washington, DC, USA

Abstract. We used Space Telescope Imaging Spectrograph (STIS) long slit medium-resolution G430M and G750M spectra to analyze the extended [O III] λ5007 emission in a sample of twelve QSO2s from Reyes *et al.* (2008). The purpose of the study was to determine the properties of the mass outflows and their role in AGN feedback. We measured fluxes and velocities as functions of deprojected radial distances. Using photoionization models and ionizing luminosities derived from [O III], we were able to estimate the densities for the emission-line gas. From these results, we derived masses, mass outflow rates, kinetic energies and kinetic luminosity rates as a function of radial distance for each of the targets. Masses are several times 10^3-10^7 solar masses, which are comparable to values determined from a recent photoionization study of Mrk 34 (Revalski *et al.* 2018). Additionally, we are studying the possible role of X-ray winds in these QSO2s.

Keywords. galaxies: active, galaxies: QSO2, galaxies: kinematics and dynamics

1. Introduction

Accreting supermassive black holes (SMBHs) are believed to be the central engines that power luminous AGNs. The ionizing radiation released by the SMBH interacts with the interstellar medium of the host galaxy which may regulate the SMBH accretion rate and evacuate star-forming gas from the host galaxy bulge, i.e. AGN feedback. AGN winds are present in most AGNs (Mullaney *et al.* 2013). Recent studies (Fischer *et al.* 2018) question whether these winds produce efficient feedback. Determining this requires characterizing their physical properties, such as mass, mass outflow rates, kinetic energies and kinetic luminosity rates.

2. Sample, Observations and Calculations

In order to study the physical properties of these QSO2s, we have obtained Hubble Space Telescope (*HST*) imaging and spectroscopy of 12 of the 15 most luminous targets from Reyes *et al.* (2008) sample under z = 0.12 (Fischer *et al.* 2018) to map [O III] velocities and widths as a function of radial distance and determine physical properties of each system. To analyze and determine the number of significant kinematic components for each emission line, we used a fitting technique (Fischer *et al.* 2017), which gave

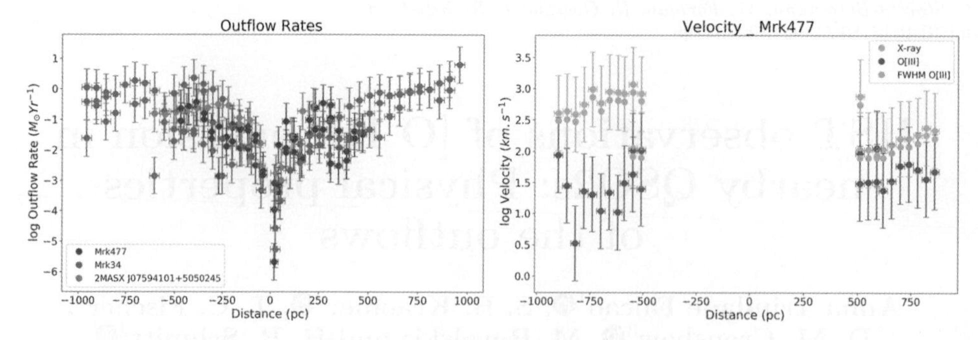

Figure 1. The left panel shows the mass outflow rates in units of $M_\odot\ yr^{-1}$ for Mrk 477, Mrk 34 and 2MASX J075941 as a function of distance from the nucleus. The right panel shows a comparison between the velocity profiles of the [O III] winds and X-ray winds, obtained using our method described in section 3, and the FWHM, for Mrk 477.

us [O III] fluxes and velocities (v) as a function of position. We used Cloudy models, a software designed to simulate conditions in interstellar matter under a broad range of conditions (Ferland *et al.* 2017), to derive the column densities and gas densities, at each de-projected distance from the nucleus (Trindade Falcao et al. in preparation).

Along the *HST* STIS slit, we calculated the gas mass as a function of distance from the nucleus using the observed [O III] fluxes as compared to those of the Cloudy model. The ionized gas mass as a function of distance from the nucleus, and M_{ion}, the total mass of ionized gas derived from the [O III] images, were calculated as described in Crenshaw *et al.* (2015). Using M_{ion} and v, we calculated the mass outflow rates ($\dot{M}_{out}$), kinetic energies (E), and kinetic luminosity rates ($\dot{E}$) (Figure 1, left panel). Among the QSO2s in our sample, the outflow region contains a total M_{ion} ranging from $5 \times 10^3 M_\odot$ to $1.10 \times 10^7 M_\odot$. The maximum E for the targets varies from 4.1×10^{50} to 1.96×10^{54} erg and the $\dot{M}_{out}$ peaks vary from $7 \times 10^{-3} M_\odot\ yr^{-1}$ to $4.15\ M_\odot\ yr^{-1}$. The peak $\dot{E}$ ranges from 1.00×10^{38} to $1.51 \times 10^{42}\ erg\ s^{-1}$.

3. Discussion and Conclusions

The maximum kinetic luminosity of the outflow for our sample reaches 6.56×10^{-9} to 8.04×10^{-4} of the AGN L_{bol}, which does not approach the 0.5%−5% range used in some models as providing efficient feedback (Di Matteo *et al.* 2005; Hopkins & Elvis 2010), **indicating that the [O III] winds are not an efficient feedback mechanism**, based on this criterion. Fischer *et al.* (2018) show that the outer regions of these QSO2s present disturbed gas with high FWHM. Assuming that the FWHM is due only to motion within the [O III] emission-line gas and to explore the possible causes of this disturbance, we calculated the kinetic energy density, U_{KE}, of the disturbed gas. If the disturbance is due to the impact of X-ray winds, they would have the same U_{KE} as the [O III] gas (Trindade Falcao *et al.* in preparation). This analysis made it possible for us to construct a velocity profile of these X-ray winds (Figure 1, right panel).

References

Crenshaw, *et al.* 2015, *ApJ*, 799, 83
Di Matteo, *et al.* 2005, *Nature*,433, 604
Ferland, *et al.* 2017, *RMxAA*, 49, 1379
Fischer, *et al.* 2017, *ApJ*, 834, 30
Fischer, T. C., Kraemer, S. B., Schmitt, H. R., *et al.* 2018, *ApJ*, 856, 102

Hopkins & Elvis 2010, *MNRAS*, 401, 7
Kraemer, *et al.* 2000, *ApJ*, 532, 256
Mullaney, *et al.* 2013, *MNRAS*, 433, 622
Revalski, M., Crenshaw, D. M., Kraemer, S. B., *et al.* 2018, *ApJ*, 856, 46
Reyes, *et al.* 2008, *AJ*, 136, 2373

Galaxy Evolution and Feedback across Different Environments
Proceedings IAU Symposium No. 359, 2020
T. Storchi-Bergmann, W. Forman, R. Overzier & R. Riffel, eds.
doi:10.1017/S1743921320002082

The two-phase gas outflow in the Circinus Galaxy

M. A. Fonseca-Faria[1] and **A. Rodríguez-Ardila**[1,2]

[1]Instituto Nacional de Pesquisas Espaciais, Av. dos Astronautas, CEP 12227-010, São José dos Campos - SP, Brazil
email: marcosfonsecafaria@gmail.com

[2]Laboratório Nacional de Astrofísica, R. dos Estados Unidos, CEP 37504-364, Itajubá - MG, Brazil

Abstract. We employ Multi Unit Spectroscopic Explorer (MUSE) data to study the ionized and very ionized gas phase of the feedback in Circinus, the closest Seyfert 2 galaxy. The analysis of the nebular emission allowed us to detect a remarkable high-ionization gas outflow, out of the galaxy plane, traced by the coronal lines [Fe VIII] 6089Å and [Fe X] 6374Å, extending up to 700 parsecs north-west from the nucleus. The gas kinematics reveal expanding gas shells with velocities of a few hundred km s^{-1}, spatially coincident with prominent hard X-ray emission detected by *Chandra*. Density and temperature sensitive line ratios show that the extended high-ionization gas is characterized by a temperature of up to 18000 K and a gas density of $n_e > 10^2$ cm^{-3}. We propose two scenarios consistent with the observations to explain the high-ionization component of the outflow: an active galactic nuclei (AGN) ejection that took place $\sim 10^5$ yr ago or local gas excitation by shocks produced by the passage of a radio jet.

Keywords. active galaxy nuclei, Seyfert, emission lines, spectroscopic.

1. Introduction

Active Galactic nuclei (AGN) are extremely important objects for understanding the formation and evolution of galaxies. Results presenting a correlation between the mass of the black hole and the mass of the galaxy's bulge (Kormendy & Richstone 1995) as well as a correlation between the luminosity of the galaxy and the stellar velocity dispersion with the mass of the black hole (Kormendy & Ho 2013) were fundamental to suggest a co-evolution of the black hole and its host galaxy. In this context, the detection of outflows is key to understand the role of AGNs at controlling the growth of their host galaxy.

Traditionally, the identification of outflows in the warm ionized phase has been done by means of the [O III] 5007 Å line. In combination with integral field spectroscopy, critical insights about the the geometry, structure, extension, and physical conditions of the outflowing gas can be derived. However, because [O III] is also emitted in the galaxy disk and star-forming regions, isolating the contribution due to outflows is not straightforward. In this respect, Rodríguez-Ardila *et al.* (2006) showed that high-ionization lines are excellent tracers of the ionized component of the outflows. The energy required for their production rules out stellar or galactic origin.

Here, using data from the Multi Unit Spectroscopic Explorer (MUSE) spectrograph, we report optical observations of the Circinus Galaxy focusing of the detection of extended coronal gas, at scales of hundreds of parsecs from the AGN.

2. Observations

The data used here were obtained by the MUSE and retrieved from the European Southern Observatory (ESO) science portal. The integral field unit (IFU) cube is fully reduced, including calibration in flux (in absolute units) and wavelength. Details of the observation and data reduction is provided in Mingozzi *et al.* (2019).

The datacube was analyzed making use of a set of custom PYTHON scripts developed by us as well as software publicly available in the literature. First, we rebinned the cube, reducing the total number of spaxels to ∼10000. We then removed the stellar continuum across the whole spectral range of MUSE (4700–9100 Å) using STARLIGHT (Cid-Fernandes *et al.* 2005) and the Bruzual & Charlot (2003) stellar population models.

3. Results and Conclusions

We constructed maps of the flux distribution for the most important lines such as [O III] λ5007, Hβ, Hα and [Fe VIII] λ6087. That information revealed the largest outflow of high-ionization gas ever observed in Circinus. Because of its close proximity, we carried out a detailed analysis of the gas morphology and kinematics. The high-ionized outflow, detected by means of the [Fe VIII] λ6087 line, extends up to a distance of 700 pc from the AGN and runs along the radio jet axis. The gas emission appears clumpy, with several knots of emission within a region of $\sim 400 \times 300$ pc^2.

The kinematics of the high-ionization gas is complex, revealing split line profiles with full width at half maximum (FWHM) that implies gas velocities of 200–350 km s^{-1}. The simultaneous detection of approaching and receding components suggests expanding gas shells, produced either by nuclear ejections or inflated by the passage of the radio jet.

The gas emitting [Fe VII] has a density $n_e > 10^2$ cm^{-3}, a temperature $\sim 1.8 \times 10^4$ K and is little affected by dust extinction. Its spatial coincidence with extended, thermal X-ray emission, and its orientation along the radio jet axis, points out that it is likely the remnant of shells inflated by the passage of the radio-jet. The gas velocity dispersion of ∼300 km s^{-1} supports this hypothesis.

We found evidence of a scenario where the extended [Fe VII] emission cannot be driven by photoionization from the AGN. The gas velocity dispersion observed is consistent with the shock velocities needed to produce that line. Models of Contini & Viegas (2001) show that shock velocities of 300 km s^{-1} or larger produce [Fe VII]/Hα ratios $> 10^{-1}$, in agreement with the observations. The gas density measured along the region emitting [Fe VII], of a few hundred cm^{-3} is also consistent with the density required to power the coronal emission.

Our results highlight the relevance of the kinetic channel as a major way of releasing nuclear energy to the ISM in low-luminosity AGN (Wylezalek & Morganti 2018). Due to its proximity, Circinus is a showcase that demands further investigation of that scenario.

References

Bruzual, G. & Charlot, S. 2003, *MNRAS*, 344, 1000
Cid-Fernandes, R., Mateus, A., Sodré, L., *et al.* 2005, *MNRAS*, 358, 363
Contini, M. & Viegas, S. M. 2001, *Apj*, 132, 211
Kormendy, J. & Richstone, D. 1995, *ARAA*, 33, 581
Kormendy, J. & Ho, L. C. 2013, arXiv:1308.6483
Mingozzi, M., Cresci, G., Venturi, G., *et al.* 2019, *A&A*, 622, A146
Rodríguez-Ardila, A., Prieto, M. A., Viegas, S., *et al.* 2006, *Apj*, 6
Wylezalek, D. & Morganti, R. 2018, *Nature Astronomy*, 2, 181

Galaxy Evolution and Feedback across Different Environments
Proceedings IAU Symposium No. 359, 2020
T. Storchi-Bergmann, W. Forman, R. Overzier & R. Riffel, eds.
doi:10.1017/S1743921320001854

Outflow signatures in Gemini GMOS-IFU observations of 5 nearby Seyfert 2 galaxies

Izabel C. Freitas[1], Rogemar A. Riffel[2] and Thaisa Storchi-Bergmann[3]

[1]Universidade Federal de Santa Maria, Colégio Politécnico, 97105-900, Santa Maria, RS, Brazil
email: izabelfisica@gmail.com

[2]Universidade Federal de Santa Maria, Departamento de Física, Centro de Ciências Naturais e Exatas, 97105-900, Santa Maria, RS, Brazil

[3]Instituto de Física, Universidade Federal do Rio Grande do Sul, Av. Bento Gonçalves 9500, 91501-970, Porto Alegre, RS, Brazil

Abstract. We use Gemini Multi-Object Spectrograph (GMOS) Integral Field Unit (IFU) observations of a sample of 5 bright nearby Seyfert galaxies to map their emission-line flux distributions and kinematics at a spatial resolution ranging from 110 to 280 pc. For all galaxies, the gas kinematics show two components: a rotation and an outflow component.

Keywords. galaxies: Seyfert, galaxies: kinematics, galaxies: nuclei

1. Introduction

We present a study of the Narrow Line Region (NLR) of 5 nearby Seyfert galaxies: Mrk 6, Mrk 79, Mrk 348, Mrk 607 and Mrk 1058. We have mapped the gas emission-line flux distributions and kinematics, as well as of the stellar kinematics using Integral Field Spectroscopy (IFS) obtained at the Gemini North Telescope, using the Gemini Multi-Object Spectrograph (GMOS). The data cover the spectral range from 4300 Å to 7100 Å, that includes the strongest emission-lines from the NLR of AGNs: Hβ, [O III] $\lambda\lambda$4959,5007, [O I] λ6300, Hα, [N II] $\lambda\lambda$6548,83 and [S II] $\lambda\lambda$6716,31.

We fit the emission-line profiles with Gaussian curves in order to map the emission-line flux distributions and gas kinematics.

2. Results

In Figure 1, we present the resulting maps for the flux distributions, line ratios and gas and stellar kinematics for Mrk 348. The highest blueshifts and redshifts, seen at 1.5″ NE and 1.5″ SW, respectively, associated to high velocity dispersion values are interpreted as being due to ionized outflows. Mrk 79, Mrk 348 and Mrk 1058 show bipolar outflows, while in Mrk 6 and Mrk 607 the geometry of the outflows are not fully constrained by our data.

3. Conclusions

Our main conclusions are: (i) The highest gas densities ($N_e \approx 1000$–$2000\,cm^{-3}$) are usually observed at the nucleus, in a few cases also extending towards regions of highest excitation. A particular case is Mrk 1058 that seems to show a circumnuclear ring of high-density gas at ≈1.7 arcsec (592 pc) from the nucleus. (ii) The average nuclear chemical

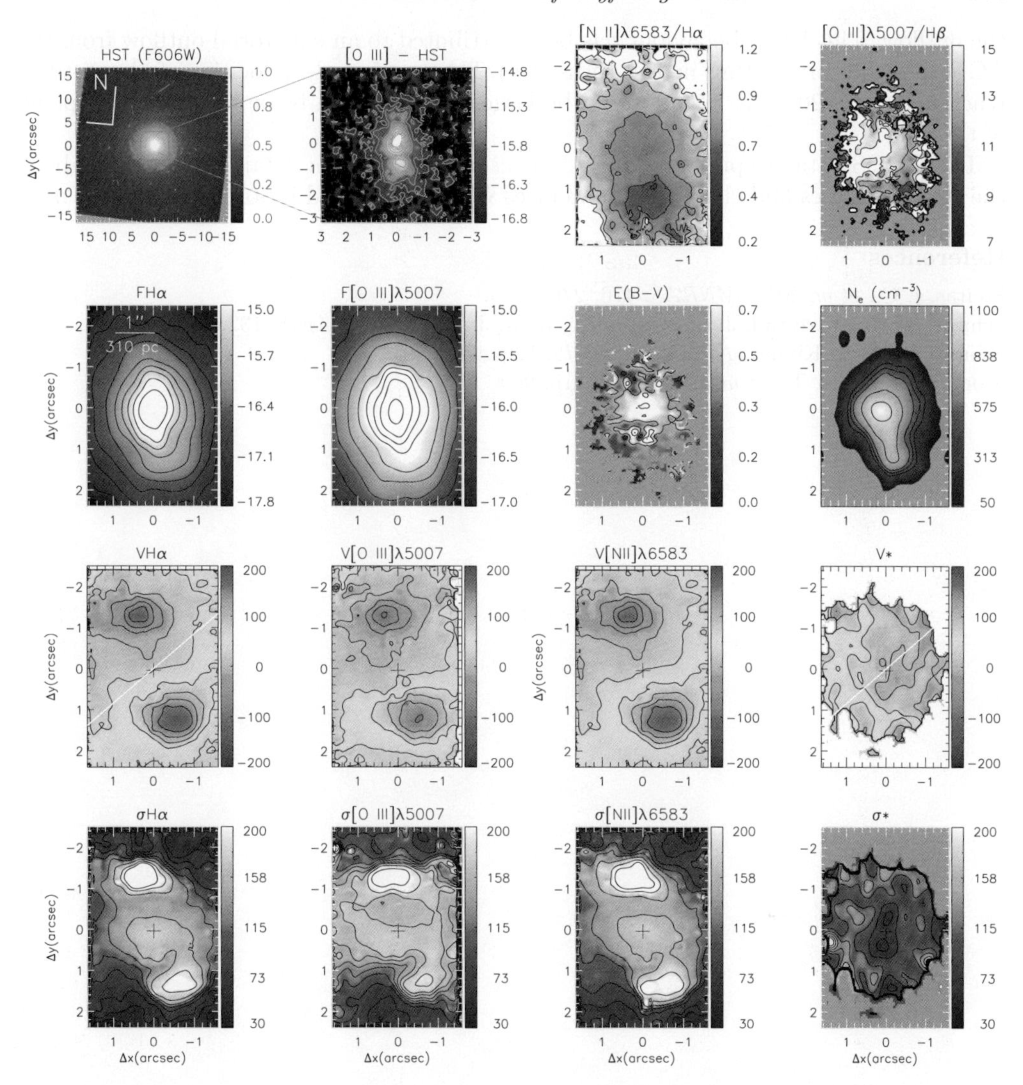

Figure 1. Two-dimensional maps for Mrk 348. The central cross marks the location of the nucleus, defined as the position of the peak of continuum emission and the spatial orientation is shown at the top left corner of the large-scale image. The contours overplotted to the [O III] λ5007 flux map (in green), velocity field (in white), and σ map (green) are from the 3.6 cm radio image of Schmitt *et al.* (2001). The white line shown in the Hα velocity field represents the major axis of the large-scale disk, measured using I-band images by Schmitt & Kinney (2000). Gray regions in the flux, ratio and velocity dispersion maps and white regions in the velocity fields correspond to locations where the signal-to-noise ratio was not high enough to obtain a good fit of the line profiles. Figure from Freitas *et al.* (2018).

abundance of the galaxies of our sample is $\langle 12 + \log O/H \rangle = 8.66 \pm 0.17$, measured within an aperture of 0.25×0.25 arcsec^2 using the strong lines method (Storchi-Bergmann *et al.* (1998)). (iii) The gas kinematics show a distorted rotation pattern that can be attributed to a combination of emission from gas in rotation in the galaxy plane and outflows. (iv) The rotation component of the gas is similar to that obtained from the stellar kinematics, except for Mrk 607 in which the gas is counter-rotating relative to the stars. (v) The gas velocity dispersion shows two typical patterns: it is enhanced at the location of the outflows or surrounding the nucleus in an elongated structure perpendicular to the outflow. This latter behaviour has been attributed to an equatorial outflow from the AGN, possibly originating in the torus. (vi) The (projected) velocities of the outflow reach at most $\approx$200 kms^{-1}, but could be larger as they seem to be mostly in the plane of the sky.

The GMOS data are presented in Freitas *et al.* (2018) and a detailed discussion about the gas kinematics and chemical abundances will be presented in forthcoming papers.

References

Freitas, I. C., *et al.* 2018, *MNRAS*, 476, 2760
Schmitt, H. R., Ulvestad, J. S., Antonucci, R. R. J., *et al.* 2001, *ApJS*, 132, 199
Schmitt, H. R. & Kinney, A. L. 2000, *ApJS*, 128, 479
Storchi-Bergmann, T., *et al.* 1998, *ApJ*, 115, 909

Galaxy Evolution and Feedback across Different Environments
Proceedings IAU Symposium No. 359, 2020
T. Storchi-Bergmann, W. Forman, R. Overzier & R. Riffel, eds.
doi:10.1017/S174392132000229X

The role of internal feedback in the evolution of the dwarf spheroidal galaxy Leo II

Roberto Hazenfratz[1], Gustavo A. Lanfranchi[1] and Anderson Caproni[1]

[1]Núcleo de Astrofísica, Universidade Cidade de São Paulo, Zip Code 01506-000, Rua Galvão Bueno 868, São Paulo, Brasil
email: robertohm@usp.br

Abstract. This work aims to explore the different processes of formation and evolution of dwarf spheroidal galaxies in the Local Group analyzing internal and external feedbacks, taking Leo II as a model of parametrization due to its adequate large distance to the Milky Way, in order to minimize potential external effects. We present a discussion of the first results regarding the processes of formation and galactic evolution from the gas hydrodynamics. Combined with previous studies for other similar systems, such results have the potential to establish strong links for the elaboration of a consistent and coherent scenario of formation and evolution of the dwarf spheroidal galaxies in the Local Group.

Keywords. Dwarf Galaxies, Internal Feedback, Hydrodynamic Simulations, Leo II

1. Introduction

The dwarf spheroidal galaxy (DSG) Leo II (Harrington & Wilson 1950) is one of the most distant Milky Way satellite galaxies (233 ± 15 kpc) and therefore suitable for studying the role of internal feedback in the hydrodynamic evolution of its gaseous content. Like all other local dwarf spheroidal galaxies, there is no sign of neutral gas and the detailed mechanisms responsible for its material loss over time are still unknown. This work has as its main objective the hydrodynamic study of the gaseous content of the DSG Leo II, using chemical evolution models combined with a three-dimensional hydrodynamic simulation code as main tools.

Leo II has a total mass of $\approx 4.3 \times 10^7 M_{sun}$ Walker *et al.* (2009), a stellar population with a mean age of 9 Gyr, formed between approximately 14 and 7 Gyr ago, whose average metallicity is [Fe/H] $= -1.59$ Kirby *et al.* (2011) and particular chemical abundance patterns. Its gaseous content could have been depleted by galactic winds due to stellar feedback (internal mechanisms) or removed by ram pressure or tidal forces (external mechanisms). Such a gas loss, in turn, would directly influence the observed patterns of chemical abundances and other properties of the galaxy.

2. Methods

The hydrodynamic simulations of galaxies, for the classical and non-magnetic regime, can be expressed in the conservative formulation of fluid dynamics equations. The hydrodynamic code used in this work was PLUTO v4.2, assuming a single star formation episode during 7 Gyr. The maximum resolution used in this simulation was 30 pc/cell.

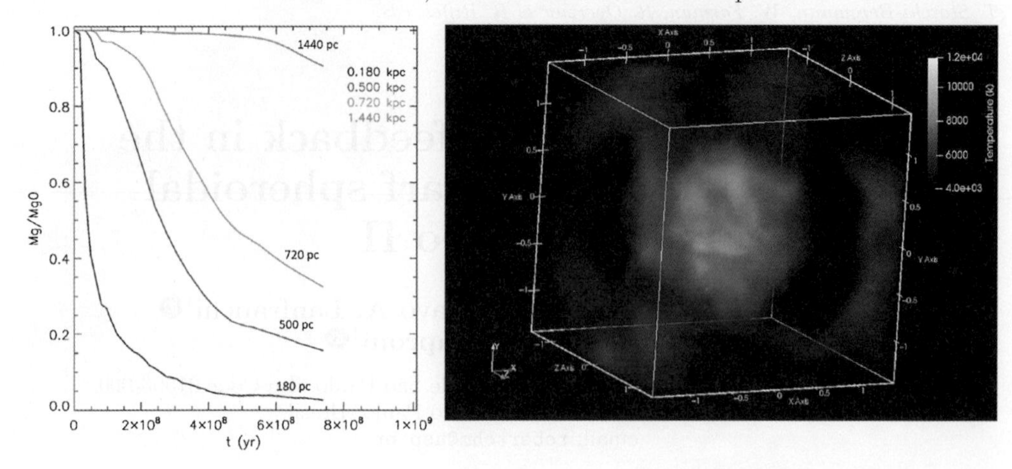

Figure 1. Residual gas mass over time and a 3D-temperature map of Leo II for ≈400 Myr.

Cosmological effects were not considered in a static cartesian grid over time. Further details can be found in Caproni *et al.* (2015, 2017).

3. Preliminary results and Perspectives

The induced gas loss by supernovae (SNe) feedback varies with respect to the galactic radius and time (Fig. 1). The gas loss is the most intense in the first 200 Myr of simulation, considering the radius of 180 pc (core radius of the galaxy), and in the first 400 Myr, considering the radius of 500 and 720 pc. It was also observed that the gas loss was higher in the radius of 180 pc (core radius), for which the residual gas mass achieved the minimum value of ≈5% around 400 Myr and ≈35% for 720 pc around 750 Myr. The latter radius represents a value that contains the tidal radius of the galaxy, estimated as 632 ± 32 Coleman *et al.* (2007).

The location and distribution of supernovae over space and time were observed by spatial distribution histograms. It was observed that SNe Ia explode in more peripheral regions since the beginning of the gas evolution. On the other hand, SNe II explode initially only in the central and denser region. However, the explosions occur in progressively larger radius over time, due to the feedback of the first supernovae, which spread the gas accross the galaxy. The difference in the patterns regarding the two types of supernovae can be explained by the fact that the progenitors of SNe II are stars of higher masses ($> 8M_{sun}$), which occur with higher frequency in high-density regions. In turn, the SNe Ia explosions are more randomly distributed due to the fact that their progenitors are stars of lower mass, which do not necessarily need high-density regions to form.

Regarding the future simulations, the resolution will be increased in order to model the detailed hydrodynamics of Leo II in higher accuracy; the chemical evolution of Leo II will be remodeled with new available data (metallicities, bynary fraction); the mechanical feedback of an intermediate mass black hole will be parameterized; and the time of simulation will be increased to comprise the main star formation duration (7 Gyr).

References

Caproni, A., Lanfranchi, G. A., da Silva, A. L., *et al.* 2015, *ApJ*, 805, 109
Caproni, A., Lanfranchi, G. A., da Silva, A. L., *et al.* 2017, *ApJ*, 838, 99

Coleman, M. G., Jordi, K., Rix, H. W., *et al.* 2007, *AJ*, 134, 1938
Harrington, R. G. & Wilson, A. G. 1950, *PASP*, 62, 118
Kirby, E. N., Lanfranchi, G. A., Simon, J. D., *et al.* 2011, *ApJ*, 727, 78
Lanfranchi, G. A. & Matteucci, F. 2010, *A&A*, 512, A85
Walker, M. G., Mateo, M., Olszewski, E. W., *et al.* 2009, *ApJ*, 704(2), 1274

Galaxy Evolution and Feedback across Different Environments
Proceedings IAU Symposium No. 359, 2020
T. Storchi-Bergmann, W. Forman, R. Overzier & R. Riffel, eds.
doi:10.1017/S174392132000215X

Effects of supernovae feedback and black hole outflows in the evolution of Dwarf Spheroidal Galaxies

Gustavo Amaral Lanfranchi, Anderson Caproni, Jennifer F. Soares and Larissa S. de Oliveira

Núcleo de Astrofísica, Universidade Cidade de São Paulo, R. Galvão Bueno 868, Liberdade, São Paulo, SP, 01506-000, Brazil
email: gustavo.lanfranchi@cruzeirodosul.edu.br

Abstract. The gas evolution of a typical Dwarf Spheroidal Galaxy is investigated by means of 3D hydrodynamic simulations, taking into account the feedback of type II and Ia supernovae, the outflow of an Intermediate Massive Black Hole (IMBH) and a static cored dark matter potential. When the IMBH's outflow is simulated in an homogeneous medium a jet structure is created and a small fraction of the gas is pushed away from the galaxy. No jet structure can be seen, however, when the medium is disturbed by supernovae, but gas is still pushed away. In this case, the main driver of the gas removal are the supernovae. The interplay between the stellar feedback and the IMBH's outflow should be taken into account.

Keywords. hydrodynamics, ISM: jets and outflows, galaxies: dwarf, evolution, Local Group

1. Introduction

A common feature to local Dwarf Spheroidal Galaxies is the absence of neutral gas, that could be removed by ram pressure or tidal stripping. Hydrodynamic simulations, however, suggest that galactic winds are very efficient in expelling the gas out of the galaxy (Caproni *et al.* 2015, 2017). A physical process, not yet considered, is the outflow from a central black hole in these galaxies.

2. Code setup

The effects of an IMBH's outflow and the stellar feedback in the internal dynamics of a classical dSph galaxy are analyzed by means of 3D hydrodynamic simulations. The initial setup is the same as described in Caproni *et al.* (2017). The galaxy is simulated for 1 Gyr inside a computacional cube of 40^3 to 200^3 cells, twice as large as Ursa Minor's tidal radius (950 pc). An initial gas mass of 2.94×10^8 $M_\odot$ is in hydrostatic equilibrium with a dark matter halo mass of 1.51×10^9 $M_\odot$. An outflow is created by inserting a density in the central cell with a velocity in the z axis at t = 0 yr. The energy of the SNe is injected in the medium following the procedure of Caproni *et al.* (2017).

3. Results

3.1. *Stellar feedback*

Two scenarios were considered: one with SNe II only and other only with SNe Ia. Initially, SNe II occur at the center, creating a shock wave that pushes the gas to outer regions. The shocks mix the gas on the galaxy, creating "bubbles" of hot gas and high density that carry material out of the system. The mass fraction inside a spherical region

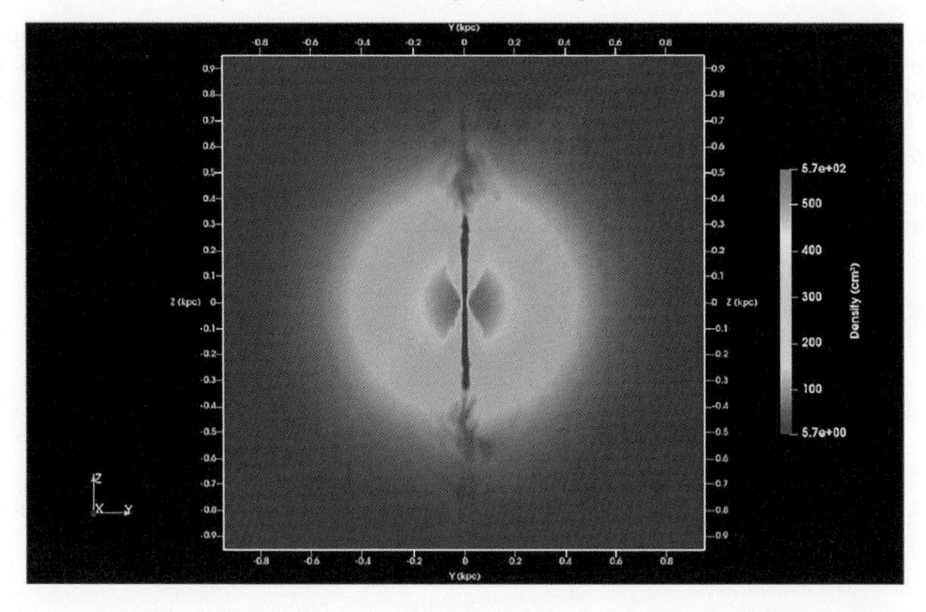

Figure 1. Density profile cut in the yz plane for the simulation with an IMBH's outflow with $\rho = 0.027$ cm^{-3}, v = 1000 km/s, $t_0 = 0$ yr.

of 300 pc radius drops slow to 85% of the inital value, but increases later to ~90%, remaining constant until the end of the simulation. Inside 950 pc radius, there is a very small decrease (~1%). In the second scenario, interactions among the SNe Ia remnants do not create a distinguished shock wave and gas is easier removed from the galaxy. The mass fraction starts decreasing fast at ~150 Myr, reaching ~80%; then the gas loss becomes slower reaching ~78% in outer regions. Compared to the SNe II case, SNe Ia removes more gas in outer than in inner regions and the decrease in the gas mass begins later and is slower.

3.2. *BH outflow*

Outflows with different initial density and the same initial velocity (1000 km.s^{-1}) are tested first in a medium where the density varies only radially. In the case $\rho = 0.008$ cm^{-3} the outflow does not have energy to break into the ISM, but gas is pushed away from the tidal radius. When the initial density is $\rho = 0.027$ cm^{-3}, the outflow creates a jet feature (Figure 1), pushing the gas and leaving behind a stream of very low density. When the medium is disturbed by SNe there is no jet feature. The shock waves created by the SNe give rise to regions of pressure and velocity much higher than the ones created by the outflow, making its propagation more difficult. Even though, the outflow pushes away gas from the galaxy. The fraction of initial mass left inside different radii is lower when the outflow's initial density is higher. In the case with an outflow beginning 30 Myr after the simulation is started, SNe lower the density of the central region , the outflow can develop easier, and the gas loss is higher in outer regions (950 pc).

4. Conclusions

SNe Ia are more efficient in removing gas, whereas SNe II affect only the central region. The BH's outflow effects depend on the conditions of the medium. These two types of feedback should be considered togheter in future studies, due to their strong interplay.

G. A. L. thanks FAPESP grants 2017/25779-2 and 2014/11156-4. The authors acknowledge the National Laboratory for Scientific Computing (LNCC/MCTI, Brazil) for providing HPC resources of the SDumont supercomputer. URL: http://sdumont.lncc.br.

References

Caproni, A., Lanfranchi, G. A., Luiz da Silva, A., *et al.* 2015, *ApJ*, 805, 109
Caproni, A., Lanfranchi, G. A., Campos Baio, G. H., *et al.* 2017, *ApJ*, 838, 99

Galaxy Evolution and Feedback across Different Environments
Proceedings IAU Symposium No. 359, 2020
T. Storchi-Bergmann, W. Forman, R. Overzier & R. Riffel, eds.
doi:10.1017/S174392132000201X

On the path toward a universal outflow mechanism in light of NGC 4151 and NGC 1068

D. May[1], J. E. Steiner[1] and R. B. Menezes[2]

[1]Instituto de Astronomia, Geofísica e Ciências Atmosféricas, Universidade de São Paulo, 05508-090, São Paulo, SP, Brazil
email: dmay.astro@gmail.com

[2]Centro de Ciências Naturais e humanas, Universidade Federal do ABC, Santo André, 09210-580 SP, Brazil

Abstract. We use near-infrared Integral Field Unit (IFU) data to analyze the galaxies NGC 4151 and NGC 1068, which have very different Eddington ratios - ∼50 times lower for NGC 4151. Together with a detailed data cube treatment methodology, we reveal remarkable similarities between both AGN, such as the detection of the walls of an "hourglass" structure for the low-velocity [Fe II] emission with the high-velocity emission within this hourglass; a molecular outflow - detected for the first time in NGC 4151; and the fragmentation of an expanding molecular bubble into bullets of ionized gas. Such observations suggest that NGC 4151 could represent a less powerful and more compact version of the outflow seen in NGC 1068, suggesting a universal feedback mechanism acting in quite different AGN.

Keywords. galaxies: active, galaxies: individual (NGC 1068, NGC 4151)

1. Context

The primary motivation of this work is to analyse the presence of an "hourglass" wall structure in NGC 1068 (May & Steiner 2017) and NGC 4151, as seen in the low-velocity [Fe II] emission of these galaxies, as well as the high-velocity emission that fills in the hourglass volume. The combination with observations having excellent seeing and our routine of data treatment has revealed a new scenario for both galaxies, which is summarized in Figs. 1 and 2.

2. Results and Conclusions

For NGC 1068, we propose that a strong secondary wind is formed in the northeast cone where the jet hits one of the H_2 molecular clouds (the primary wind comes from the accretion disc). This wind probably changes the direction of the accelerating [Si VI] emitting blobs. Furthermore, the north-south asymmetry in the ionized gas emission probably arises from this interaction. The same process likely occurs in the southwest cone, but without a molecular barrier so close to the AGN. In the northeast and southwest cones, this wind would be responsible for accelerating the narrow line region (NLR) blobs and inflating the bubble. For NGC 4151, in turn, we propose that the ionized outflow is mostly a consequence of a molecular fragmentation process, leading to the formation of bullets of gas. Furthermore, both the low- and high-velocity H_2 structures are connected

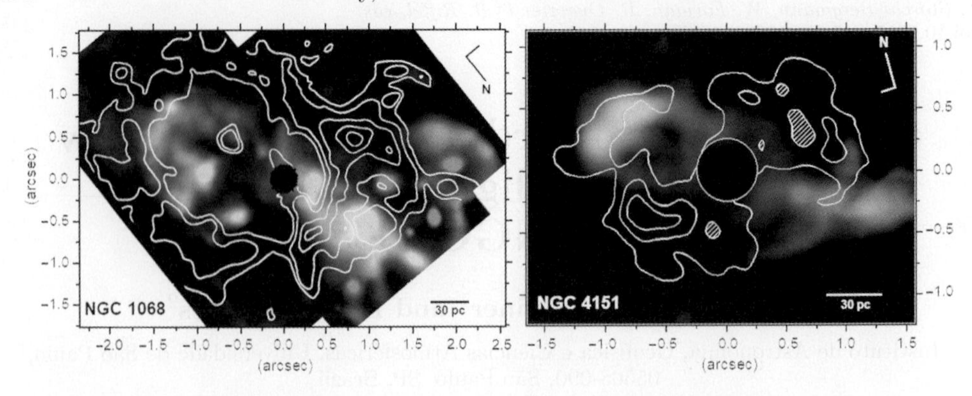

Figure 1. High-velocity [Fe II] emission and the H_2 molecular structure (contours) for NGC 1068 (left panel), and NGC 4151 (right panel).

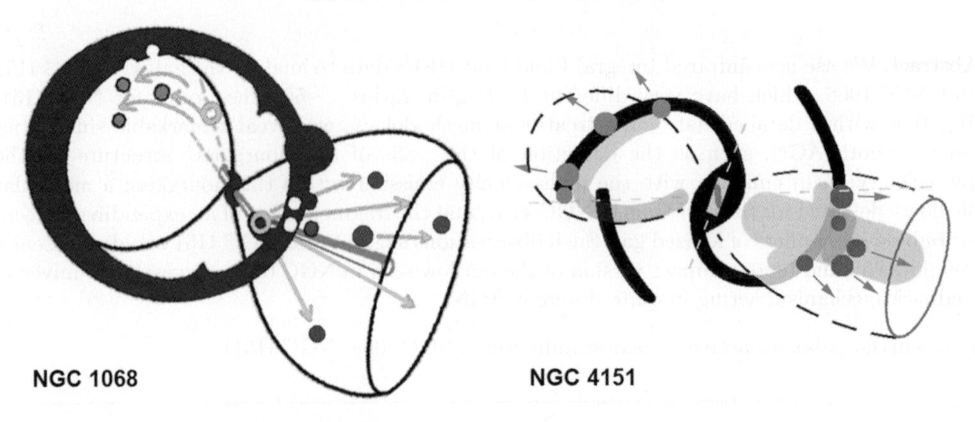

Figure 2. Left panel: a sketch of the interaction between the jet and the molecular bubble for NGC 1068. The production of the secondary wind is marked by two open circles and the arrows illustrate their directions. The molecular cavity (in black) remains intact in one side of the cone and is disrupted on the side where the blobs of gas are being blown away after the jet-cloud interaction. Right panel: sketch of the scenario proposed for the NLR of NGC 4151. The discontinuous molecular (H_2) walls are shown in black together with the dashed contours of the hourglass structure. The filled circles represent the bullets of ionized and molecular gas, with the arrows denoting their radial motion, while the light gray inside the hourglass illustrates the main regions of the high-velocity [Fe II] emission. The straight dashed lines denote the PA of the jet and the torus (central rectangle), while the accretion disc is shown inside the torus (not to scale). See a colour version of the presented panels in May & Steiner 2017 and May *et al.* 2020 (accepted).

to the [Fe II] emission inside and outside the cones. Our observations support the scenario that an expanding molecular bubble is being inflated and disrupted by the AGN and also by the jet.

References

May, D. & Steiner, J. E. 2017, *MNRAS*, 469, 994
May, D., Steiner, J. E., Menezes, R. B., *et al.* 2020, *MNRAS*, 496, 1488

Galaxy Evolution and Feedback across Different Environments
Proceedings IAU Symposium No. 359, 2020
T. Storchi-Bergmann, W. Forman, R. Overzier & R. Riffel, eds.
doi:10.1017/S1743921320002185

Identifying the extent of AGN outflows using spatially resolved gas kinematics

Beena Meena[1], D. M. Crenshaw[1], T. C. Fischer[2], Henrique R. Schmitt[3], M. Revalski[2] and G. E. Polack[1]

[1]Department of Physics and Astronomy, Georgia State University, Atlanta, GA, USA
email: bmeena@astro.gsu.edu

[2]Space Telescope Science Institute, Baltimore, MD, USA

[3]Naval Research Laboratory, Washington D.C., DC, United States

Abstract. We present spatially resolved kinematics of ionized gas in the narrow-line region (NLR) and extended narrow-line region (ENLR) in a sample of nearby active galaxies. Utilizing long-slit spectroscopy from Apache Point Observatory (APO)'s ARC 3.5 m Telescope and *Hubble Space Telescope* (*HST*) we analyzed the strong λ5007 Å [O III] emission line profiles and mapped the radial velocity distribution of gas at increasing radii from the center. We identified the extents of Active Galactic Nuclei (AGN) driven outflows in our sample and determined the distances at which the observed gas kinematics is being dominated by the rotation of the host galaxy. We also measured the effectiveness of radiative driving of the ionized gas using mass distribution profiles calculated with two-dimensional modeling of surface brightness profiles in our targets. Finally, we compared our kinematic results of the outflow sizes with the maximum distances at which the gas is being radiatively driven to investigate whether these outflows are capable of disrupting or evacuating the star-forming gas at these distances.

Keywords. galaxies: active, AGN, NLR, kinematics, feedback, outflows

1. Introduction, Observations & Analysis

The AGN driven outflows observed in the NLRs and ENLRs are thought to play a crucial role in regulating the growth of supermassive black holes and the bulges of the host galaxies. It is important to quantify the significance of these outflows in affecting the star forming gas in the bulge. Our goal is to measure the sizes of AGN driven outflows in a sample of nearby Seyfert galaxies ($z < 0.05$) and determine the fractions of bulges that are being evacuated by these outflows. In this work, we present preliminary results for two of the galaxies in our sample. The long slit observations were taken using the Dual Imaging Spectrograph (DIS) at APO with a resolving power of ∼4000−5500. We employed a multiple component gaussian fitting routine (Fischer *et al.* 2017) to determine the wavelength centroids, widths and peak fluxes of the spatially resolved λ5007 Å [O III] emission lines along the slit. The gaussian parameters are used to map the kinematics, velocity dispersion and flux distributions of the ionized gas for different components. The left panel in Figure 1 shows the composite images of NGC 3227 and NGC 4051 with the APO slit positions. The pseudo-IFU diagrams (middle panel) show the velocity maps along the slits for the highest flux components.

2. Results

NGC 3227: We observed high blueshift velocities ∼900 km/s to the NE within 1″ (72 pc) of the nucleus similar to Barbosa *et al.* (2009). There is evidence of faint

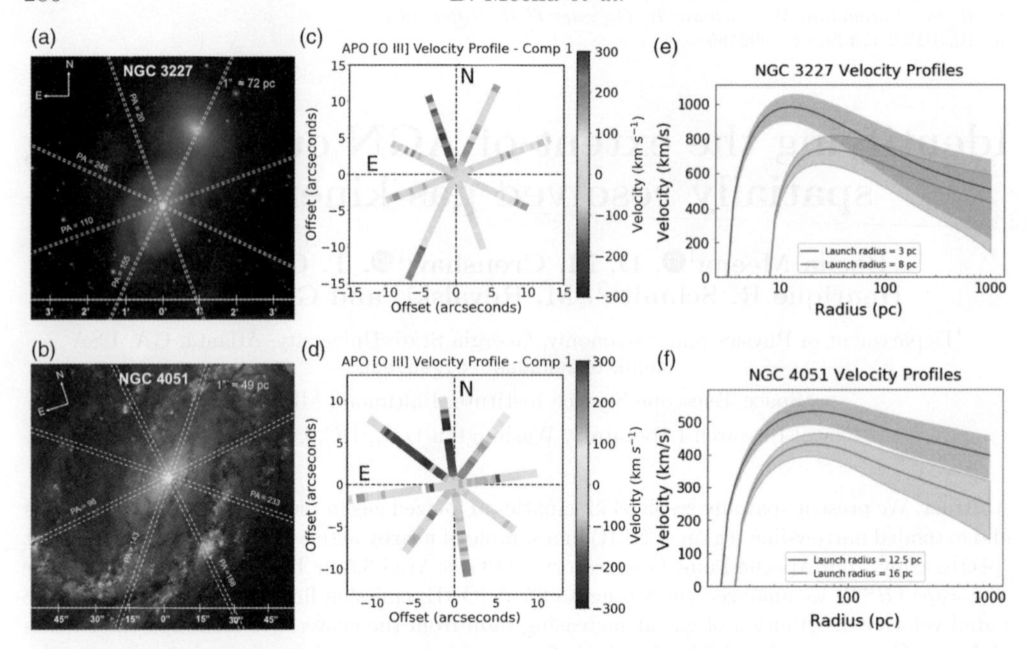

Figure 1. Left panel: (a) NGC 3227 (Sloan Digital Sky Survey) and (b) NGC 4051 (ESA/Hubble & NASA, D. Crenshaw and O. Fox) with the APO-DIS slit positions. Middle panel: Pseudo IFU diagrams for highest flux component. Right Panel: Model outflow velocity profiles calculated using the galaxy mass distribution and radiative driving pressure.

outflows up to $\pm 7''$ ($\sim$500 pc) to the NE and SW sides of the suggested bicone (Fischer *et al.* 2013). The kinematics along position angle (PA) of 155° shows redshift velocity of $\sim$200 km/s to the SW which is similar to stellar and CO velocities shown in Schinnerer *et al.* (1999).

NGC 4051: Blueshift outflows with velocities of $\sim$400 km/s up to $8''$ ($\sim$400 pc) were observed to the NE for PA 188° and PA 233°. That indicates front face of the bicone as seen in STIS (Fischer *et al.* 2013) and GMOS observations (Barbosa *et al.* 2009). Kinematics in PA 143° shows rotational velocity of 100 km/s outside the inner $\pm 5''$ with the NW (receding) and SE (approaching) similar to H_2 kinematics (Riffel *et al.* 2008).

Radiative Driving: We compared the observed velocities with velocities predicted via radiative driving models which was calculated from mass distribution M(r) and radiative acceleration (Das *et al.* 2007). The velocity profiles are plotted in the right panel of Figure 1. The velocities increase with decreasing launch radii and an increasing value of the force multiplier. The uncertainties are propagated considering a range of force multipliers from 2700 to 3300. Mass profiles to calculate the velocities were generated using 2D modeling of surface brightness profiles produced by Bentz *et al.* (2009). The effective radii of the bulges are 1200 pc and 860 pc for NGC 3227 and NGC 4051 respectively.

3. Conclusions

We identified the extents of outflows and found rotational signatures from the observed [O III] gas kinematics in NGC 3227 and NGC 4051. Preliminary velocities derived using radiative acceleration and enclosed mass profiles are consistent with the observed outflow velocities (within a few hundred pc). We observed that radiatively driven outflows terminate inside the bulges of these galaxies. This indicates that only a fraction of the bulges are being affected by AGN driven outflows as seen in Mrk 573 (Fischer *et al.* 2017).

References

Barbosa, F. K. B., *et al.* 2009, *MNRAS*, 296, 1
Bentz, M. C., *et al.* 2009, *ApJ*, 697,1
Das, V., *et al.* 2007, *ApJ*, 656, 2
Fischer, T. C., *et al.* 2013, *ApJ*, 209, 1
Fischer, T. C., *et al.* 2017, *ApJ*, 834, 30
Riffel, R. A., *et al.* 2008, *MNRAS*, 385,3
Schinnerer, E., *et al.* 1999, *ApJ*, 533, 2

Galaxy Evolution and Feedback across Different Environments
Proceedings IAU Symposium No. 359, 2020
T. Storchi-Bergmann, W. Forman, R. Overzier & R. Riffel, eds.
doi:10.1017/S1743921320002288

Mapping the inner kpc of the interacting Seyfert galaxy NGC 2992: Stellar populations and gas kinematics

Muryel Guolo-Pereira and Daniel Ruschel-Dutra

Universidade Federal de Santa Catarina, 88040-900 Florianópolis, SC, Brazil
email: muryel@astro.ufsc.br

Abstract. We present Gemini Multi-Object Spectrograph Integral Field Unit (GMOS-IFU) observations of the inner 1.1 kpc of the interacting Seyfert galaxy NGC 2992. From full spectral synthesis we found that the stellar population is mainly (up to 80 per cent of the total light) composed by an old (t $\geqslant$ 1.4 Gyr) metal-rich ($Z \geqslant 2.0$ Z$_\odot$) populations with a smaller but considerable contribution (up to 30 per cent) from young (t $\leqslant$ 100 Myr) metal-poor ($Z \leqslant 1.0$ Z$_\odot$) populations. The gas kinematics presents two main components: one from gas in orbit in the galaxy disk and an outflow with mass outflow rate of $\sim$2 M$_\odot$ yr^{-1} and a kinematic power of $\sim 2 \times 10^{40}$ erg s^{-1}.

Keywords. galaxies: individual (NGC2992, Arp 245), galaxies: active, galaxies: interactions

---

1. Introduction and Observations

NGC2992 is a nearby (z = 0.0077) interacting Seyfert galaxy, that together with NGC 2993 forms the system Arp 245. The pericentre passage between NGC2992 and NGC2993 is predicted to have occurred $\sim$100 Myr ago (Duc *et al.* 2000). In order to investigate the role the active galaxy nucleus (AGN) and the interaction in both circumnuclear stellar and gaseous content, we obtained IFU observations of the inner 1.1 kpc of the galaxy, with the Gemini instrument GMOS-IFU. The final reduced data cube has a field-of-view (FoV) of $6.0'' \times 6.1''$, with $\sim$2500 spaxels, and wavelength range $\sim$4400$-$6800Å. The spatial resolution is $\sim 0.8''$(120 pc) and the spectral is $\sim$45 km s^{-1}.

2. Stellar populations

We performed stellar population synthesis using the STARLIGHT code (Cid Fernandes *et al.* 2005). We use Bruzual & Charlot (2003) simple stellar populations (SSP) models. The base consists of 45 SSPs with 15 ages and 3 metallicities (0.2 Z$_\odot$, 1.0 Z$_\odot$, 2.5 Z$_\odot$) in addition to a power law featureless continuum (FC) with the form of Flux(λ) $\propto \lambda^{-1.7}$. We applied Cappellari & Copin (2003) Voronoi binning technique with a target S/N of 20. We also divided the SSPs in 3 age groups: a young one (x_y) with age range from 1 Myr to 100 Myr, an intermediate one (x_I) with age from 0.1 Gyr to 1.4 Gyr and a old one (x_O) with age from 1.5 to 13 Gyr, where the quantities x_y, x_I and x_O are respectively the percentage contribution to the total light from young, intermediaries and old SSPs.

The stellar population synthesis shows that the stellar population in the inner 1.1 kpc of NGC 2992 is mainly composed (60 percent $\leqslant x_O \leqslant$ 80 per cent) by old (t $\geqslant$ 1.4 Gyr) metal rich ($\langle Z \rangle_O \geqslant 2.0Z_\odot$) populations with a smaller, but considerable, contribution (10 per cent $\leqslant x_Y \leqslant$ 30 percent) by young (t $\leqslant$ 100 Myr) metal poor ($\langle Z \rangle_Y \leqslant 1.0Z_\odot$)

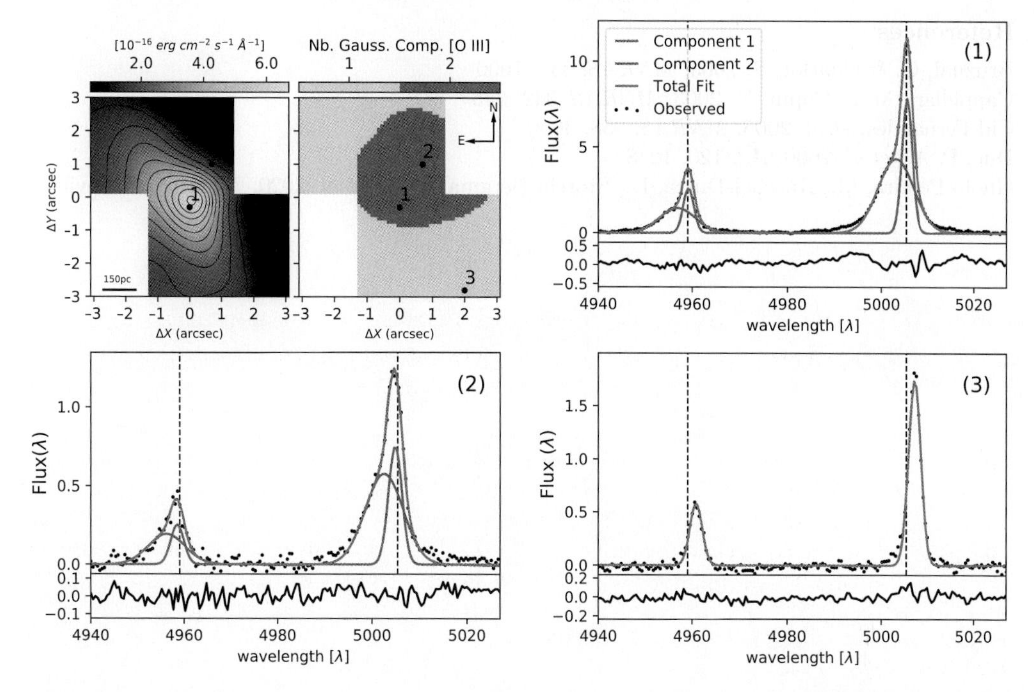

Figure 1. Top Left: Continuum map alongside with the map with the number of Gaussian component fitted to the [O III] emission lines. Other panel: Examples of fits profiles in three distinct regions of the FoV. Vertical axis units in the panels showing the emission-line profiles are 10^{-16}erg cm^{-2} s^{-1}.

populations, trough the entire FoV. A contribution up to 20 per cent of the total light from the FC component is also found in the innermost spaxels.

The pericentre passage of the galaxies is estimated to have occurred ∼100 Myr ago (Duc *et al.* 2000), thus a possible scenario is that metal-poor gas inflow has let to interaction-driven circumnuclear star formation which can explain the presence of such young metal-poor stellar population in the nucleus. Such inflows could also be responsible to trigger the nuclear activity. For numerical simulations on the AGN-merger link in Arp 245 see abstract by Lösch & Ruschel-Dutra in this same IAU Proceeding Volume.

3. Gas kinematics

We use gaussian profiles to fit the ionized gas emission lines spaxel by spaxel, performed using the IFSCUBE code (https://github.com/danielrd6/ifscube). Examples of the fits in distinct portions of the FoV can be seen in Figure 1. Two main kinematic components were adjusted to [O III]λ5007. The first one was present in the entire FoV, showing a clear rotation pattern, interpreted as originating in the galaxy disk in rotation. The second one was present from the central to the NW region, blueshifted by ∼150−250 km s_{-1} and interpreted as due to a nuclear outflow. For the outflow component we estimated both its mass outflow rate and kinetic power, combining the velocities from [O III]λ5007, the luminosity from H αand the electron density from the [S II]λ6717/λ6731 ratio. The values found were $\dot{M}_{out} \approx 1.6 \pm 0.6$ M$_\odot$ yr^{-1} and $\dot{E}_{out} = 2.2 \pm 0.3 \times 10^{40}$ erg s^{-1}. We do not find any clear evidence of direct influence of the interaction in the kinematics of the circumnuclear region of NGC 2992.

More details of this study can be found in the paper Guolo-Pereira *et al.* (2020).

References

Bruzual, G. & Charlot, S. 2003, *MNRAS*, 344, 1000
Cappellari, M. & Copin, Y. 2003, *MNRAS*, 342, 345
Cid Fernandes, *et al.* 2005, *MNRAS*, 358, 36
Duc, P. A., *et al.* 2000, *AJ*, 120, 1238
Guolo-Pereira, M., Ruschel-Dutra, D., Storchi-Bergmann, T., *et al.* 2020, *MNRAS*, in Review

Galaxy Evolution and Feedback across Different Environments
Proceedings IAU Symposium No. 359, 2020
T. Storchi-Bergmann, W. Forman, R. Overzier & R. Riffel, eds.
doi:10.1017/S1743921320001593

A MUSE study of NGC 7469: Spatially resolved star-formation and AGN-driven outflows

A. C. Robleto-Orús[1], J. P. Torres-Papaqui[1], A. L. Longinotti[2], R. A. Ortega-Minakata[3], S. F. Sánchez[4], Y. Ascasibar[5], E. Bellocchi[5], L. Galbany[6], M. Chow-Martínez[7,1], J. J. Trejo-Alonso[8], A. Morales-Vargas[1], F. J. Romero-Cruz[1,9], K. A. Cutiva-Alvarez[1] and R. Coziol[1]

[1]Departamento de Astronomía, Universidad de Guanajuato, Mexico
[2]Instituto Nacional de Astrofísica, Óptica y Electrónica, Mexico
[3]Instituto de Radioastronomía y Astrofísica, Univ. Nac. Aut. de México, Mexico
[4]Instituto de Astronomia, Universidad Nacional Autónoma de México, Mexico
[5]Departamento de Física Teórica, Universidad Autónoma de Madrid, Spain
[6]Departamento de Física Teórica y del Cosmos, Universidad de Granada, Spain
[7]Instituto de Geología y Geofísica, Universidad Nacional Autónoma de Nicaragua, Nicaragua
[8]Facultad de Ingeniería, Universidad Autónoma de Querétaro, Mexico
[9]Tecnológico de Monterrey, Campus Irapuato, Mexico

Abstract. NGC 7469 is a well-known type 1 AGN with a cirumnuclear star formation ring. It has previous detections of X-rays warm absorbers and an infrared biconical outflow. We analysed archival MUSE/VLT observations of this galaxy in order to look for an optical counterpart of these outflows. We report spatially resolved winds in the [O III]λ5007 emission line in two regimes: a high velocity regime possibly associated with the AGN and a slower one associated with the massive star formation of the ring. This slower regime is also detected with Hβ.

Keywords. Galaxies: Seyfert, nuclei

---

1. Introduction

NGC 7469 is a luminous infrared galaxy (LIRG) hosting a Seyfert 1 active galactic nucleus (AGN). It shows a starburst, likely caused by interactions with galaxy IC 5283, concentrated in a circumnuclear star formation ring (hereafter: CSFR) between 0.4 and 1.6 kpc from the centre, producing 80% of the bolometric flux (Genzel *et al.* 1995). Stellar populations within the ring are very young with ages <20 Myr (Diaz-Santos *et al.* 2007).

Nuclear winds have been reported in X-rays (warm absorbers) and ultraviolet by Blustin *et al.* (2007), with line-of sight velocities (LoSV) from -500 to -2000 km s^{-1}, which have been associated with the AGN. A biconical outflow was reported by Müller-Sánchez *et al.* (2011), in scales of 10^2 pc, using infrared integral field spectroscopy (IFS). Here we present a search for optical counterparts of this outflow using archival IFS data of the Multi Unit Spectroscopic Explorer (MUSE) instrument at the Very Large Telescope (VLT, European Southern Observatory, Chile).

2. Methodology

We used archival data from August 19, 2014 from the MUSE science verification run. The data reduction followed the standard procedures, with the REFLEX software (Freudling *et al.* 2013) and MUSE pipelines (Weilbacher *et al.* 2014).

Correction for beam smearing was applied (to account for contamination by extended emission from the non-resolved AGN due to seeing) using QDEBLEND3D, described in Husemann *et al.* (2014). This yields a datacube containing stellar and nebular emission from the host galaxy, as well as the emission from the extended narrow line region (NLR), where evidence of outflows could be found. In this way, contamination by the broad line region (BLR), the inner NLR and the AGN continuum were subtracted.

A synthetic stellar continuum (obtained with STARLIGHT, Cid Fernandes *et al.* 2005) using the MILES spectral libraries (a 2016 update of the ones by Bruzual & Charlot 2003) was subtracted from each spaxel. Afterwards, three Gaussian components were fitted to the [O III]λ5007 and Hβ emission lines. Following the non-parametric approach (Harrison *et al.* 2014), we measured the full width at half maximum ($FWHM$) and the width at 80% of the flux (W_{80}), with the ratio $W_{80}/FWHM > 1.2$ criterion corresponding to a line profile so broadened at its base that it cannot be fitted with a single Gaussian. The velocity offset Δv (i.e. the mean of the velocities at the 5th and 95th percentiles) was estimated, which is related to the asymmetry of the line. Large values of these parameters are evidence of the presence of outflows.

3. Results

For the [O III]λ5007 line we found a broadening ($2 \leqslant W_{80}/FWHM \leqslant 4$) and blue-shifted asymmetry ($\Delta_v \sim -300$ km s^{-1}) at $\sim$200 pc in the north-west direction between the AGN and the CSFR. This is consistent with an extended outflow moving towards the observer. Lower velocity offsets ($\Delta v \sim -200$ to -100 km s^{-1}) extend across the north east, north west and south west regions, covering part of the CSFR. This lower velocity outflow is consistent with what we observe in Hβ, and we propose that it is associated with the massive star forming regions of the CSFR. Nevertheless, the higher velocity [O III]λ5007 outflow was not detected in Hβ, for which we propose and AGN origin. More details will be presented in Robleto-Orús *et al.* (in preparation).

References

Blustin, A. J., Kriss, G. A., Holczer, T., *et al.* 2007, *A&A*, 466, 1, 107–118
Bruzual, G. & Charlot, S. 2003, *MNRAS*, 344, 1000–1028
Cid Fernandes, R., Mateus, A., Sodré, L., *et al.* 2005, *MNRAS*, 358, 363–378
Diaz-Santos, T., Alonso-Herrero, A., Colina, L., *et al.* 2007, *ApJ*, 661, 1, 149–164
Freudling, W., Romaniello, M., Bramich, D. M., *et al.* 2013, *A&A*, 559, A96
Genzel, R., Weitzel, L., Tacconi-Garman, Blietz, M., *et al.* 1995, *ApJ*, 444, 129–145
Harrison, C. M., Alexander, D. M., Mullaney, J. R., *et al.* 2014, *MNRAS*, 441, 4, 3306–3347
Husemann, B., Jahnke, K., Sánchez, S. F., *et al.* 2014, *MNRAS*, 443, 1, 755–783
Müller-Sánchez, F., Prieto, M. A., Hicks, E. K. S., *et al.* 2011, *ApJ*, 739, 2, 69
Weilbacher, P. I., Streicher, O., Urrutia, T., *et al.* 2014, *ASP-CS*, 485, 451

Giacomo Venturi

Frederik Hamman

Alberto Rodriguez Ardila

Session 5: Secular evolution and internal processes: Mechanisms for fueling star formation and AGN

Galaxy Evolution and Feedback across Different Environments
Proceedings IAU Symposium No. 359, 2020
T. Storchi-Bergmann, W. Forman, R. Overzier & R. Riffel, eds.
doi:10.1017/S1743921320004251

INVITED LECTURE

Nature of inflows and outflows in AGNs

Keiichi Wada[1], Yuki Kudoh[1,3], Naomichi Yutani[1] and Nozomu Kawakatu[2]

[1]Kagoshima University, Kagoshima 890-0065, Korimoto 1-21-35, Kagoshima, Japan
email: wada@astrophysics.jp

[2]National Institute of Technology, Kure Collge, Kure, Japan

[3]National Astronomical Observatory of Japan, Mitaka, Japan

Abstract. Despite many theoretical studies and observations, we still do not fully understand the feeding mechanism in AGNs even in nearby galaxies, and how feedback from AGNs affects the gas dynamics itself in the galactic central regions. In this article, we summarize our recent theoretical studies and preliminary results in terms of the mass inflow and outflows on sub-parsec to 100 parsecs scales around AGNs. We introduce different studies: 1) How do galaxy-galaxy mergers trigger AGN activity and obscuration?, 2) How do the radiative feedback affect formation of outflows and obscuration of the nucleus? and 3) How does the AGN plus starburst feedback contribute to the obscuration?

Keywords. merger, obscuration, feedback

1. Introduction

Nuclear activity in galaxies, such as the Active Galactic Nuclei (AGNs) and the nuclear starburst are believed to be triggered by mass inflow to the galactic central region. The underlying physics here is clear: transfer or redistribution of the angular momentum of the interstellar medium. This "mass transfer-induced activity" in galaxies has been a long-standing problem for more than three decades (e.g., Shlosman (1993)), yet not fully understood. This is partly because the fueling and feedback of AGNs span spatially many orders of magnitude. An important open question is how the energy, momentum and radiative feedback from the AGNs and the mass transfer to the central region coexist. Strong feedback may prevent from accreting the material, but if this is the case, AGN activity eventually die. The non-spherical mass accretion and outflows driven by the radiation from the accretion disc could be a key mechanism (e.g. Wada 2012, 2015, 2016, 2018a,b; see also Williamson in this volume), but episodic mass accretion and feedback during the AGN lifetime may also explain the coexistence.

Gas supply to the central 100 pc from the galactic scale would not be a big problem. As many numerical simulations have revealed, the stellar bars, galaxy-galaxy encounters or major/minor mergers may remove the angular momentum of the gas. However, the gas dynamics from several tens parsecs to the accretion disc scale is barely understood. Observationally, the structures and dynamics of the gas in the central tens of parsecs of external galaxies were not well mapped and sampled, but thanks to the ALMA, spatial structures and dynamics of the molecular gas in the central regions in some nearby AGNs have been recently resolved (Combes *et al.* (2019); Izumi *et al.* (2018); see also Garcia-Burillo in this volume). The X-ray observations may also reveal the structures of the ISM in the vicinity of the AGN. For example, Buchner *et al.* (2015) analyzed X-ray selected

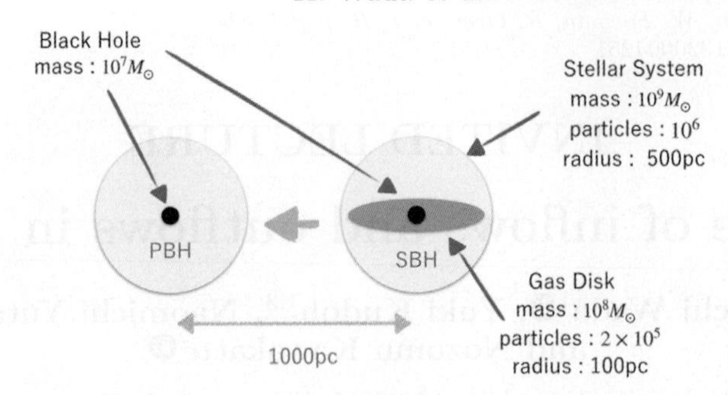

Figure 1. Schematic picture of initial setup of the fiducial model.

2000 AGNs, and they found that a large fraction of the AGNs is obscured by dense gas with $N_H > 10^{22}$ cm^{-2}. This obscured nature of AGNs should also be related with the fueling and feedback processes in the region of $r \sim$ sub-pc to several 100 pc.

In this article, we summarize our recent theoretical studies on the mass inflow and outflows from three different points of view: 1) How do galaxy-galaxy mergers trigger AGN activity and obscuration (Kawaguchi *et al.* (2020); Yutani & Wada in prep.), 2) How do the radiative feedback affect formation of outflows and obscuration of the nucleus? (Kudoh & Wada in prep.), and 3) How does the AGN plus starburst feedback contribute to the obscuration? (Kawakatu *et al.* (2020)).

2. Triggering AGN activity and obscuration during mergers

It is widely believed that galaxy mergers contribute to the growth of supermassive black holes (SMBH) in galaxies, and nuclear activity can be triggered by major mergers. Mergers of two or more nucleated galaxies result in the formation of binary BHs. Governato *et al.* (1994) studied the orbital decay of binary black holes (BBH) during galaxy mergers using N-body experiments. It has also been observed that the interstellar medium (ISM) around binary BHs may affect the orbital decay of BBHs. Because of the interaction between BBHs and the ISM, a nuclear starburst could be enhanced (Taniguchi & Wada (1996)). More recently, Prieto *et al.* (2017) studied mass transport in high-z galaxies and BH growth using cosmological hydrodynamic simulations.

In Kawaguchi *et al.* (2020), we studied the interactions between a BBH (and triple BHs) and the interstellar medium in the central sub-kpc region, using the N-body/SPH code ASURA (Saitoh *et al.* (2008, 2009); Saitoh & Makino (2013)). The numerical experiments aimed to understand the fate of the gas supplied by mergers of two or more galaxies with SMBHs, and the efficiency of the BH growth resulting from the gas supply to the galactic central region by mergers.

The model setup is schematically shown in Fig. 1. The mass resolutions are $10^3 M_\odot$ for stars and $500 M_\odot$ for gas. The gravitational softening radii are 0.5 pc for both SPH and star particles. This secondary BH (SBH) system falls toward the primary BH (PBH) system, and they merge (Fig. 2).

We found that the mass accretion rate to one SMBH exceeds the Eddington rate as the distance of two BHs rapidly decreases (see also Fig. 4). However, this rapid accretion phase does not last more than 10 Myrs, and it decreases to a sub-Eddington value ($\sim$10% of the Eddington mass accretion rate). The rapid accretion is caused by the angular momentum transfer from the gas to the stellar component, where gravitational torque dominates the torque created by the turbulent pressure gradient. The rapid accretion phase is followed by a quasi-steady accretion phase where the angular momentum is redistributed not only by the gravitational torque, but also by the turbulent viscosity in the gas disc.

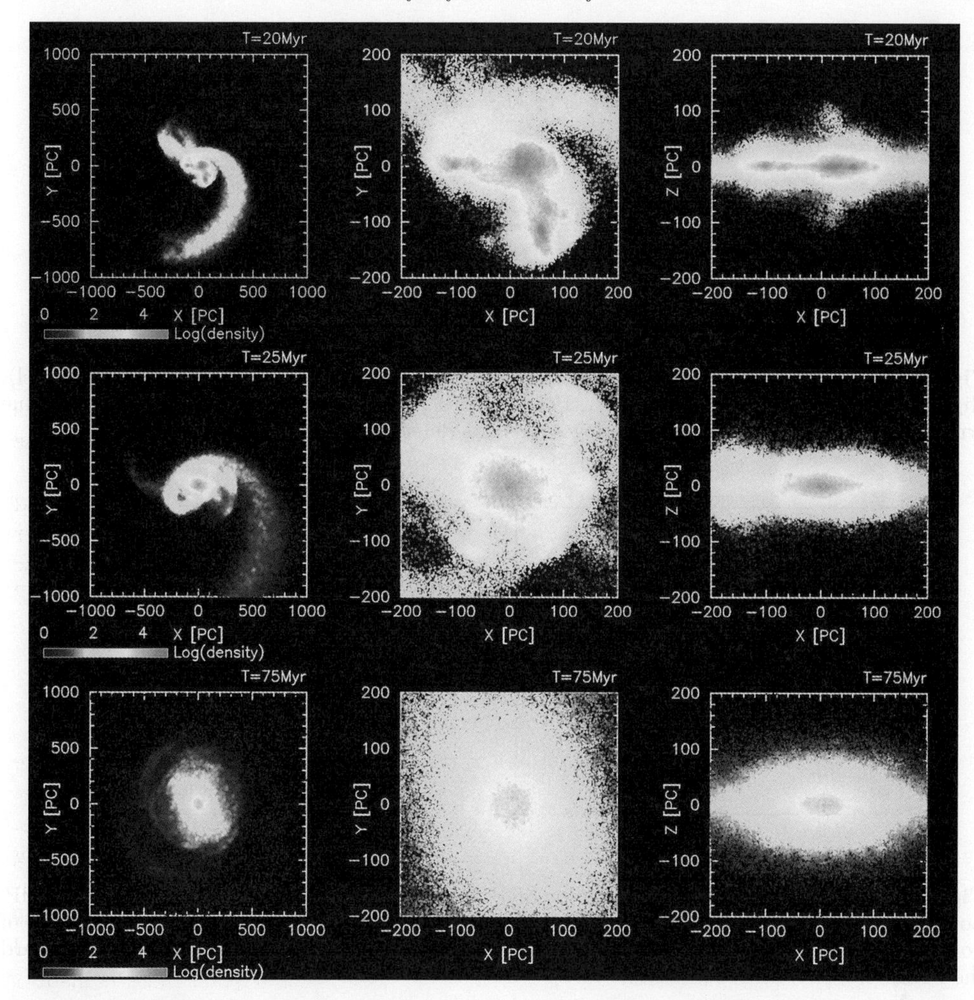

Figure 2. Evolution of gas density in the fiducial model. Three snapshots, at $t = 20, 25$, and 75 Myr, are shown. The left panels show distribution projected onto the x-y plane ($2\,\text{kpc} \times 2\,\text{kpc}$), and the right two columns are close-ups for the x-y and x-z planes ($0.4\,\text{kpc} \times 0.4\,\text{kpc}$). The axis units are in parsecs. The color bar represents log-scaled density ($M_\odot$ pc^{-3}).

We now examine how the energy feedback from AGNs affect the mass accretion processes. Figure 3 is a close-up of the density field in the central 200 pc $\times$ 200 pc for a model without AGN feedback and two models with AGN feedback. We found that the AGN feedback does not change the large-scale morphology (cf. Fig. 2), but it makes the gas diffuse around the BHs. In Model HighAGN, where the feedback energy rate is calculated as $0.02\dot{M}c^2$, at a mass accretion rate ($\dot{M}$) to $r = 1$ pc, almost no high-density gas remains around the BHs. This result is in contrast to Models AGN (LowAGN), where the energy conversion efficiency is 1/10 (1/100) of that in Model HighAGN.

Figure 4 (left) compares the time evolution of the mass accretion rates to SBH in the three feedback models (Models AGN, HighAGN, and LowAGN) with that of the fiducial model (hereafter, Model hP). There is no significant difference between the mass accretion rates of Model hP and Model LowAGN, and the accretion rate in Model AGN is slightly smaller than those in the other two models after $t \sim 50$ Myr. All three models show a peak of mass accretion, $\dot{M} \simeq 3$–$4 M_\odot$ yr^{-1} at approximately $t \sim 20$ Myr. However, the mass accretion rate at the peak is sub-Eddington ($0.7 M_\odot$ yr^{-1}) in model

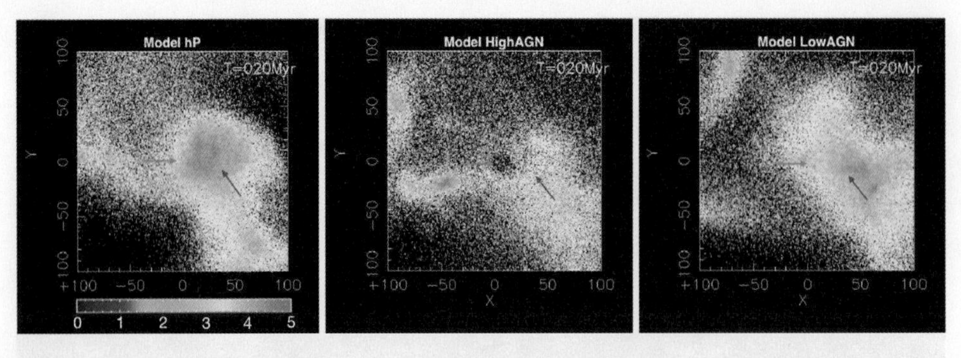

Figure 3. Gas density distributions of central 100-pc regions in Models hP (fiducial model), HighAGN, and LowAGN at $t = 20$ Myr. Positions of PBHs and SBHs are shown by red and blue arrows. The color bar represents the log-scaled gas density in $M_\odot$ pc^{-3}.

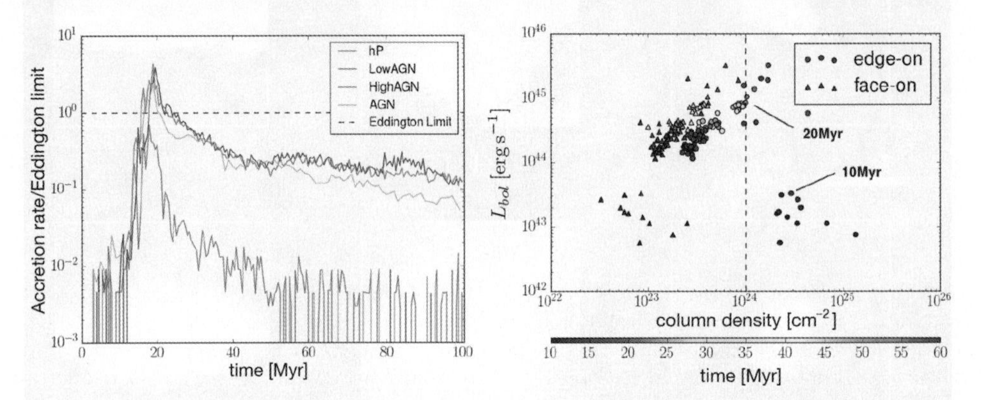

Figure 4. (left) Mass accretion rates for SBH normalized by Eddington rate for Models hP, LowAGN, HighAGN, and AGN. (right) Evolutionary track in model AGN on the plane of bolometric luminosity, based on the mass accretion to the PBH, and column densities toward the PBH.

HighAGN, and rapidly decreases to ∼1/10 in Model hP and Model LowAGN in the late accretion phase ($t > 20$ Myr). In Model HighAGN, 2% of $\dot{M}c^2$ at $r = 1$ pc is supplied to the the circumnuclear gas. This result shows that AGN feedback and black hole growth (i.e., mass accretion) can coexist in galaxy merger simulations if the feedback efficiency at 1 pc is ∼0.02−0.2%. Kawaguchi *et al.* (2020) also found that the luminous phase of the AGN ($L_{bol} > 10^{45}$ erg s^{-1}) during the merger events is heavily obscured ($N_H > 10^{24}$ cm^{-2}) by the supplied gas, and the moderate AGN feedback does not alter this property (Fig. 4(right)).

The fraction of the gas that accretes to each BH is approximately 5–7% of the supplied total gas mass ($10^8 M_\odot$), and 15–20% of the gas forms a circumnuclear gas within 100 pc of the BH. Star formation consumes approximately 15% of the gas supplied by mergers, and the rest forms a circumnuclear gas. Only 1/10 of the supplied gas accumulates to the central 1 pc and it is used to grow the BHs in each event. This idealized situation implies that frequent mergers are necessary for the continuous growth of BHs.

The gas inside $r < 100$ pc mostly contributes to the large column density. Ricci *et al.* (2017a) have studied 52 galactic nuclei in infrared-selected local Luminous and Ultra-luminous infrared galaxies (ULIRG) in different merger stages in the hard X-ray band, and found that the fraction of Compton-thick AGN in late merger galaxies are higher than in local hard X-ray selected AGN. They suggested that the material is most effectively

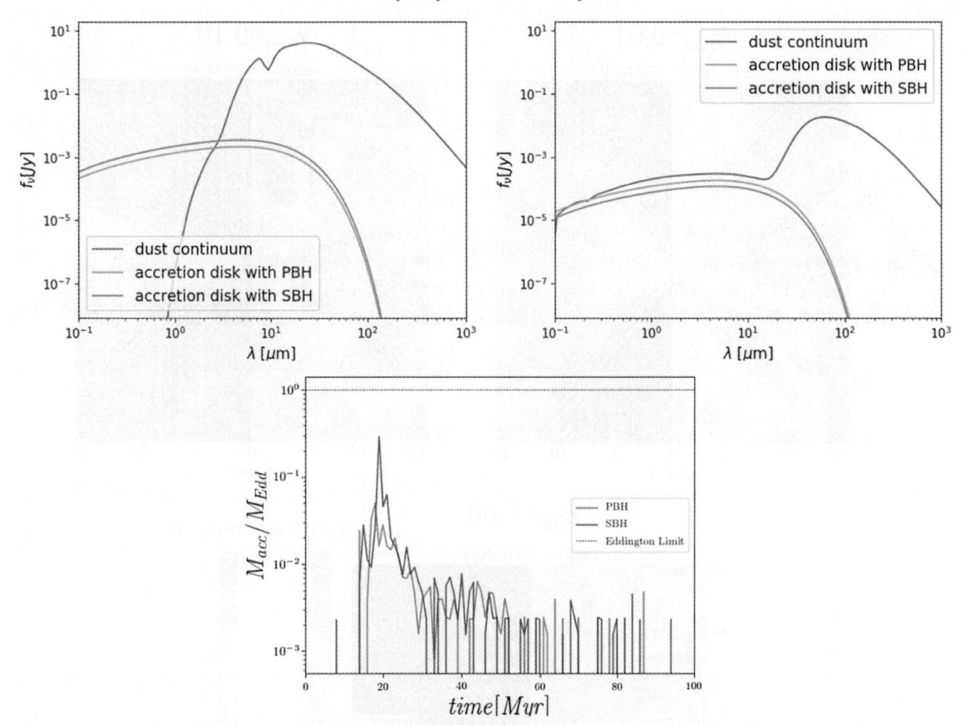

Figure 5. SED evolution of mergers with $M_{BH} = 10^7 M_\odot$, $M_{gas} = 2 \times 10^8 M_\odot$ and $M_{star} = 1 \times 10^9 M_\odot$, $R_{gas} = 100$ pc. Top left is at $t = 20$ Myr and top right is at $t = 70$ Myr. The distance of the observer is 100 Mpc. The bottom panel shows time evolution of the mass accretion rate. When two systems are merged ($t \sim 20$ Myr), the accretion rates become the maximum value ~0.3 for PBH and 0.06 of the Eddington rate. The mass resolution is 1000 $M_\odot$ for both gas and stars. (Yutani, & Wada in prep.)

funnelled to the inner tens of parsecs during the late stages of galaxy mergers. Our results above are qualitatively consistent with Ricci *et al.* (2017a).

Our model can also be compared with the recent findings of a high fraction of Compton-thick AGNs in merging systems and AGNs in dust-obscured galaxies (DOGs) (Dey *et al.* (2008); Fiore *et al.* (2008); Toba *et al.* (2017); Riguccini *et al.* (2019)). In Fig. 5, we show preliminary results of the SED evolution during the mergers of two gas-rich systems with BHs, which is slightly different from the model set-up in Kawaguchi *et al.* (2020). Here the AGN feedback efficiency is assumed to be 0.2%. At $t = 20$ Myr, when the mass accretion rate to the PBH and SBH is 30% of the Eddington rate, the UV light from the nucleus is attenuated. At $t \sim 70$ Myr after the two systems are merged, the nucleus becomes "bluer" and less obscured (Fig. 5, top right).

3. How does the radiative feedback affect outflows formation and obscuration?

Using a systematic multi-wavelength survey of hard X-ray-selected black holes, Ricci *et al.* (2017b) suggested that radiation pressure on dusty gas is the main physical mechanism regulating the distribution of the circumnuclear material. We are trying to confirm this observational result by using a grid-based MHD code, CANS+ (Matsumoto *et al.* (2019)). Figure 6 shows a preliminary result, where density and temperature distributions of three models with the Eddington ratios of 0.01, 0.1 and 1.0 in the central 16 pc are shown. The model is axisymmetric, and the grid cell size is 0.01 pc. The black hole mass is $10^7 M_\odot$. The non-spherical radiation feedback (the radiation pressure for dust

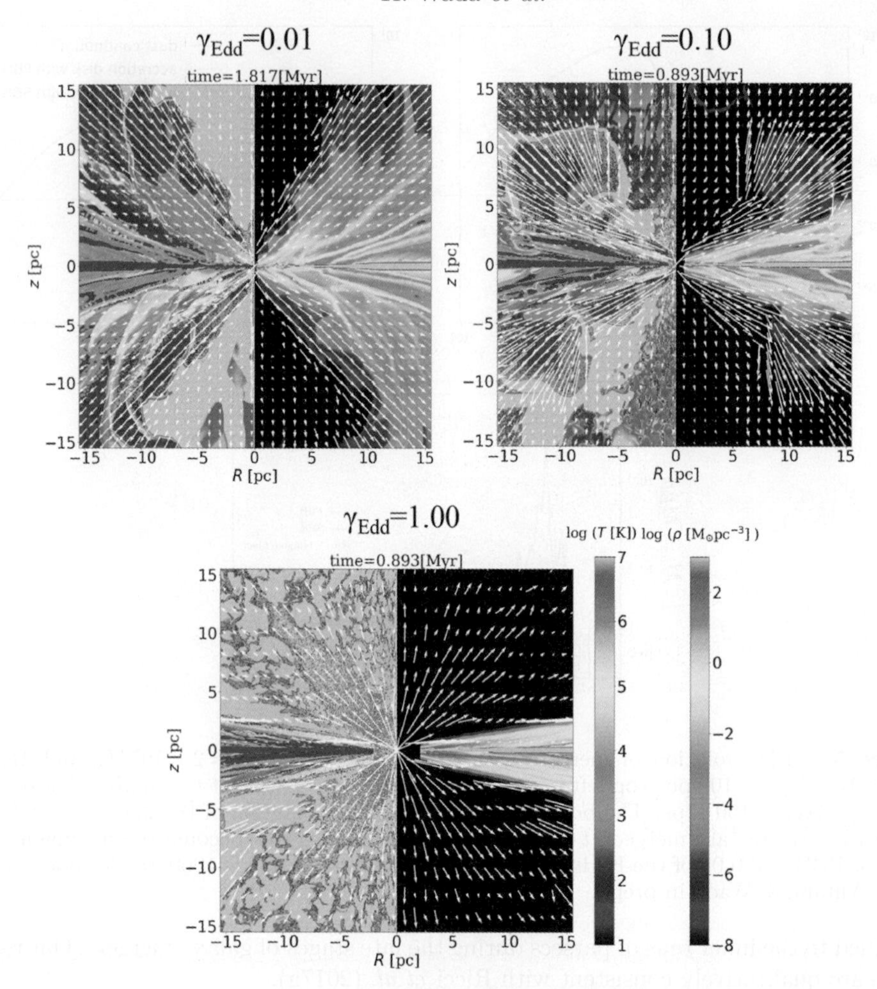

Figure 6. The dependence of the outflow properties around AGN on the Eddington ratio (γ_{Edd}). Density (right panels) and temperature (left panels) are shown (Kudoh, & Wada in prep.)

and the X-ray heating) is considered. As suggested by the observations, the scale height of the disc becomes smaller for larger the Eddington ratio. However, for much smaller γ_{Edd}, the radiation-driven outflows themselves are not formed; therefore it is difficult to explain the high obscured fraction in AGNs with $\gamma_{Edd} \ll 0.01$ only by the circumnuclear region on this scale.

4. How do the AGN and starburst feedback contribute to obscuration?

In Kawakatu *et al.* (2020), we investigated the structure at 10 pc scale obscuring the circumnuclear discs (CNDs) by considering the SN feedbacks from nuclear starburst and the effect of anisotropic radiation pressure. We explored how structures of 1–10 pc dusty CNDs depend on the BH mass ($M_{\rm BH}$), AGN luminosity ($L_{\rm AGN}$), and physical properties of CNDs. The model is based on Wada & Norman (2002); Kawakatu & Wada (2008), and Kawakatu & Wada (2009).

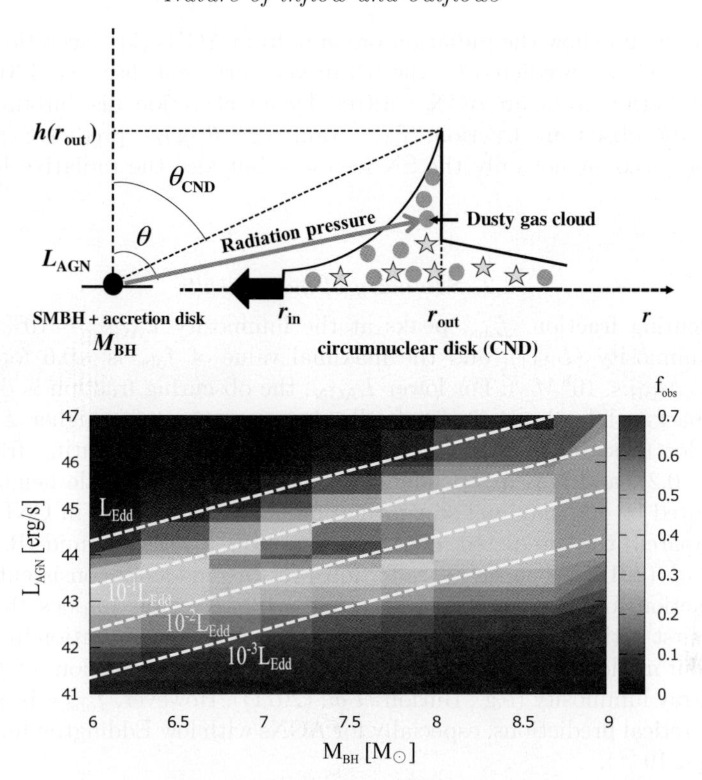

Figure 7. (top) A schematic picture of the CND with nuclear starbust irradiated by the radiation from the accretion disc. (bottom) Obscured fractions as a function of the AGN luminosity and the black hole mass.

4.1. *Structure of circumnuclear discs with the radiative feedback from the AGN*

We assume that the vertical structure of CNDs is in hydrostatic equilibrium (see details in Wada & Norman 2002). We suppose that the turbulent pressure associated with SN explosions is balanced with gravitation in the vertical direction, i.e.,

$$\rho_{\rm g} v_{\rm t}^2 = \rho_{\rm g} g h, \tag{4.1}$$

where $\rho_{\rm g}$, $v_{\rm t}$, and h are the gas density, turbulent velocity, and scale height of the disc, respectively. Under the energy between supernova feedback $E_{\rm in}$ and the turbulent dissipation $E_{\rm out}$, the turbulent velocity $v_{\rm t}$ and scale height h are expressed as

$$v_{\rm t} = \left(\frac{GM_{\rm BH}}{r^3}\right)^{1/2} h, \tag{4.2}$$

$$h = \left(\frac{GM_{\rm BH}}{r^3}\right)^{-3/4} (\eta E_{\rm SN} C_*)^{1/2},$$
$$= 14\,{\rm pc} \left(\frac{C_*}{10^{-8}\,{\rm yr}^{-1}}\right)^{1/2} \left(\frac{M_{\rm BH}}{10^7 M_\odot}\right)^{-3/4} \left(\frac{r}{30\,{\rm pc}}\right)^{9/4}. \tag{4.3}$$

Here C_* is the star-formation efficiency and E_{SN} is the ejecta energy of a single supernova. The model predicts that the disc will have a concave structure due to the SN-driven turbulence, i.e., $h \propto r^{9/4}$ (see Fig. 7 top).

In order to examine how the radiation pressure from AGNs (i.e., accretion disc) affects the structure of CNDs predicted by the SN-driven turbulent disc (eq. 4.3), we consider anisotropic radiation from an AGN emitted by an accretion disc around an SMBH. We evaluate the obscuring fraction, $f_{\rm obs} \equiv \tan(\pi/2 - \theta_{CND})$, predicted by the model that takes into account not only the SN feedback but also the radiative feedback from the AGN.

4.2. *Summary of main results*

• The obscuring fraction, $f_{\rm obs}$, peaks at the luminosity $L_{\rm AGN,p} \sim 10\%$ of the AGN Eddington luminosity ($L_{\rm Edd}$), and the maximal value of $f_{\rm obs}$ is ~0.6 for less massive SMBHs (e.g., $M_{\rm BH} < 10^8 M_\odot$). For lower $L_{\rm AGN}$, the obscuring fraction is determined by the SN feedback, while the radiative feedback is important for higher $L_{\rm AGN}$. On the other hand, for massive SMBHs (e.g., $M_{\rm BH} > 10^8 M_\odot$), the obscuring fraction $f_{\rm obs}$ is always low (<0.2), and it is independent of $L_{\rm AGN}$ because the scale height of CNDs is mainly regulated by the maximal star-formation efficiency, $C_{*.\rm max}$, in CNDs.

• We compared the predicted obscuring fraction $f_{\rm obs}$ with mid-IR observations (Ichikawa *et al.* (2019)). The SN + radiation pressure model is consistent with the IR obscuring fraction for massive BHs with $M_{\rm BH} = 10^8 M_\odot$. This implies that an intense nuclear starburst with $C_{*,\rm max} = 10^{-7}\ {\rm yr}^{-1}$ contributes to the obscuration in these objects. In addition, our model can qualitatively explain the observed behaviour of $f_{\rm obs}$ as a function of the X-ray luminosity (e.g., Burlon *et al.* (2011)). However, $f_{\rm obs,X}$ is always greater than our theoretical predictions, especially for AGNs with low Eddington luminosity ratio ($L_{\rm AGN}/L_{\rm Edd} < 10^{-2}$).

References

Buchner, J., Georgakakis, A., Nandra, K., *et al.* 2015, *ApJ*, 802, 89
Burlon, D., Ajello, M., Greiner, J., *et al.* 2011, *ApJ*, 728, 58
Combes, F., García-Burillo, S., Audibert, A., *et al.* 2019, *A&Ap*, 623, A79
Debuhr, J., Quataert, E., & Ma, C.-P. 2011, *MNRAS*, 412, 1341
Dey, A., Soifer, B. T., Desai, V., *et al.* 2008, *ApJ*, 677, 943
Fiore, F., Grazian, A., Santini, P., *et al.* 2008, *ApJ*, 672, 94
Governato, F., Colpi, M., & Maraschi, L. 1994, *MNRAS*, 271,
Kawaguchi, T., Yutani, N., & Wada, K. 2020, *ApJ*, 890, 125
Kawakatu, N., Wada, K., & Ichikawa, K. 2020, *ApJ*, 889, 84
Kawakatu, N. & Wada, K. 2008,*ApJ*, 681, 73
Kawakatu, N. & Wada, K. 2009, *ApJ*, 706, 676
Izumi, T., Wada, K., Fukushige, R., *et al.* 2018, *ApJ*, 867, 48
Ichikawa, K., Ricci, C., Ueda, Y., *et al.* 2019, *ApJ*, 870, 31
Matsumoto, Y., *et al.* "Magnetohydrodynamic Simulation Code CANS+: Assessments and Applications", *Publ. Astron. Soc.* Japan, doi:10.1093/pasj/psz064
"Mass-transfer induced activity in galaxies", ed. by I. Shlosman, 1994, Cambridge Univ. Press, NY
Prieto, J., Escala, A., Volonteri, M., & Dubois, Y. 2017, *ApJ*, 836, 216
Ricci, C., Bauer, F. E., Treister, E., *et al.* 2017a, *MNRAS*, 468, 1273
Ricci, C., Trakhtenbrot, B., Koss, M. J., *et al.* 2017b, *Nature*, 549, 488
Riguccini, L. A., Treister, E., Menéndez-Delmestre, K., *et al.* 2019, *AJ*, 157, 233
Saitoh, T. R., Daisaka, H., Kokubo, E., *et al.* 2008, *PASJ*, 60, 667
Saitoh, T. R., Daisaka, H., Kokubo, E., *et al.* 2009, *PASJ*, 61, 481
Saitoh, T. R. & Makino, J. 2013, *ApJ*, 768, 44
Taniguchi, Y. & Wada, K. 1996, *ApJ*, 469, 581

Toba, Y., Nagao, T., Kajisawa, M., *et al.* 2017, *ApJ*, 835, 36
Wada, K. & Norman, C. A. 2002, *ApJL*, 566, L21
Wada, K., Yonekura, K., & Nagao, T. 2018a, *ApJ*, 867, 49
Wada, K., Fukushige, R., Izumi, T., *et al.* 2018b, *ApJ*, 852, 88
Wada, K., Schartmann, M., & Meijerink, R. 2016, *ApJL*, 828, L19
Wada, K. 2015, *ApJ*, 812, 82
Wada, K. 2012, *ApJ*, 758, 66

Hekatelyne Carpes

Sandra Raimundo

Galaxy Evolution and Feedback across Different Environments
Proceedings IAU Symposium No. 359, 2020
T. Storchi-Bergmann, W. Forman, R. Overzier & R. Riffel, eds.
doi:10.1017/S1743921320002239

ORAL CONTRIBUTIONS

Feeding and feedback in nuclei of galaxies†

Anelise Audibert[1], Françoise Combes[2], Santiago García-Burillo[3] and Kalliopi Dasyra[4]

[1]IAASARS, National Observatory of Athens, Penteli, Greece
email: anelise.audibert@noa.gr

[2]Observatoire de Paris, LERMA, Collège de France, CNRS, PSL University, UPMC, Paris

[3]Observatorio Astronómico Nacional (OAN-IGN)- Observatorio de Madrid, Madrid, Spain

[4]Department of Astrophysics, Astronomy & Mechanics, University of Athens, Athens, Greece

Abstract. Our aim is to explore the close environment of Active Galactic Nuclei (AGN) and its connection to the host galaxy through the morphology and dynamics of the cold gas inside the central kpc in nearby AGN. We report Atacama Large Millimeter/submillimeter Array (ALMA) observations of AGN feeding and feedback caught in action in NGC613 and NGC1808 at high resolution (few pc), part of the NUclei of GAlaxies (NUGA) project. We detected trailing spirals inside the central 100 pc, efficiently driving the molecular gas into the SMBH, and molecular outflows driven by the AGN. We present preliminary results of the impact of massive winds induced by radio jets on galaxy evolution, based on observations of radio galaxies from the ALMA Radio-source Catalogue.

Keywords. galaxies: active, galaxies: kinematics and dynamics, ISM: jets and outflows

---

1. Overview: NUGA project

The key elements in galaxy evolution are the interplay of the fuelling of SMBH at the center of galaxies and the subsequent feedback from their AGN. Gas inflows into the center of galaxies can fuel the SMBH and the energy input by the AGN can trigger subsequent feedback. One of the outstanding problems is to identify the mechanism that drives gas from the disk towards the nucleus, removing its large angular momentum (as discussed in Wada (2004) and Jogee (2006), for instance). Feedback processes can be responsible of regulating the SMBH growth (Croton *et al.* 2006) and explain the co-evolution of SMBH and their host galaxies, which is now well established by the tight M-σ relation (Magorrian *et al.* 1998; McConnell & Ma 2013). Recent discoveries of massive molecular outflows (e.g., Fiore *et al.* 2017; Fluetsch *et al.* 2019) have been promoting the idea that winds may be major actors in sweeping the gas out of galaxies, in agreement with theoretical predictions of AGN-driven wind models (see Faucher-Giguère & Quataert 2012; Zubovas & King 2012).

Nearby low luminosity AGN (LLAGN) are ideal laboratories to explore the details of outflow and inflowing gas mechanisms. In the NUGA project, we have performed high resolution observations ($\lesssim 0.1''$) of the CO(3-2) and dense gas tracers emission with ALMA in a sample of 7 nearby LLAGN. The sample spans more than a factor of 100 in AGN power (X-ray and radio luminosities), a factor of 10 in star formation rate (SFR),

† This project has received funding from the Hellenic Foundation for Research and Innovation (HFRI) and the General Secretariat for Research and Technology (GSRT), under grant agreement No 1882.

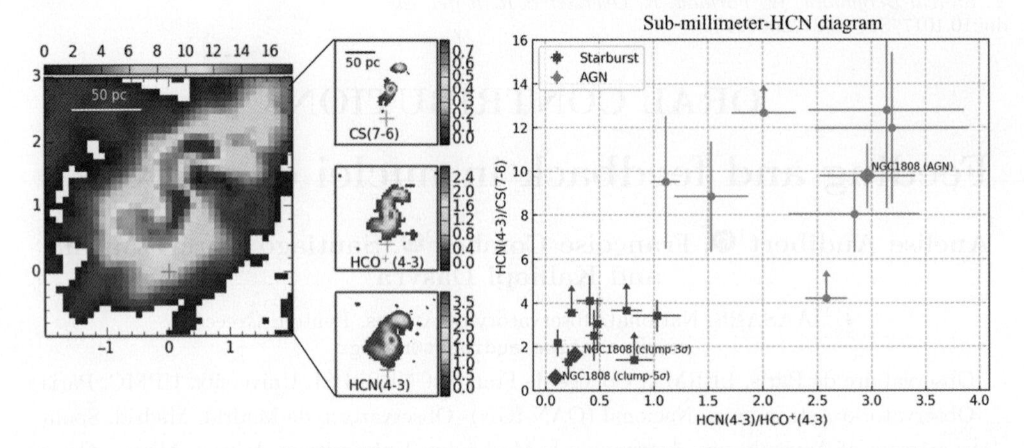

Figure 1. *Left*: a zoomed $4'' \times 4''$ region of CO(3-2) intensity map for NGC1808 and dense gas tracers CS(7-6), HCO^+(4-3) and HCN(4-3). *Right:* submillimetre-HCN diagram (Izumi *et al.* 2016) using the line intensity ratios R_{HCN/HCO^+} and $R_{HCN/CS}$. We include the line ratios of NGC1808 (diamonds) measured at the centre, or "AGN", and in a clump detected 140 pc north-west of central position in the dense tracers.

and a wide range of galaxy inner morphology (Combes *et al.* 2019). Our goal is to probe feeding and feedback phenomena in these LLAGN, through the study of the morphology and kinematics of the cold molecular gas in galaxy disks and the characterization of the mechanisms driving gas inflows and/or outflows.

We mapped the CO(3-2) and HCN(4-3), HCO^+(4-3) and CS(7-6) emission and compared the morphology of the cold gas to optical images from HST and ionised and warm molecular gas observed in the near-infrared (NIR) with SINFONI. We derived the rotation curves and have modelled the observed velocity field of the CO(3-2) line emission in the galaxy disks in order to find patterns of non-circular motions that could be associated to streaming motions of inflowing gas and/or outflow signatures. To estimate the fuelling efficiency, we have computed the gravitational potential from the stars within the central kpc, from the HST images. Weighting the torques on each pixel by the gas surface density observed in the CO(3-2) line has allowed us to estimate the sense of the angular momentum exchange and its efficiency. In this work, we focus on the study of two individual objects: NGC 1808 and NGC 613.

2. Nuclear trailing spiral in NGC1808

The "hot spot" H II/Sy galaxy NGC 1808 was studied using ALMA Cycle 3 observations at 12 pc spatial resolution. The CO(3-2) is distributed in a patchy ring at a radius 350 pc, that is most prominent in the south part and another broken ring at 180 pc. They are connected by multiple spiral arms. Inside the star-forming ring, a 2-arm spiral structure is clearly detected at ~50 pc radius (left panel of Fig. 1), as presented in Audibert *et al.* (2017). The nuclear spiral region corresponds to the peak of the velocity dispersions ($\sigma \gtrsim 100$ km/s). The CO morphology shows a remarkable resemblance between the ionised and warm molecular gas along the star forming ring at $\sim 4''$ radius, traced by the Pa α and H_2 emission with SINFONI (Busch *et al.* 2017). We found that the nuclear spiral is kinematically decoupled from the larger disk, the position angle being tilted from 323° to close to 270°.

Previous CO(1-0) ALMA observations reported a molecular outflow in the central ~250 pc (Salak *et al.* 2016), but we did not detect outflow signatures in our

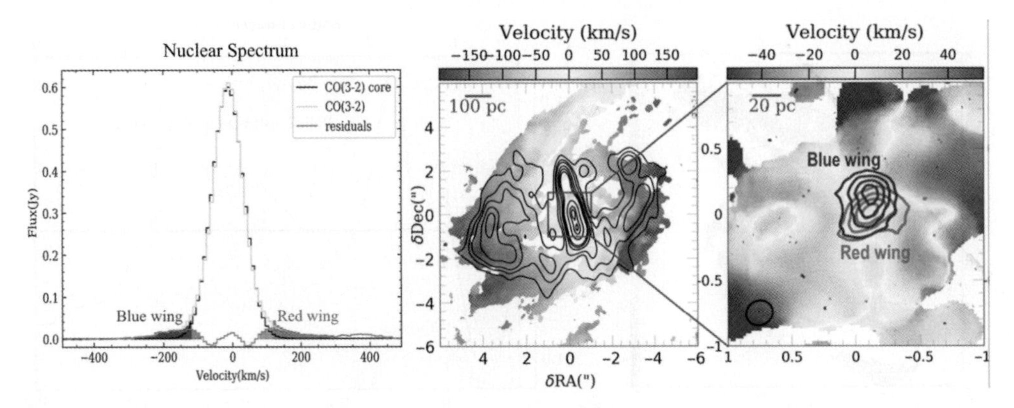

Figure 2. *Left:* the nuclear CO(3-2) spectrum extracted in a 0.28″ region. The blue (−400 to −120 km/s) and red (120 to 300 km/s) wings are associated to the outflow. *Middle and right:* the velocity distribution of the CO(3-2) emission with the VLA radio contours at 4.86 GHz and a 2″ × 2″ zoom of the velocity distribution and the contours of the blue and red wings emission.

high-resolution observations. The velocities are mainly due to circular rotation and some perturbations from coplanar streaming motions along the spiral arms.

We confirm the HCN enhancement in circumnuclear molecular gas around AGN, by measuring the HCN(4-3)/HCO$^+$(4-3) and HCN(4-3)/CS(7-6) intensity ratios in the submillimetre diagram (Izumi *et al.* 2016). We find that the nuclear region of NGC 1808 presents line ratios that indicate excitation conditions typical of X-ray dominated regions in the vicinity of AGN (Fig. 1). What is remarkable in our observations, is that the nuclear trailing spiral is even more contrasted in the dense gas tracers. The two-arm spiral structure is also detected in the residual maps in the NIR by Busch *et al.* (2017), supporting the scenario of gas inflow towards the nucleus of NGC 1808.

3. Nuclear trailing spiral and molecular outflow in NGC613

In the Seyfert/nuclear starburst galaxy NGC 613, we have combined ALMA Cycles 3 and 4 observations at a spatial resolution of 17 pc (Audibert *et al.* 2019). The morphology of CO(3-2) line emission reveals a 2-arm trailing nuclear spiral at r $\lesssim$ 100 pc and a circumnuclear ring at ∼350 pc radius, that is coincident with the star-forming ring seen in the optical images. The molecular gas in the galaxy disk is in a remarkably regular rotation, however, the kinematics in the nuclear region is very skewed. We find broad wings in the nuclear spectra of CO and dense gas tracers, with velocities reaching up to ±300 km/s, associated with a molecular outflow emanating from the nucleus ($r \sim 25$ pc, Fig. 2).

We derive a molecular outflow mass $M_{out} = 2 \times 10^6 M_\odot$ and a mass outflow rate of $\dot{M}_{out} = 27\, M_\odot yr^{-1}$. The molecular outflow energetics exceed the values predicted by AGN feedback models: its kinetic power corresponds to $P_{K,out} = 20\% L_{AGN}$ and the momentum rate is $\dot{M}_{out} v \sim 400 L_{AGN}/c$. The outflow is mainly boosted by the AGN through entrainment by the radio jet, but given the weak nuclear activity of NGC 613, we proposed that we might be witnessing a *fossil outflow*, resulted from a strong past AGN that now has already faded. From 25 to 100 pc, the nuclear trailing spiral observed in CO emission inside the Inner Lindblad Resonance (ILR) ring is efficiently driving gas towards the center. The gravitational torques exerted in the gas show that the gas loses its angular momentum in a rotation period, i.e., in ∼10 Myr dynamical timescale (Fig. 3). NGC 613 is a remarkable example of the complexity of fuelling and feedback mechanisms in AGN: given the relative short flow timescale, $t_{flow} \sim 10^4$ yr, the molecular outflow could be a response of the inflowing gas, and eventually acts to self-regulate the gas accretion.

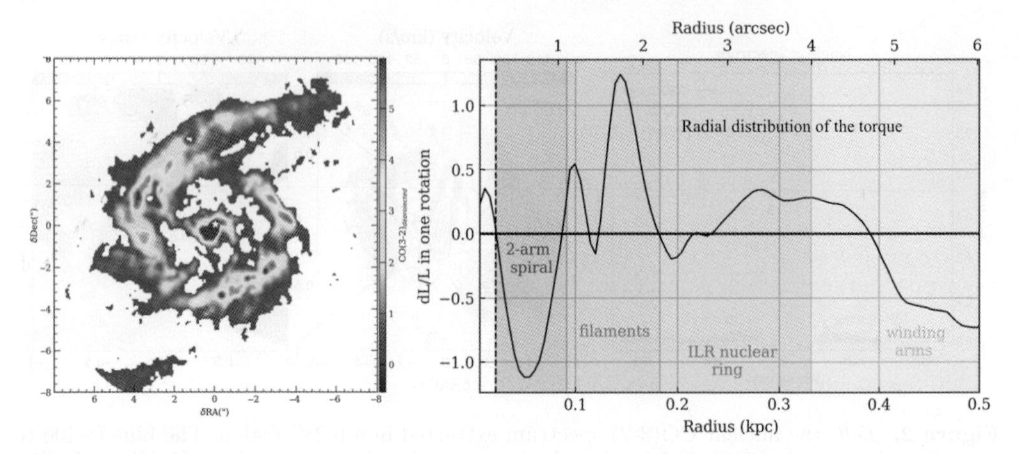

Figure 3. *Left:* deprojected image of the CO(3-2) emission of NGC 613 with the morphological features in the *right* panel, showing the radial distribution of the torque, quantified by the fraction of the angular momentum transferred from the gas in one rotation, dL/L.

4. Summary

Among the total NUGA sample of 8 galaxies (including the prototypical Sy 2 galaxy NGC 1068 studied by our group), there is evidence of outflows in half of the sample, namely, NGC 613, NGC 1433 (Combes *et al.* 2013), NGC 1068 (García-Burillo *et al.* 2014, 2019) and NGC 1808 (detected in CO(1-0) by Salak *et al.* (2016), but not confirmed in our high resolution analysis in CO(3-2) emission). The mass outflow rates range from $\sim$1-70 $M_{\odot} yr^{-1}$, and we confirm the expectations from theoretical models, even in the case of LLAGN: the mass load rates of the outflows increase with the radio power and the AGN luminosity. In the case of the H II/Sy 2 galaxy NGC 1808, the weakest active object among the detections, the outflow is most likely to be starburst-driven. However, in the other three galaxies, the nuclear SFRs are not able to drive the observed outflows and the properties of the flow require an AGN contribution. Therefore we favour the AGN-driven scenario, in particular the radio mode, where the molecular flow is entrained by the interaction between the radio jet and the interstellar medium (ISM). At the same time, the observed outflows could regulate gas accretion in the CND and in short timescales quench the star formation in the nuclear rings, maintaining the balance between gas cooling and heating.

The molecular galaxy disk morphologies reveal the presence of contrasted nuclear rings in the totality of the sample. These rings are quite often the spots of nursery of stars, i.e. usually associated with high SFRs and young star formation, most commonly observed in the optical and NIR. The nuclear rings detected in CO(3-2) emission are usually at the ILR, and in a few cases located at the inner ILR of the nuclear bar, with radius varying from $\sim$170 to 800 pc. Since all galaxies in the sample are barred, with different bar strengths, the detection of molecular rings provides evidence of the efficiency of torques due to the bar, driving and piling up the cold gas in rings to eventually form new stars. Although bars are very efficient to drive the gas to a few hundreds of pc scales, an additional mechanism is necessary to bring the gas to the very center and feed the modest black holes at the center of these LLAGN. We find clear evidence of nuclear trailing spirals in 3 galaxies inside the ILR or inner ILR: NGC 613, NGC 1808 and NGC 1566 (presented in Combes *et al.* 2014). Previous works have computed the torques in NGC 1365 and NGC 1433. In the case of NGC 1365, it was possible to show that the gas is inflowing to the center, driven by the bar, on a timescale of 300 Myr (Tabatabaei *et al.* 2013). For the Sy 2 galaxy NGC 1433, the gas is driven towards a nuclear ring of 200 pc radius, at the inner ILR of the nuclear bar, and viscous torques could drive the gas infall towards the very center (Combes *et al.* 2013; Smajić *et al.* 2014).

The project will notably benefit from the improving in the statistics by joining forces with the Galactic Activity, Torus and Outflow Survey (GATOS: `gatos.strw.leidenuniv.nl`). GATOS is also mapping the CO(3-2) and HCO^+(4-3) emission with ALMA in the circumnuclear disks of 20 Seyfert galaxies, selected from a ultra-hard X-ray sample, with similar spatial resolution of 0.1″. Together, NUGA and GATOS will provide a wider range of AGN luminosities and Eddington ratios to explore the connection of inflowing/outflowing gas and molecular tori properties to the host galaxies.

5. The ALMA Radio-source Catalogue

The importance of radio jets in shaping the galaxy evolution have been highlighted in this IAU Symposium. The interaction between radio jets with the ISM has been revealed that relativistic jets can drive molecular and atomic gas outflows, as in the case of the radio bright Seyfert IC 5063 (Morganti *et al.* 2015; Dasyra *et al.* 2016). ALMA observations have even revealed previously unknown jets thanks to collimated molecular outflows detected in CO (e.g. in NGC 1377 and ESO 420-G13, Aalto *et al.* 2016; Fernández-Ontiveros *et al.* 2020, respectively). To quantify the impact of radio jets on host galaxies, we built a representative sample of radio galaxies observed with ALMA, the ALMA Radio-source Catalogue, even exploring calibrators. New CO detections, even at high velocities are discovered in this sample.

References

Aalto, S., Costagliola, F., Muller, S., *et al.* 2016, *A&A*, 590, A73
Audibert, A., Combes, F., García-Burillo, S., *et al.* 2017, *Frontiers in Astronomy and Space Sciences*, 4, 58
Audibert, A., Combes, F., García-Burillo, S., *et al.* 2019, *A&A*, 632, A33
Busch, G., Eckart, A., Valencia-S., M., *et al.* 2017, *A&A*, 598, A55
Combes, F., García-Burillo, S., Casasola, V., *et al.* 2013, *A&A*, 558, A124
Combes, F., García-Burillo, S., Casasola, V., *et al.* 2014, *A&A*, 565, A97
Combes, F., García-Burillo, S., Audibert, A., *et al.* 2019, *A&A*, 623, A79
Croton, D. J., Springel, V., White, S. D. M., *et al.* 2006, *MNRAS*, 365, 11
Dasyra, K. M., Combes, F., Oosterloo, T., *et al.* 2016, *A&A*, 595, L7
Faucher-Giguère, C.-A. & Quataert, E. 2012, *MNRAS*, 425, 605
Fernández-Ontiveros, J. A., Dasyra, K. M., Hatziminaoglou, E., *et al.* 2020, *A&A*, 633, A127
Fiore, F., Feruglio, C., Shankar, F., *et al.* 2017, *A&A*, 601, A143
Fluetsch, A., Maiolino, R., Carniani, S., *et al.* 2019, *MNRAS*, 483, 4586
García-Burillo, S., Combes, F., Usero, A., *et al.* 2014, *A&A*, 567, A125
García-Burillo, S., Combes, F., Ramos Almeida, C., *et al.* 2019, *A&A*, 632, A61
Izumi, T., Kohno, K., Aalto, S., *et al.* 2016, *ApJ*, 818, 42
Jogee, S. 2006, *Physics of Active Galactic Nuclei at All Scales*, 143
Magorrian, J., Tremaine, S., Richstone, D., *et al.* 1998, *AJ*, 115, 2285
McConnell, N. J. & Ma, C.-P. 2013, *ApJ*, 764, 184
Morganti, R., Oosterloo, T., Oonk, J. B. R., *et al.* 2015, *A&A*, 580, A1
Salak, D., Nakai, N., Hatakeyama, T., *et al.* 2016, *ApJ*, 823, 68
Smajić, S., Moser, L., Eckart, A., *et al.* 2014, *A&A*, 567, A119
Tabatabaei, F. S., Weiß, A., Combes, F., *et al.* 2013, *A&A*, 555, A128
Wada, K. 2004, *Coevolution of Black Holes and Galaxies*, 186
Zubovas, K. & King, A. 2012, *ApJ*, 745, L34

Galaxy Evolution and Feedback across Different Environments
Proceedings IAU Symposium No. 359, 2020
T. Storchi-Bergmann, W. Forman, R. Overzier & R. Riffel, eds.
doi:10.1017/S1743921320001544

Circum-nuclear molecular disks: Role in AGN fueling and feedback

Francoise Combes
Observatoire de Paris, LERMA, Collège de France, CNRS, PSL University,
Sorbonne University, UPMC, Paris
email: francoise.combes@obspm.fr

Abstract. Gas fueling AGN (Active Galaxy Nuclei) is now traceable at high-resolution with ALMA (Atacama Large Millimeter Array) and NOEMA (NOrthern Extended Millimeter Array). Dynamical mechanisms are essential to exchange angular momentum and drive the gas to the super-massive black hole. While at 100pc scale, the gas is sometimes stalled in nuclear rings, recent observations reaching 10pc scale (50mas), may bring smoking gun evidence of fueling, within a randomly oriented nuclear gas disk. AGN feedback is also observed, in the form of narrow and collimated molecular outflows, which point towards the radio mode, or entrainment by a radio jet. Precession has been observed in a molecular outflow, indicating the precession of the radio jet. One of the best candidates for precession is the Bardeen-Petterson effect at small scale, which exerts a torque on the accreting material, and produces an extended disk warp. The misalignment between the inner and large-scale disk, enhances the coupling of the AGN feedback, since the jet sweeps a large part of the molecular disk.

Keywords. galaxies: active, galaxies: general, galaxies: nuclei, galaxies: Seyfert, galaxies: spiral

---

1. Introduction

It is now well established that there exists a tight relation between the mass of the central black hole, and the bulge mass, or central velocity dispersion, which has been interpreted as a co-evolution of galaxies and black holes (e.g., Kormendy & Ho 2013; Heckman & Best 2014). This co-evolution might be due to a common feeding mechanism, either through mergers or cosmic gas accretion followed by secular evolution, as recenty reviewed by Storchi-Bergmann & Schnorr-Müller (2019), and/or to AGN feedback mechanisms, regulating the star formation in the galaxy host (e.g., Fabian 2012; Morganti & Oosterloo 2018).

The new frontier in this domain is to understand in more details the feeding and feedback mechanisms at the highest resolution, in the complex circumnuclear region, surrounding the black hole, with the help of multi-wavelength observations (Ramos-Almeida & Ricci 2017). A new view is emerging, where the absorbing material is not due to the long-expected dusty torus (Hönig 2019). VLT (Very Large Telescope) Interferometer (VLTI) observations showed that the dust on parsec scales is not mainly in a thick torus, but instead in a polar structure, forming like a hollow cone, perpendicular to a thin disk (e.g., Asmus *et al.* 2016; Asmus 2019). The circumnuclear region, as a transition between the Broad Line Region (BLR) of the accretion disk, and the Narrow Line Region (NLR), is complex and clumpy, and contains both inflowing material in a thin disk, where millimeter lines and H_2O masers are found, and an outflowing component, in the perpendicular direction (e.g., Cicone *et al.* 2014; Garcia-Burillo *et al.* 2016).

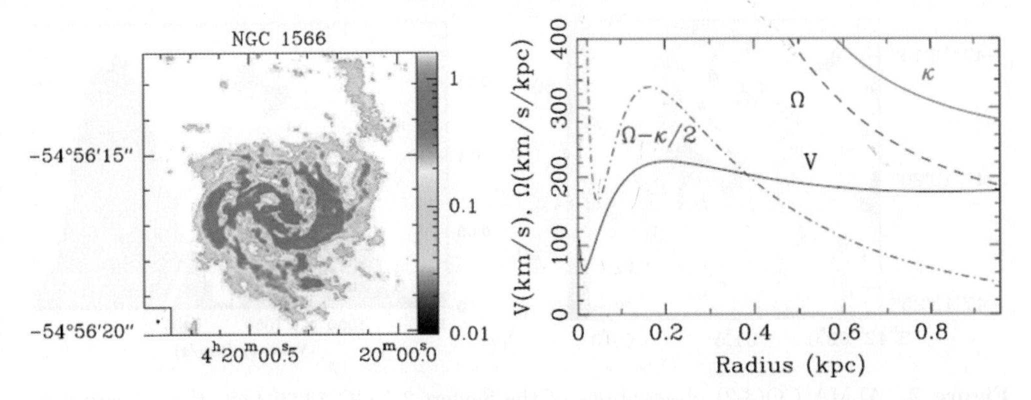

Figure 1. ALMA observations of the Seyfert-1 galaxy NGC 1566. Left is a zoomed 8" × 8" region of the CO(3-2) intensity map, showing the nuclear trailing spiral (1" = 35pc). Right, is the model rotation curve from NIR images, represented schematically, with the circular velocity (black) epicyclic frequency κ (red), and corresponding $\Omega - \kappa/2$ curve (blue) within the central kpc. The contribution of a super-massive black hole in the nucleus with MBH = 8.3×10^6 $M_\odot$ has been included. From Combes *et al.* (2014, 2019).

In the following, I review ALMA observations at high-angular resolution of the molecular gas, revealing nuclear trailing spiral features, that explain the feeding of the central black hole, through exchange of angular momentum. ALMA observations have also revealed outflows, some being extremely collimated in a molecular jet. These outflows must be due to the radio mode of AGN feedback, even when no radio jet has yet been detected. ALMA has also revealed in most nearby Seyferts the existence of molecular circum-nuclear disks, misaligned with the large-scale disks, and with decoupled kinematics. We identify these parsec-scale structures as molecular tori, able to obscure the central accretion disks. Several mechanisms are reviewed to explain the misalignments.

2. Feeding the monster

To fuel the central black hole, the main problem is to transfer the angular momentum of the gas outwards. This can be done by the gravity torques exerted by bars on the gas in spiral arms (Garcia-Burillo *et al.* 2005). Torques are positive outside corotation, and the gas is driven outward to accumulate in a ring at the Outer Lindblad Resonance (OLR). Inside corotation, torques are negative, and gas is driven inward, to pile up in a nuclear ring at the Inner Lindblad resonance (ILR).

What happens inside the ILR depends on the winding sense of orbits there. The gas is orbiting in elliptic streamlines, which gradually tilt by 90° at each resonance and wind up in spiral structures. The precession rate of these elliptical orbits is equal to $\Omega - \kappa/2$, with Ω the rotation frequency = V/r, and κ the epicyclic frequency. Usually, inside ILR, and far from the black hole, $\Omega - \kappa/2$ increases with radius, and the spiral is leading. The torque of the bar is positive, and the gas is driven back to the ILR. But near the massive black hole, the precessing frequency $\Omega - \kappa/2$ is decreasing with radius, and the spiral is trailing. The gas can then fuel the AGN (Buta & Combes 1996).

ALMA has the resolution to enter the sphere of influence of the black hole, and a trailing nuclear spiral was first seen in NGC 1566 (Combes *et al.* 2014). This nuclear spiral is located well inside the r = 400pc ring, corresponding to the ILR of the bar. Fig. 1 shows the nuclear spiral in the CO(3-2) line (left), and at right the rotation curve and corresponding frequencies, derived from the stellar potential traced by near-infrared images. The precessing rate $\Omega - \kappa/2$ increases towards the center, inside 50pc, due to a

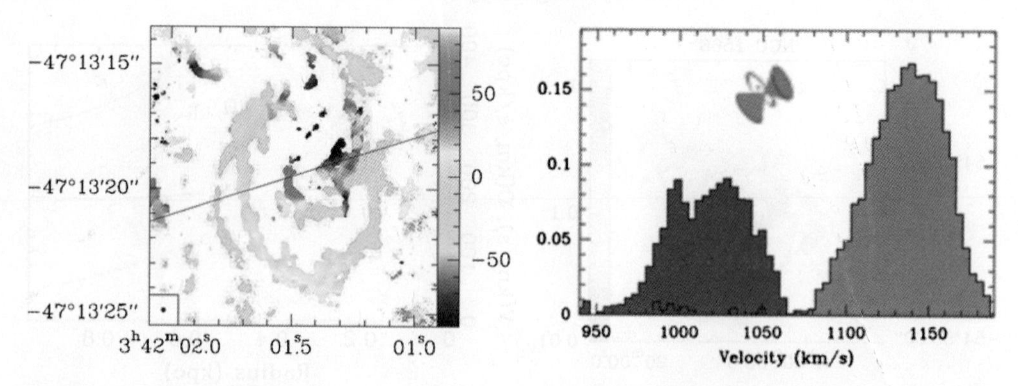

Figure 2. ALMA CO(3-2) observations of the Seyfert-2 NGC 1433: Left, the velocity field, with the color bar labelled in km/s; The thin line indicates the minor axis (PA = 109°). Right, spectrum summing the blue and red-shifted components close to the center, along the minor axis. The systemic velocity is $V_{sys} = 1075$ km/s. The mass in the outflow is 3.6 10^6 $M_\odot$. From Combes *et al.* (2013, 2019).

black hole of mass 8.3×10^6 $M_\odot$. Such trailing nuclear spirals have been found also in NGC 613 (Audibert *et al.* 2019) and in NGC 1808 (Audibert et al. 2020, in prep.).

3. AGN feedback: jets and winds

Molecular outflows are now commonly observed as AGN feedback (Cicone *et al.* 2014). If NGC 1566 does not reveal any outflow, both inflow and outflow can be observed simultaneously, as in NGC 613, where a very short (23pc) and small velocity (300km/s) outflow is detected on the minor axis, parallel to the VLA radio jet (Audibert *et al.* 2019). A very small molecular outflow is also seen in NGC 1433, along the minor axis, cf Fig. 2. This might be the smallest molecular outflow in a nearby Seyfert galaxy, and could be associated to a past radio jet (e.g., Combes *et al.* 2013; Smajic *et al.* 2014).

In these nearby low-luminosity AGN, which accrete far below the Eddington limit, the main mechanism to drive molecular outflows is the radio mode, i.e. entrainement by the radio jets. In some more luminous cases, where L approaches $L_{Edd}/100$, there could be both the radio mode, and winds generated by radiation pressure (either in the ionized gas, or on dust). This might be the case of the prototypical Seyfert-2 NGC 1068, where there is clearly a molecular outflow parallel to the radio jet, sweeping part of the galactic disk (Garcia-Burillo *et al.* 2014). The jet is not perpendicular to the plane, due to the misalignment of the accretion disk with the plane. The ALMA observations of the various CO rotational lines reveal clearly a molecular torus, almost edge-on, and a molecular flow in the perpendicular direction, aligned with the polar dust. The molecular disk appears warped and tilted with respect to the H_2O maser disk (Garcia-Burillo *et al.* 2016).

The lenticular galaxy NGC 1377 is an exceptional case, with a very thin and highy collimated molecular outflow, in the absence of any detected radio jet (Aalto *et al.* 2016). The molecular outflow changes sign along the flow, on each side of the galaxy. This means that the jet is almost in the plane of the sky, and that a slight precession of only 10° is able to tip the jet from redshifted to blue-shifted and back. Such a precession is observed in micro-quasars jets in the Milky Way, for instance SS433 (Mioduszewski *et al.* 2005). But this can be attributed to the companion star. Here there must exist another origin of the precesion, which could be relativistic (see next section).

A precessing molecular outflow model is compatible with the data (Aalto *et al.* 2016). The flow is launched close to the center (r < 10pc). A radio jet must exist at a low level, or has existed in a recent past.

4. Molecular tori: misalignment

With the high spatial resolution of ALMA, it was possible to unveil circumnuclear disks in the CO emission, towards nearby Seyferts. These happen to be misaligned to the large-scale disks, and kinematically decoupled. We call them molecular tori, they exist in 7 out of the 8 cases observed (Combes *et al.* 2019). The average radius of the molecular tori is 18 pc, with a median at 21 pc. Their average molecular mass is $M(H_2) = 1.4 10^7 M_\odot$, and on average their inclination relative to the plane of the sky differs by 29° from that of their galactic disk.

These molecular tori are frequently within the sphere of influence of their black holes, and can serve to measure their mass, provided that their inclination is sufficient (Combes *et al.* 2019). This has been done also for more massive early-type galaxies, by the WISDOM project (Davis *et al.* 2018).

We can invoke at least three mechanisms of misalignment between the large-scale galactic disks and the molecular tori and/or accretion disks. One of them is the radiation-driven warping instability (Pringle 1996). A tilted optically thick disk, which absorbs the radiation from the central AGN, receives in each point some momentum from the radiation, but no torque, because of the radial direction. But then it re-radiates perpendicularly to its orientation, and this produce torques, which maintain and amplify the warping. Assuming the luminosity is powered by accretion eliminates the unknown viscosity parameter α. The instability occurs for radii $R > 0.1 pc\ M_{BH}/(10^8 M_\odot)$. The efficiency of the mechanism was tested by simulations, both in the case of retrograde and prograde precession with respect to the disk rotation (Maloney & Begelmann 1997). A second meschanism is the magnetic instability, and consequent torques but compatibility with AGN observations is contrived, it is more adapted to accretion disks around magnetic stars (Pfeiffer & Lai 2004).

A third mechanism uses the Bardeen-Petterson effect (Bardeen & Petterson 1975), due to Lense-Thirring precession. The accretion disk has a random orientation, generally not aligned with the black hole spin. The relativistic frame dragging effect induces a precession, which tends to align the inner parts of the accretion disk with the black hole equator. The disk develops a warping up to distances 10^2 to 10^4 Schwarzschild radius R_s. The precession of the disk and its warp can be seen from inner to outer disk, up to $1\ pc\ M_{BH}/(10^9 M_\odot)$. According to the amplitude of viscosity, one can distinguish two regimes: the diffusion, when $\alpha > H/R$, where H is the height of the disk, and the regime of bending waves, when $\alpha < H/R$ (Papaloizou & Pringle 1983). In the first case, the disk is warping smoothly and continuously, while in the second case, the disk can break in several rings, with different inclinations and precessing rates. Then the various rings, with differential precession, collide, and drive the gas to fuel the AGN more quickly. This regime has been simulated by Nealon *et al.* (2015). Some works found that the alignment of accretion disks with the black hole equator, through the Bardeen-Petterson effect, was inefficient (Zhuravlev *et al.* 2014; Banerjee *et al.* 2019a,b). More precisely, according to some viscous parameters (parallel of perpendicular to the disk), and viscosity generated by magnetized turbulence, the disk near the black hole can retain its inital inclination, instead of aligning, cf. Fig. 3. Then precessing jets can be launched, perpendicular to the disk, but not aligned to the black hole spin (Liska *et al.* 2018, 2019).

A manifestation of these warping instabilities is the observation of the warped maser disks. Water masers in the prototypes NGC 4258 (Herrnstein *et al.* 1999) and NGC 1068 (Gallimore *et al.* 2004) are detected on 0.3-0.8 pc warped discs. These observations are best represented by the Lense-Thirring effect and/or the radiation driven warps. These perturbations also heat the disk. The fitting of the observations has been done for NGC 4258 by Martin (2008), and for NGC 1068 by Caproni *et al.* (2006). In NGC 1068,

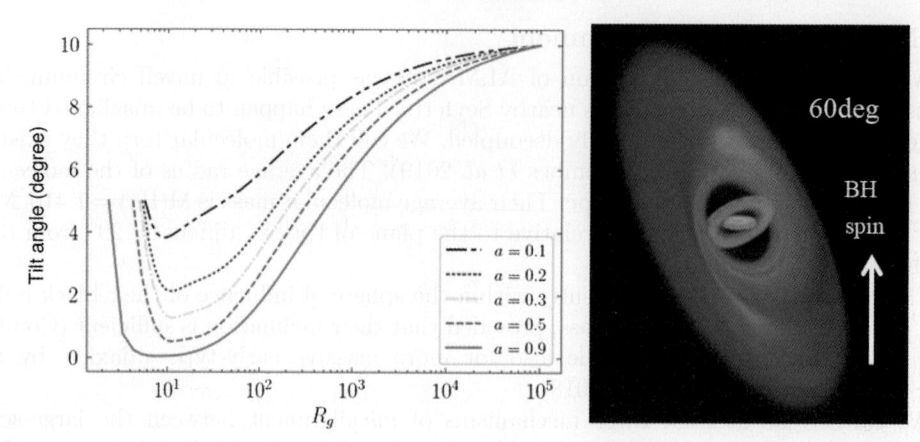

Figure 3. Misalignment of the accreting material near a black hole: Left: radial profiles of the disk tilt-angle (with respect to the black hole spin), for several values of the parameter a, dimensionless angular momentum J of the black hole M, $0 < a < 1$ (a– $cJ/(GM^2)$). The initial tilt-angle is 5°. The radius is in unit of the gravitational radius $R_g = GM/c^2$. From Banerjee *et al.* (2019a). Right: Simulation of the Bardeen-Petterson effect, in a disk initially inclined by 60° with respect to the black hole spin. The Lense-Thirring precession causes the disk to break in a few tilted rings. Adapted from Nealon *et al.* (2015).

where the Bardeen-Petterson mechanism gives the best fit with the observations, the disk is aligned with the black hole spin, until the radius $R_{BP} = 10^{-5}$ to 10^{-4} pc, then warps. For one of the best fit models the alignment time-scale is 7580 yr, the misalignment angle 40° and the velocity of the jet 0.17c. The shape and precession of the pc and kpc-scale jet is also fitted, following Wilson & Ulvestad (1987), in addition to the warped H_2O maser disk by Gallimore *et al.* (2004).

How can the gas accreted from the galactic disk be so misaligned with the disk itself? First, the potential in the center is almost spherical, and the disk very thick with respect to the parsec-scales in question, and second, star formation feedback constantly ejects some gas out of the plane, which rains down in a fountain effect at a random orientation, sometimes in a polar ring (Renaud *et al.* 2015; Emsellem *et al.* 2015).

5. Summary

Thanks to the high resolution provided by ALMA on the molecular gas, it is now possible to better understand the AGN fueling mechanisms. If at large scale, the primary bars can drive the gas towards the 100pc scales, in ILR rings, the nuclear bars act on a trailing nuclear spiral to drive the gas towards the black hole, when the circumnuclear gas enters its sphere of influence. Inside the nuclear spiral, ALMA has revealed the existence of morphologically and kinematically decoupled circum-nuclear disks, or molecular tori.

In some less frequent cases, we can see both AGN fueling, and molecular outflows, through AGN feedback. This can occur via entrainment by radio jets, or through disk winds (or both). The radio mode is distinguished by extremely thin and collimated molecular jets, sometimes precessing, as are the entraining radio jets.

To explain the precession, and the misalignment of the circum-nuclear disks, we can invoke at small scale the Bardeen-Petterson effect, which produces torques in the accreting material, to align it with the black hole spin, and then induces a warping up to a fraction of parsec-scale. The black hole is likely to be fueled via several accretion episodes, coming from the large-scale galacic disk, but also the fountain gas, ejected above the plane through supernovae feedback.

References

Aalto, S., Costagliola, F., Muller, S., *et al.* 2016, *A&A*, 590, A73
Asmus, D. 2019, *MNRAS*, 489, 2177
Asmus, D., Hönig, S. F., Gandhi, P., *et al.* 2016, *ApJ*, 822, 109
Audibert, A., Combes, F., Garcia-Burillo, S., *et al.* 2019, *A&A*, 632, A33
Banerjee, S., Chakraborty, C., Bhattacharyya, S., *et al.* 2019a, *ApJ*, 870, 95
Banerjee, S., Chakraborty, C., Bhattacharyya, S., *et al.* 2019b, *MNRAS*, 487, 3488
Bardeen, J. M. & Petterson, J. A. 1975, *ApJ*, 195, L65
Buta, R. & Combes, F. 1996, *Fund. Cosmic Phys.*, 17, 95 .
Caproni, A., Abraham, Z., Mosquera Cuesta, H. J., *et al.* 2006, *ApJ*, 638, 120
Cicone, C., Maiolino, R., Sturm, E., *et al.* 2014, *A&A*, 562, A21
Combes, F., Garcia-Burillo, S., Casasola, V., *et al.* 2013, *A&A*, 558, A124
Combes, F., Garcia-Burillo, S., Casasola, V., *et al.* 2014, *A&A*, 565, A97
Combes, F., Garcia-Burillo, S., Audibert, A., *et al.* 2019, *A&A*, 623, A79
Davis, T. A., Bureau, M., Onishi, K., *et al.* 2018, *MNRAS*, 473, 3818
Emsellem, E., Renaud, F., Bournaud, F., *et al.* 2015, *MNRAS*, 446, 2468
Fabian, A. C. 2012, *ARAA*, 50, 455
Gallimore, J. F., Baum, S. A., O'Dea, C. P., *et al.* 2004, *ApJ*, 613, 794
Garcia-Burillo, S., Combes, F., Schinnerer, E., *et al.* 2005, *A&A*, 441, 1011
Garcia-Burillo, S., Combes, F., Usero, A., *et al.* 2014, *A&A*, 567, A125
Garcia-Burillo, S., Combes, F., Ramos Almeida, C., *et al.* 2016, *ApJ*, 823, L12
Heckman, T. M. & Best, P. N. 2014, *ARAA*, 52, 589
Herrnstein, J. R., Moran, J. M., Greenhill, L. J., *et al.* 1999, *Nature*, 400, 539
Hönig, S. F. 2019, *ApJ*, 884, 171
Kormendy, J. & Ho, L. C. 2013, *ARAA*, 51, 511
Liska, M., Hesp, C., Tchekhovskoy, A., *et al.* 2018, *MNRAS*, 474, L81
Liska, M., Tchekhovskoy, A., Ingram, A., *et al.* 2019, *MNRAS*, 487, 550
Maloney, P. R. & Begelman, M. C. 1997, *ApJ*, 491, L43
Martin, R. 2008, *MNRAS*, 387, 830
Mioduszewski A. J., Dhawan, V., Rupen, M. P., *et al.* 2005, *ASPC*, 340, 281
Morganti, R. & Oosterloo, T. 2018, *A&ARv*, 26, 4
Nealon, R., Price, D. J., Nixon, C. J., *et al.* 2015, *MNRAS*, 448, 1526
Papaloizou, J. C. B. & Pringle, J. E.1983, *MNRAS*, 202, 1181
Pfeiffer, H. P. & Lai, D.2004, *ApJ*, 604, 766
Pringle, J. E. 1996, *MNRAS*, 281, 357
Ramos-Almeida, C. & Ricci, C. 2017, *NatAs*, 1, 679
Renaud, F., Bournaud, F., Emsellem, E., *et al.* 2015, *MNRAS*, 454, 3299
Smajic, S., Moser, L., Eckart, A., *et al.* 2014, *A&A*, 567, A119
Storchi-Bergmann, T. & Schnorr-Müller, A. 2019, *NatAs*, 3, 48
Wilson, A. S. & Ulvestad, J. S. 1987, *ApJ*, 319, 105
Zhuravlev, V. V., Ivanov, P. B., Fragile, P. C., *et al.* 2014, *ApJ*, 796, 104

Galaxy Evolution and Feedback across Different Environments
Proceedings IAU Symposium No. 359, 2020
T. Storchi-Bergmann, W. Forman, R. Overzier & R. Riffel, eds.
doi:10.1017/S1743921320001623

Observations of AGN feeding and feedback on Nuclear, Galactic, and Extragalactic Scales

D. Michael Crenshaw[1], C. L. Gnilka[1], T. C. Fischer[2], M. Revalski[2], B. Meena[1], F. Martinez[1], G. E. Polack[1], C. Machuca[3], D. Dashtamirova[2], S. B. Kraemer[4], H. R. Schmitt[5], R. A. Riffel[6,7] and T. Storchi-Bergmann[8]

[1]Department of Physics and Astronomy, Georgia State University, 25 Park Place, Suite 605, Atlanta, GA 30303, USA
email: crenshaw@astro.gsu.edu

[2]Space Telescope Science Institute, 3700 San Martin Drive, Baltimore, MD 21218, USA

[3]Department of Astronomy, University of Wisconsin, Madison, WI 53706, USA

[4]Institute for Astrophysics and Computational Sciences, Department of Physics, The Catholic University of America, Washington, DC 20064, USA

[5]Naval Research Laboratory, Washington, DC 20375, USA

[6]Departamento de Física, Centro de Ciências Naturais e Exatas, Universidade Federal de Santa Maria, 97105-900 Santa Maria, RS, Brazil

[7]Department of Physics & Astronomy, Johns Hopkins University, Bloomberg Center, 3400 N. Charles St, Baltimore, MD 21218, USA

[8]Departamento de Astronomia, Universidade Federal do Rio Grande do Sul, IF, CP 15051, 91501-970 Porto Alegre, RS, Brazil

Abstract. We investigate the processes of active galactic nuclei (AGN) feeding and feedback in the narrow line regions (NLRs) and host galaxies of nearby AGN through spatially resolved spectroscopy with the *Gemini* Near-Infrared Integral Field Spectrograph (NIFS) and the *Hubble Space Telescope's* Space Telescope Imaging Spectrograph (STIS). We examine the connection between nuclear and galactic inflows and outflows by adding long-slit spectra of the host galaxies from Apache Point Observatory. We demonstrate that nearby AGN can be fueled by a variety of mechanisms. We find that the NLR kinematics can often be explained by in situ ionization and radiative acceleration of ambient gas, often in the form of dusty molecular spirals that may be the fueling flow to the AGN.

Keywords. galaxies: active, galaxies: kinematics and dynamics, ISM: jets and outflows

1. Introduction

Recent work summarized by Storchi-Bergmann & Schnorr-Müller (2019) identify several ways to fuel AGN on galactic or extragalactic scales (in addition to chaotic cold accretion onto AGN in the bright center galaxies of rich clusters). 1) Major mergers, where the mass ratio of the two galaxies is $\leqslant 4$, are more likely to occur for luminous quasars at high ($z \geqslant 2$) redshifts. 2) Minor mergers with mass ratios >4, have been identified as fueling mechanisms for a number of local AGN (Martini *et al.* 2013; Fischer *et al.* 2015; Riffel *et al.* 2015; Raimundo *et al.* 2017). 3) Tidal interactions between two galaxies that are not as severe as mergers can result in an exchange of gas from one galaxy to the

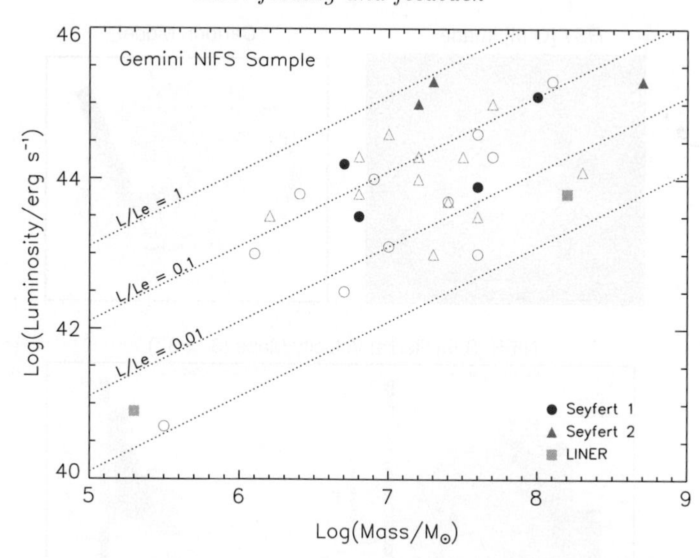

Figure 1. Fundamental properties of the AGN NIFS sample, color- and symbol-coded by AGN type. Existing and planned Z-band observations are given as filled and open symbols, respectively. The new observations will increase the sample by a factor of ∼4, revealing the dependence of outflow and fueling properties on the AGN parameters of black-hole mass, L_{bol}, and L_{bol}/L_{Edd}, as well as other galactic and extragalactic parameters including bulge size, fueling mechanism(s), and environment.

other to fuel the AGN (Davies *et al.* 2017). 4) Secular processes within a galaxy, particularly inflows along a large-scale stellar bar, can drive gas to within a few hundred to a thousand pc of the supermassive black hole (SMBH) (Shlosman *et al.* (1989); Regan *et al.* (1999). Here we present several different examples of these fueling mechanisms and their effects on nuclear ($\leqslant$1 kpc) scales.

The fueling of active galactic nuclei (AGN) and subsequent feedback via radiation and gas outflows is thought to play a critical role in the formation of large-scale structure (Scannapieco *et al.* 2004), chemical enrichment of the intergalactic medium (Di Matteo *et al.* 2010), and self-regulation of supermassive black hole (SMBH) and galactic bulge growth (Hopkins *et al.* 2010). Recent progress in understanding the detailed mechanisms of AGN feeding and feedback has come from adaptive optics observations with integral field units such as *Gemini* NIFS to map the kinematics of ionized and molecular gas at $\sim 0''.1$ resolution in nearby AGN. We present examples of outflows based on NIFS and STIS observations and discuss their possible connections to the fueling flow.

2. Sample

Our overall sample consists of 35 nearby ($z < 0.035$) AGN observed in the K-band by NIFS and present in the *Gemini* archives. The K band is important for determining the gravitational potential via the stellar CO bandheads and the kinematics of the warm molecular gas via H_2 lines. We are obtaining matching observations of the [S III] line in the Z band, because it is the brightest ionized gas line in the IR and therefore provides the best opportunity to trace multiple kinematic components of this gas in the NLR.

Figure 1 shows the fundamental properties of these AGN, which span a wide range in properties including black-hole mass (3.4 dex), bolometric luminosity (L_{bol}, 4.6 dex), and Eddington ratio (L_{bol}/L_{Edd}, 2.8 dex). The AGN host galaxies also span a wide

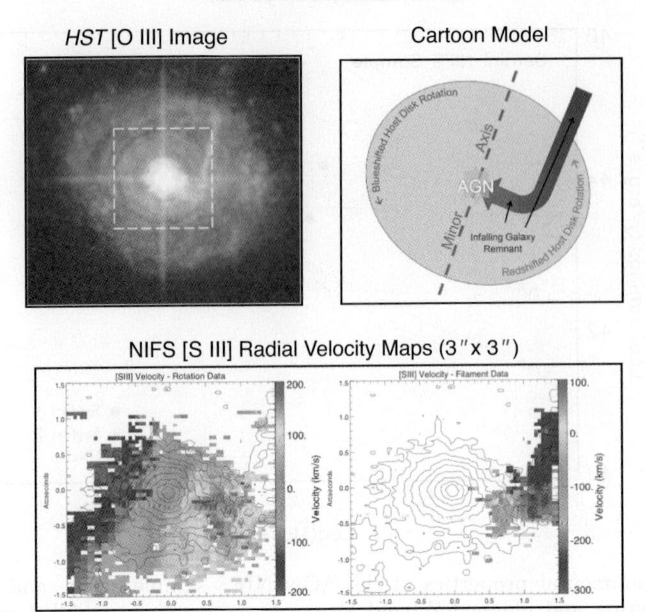

Figure 2. Kinematics of Mrk 509 from Fischer *et al.* (2015). **Top left:** *HST* [O III] image showing "check mark" feature that also appears in the optical continuum. The dashed-lined box show the $3'' \times 3''$ NIFS field of view. **Bottom:** Radial velocity maps of [S III] showing rotation (left) and a mostly blueshifted component associated with the check mark. **Top Right:** Our interpretation that this feature is due to tidal disruption of a satellite galaxy that is now fueling the host galaxy of Mrk 509 and its AGN.

range in galactic and extragalactic parameters including bulge size, fueling mechanism(s), and environment. We have obtained Z-band observations of 9 of these AGN to date, and our results from some of these are detailed below.

3. AGN feeding

An example of fueling an AGN through a minor merger is given in Figure 2, where we see a "check mark" in ionized gas and continuum emission from *HST* images of the bright Seyfert 1 Mrk 509. NIFS radial velocity maps of [S III] emission show that the kinematics of the ionized gas in the inner ∼1 kpc is dominated by rotation and blueshifted emission from the check mark. We claim this feature is the result of a minor merger, which has torn a galaxy that about the size of the Small Magellanic Cloud apart and is now fueling the central host galaxy and AGN (Fischer *et al.* 2015).

An example of AGN fueling by a tidal interaction is shown in Figure 3. A tidal tail of H I gas extends from the gas-rich galaxy UGC 3422 to Mrk 3 (UGC 3426), an S0 galaxy containing a Seyfert 2 nucleus, and beyond. This fueling results in a large-scale gas/dust disk offset from the stellar major axis in position angle by $\sim 100^{\circ}$ and in the direction of the tidal stream, which we confirmed with *APO* long-slit spectra showing its rotation curve (Gnilka *et al.* 2020). NIFS observations of the central 800 pc × 800pc region in Mrk 3 shows a nuclear stellar disk aligned with the large-scale disk, and a counter-rotating ionized and warm molecular gas disk. This finding is consistent with claims that AGN in S0 galaxies are fueled by external sources (Hicks *et al.* 2013; Davies *et al.* 2014; Raimundo *et al.* 2017).

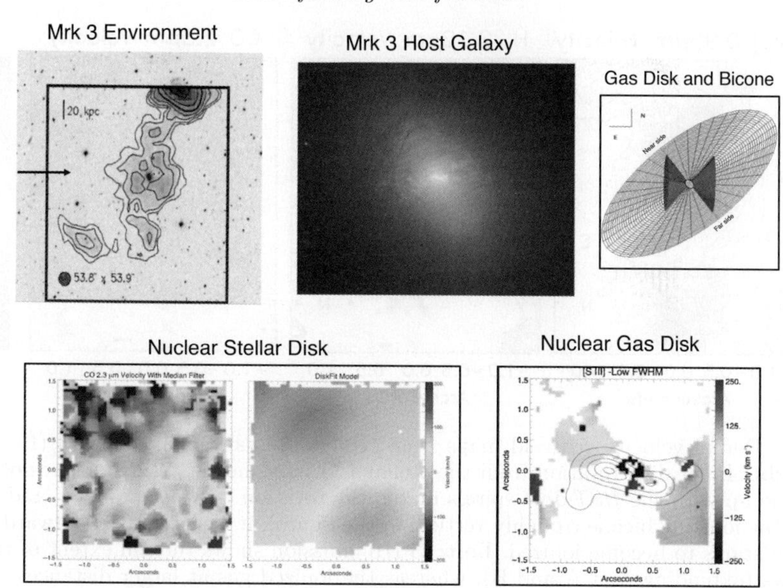

Figure 3. Fueling of the AGN Mrk 3 on different scales. **Upper Left:** Contour map of H I 21 cm emission from Noordermeer *et al.* (2005) superimposed on a DSS red image, showing a tidal tail of H I gas extending from UGC 3422 (100 kpc to the NW) to Mrk 3 and beyond. **Upper Middle:** Composite $50'' \times 50''$ (13.5 kpc $\times$ 13.5 kpc) *HST* image of Mrk 3 showing the orientation of the host galaxy (red), dust lanes (dark), and the ionized gas in the NLR (white) and ENLR (blue). **Upper Right:** Geometric model of the NLR bicone and large-scale gas/dust disk in Mrk 3, offset in position angle by 100° from the stellar major axis (Crenshaw *et al.* 2010). **Lower Left:** Radial velocity map from NIFS $3'' \times 3''$ (800 pc $\times$ 800 pc) observation of the stellar CO bandheads smoothed with a 5×5 pixel median filter and the resulting Diskfit rotation model. **Lower Right:** NIFS radial velocity map of the low width (FWHM $\leqslant$ 250 km s^{-1}) ionized gas, showing counter-rotation with respect to the stellar nuclear and galactic disks.

4. AGN feedback

An intimate connection between feeding and feedback is demonstrated in Figure 4, which shows NIFS radial velocity maps of the stellar, warm molecular, and ionized gas in the central ~1 kpc of Mrk 573. Nuclear dust/gas spirals thought to represent the fueling flows to the AGN are illuminated when they enter its bicone of ionizing radiation (Fischer *et al.* 2017). NIFS K-band observations of H_2 show that the inner dust spirals coincide with warm molecular gas that is rotating with the disk until it enters the bicone within a few hundred parsecs of the AGN where it is driven outward. NIFS Z-band observations of [S III] show that this gas is then ionized and radiatively driven to a distance of only ~600 pc, whereas AGN-ionized gas at greater distances is rotating with the galactic disk out to ~4 kpc according to our *APO* long-slit spectra. Mrk 3 shows a similar pattern of ionization and acceleration of ambient gas from the nuclear spirals (Gnilka *et al.* 2020).

We have measured spatially-resolved mass and kinetic energy outflow rates based on *HST* [O III] images, STIS long-slit spectra, and photoionization models (Crenshaw *et al.* 2015; Revalski *et al.* 2018a,b, 2021). In general we find that the mass outflow rates peak at values of $3-12$ $M_\odot$ and that the peaks and extents of the outflows appear to scale with luminosity. However, the outflow extents at these Seyfert luminosities are in the range $100-800$ pc, which are significantly less than the effective bulge radii of their host galaxies. Thus, we are extending our studies to quasars at higher redshifts (Fischer *et al.* 2018; A.L. Trindade Falcao, in preparation) to determine if there is a transition luminosity at which the NLR outflows can clear their bulges.

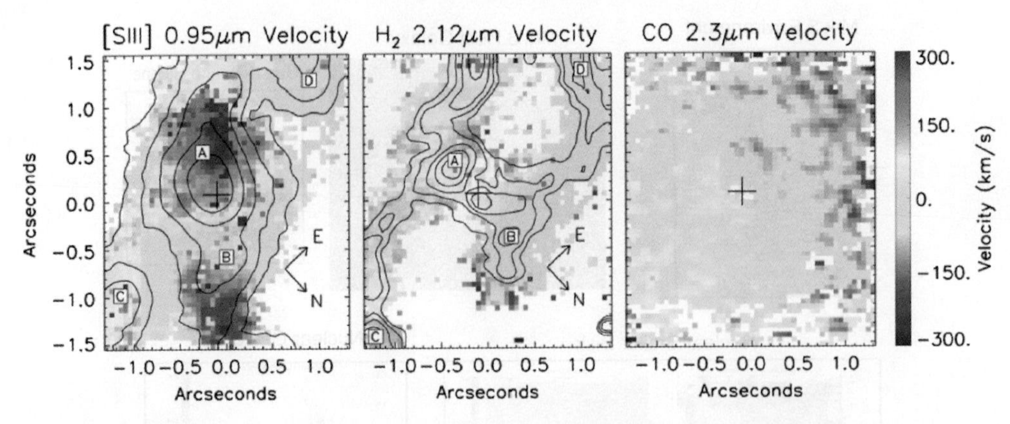

Figure 4. Radial velocity centroid maps from NIFS observations of Mrk 573 (Fischer *et al.* 2017). **Right:** Inner disk rotation from the stellar CO band heads. **Middle:** Arcs of warm H_2 gas that line up with the *HST* dust spirals match the rotation curve at points C and D. As the gas enters the ionizing bicone (roughly vertical in the figure), it is accelerated outward (points A and B) and starts to become ionized. **Left:** [S III] emission shows the full extent of the ionized gas outflows (extending past A and B), whereas the ionized gas at larger distances (beginning at C and D) is primarily in rotation.

References

Crenshaw, D. M., Kraemer, S. B., Schmitt, H. R., *et al.* 2010, *AJ*, 139, 871
Crenshaw, D. M., Fischer, T. C., Kraemer, S. B., *et al.* 2015, *ApJ*, 799, 83
Davies, R. I., Maciejewski, W., Hicks, E. K. S., *et al.* 2014, *ApJ*, 792, 101
Davies, R. I., Hicks, E. K. S., Erwin, P., *et al.* 2017, *MNRAS*, 466, 4917
Di Matteo, T., *et al.* 2010, *MNRAS*, 401, 1
Fischer, T. C., Crenshaw, D. M., Kraemer, S. B., *et al.* 2015, *ApJ*, 799, 234
Fischer, T. C., Machuca, C., Diniz, M. R., *et al.* 2017, *ApJ*, 834, 30
Fischer, T. C., Kraemer, S. B., Schmitt, H. R., *et al.* 2018, *ApJ*, 856, 102
Gnilka, C. L., *et al.* 2020, *ApJ*, in press
Hicks, E. K. S., Davies, R. I., Maciejewski, W., *et al.* 2013, *ApJ*, 768, 107
Hopkins, P. F. & Elvis, M. 2010, *MNRAS*, 401, 1
Martini, P., Dicken, D., & Storchi-Bergmann, T. 2013, *ApJ*, 766, 121
Noordermeer, E., van der Hulst, J. M., Sancisi, R., *et al.* 2005, *A&A*, 442, 137
Raimundo, S. I., Davies, R. I., Canning, R. E. A., *et al.* 2017, *MNRAS*, 464, 4227
Regan, M. W., Sheth, K., & Vogel, S. N. 1999, *ApJ*, 526, 97
Revalski, M., Crenshaw, D. M., Kraemer, S. B., *et al.* 2018a, *ApJ*, 856, 46
Revalski, M., Dashtamirova, D., Crenshaw, D. M., *et al.* 2018b, *ApJ*, 867, 88
Revalski, M., Meena, B., Martinez, F., *et al.* 2021, arXiv:2101.06270
Riffel, R. A., Storchi-Bergmann, T., & Riffel, R. 2015, *MNRAS*, 451, 3587
Shlosman, I., Frank, J., & Begelman, M. C. 1989, *Nature*, 338, 45
Scannapieco, E. & Oh, S.P. 2004, *ApJ*, 608, 62
Storchi-Bergmann, T. & Schnorr-Müller, A. 2019, *Nature Astronomy*, 3, 48

Galaxy Evolution and Feedback across Different Environments
Proceedings IAU Symposium No. 359, 2020
T. Storchi-Bergmann, W. Forman, R. Overzier & R. Riffel, eds.
doi:10.1017/S1743921320002422

Interstellar medium properties and feedback in local AGN with the MAGNUM survey

M. Mingozzi[1], G. Cresci[2], G. Venturi[3,2], A. Marconi[4,2] and F. Mannucci[2]

[1]INAF - Osservatorio astronomico di Padova, Vicolo dell'Osservatorio 5, 35122 Padova, Italy
email: matilde.mingozzi@inaf.it

[2]INAF – Osservatorio Astrofisico di Arcetri, Largo E. Fermi 5, I-50157, Firenze, Italy

[3]Instituto de Astrofísica, Pontificia Universidad Católica de Chile, Avda. Vicuña Mackenna 4860, 8970117, Macul, Santiago, Chile

[4]Dipartimento di Fisica e Astronomia, Università degli Studi di Firenze, Via G. Sansone 1, I-50019 Sesto Fiorentino, Firenze, Italy

Abstract. We investigated the interstellar medium (ISM) properties in the central regions of nearby Seyfert galaxies characterised by prominent conical or bi-conical outflows belonging to the MAGNUM survey by exploiting the unprecedented sensitivity, spatial and spectral coverage of the integral field spectrograph MUSE at the Very Large Telescope. We developed a novel approach based on the gas and stars kinematics to disentangle high-velocity gas in the outflow from gas in the disc to spatially track the differences in their ISM properties. This allowed us to reveal the presence of an ionisation structure within the extended outflows that can be interpreted with different photoionisation and shock conditions, and to trace tentative evidence of outflow-induced star formation ("positive" feedback) in a galaxy of the sample, Centaurus A.

Keywords. Galaxies: ISM, Seyfert, jets

1. Introduction

Galaxy-scale outflows driven by active galactic nucleus (AGN) activity are thought to be so powerful to sweep away most of the gas of the host galaxy, providing a mechanism for the central black hole (BH) to possibly regulate star formation (SF) activity. This mechanism, the so-called negative feedback, could potentially explain the relation between the BH mass and the galaxy bulge properties (Silk & Rees 1998; Fabian 2012). Recently, models and observations have revealed that outflows and jets can also have a positive feedback effect, triggering SF in the galaxy disc and also within the outflowing gas itself (e.g. Silk 2013; Cresci *et al.* 2015a; Maiolino *et al.* 2017). Outflows are now routinely detected in luminous active galaxies on different physical scales and in different gas phases (e.g., ionised, atomic and molecular gas; Cicone *et al.* 2018 and references therein), even though understanding their role in galaxy evolution is still a challenging task. In this context, nearby galaxies represent ideal laboratories to explore in high detail outflow properties, their formation and acceleration mechanisms, as well as the effects of SF and AGN activities on host galaxies.

Here we present the results of the Measuring AGN under MUSE microscope (MAGNUM) survey (P.I. Marconi), aimed at investigating the inner regions of a number of local AGN, all showing evidence for the presence of outflows, with the unprecedented combination of spatial and spectral coverage of the integral field spectrograph MUSE (Bacon *et al.* 2010) at the Very Large Telescope. This contribution is based on recent published papers (Cresci *et al.* 2015a; Venturi *et al.* 2017, 2018; Mingozzi *et al.* 2019) and on unpublished material from M. Mingozzi's Phd Thesis (2020, University of Bologna).

2. The MAGNUM survey

MAGNUM galaxies have been selected to be observable from Paranal Observatory and with a luminosity distance $D_L < 50$ Mpc. In Venturi *et al.* (2021, in preparation) we present our sample, explaining the selection criteria, data reduction and analysis, and investigating the kinematics of the ionised gas. Here, we show our results for the nine Seyfert galaxies analysed in Mingozzi *et al.* (2019) (M19 hereafter), namely Centaurus A, Circinus, NGC 4945, NGC 1068, NGC 1365, NGC 1386, NGC 2992, NGC 4945 and NGC 5643. The MUSE field of view (FOV) covers their central regions, spanning from 1 to 10 kpc, according to their distance. The average seeing of the observations is ∼0.6"−0.8". The datacubes were analysed with a set of custom python scripts in order to fit and subtract the stellar continuum in each single-spaxel spectrum and fit the main emission lines (i.e. Hβ, [O III]$\lambda\lambda$4959,5007, Hα, [N II] $\lambda\lambda$6548,84, [S II]$\lambda\lambda$6717,31, [S III]λ9069) with multiple Gaussians where needed. This happens in the central parts of the galaxies and in the outflowing cones. All the details about the applied procedure are given in M19.

2.1. *Gas properties: disc versus outflow*

In many works, disc and outflow are separated according to the width of the two Gaussian components used to fit the main emission lines (narrower and broader, respectively). In our analysis this approach is not feasible since the line profiles can be very complex, requiring three or four Gaussians to be fully reproduced. Therefore, we disentangle the outflow from the systemic gas by applying a novel approach, explained in detail in M19. In brief, we assume that the stellar velocity is generally a good approximation of the gas velocity in the disc, defining as *disc component* the low-velocity ionised gas rotating similarly to the stars, while the *outflow component* (i.e. the high-velocity component) is moving faster than the stellar velocity, and is partly blueshifted and partly redshifted with respect to it. As an example, Fig. 1 shows the Hα disc component flux maps, superimposing the [O III]λ5007 outflow component contours for Circinus and Centaurus A†. Indeed, the Hα emission is in general dominant in the disc, while the [O III] is enhanced in the outflow (Venturi *et al.* 2018). In Circinus (1" ∼20.4 pc), the outflow is extended on North-West in one-sided and wide-angled kpc-scale [O III] cone, first revealed by Marconi *et al.* (1994). In Centaurus A (1" ∼18.5 pc) the outflow is mainly distributed in two cones (direction north-east and south-west) in the same direction of the extended double-sided jet revealed both in the radio and X-rays (e.g. Hardcastle *et al.* 2003), and located perpendicularly with respect to the gas in the disc component.

In M19, we calculated dust extinction, gas density and ionisation parameter (i.e. a measure of the radiation field intensity, relative to gas density) for the disc and outflow components, using Hα/Hβ, [S II]λ6717/[S II]λ6731 and [S III]$\lambda\lambda$9069,9532/[S II]$\lambda\lambda$6717,31 line ratios, respectively. We found that the outflow is characterised by higher values of density and ionisation parameter ($A_V \sim 0.9$, $n_e \sim 250$ cm^{-3}, log([S III]/[S II]) ~ 0.16) than the disc component, that is instead more affected by dust extinction ($A_V \sim 1.75$, $n_e \sim 130$ cm^{-3}, log([S III]/[S II]) ~ -0.38). Interestingly, our median outflow density is lower than what is found in literature, but a more consistent value can be obtained calculating the median density weighting by the [S II] line flux (disc: $n_e \sim 170$ cm^{-3}; outflow: $n_e \sim 815$ cm^{-3}). This means that many values of outflow density found in literature could be biased towards higher densities because they are based only on the most

† Centaurus A shows a strong misalignment between stars and gas (Morganti *et al.* 2010), so we consider the global systemic velocity ($v_{\rm sys} = 547$ km/s) as a reference for the disc gas.

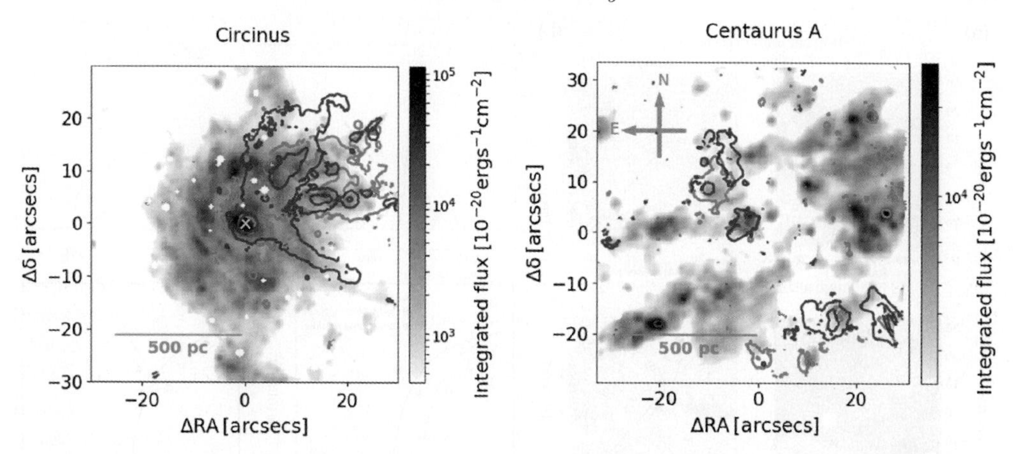

Figure 1. Circinus and Centaurus A Hα disc component maps with [O III]λ5007 blueshifted and redshifted outflow component contours superimposed (in blue and red, respectively). We show only the spaxels with a signal-to-noise S/N > 5. East is to the left. The magenta bar represents a physical scale of $\sim$500 pc. The white circular regions are masked foreground stars. The green cross marks the position of the peak of the continuum in the wavelength range 6800$-$7000 Å.

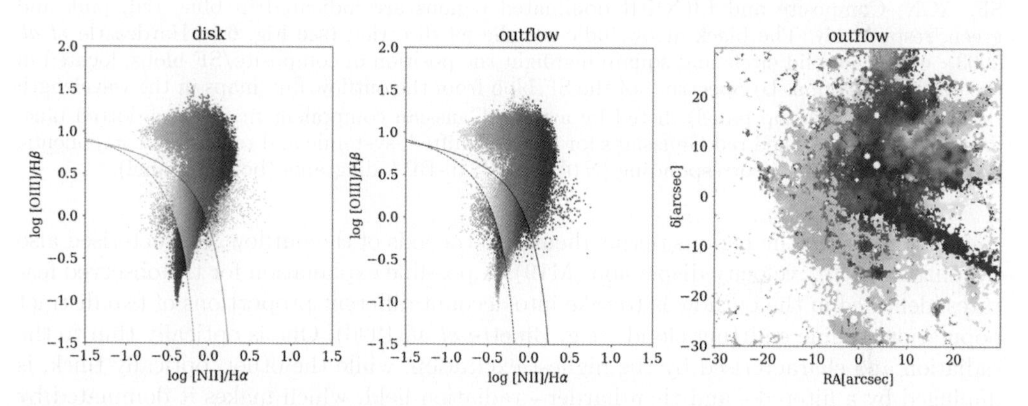

Figure 2. [N II] -BPT diagrams for the disc and outflow components of Circinus, and the corresponding outflowing gas component map. Shades of blue, pink and red denote SF, composite and AGN dominated regions, respectively (darker shade means higher [N II] /Hα). The black dashed curve is the boundary between star-forming galaxies and AGN (Kauffmann *et al.* 2003), while the black solid curve is the theoretical upper limit on SF line ratios (Kewley *et al.* 2001). The grey dots in the BPTs and the dashed grey regions in the corresponding map show the disc and outflow component together. We show only spaxels with a S/N > 5 for all the lines involved.

luminous and densest outflowing regions, characterised by a high S/N ratio. Finally, in M19 we used spatially and kinematically resolved Baldwin-Phillips-Terlevich (BPT, Baldwin *et al.* 1981) diagrams to explore the dominant contribution to ionisation in the disc and outflow components separately. The left and middle panels of Fig. 2 show the [N II] -BPT diagrams of Circinus for the disc and outflow, respectively. The corresponding position on the outflowing gas component map is shown in the right panel of Fig. 2. We noticed that the highest and lowest values of low-ionisation line ratios (LILrs; i.e. [N II] /Hα and [S II]/Hα), displayed in dark red and orange, are prominent in the AGN/LI(N)ER-dominated outflow component, while they are not observable in the disc component. These features are visible in almost all the MAGNUM sample: in general, the highest LILrs trace the inner parts along the axis of the emitting cones, where the [S III]/[S II] line ratio is enhanced (i.e. high ionisation), while the lowest LILrs follow the

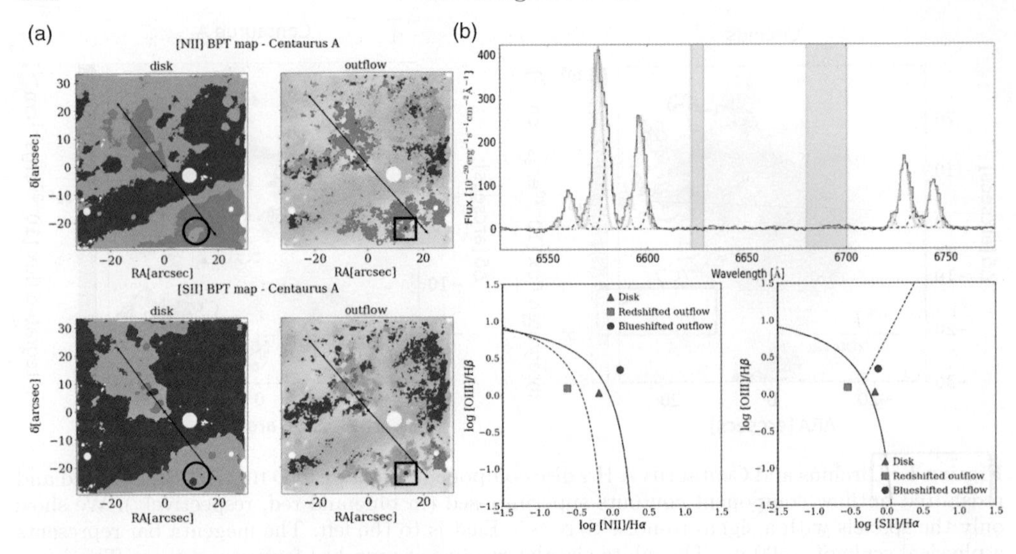

Figure 3. **a**) [N II] (top panel) and [S II] (bottom panel) BPT maps for Centaurus A disc (-150 km/s < v < $+150$ km/s) and outflow components (v > $+200$ km/s and v < -200 km/s). SF, AGN, Composite and LI(N)ER dominated regions are indicated in blue, red, pink and green, respectively. The black arrow indicates the jet direction (see Fig. 6 in Hardcastle *et al.* 2003), while the solid circle and square highlight the position of composite/SF blobs, located in the outflow direction. **b**) Spectrum of the SF blob from the outflow flux maps in the wavelength range 6500–6650 Å(top panel), fitted by a three Gaussian component fit (dashed-dotted blue, dotted green and dashed red Gaussians for the blueshifted, systemic and redshifted components, respectively) and their corresponding [N II] - and [S II]-BPT diagrams (bottom panel).

cone edges and/or the regions perpendicular to the axis of the outflow, characterised also by a higher [O III] velocity dispersion (M19). A possible explanation for the observed features identified in the outflow is to take into account different proportions of two distinct populations of line emitting clouds (e.g. Binette *et al.* 1996). One is optically thin to the radiation and characterised by the highest excitation, while the other, optically thick, is impinged by a filtered - and then harder - radiation field, which makes it dominated by low-excitation lines and characterised by lower [S III]/[S II] line ratios. The highest LILrs may be due to shocks and/or to a hard-filtered radiation field from the AGN (M19).

2.2. *Centaurus A: a local laboratory to study AGN positive feedback*

Centaurus A shows the best example of a radio jet emitted by the central AGN interacting with the ISM (e.g. Santoro *et al.* 2016 and references therein). In this context, we investigated its central region, exploiting our approach of disentangling the disc from the outflow component to obtain an independent classification of their ionisation sources. Fig. 3a shows the [N II] - and [S II]-BPT maps for the disc (on the left) and the outflow (on the right). The galaxy FOV South-West portion hosts a blob (solid circle) with a velocity consistent with the gas disc and by Composite and SF dominated ionisation (in [N II] - and [S II]-BPT, respectively), that can be interpreted as SF triggered in the ionisation cone due to compression of the galaxy ISM by the outflow, as already revealed in another galaxy of the MAGNUM survey, NGC 5643, by Cresci *et al.* (2015a). Moreover, a nearby clump (solid square) appears to have SF ionisation, but velocities consistent with the gas in the outflow, suggesting that newborn stars could be forming directly in outflowing gas (see Maiolino *et al.* 2017). The top panel of Fig. 3b shows the spectrum of the star-forming blob in the outflow (solid square in Fig. 3a). The asymmetric line profiles can be reproduced by a fit with three Gaussian components, whose corresponding positions in [N II] - and [S II]-BPT diagrams are shown in the bottom panel of Fig. 3b: the redshifted Gaussian component (red) has a very strong Hα emission (star formation rate SFR $\sim 8 \times 10^{-3}$ $M_\odot$/yr, using Lee *et al.* 2009 calibration) and SF ionisation.

Overall, the two blobs have a total SFR ~ 0.01 $M_\odot$/yr, which is $\sim$3% of the global value of the galaxy and could represent the first evidence of both the two modes of positive feedback – triggered SF both in the galaxy disc and within the outflowing gas – operating in the nuclear region of this galaxy. Hence, we asked for X-SHOOTER observations (0102.B-0292, P.I. Mingozzi), that we will discuss in a forthcoming paper (Mingozzi *et al.* 2021, in preparation) to undoubtedly identify SF signatures in the outflow, exploiting IR diagnostics to discard AGN and shock ionisation, and stellar absorption lines to investigate the presence of newborn stars (see Maiolino *et al.* 2017).

3. Conclusions

The MAGNUM survey is exploring gas properties and ionisation sources of the outflowing gas in the central regions of nearby Seyfert galaxies. We found that the gas in the outflowing cones of our galaxies is set up in clumpy clouds characterised by higher density and ionisation with respect to disc gas. The cone innermost regions are generally highly ionised and directly heated by the AGN. The cone edges and the regions perpendicular to the outflow axis could instead be dominated by shocks due to the interaction between the outflow and the ISM. Alternatively, these regions, generally characterised by low ionisation, could be impinged by an ionising radiation filtered by clumpy, ionised absorbers. Separating the outflow and disc components allowed us also to detect in one of the sources, Centaurus A, two blobs dominated by SF ionisation apparently embedded in the AGN ionisation cone, possibly tracing positive feedback (both in the disc and in the outflow) and accounting for $\sim$3% of the galaxy global SFR. If the new X-SHOOTER data confirmed it, this would be the first example of the two modes of positive feedback coexisting in the same object. The contribution to the total SFR might seem irrelevant, but it remains that Centaurus A could be a local test bench to explore in detail this phenomenon. Positive feedback may play a significant role in the formation of galaxy spheroidal component at high redshift, where AGN-driven outflows are more prominent and the associated SF inside those very massive outflows possibly far higher (Gallagher *et al.* 2019; Rodríguez del Pino *et al.* 2019).

References

Bacon, R., Accardo, M., Adjali, L., *et al.* 2010, *Ground-based and Airborne Instrumentation for Astronomy III*, Proc. SPIE, 7735, 773508

Baldwin, J. A., Phillips, M. M., & Terlevich, R. 1981, *Publications of the ASP*, 78, 20

Binette, L., Wilson, A. S., & Storchi-Bergmann, T. 1996, *A&A*, 78, 20

Cicone, C., Brusa, M., Ramos Almeida, C., *et al.* 2018, *Nature Astronomy*, 2, 176 6

Cresci, G., Marconi, A., Zibetti, S., *et al.* 2015a, *A&A*, 582, A63

Fabian, A. C. 1995, *ARA&A*, 50, 455

Gallagher, R., Maiolino, R., Belfiore, F., *et al.* 2019, *MNRAS*, 485, 3409

Hardcastle, M. J., Worrall, D. M., Kraft, R. P., *et al.* 2003, *ApJ*, 593, 169

Kauffmann, G., Heckman, T. M., Tremonti, C., *et al.* 2003, *MNRAS*, 346, 1055

Kewley, L. J., Dopita, M. A., Sutherland, *et al.* 2001, *ApJ*, 78, 20

Lee, J. C., Gil de Paz, A., Tremonti, C., *et al.* 2009, *ApJ*, 706, 599

Maiolino, R., Russell, H. R., Fabian, A. C., *et al.* 2017, *Nature*, 544, 202 19, 71, 75, 165

Marconi, A., Moorwood, A. F. M., Origlia, L., *et al.* 2019, *The Messenger*, 78, 20

Mingozzi, M., Cresci, G., Mannucci, F., *et al.* 2019, *A&A*, 622, A146

Morganti, R. 2010, *PASA*, 27, 463
Rodríguez del Pino, B., Arribas, S., Piqueras López, *et al.* 2019, *MNRAS*, 486, 344
Santoro, F., Oonk, J. B. R., Morganti, R., *et al.* 2016, *A&A*, 590, A37
Silk, J. & Rees, M. J. 1998, *A&A*, 331, L1
Silk, J. 2013, *ApJ*, 772, 112
Venturi, G., Marconi, A., Mingozzi, M., *et al.* 2017, *Front. Astron. Space Sci.*, 4, 46
Venturi, G., Nardini, E., Marconi, A., *et al.* 2018, *A&A*, 619, A74

Galaxy Evolution and Feedback across Different Environments
Proceedings IAU Symposium No. 359, 2020
T. Storchi-Bergmann, W. Forman, R. Overzier & R. Riffel, eds.
doi:10.1017/S1743921320003981

The first AI simulation of a black hole

Rodrigo Nemmen[1], Roberta Duarte[1] and João P. Navarro[2]

[1]Universidade de São Paulo, Instituto de Astronomia, Geofísica e Ciências Atmosféricas, Departamento de Astronomia, São Paulo, SP 05508-090, Brazil
email: rodrigo.nemmen@iag.usp.br

[2]NVIDIA

Abstract. We report the results from our ongoing pilot investigation of the use of deep learning techniques for forecasting the state of turbulent flows onto black holes. Deep neural networks seem to learn well black hole accretion physics and evolve the accretion flow orders of magnitude faster than traditional numerical solvers, while maintaining a reasonable accuracy for a long time.

Keywords. Black hole physics, astrostatistics, active galactic nuclei

---

1. Introduction

My presentation was supposed to be about the new constraints on the spin of the supermassive black hole (SMBH) in M87. Here is the bottom line from that work: we have constrained the spin parameter to be $|a_*| > 0.4$ (Nemmen 2019). The spin is the second fundamental parameter of black hole (BH) spacetimes. This constraint should set expectations for future estimates of the M87* spin with the Event Horizon Telescope and other observatories.

Instead, I will present some early exciting results from our pilot investigation of artificial intelligence (AI) methods as tools to accelerate numerical simulations of BH accretion flows. Here, we address two inter-related questions: Can we make the models faster while maintaining an accuracy comparable to explicit solvers of the fluid conservation equations? Can deep neural networks learn fluid dynamics?

2. Deep learning

Let me begin with the fundamental problem of BH astrophysics: to figure out the function

$$\mathrm{AGN}(t) = f(M, a_*, \dot{M}) \tag{2.1}$$

which quantifies the complete time-evolution of "weather" around SMBHs, where M and $\dot{M}$ are the BH mass and mass accretion rate, respectively. The challenge is that BH weather is a complex process, requiring the solution of nonlinear, multidimensional partial differential equations which are very time-consuming (e.g. Porth *et al.* 2019). Ideally, we want those simulations to have a duration much shorter than a typical PhD thesis timescale, so we need to make them as fast as possible.

Here, we are investigating the use of deep learning (DL) techniques for that purpose. DL consists of using deep neural networks inspired by the way the brain works, with a large number of layers and parameters ("neurons") (Goodfellow, Bengio & Courville 2016). Deep neural networks are good approximators for empirical functions which are too complex to be have an analytical form (e.g. Cybenko 1989). It is not an exaggeration to say that a considerable fraction of AI work today consists of applications—and improvements upon—DL.

In practice, DL algorithms "learn from experience": instead of explicitly coding the instructions in the code, one trains the machine by showing a lot of examples and comparing the output of trained algorithm to a test dataset. Then, by tweaking the network architecture and its hyperparameters, one arrives at a trained deep net (LeCun 2015). DL is leading to several breakthroughs in many fields (e.g. Mnih *et al.* 2015; Silver *et al.* 2016; Krizhevsky, Sutskever & Hinton 2017), including astronomy (e.g. Hausen & Robertson 2020; Zhang *et al.* 2019). The downsides of DL is that it is data-hungry (needs a lot of data to be effective) and computationally expensive to train. Once trained, however, the algorithm deploys answers very quickly.

Here, we devise the prediction challenge as a computer vision problem and infer the evolution of the system from the sequence of input data cubes that comprise previous states of an accreting BH. This is a data-driven, equation-free approach in which a model learns to approximate the relevant physics from the training examples alone and not by incorporating a priori knowledge about the equations underlying the processes. This is similar to the work of Jaeger & Haas (2004); Tompson *et al.* (2016).

3. Teaching a machine about BH accretion

The training data set consists of the hydrodynamical BH accretion simulations performed by Almeida & Nemmen (2020). This is a set of long-duration (durations $>10^5\ GM/c^3$), 2D models designed to explore the winds produced by low-luminosity active galactic nuclei, where a Schwarzschild BH is fed by a hot, geometrically thick accretion flow. The specific model we chose is PNSS3, which has one of the longest durations among the models computed by Almeida & Nemmen (2020). The numerical simulation computes the spacetime distribution of gas densities around the BH which we feed to the DL model.

For the learning algorithm, we use the well-known U-Net convolutional neural network (ConvNet) architecture, which is commonly used to extract information from datasets which involve spatial and temporal coherence (e.g. Karpathy *et al.* 2014). For the training, we divide the time series data into 67.5% training, 12.5% cross-validation and 20% testing phases. How well does the trained DL model predict the future state of the density field around the BH?

4. Preliminary results

We quantify the performance of the DL approach by comparing a number of indicators with those obtained from the explicit solution to the conservation equations (Duarte, Nemmen & Navarro, in preparation). Here, we present some early, preliminary results.

The left panel of Figure 1 displays the density difference between the numerical simulations (i.e. the target) and the trained ConvNet (the prediction) as a function of time. What we mean by error here is defined in the lower left corner of Figure 1. We see that the error gradually builds up over time, as if the predictions drifted from the ground truth solutions of the conservation equations. The middle panel shows the target density map, which results from solving the fluid conservation equations. The right panel corresponds to the inferred deep learning model—the neural network's imagination.

Figure 2 summarizes the main result of this presentation. Each panel displays the density marginalized over the polar angle. Think of this as the average density in spherical shells of increasing radii. On a first inspection, we can see that the DL forecast and the data are virtually identical. Only after analyzing the residuals we realize that

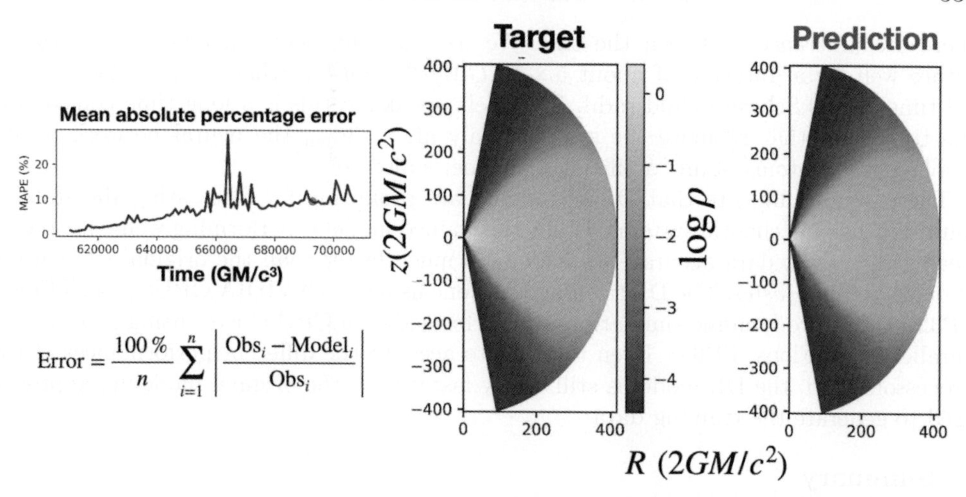

Figure 1. Comparison between the deep neural network's "imagination" (indicated as *prediction*) and the actual numerical solutions from the fluid conservation laws (the *target*), for a black hole surrounded by a hot accretion flow. Left panel: time evolution of the difference between the prediction and target (i.e. the error of the trained model). The lengths are expressed in terms of the Schwarzschild radius. Time in gravitational units.

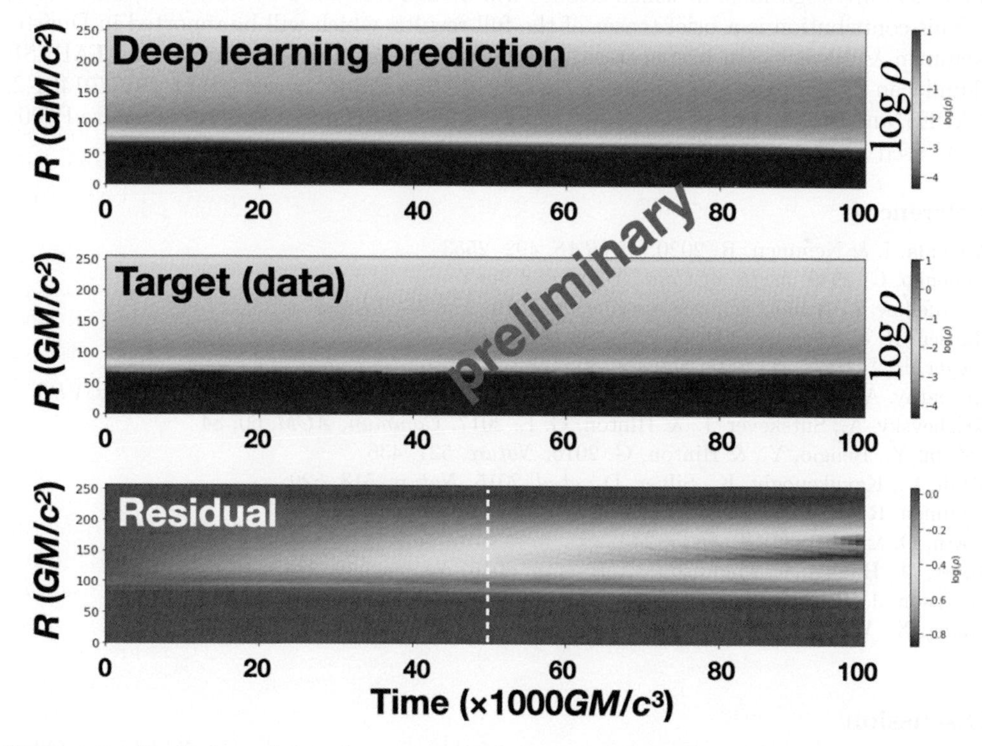

Figure 2. Prediction of deep learning model compared with the data from the numerical simulations of BH accretion. The colors map the logarithm of the gas density. The density is averaged over the polar angle. The residuals are the difference between the logarithms. The vertical white line indicates the time at which the deep nets begin to drift away from the ground truth.

there are differences between the two. We see that the neural network imagines the future well for a duration of about $5 \times 10^4 GM/c^3 \approx 33 t_{\rm dyn}$, where $t_{\rm dyn}$ is the dynamical time at 100 Schwarzschild radii. Relatively speaking, this is a long time considering the timescales that regulate the system. Only after $33 t_{\rm dyn}$ the neural network begins displaying symptoms of an "artificial Alzheimer's disease".

These results indicate that DL techniques are promising for forecasting the state of multidimensional chaotic systems. While the trained model is performing well—i.e. reproducing the target data accurately—it does so much faster than the original simulation: about *600 times faster*. The DL training was done using two NVIDIA GPUs ($\approx$40 TFlops, FP32); the hydrodynamic simulation was performed on a CPU cluster using 200 cores in parallel ($\sim$3 TFlops, FP64). Even taking into account the different performances of the processors used, the DL model is still vastly faster than the original modeling approach used to generate the training data.

5. Summary

This work is a pilot study of the performance of deep learning in forecasting the state of large spatiotemporally chaotic systems comprised by turbulent flows onto black holes. So far, we have obtained promising preliminary results. Not only deep neural networks seem to learn well black hole accretion physics: they evolve the accretion flow $\sim$1000 times faster than traditional numerical solvers. The future seems bright for black hole numerical investigations, in which science will be less restricted by hardware limitations.

This contribution is a brief teaser of the full results, which will be reported in Duarte, Nemmen & Navarro (in preparation). We gratefully acknowledge support by FAPESP (Fundação de Amparo à Pesquisa do Estado de São Paulo) under grant 2017/01461-2, CAPES, and the support of NVIDIA Corporation with the donation of the Quadro P6000 GPU used for this research.

References

Almeida, I. & Nemmen, R. 2020, *MNRAS*, 492, 2553
Cybenko, G. 1989, *Math. of Cont., Sign. and Sys.*
Goodfellow, I., Bengio, Y., & Courville, A. 2016, *The MIT Press*
Hausen, R. & Robertson, B. E. 2020, *ApJS*, 60, 84
Jaeger, H. & Haas, H. 2004, *Science*, 304, 78
Karpathy, A., Toderici, G., Shetty, S., Leung, T., Sukthankar, R., & Fei-Fei, L. 2014 *CVPR*
Krizhevsky, A., Sutskever, I., & Hinton, G. E. 2017, *Commun. ACM*, 60, 84
LeCun, Y., Bengio, Y., & Hinton, G. 2015, *Nature*, 521, 436
Mnih, V., Kavukcuoglu, K., Silver, D., *et al.* 2015, *Nature*, 518, 529
Nemmen, R. 2019, *ApJL*, 880, L26
Porth, O. & others 2019, *ApJS*, 243, 26
Silver, D., Huang, A., Maddison, C. J., *et al.* 2016, *Nature*, 529, 484
Tompson, J., Schlachter, K., Sprechmann, P., *et al.* 2016, arXiv:1607.03597
Zhang, X., Wang, Y., Zhang, W., *et al.* 2019, arXiv:1902.05965

Discussion

WEINBERGER: From the movie you showed us, it seems that the accretion flow is almost stationary. I do not see much turbulence going on.

NEMMEN: We have chosen for the training data a model from Almeida & Nemmen (2020) that shows little variability, because this was the longest model we had (more data improves the performance of the training). We are also investigating and generating data with much more variability and comparable durations.

Sanchez: I heard that there are also some GPU codes that solve fluid dynamics and are faster than CPU solvers. Did you compare the speedups of your deep learning approach with those GPU solvers?

Nemmen: I only know of one GPU solver for black hole accretion simulations, H-AMR developed by Liska et al. In my understanding, H-AMR is obtaining speedups of about 10x when comparing GPUs and multicore CPUs. Our early results indicate a speedup of about 1000x using a good GPU of the kind that gamers use. So it seems that a DL approach would offer a great cost-benefit. But much remains to be investigated yet.

Wylezalek: Can you trust the predictions of an algorithm that does not solve the physical laws? In other words, can you really do physics without doing physics? You don't know what the AI algorithm is really doing, how it arrives at a given answer.

Nemmen: This is a big issue in AI nowadays: the issue of explainability—explaining how the neural networks arrive at a given output. I do not have an answer for that. But I do know that there is progress in this front by some AI research groups. Using AI for the purpose described in this presentation involves a shift in the underlying philosophy of the numerical simulations. What if it does indeed work well for forecasting while being much faster than current simulations? I think this is a very interesting question.

Galaxy Evolution and Feedback across Different Environments
Proceedings IAU Symposium No. 359, 2020
T. Storchi-Bergmann, W. Forman, R. Overzier & R. Riffel, eds.
doi:10.1017/S1743921320001672

Gas flows in a changing-look AGN

Sandra I. Raimundo

DARK, Niels Bohr Institute, University of Copenhagen, Lyngbyvej 2,
DK-2100 Copenhagen, Denmark
email: sandra.raimundo@nbi.ku.dk

Abstract. The galaxy Mrk 590 is one of the few known 'changing-look' Active Galactic Nuclei (AGN) to have transitioned between states twice, having just increased its flux after a period of ∼10 years of low activity. In addition to the increase in flux, the optical broad emission lines have reappeared but show a different profile than what was observed before they disappeared. The gas motions in the host galaxy of this changing-look AGN show outflows and dynamical structures able to drive gas to the nucleus, suggesting an interplay between inflow and outflow in the centre of the galaxy.

Keywords. galaxies:active, galaxies:nuclei, galaxies:individual(Mrk 590)

---

1. Introduction

Optical changing-look Active Galactic Nuclei (AGN) transition between types (type 1 to type 2 or vice versa) in a matter of years or decades. This fast transition is characterised by a dramatic change in flux and by the appearance or disappearance of broad emission lines. While the first changing-look AGN were discovered serendipitously, more recent studies have employed systematic searches to find more examples of this class of objects (e.g. Runco *et al.* 2016; Ruan *et al.* 2016). AGN that show extreme levels of variability (>1 mag in a timescale of years or decades) seem to have lower Eddington ratios than their less variable counterparts (Rumbaugh *et al.* 2018). Out of these highly variable AGN, 30% of them are expected to show changing-look characteristics (MacLeod *et al.* 2018), suggesting that there are still a significant number of changing-look AGN that we have not discovered yet. Most of the changing-look AGN found are not caused by obscuration but by a change in the black hole mass accretion rate, possibly driven by a rapid change in the accretion physics. Strong evidence of this comes from observations in the mid-infrared and radio (e.g. Stern *et al.* 2018; Koay *et al.* 2016b), which are wavebands less affected by obscuration but that also show changing-look transitions. Understanding dramatic mass accretion rate changes occurring in a matter of years or decades is a challenge for this field. Several options to modify our accretion disc paradigm have been put forward (e.g. Ross *et al.* 2018; Noda & Done 2018), but there is still some difficulty in matching the short timescales observed. The appearance and disappearance of the broad emission lines is likely associated with the change in mass accretion rate, suggesting the existence of a threshold value in luminosity or a combination of luminosity and black hole mass (e.g. Elitzur *et al.* 2014), for the formation of broad emission lines.

2. Mrk 590

Mrk 590 is one of the few changing-look AGN that has gone through a type transition (at least) twice (Raimundo *et al.* 2019a). In the 1970's and 1980's spectra of Mrk 590 showed a broad Hα line and weak Hβ broad component. In the following decade the

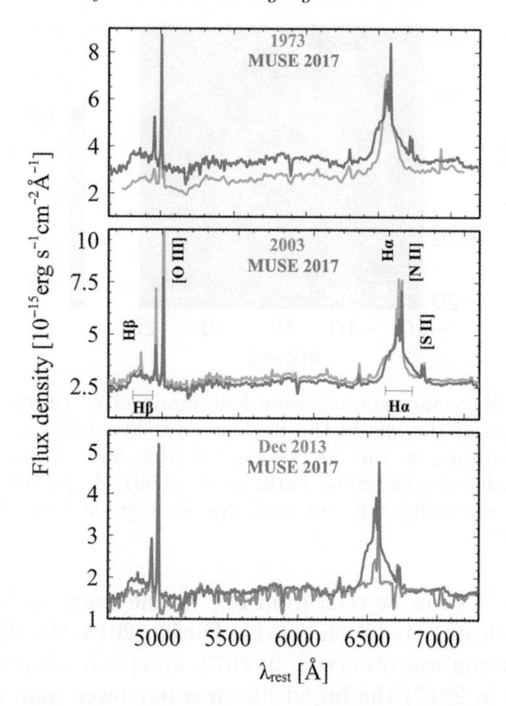

Figure 1. Comparison between spectral features observed in Mrk 590. The grey solid lines show spectra measured at different dates, from top to bottom: 1973, 2003 and 2013 (Denney *et al.* 2014). The blue solid lines are spectra from 2017, obtained with the Multi Unit Spectroscopic Explorer (MUSE) in the Very Large Telescope (Raimundo *et al.* 2019a), and extracted using artificial apertures to match each of the grey spectra. To note the asymmetric shape of the broad Hα line observed in 2017, significantly different from the shape observed in 2003 and 1973.

AGN flux increased until Mrk 590 became a typical Seyfert type 1 galaxy in the 1990's, accreting with an Eddington ratio of $\sim$0.1 (e.g. Peterson*et al.* 2004; Koay *et al.* 2016a), and showing a strong AGN continuum and broad Balmer emission lines in the optical (Peterson *et al.* 1998). Since then, the AGN continuum flux has decreased by a factor of more than 100 across all wavebands and sometime between 2006 and 2012 the broad Hβ emission line disappeared (Denney *et al.* 2014). Raimundo *et al.* (2019a) found that in 2017 the optical broad lines (Hβ and Hα) had reappeared after a period of absence of $\sim$10 years. The AGN optical continuum flux has not increased to the highest levels observed and it is still $\sim$10 times lower than that in the 1990s. According to the model of Elitzur *et al.* (2014), the small increase in continuum luminosity in Mrk 590 would have been enough to cross the luminosity/black hole mass ratio threshold for broad line production. Since UV photons are known to be present (Mathur *et al.* 2018) to produce the broad emission lines, a lack of optical photons suggests that despite producing broad emission lines, Mrk 590 has not fully built its accretion disc yet (Raimundo *et al.* 2019a).

2.1. *Evolution of the spectral shape*

Fig. 1 compares the spectral shape of Mrk 590 at different epochs. The 2017 spectrum in each panel was extracted from the MUSE/VLT datacube shown in Raimundo *et al.* (2019a) using an artificial aperture to match the slits used in the archival spectra. A compilation of archival spectra for Mrk 590 can be found in Denney *et al.* (2014). Here we only show the epochs from 1973, 2003 and 2013 to compare the shape of the Hα broad emission line. The flux calibration is uncertain for some epochs so we focus our discussion

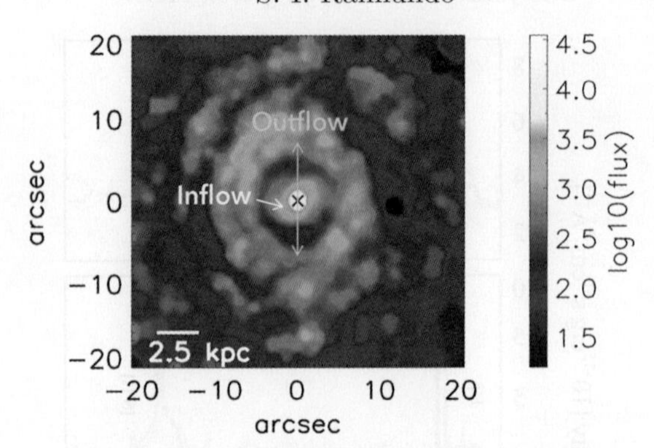

Figure 2. Ionised gas flux map adapted from Raimundo *et al.* (2019a). The arrows illustrate the dynamical properties of the gas in the nucleus and the black cross indicates the position of the AGN. Nuclear spirals, as the one shown for Mrk 590, are candidate mechanisms to drive gas inflows (e.g. Maciejewski 2004; Fathi *et al.* 2006). A gas outflow in the north-south projected direction and extending out to r$\sim$1.5 kpc is detected from the [O III] emission - see Raimundo *et al.* (2019a).

on the spectral shape. As can be seen from Fig. 1, the shape of the Hα broad emission line has changed significantly between epochs. During 2013, the AGN was in a low state and no broad Hα emission was observed. In 1973, an epoch where the AGN was increasing its luminosity (as in 2017) the broad line was narrower than that observed in 2017 and with a 'red shoulder'. The most striking difference is between the 2003 and 2017 line shapes. The broad Hα line from 2003 is stronger in the blue side of the peak while in 2017 the opposite happens, with a strong red tail appearing in the line. The gas producing this line originates from within light-days distance from the black hole, with gas motions away from, or towards the observer contributing to the redshifted or blueshifted components of the broad line (e.g. Raimundo *et al.* 2019b). Such a strong difference in the line shapes may be associated with changes occurring within the dynamical timescale of the broad line region or associated with the process of starting the production of broad emission lines after a quiescent period. Since broad line shapes contain information on the gas geometry and dynamics in the broad line region (e.g. Raimundo *et al.* 2020), more observations of AGN going through this transition may shed light on the physics behind the production of broad emission lines.

2.2. *Host galaxy properties*

The host galaxy shows the presence of ionised and molecular gas in the nucleus, surrounded by a nuclear spiral (clearly observed between r$\sim$0.5−2 kpc) and a star-forming ring (r$\sim$4.5 kpc) (Raimundo *et al.* 2019a). The gas observed indicates that the AGN in Mrk 590 will not run out of gas to fuel its activity but that there may be a cyclical mechanism (likely at the accretion disc scales) responsible for the change in black hole accretion rate. At the hundreds of parsecs scales the gas dynamics in Mrk 590 suggest a balance between gas inflow and outflow (Fig. 2), that can replenish or remove gas from the nucleus, modulating the gas available to the black hole.

3. Conclusions and Outlook

The fast transitions observed in changing-look AGN allow us to probe changes in supermassive black hole accretion physics within a human lifetime. While explaining such transitions is currently challenging, understanding why black holes transition between states can shed light on how and why black holes become active or quiescent. To make progress in this field we need to study both the spectral changes and host galaxy properties (as in Mrk 590) but with a high cadence monitoring to observe changes as they happen. We also need to expand the sample of known changing-look AGN for population studies. Current and new time domain surveys will play a major role in this, as hundreds of new changing-look AGN are expected to be found each year in surveys such as e.g. Zwicky Transient Facility (Graham *et al.* 2019), the Young Supernova Experiment (Jones *et al.* 2019, Jones *et al.* in prep.) and the upcoming Vera C. Rubin Observatory Legacy Survey of Space and Time (LSST).

References

Denney, K. D., *et al.* 2014, *ApJ*, 796, 134
Elitzur, M., Ho, L. C., Trump, J. R., *et al.* 2014, *MNRAS*, 438, 3340
Fathi, K., Storchi-Bergmann, T., Riffel, R. A., *et al.* 2006, *ApJL*, 641, L25
Graham, M. J., *et al.* 2019, *PASP*, 131, 078001
Jones, D. O., *et al.* 2019, *The Astronomer's Telegram*, 13330, 1
Koay, J. Y., Vestergaard, M., Casasola, V., *et al.* 2016a, *MNRAS*, 455, 2745
Koay, J. Y., Vestergaard, M., Bignall, H. E., *et al.* 2016b, *MNRAS*, 460, 304
MacLeod, C. L., *et al.* 2018, *AJ*, 155, 6
Maciejewski, W. 2004, *MNRAS*, 354, 892
Mathur, S., *et al.* 2018, *ApJ*, 866, 123
Noda, H. & Done, C., 2018, *MNRAS*, 480, 3898
Peterson, B. M., Wanders, I., Bertram, R., *et al.* 1998, *ApJ*, 501, 82
Peterson, B. M., *et al.* 2004, *ApJ*, 613, 682
Raimundo, S. I., Vestergaard, M., Koay, J. Y., *et al.* 2019a, *MNRAS*, 486, 123
Raimundo, S. I., Pancoast, A., Vestergaard, M., *et al.* 2019b, *MNRAS*, 489, 1899
Raimundo, S. I., Vestergaard, M., Goad, M. R., *et al.* 2020, *MNRAS*, 493, 1227
Ross, N. P., *et al.* 2018, *MNRAS*, 480, 4468
Ruan, J. J., *et al.* 2016, *ApJ*, 826, 188
Rumbaugh, N., *et al.* 2018, *ApJ*, 854, 160
Runco, J. N., *et al.* 2016, *ApJ*, 821, 33
Stern, D., *et al.* 2018, *ApJ*, 864, 27

Michael Crenshaw, Travis Fisher, Henrique Schmitt, Beena Meena, Steven Kraemer and Thaisa

Gabriel Roier

Galaxy Evolution and Feedback across Different Environments
Proceedings IAU Symposium No. 359, 2020
T. Storchi-Bergmann, W. Forman, R. Overzier & R. Riffel, eds.
doi:10.1017/S1743921320002094

POSTERS

The relation between the environment and nuclear activity in nearby QSOs: Defining a control sample

Bruna L. C. Araujo[1], **Thaisa Storchi-Bergmann**[2] **and Sandro B. Rembold**[3]

[1]Departamento de Física, CCN, Universidade Federal do Piauí, 64049-550, Teresina, PI, Brazil
email: araujo.brunalc@gmail.com

[2]Instituto de Física, Universidade Federal do Rio Grande do Sul, CP 15051, Porto Alegre, RS, 91501-970, Brazil

[3]Departamento de Física, CCNE, Universidade Federal de Santa Maria, 97105-900, Santa Maria, RS, Brazil

Abstract. In this study, we aim to investigate the relation between nuclear activity and the environment for luminous (L[O III] $> 7.63 \times 10^{41}$ erg s^{-1}) Active Galactic Nuclei (AGN) - that, at these luminosities are classified as quasi-stellar objects (QSOs) - using a sample of 436 type 2 QSOs. Recent studies suggest that there is an excess of interacting hosts in luminous AGN, indicating that interactions trigger the nuclear activity. In order to examine this, it is necessary to select a control sample of non-active galaxies, matched to the active ones by the properties of the host galaxies, such as distance and stellar mass. We present here the results of the search for such a control sample.

Keywords. galaxies: active, galaxies: interactions; quasars

1. Introduction

The QSO 2 sample we used is a subset of that of Reyes *et al.* (2008), selected according to their spectra in the Sloan Digital Sky Survey (SDSS) for being luminous but sufficiently nearby ($z < 0.3$) for resolving the morphology of their host galaxies, as done previously in Storchi-Bergmann *et al.* (2018). The small sample of Storchi-Bergmann *et al.* (2018) indicated an excess of interacting galaxies. Our goal is to investigate the incidence of companions in a larger, statistically significant sample.

To achieve this objective, we needed to define a control sample. The first strategy to build the control sample was to use SDSS images to match the continuum luminosity of the QSOs with that of non-active galaxies at similar distances. The goal was to find a continuum band that was free of emission lines in the QSOs, so that we could select a control based on the luminosity of that particular band.

The second strategy was to define the control sample based on the stellar mass of the galaxy and its redshift. For this we needed to derive the stellar mass of the QSO host galaxies, which we did through synthesis of their stellar populations. As the stellar continuum flux is weak in the QSO spectra, in order to have enough signal-to-noise ratio in the continuum we separated our sample in 10 redshift bins, with 43 or 44 QSOs each. We then obtained the median spectra of the QSOs in each bin, after correcting the individual

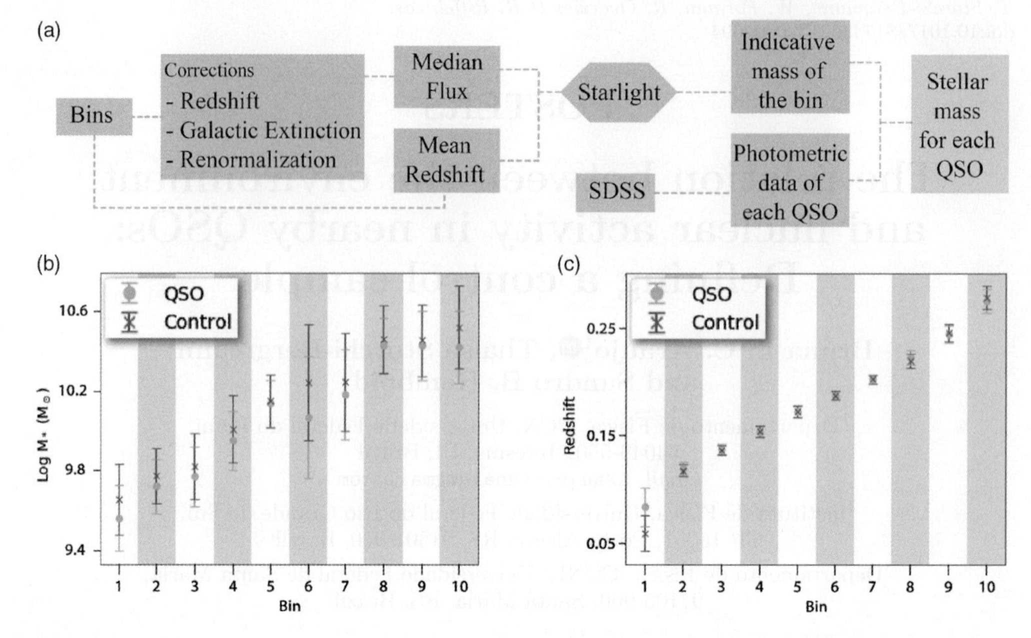

Figure 1. (a) Summary of the process to obtain the stellar mass of the QSOs. The bottom panels show the comparison between the QSOs (orange) and possible control galaxies (green) in terms of (b) mean stellar masses and (c) mean redshifts. The error bars are the standard deviation of the mean.

spectra for redshift and galactic extinction and renormalizing them. We were able to obtain the corresponding stellar masses for each bin from stellar population synthesis of the continuum spectrum using the program STARLIGHT (Cid Fernandes *et al.* 2014; Mateus *et al.* 2006). Then, we calculated the approximate mass for all QSOs using the indicative mass for each bin and photometric data for each QSO. Figure 1a shows a diagram summarizing this process.

2. Results

When inspecting the spectra of the 436 QSOs, we noticed that the majority did not have any SDSS band (u, g, r, i, z) free from emission lines. None of the spectra possess a full coverage of the u-band, and only a few extend to the end of z-band, thus, only the bands g, r, and i could be used. For the majority of the objects, these bands were contaminated by the emission lines: [O III]λ5007, [O II]λ3727, [N III]λ3868, Hα, [O I]λ6300, [N II]λ6583 and [S II]λ6716 + λ6731. Therefore, we were unable to select a control sample via the continuum luminosity in the SDSS image bands.

Our next strategy was to match the control galaxies to each QSO via stellar masses. After obtaining an estimate for the stellar mass of each QSO we calculated the mean and standard deviation of the stellar masses for each bin. We thus selected 150 control galaxies that matched the redshift range and with stellar masses within two standard deviations of the mean value. Figure 1b shows the mean stellar mass of each bin for the QSOs as compared to the mean stellar mass of the 150 possible control galaxies. Figure 1c compares the mean redshifts.

We have now shown that we are able to create an appropriate control sample. The next step in our program is to select the best galaxies among the 150 of each bin to be the controls for each QSO of our sample, according to the stellar mass and redshift.

References

Cid Fernandes, R., González Delgado, R. M., García Benito, R., *et al.* 2014, *A&A*, 561, A130
Mateus, A., Sodré, L., Cid Fernandes, R., *et al.* 2006, *MNRAS*, 370, 721
Reyes, R., Zakamska, N. L., Strauss, M. A., *et al.* 2008, *AJ*, 136, 2373
Storchi-Bergmann, T., Dall'Agnol de Oliveira, B., Longo Micchi, L. F., *et al.* 2018, *ApJ*, 868, 14

Galaxy Evolution and Feedback across Different Environments
Proceedings IAU Symposium No. 359, 2020
T. Storchi-Bergmann, W. Forman, R. Overzier & R. Riffel, eds.
doi:10.1017/S1743921320001763

The relation between dust amount and galaxy mass across the cosmic time

J. H. Barbosa-Santos and G. B. Lima Neto

Instituto de Astronomia, Geofísica e Ciências Atmosférica, USP, Rua do Matão, 1226 - Cidade Universitária, 05508-090, São Paulo, SP, Brazil
email: `jullian.santos@usp.br`

Abstract. Dust Obscured Galaxies (DOGs) are observed as far as the reionization epoch. Their cosmic density peaks together with the star formation rate. DOGs also rule the star formation in high stellar mass galaxies. In this work we used a chemodynamical model to evolve the amount of dust in galaxies. We ran forty models varying initial mass and both dust formation efficiency and dust production. We find that for high star formation rate systems the accretion dominates the dust evolution and it explains high-z DOGs. Low star formation rate systems are better suited to investigate dust production. Also, we find that a $M_{\rm Dust}/M_{\rm Gas}$ versus $M_{\rm Dust}/M_*$ diagram is a good tracer of galaxy evolution.

Keywords. galaxies: evolution, galaxies: ISM, galaxies: high-redshift, (ISM:) dust, extinction

1. Introduction

Dust obscured galaxies (DOGs) are objects with almost all their light obscured by dust. They are the most luminous galaxies and the most intense stellar nurseries in the Universe. The observation of dusty evolved galaxies at $z \gtrsim 6$, in the reionization era, constrains the maximum time taken by dust production Knudsen *et al.* 2017.

High mass galaxies (even normal star forming ones) have most of their star formation (SF) obscured by dust, reaching ~90% in galaxies with $\log(M/{\rm M}_\odot) = 10.5$, while low mass ones tend to have most of the SF unobscured (Whitaker *et al.* 2017). This pattern seems to be present in galaxies with $z \lesssim 2.5$–3.0 (Whitaker *et al.* 2017; Magdis *et al.* 2017). In this work, we investigate the dependence of dust and star formation rates (SFR) on the build-up of dust in galaxies. We also investigated the dominant processes to produce dust during the reionization epoch.

2. Simulation

We used the Friaça & Terlevich (1998) chemodynamical model to investigate dust amount evolution. We assumed a Salpeter initial mass function and a specific SF law, $\nu_{\rm SF} \propto \rho^{1/2}$, as in Friaça & Barbuy (2017). We simulated five initial galaxy masses, $M_{G,0}$, in the range 5×10^7 and 2×10^{12} ${\rm M}_\odot$ (see Fig. 1). The galaxy is initially composed by pristine gas and dark matter with the ratio of $M_{DM}/M_{G,0} = 5.6$. The SF is characterized by a star formation efficiency, ν_0, varying between 0.1 and 10.0 Gyr^{-1} (see Fig. 1).

The dust production formulation "Case A" is the same as in Dwek (1998), while "Case B" has lower grain condensation efficiency, $\delta^X(A)$, set as 0.1 for stellar winds and type II supernova (SN), and 0.0 for SN Ia. The accretion in the cold interstellar medium

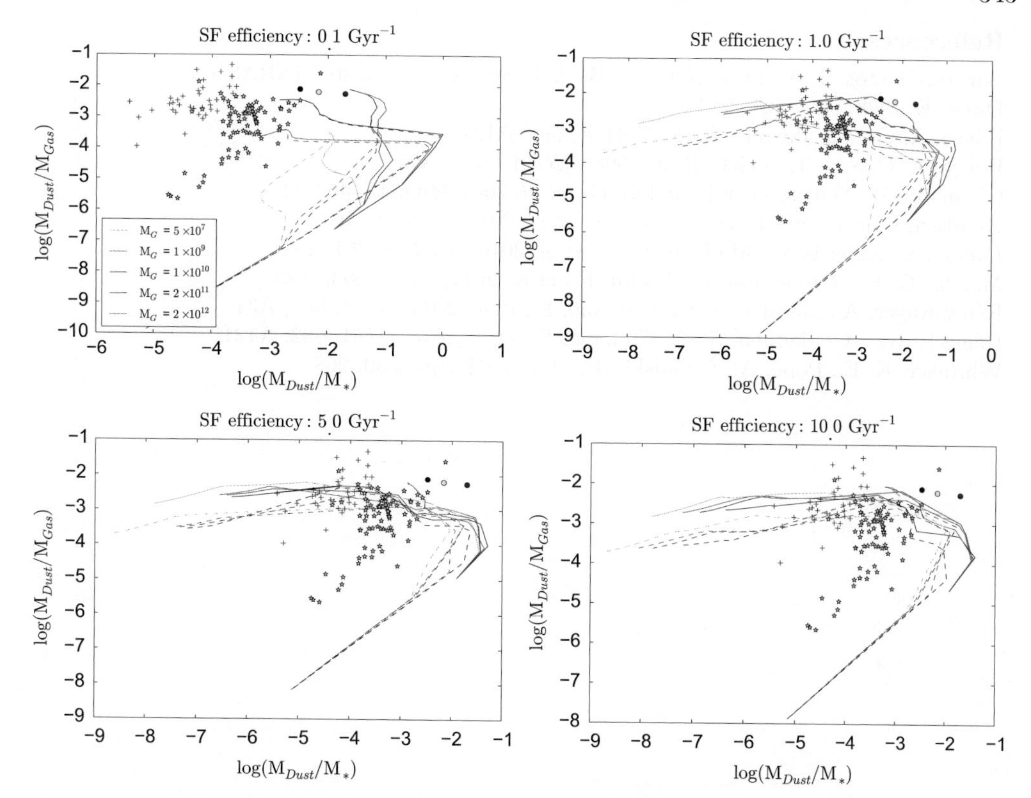

Figure 1. Dust-to-gas vs. dust-to-star mass ratio predicted by the models. Each panel corresponds to a star formation efficiency of the model and the model tracks are color-coded according to the initial galaxy mass. Solid and dashed lines stand for Case A and Case B dust production formulation, respectively. Rémy-Ruyer *et al.* (2014, 2015) are tagged as yellow and pink stars, Lianou *et al.* (2016) [elliptical galaxies] as blue cross, Magdis *et al.* 2017 D49 as yellow, and M28 as black, and Knudsen *et al.* (2017), A1689-zD1 as a blue big dot.

follows Gioannini *et al.* (2017). We considered grains formed by C and Si. We combine all $M_{G,0}$ values, with all ν_0 and with both $\delta^X(A)$ formulations, resulting in forty different models.

3. Results and Conclusion

High SFR models build the bulk of their dust mass in ~0.6 Gyr, nearly the age of the Universe at the reionization. Their dust mass is almost insensitive to $\delta^X(A)$ during star formation peak, due to grain accretion. Low SFR models take a few Gyr to build the bulk of their dust mass. The $M_{G,0} = 1 \times 10^{10}$ $M_\odot$ and $\nu_0 = 10.0$ Gyr^{-1} models need ~0.4 Gyr to reach the equivalent of A1689-zD1 (at $z \sim 7.5$) dust-to-gas ratio. Our result also points to a constant obscuration rate in galaxies with $z \gtrsim 3$.

To balance evolutionary effects, we propose a $M_{\rm Dust}/M_{\rm Gas}$ versus $M_{\rm Dust}/M_*$ diagram (Fig. 1). In this figure, the star forming galaxies follow a clear path, while the elliptical galaxies lie at higher $M_{\rm Dust}/M_{\rm Gas}$ and $M_{\rm Dust}/M_*$ locus. The high-z sample does not exhibit a distinguished pattern. The $M_{\rm Dust}/M_{\rm Gas}$ versus $M_{\rm Dust}/M_*$ diagram is thus a powerful tool to study the interplay between obscuration and galaxy evolution. Further discussion can be found in Barbosa-Santos *et al.* (2020).

References

Barbosa-Santos, J. H., Lima Neto, G. B., & Friaça A. C. S. 2020, MNRAS *submitted*
Dwek, E. 1998, *ApJ*, 501, 643
Friaça, A. C. S. & Barbuy, B. 2017 *A&A*, 598, A121
Friaça, A. C. S. & Terlevich, R. J. 1998 *MNRAS*, 298, 399
Gioannini, L., Matteucci, F., Vladilo, G., *et al.* 2017 *MNRAS*, 464, 985
Knudsen, K. K., Watson, D., Frayer, D., *et al.* 2017 *MNRAS*, 466, 138
Lianou, S., Xilouris, E., Madden, S. C., *et al.* 2016 *MNRAS*, 461, 2856
Magdis, G. F., Rigopoulou, D., Daddi, E., *et al.* 2017, *A&A*, 603, A93
Rémy-Ruyer, A., Madden, S. C., Galliano, F., *et al.* 2014, *A&A*, 563, A31
Rémy-Ruyer, A., Madden, S. C., Galliano, F., *et al.* 2015, *A&A*, 582, A121
Whitaker, K. E., Pope, A., Cybulski, R., *et al.* 2017 *ApJ*, 850, 208

Galaxy Evolution and Feedback across Different Environments
Proceedings IAU Symposium No. 359, 2020
T. Storchi-Bergmann, W. Forman, R. Overzier & R. Riffel, eds.
doi:10.1017/S1743921320001982

Kinematics of the parsec-scale jet of the blazar AO 0235+164

Flávio Benevenuto da Silva Junior and Anderson Caproni

Núcleo de Astrofísica, Universidade Cidade de São Paulo, R. Galvão Bueno 868,
Liberdade, São Paulo, SP, 01506-000, Brazil
email: flavio.ben@outlook.com

Abstract. Radio interferometric maps of the blazar AO 0235+164 show the existence of a stationary core, and a compact jet composed of multiple receding components. In this work, we determined the structural characteristics of these jet components (core-component distance, position angle, flux density, etc.) using the statistical method for global optimization Cross-Entropy (CE). The images we analyzed were extracted from public databases, totaling 41 images at 15 GHz and 128 images at 43 GHz. Using criteria such as the value of the CE merit function, and mean residuals, we determined the optimum number of components in each map analyzed in this work. We found that jet components are distributed across all four quadrants on the plane of the sky, indicating a possible non-fixed jet orientation during the monitoring interval. The time evolution of the equatorial coordinates of the jet components were used to determine their respective speeds, ejection epochs, and mean position angles on the plane of the sky. We have identified more than 20 components in the jet of AO 0235+164, with their apparent speeds ranging roughly from 2c to 40c, and distributed across all four quadrants on the plane of the sky. From the kinematics of these jet components we could derive a lower limit of about 39 for its bulk jet Lorentz factor and an upper limit of approximately 42 degrees for its jet viewing angle.

Keywords. BL Lacertae objects: individual: AO 0235+164 — galaxies: active-galaxies: jets — methods: data analysis — radio continuum: galaxies — techniques: interferometric

---

1. Introduction

As usual in its active galactic nuclei (AGN) class, the BL Lac object AO 0235+164 exhibits violent variability across the electromagnetic spectrum on time-scales from hours to years (e.g., Raiteri *et al.* 2008; Ackermann *et al.* 2012).

Radio interferometric images of AO 0235+164 ($z = 0.94$ e.g., Cohen *et al.* (1987)) show a very compact source ($\lesssim$1 mas) with a stationary component identified as the core region and a few receding jet components (e.g., Chu *et al.* 1996; Jorstad *et al.* 2001; Lee *et al.* 2008; Lister *et al.* 2009; Kutkin *et al.* 2018).

2. Methodology and Results

We extended previous studies analyzing 128 43-GHz interferometric images obtained from the public archive VLBA-BU-BLAZAR Multi-Wavelength Monitoring Program (Jorstad & Marscher 2016; Jorstad *et al.* 2017), and 41 15-GHz maps gathered from the MOJAVE/2cm Survey Data Archive (Lister *et al.* 2009). Structural parameters of the elliptical Gaussian components were determined via Cross-Entropy global optimization techniques (e.g., Rubinstein 1997; Caproni *et al.* 2009, 2014, 2017).

We found that jet component motions are compatible with non-accelerated trajectories residing in all quadrants on the plane of sky. Their ejections coincide closely in time with flares seen in the core and gamma-ray light curves. The minimum value for the Lorentz factor was obtained from the maximum apparent speed, $\beta_{\rm app,max}$, determined in this work: $\gamma_{min} = \sqrt{1+\beta^2_{\rm app,max}} \geqslant 39.2 \pm 11.8$. The maximum jet viewing angle comes from the minimum apparent speed among jet components, $\beta_{\rm app,min}$, and taking the limit $\beta \to 1$, where β is the bulk jet speed: $\theta_{max} = \arccos\left[\left(\beta^2_{\rm app,min} - 1\right) / \left(\beta^2_{\rm app,min} + 1\right)\right] = 42\overset{\circ}{.}1 \pm 19\overset{\circ}{.}8$.

3. Conclusions

We were able to identify 27 components that have different apparent speeds, ranging from 2c to 40c. It allowed us to estimate a lower limit for the Lorentz factor (∼39), as well as an upper limit for the jet viewing angle (∼42°), in agreement with previous estimates (e.g., Zhang *et al.* 1998; Volvach *et al.* 2015).

We also found that the position angles of the jet components varied roughly from −10° to −345°. This very broad range indicates that the jet direction has not remained fixed over time. A possible explanation for such dispersion could be the jet precession phenomenon, which will be pursued in future work.

Acknowledgements

F.B.S.J. thanks the Brazilian agency CAPES and the Brazilian Astronomical Society (SAB) for financial support. A.C. thanks the Brazilian agency FAPESP (grants 2017/25651-5 and 2014/11156-4). This research has made use of data from the MOJAVE database that is maintained by the MOJAVE team (Lister *et al.* 2018, *ApJS*, 232, 12). This study makes use of 43 GHz VLBA data from the VLBA-BU Blazar Monitoring Program (VLBA-BU-BLAZAR; http://www.bu.edu/blazars/VLBAproject.html), funded by NASA through the Fermi Guest Investigator Program. The VLBA is an instrument of the National Radio Astronomy Observatory. The National Radio Astronomy Observatory is a facility of the National Science Foundation operated by Associated Universities, Inc.

References

Ackermann, M. 2012, *ApJ*, 751, 159
Caproni, A., Monteiro, H., Abraham, Z., *et al.* 2009, *MNRAS*, 399, 1415
Caproni, A., Tosta e Melo, I., Abraham, Z., *et al.* 2014, *MNRAS*, 441, 187
Caproni, A., Abraham, Z., Motter, J. C., *et al.* 2017, *ApJL*, 851, L39
Chu, H. S., Baath, L. B., Rantakyroe, F. T., *et al.* 1996, *A&A*, 307, 15
Cohen, R. D., Smith, H. E., Junkkarinen, V. T., *et al.* 1987, *ApJ*, 318, 577
Jorstad, S. G., Marscher, A. P., Mattox, J. R., *et al.* 2001, *ApJS*, 134, 181
Jorstad, S. G. & Marscher, A. P. 2016, *Galaxies*, 4, 47
Jorstad, S. G., Marscher, A. P., Morozova, D. A., *et al.* 2017, *ApJ*, 846, 98
Kutkin, A. M., Pashchenko, I. N., Lisakov, M. M., *et al.* 2018, *MNRAS*, 475, 4994
Lee, S., Lobanov, A. P., Krichbaum, T. P., *et al.* 2008, *ApJ*, 136, 159
Lister, M. L., Aller, H. D., Aller, M. F., *et al.* 2009, *AJ*, 137, 3718
Raiteri, C. M., Villata, M., Larionov, V. M., *et al.* 2008, *A&A*, 480, 339
Rubinstein, R. Y. 1997, *European Journal of Operational Research*, 99, 89
Volvach, A. E., Larionov, M. G., Volvach, L. N., *et al.* 2015, *Astronomy Reports*, 59, 145
Zhang, F. J., Chen, Y. J., Zhu, H. S., *et al.* 1998, *Chin. Astron. Astrophys.*, 22, 138

Galaxy Evolution and Feedback across Different Environments
Proceedings IAU Symposium No. 359, 2020
T. Storchi-Bergmann, W. Forman, R. Overzier & R. Riffel, eds.
doi:10.1017/S1743921320004317

Multiwavelength analysis of OH Megamaser galaxies: The case of IRAS11506-3851

Carpes P. Hekatelyne and Thaisa Storchi-Bergmann

Departamento de Astronomia, Universidade Federal do Rio Grande do Sul,
91501-970 Porto Alegre, Rio Grande do Sul, Brazil
email: hekatelyne.carpes@gmail.com

Abstract. We present Multi-Object Spectrograph (GMOS) Integral Field Unit (IFU), Hubble Space Telescope (HST) and Very Large Array (VLA) observations of the inner kpc of the OH Megamaser galaxy IRAS 11506-3851. In this work we discuss the kinematics and excitation of the gas as well as its radio emission. The HST images reveal an isolated spiral galaxy and the combination with the GMOS-IFU flux distributions allowed us to identify a partial ring of star-forming regions surrounding the nucleus with a radius of ≈500 pc. The emission-line ratios and excitation map reveal that the region inside the ring present mixed/transition excitation between those of Starbursts and Active Galactic Nuclei (AGN), while regions along the ring are excited by Starbursts. We suggest that we are probing a buried or fading AGN that could be both exciting the gas and originating an outflow.

Keywords. masers; galaxies: active; galaxies: interactions; galaxies: ULIRGs

---

1. Introduction

Ultra Luminous Infrared Galaxies (ULIRGs; $L \geqslant 10^{12}$ $L_\odot$) are known to be gas-rich galaxies undergoing strong interactions and associated with advanced mergers (Sanders & Mirabel 1996). Approximately 20% of LIRGs present masers at 1665 and 1667 MHz with luminosities of about 10^{2-4} $L_\odot$ (Lo 2005). These galaxies are called OH Megamaser galaxies (hereafter OHM galaxies). The OH maser emission is the result of stimulated emission, being produced by the amplification of the nuclear radio continuum by foreground molecular gas, with the inverted level populations being pumped by far infrared radiation that could be due to a Starburst or its combination with an obscured Active Galactic Nuclei (AGN) (Parra *et al.* 2005).

There is evidence that many of these OHM galaxies host an AGN that is still immersed in dense layers of dust and gas, suggesting that they could represent a key stage in galaxy evolution in which the AGN is being triggered by the accretion of matter to the central supermassive black hole (Darling & Giovanelli 2000). The investigation of the nature of OHM galaxies can provide information about the evolution of gas-rich mergers and about how the OH maser emission is related to the nature of the dominant heating source.

This study is part of an ongoing project aimed to investigate the nature of OHM galaxies using multi-wavelength analysis (Sales *et al.* 2015, 2019; Hekatelyne *et al.* 2018a,b). The main result, so far, is that the four galaxies already studied present signatures of an embedded AGN surrounded by star-forming regions. In addition, it is possible to identify signatures of gas non-circular motions that are attributed to inflows and outflows. Although we find evidence of the presence of AGNs, the number of objects is still too small for a more definitive conclusion about the nature of OHM galaxies.

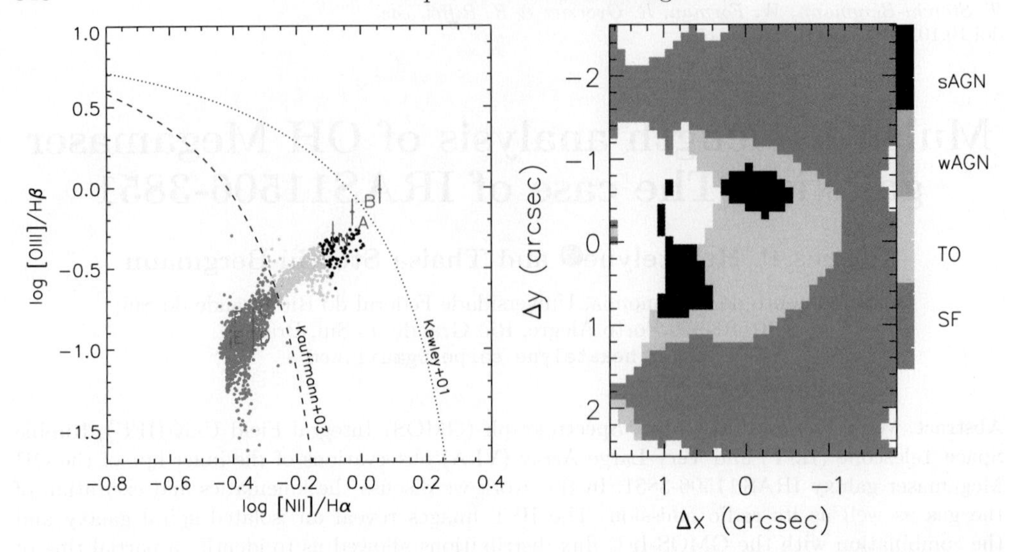

Figure 1. [O III]λ5007/H_β versus [N II]/H_α diagnostic diagram (left). The dotted and dashed lines represent the Kewley and Kauffmann criteria respectively. Excitation map identifying the regions within the FoV presenting distinct excitation mechanisms: strong AGN (sAGN), weak AGN (wAGN), Transition object (TO) and star-forming (SF) (right).

2. Results

We report a multi-wavelength data analysis of IRAS 11506-3851. The GMOS-IFU observations cover the inner $720 \times 1200\,\mathrm{pc}^2$ at a spatial resolution of 193 pc and spectral resolution of 1.8 Å. Our main results are:

The HST images reveal an isolated spiral galaxy and several knots that in combination with the GMOS-IFU data can be attributed to star-forming regions in a 500 pc radius circumnuclear ring. Moreover, the VLA images reveal a steep continuum spectrum kpc-scale bipolar structure in the east-west direction, consistent with an AGN outflow.

Inside the circumnuclear ring the excitation is due to both Starburst and AGN activity (see Fig. 1). On the other hand, the excitation in the external regions (ring) it is due only to Starburst activity. Besides, the presence of radio emission indicates another source of excitation in the nuclear region. From VLA data we also conclude that the gas in the inner 240 pc is partially ionized by an embedded AGN.

We fitted and subtracted a rotating-disk model to the gas kinematics and find deviations from pure rotation within the inner 240 pc that are co-spatial to an outflow in neutral gas that was previously detected (Cazzoli *et al.* 2016). Moreover, these regions present the highest velocity dispersion and are coincident with the radio emission. We argue that the nuclear region is unveiling a new plasma bubble ejection that is pushing the surrounding gas and either shocks or escape from the AGN radiation are increasing the gas excitation in these regions.

References

Cazzoli, S., Arribas, S., Maiolino, R., *et al.* 2016, *A&A*, 590, A125
Darling, J. & Giovanelli, R. 2000, *AJ*, 119, 3003
Hekatelyne, C., Riffel, R. A., Sales, D., *et al.* 2018a, *MNRAS*, 474, 5319
Hekatelyne, C., Riffel, R. A., Sales, D., *et al.* 2018b, *MNRAS*, 479, 3966

Lo, K. Y. 2005, *ARA&A*, 43
Parra, R., Conway, J. E., Elitzur, M., *et al.* 2005, *A&A* 443, 383
Sales, D. A., Robinson, A., Axon, D. J., *et al.* 2015, *ApJ*, 799, 25
Sales, D., Robinson, A., Riffel, R. A., *et al.* 2019, *MNRAS*, 486, 3350
Sanders & Mirabel 1996, *ARA&A*, 34, 749

Galaxy Evolution and Feedback across Different Environments
Proceedings IAU Symposium No. 359, 2020
T. Storchi-Bergmann, W. Forman, R. Overzier & R. Riffel, eds.
doi:10.1017/S1743921320001799

Galactic nuclear off-centerings: the innermost accretion mechanism?

Gaia Gaspar[1], **Rubén Díaz**[1,2], **Damían Mast**[1] **and María P. Agüero**[1]

[1]Observatorio Astronómico de Córdoba, Universidad Nacional de Córdoba,
Postcode 5000, Córdoba, Argentina
email: gaiagaspar@unc.edu.ar

[2]Gemini Observatory, La Serena, Chile
email: rdiaz@gemini.edu

Abstract. In the current scenario of galaxy evolution, supermassive black holes (SMBH) are present in almost all galaxies. To trigger nuclear activity, large amounts of material have to fall from kpc to pc and even smaller scales. Hence, an efficient angular momentum removal mechanism is needed. A growing black hole could still not be fixed in the gravitational potential well of the galaxy. This can be observed as a break in the symmetry between the global structure of the galaxy and the central source and could be part of the mechanism that drives material from the last hundred parsecs onto accretion in the SMBH. We present spatial profile decomposition of 16 galaxies observed with GNIRS (Gemini North) in the K_{long} band. We have been able to measure off-centerings in 3 of 16 galaxies. We found a possible correlation between the presence of an off-centering and the SMBH mass.

Keywords. galaxies: nuclei, galaxies: evolution, galaxies: Seyfert, techniques: spectroscopic

1. Context

Numerical simulations have shown that there exists a phase in Active Galactic Nuclei (AGN) growth where the SMBH is decoupled from the global gravitational potential of the host galaxy (Miller & Smith (1992); Taga & Iye (1998); Emsellem *et al.* (2015)). Observational works have presented evidence of AGN that do not reside in the center of the host galaxies (Díaz, *et al.* (1999); Côté *et al.* (2006); Combes *et al.* (2019)).

The frequency of off-centerings in galactic nuclei and the correlation with other nuclei parameters such as SMBH mass, activity type and X-ray luminosity could shed light onto the sub-10 pc accretion mechanism. It is important to separate the underlying red stellar population, a good tracer of the distribution of mass, from the discretized components such as ionized, atomic, and molecular gas, young star clusters and massive star clumps. This characteristic is achieved in the K band of the Near Infrared (NIR).

The fitting

In this work we used spectra obtained by Mason *et al.* (2015), taken with GNIRS in the K_{long} band. In order to perform the spatial component fitting we extracted continuum spatial profiles in two wavelength ranges: $\lambda 2.097 - 2.104$ μm and $\lambda 2.247 - 2.254$ μm. To model the profiles we used 2 components: a *gaussian* for the nuclear source and a Sérsic profile to represent the larger structure leaving as a free parameter the off-center between them. In Table 1 we present the results of the fitting. Two of the off-centered nuclei found were discarded due to obscuration near the slit as seen in HST/NICMOS images.

Table 1. Parameters of the spatial profile fitting for the 16 galaxies of the sample. Column 2 lists the *fwhm* of the nuclear *gaussian* function, column 3 corresponds to the scalelength radius of the Sérsic component and column 4 is the Sérsic index. Columns 5 and 6 present the Off-centerings in arc seconds and parsecs respectively.

Galaxy	fw_g ["]	R_e["]	n	Off-center ["]	Off-center [pc]
NGC 2273	0.67	2.25	1.5	–	–
NGC 3031	0.92	6.3	1.6	–	–
NGC 3718	1.13	3.45	1.6	0.06	4.95
NGC 3998	0.56	4.5	2.2	–	–
NGC 4258	0.77	6.9	1.59	–	–
NGC 4388	0.56	6.75	2.8	–	–
NGC 4450	0.78	4.35	2	–	–
NGC 4548	0.6	4.2	2.09	–	–
NGC 4565	0.91	3.45	1	–	–
NGC 4594	–	10.5	2.8	–	–
NGC 4725	1.23	4.27	1.2	–	–
NGC 5005	0.7	2.1	0.79	–	–
NGC 5033	0.7	7.8	2.2	0.21	19.62
NGC 5194	1.06	49.35	1	–	–
NGC 7331	0.63	21.45	2.9	0.11	7.26
NGC 7743	0.7	5.25	1.4	–	–

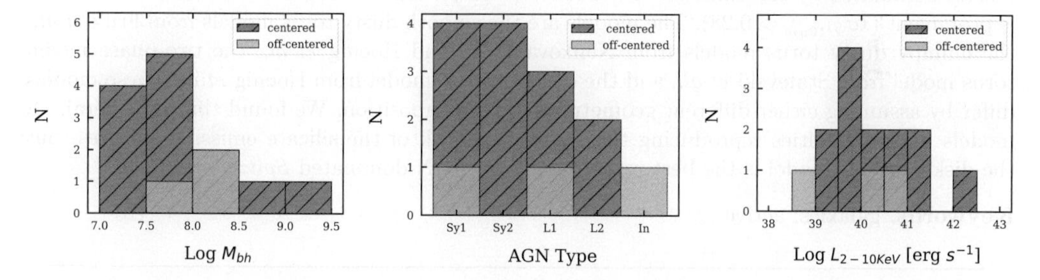

Figure 1. Histograms of 3 nuclei parameters for the 16 galaxies of the sample: black hole mass in the left panel, AGN type in the middle panel and X-ray luminosity in 2-10 keV in the right panel. Blue represents the galaxies with centered nuclei, in pink galaxies with off-centered nuclei are differentiated

2. Results

We have found off-centerings in 3 of the 16 galaxies. In Figure 1 we present three parameter distributions of the sample. The presence of an off-centering does not correlate with X-ray luminosity nor AGN type for this sample. But, there is a suggestion that it could correlate with the SMBH mass. The three off-centered nuclei fall in the range of intermediate mass (Log(M/Msun) = 7.9, 8 and 8.1 respectively). This result could indicate that more massive nuclei are already fixed to or dominating the gravitational potential of the galaxy and therefore are no longer causing an asymmetry in the circumnuclear material distribution.

References

Combes, F., *et al.* 2019, *A&A*, 623, A79

Côté, P., *et al.* 2006, *ApJS*, 165, 57

Díaz, R., Carranza, G., Dottori, H., *et al.* 1999, *ApJ*, 512, 623

Emsellem, E., Renaud, F., Bournaud, F., *et al.* 2015, *MNRAS*, 446, 2468

Mason, R. E., *et al.* 2015, *ApJS*, 217, 13

Miller, R. H. & Smith, B. F. 1992, *ApJ*, 393, 508

Taga M. & Iye M. 1998, *MNRAS*, 299, 111

Galaxy Evolution and Feedback across Different Environments
Proceedings IAU Symposium No. 359, 2020
T. Storchi-Bergmann, W. Forman, R. Overzier & R. Riffel, eds.
doi:10.1017/S1743921320001635

Modelling the silicate emission features in type 1 AGNs: Dusty torus and disk+outflow models

M. Martínez-Paredes[1] and I. Aretxaga[2]

[1]Korea Astronomy & Space Science Institute, Daejeon, South Korea
email: mariellauriga@kasi.re.kr

[2]Instituto de Astrofísica, Óptica y Electrónica, Tonantzintla, Puebla, Mexico

Abstract. We investigated how the most common dusty torus models reproduce both the 10 and 18μm silicate emission features observed in the nuclear infrared (IR) *Spitzer* spectrum of type 1 active galactic nuclei (AGN). We use a sample of type 1 AGN for which the *Spitzer* spectrum is mostly dominated by the emission of the AGN (>80%), and the 10μm silicate emission feature is prominent ($1\sigma_{Si_{10\mu m}} > 0.28$). The models are the smooth dusty torus models from Fritz *et al.*, the clumpy dusty torus models from Nenkova *et al.* and Hoenig *et al.*, the two-phase media torus model from Stalevski *et al.*, and the disk+outflow model from Hoenig *et al.* These models differ by assuming either different geometry or dust composition. We found that in general, all models have difficulties reproducing the shape and peak of the silicate emission features, but the disk+outflow model is the best reproducing the AGN-dominated *Spitzer* spectrum.

Keywords. galaxies: active

1. Introduction

The emission of both 10 and 18 μm silicate features observed in the nuclear spectrum of type 1 AGN, obtained with the Infrared Spectrograph (IRS) on the *Spitzer* Space Telescope, is intimately linked to the physical properties of the dust in the nucleus of these galaxies, where is located the "putative" dusty torus that surrounds the central engine (super massive black hole, accretion disk and, broad line region) of the AGN. The dust in the torus absorbed the UV and optical emission produced by the accretion disk and re-emit it in the IR. In type 1 AGNs is possible to observe the emission from the dust directly heated by the AGN, while in type 2 AGNs, this emission is blocked by the outermost cold dust located along the line of sight. Therefore, to investigate how the most used dusty torus models in the literature reproduce the silicate emission features observed in type 1 AGNs, we selected a sample of local ($z < 0.1$) type 1 AGNs, for which the IRS/*Spitzer* spectrum, between $\sim 5 - 30$ μm, is mostly dominated (>80%) by the emission of the dust heated by the AGN, between them, we chose those in which the 10 μm silicate emission feature is prominent, i.e, those with the strongest 10 μm silicate strength ($1\sigma_{Si_{10\mu m}} > 0.28$, 10 objects). We used four dusty torus models, the smooth (Fritz *et al.* 2006) torus model, two clumpy (Nenkova *et al.* 2008a and Hönig *et al.* 2010) torus models, the two-phase media (Stalevski *et al.* 2016) torus model, and the disk plus outflow (Hönig & Kishimoto 2017) model, to investigate which of these models better reproduce the shape and peak of both silicate emission features since these models assume different geometries and/or dust composition (see Martínez-Paredes *et al.* 2020).

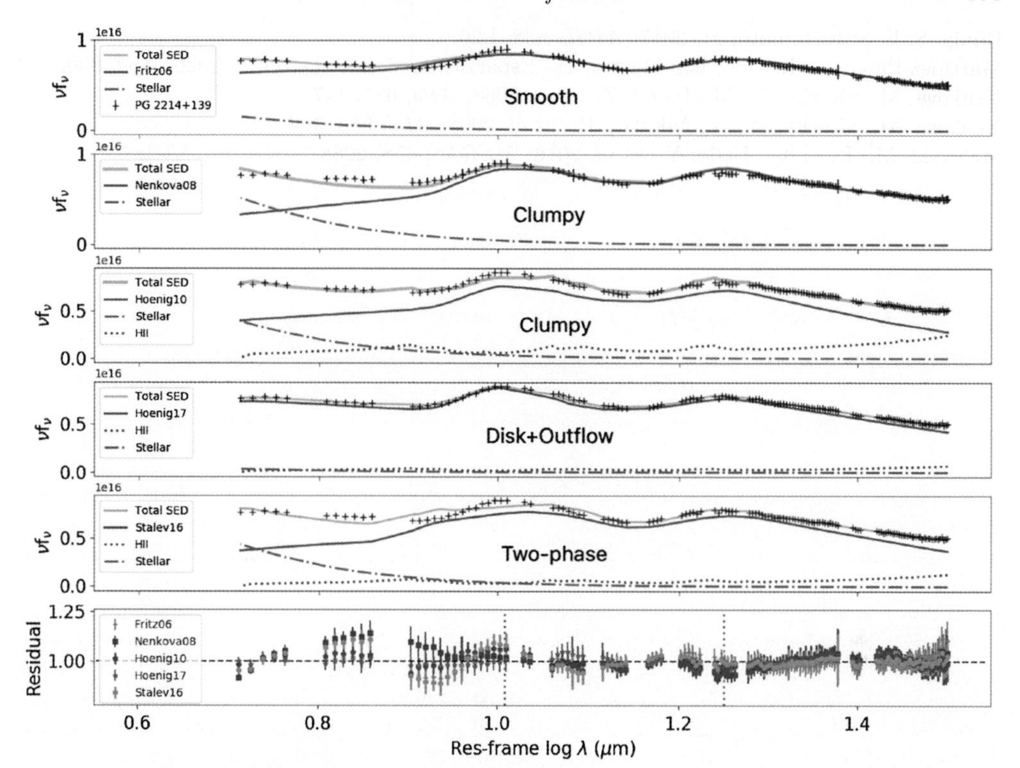

Figure 1. Modelling of the AGN-dominated IRS/*Spitzer* spectrum (in black) of the type 1 AGN PG 2214+139. From top to bottom is the modelling assuming the smooth dusty torus from Fritz *et al.* (2006), the clumpy dusty torus from Nenkova *et al.* (2008a,b), the clumpy dusty torus from Hönig *et al.* (2010), the disk+outflow model from Hönig & Kishimoto (2017), and the two-phase dusty torus model from Stalevski *et al.* (2016). The last panel shows the residuals. This figure is from Martínez-Paredes *et al.* (2020).

2. Results

We measured, for models and observations, both the 10 and 18 μm silicate strength ($Si_{\lambda_p} = ln f_{\lambda_p}(spectrum)/f_{\lambda_p}(continuum)$) at the wavelength where they peak, as well as the near- ($5.5 - 7.5\mu$m) and mid-IR ($7.5 - 14\mu$m) spectral indexes. Comparing the synthetic and observational values, we found that in general, the models predict the 10 and 18 μm silicate strength values observed in these objects. However, when we compare the 10 and 18 μm central wavelengths, we found that only the values observed in the objects (2) with the lowest bolometric luminosity ($L_{bol} < 10^{42}$erg s^{-1} cm^{-2}) are poorly sampled by all models. In Figure 1 we show the fitting of the AGN-dominated IRS/*Spitzer* spectrum of the object PG 2214+139. In general, we found that clumpy models better reproduce the 10 and 18 μm silicate emission features than smooth models. However, on average, we noted that the disk+outflow model better reproduces the AGN-dominated IRS/*Spitzer* spectrum between $\sim 5 - 30\,\mu$m, specially for objects with the highest luminosity ($\sim 10^{44} - 10^{46}$ erg s^{-1}cm^{-2}).

References

Fritz, J., Franceschini, A., & Hatziminaoglou, E. 2006, *MNRAS*, 366, 767
Hönig, S. F., Kishimoto, M., Gandhi, P., *et al.* 2010, *AAP*, 515, A23

Hönig, S. F. & Kishimoto, M. 2017, *APJL*, 838, L20
Martínez-Paredes, M., González-Martín, O., Esparza-Arredondo, D., *et al.* 2020, *ApJ*, 890, 152
Nenkova, M., Sirocky, M. M., Ivezić, Ž., *et al.* 2008, *APJ*, 685, 147
Nenkova, M., Sirocky, M. M., Nikutta, R., *et al.* 2008, *APJ*, 685, 160
Stalevski, M., Ricci, C., Ueda, Y., *et al.* 2016, *MNRAS*, 458, 2288

Galaxy Evolution and Feedback across Different Environments
Proceedings IAU Symposium No. 359, 2020
T. Storchi-Bergmann, W. Forman, R. Overzier & R. Riffel, eds.
doi:10.1017/S1743921320002033

NIR–IFU observations of the merger remnant NGC 34

Juliana C. Motter[1], Rogério Riffel[1], Tiago V. Ricci[2], Natacha Z. Dametto[3], Luis G. Dahmer-Hahn[4], Marlon R. Diniz[5], Rogemar A. Riffel[5], Miriani G. Pastoriza[1], Alberto Rodríguez-Ardila[4], Thaísa Storchi-Bergmann[1] and Daniel Ruschel-Dutra[6]

[1]Departamento de Astronomia, Universidade Federal do Rio Grande do Sul, 91501-970, Porto Alegre, RS, Brazil
email: juliana.motter@ufrgs.br

[2]Universidade Federal da Fronteira Sul, Campus Cerro Largo, 97900-000, Cerro Largo, RS, Brazil

[3]Centro de Astronomía (CITEVA), Universidad de Antofagasta Avenida, Angamos 601, Antofagasta, Chile

[4]Laboratório Nacional de Astrofísica, 37500-000, Itajubá, MG, Brazil

[5]Departamento de Física, Universidade Federal de Santa Maria, 97105-900, Santa Maria, RS, Brazil

[6]Departamento de Física, Universidade Federal de Santa Catarina, 88040-900, Florianópolis, SC, Brazil

Abstract. Understanding the interplay between the phenomena of active galactic nuclei (AGN) and starbursts remains an open issue in studies of galaxy evolution. The galaxy NGC 34 is the remnant of the merger of two former gas-rich disc galaxies and it also hosts a strong nuclear starburst. In this work, we map the ionized and molecular gas present in the nuclear regions of the galaxy NGC 34 using adaptive optics (AO) assisted near infrared (NIR) integral field unity (IFU) observations. Our main goals are to better constrain the energy source of this object and to use NGC 34 as a laboratory to probe the AGN-starburst connection in the context of galaxy evolution and AGN feeding and feedback processes.

Keywords. galaxies: interactions, galaxies: active, galaxies: starburst, infrared: galaxies

1. Introduction

The phenomenon of active galactic nuclei (AGN) represents a critical phase in galaxy evolution, since AGN feedback may impact star formation (SF) over galactic scales (Storchi-Bergmann & Schnorr-Muller 2019). However, understanding the feeding and feedback of AGN becomes a challenging task for objects that seem to host not only an AGN but also a nuclear starburst. Another problem involves quantifying feedback effects in high redshift objects, which can be addressed through studies of Local Universe analogs.

In this context, the galaxy NGC 34, $z = 0.01962$ (Rothberg & Joseph 2006), a local luminous infrared galaxy (LIRG), is an ideal laboratory for such studies. It hosts a strong nuclear starburst and tidal tails indicative of the merger of two former gas-rich disc galaxies (Schweizer & Seitzer 2007). Although X-ray observations provide compelling evidence for the presence of an AGN in its central regions (Esquej *et al.* 2012), the

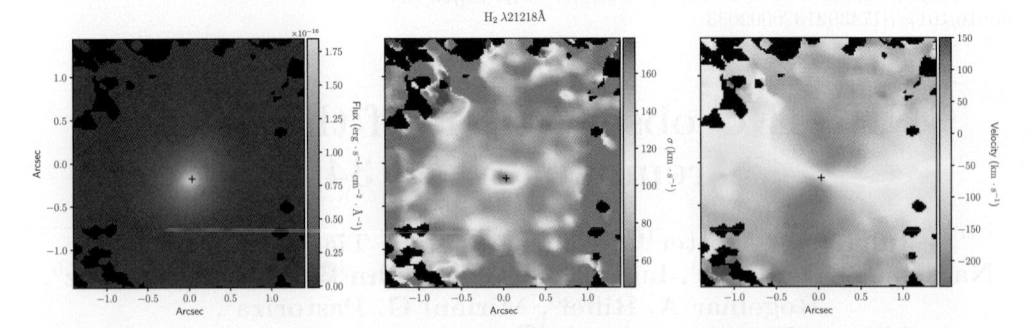

Figure 1. Flux, velocity dispersion σ and centroid velocity maps for the H_2 λ21218Å emission line. The black cross indicates the galaxy center. Black pixels were not considered in the fit.

nature of its nuclear emission line spectrum is still highly controversial. In this work, we use near infrared (NIR) integral field unity (IFU) observations to map the distribution of the ionized and molecular gas present in the central regions of NGC 34 in order to better constrain the energy source powering this object and to use NGC 34 as a local laboratory to probe the AGN-starburst connection in the context of galaxy evolution.

2. Observations and Data analysis

We have taken adaptive optics (AO) assisted Gemini North Near-Infrared Integral Field Spectrograph (NIFS) data cubes in September 2011 in the J and K_l bands. The data were reduced using IRAF and standard reduction scripts made available by the Gemini team. We applied the treatment techniques suggested by Menezes *et al.* (2014). Our final data cubes have fields–of–view (FoV) of $\approx$3.0 arcsec $\times$ 3.0 arcsec, corresponding to 1.2 kpc $\times$ 1.2 kpc at the galaxy, and the spatial resolution is 0.17 arcsec ($\approx$70 pc) for both bands. We subtracted the stellar component from the observed spectra of the galaxy using the IRTF Spectral Library (Cushing *et al.* 2005; Rayner *et al.* 2009) and the Penalized Pixel-Fitting (PPXF) code (Cappellari 2017). The fitting of the profiles of the emission lines is being carried out using the package IFSCUBE.

3. Preliminary results and Ongoing analysis

The main NIR emission lines that can be seen in our spectra are [P II] λ11470Å, [P II] λ11886Å, [Fe II] λ12570Å and Paβ in the J band, and H_2 λ21218Å, Brγ, H_2 λ22230Å and H_2 λ22470Å in the K band. The flux distribution and velocity field maps for H_2 λ21218Å are shown in Fig. 1. The map of velocities shows a rotation signature that resembles a disc with a northern receding and a southern approaching side. Our next steps include obtaining spatially resolved maps of the emission line ratios H_2 $\lambda 2.121\mu$m/Brγ and [FeII]$\lambda 1.257\mu$m/Paβ to be used in NIR diagnostic diagrams that will allow us to better constrain the excitation mechanisms of the multi–phase gas in NGC 34.

References

Cappellari, M. 2017, *MNRAS*, 466, 798
Cushing, M. C., Rayner, J. T. & Vacca, W. D. 2005, *ApJ*, 623, 1115
Esquej, P., Alonso-Herrero, A., Pérez-García, A. M., *et al.* 2012, *MNRAS*, 423, 185
Menezes, R. B., Steiner, J. E., & Ricci, T. V. 2014, *MNRAS*, 438, 2597
Rayner, J. T., Cushing, M. C. & Vacca, W. D. 2009, *ApJS*, 185, 289
Rothberg, B. & Joseph, R. D. 2006, *AJ*, 131, 185
Schweizer, F. & Seitzer, P. 2007, *AJ*, 133, 2132
Storchi-Bergmann, T. & Schnorr-Muller, A. 2019, *Nat. Astron.*, 3, 48

Galaxy Evolution and Feedback across Different Environments
Proceedings IAU Symposium No. 359, 2020
T. Storchi-Bergmann, W. Forman, R. Overzier & R. Riffel, eds.
doi:10.1017/S1743921320001714

Optical properties of CSS/GPS sources

Raquel S. Nascimento[1], Alberto Rodríguez-Ardila[1], Marcos F. Faria[2], Murilo Marinello[1] and Luis G. Dahmer-Hahn[1]

[1]Laboratório Nacional de Astrofésica, Itajubá-MG, Brazil
email: rnascimento@lna.br

[2]Instituto Nacional de Pesquisas Espaciais, São José dos Campos-SP, Brazil

Abstract. In this work, we study the optical properties of 58 CSS/GPS radio sources selected from the literature in order to determine the impact of the radio-jet in the circumnuclear environment of these objects. We obtained optical spectra for all sources from SDSS-DR12 and performed a stellar population synthesis using the Starlight code. Our results indicate that the sample is dominated by intermediate to old stellar populations and there is no strong correlation between optical and radio properties of these sources.

Keywords. galaxies:active, galaxies:ISM, galaxies:jet

1. Introduction

Compact Steep Spectrum (CSS) and GHz Peaked Spectrum (GPS) objects are powerful extragalactic radio sources with projected linear sizes of the order of ~20 kpc. They are considered young active galactic nuclei (AGN) with ages varying between 10^2 and 10^3 years. Also, in these sources, the jet is still crossing the Interstellar Medium (ISM) and the interaction with the ISM is stronger than in large radio sources. Such features make CSS/GPS objects very important to our understanding of the jet evolution and the interactions between the radio source and the ISM. In this work we investigate the stellar population properties of a sample of 58 CSS/GPS obtained from catalogues publicly available in the literature (Spencer *et al.* 1989; Fanti *et al.* 1990, 2001; Kunert *et al.* 2002; Hancock *et al.* 2010; Snellen *et al.* 1998, 2004; Kunert-Bajraszewska & Labiano 2010; Stanghellini *et al.* 1998; Tengstrand *et al.* 2009; Son *et al.* 2012; Jeyakumar 2016; and Peck & Taylor 2000). The main goal is to determine the impact of the radio-jet on the circumnuclear environment of these objects.

2. Sample selection and Stellar population synthesis

Initially we found 204 CSS/GPS sources with spectroscopic redshift $z < 1$. We then searched for optical counterparts of these objects in the Sloan Digital Sky Survey (SDSS) Data Release 12 and found 75 objects with signal-to-noise ratio (SNR) greater than 3 of which 58 are clearly dominated by a strong stellar-like continuum. For these, we performed a stellar population synthesis using the STARLIGHT spectral synthesis code (Cid Fernandes *et al.* 2005a,b; Asari *et al.* 2007) following the procedure described in Riffel *et al.* (2009) and Dametto *et al.* (2014). The figure 1 show the comparison between the stellar properties derived with STARLIGHT, redshift, radio luminosity and optical morphology for the sample of 58 CSS/GPS sources with good quality spectra. The red, blue, purple and green symbols represent, respectively, the values for elliptical, spiral, irregular/merger and point sources. For each graph, we show in black the linear regression model, assuming the x axis as the independent variable. Our results indicate that our

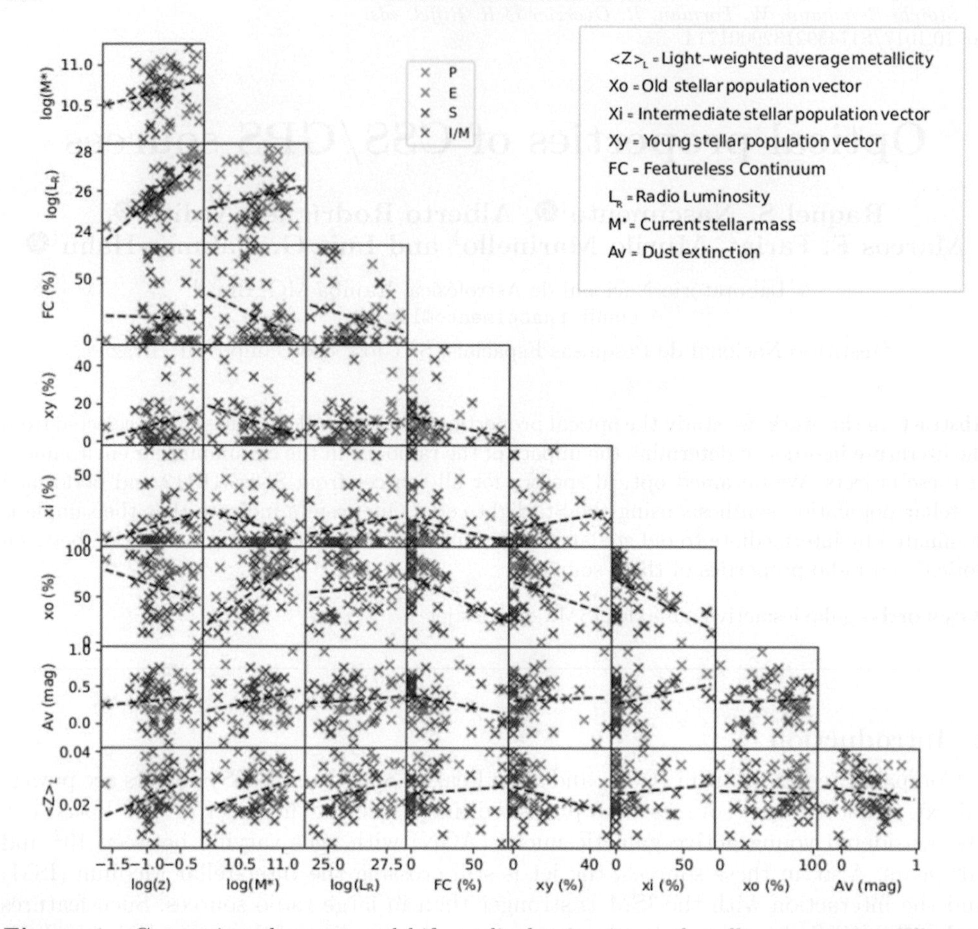

Figure 1. Comparison between redshift, radio luminosity and stellar properties. Elliptical galaxies are denoted red, spirals are blue, irregular/merger purple and point sources green. For each graph, we show in black the linear regression model, assuming the x axis variable as the independent one.

sample is dominated by intermediate to old stellar populations and there is no strong correlation between optical and radio properties of these sources.

References

Asari, N. V., Cid Fernandes, R., Stasińska, G., *et al.* 2007, *MNRAS*, 381, 263

Cid Fernandes, R., González Delgado, R. M., Storchi-Bergmann, T., *et al.* 2005a, *MNRAS*, 356, 270

Cid Fernandes, R., Mateus, A., Sodré L., *et al.* 2005b, *MNRAS*, 358, 363

Dametto, N. Z., Riffel, R., Pastoriza, M. G., *et al.* 2014, *MNRAS*, 443, 1754

Fanti, R., Fanti, C., Schilizzi, R. T., *et al.* 1990, *A&A*, 231, 333

Fanti, C., Pozzi, F., Dallacasa, D., *et al.* 2001, *A&A*, 369, 380

Jeyakumar, S. 2016, *MNRAS*, 458, 3786

Kunert, M., Marecki, A., Spencer, R. E., *et al.* 2002, *A&A*, 391, 47

Kunert-Bajraszewska, M. & Labiano A. 2010, *MNRAS*, 408, 2279

Paul, J. Hancock, P. J., Sadler, E. M., *et al.* 2010, *MNRAS*, 408, 1187

Peacock, J. A. & Wall, J. V. 1982, *MNRAS*, 198, 843

Peck, A. B. & Taylor, G. B. 2000, *ApJ*, 534, 90

Riffel, R., Pastoriza, M. G., Rodrguez-Ardila, A., *et al.* 2009, *MNRAS*, 400, 273
Snellen, I. A. G., Mack, K.-H., Schilizzi, R. T., *et al.* 2004, *MNRAS*, 348, 227
Snellen, I. A. G., Schilizzi, R. T., de Bruyn, A. G., *et al.* 1998, *A&AS*, 131, 435
Son, D., Woo, J.-H., Kim, S. C., *et al.* 2012, *ApJ*, 757, 140
Spencer, R. E., McDowell, J. C., Charlesworth, M., *et al.* 1989, *MNRAS*, 240, 657
Stanghellini, C., O'Dea, C. P., Dallacasa, D., *et al.* 1998, *A&AS*, 131, 303
Tengstrand, O., Guainazzi, M., Siemiginowska, A. *et al.* 2009, *A&A*, 501, 89

Galaxy Evolution and Feedback across Different Environments
Proceedings IAU Symposium No. 359, 2020
T. Storchi-Bergmann, W. Forman, R. Overzier & R. Riffel, eds.
doi:10.1017/S1743921320002057

The excitation mechanisms of X-ray oxygen emission-lines

Victoria Reynaldi[1], Matteo Guainazzi[2], Stefano Bianchi[3], Ileana Andruchow[1,4], Federico García[1,5], Iván López[1] and Nicolás Salerno[1]

[1]Facultad de Ciencias Astronómicas y Geofísicas - Universidad Nacional de La Plata
Paseo del Bosque s/n. La Plata, Argentina
email: vreynaldi@fcaglp.unlp.edu.ar

[2]ESA/European Space Technology and Research Centre (ESTEC)
D-SRE, Keplerlaan 1, 2200 AG, Noordwijk, The Netherlands

[3]Dipartimento di Matematica e Fisica, Università degli Studi Roma Tre
via della Vasca Navale 84, I-00146 Roma, Italy

[4]Instituto de Astrofísica de La Plata, Paseo del Bosque s/n. La Plata, Argentina

[5]Kapteyn Astronomical Institute, University of Groningen, The Netherlands

Abstract. We present the Catalogue of High REsolution Spectra of Obscured Sources (CHRESOS) from the *XMM-Newton* Science Archive. It comprises soft X-ray emission-lines from C to Si and the Fe 3C and Fe 3G L-shell transitions. Here, we concentrate on the oxygen emission-lines O VII(f) and O VIII Lyα to shed light onto the physical processes with which their formation can be related to: active galactic nucleus vs. star-forming regions. We are analysing the relationships between the oxygen lines and the luminosities of: [OIII]λ5007, [OIV]25.89μm, MIR-12μm, FIR-60μm, FIR-100μm, and hard X-rays continuum bands.

Keywords. line: formation, plasmas, galaxies: active, galaxies: Seyfert

---

1. The data and the astrophysical question

The Catalogue of High REsolution Spectra of Obscured Sources (CHRESOS) arises from RGS (Reflection Grating Spectrometer onboard *XMM-Newton*; den Herder *et al.* 2001) data of 62 nearby ($z < 0.07$), Seyfert-type active galactic nuclei (AGNs). The data were obtanined from the *XMM-Newton* Science Archive (XSA). CHRESOS gathers for the first time the soft X-ray emission-line luminosities of H-like and He-like transitions from C to Si, and the Fe 3C and Fe 3G L-shell transitions. We focus our analysis on two important oxygen emission-lines: O VII(f) (0.561 keV) and O VIII Lyα (0.654 keV). We are currently analysing them with multiwavelength (MW) nuclear data: Continuum Fluxes in 14-195 keV, 2-10 keV, Mid-Infrared (MIR)-12μm, Far Infrared (FIR)-60μm, and FIR-100μm continua, and fluxes of two other important oxygen lines: [OIII]λ5007 in the optical, and [OIV]25.89μm in the IR. Since we are probing into the formation mechanism of O VII(f) and O VIII Lyα, the MW data were chosen because of their known relationships with the two scenarios from where these emission-lines can emerge: the AGN and the (nuclear/near nuclear) star-forming regions, or starbursts (SB).

We show some preliminary results in Figure 1. The two diagrams represent the luminosity of O VII(f) and O VIII Lyα against that of [OIII]λ5007 and FIR-100μm, respectively. The former as a proxy of the AGN ionizing power in the Narrow-Line Region (NLR; e.g. Bassani *et al.* 1999; Schmitt *et al.* 2003; Heckman *et al.* 2005; LaMassa *et al.* 2010;

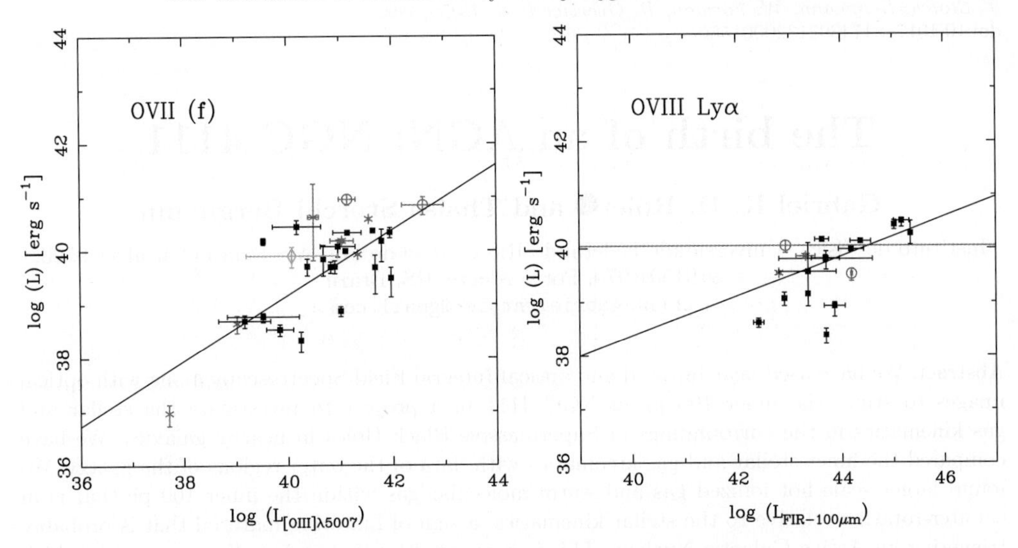

Figure 1. *Left*: diagram of O VII(f) luminosity vs. [OIII]λ5007 (NLR, AGN ionizing power). *Right*: luminosity of O VIII Lyα vs. FIR-100μm continuum (star formation indicator). Seyfert 1-1.2 sources are plotted as empty (red) circles; Seyfert 1.5-1.9 as (blue) asterisks, and Seyfert 2 as filled (black) squares; unclassified sources were drawn as empty (orange) diamonds.

Zhang & Feng 2017), and the the latter as that of star-forming regions (Rodriguez-Espinosa *et al.* 1986, 1987; Mouri & Taniguchi 1992; Hatziminaoglou *et al.* 2010). The sources are described in Fig 1. The continuous line in each diagram represents the best linear fitting (in log space). The data of [OIII]λ5007 were collected from Schmitt *et al.* (2003); Heckman *et al.* (2005) and Panessa *et al.* (2006); and that from FIR-100μm continuum was obtained from Sanders *et al.* (1989, IRAS).

So far, we have found that O VII(f) and O VIII Lyα luminosities are strongly related to [OIII]λ5007, [OIV]25.89μm, MIR-12μm, and the two primary continuum X-ray bands. However, the relationships that point to an origin from the star-forming regions are also meaningful. The two relationships shown in Fig. 1 are statistically significant (90% confidence level) in spite of the small sample size in each diagram (28 and 17 data points, respectively). We continue analysing these relationships and their statistical significance, in order to disentangle the main ionizing mechanisms taking place in the soft X-ray emitting gas.

References

Bassani, L., Dadina, M., Maiolino, R., *et al.* 1999, *ApJS*, 121, 473
den Herder, J. W., *et al.* 2001, *A&A*, 365, L7
Hatziminaoglou, E., *et al.* 2010, *A&A*, 518, L33
Heckman, T. M., Ptak, A., Hornschemeier, A., *et al.* 2005, *ApJ*, 634, 161
LaMassa, S. M., Heckman, T. M., Ptak, A., *et al.* 2010, *ApJ*, 720, 786
Mouri, H., & Taniguchi, Y. 1992, *ApJ*, 386, 68
Panessa, F., Bassani, L., Cappi, M., *et al.* 2006, *A&A*, 455, 173
Rodriguez-Espinosa, J. M., Rudy, R. J., *et al.* 1986, *ApJ*, 309, 76
Rodriguez Espinosa, J. M., Rudy, R. J., & Jones, B. 1987, *ApJ*, 312, 555
Sanders, D. B., Phinney, E. S., Neugebauer, G., *et al.* 1989, *ApJ*, 347, 29
Schmitt, H. R., Donley, J. L., Antonucci, R. R. J., *et al.* 2003, *ApJS*, 148, 327
Zhang, X. G. & Feng L. L. 2017, *MNRAS*, 468, 620

Galaxy Evolution and Feedback across Different Environments
Proceedings IAU Symposium No. 359, 2020
T. Storchi-Bergmann, W. Forman, R. Overzier & R. Riffel, eds.
doi:10.1017/S1743921320001556

The birth of an AGN: NGC 4111

Gabriel R. H. Roier and Thaisa Storchi-Bergmann

Instituto de Física, Universidade Federal do Rio Grande do Sul, Av. Bento Gonçalves 9500, 91501-970, Porto Alegre, RS, Brazil
email: gabrielrhroier@gmail.com

Abstract. We have used near-infrared and optical Integral Field Spectroscopy along with optical images to study the inner 100 pc of NGC 4111 in a project to investigate the stellar and gas kinematics in the surroundings of Supermassive Black Holes in nearby galaxies. We have compared the inner stellar and gas kinematics with data of the outer regions of the galaxy. We found larger scale hot ionized gas and warm molecular gas within the inner 100 pc that is in counter-rotation relative to the stellar kinematics, a sign of inflowing material that is probably triggering an Active Galactic Nucleus. This is supported by the nuclear X-ray emission which is heating the molecular gas and causing it to emit. The presence of large amounts of dust in a polar ring suggests that this is a fairly recent event probably due to the capture of a dwarf galaxy.

Keywords. galaxies: active, galaxies: individual: NGC 4111, galaxies: kinematics and dynamics

1. Introduction

NGC 4111 is an S0 edge-on galaxy from the Ursa Major galaxy group, at a distance of 15 Mpc, has 4 companion galaxies located within a distance of 250 kpc (Pak *et al.* 2014) and seems to have suffered a merger with a companion dwarf galaxy. Hubble Space Telescope (HST) optical images show an extended dust-rich polar ring – perpendicular to the galactic plane – which is also associated with H I filaments probably originating in tidal-stripping gas from nearby galaxies (Pak *et al.* 2014). Chandra X-ray data (González-Martín *et al.* 2009) shows that NGC 4111 has an extended soft X-ray ($0.5 - 2.0$ keV) emission along the galaxy plane due to stellar contribution and a hard X-ray ($2.0 - 10.0$ keV) emission located at the nucleus due to supermassive black hole activity. Our and previous data suggest that NGC 4111 has a fairly extincted Active Galactic Nucleus (AGN) embedded in the dusty polar ring. Further evidence for this is given by its classification as a LINER in diagnostic diagrams.

To investigate the inner 100 pc kinematics, we used the Gemini North Near-Infrared Integral Field Spectrometer (NIFS) data in the K-band, which has a Field-of-View (FOV) of $3'' \times 3''$, that corresponds to 219 pc $\times$ 219 pc at the galaxy. The near infrared spectra shows abundance of H_2 emission lines and an absence of the Brγ recombination line. To further explain the nature of these emissions, we also used HST F475W and F814W images and data from the Spectrographic Areal Unit for Research on Optical Nebulae survey (SAURON) to study the gas and stellar kinematics of the outer regions of the galaxy and see how it links to the central region small scale kinematics. We also investigate the physical properties of the gas in these small and large scales.

2. Results

Figure 1 shows a HST composite image in the visible with the NIFS FOV as a white dashed box. The right panels show the stellar (upper panels) and the H_2 gas (lower panels) kinematics from NIFS. The right panels show that the molecular gas is counter

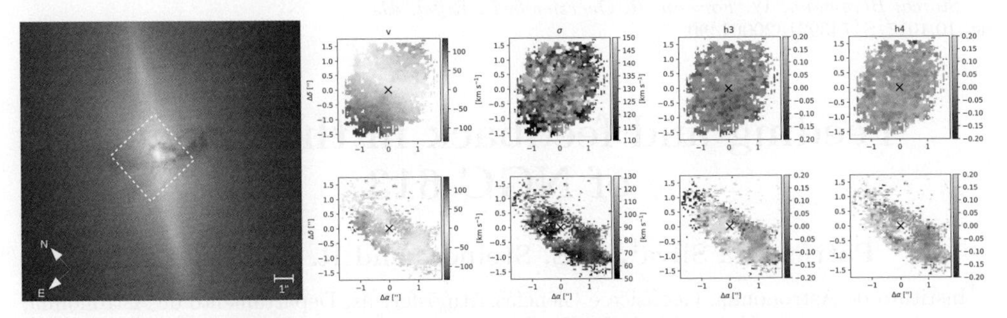

Figure 1. Left: HST composite image with the NIFS FOV shown as a white dashed box. **Right:** Stellar and Gas kinematics obtained from NIFS.

rotating with respect to the stellar component of the galaxy, and it is located in a ring-like structure not centered in the nucleus which is associated with the dusty polar ring. For further analysis, we created channel maps to investigate the 3D geometry of this molecular ring in search for inflows and/or outflows. The stellar kinematics shows a σ drop in the nuclear region, which may be due to young stellar population which hasn't entered the same velocity field of the older stars yet. SAURON data confirms the presence of a young stellar population in the central region, that may be due to inflowing gas from the dusty polar ring.

The HST F475W and F814W images show the presence of a dusty polar ring which has an extension of 458 pc and 129 pc width. SAURON data shows the outer gas kinematics follows a rotation-like pattern along the polar ring, but the subtraction of the stellar velocity field from that of the gas shows a spiral-like pattern in the gas kinematics with the central portion similar to the NIFS gas kinematics, and comparison with the channel maps show consistent signs of inflowing material.

Therefore, we conclude that the polar ring structure is inflowing towards the nucleus and then forming new stars on the way in and fueling the central black hole, also causing the hard X-ray emission seen at the nucleus. However, as the dust polar ring is still present, it suggests that the nuclear activity is fairly recent, as there wasn't enough time to heat the medium enough to sublimate the dust. The measured temperature of the H_2 gas is 2704K (following Wilman, Edge & Johnstone 2005) and the molecular emission is caused by thermal processes, as shown by the H_2 $\lambda 2.2477$ μm/H_2 $\lambda 2.1218$ μm line ratio (Storchi-Bergmann *et al.* 2009). We believe it is been heated by the embedded X-ray emission, which is consistent with the absence of the Brγ line in the spectrum, as it isn't caused by stellar ionization.

References

González-Martín, O., Masegosa, J., Márquez, I., *et al.*2009, *A&A*, 506, 1107
Pak, M., Rey, S.-C., Lisker, T., *et al.* 2014, *MNRAS*, 445, 630
Storchi-Bergmann, T., McGregor, P. J., Riffel, R. A., *et al.*2009, *MNRAS*, 394, 1148
Wilman, R. J., Edge, A. C., & Johnstone, R. M. 2005, *MNRAS*, 359, 755

Galaxy Evolution and Feedback across Different Environments
Proceedings IAU Symposium No. 359, 2020
T. Storchi-Bergmann, W. Forman, R. Overzier & R. Riffel, eds.
doi:10.1017/S1743921320001490

Feeding and feedback in the nucleus of NGC 613

Patrícia da Silva[1], J. E. Steiner[1] and R. B. Menezes[2]

[1]Instituto de Astronomia, Geofísica e Ciências Atmosféricas, Departamento de Astronomia, Universidade de São Paulo, 05508-090, SP, Brazil
emails: p.silva2201@gmail.com, joao.steiner@iag.usp.br

[2]Instituto Mauá de Tecnologia, Praça Mauá 1, 09580-900, São Caetano do Sul, SP, Brazil
email: roberto.menezes@maua.br

Abstract. Active Galactic Nuclei (AGN) are objects in which a supermassive black hole is fed by gas and, as this generates energy, can ionise the environment and interact with it by jets and winds. This work is focused on the processes of feeding and feedback in the nucleus of NGC 613. This object is a case in which both phenomena can be studied in some detail. The kinematics and morphology of the molecular gas trace the feeding process while the ionization cone, seen in [OIII]λ5007 and soft X-rays, as well as the radio jet and wind/outflows are associated with feedback processes. In addition, we see 10 HII regions, associated with nuclear and circumnuclear young stellar populations, dominant in the optical, that makes the analysis complicated, though more interesting. For all these phenomena, NGC 613 nucleus is a vibrant example of the interplay between the AGN and the host galaxy.

Keywords. galaxies: active, galaxies: individual: NGC 613, galaxies: nuclei

---

1. Introduction

NGC 613 is an SB(rs)bc galaxy located at 26 ± 5 Mpc (Nanosova *et al.* 2011). It contains a very rich nucleus, composed by a circumnuclear ring of star formation (Hummel & Jorsater 1992; Böker *et al.* 2008; Fálcon-Barroso *et al.* 2014), a nuclear spiral of molecular gas (Audibert *et al.* 2019) and an obscured Active Galactic Nucleus (AGN, Castangia *et al.* 2013; Asmus *et al.* 2015) with a Low Ionization Emission-Line Region (LINER) emission (Audibert *et al.* 2019).

For more details about this work, see da da Silva *et al.* (2020a).

2. Metodology

The observations were taken with the Integral Field Unit (IFU) of Gemini Multi-Object Spectrograph (GMOS) of Gemini South Telescope and with the SOAR Integral Field Spectrograph (SIFS) from the SOAR telescope. We also analysed the data of public archives from the Chandra Space Telescope, the Hubble Space Telescope (*HST*) and the Atacama Large Millimeter/Submillimeter Array (ALMA) and SINFONI from the Very Large Telescope (VLT).

The treatment of SIFS and GMOS data cubes was performed using scripts developed by our group and its reduction and observational information is described in da Silva *et al.* (2020a). For more details on treatment processes, see Menezes *et al.* (2014, 2015) and Menezes *et al.* (2019).

3. Results

The difference of F475W–F814W filters from *HST* (*B-I* in magnitude scale) shows the inner morphology of the bar of NGC 613 (Fig. 1). The streams of gas and dust

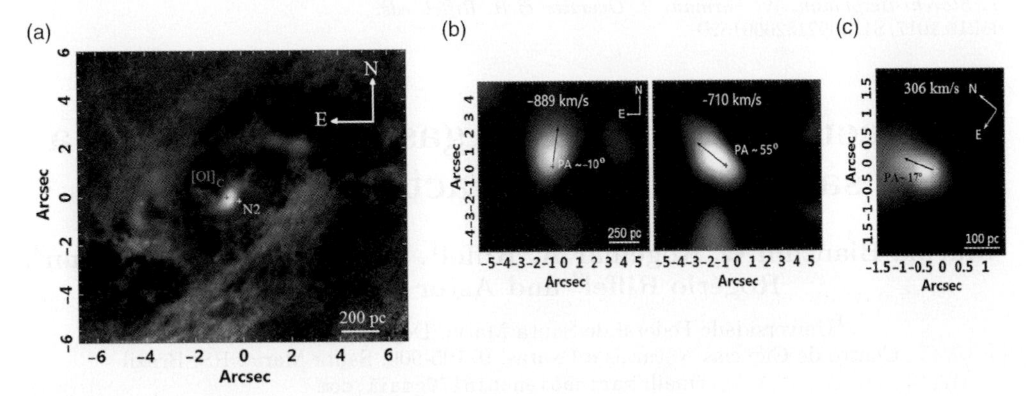

Figure 1. (a) RB composition of F475W–F814W filters from *HST* (*B-I* in magnitude scale) in red and CO(3-2) from the ALMA data cube in green. The purple and cyan crosses represent the position of the centroid of the [OI]λ6300 emission line and the center of N2 (referential point for the superposition of the data –see da Silva *et al.* 2020a for more details), respectively, and its size the uncertainty of 3σ, considering the size of the pixel of the HST images. Outflows of [OIII]λ5007 (b) and Hα (c) observed in the GMOS data cube. The cross reprents the position of the AGN.

are connected to the circumnuclear ring in two opposite points. This is compatible with the findings of Böker *et al.* (2008), that suggest that the ring is being fed by the bar through these two points. By its turn, the nuclear spiral is also connected to the bar in two nearby points. The kinematics of the nuclear spiral (da Silva *et al.* 2020b) reveals that this structure is directly injecting gas and dust to the center.

The AGN, which is located in the center of the nuclear spiral (see da Silva *et al.* 2020a), is highly obscured and the nuclear spiral is responsible for this obscuration, since the optical image of the nucleus is composed by two sources (which we called N1 and N2, see Fig. 9 of da Silva *et al.* 2020a) separated by the nuclear spiral. The Chandra data present an extended soft X-ray emission, which delineates the ionization cone observed in [Oiii]λ5007 (see Fig. 15 of da Silva *et al.* 2020a). Two outflows were detected in [Oiii]λ5007, one of them in blueshift (v~ -710 km s^{-1}) with a PA compatible with the one of the ionization cone axis. The Hα emission also presented an outflow with redshift emission (v~ 306 km s^{-1}) with PA compatible with the one of the radio jet observed by Hummel & Jorsater (1992).

The AGN interacts with the environment by outflows of gas and the ionization cone, and the galaxy interacts with it by feeding it through the nuclear spiral that is connected to its bar.

References

Asmus, D., Gandhi, P., Honig, S. F., *et al.* 2015, *MNRAS*, 454, 766
Audibert, A., Combes, F., García-Burillo, S. 2019, *A&A*, 632, 33
Böker, T., Fálcon-Barroso, J., Schinnerer, E., *et al.* 2008, *AJ*, 135, 479
Castangia, P., Panessa, F., Henkel, C., *et al.* 2013, *MNRAS*, 436, 3388
da Silva, P., Menezes, R. B., Steiner, J. E. 2020a, *MNRAS*, 492, 5121
da Silva, P., Menezes, R. B., Steiner, J. E. 2020b, *MNRAS*, 496, 943
Fálcon-Barroso, J., Ramos Almeida, C., Boker, T., *et al.* 2014, *MNRAS*, 438, 329
Hummel, E. & Jorsater S. 1992, *A&A*, 261, 85
Menezes, R. B., Steiner, J. E., Ricci, T. V., *et al.* 2014, *MNRAS*, 438, 2597
Menezes, R. B., da Silva, P., Ricci, T. V. *et al.* 2015, *MNRAS*, 450, 369
Menezes, R. B., Ricci, T. V., Steiner, J. E. *et al.* 2019, *MNRAS*, 483, 3700
Nanosova, O. G., de Freitas Pacheco, J. A., Karachentsev, I. D., *et al.* 2011, *A&A*, 532, A104

Galaxy Evolution and Feedback across Different Environments
Proceedings IAU Symposium No. 359, 2020
T. Storchi-Bergmann, W. Forman, R. Overzier & R. Riffel, eds.
doi:10.1017/S1743921320001520

Molecular and ionised gas kinematics in a sample of nearby active galaxies

Marina Bianchin[1], Rogemar A. Riffel[1], Thaisa Storchi-Bergmann[2], Rogério Riffel[2] and Astor J. Schonell[3]

[1]Universidade Federal de Santa Maria, Departamento de Física, Centro de Ciências Naturais e Exatas, 97105-900, Santa Maria, RS, Brazil
email: marinabianchin17@gmail.com

[2]Instituto de Física, Universidade Federal do Rio Grande do Sul, Av. Bento Gonççalves 9500, 91501-970 Porto Alegre, RS, Brazil

[3]Instituto Federal de Educação, Ciência e Tecnologia Farroupilha, BR287, km 360, Estrada do Chapadão, 97760-000 Jaguari, RS, Brazil

Abstract. We used Gemini NIFS integral field spectroscopy to analyse the molecular and ionised gas kinematics of six nearby ($z \leqslant 0.015$) Seyfert galaxies with a spatial coverage of 0.1–0.6 kpc^2. By fitting the emission-line profiles using multiple Gaussian components we determined that the ionised and hot molecular gas kinematics are dominated by gas outflows and rotation, respectively, even though three objects also present molecular outflows.

Keywords. galaxies: Seyfert, galaxies: kinematics and dynamics, galaxies: nuclei

1. Introduction

Feeding and feedback processes in active galaxies are one of the keys to constrain galaxy evolution models. Only in the local Universe it is possible to achieve spatial resolutions of tens of parsecs allowing us to resolve the gas kinematics in the vicinity of supermassive black holes. The gas distribution and kinematics in the inner kpc of the host galaxies of Active Galactic Nuclei (AGN) can provide insightful information about the physical processes involved in AGN feeding and feedback. The feeding, that occurs through gas accretion, is the source of the AGN power, while the feedback is required to constrain galaxy evolution models. Understanding these physical processes in high spatial and spectral resolution is important to such theoretical studies.

In this work we studied the gas kinematics of six nearby Seyfert galaxies observed with integral field spectroscopy in the near-infrared. These galaxies have been previously studied by Schönell *et al.* (2019), who mapped the gas and mass distribution concluding that the molecular and ionised gas present distinct flux distributions and kinematics. They estimated the mass of the former to be 10^{3-4} times smaller than that of the latter.

2. The data and measurements

Our sample is composed of six nearby Seyfert galaxies that are part of the AGNIFS sample (Riffel *et al.* 2018). This set of galaxies was built according to the following criteria: (i) $L_X \geqslant 10^{41.5} L_\odot \mathrm{erg\,s^{-1}}$ in the 105 month 14-195 keV Swift catalogue; (ii) $z \leqslant 0.015$; (iii) $-30^\circ < \delta < 73^\circ$ and (iv) extended [O III] emission. The total sample is composed of 24 galaxies, but here we studied: NGC 5899, Mrk 607, NGC 788, NGC 3227, NGC 3516 and NGC 5506 Schönell *et al.* (2019).

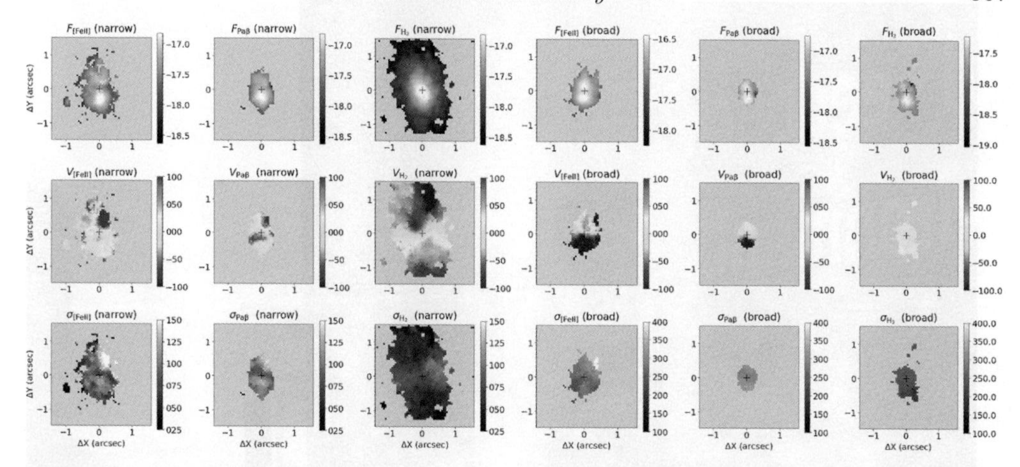

Figure 1. 2D maps for NGC 5899. Top: the flux (F) distribution for the narrow and broad components of the [FeII], Paβ and $H_2\lambda2.12\mu$m. Middle and bottom: velocity (V) and velocity dispersion (σ) in the same order as the top panels. The H_2 narrow component is dominated by emission of gas in the disk, while the ionised gas emission is possibly dominated by outflows.

We used Gemini NIFS (Near-intrared Integral Field Spectrograph) to map the gas and stellar kinematics in the inner kpc of the galaxies of our sample, at a velocity resolution of $\sim$25 km s^{-1} and a spatial resolution of 12–35 pc at the distance of the galaxies. Details about the data reduction can be found in Schönell *et al.* (2019).

In order to measure the fluxes, velocities and velocity dispersions (σ) of the J and K band emission lines, we used the IFSCUBE code† to fit each line profile with up to three Gaussian functions. The broad components ($\sigma \gtrsim$150 km s^{-1}) are usually associated with gas outflows and the narrow components ($\sigma \lesssim$150 km s^{-1}) with the gas rotating in a disk, the details will be analysed for each individual galaxy.

3. Results

We constructed flux, velocity and velocity dispersion maps for [Fe II]λ1.2570 μm, Paβ and $H_2\lambda2.1218\,\mu$m for all galaxies of our sample. As an example, Figure 1 shows the corresponding maps for NGC 5899. To represent the line profiles observed in this galaxy two-Gaussian components are required. A clear rotation pattern is observed for the narrow component of $H_2\lambda2.1218\,\mu$m, while for the ionised gas, the narrow component seems to be tracing a more disturbed kinematics, besides the emission from the disk. The broad components are interpreted as being due to emission within a bicone, with the near side of the cone seen to the south of the nucleus and its far side to the north of it. The broad component for the molecular gas seems to be tracing the interaction of the outflows with the gas of the disk of the galaxy, as indicated by the low velocities.

All galaxies of our sample present outflows in ionised gas, and at least three of them also show molecular outflows. Although for some galaxies (such as NGC 5899 and NGC 5506) the outflows seem to be observed within conical structures, the geometry of the outflows must be further constrained by modeling the gas kinematics. This will be presented in Bianchin *et al.* (in preparation).

References

Riffel, R. A., *et al.* 2018, *MNRAS*, 474, 1373

Schönell, A. J., *et al.* 2019, *MNRAS*, 485, 2054

† https://github.com/danielrd6/ifscube

Francoise Combes

Session 6: The present-day Universe: Spatially resolved studies of stellar and gas content, excitation and metallicity

Galaxy Evolution and Feedback across Different Environments
Proceedings IAU Symposium No. 359, 2020
T. Storchi-Bergmann, W. Forman, R. Overzier & R. Riffel, eds.
doi:10.1017/S1743921320002124

INVITED LECTURES

The importance of the diffuse ionized gas for interpreting galaxy spectra

Natalia Vale Asari[1,2]† and **Grażyna Stasińska**[3]

[1]Departamento de Física - CFM - Universidade Federal de Santa Catarina,
Florianópolis, SC, Brazil
email: natalia@astro.ufsc.br

[2]School of Physics and Astronomy, University of St Andrews, North Haugh,
St Andrews KY16 9SS, UK

[3]LUTH, Observatoire de Paris, PSL, CNRS 92190 Meudon, France

Abstract. Diffuse ionized gas (DIG) in galaxies can be found in early-type galaxies, in bulges of late-type galaxies, in the interarm regions of galaxy disks, and outside the plane of such disks. The emission-line spectrum of the DIG can be confused with that of a weakly active galactic nucleus. It can also bias the inference of chemical abundances and star formation rates in star forming galaxies. We discuss how one can detect and feasibly correct for the DIG contribution in galaxy spectra.

Keywords. ISM: abundances, galaxies: abundances, galaxies: ISM

---

1. Introduction

A lot can be learned from studying the integrated spectra of galaxies. The power is in the numbers: a large statistical sample tells us about trends in astrophysics, but also about dispersions in those trends. Empirical relations thus constructed can provide useful guidance for chemical evolution models.

For the sake of the argument let us focus on some empirical laws using the Sloan Digital Sky Survey (SDSS, York *et al.* 2000) data. One is the stellar mass–nebular metallicity relation (Tremonti *et al.* 2004), which informs us about the history of chemical enrichment of galaxies. The metallicity, Z, depends not only on the yields and on star-formation histories, but also on the inflow and outflow of gas with chemical compositions different from that of the galaxy.

Another important relation is the stellar mass–star formation rate ($M_\star$–SFR) relation (Brinchmann *et al.* 2004). It shows that larger galaxies are also forming more stars. This relation has been later wrongly extrapolated, Hα being carelessly transformed into SFR to reveal a 'quiescent sequence'. This quiescent sequence is nothing more than a sequence of retired galaxies and has nothing to do with star formation (see Section 3 below).

One of the most popular empirical relations nowadays is the $M_\star$–Z–SFR relation (e.g. Mannucci *et al.* 2010). In the representation by Mannucci *et al.* galaxies in different mass bins show different trends in the Z versus SFR plane. Low-mass bins show an anticorrelation of Z with SFR, whereas high-mass bins show no trend at all.

As a matter of fact, whereas the concept of a galaxy's total stellar mass and global star formation rate make sense and are intrinsically related to the galaxy as a whole,

† Royal Society–Newton Advanced Fellowship

the 'metallicity' is more a fraught term, because the methods to measure the galaxy's 'metallicity' have actually been developed for (giant) H II regions. This ignores the fact that the line-emission regions in a galaxy comprise compact H II regions, giant H II regions of diverse morphologies, and diffuse ionized regions.

Several biases may permeate the results in the above and similar papers. As a whole, one needs to care about sample selection and aperture effects. For the SFR, one also needs to deal with dust correction (see Vale Asari *et al.* in prep.), with the calibration used for the SFR, and with the contamination by the diffuse ionized gas (DIG). For the determination of the metallicity, it is well-known that the method, indices and calibration used may change the results dramatically (see e.g. Maiolino & Mannucci 2019; Kewley *et al.* 2019). So far, the influence of the DIG has not been studied in detail (except by a few like Kumari *et al.* 2019, Poetrodjojo *et al.* 2019, and Vale Asari *et al.* 2019, hereafter VA19) but it could be important.

There is another domain of galaxy research where the DIG is relevant: this is the field of active galactic nuclei (AGN). Weak line emission in the integrated spectra of galaxies has been traditionally interpreted as due to low-level activity linked to accretion onto a supermassive black hole (Kauffmann *et al.* 2003; Kewley *et al.* 2006). However, it has been shown that, in galaxies which have stopped forming stars, dubbed 'retired' galaxies, ionization by hot low-mass evolved stars (HOLMES) is able to explain both the observed emission line-ratios and their luminosities (Stasińska *et al.* 2008; Cid Fernandes *et al.* 2011).

2. A condensed history of the DIG

The DIG was first discovered as a faint extraplanar emission in the Milky Way (Reynolds 1971, 1989) and in edge-on galaxies (Dettmar 1990; Hoopes *et al.* 1996, 1999). In the context of our Galaxy it is often referred to as the warm ionized medium or diffuse ionized medium. It has later been found in interarm regions, where the H II regions do not outshine it, or the density of the gas is smaller (Walterbos & Braun 1994; Wang *et al.* 1999; Zurita *et al.* 2000). Some studies find that 30 to 60 per cent of the total Hα in emission in a spiral galaxy may be due to the DIG (e.g. Oey *et al.* 2007). Warm ionized gas has also been detected in early-type galaxies (Phillips *et al.* 1986; Martel *et al.* 2004; Jaffé *et al.* 2014; Johansson *et al.* 2016).

The recent development of integral field spectroscopy (IFS) boosted studies of resolved properties of nearby galaxies in which the properties of bona fide H II regions and DIG can be separated (e.g. Blanc *et al.* 2009; Kaplan *et al.* 2016; Kreckel *et al.* 2016; Poetrodjojo *et al.* 2019).

Already several decades ago it was found that the DIG has a lower electron density, higher electron temperature, and enhanced collisionally-excited to recombination emission line ratios ([N II]λ6584/Hα, [S II]λ6716 Hα, also usually [O III]λ5007/Hβ) as compared to H II regions (Galarza *et al.* 1998). This suggests that the DIG is ionized by a mechanism other than photoionization by OB stars. Propositions include cosmic rays (Reynolds & Cox 1992; Vandenbroucke *et al.* 2018), photoionization by old supernova remnants (Slavin *et al.* 2000), dissipation of turbulence (Minter & Spangler 1997; Minter & Balser 1997; Binette *et al.* 2009), contribution of dust-scattered light (Wood & Reynolds 1999), shocks from supernova winds (Collins & Rand 2001), ionization by photons leaking from star-forming (SF) regions (Domgorgen & Mathis 1994; Haffner *et al.* 2009; Weilbacher *et al.* 2018), and photoionization by HOLMES (Binette *et al.* 1994; Stasińska *et al.* 2008; Athey & Bregman 2009; Flores-Fajardo *et al.* 2011; Yan & Blanton 2013).

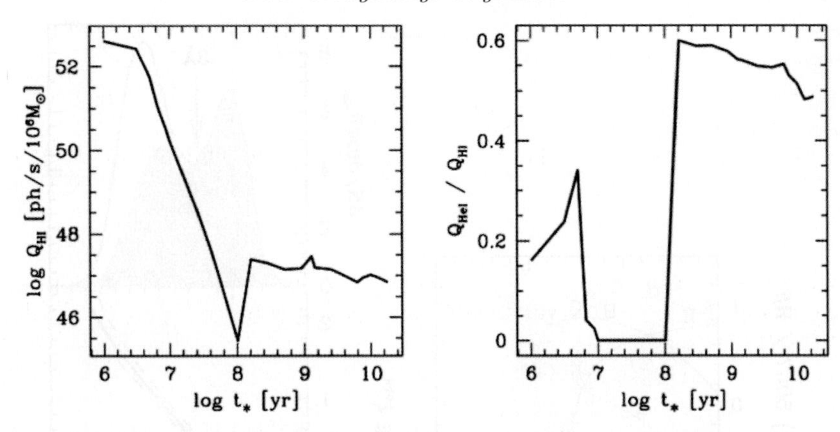

Figure 1. *Left:* The rate $Q_{\rm H}$ of photons capable of ionizing a hydrogen atom for a simple stellar population from Bruzual & Charlot (2003). *Right:* The hardness $Q_{\rm He}/Q_{\rm H}$ of the ionizing radiation field as a function of age for the same SSP.

3. The DIG in early-type galaxies and in bulges

Emission-line ratios can serve to distinguish the main ionization mechanism in galaxies. The most famous diagram, [N II]λ6584/Hα versus [O III]λ5007/Hβ (Baldwin *et al.* 1981; BPT) drew an empirical line to separate giant H II regions from planetary nebulae and objects ionized by a power-law spectrum or excited by shocks.

With the advent of the SDSS and its wealth of spectroscopic data, the separation of the BPT plane in several zones became much clearer (Kauffmann *et al.* 2003; Kewley *et al.* 2006). It has been commonly said that SF galaxies lie in the same region as giant H II regions, and objects on the right-hand side of the diagram are AGNs, subdivided into Seyfert and LINERs. Note that the acronym LINER stands for low-ionization *nuclear* emission regions (Heckman 1980), and a priori does not apply to SDSS spectral observations, which where made through 3-arcsec fibers and covered a significant portion of the galaxies – except for the nearest ones. So the part of the diagram where SDSS galaxies with LINER-like spectra lie cannot all be populated by *bona fide* LINERs.

Stasińska *et al.* (2008) proposed that HOLMES† could be responsible for the observed LINER-like emission-line ratios. Fig. 1 shows the reasoning behind it. The panel on the left shows the rate $Q_{\rm H}$ of photons capable of ionizing a hydrogen atom as a function of time, for a simple stellar population (SSP) from Bruzual & Charlot (2003, BC03). Here a Chabrier (2003) initial-mass function (IMF) and solar metallicity are used. $Q_{\rm H}$ is seen to fall by 5 orders of magnitude between 10 Myr and 100 Myr. However, there is a continuous production of ionizing photons for ages larger than 100 Myr. These arise from stars that have evolved off the asymptotic giant branch, have become very hot (some of them may reach 200,000 K) and are on the way of becoming degenerate stars of freshly-formed white dwarfs. Such stars are faint in comparison to OB stars, but their huge numbers from the IMF make up for their faintness.

More importantly, the right panel of Fig. 1 shows the hardness of the ionizing radiation field as a function of age for the same SSP, where $Q_{\rm He}$ is the rate of photons with energies >24.6 eV. The ionizing photons from HOLMES are on average more energetic than those from young stars, which means that on average the electron kinetic energy will be greater

† Stasińska *et al.* (2008) did not use the HOLMES terminology at that time, but the expression 'post-AGB stars', which has a double meaning in astrophysics.

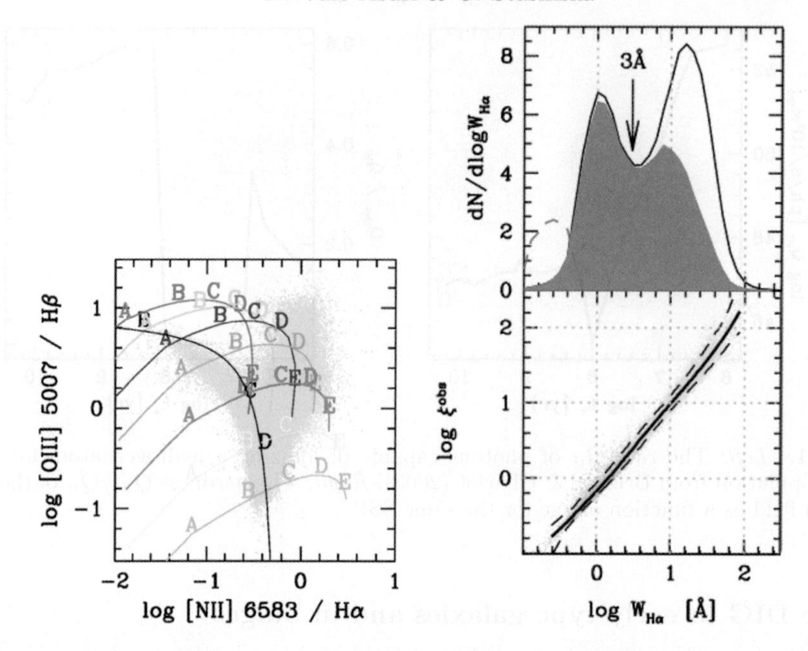

Figure 2. *Left*: Sequences of photoionization models (of given metallicity and varying ionization parameter) where the ionization source are HOLMES in the BPT plane plotted over observed SDSS galaxies (Figure from Stasińska *et al.* 2008). *Right*: Top panel shows the $W_{\mathrm{H}\alpha}$ histogram for SDSS galaxies. The bottom panel shows ξ (the ratio of the total observed Hα luminosity to the expected Hα luminosity due to ionization by HOLMES) versus $W_{\mathrm{H}\alpha}$ for the same galaxies, where the solid (dashed) line is the median (10 and 90 percentile) relation. The 3 Å arrow delimits the separation between galaxies solely explained by ionization by HOLMES from galaxies where an extra ionization source is needed. (Figure adapted from Cid Fernandes *et al.* 2011)

in a gas ionized by HOLMES than by OB stars. This implies that the collisionally-excited lines emitted by this gas will be stronger with respect to recombination lines than in the case of H II regions.

The first *ab initio* models for retired galaxies ionized by HOLMES were made by Stasińska *et al.* (2008). Fig. 2 (left) shows an example of line ratios calculated from photoionization models using as an ionization source the spectrum obtained by the spectral synthesis code STARLIGHT (Cid Fernandes *et al.* 2005) for galaxies in the LINER-like region of the BPT. The stellar populations were modeled in the optical region to reproduce the SDSS spectra, and the ionizing part of the spectrum was extrapolated from the BC03 stellar population models. The photoionization models show that radiation from HOLMES can explain the whole BPT plane, except for the rightmost tip where Seyfert galaxies live.

Fig. 2 (right), from Cid Fernandes *et al.* (2011), shows that not only line ratios for galaxies with LINER-like spectra can be explained only by HOLMES, but also the budget of their ionizing photons. The parameter ξ is the total observed Hα luminosity divided by the expected Hα luminosity assuming that all the photons from HOLMES (>100 Myr) are absorbed by the gas. This parameter ξ is actually very well correlated with the Hα equivalent width ($W_{\mathrm{H}\alpha}$). The sample shown in this figure is a mixture of all SDSS galaxies. The bimodality shown in the $W_{\mathrm{H}\alpha}$ distribution separates galaxies where HOLMES can be the sole ionization mechanism (below 3 Å), and other galaxies where extra sources are needed to explain the enhanced Hα luminosity (SF, AGN, etc.).

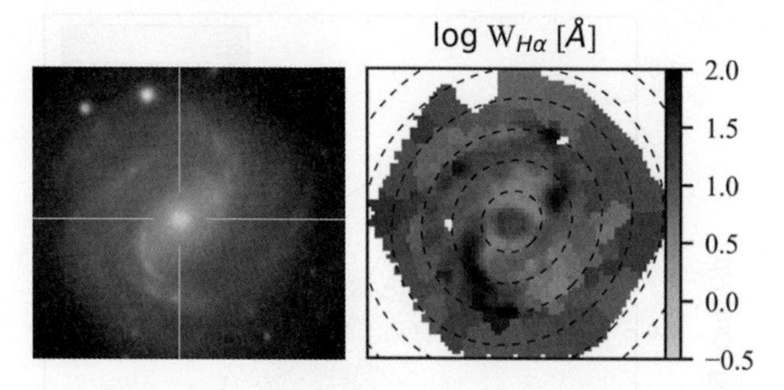

Figure 3. SDSS image and $W_{\mathrm{H}\alpha}$ map for the CALIFA galaxy 0073. (Figure adapted from Lacerda *et al.* 2018.)

4. The DIG in face-on late-type galaxies

Many of the problems found in the SDSS studies have come about because we are unable to separate the ionizing sources in a galaxy. Several studies based on IFS (e.g. Sarzi *et al.* 2010; Belfiore *et al.* 2016; Gomes *et al.* 2016) have found evidence of real LINERS in the nuclei of some galaxies, and of LINER-like emission *outside* the nuclei of late-type galaxies (dubbed 'LIER', where the *N* for *nuclear* has been dropped). Recent IFS studies of SF galaxies have found that the DIG has kinematic properties which are different from those of H II regions, having been found to come from a thicker layer (den Brok *et al.* 2020) and to sustain more turbulence (Della Bruna *et al.* 2020) In the following we show how DIG biases metallicity measurements in SF galaxies based on emission line ratios, and present ways to mitigate those biases using IFS data.

4.1. *Digging out the DIG with integral field spectroscopy*

There are several ways to identify regions where the DIG emission is important. One of them is based on line ratios, such as [S II]λ6716/Hα (e.g. Kreckel *et al.* 2016). However, if we would ultimately like to quantify the contribution of the DIG to line ratios, tagging DIG regions using this criterion would make our analysis circular.

A second method uses the Hα surface brightness (e.g. Zhang *et al.* 2017). This criterion may break down due to a simple geometrical effect, e.g. in the bulge of late-type galaxies the column density of the gas in the line-of-sight is larger, which makes Hα brighter. Therefore, diffuse emission from bulges would be missed with a simple Hα surface brightness cut.

A third criterion, based on $W_{\mathrm{H}\alpha}$, is the one we favour. Fig. 3 shows a SDSS colour-image and a $W_{\mathrm{H}\alpha}$ map from the Calar Alto Legacy Integral Field Area (CALIFA, Sánchez *et al.* 2016) survey for the same galaxy; one may note that high $W_{\mathrm{H}\alpha}$ regions trace the spiral arms quite well. Lacerda *et al.* (2018) have thus proposed a classification built upon of the bimodality of the $W_{\mathrm{H}\alpha}$ distribution, which is found for both SDSS spectra and CALIFA spaxels. CALIFA spaxels have been sorted into three classes: (1) HOLMES DIG (hDIG) where $W_{\mathrm{H}\alpha} < 3$ Å, (2) mixed DIG (mDIG) where $3 < W_{\mathrm{H}\alpha} < 14$ Å, (3) SF complexes (SFc) where $W_{\mathrm{H}\alpha} > 14$ Å.

Spaxel sizes in CALIFA are ~1-kpc wide, so hDIG and mDIG regions may still contain buried-in H II regions. SFc spaxels, on the other hand, are not classical H II regions (which usually have $W_{\mathrm{H}\alpha} \sim 100$–1000 Å), but may encompass many H II regions and some DIG.

Fig. 4 from Lacerda *et al.* (2018) shows only spaxels classified as mDIG for CALIFA galaxies on the BPT plane, colour-coded by $W_{\mathrm{H}\alpha}$. Smaller $W_{\mathrm{H}\alpha}$ values (i.e. where larger DIG contribution is expected) correspond to larger [N II] /Hα and [O III]/Hβ line ratios.

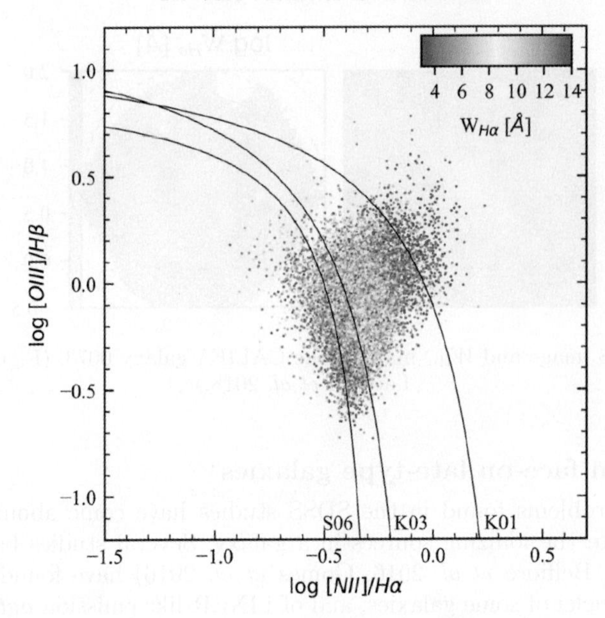

Figure 4. BPT diagram for CALIFA regions with $W_{\mathrm{H}\alpha}$ in the 3−14 Å range, coloured according to $W_{\mathrm{H}\alpha}$, and excluding zones inwards of one half light radius. (Figure from Lacerda *et al.* 2018.)

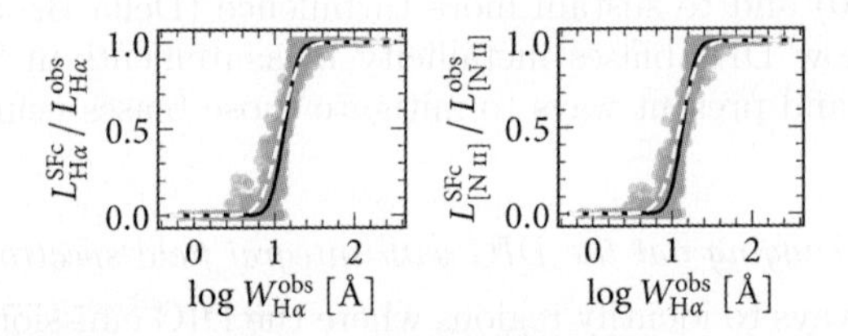

Figure 5. Correction for DIG emission appropriate for SDSS observations, based on 1409 MaNGA star-forming galaxies. Panels show the Hα and [N II]λ6584 line luminosities in SFc spaxels normalised by the total luminosity versus total $W_{\mathrm{H}\alpha}$, where all measurements were taken within a circular $0.7R_{50}$-diameter aperture. The solid line shows a fit to the data; the dashed line shows a fit for measurements made in $2.0R_{50}$-diameter apertures. (Figure adapted from VA19.)

This means that spaxels with more DIG emission do have systematically different emission line ratios, which must bias studies using those ratios as proxies for gas-phase metallicity.

4.2. *A method to remove the DIG contribution in the integrated spectrum of a galaxy*

To be able to extract the contribution of *bona fide* H II regions from an observed emission-line spectrum, one needs an empirical method. VA19 developed a method using Mapping Nearby Galaxies at APO (MaNGA, Blanton *et al.* 2017) IFS data. Similarly to CALIFA these data have ∼1 kpc resolution and the contamination of the DIG to SFc spaxels still holds true, the advantage of using MaNGA being essentially a larger sample of galaxies.

Fig. 5 shows the correction proposed for [N II]λ6584 and Hα emission lines (other lines are in VA19). The abscissa for both panels is the global observed $W_{\mathrm{H}\alpha}$, i.e. measured

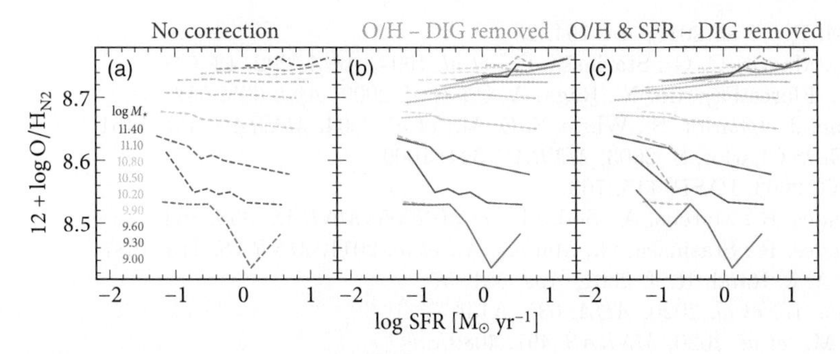

Figure 6. O/H as a function of SFR for ∼10, 000 SDSS star-forming galaxies in stellar mass bins (whose centres are given in the left handside panel). O/H has been calculated using the [N II]λ6584/Hα index. (a) M–Z–SFR relation with no correction. This is repeated (as dashed translucent lines) in the two other panels for comparison. (b) Correcting [N II]λ6584 and Hα for the DIG contamination (using the fits shown in Fig. 5) prior to calculating O/H. (c) Correcting Hα prior to obtaining the SFR as well. (Figure adapted from VA19.)

in circular $0.7R_{50}$-diameter apertures. The ordinate shows the ratio $L_{\rm SFc}/L_{\rm obs}$, where $L_{\rm obs}$ is the total luminosity in a line, and $L_{\rm SFc}$ is the line luminosity adding up only spaxels tagged as SFc (i.e. removing hDIG and mDIG spaxels). The left panel concerns the Hα line and the right one [N II] . This ratio increases from zero – where there is no contribution from SFc to the total spectra – to one – the whole emission is in SFc spaxels. Crucially, these curves are slightly different for each emission line.

Fig. 6 shows the M–Z–SFR relation for SDSS galaxies, where the oxygen abundance is calculated using the [N II]/Hα line ratio. Panel (a) shows the uncorrected relation, which is repeated in translucent dashed lines in the other panels. Panel (b) overplots, in solid lines, the relation where O/H has been recalculated by removing the contribution from the DIG to [N II] and Hα using the fit from Fig. 5. High-mass bins are the most affected, now featuring a correlation which was absent in panel (a). Panel (c) shows the effect of also removing the DIG contribution to Hα prior to computing the SFR.

Changes to the M–Z–SFR are small because, as mentioned before, for MaNGA observations even SFc spaxels still contain non-negligible contribution from the DIG. A similar approach using a large sample of data obtained with the Multi Unit Spectroscopic Explorer (MUSE, Bacon *et al.* 2010) should give a larger difference and allow a more reliable correction. Note that this method is purely empirical and does not rely on any assumption regarding the source of ionization of the DIG.

Acknowledgements

NVA acknowledges support of FAPESC and CNPq, and of the Royal Society–Newton Advanced Fellowship award (NAF\R1\180403). GS acknowledges a CNPq visiting professor grant.

References

Athey, A. E. & Bregman, J. N., 2009, *ApJ*, 696, 681
Bacon, R., *et al.* 2010, SPIE, 7735, 773508, SPIE.7735
Baldwin, J. A., Phillips, M. M., & Terlevich, R. 1981, *PASP*, 93, 5
Belfiore, F., *et al.* 2016, *MNRAS*, 461, 3111
Blanc, G. A., Heiderman, A., Ge bhardt, K., *et al.* 2009, *ApJ*, 704, 842

Blanton, M. R., *et al.* 2017, *AJ*, 154, 28
Binette, L., Magris, C. G., Stasińska, G., *et al.* 1994, *A&A*, 292, 13
Binette, L., Flores-Fajardo, N., Raga, A. C., *et al.* 2009, *ApJ*, 695, 552
Brinchmann, J., Charlot, S., White, S. D. M., *et al.* 2004, *MNRAS*, 351, 1151
Bruzual, G. & Charlot, S. 2003, *MNRAS*, 344, 1000
Chabrier, G. 2003, PASP, 115, 763
Cid Fernandes, R., Mateus, A., Sodré L., *et al.* 2005, *MNRAS*, 358, 363
Cid Fernandes, R., Stasińska, G., Mateus, A., *et al.* 2011, *MNRAS*, 413, 1687
Collins, J. A. & Rand, R. J. 2001, *ApJ*, 551, 57
Della Bruna, L., *et al.* 2020, *A&A*, 635, A134
den Brok, M., *et al.* 2020, *MNRAS*, 491, 4089
Dettmar, R.-J. 1990, *A&A*, 232, L15
Domgorgen, H. & Mathis, J. S. 1994, *ApJ*, 428, 647
Dopita, M. A., Sutherland, R. S., Nicholls, D. C., *et al.* 2013, *ApJS*, 208, 10
Flores-Fajardo, N., Morisset, C., Stasińska, G., *et al.* 2011, *MNRAS*, 415, 2182
Galarza, V. C., Walterbos, R. A. M., & Braun, R. 1998, *A&AS*, 192, 40.07
Gomes, J. M., *et al.* 2016, *A&A*, 588, A68
Haffner, L. M., Dettmar, R.-J., Beckman, J. E., *et al.* 2009, *Reviews of Modern Physics*, 81, 969
Heckman, T. M. 1980, *A&A*, 500, 187
Hoopes, C. G., Walterbos, R. A. M., & Rand, R. J. 1999, *ApJ*, 522, 669
Hoopes, C. G., Walterbos, R. A. M., & Greenwalt, B. E. 1996, *AJ*, 112, 1429
Jaffé, Y. L., *et al.* 2014, *MNRAS*, 440, 3491
Johansson, J., Woods, T. E., Gilfanov, M., *et al.* 2016, *MNRAS*, 461, 4505
Kaplan, K. F., *et al.* 2016, *MNRAS*, 462, 1642
Kauffmann, G., Heckman, T. M., Tremonti, C., *et al.* 2003, *MNRAS*, 346, 1055
Kewley, L. J., Dopita, M. A., Sutherland, R. S., *et al.* 2001, *ApJ*, 556, 121
Kewley, L. J., Groves, B., Kauffmann, G., *et al.* 2006, *MNRAS*, 372, 961
Kewley, L. J., Nicholls, D. C., Sutherland, R. S., *et al.* 2019, *ARAA*, 57, 511
Kreckel, K., Blanc, G. A., Schinnerer, E., *et al.* 2016, *ApJ*, 827, 103
Kumari, N., Maiolino, R., Belfiore, F., *et al.* 2019, *MNRAS*, 485, 367
Lacerda, E. A. D., *et al.* 2018, *MNRAS*, 474, 3727
Maiolino, R. & Mannucci, F. 2019, *AARv*, 27, 3
Mannucci, F., Cresci, G., Maiolino, R., *et al.* 2010, *MNRAS*, 408, 2115
Martel, A. R., *et al.* 2004, *AJ*, 128, 2758
Minter, A. H. & Balser, D. S. 1997, ApJL, 484, L133
Minter, A. H. & Spangler, S. R. 1997, *ApJ*, 485, 182
Oey, M. S., Meurer, G. R., Yelda, S., *et al.* 2007, *ApJ*, 661, 801
Phillips, M. M., Jenkins, C. R., Dopita, M. A., *et al.* 1986, *AJ*, 91, 1062
Poetrodjojo, H., *et al.* 2019, *MNRAS*, 487, 79
Reynolds, R. J. 1971, *Ph.D. Thesis*
Reynolds, R. J. 1989, *ApJL*, 339, L29. doi: 10.1086/185412
Reynolds, R. J. & Cox, D. P. 1992, *ApJl*, 400, L33
Sánchez, S. F., *et al.* 2016, *A&A*, 594, A36
Sarzi, M., *et al.* 2010, *MNRAS*, 402, 2187
Slavin, J. D., McKee, C. F., Hollenbach, D. J., *et al.* 2000, *ApJ*, 541, 218
Stasińska, G., Cid Fernandes, R., Mateus, A., *et al.* 2006, *MNRAS*, 371, 972
Stasińska, G., Vale Asari, N., Cid Fernandes, R., *et al.* 2008, *MNRAS*, 391, L29
Tremonti, C. A., *et al.* 2004, *ApJ*, 613, 898
Vale Asari, N., *et al.* 2019, *MNRAS*, 489, 4721 (VA19)
Vandenbroucke, B., Wood, K., Girichidis, P., *et al.* 2018, *MNRAS*, 476, 4032
Walterbos, R. A. M. & Braun, R. 1994, *ApJ*, 431, 156
Wang, J., Heckman, T. M., Lehnert, M. D., *et al.* 1999, *ApJ*, 515, 97

Weilbacher, P. M., Monreal-Ibero, A., Verhamme, A., *et al.* 2018, *A&A*, 611, A95
Wood, K. & Reynolds, R. J. 1999, *ApJ*, 525, 799
Yan, R. & Blanton, M. R. 2013, *IAUS*, 295, 328,
York, D. G., *et al.* 2000, *AJ*, 120, 1579
Zhang, K., *et al.* 2017, *MNRAS*, 466, 3217
Zurita, A., Rozas, M., Beckman, J. E. *et al.* 2000, *A&A*, 363, 9

Sebastian Sanchez

Natalia Vale Asari

Galaxy Evolution and Feedback across Different Environments
Proceedings IAU Symposium No. 359, 2020
T. Storchi-Bergmann, W. Forman, R. Overzier & R. Riffel, eds.
doi:10.1017/S1743921320004020

ORAL CONTRIBUTIONS

Reconstructing the mass accretion histories of nearby red nuggets with their globular cluster systems

Michael A. Beasley[1,2], Ryan Leaman[3], Ignacio Trujillo[1,2], Mireia Montes[4], Alejandro Vazdekis[1,2], Núria Salvador Rusiñol[1,2], Elham Eftekhari[1,2], Anna Ferré-Mateu[5] and Ignacio Martín-Navarro[1,2]

[1]Instituto de Astrofísica de Canarias, c/ Vía Láctea s/n, E-38250, La Laguna, Tenerife, Spain
email: beasley@iac.es

[2]Departamento de Astrofísica, Universidad de La Laguna, E-38205, Tenerife, Spain

[3]Max-Planck Institut für Astronomie, Königstuhl 17, D-69117 Heidelberg, Germany

[4]School of Physics, University of New South Wales, 2052, Sydney, Australia

[5]Institut de Ciéncies del Cosmos (ICCUB), Universitat de Barcelona (IEEC-UB), Barcelona 08028, Spain

Abstract. It is generally recognized that massive galaxies form through a combination of *in-situ* collapse and *ex-situ* accretion. The *in-situ* component forms early, where gas collapse and compaction leads to the formation of massive compact systems (blue and red "nuggets") seen at $z > 1$. The subsequent accretion of satellites brings in *ex-situ* material, growing these nuggets in size and mass to appear as the massive early-type galaxies (ETGs) we see locally. Due to stochasticity in the accretion process, in a few rare cases a red nugget will evolve to the present day having undergone little *ex-situ* mass accretion. The resulting massive, compact and ancient objects have been termed "relic galaxies". Detailed stellar population and kinematic analyses are required to characterise these systems. However, an additional crucial aspect lies in determining the fraction of *ex-situ* mass they have accreted since their formation. Globular cluster systems can be used to constrain this fraction, since the oldest and most metal-poor globular clusters in massive galaxies are primarily an accreted, *ex-situ* population. Models for the formation of relic galaxies and their globular cluster systems suggest that, due to their early compaction and limited accretion of dark-matter dominated satellites, relic galaxies should have characteristically low dark-matter mass fractions compared to ETGs of the same stellar mass.

Keywords. galaxies: massive, accretion, evolution, dark matter, globular clusters

---

1. Introduction

The most massive ($\log (M_*/M_\odot) > 11$) galaxies in the nearby Universe are early-type galaxies (ETGs; e.g. Kauffmann *et al.* 2003; Kelvin *et al.* 2014). These galaxies have large effective radii ($R_e > 5$ kpc) and generally show extended stellar envelopes (e.g. Caon *et al.* 1993; Spavone *et al.* 2017). The inner (∼1 kpc) of ETGs are extremely old, metal-rich and α-element enhanced, suggestive of a rapid and early dissipational collapse (Trager *et al.* 2000; Thomas *et al.* 2005). In addition, the central regions of ETGs appear to be dominated by a dwarf-rich "bottom-heavy" stellar initial mass function (IMF) (Cenarro *et al.* 2003; Conroy & van Dokkum 2012; La Barbera *et al.* 2013)

In contrast, mass-matched quiescent galaxies at $z \sim 2$ do not appear extended, but are in fact extremely compact systems (Re $\sim$ 1 kpc) (Trujillo *et al.* 2007; van Dokkum *et al.* 2008; Damjanov *et al.* 2009). These "red nuggets" (Damjanov *et al.* 2009) show properties remarkably similar to the central regions of ETGs (Trujillo *et al.* 2014; Martín-Navarro *et al.* 2015; Toft *et al.* 2017; Newman *et al.* 2018). In order to reconcile the high- and low-redshift populations of massive galaxies, red nuggets must grow in effective radius by a factor of ~4 without further significant star formation (Trujillo *et al.* 2007; van Dokkum *et al.* 2008). These observations, in conjunction with models of galaxy evolution (e.g., Oser *et al.* 2010; Shankar *et al.* 2013; Ceverino *et al.* 2015) have led to a "two-phase" picture of massive galaxy formation. In the first phase an initial gas collapse, compaction and quenching event(s) forms a passively evolving red nugget. In the second phase the red nugget grows in size (and mass) via mergers and satellite accretions to form what we see as a present-day ETG. In this basic picture, the innermost regions of ETGs comprise *in-situ* material (stars, gas and dark matter), while the outer envelope is composed of *ex-situ* material originating from accreted satellites (e.g., Martín-Navarro *et al.* 2019).

2. Relic galaxies

The individual accretion histories of massive galaxies are expected to vary quite significantly, and this stochasticity in part drives the scatter seen in $M_* - M_{\rm halo}$ relations at the high mass end of the galaxy mass function (Moster *et al.* 2013; Behroozi *et al.* 2013). This stochasticity leads to the prediction that a few, rare objects may essentially skip the second phase of ETG formation entirely such that some red nuggets may reach us from the high redshift universe essentially unaltered (Quilis & Trujillo 2013). Studying these nearby "relic galaxies"† will allow us to understand the stellar populations, kinematics, environments, central black-holes and dark matter content resulting from the first phase of galaxy formation at the highest spatial resolutions and signal to noise.

The first, best sample of nearby relic galaxies was discovered serendipitously in the HETMGS survey (van den Bosh *et al.* 2015). HETMGS searched for suitable galaxies to measure black hole masses based on sphere-of-influence arguments, which automatically selected for compact galaxies. Besides compactness, relic galaxies are observed to be old, metal-rich and have bottom-heavy IMFs at all radii (e.g. Martín-Navarro *et al.* 2015). Kinematically, they are found to appear as "hot discs" similar to the red nuggets at $z \sim 2$ (Toft *et al.* 2017; Newman *et al.* 2018) indicative of a predominantly dissipational collapse. Follow-up observations of the most compact ($R_e < 3$ kpc) and massive (log $(M_*/M_\odot > 11)$ galaxies in HETMGS have produced a sample of 14 galaxies with the above properties (Trujillo *et al.* 2014; Yıldırım *et al.* 2017; Ferré-Mateu *et al.* 2017).

Beyond using the structural, chemical and dynamics properties to identify true relic galaxies, ideally one wants to have a handle on their mass accretion histories. By constraining the amount of *ex-situ* material in massive galaxies, a clean sample of relic galaxies can be defined.

3. Globular clusters in relic galaxies

Similar to developments in the field of massive galaxy formation, the presently favoured picture for the formation of globular cluster (GC) systems around massive galaxies is also a two phase one. GC systems in massive ETGs show complex colour and metallicity distributions functions (e.g., Peng *et al.* 2006; Harris *et al.* 2017). The most metal-rich "red" GCs are predominantly an *in-situ* population likely formed during the galaxies' initial collapse. Metal-poor "blue" GCs are found in all but the lowest mass dwarfs,

† Also sometimes called "naked red nuggets" or massive ultracompact galaxies (MUGs).

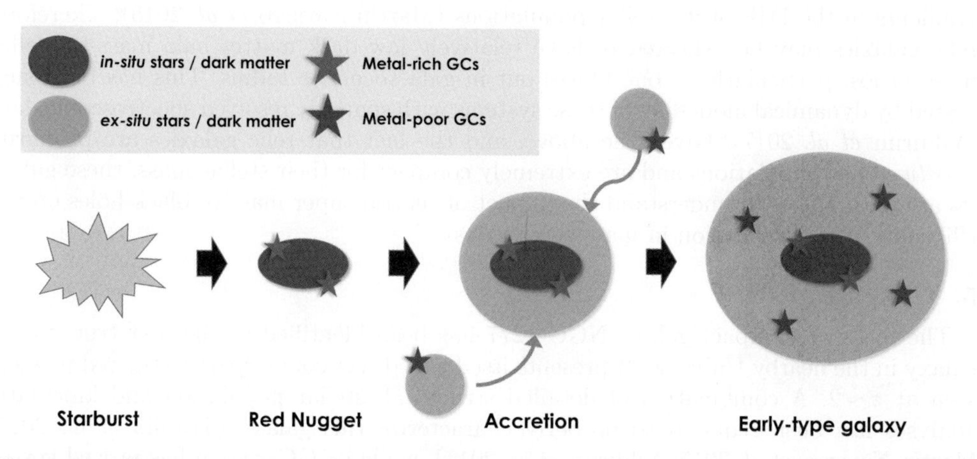

Figure 1. Cartoon overview of the possible formation and co-evolution of massive galaxies and their globular cluster systems. After an initial gas-compaction, star-burst and subsequent quenching, the resulting red nugget is a compact stellar system comprising *in-situ* stars with a retinue of metal-rich globular clusters. Subsequent accretion and merging grows the red nugget in size and mass to form a present-day ETG. The galaxy halo and metal-poor globular clusters come primarily from accreted *ex-situ* material.

but in massive galaxies are regarded as a largely *ex-situ* population brought in via satellite accretion†. This general model finds extensive observational and theoretical support (e.g., Côté *et al.* 1998; Beasley *et al.* 2002; Tonini 2013; Leaman *et al.* 2013; Mackey *et al.* 2019; Kruijssen *et al.* 2019). A cartoon of the basic scheme is shown in Fig. 1.

The colour and metallicity distributions of GC systems can be readily observed and compared to expectations from hierarchical merger models. This exercise has been performed for the archetypical relic galaxy NGC 1277 in the Perseus galaxy cluster (Beasley *et al.* 2018). Initially identified as a massive, compact galaxy with an "over-massive" black hole for its stellar mass (van den Bosch *et al.* 2012), the properties of NGC 1277 have been subsequently shown to be in strikingly good agreement with expectations for a $z \sim 2$ red nugget (Trujillo *et al.* 2014; Martín-Navarro *et al.* 2015; Yıldırım *et al.* 2017). In terms of its GC system, Beasley *et al.* (2018) found that NGC 1277 has very few, if any metal-poor GCs. This observation suggests that the galaxy contains very little *ex-situ* material. By connecting distinct GC subpopulations with *in-situ* and *ex-situ* origins, a picture of the accreted sub-halo mass function of massive galaxies can be constructed. The colour distributions and total population of GCs in "normal" ETGs suggests accreted mass fractions of 40–70% (Beasley *et al.* 2018). This is in broad agreement with other observational approaches (e.g., Spavone *et al.* 2017). By contrast, the inferred accretion fraction for NGC 1277 – based on its GC system – is <12%. This is the first relic galaxy for which this technique has been applied, but promises to be a useful approach for identifying true relic galaxies by quantification of their *ex-situ* mass fractions.

4. Dark matter and the baryon fraction

An additional, interesting area for investigation concerns the dark matter content of relic galaxies. As a consequence of their extremely modest *ex-situ* fractions, relic galaxies accrete few low stellar mass, dark matter-dominated satellites during their second

† The precise mapping between blue and red GCs, and *ex-situ* and *in-situ* components is more complicated that the picture outlined here (see e.g. Fahrion *et al.* 2020).

phase of evolution. Again, this is evidenced by the lack of very metal poor GCs in the GC system of NGC 1277 (Beasley *et al.* 2018), in addition to a lack of noticeable radial gradients in the IMF of its stellar populations (Martín-Navarro *et al.* 2015). Therefore, relic galaxies may be expected to have relatively low dark matter halo mass-to-stellar mass ratios, particularly as one moves out in galactocentric radius. This has been suggested by dynamical modelling of these systems with spatially resolved spectroscopic data (Yıldırım *et al.* 2017). Given the above, and the fact that relic galaxies are primarily *in-situ* stellar populations and are extremely compact for their stellar mass, these galaxies are ideal places to understand the impact of central super-massive black holes on the efficiency of star formation in massive galaxies.

5. Conclusions

The massive, compact galaxy NGC 1277 has been identified as the first true "relic" galaxy in the nearby Universe. It presents itself as a direct counterpart to the red nuggets seen at $z \sim 2$. A combination of detailed structural, stellar population and kinematic analyses has been required to properly characterize this galaxy (Trujillo *et al.* 2014; Martín-Navarro *et al.* 2015; Yıldırım *et al.* 2017), while its GC system has proved key to constrain its accretion history (Beasley *et al.* 2018). A detailed exploration of the X-ray properties of NGC 1277 (c.f., Buote & Barth 2019) may bring important constraints on its dark matter halo, as will dynamical studies at large radii via kinematics of the GC system with *James Webb Space Telescope.* Upcoming analysis of its stellar populations in the UV will be used to search for possible star formation even at the sub- one percent level (e.g., Salvador-Rusiñol *et al.* 2019), and infrared spectroscopy will bring further insights into the properties of its stellar populations.

The presence of NGC 1277 in the Perseus cluster places a lower limit on the space densities of relic galaxies, $\rho_{\rm relic} > 7 \times 10^{-7}$ Mpc^{-3}, and we note that several more relics have now been identified (Ferré-Mateu *et al.* 2017; Yıldırım *et al.* 2017). A study of the GC populations of these galaxies will be crucial for constraining their accretion fractions. The identification and study of relic galaxies provide the opportunity to study a population normally only accessible at $z > 2$, and gain a detailed understanding of the earliest phases of massive galaxy formation.

References

Beasley, M. A., Baugh, C. M., Forbes, D. A., *et al.* 2002, *MNRAS*, 333, 383
Beasley, M. A., Trujillo, I., Leaman, R., *et al.* 2018, *Nature*, 555, 483
Behroozi, P. S., Wechsler, R. H., & Conroy, C. 2013, *ApJ*, 770, 57
Buote, D. A. & Barth, A. J. 2019, *ApJ*, 877, 91
Caon, N., Capaccioli, M., & D'Onofrio, M. 1993, *MNRAS*, 265, 1013
Cenarro, A. J., Gorgas, J., Vazdekis, A., *et al.* 2003, *MNRAS*, 339, L12
Ceverino, D., Dekel, A., Tweed, D., *et al.* 2015, *MNRAS*, 447, 3291
Conroy, C. & van Dokkum, P. G. 2012, *ApJ*, 760, 71
Côté, P., Marzke, R. O., & West, M. J. 1998, *ApJ*, 501, 554
Damjanov, I., McCarthy, P. J., Abraham, R. G., *et al.* 2009, *ApJ*, 695, 101
Fahrion, K., Lyubenova, M., Hilker, M., *et al.* 2020, *A&A*, 637, A27
Ferré-Mateu, A., Trujillo, I., Martín-Navarro, I., *et al.* 2017, *MNRAS*, 467, 1929
Harris, W. E., Ciccone, S. M., Eadie, G. M., *et al.* 2017, *ApJ*, 835, 101
Kauffmann, G, Heckman, T. M., White, S. D. M., *et al.* 2003, *MNRAS*, 341, 54
Kelvin, L., *et al.* 2014, *MNRAS*, 444, 1647
Kruijssen, J. M. D., Pfeffer, J. L., Reina-Campos, M., *et al.* 2019, *MNRAS*, 486, 3180
La Barbera, F., Ferreras, I., Vazdekis, A., *et al.* 2013, *MNRAS*, 433, 3017
Leaman, R., VandenBerg, D. A., Mendel, J. T., *et al.* 2013, *MNRAS*, 436, 122
Mackey, D., Lewis, G. F., Brewer, B. J., *et al.* 2019, Nature, 574, 69

Martín-Navarro, I., La Barbera, F., Vazdekis, A., *et al.* 2015, *MNRAS*, 451, 1081
Martín-Navarro, I., van de Ven, G., & Yıldırım, A. 2019, *MNRAS*, 487, 4939
Moster, B. P., Naab, T., & White, S. D. M. 2013, *MNRAS*, 428, 3121
Newman, A. B., Belli, S., Ellis, R. S., *et al.* 2018, *ApJ*, 862, 126
Oser, L., Ostriker, J. P., Naab, T., *et al.* 2010, *ApJ*, 725, 2312
Peng, E. W., Jordán, A., Côté, P., *et al.* 2006, *ApJ*, 639, 95
Quilis, V. & Trujillo, I. 2013, *ApJL*, 773, L8
Salvador-Rusiñol, N., Vazdekis, A., La Barbera, F., *et al.* 2019, *Nature Astronomy*, 4, 252
Shankar, F., Marulli, F., Bernardi, M., *et al.* 2013, *MNRAS*, 428, 109
Spavone, M, Capaccioli, M, Napolitano, *et al.* *A&A*, 603, 38
Tonini, C. 2013, *ApJ*, 762, 39
Toft, S., Zabl, J., Richard, J., *et al.* 2017, *Nature*, 546, 510
Thomas, D., Maraston, C., Bender, R., *et al.* 2005, *ApJ*, 621, 673
Trager, S. C., Faber, S. M., Worthey, G., *et al.* 2000, *AJ*, 119, 1645
Trujillo, I., Conselice, C. J., Bundy, K., *et al.* 2007, *MNRAS*, 382, 109
Trujillo, I., Ferré-Mateu, A., Balcells, M., *et al.* 2014, *ApJL*, 780, L20
van den Bosch, R. C. E., Gebhardt, K., Gültekin, K., *et al.* 2015, *ApJS*, 28, 10
van den Bosch, R. C. E., Gebhardt, K., Gültekin, K., *et al.* 2012, *Nature*, 491, 729
van Dokkum, P. G., Franx, M., Kriek, M., *et al.* 2008, *ApJL*, 677, L5
Yıldırım, A., van den Bosch, R. C. E., van de Ven, G., *et al.* 2017, *MNRAS*, 468, 4216

Galaxy Evolution and Feedback across Different Environments
Proceedings IAU Symposium No. 359, 2020
T. Storchi-Bergmann, W. Forman, R. Overzier & R. Riffel, eds.
doi:10.1017/S1743921320001647

Recovering the star formation history of galaxies through spectral fitting: Current challenges

Lucimara P. Martins

Núcleo de Astrofísica/Universidade Cidade de São Paulo/Universidade Cruzeiro do Sul
Rua Galvão Bueno, 868, São Paulo, SP, Brazil, 01506-000
email: lucimara.martins@cruzeirodosul.edu.br

Abstract. With the exception of some nearby galaxies, we cannot resolve stars individually. To recover the galaxies star formation history (SFH), the challenge is to extract information from their integrated spectrum. A widely used tool is the full spectral fitting technique. This consists of combining simple stellar populations (SSPs) of different ages and metallicities to match the integrated spectrum. This technique works well for optical spectra, for metallicities near solar and chemical histories not much different from our Galaxy. For everything else there is room for improvement. With telescopes being able to explore further and further away, and beyond the optical, the improvement of this type of tool is crucial. SSPs use as ingredients isochrones, an initial mass function, and a library of stellar spectra. My focus are the stellar libraries, key ingredient for SSPs. Here I talk about the latest developments of stellar libraries, how they influence the SSPs and how to improve them.

Keywords. atomic data, stars: fundamental parameters, galaxies: stellar content

1. Introduction

Most of the light in galaxies comes from stars. The star formation history (SFH) of galaxies contains information of how they were shaped and evolved. Our knowledge of stars and stellar evolution improved significantly over the last decades, but is based mainly on observing, modelling and interpreting stars of our solar vicinity. However, with the exception of a few nearby galaxies, we cannot resolve stars individually down to the turn-off and below. This means that the information about the star formation history of a given galaxy will be encoded in its integrated spectrum. Extracting physical, chemical and evolutionary information about galaxies from this type of spectrum is one of the major challenges astronomers face today.

Several techniques are available to extract information from integrated spectra, mostly involving the comparison of the observations with stellar population model libraries with a wide range of ages and metallicities (e.g. Cid Fernandes *et al.* 2005; Ocvirk *et al.* 2006; Walcher *et al.* 2006; Koleva *et al.* 2008). This is called full spectral fitting technique. The simplest models used in this technique are the simple stellar population (SSP) models, which are spectra built theoretically using as ingredients isochrones, an initial mass function (IMF), and a library of stellar spectra (e.g. Bruzual 1983; Bressan *et al.* 1994; Worthey 1994; Leitherer *et al.* 1999; Bruzual & Charlot 2003; González-Delgado *et al.* 2005; Maraston 2005; Conroy & Gunn 2010; Vazdekis*et al.* 2010; Meneses-Goytia *et al.* 2015).

Most spectral fitting codes, using different SSPs, produce similar results, but for a minimal S/N, in the optical spectral range and for metallicities close to the solar value. For everything else there is a lot of room for improvement. All the ingredients and codes involved in the creation of SSPs used for the spectral fitting technique had major developments in the last decades, which greatly improved our capacity to interpret integrated spectra. However in all of them there are approximations and imprecisions that might affect the final result. For example, our knowledge of the IMF is based on our interpretation of observations and many assumptions, such as that the IMF is universal and constant in time. Doubts about these assumptions are still in debate today and many studies try to explore different possibilities. (Chiappini *et al.* 2000; Chieffi *et al.* 2002; Bastian *et al.* 2010; Bonatto *et al.* 2012). In the case of the isochrones, there are many aspects of stellar evolution that we still cannot model. Evolutionary effects of chemical variations like α-enhancement (e.g. Salasnich *et al.* 2000; Pietrinferni *et al.* 2009) or individual element variations (e.g. Dotter *et al.* 2007) have been investigated. Problems exist in this field, with different treatments by different groups leading to different evolutionary tracks, even when using the same input parameters (Walche *et al.* 2011; Martins *et al.* 2013).

In the case of stellar libraries, they can be either empirical or theoretical. Empirical libraries are based on observational data, which implies that all features contained in the resulting SSP spectra will be real. The disadvantage, however, is that these libraries are biased towards the star formation and chemical enrichment histories of the solar neighbourhood, the Small and Large Magellanic Clouds or Galactic Globular Clusters (GCs), limiting the coverage and sampling of the Hertzsprung–Russell (HR) diagram. Jain *et al.* 2020 shows that a limited coverage in HR might produce different results when using different empirical libraries.

Theoretical libraries do not have this setback, since it is possible to generate stellar spectra with virtually any temperature or metallicity desired, in any wavelength range, covering the whole parameter space. Of course, this also comes with a limitation, since they are build from models which are always based on physics approximations and simplifications (Bessell *et al.* 1998; Kurucz 2006; Martins & Coelho 2007; Coelho *et al.* 2009; Sansom *et al.* 2013; Kitamura *et al.* 2017). Despite that, theoretical libraries have improved over the years, and it has been shown that their performance is not much worst than that of the empirical libraries (Martins *et al.* 2019). Due to the importance of this ingredient to the quality of SSPs, many groups have been dedicated to create better libraries.

2. Improving theoretical libraries

Martins *et al.* (2019) showed that modern theoretical libraries are capable of reproducing the integrated spectra of globular clusters (GCs) almost as well as empirical libraries. In this work they used a sample of GCs for which there was integrated spectra and CMDs available. For each star in the CMD they associated a spectrum from a given library (they tested two empirical and two synthetic libraries) and with that created an integrated spectrum without the need of an IMF or isochrones. With that they could directly access the quality of each stellar library. Figure 1 shows an example of their results, for the GC NGC 1904. For individual signatures and details of the spectrum, empirical libraries are still better than theoretical ones. But taking the overall shape of the continuum into account, theoretical libraries outperformed the empirical libraries in 13 out of 18 of the cases.

Martins & Coelho (2007) showed that, only by improving the atomic line list used to generate stellar synthetic spectra, it is possible to greatly improve their quality. In general, one half of the discernible lines in observed stellar spectra are missing from the

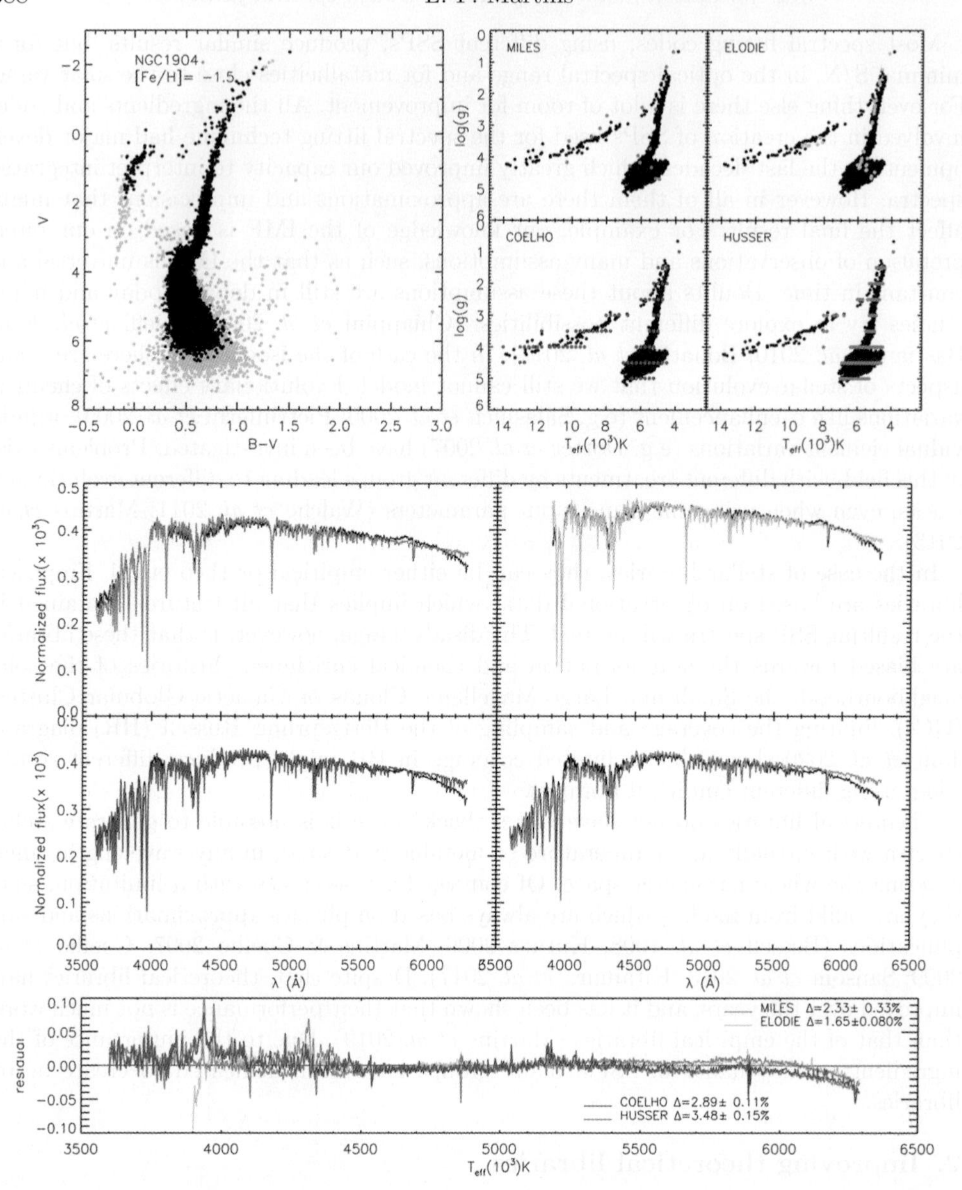

Figure 1. Result of the synthetic spectrum creation for NGC 1904. The top left panel shows the clean CMD used in this work (in black), on top of the original CMD (in gray). The four figures in the top right part of the figure show the $\log g$ vs. $T_{\rm eff}$ diagram, where in black are the stars of the GC and in red, orange, blue and magenta are the selected stars from MILES, ELODIE, COELHO and HUSSER libraries respectively. In the middle panel we show the synthetic spectra created for the GC for each of these libraries. In the bottom panel we show the residual difference (observed - synthetic spectra) for each of the libraries.

line lists with good wavelengths (Kurucz 2011). Efforts are underway to reduce transition probability uncertainties of selected lines (e.g. Fuhr & Wiese 2006; Safronova & Safronova 2010; Wiese, Fuhr & Bridges 2011; Civiš *et al.* 2012) and accurately compute broadening parameters. It has also been shown in the literature (e.g. Barbuy *et al.* 2003; Martins & Coelho 2007) that even empirical calibrations of some specific lines can produce significant improvement on the synthetic spectra generated. The empirical calibration is done by changing the values of the parameters on the line list, generating models and comparing them with observations of very well known stars (like the Sun or Arcturus, for example). This process is repeated until the results are adequate (Barbuy *et al.* 2003). Franchini *et al.* (2018) is a recent example of this approach. They manually derived oscillator strengths for 2229 lines, an unprecedented effort that greatly improved the quality of their synthetic stellar library, although for a very small wavelength range.

These approaches tend to improve the quality of a selective group lines, mostly those that were considered more suitable to chemical abundance measurements, where relatively weak lines in the linear part of the curve of growth are favoured. On the other hand, the strong lines close to saturation and blended features are the ones which dominate spectral indices in integrated spectra of stellar populations. While these efforts are improving the parameters for thousands of lines, tens of millions of lines are estimated to be needed to compute, say, a complete stellar grid with a good range of atmospheric parameters. Therefore, an innumerous amount of lines remain poorly characterised. For the ultimate goal of computing a large grid of theoretical stellar spectra for further use in automatic classification of stellar spectroscopic surveys and stellar population modelling, this is rather limiting.

Recently, an automatic method has been developed to overcome limitations of the atomic line lists: ALICCE (Atomic Lines Calibration using the Cross-Entropy Algorithm, Martins *et al.* 2014) is a code developed to automatically calibrate atomic lines using the cross-entropy method. The cross-entropy method is a general Monte Carlo approach to combinatorial and continuous multi-extremal optimisation and importance sampling, which is a general technique for estimating properties of a particular distribution using samples generated randomly from a different statistical distribution rather than the distribution of interest (Rubinstein 1997, 1999). The first application of the code was made by Kitamura *et al.* (2017), to calibrate missing lines in the Solar spectrum. We hope, in the near future, to have improved line lists that will be used to generate more accurate synthetic stellar spectra, improving the quality of theoretical stellar libraries.

3. Conclusion

When using SSP models to interpret the integrated spectra of stellar systems, the user is not always aware of the limitations in these models due its ingredients. It is always important for the stellar population modelling community to try to keep users well informed of the limitations of these models. The accuracy of the techniques using these models have greatly improved over the years, but there is yet a lot of room for improvement. One of the ways stellar population models can be improved is by perfecting one of its main ingredients: the stellar spectral libraries. We developed a code to calibrate the atomic line lists used to generate synthetic stellar spectra, which we believe will be used to create stellar synthetic spectra with unprecedented accuracy. This will certainly expand the applications of SSPs from what we have today.

References

Barbuy, B., Perrin, M.-N., Katz, D., *et al.* 2003, *A&A*, 404, 661
Bastian, N., Covey, K. R., Meyer, M. R., *et al.* 2010, *ARA&A*, 48, 339
Bessell, M. S., Castelli, F., & Plez, B., 1998, *A&A*, 333, 231
Bonatto, C., Bica, E., *et al.* 2012, *MNRAS*, 423, 1390
Bressan, A., Chiosi, C., Fagotto, F., *et al.* 1994, *ApJS*, 94, 63
Bruzual, A. G. 1983, *ApJ*, 273, 105
Bruzual, G. & Charlot, S. 2003, *MNRAS*, 344, 1000
Chiappini, C., Matteucci, F., Padoan, P., *et al.* 2000, *ApJ*, 528, 711

Chieffi, A., Limongi, M., *et al.* 2002, *ApJ*, 577, 281
Coelho, P., Mendes de Oliveira, C., & Cid Fernandes, R. 2009, *MNRAS*, 396, 624
Cid Fernandes, R., Mateus, A., Sodré, L., *et al.* 2005, *MNRAS*, 358, 363
Civiš, S., Ferus, M., Kubelík, P., *et al.* 2012, *A&A*, 542, 35
Conroy, C. & Gunn, J. E. 2010, *ApJ*, 712, 833
Dotter, A., Chaboyer, B., Jevremović, D., *et al.* 2007, *AJ*, 134, 376
Franchini, M., Morossi, C., Di Marcantonio, P., *et al.* 2018, *ApJ*, 862, 146
Fuhr, J. R. & Wiese, W. L. 2006, *Journal of Physical and Chemical Reference Data*, 35, 1669
González-Delgado, R. M., Cerviño, M., Martins, L. P., *et al.*2005, *MNRAS*, 357, 945
Jain, R., Prugniel, P., Martins, L., *et al.* 2020, *A&A*, 635, 161
Kitamura, J. R., Martins, L. P., Coelho, P., *et al.* 2017, *A&A*, 600, 11
Koleva, M., Prugniel, P., Ocvirk, P., *et al.* 2008, *MNRAS*,385, 1998
Kurucz, R. L. 2006, in Stee, P., ed., EAS Publications Series Vol. 18, EAS Publications Series. pp 129–155, doi: 10.1051/eas:2006009
Kurucz, R. L. 2011, *Canadian Journal of Physics*, 89, 417
Leitherer, C., *et al.* 1999, *ApJS*, 123, 3
Maraston, C. 2005, *MNRAS*, 362, 799
Martins, L. P. & Coelho, P. 2007, *MNRAS*, 381, 1329
Martins, L. P., Rodríguez-Ardila, A., Diniz, S., *et al.* 2013, *MNRAS*, 435, 2861
Martins, L. P., Coelho, P., Caproni, A., *et al.* 2014, *MNRAS*, 442, 1294
Martins, L. P., Lima-Dias, C., Coelho, P. R. T., *et al.* 2019, *MNRAS*, 484, 2388
Meneses-Goytia, S., Peletier, R. F., Trager, S. C., *et al.* 2015, *A&A*, 582, A97
Ocvirk, P., Pichon, C., Lançon, A., *et al.* 2006, *MNRAS*, 365, 46
Pietrinferni, A., Cassisi, S., Salaris, M., *et al.* 2009, *ApJ*, 697, 275
Rubinstein, R. Y. 1997, *European Journal of Operational Research*, 99, 89
Rubinstein, R. Y. 1999, *Methodology and Computing in Applied Probability*, 2, 127
Safronova, U. I., Safronova, A. S., & Johnson, W. R. 2010, *Journal of Physics B Atomic Molecular Physics*, 43, 144001
Salasnich, B., Girardi, L., Weiss, A., *et al.* 2000, *A&A*, 361, 1023
Sansom, A. E., Milone, A. de C., Vazdekis, A., *et al.* 2013, *MNRAS*, 435, 952
Vazdekis, A., Sánchez-Blázquez, P., Falcón-Barroso, J., *et al.* 2010, *MNRAS*, 404, 1639
Walcher, C. J., Böker, T., Charlot, S., *et al.* 2006 *ApJ*, 649, 692
Walcher, J., Groves, B., Budavári T., *et al.* 2011, *Ap&SS*, 331, 1
Wiese, W. L., Fuhr, J. R., & Bridges, J. M., *et al.* 2011, in *2010 NASA Laboratory Astrophysics Workshop*, 16
Worthey, G. 1994, *ApJS*, 95, 107

Galaxy Evolution and Feedback across Different Environments
Proceedings IAU Symposium No. 359, 2020
T. Storchi-Bergmann, W. Forman, R. Overzier & R. Riffel, eds.
doi:10.1017/S1743921320001568

From global to local scales in galaxies

Sebastian F. Sánchez and Carlos Lopez Cobá

Instituto de Astronomía, Universidad Nacional Autonóma de México,
A.P. 70-264, 04510 México, D.F., Mexico
email: sfsanchez@astro.unam.mx

Abstract. We summarize here some of the results reviewed recently by Sanchez (2020) comprising the advances in the comprehension of galaxies in the nearby universe based on integral field spectroscopic galaxy surveys. In particular we explore the bimodal distribution of galaxies in terms of the properties of their ionized gas, showing the connection between the star-formation (quenching) process with the presence (absence) of molecular gas and the star-formation efficiency. We show two galaxy examples that illustrates the well known fact that ionization in galaxies (and the processes that produce it), does not happen monolitically at galactic scales. This highlight the importance to explore the spectroscopic properties of galaxies and the evolutionary processes unveiled by them at different spatial scales, from sub-kpc to galaxy wide.

Keywords. galaxies: evolution, galaxies: star-formation, galaxies: ISM

1. Integrated properties of galaxies

Along the last years our knowledge of nearby galaxies ($z < 0.1$) and the comprehension of the evolutionary processes that have produced them have been considerably increased due to the advent of large and statistical significant imaging and spectroscopic galaxy surveys (e.g., Sloan Digital Sky Survey, SDSS, Galaxy and Mass Assembly survey, GAMA York *et al.* 2000; Driver *et al.* 2009). They have allowed us to explore the multi-band photometry and imaging of millions of galaxies, providing with single aperture spectroscopic observations of hundred of thousand of them. A recent review by Blanton & Moustakas (2009) have summarized the main results derived from these surveys, including their global distributions in colors, luminosities, stellar and atomic gas masses and fractions and their corresponding mass/luminosity functions, for different galaxy types and environments.

The comparison of the properties of galaxies at $z \sim 0$ explored by these surveys with those derived at different redshifts have demonstrated that all galaxies formed stars in the past, at a much higher rate than today (e.g. Heavens *et al.* 2004; Speagle *et al.* 2014; Madau & Dickinson 2014). However, in the last $\lesssim$8 Gyr galaxies present a clear bimodal distribution in terms of their star-formation activity, with two well distinguished populations: (i) disk-dominated, dynamically supported by rotation late-type galaxies, actively forming stars and (ii) bulge-dominated, with a large fraction of hot and warm stellar orbits, early-type galaxies, currently retired from the star-formation budget. These two groups are well separated in different diagrams, like the color-magnitude (or age-mass), or the SFR-M_* (e.g. Renzini & Peng 2015), where the distinction is stronger than in any other one. This separation is evident when using the average EW(Hα), a parameter strongly correlated with the sSFR (i.e., $\frac{SFR}{M_*}$, the two parameters involved in the SFR-M_* diagram). Retired galaxies (RGs) in general present low EW(Hα) (<3Å, Stasińska *et al.* 2008), exhibiting in the majority of the cases a weak emission of ionized gas, with line

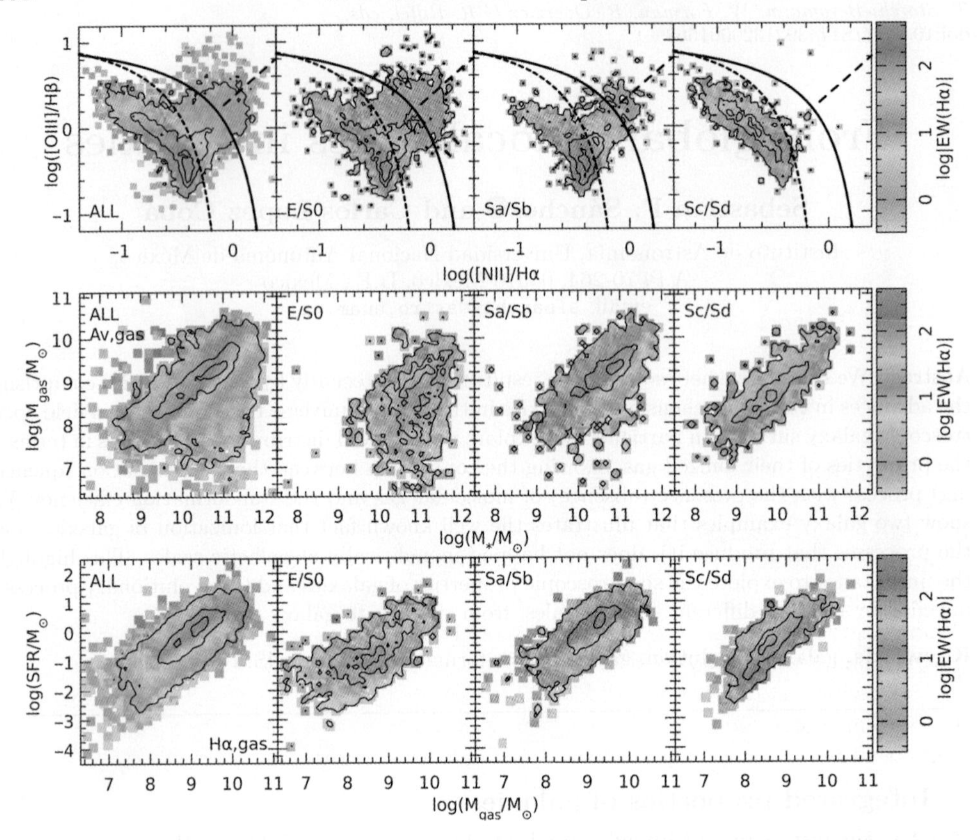

Figure 1. *Top panels:* BPT diagnostic diagram for the emission lines of the ionized gas observed in the considered sample of galaxies, averaged in a ring at one effective radius. The left-most panel show the distribution for all the galaxies, with similar distributions segregated by morphology shown in the subsequent panels, from earlier types to later ones from left to right. *Middle panels:* Distribution of the gas mass estimated from the dust attenuation derived by the Hα/Hβ line ratios, along the stellar masses of the galaxies in the same sample. *Bottom panels:* Similar figure showing the distribution of the integrated SFR derived from dust-corrected Hα luminosity along the gas mass estimated for each galaxy. Along all panels the density distributions are shown as successive contours including a 95%, 50% and 10% of the points, while the colormaps indicate the average logarithm of the EW(Hα) at a particular location.

ratios indicating the presence of a hard ionizing source. Star-forming galaxies (SFGs), present in general higher values of the EW(Hα) (>10Å Lacerda *et al.* 2018), with stronger ionized emission, and line ratios compatible with a softer ionizing source. On top of that, strong AGNs present strong emission lines, with large values of EW(Hα) (> 6Å, Cid Fernandes *et al.* 2010), and hard ionizing sources. In general, RGs are associated with early-type galaxies, while SFGs are dominated by late-type ones. Figure 1 illustrates this segregation, showing clear differences in the distribution of galaxies of different morphologies across the classical diagnostic diagrams (e.g. Baldwin *et al.* 1981, BPT diagram), and their average EW(Hα). For this particular diagram we use the compilation of optical integral field spectroscopic (IFS) data recently presented by Sanchez (2020), comprising more than 8000 galaxies observed by different IFS galaxy surveys in the nearby universe: CALIFA (Sánchez *et al.* 2012), MaNGA (Bundy *et al.* 2015), SAMI (Croom *et al.* 2012) and AMUSSING++ (López-Cobá *et al.* 2020). All data have been processed using the same pipeline, (Pipe3D Sánchez *et al.* 2016), in order to homogenize at maximum this heterogeneous compilation.

The bimodal distribution in the different described diagrams indicates that the necessary transition between SFGs (all galaxies at very high-redshift), and the current observed population of RGs (which increases towards $z \sim 0$, e.g., Muzzin *et al.* 2013) should be rather fast (compared to the Hubble type, i.e., $\lesssim$1 Gyr, e.g., Sánchez *et al.* 2019). This transition involves a morphological transformation, and therefore should be somehow connected to the dynamical stage of galaxies (e.g. Martig *et al.* 2009). It may involve a highly and short-lived process that injects energy to the system heating or removing gas, i.e., the main ingredient of the SF process (an AGN?, e.g. Hopkins *et al.* 2009). The presence of AGN hosts in the transition regime (Green-Valley) between the SFGs and RGs in different diagrams seems to support this hypothesis (e.g. Schawinski *et al.* 2014; Sánchez *et al.* 2018). Whatever is the triggering mechanism of the quenching, what it is clear is that RGs present a general deficit of cold gas (molecular gas in particular, Saintonge *et al.* 2016; Sánchez *et al.* 2018). This is clearly observed in the distribution of integrated gas mass (M_{gas}) along the stellar mass (M_*) seen in Fig. 1, middle panel. Indeed, lack of (molecular) gas, either removed or heated, seems to be the primary driver (or connection) with the observed decrease of SF in quenched/RGs. Actually, the segregation is stronger regarding the amount of (molecular) gas, than regarding the efficiency in how this gas is transformed to stars (defined as SFE= $\frac{SFR}{M_{gas}}$). Fig. 1, bottom panel, shows the integrated SFR along M_{gas}, i.e., the integrated version of the Schmidt-Kennicutt law (Kennicutt 1998), segregated by morphology, and highlighted by the EW(Hα). Retired or partially retired galaxies form stars at a slightly lower rate than SF ones for the same amount of (molecular) gas. However, they present a much lower amount of SFR because they have a much lower amount of gas. Among SFGs, SFE and gas fraction seems to equally compete to explain the distribution along the main trends (e.g., Sa/Sb galaxies).

2. Local processes in galaxies

Despite of the tremendous improvement in our understanding of galaxies provided by the quoted surveys, they still provide with an incomplete picture of the evolution processes in galaxies. While the information provided by imaging survey allow us to explore the spatial resolved distribution of light, stellar mass, and the morphological properties of galaxies, the spectroscopic information is in general limited to a single aperture that, in many cases, samples different physical extensions of the objects depending on their redshift (and in most cases biased towards their central regions). In general, the studies based on these surveys consider galaxies as unresolved entities, being either SF or retired, and assuming that any process happens simultaneously across their full extension. Therefore, they cannot provide with a real description of processes that occur at kpc or sub-kpc scales.

For instance, the condensation of diffuse atomic gas into molecular clouds, and the dynamical collapse of that clouds to give birth to stars are not processes that happen at a galactic scale. This is evident when exploring the distribution of recently formed young stars in galaxies, traced by the H II regions. Figure 2, top panel, shows both a three-color continuum image (u-, g- and r-band) and an emission line image ([O III], Hα and [N II]), of the late-type galaxy 2MASXJ09534925+0911377. This latter image is mapped into a spatial resolved BPT diagnostic diagram. The central/bulge dominated regions of the galaxy present by a weak, diffuse, hard ionization that exhibits low values of the EW(Hα), located at the right-hand side of the BPT diagram (in the LINER-like regime). This ionization is produced by either hot-evolved/post-AGB stars (e.g. Binette *et al.* 1994; Singh *et al.* 2013) or slow velocity shocks (e.g. Dopita *et al.* 1996). On the contrary, the disk of this galaxy presents a series of clumpy ionized regions with strong, but softer ionization, that have large values of the EW(Hα), located in the BPT diagram at the

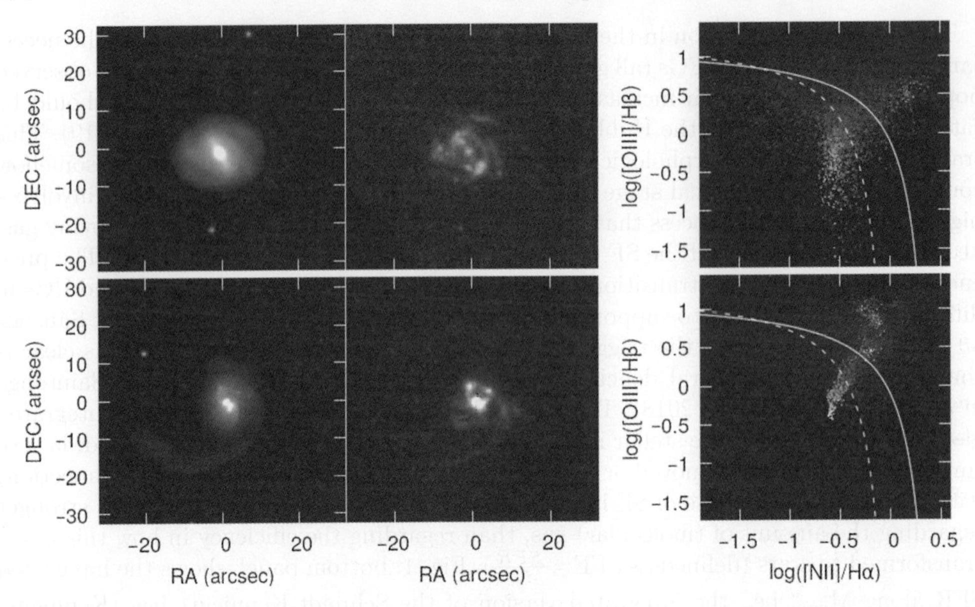

Figure 2. *Left panels:* Three-color continuum images created using g (blue),r (green) and i-band (red) images extracted from the MUSE datacube of 2MASXJ09534925+0911377 (top panels) and ESO253-G003 (bottom panels) galaxies,. *Middle panels:* Emission line image created using the [O III] (blue), Hα (green) and [N II] (red) emission line maps extracted from the same datacube using PIPE3D. *Right panels:* Spatial resolved classical BPT diagnostic diagram involving the [O III]/Hβ and [N II]/Hα line ratios, color-coded by the values in the emission line image shown in the middle panel for each spaxel. The solid- and dashed-lines represent the location of the Kewley *et al.* (2001) and Kauffmann *et al.* (2003) demarcation lines, respectively.

classical location of H II regions (e.g., Osterbrock 1989). Those are the signatures of star-forming regions (e.g. Espinosa-Ponce *et al.* 2020). Thus, this figure illustrates that neither star-formation (located in clumpy regions across the disk), not quenching (evident in the central regions), happens simultaneously through the optical extension of galaxies. This is also true for any other ionization process observed in galaxies. Fig. 2, bottom panel, shows similar plots for the galaxy merger ESO253-G003. Like in the previous case, this galaxy presents strong star-forming regions, seen as green nodules in the emission line combined image, in particular two gigantic ones at the north of the two cores of the merging galaxies. However, contrary to the previous case, this galaxy has a strong, hard ionization and well peaked emission in the central regions, with high values of the EW(Hα). This is the clear signature of an AGN (Cid Fernandes *et al.* 2010; Gomes *et al.* 2016). On top of that, the hard ionization is distributed along a set of filamentary/clumpy structures emanating from the the central regions and distributed following a more or less biconical structure at both the east and west side of the interacting galaxies. Those are clear signatures of galactic scale outflows (e.g. López-Cobá *et al.* 2020). Therefore, it is evident that the physical processes that produce the ionization in galaxies, that are related with the evolutionary processes in galaxies (star-formation, retirement, AGN activity, outflows, quenching...), does not happens monolitically and simustaneously at a galactic scale. Despite that this is a rather simple, evident and accepted known fact, all these processes have been widely explored and explained as if this was not the case.

The advent of wide-field and multiplexed Integral Field Units (IFUs) in the last decade has allowed to adopt a completely different approach in the exploration of galaxies (in general), and those better resolved (i.e., the ones at the nearby universe). The application of this technique over large and well defined statistical samples of galaxies have allowed us to uncover local/spatial resolved relations and patterns. Among them, the most relevant ones show that: (i) ionization processes in galaxies happen at local scales, and can be understood only by the combined exploration of line-ratios, EW(Hα), morphology/shape of the ionized structures and kinematic analysis (e.g., Fig. 2 López-Cobá *et al.* 2020); (ii) the global scaling relations observed among SFGs, like the SFMS, the Mgas-M$_*$, the SK-law or the Mass-Metallicity relation, have local counter-parts that indeed explain both the global ones and the radial gradients observed in many properties across the optical extension of galaxies; (iii) galaxies seem to grow from the inside-out, at least those more massive than $10^{9.5}$M$\odot$, presenting a local/resolved downsizing, meaning that more massive regions (in terms of Σ_*) in galaxies evolves faster than less massive ones; (iv) finally, quenching happens too from the inside-out, strongly connected with bulge growth (and thus, with the presence of hot orbits). The results of all these recent IFS-GS have been reviewed recently in Sanchez (2020), including a detailed explanation of the data used along this manuscript.

References

Baldwin, J. A., Phillips, M. M., & Terlevich, R. 1981, *PASP*, 93, 5
Binette, L., Magris, C. G., Stasińska, G., & Bruzual, A. G. 1994, *A&A*, 292, 13
Blanton, M. R. & Moustakas, J. 2009, *ARA&A*, 47, 159
Bundy, K. Bershady, M. A., Law, D. R., *et al.* 2015, *ApJ*, 798, 7
Cid Fernandes, R., Stasińska, G., Schlickmann, M. S., *et al.* 2010, *MNRAS*, 403, 1036
Croom, S. M., Lawrence, J. S., Bland-Hawthorn, J., *et al.* 2012, *MNRAS*, 421, 872
Dopita, M. A., Koratkar, A. P., Evans, I. N., *et al.* 1996, in Astronomical Society of the Pacific Conference Series, Vol. 103, The Physics of Liners in View of Recent Observations, ed. M. Eracleous, A. Koratkar, C. Leitherer, & L. Ho, 44
Driver, S. P., Norberg, P., Baldry, I. K., *et al.* 2009, *Astronomy and Geophysics*, 50, 5.12
Espinosa-Ponce, C., Sánchez, S. F., Morisset, C., *et al.* 2020, *MNRAS*, 494, 1622
Gomes, J. M., Papaderos, P., Vílchez, J. M., *et al.* 2016, *A&A*, 585, A92
Heavens, A., Panter, B., Jimenez, R., *et al.* 2004, *Nature*, 428, 625
Hopkins, P. F., Cox, T. J., Younger, J. D., *et al.* 2009, *ApJ*, 691, 1168
Kauffmann, G., Heckman, T. M., Tremonti, C., *et al.* 2003, *MNRAS*, 346, 1055
Kennicutt, Jr., R. C. 1998, *ApJ*, 498, 541
Kewley, L. J., Dopita, M. A., Sutherland, R. S., *et al.* 2001, *ApJ*, 556, 121
Lacerda, E. A. D., Cid Fernandes, R., Couto, G. S., *et al.* 2018, *MNRAS*, 474, 3727
López-Cobá, C., Sánchez, S. F., Anderson, J. P., *et al.* 2020, *AJ*, 159, 167
Madau, P., & Dickinson, M. 2014, *ARA&A*, 52, 415
Martig, M., Bournaud, F., Teyssier, R., *et al.* 2009, *ApJ*, 707, 250
Muzzin, A., Marchesini, D., Stefanon, M., *et al.* 2013, *ApJ*, 777, 18
Osterbrock, D. E. 1989, Astrophysics of gaseous nebulae and active galactic nuclei (University Science Books)
Renzini, A., & Peng, Y.-j. 2015, *ApJL*, 801, L29
Saintonge, A., Catinella, B., Cortese, L., *et al.* 2016, *MNRAS*, 462, 1749
Sanchez, S. F. 2020, *ARA&A*, 58, 99
Sánchez, S. F., Kennicutt, R. C., Gil de Paz, A., *et al.* 2012, *A&A*, 538, A8
Sánchez, S. F., Pérez, E., Sánchez-Blázquez, P., *et al.* 2016, *Rev. Mexicana Astron. Astrofis.*, 52, 171
Sánchez, S. F., Avila-Reese, V., Hernandez-Toledo, H., *et al.* 2018, *Rev. Mexicana Astron. Astrofis.*, 54, 217
Sánchez, S. F., Avila-Reese, V., Rodríguez-Puebla, A., *et al.* 2019, *MNRAS*, 482, 1557
Schawinski, K., Urry, C. M., Simmons, B. D., *et al.* 2014, *MNRAS*, 440, 889
Singh, R., van de Ven, G., Jahnke, K., *et al.* 2013, *A&A*, 558, A43
Speagle, J. S., Steinhardt, C. L., Capak, P. L., *et al.* 2014, *ApJS*, 214, 15
Stasińska, G., Vale Asari, N., Cid Fernandes, R., *et al.* 2008, *MNRAS*, 391, L29
York, D. G., Adelman, J., Anderson, Jr., J. E., *et al.* 2000, *AJ*, 120, 1579

Galaxy Evolution and Feedback across Different Environments
Proceedings IAU Symposium No. 359, 2020
T. Storchi-Bergmann, W. Forman, R. Overzier & R. Riffel, eds.
doi:10.1017/S1743921320001866

Radio galaxies with and without emission lines

Grażyna Stasińska[1], Natalia Vale Asari[2] and Dorota Kozieł-Wierzbowska[3]

[1]LUTH, Observatoire de Paris, PSL, CNRS 92190 Meudon, France
email: grazyna.stasinska@obspm.fr

[2]Departamento de Física - CFM - Universidade Federal de Santa Catarina, Florianópolis, SC, Brazil

[3]Astronomical Observatory, Jagiellonian University, ul. Orla 171, PL-30244 Krakow, Poland

Abstract. Using the recent ROGUE I catalogue of galaxies with radio cores (Kozieł-Wierzbowska *et al.* 2020) and after selecting the objects which are truly radio active galactic nuclei, AGNs, (which more than doubles the samples available so far), we perform a thorough comparison of the properties of radio galaxies with and without optical emission lines (galaxies where the equivalent width of Hα is smaller than 3Å are placed in the last category). We do not find any strong dichotomy between the two classes as regards the radio luminosities or black hole masses. The same is true when using the common classification into high- and low-excitation radio galaxies (HERGs and LERGs respectively).

Keywords. radio continuum: galaxies, galaxies: active, catalogs

---

1. Introduction

Recently, the first part of a catalogue of Radio sources associated with Optical Galaxies and having Unresolved or Extended morphologies (ROGUE I) has been published by Kozieł-Wierzbowska *et al.* (2020). It contains 32,616 spectroscopically selected galaxies from the Sloan Digital Sky Survey (SDSS; York *et al.* 2000) that have core identifications in the First Images of Radio Sky at Twenty Centimetre survey (FIRST: Becker *et al.* 1995, http://sundog.stsci.edu). The ROGUE II catalogue which will contain galaxies *without* cores is still in preparation, so here we present results based only on ROGUE I. This is by far the largest handmade catalog of this kind. Here we use it to revisit the differences between HERGs and LERGs previously discussed by many authors (eg. Best & Heckman 2012 and references therein). In Section 2 we briefly present the ROGUE I catalogue, in Section 3 we show how we extracted the radio AGNs. In Section 4 we recall several definitions of HERGs and LERGs and propose our own. In Sections 5 and 6 we compare the properties of the two classes.

2. The ROGUE I catalogue

The master sample is drawn from the 7th SDSS data release (Abazajian *et al.* 2009). It consists of the 662,531 galaxies from the Main Galaxy Sample and the Red Galaxy Sample for which the spectra have a signal-to-noise (S/N) ratio in the continuum at 4020Å larger than 10. Such a condition allows reliable determinations of parameters such as stellar masses, $M_{\star}$, black hole masses, $M_{\rm BH}$, emission-line intensities etc.

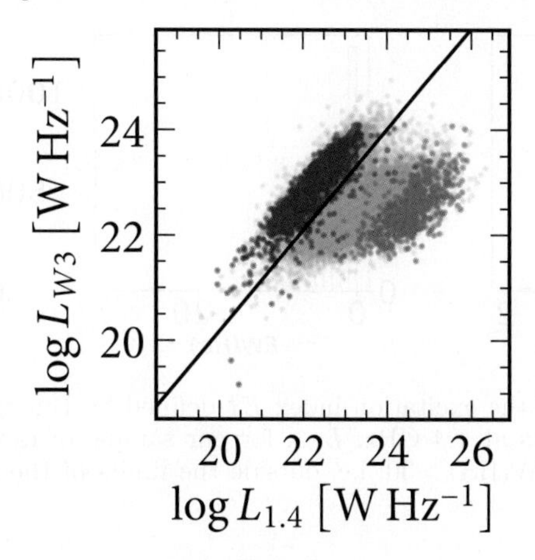

Figure 1. Diagram separating radio-AGNs from galaxies whose radio emission is related to star formation only. Blue points are star-forming galaxies according to the BPT diagram, red points are extended radio galaxies.

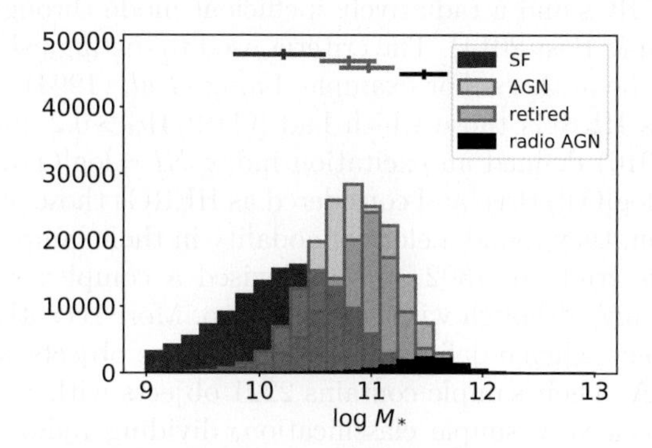

Figure 2. Histograms of stellar masses for the various galaxy types in the master sample.

The main products of the ROGUE I catalog are the identifications and redshifts of 32,616 radio sources associated to galaxies, the overlay maps, the total and core radio fluxes at 1.4 GHz, and a morphological classification in both the radio and the optical.

3. Extracting radio AGNs

ROGUE I contains both radio AGNs and sources whose radio emission is related to star formation only. It has been found (Kozieł-Wierzbowska *et al.* 2021, in preparation) that plotting the luminosities in the mid-infrared band W3, L_{W3}, versus the radio luminosities at 1.4 GHz, $L_{1.4}$, separates the two categories of radio sources very neatly (see Figure 1). Out of the 32,616 radio-sources in ROGUE I, 22,918 have a redshift $z < 0.3$ and can be plotted in this diagram. Out of those, 10,826 are radio-AGNs while 12,092 are in the star-forming branch (which also contains optical AGNs).

Figure 2 shows that, as expected, radio-AGNs are at the tip of the distribution of stellar masses in the ROGUE I master sample.

4. HERG and LERG definitions

Hines & Longair (1979), studying the spectra of 72 3CR galaxies, distinguished galaxies with strong emission-line spectra and those without strong emission lines. Later studies of the radio galaxies divided them into high- and low-excitation objects (HERGs and

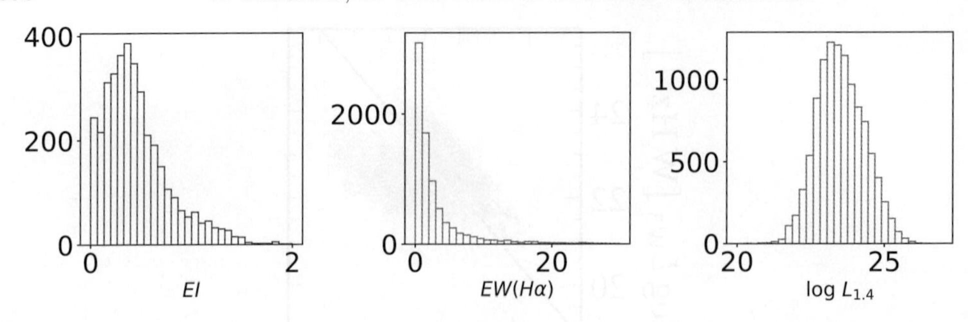

Figure 3. Histograms of the excitation index EI defined by Buttiglione, of EW(Hα), and of the total radio luminosities at 1.4 GHz, $L_{1.4}$, for our sample of radio galaxies. Note that the number of objects with EW(Hα) > 30, i.e. outside the limits of the histogram for EW(Hα), is around 200 only.

LERGs; see Padovani *et al.* 2017 for a review). These two classes have been suggested to correspond to different fueling modes of the AGN: a radiatively efficient mode through an accretion disk for HERGs and a radiatively inefficient mode through spherical accretion for LERGs (Heckman & Best 2014). The criteria used to distinguish HERGs and LERGs vary slightly among the authors. For example, Laing *et al.* (1994), studying a sample of 88 objects defined as HERGs those which had [O III]/H$\alpha > 0.2$ and EW([O III]) > 3 Å. Buttiglione *et al.* (2010) defined an excitation index $EI = \log$[O III]/H$\beta - 1/3(\log$[N II]/H$\alpha + \log$[S II]/H$\alpha + \log$[O I]/Hα), and considered as HERGs those objects having $EI > 1$. With such a definition, they found a clear bimodality in their sample of 113 objects. Best & Heckman for their study of 7302 objects devised a complex scheme and found an observational dichotomy, although with some overlap. More recently, Pracy *et al.* (2016) came back to a simpler scheme defining as HERGs those objects with S/N ([O III]) > 3 and EW([O III]) > 5 Å. Their sample contains 2221 objects with $z < 0.3$.

We propose to use a very simple classification, dividing radio galaxies into objects with and without emission lines. Among objects 'without' emission lines, we count those which have EW(Hα) < 3Å, since in this case the emission lines are likely due to ionization by hot low-mass evolved stars (HOLMES, Cid Fernandes *et al.* 2011). Although in the following, we still use the common HERGs/LERGs nomenclature, the real meaning of the division we propose is radio galaxies with or without an optical AGN.

5. HERGs versus LERGs in the ROGUE I sample

Fig. 3 shows the distribution of the excitation index EI defined by Buttiglione *et al.* (2010), of EW(Hα), and of the total radio luminosities at 1.4 GHz, $L_{1.4}$, for the 10,826 radio-AGNs of our ROGUE I sample. No dichotomy is apparent in our sample for any of those parameters.

It has been claimed that HERGs have larger radio luminosites than LERGs. The histograms plotted in Fig. 4 together with the values of the medians and quartiles show that the distributions of radio luminosities are indistinguishable between HERGs and LERGs, whatever definition of these categories is adopted.

The same applies for the black hole masses, as seen in Fig. 5.

In fact, among all the parameters tested the ones which show the largest difference between HERGs and LERGs are the stellar extinction, A_V (obtained by spectral synthesis fitting of the optical continua using the code *STARLIGHT* of Cid Fernandes *et al.* 2005) and D_{4000}, the discontinuity at 4000Å in the optical spectra. Both parameters are related to star formation: A_V indicates the presence of dust, implying the existence of cold or warm gas, while D_{4000} is an indicator of the mean age of the stellar population, being

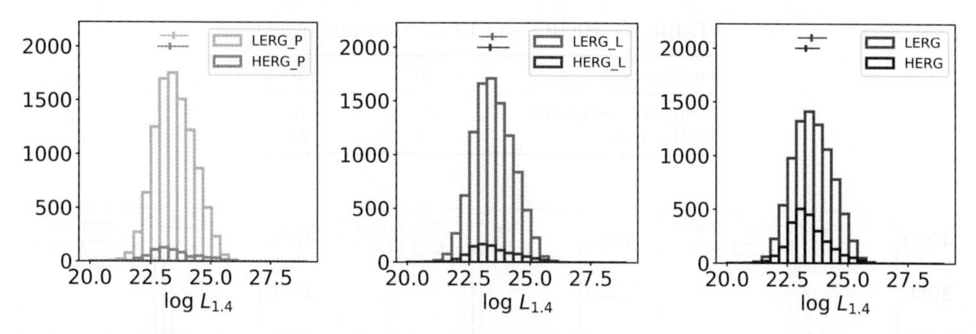

Figure 4. Histograms of total (i.e. including the lobes in case of extended sources) radio luminosities at 1.4 GHz for different definitions of HERGs (blue) and for LERGs (red). *Left panel:* definition of Pracy *et al.* (2016); *middle:* definition of Laing *et al.* (1994); *right:* our definition. The segments on top indicate the values of the medians and quartiles.

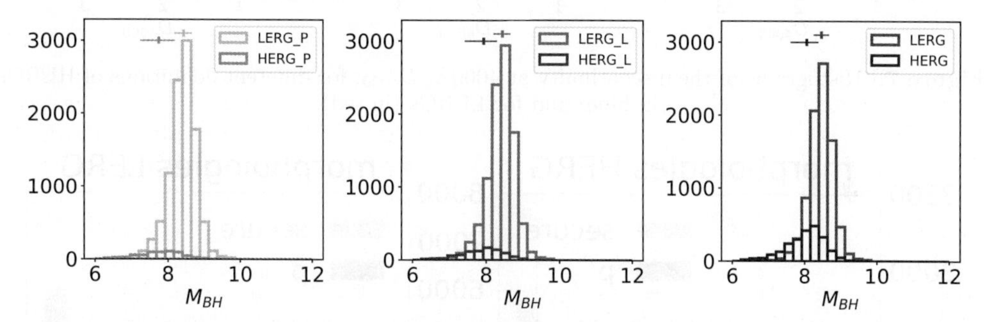

Figure 5. Histograms of the black hole masses for different definitions of HERGs (in blue) and for LERGs (in red).

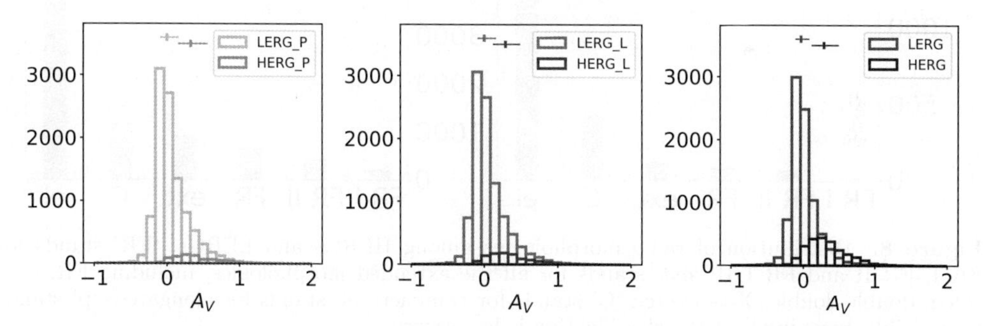

Figure 6. Histograms of the stellar extinction, A_V for different definitions of HERGs (in blue) and for LERGs (in red).

smaller in the presence of young stellar populations. The difference seen between HERGs and LERGs from these two parameters (see Figs. 6 and 7) indicates a larger amount of low-level star formation for HERGs. It must be noted, however, that the overlap is quite important.

6. HERGs and LERGs radio morphologies

Table 1 summarizes the radio properties and Figure 8 shows the distribution of radio morphologies in the HERG and LERG samples (using our definition of HERG and LERG). The vast majority of the objects are compact (ie. unresolved), or elongated (ie. one-component source with one deconvolved dimension larger than zero).

Table 1. HERG and LERG morphologies.

number ratio	HERGs	LERGs
compacts / extended	23	10
FRII / FRI	6.2	2.4

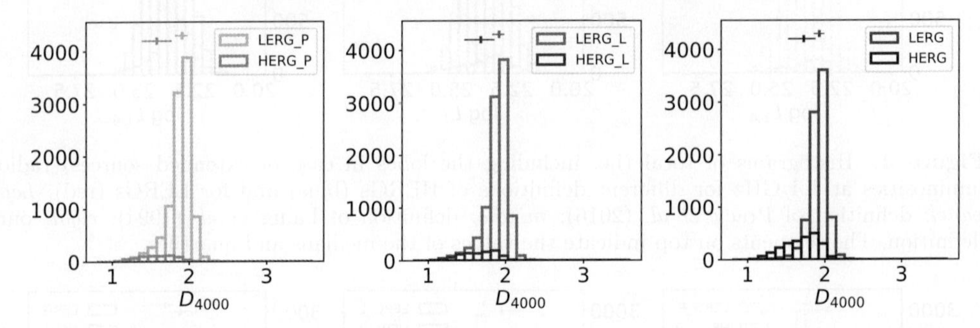

Figure 7. Histograms of the discontinuity at 4000Å, D_{4000}, for different definitions of HERGs (in blue) and for LERGs (in red).

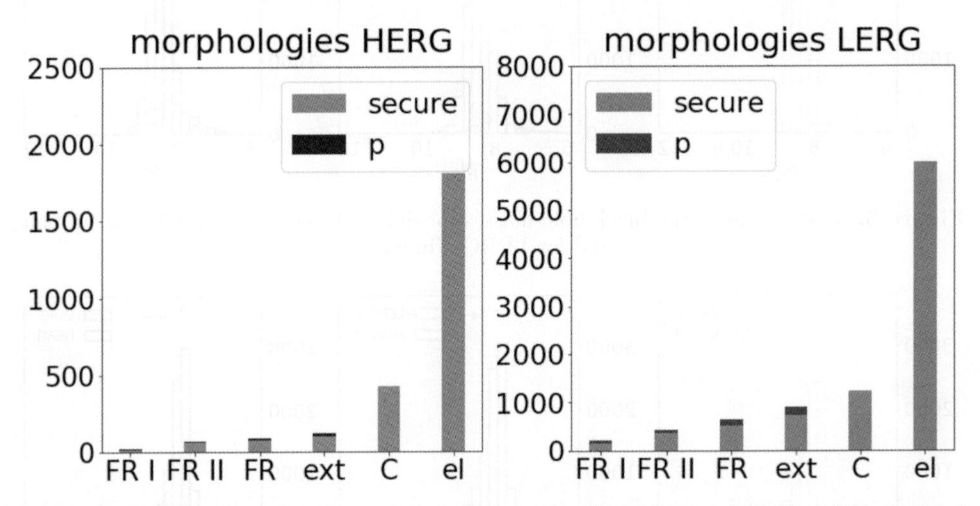

Figure 8. Distribution of radio morphologies among HERGs and LERGs. 'FR' stands for FR I, FR II and FR I/II; 'ext' stands for all the extended morphologies, including FR, one-sided, double-double, X-type etc.; 'C' stands for compact; 'el' stands for elongated; 'p' stands for possible, meaning that the classification is less secure.

The proportion of extended radio sources is larger for LERGs than HERGs, as is the proportion of FR I with respect to FR II.

A detailed version of this work will be published including the results from ROGUE II, the catalogue of radio sources without radio cores (in preparation).

References

Abazajian, K. N., *et al.* 2009, *ApJS*, 182, 543
Becker, R. H., White, R. L., Helfand, D. J., *et al.* 1995, *ApJ*, 450, 559
Best, P. N. & Heckman, T. M. 2012, *MNRAS*, 421, 1569
Buttiglione, S., Capetti, A., Celotti, A., *et al.* 2010, *A&A*, 509, A6
Cid Fernandes, R., Mateus, A., Sodré, L., *et al.* 2005, *MNRAS*, 358, 363
Cid Fernandes, R., Stasińska, G., Mateus, A., *et al.* 2011, *MNRAS*, 413, 1687

Heckman, T. M. & Best, P. N. 2014, *ARAA*, 52, 589
Hine, R. G. & Longair, M. S. 1979, *MNRAS*, 188, 111
Kozieł-Wierzbowska, D., Goyal, A., & Żywucka, N. 2020, *ApJS*, 247, 53
Laing, R. A., Jenkins, C. R., Wall, J. V., *et al.* 1994, *ASPC*, 54, 201
Padovani, P., *et al.* 2017, *AARv*, 25, 2
Pracy, M. B., Ching, J. H. Y., Sadler, E. M., *et al.* 2016, *MNRAS*, 460, 2
York, D. G., *et al.* 2000, *AJ*, 120, 1579

Galaxy Evolution and Feedback across Different Environments
Proceedings IAU Symposium No. 359, 2020
T. Storchi-Bergmann, W. Forman, R. Overzier & R. Riffel, eds.
doi:10.1017/S1743921320001817

The DIVING^{3D} Survey - Deep IFS View of Nuclei of Galaxies

J. E. Steiner[1], R. B. Menezes[2], T. V. Ricci[3] and DIVING^{3D} team

[1]Instituto de Astronomia, Geofísica e Ciências Atmosféricas, Universidade de São Paulo, 05508-090, São Paulo, SP, Brasil
email: joao.steiner@iag.usp.br

[2]Instituto Mauá de Tecnologia, Praça Mauá 1, 09580-900, São Caetano do Sul, SP, Brasil

[3]Universidade Federal da Fronteira Sul, Campus Cerro Largo, RS, 97900-000, Brasil

Abstract. The DIVING^{3D} Survey (Deep Integral Field Spectrograph View of Nuclei of Galaxies) aims to observe, with high signal/noise and high spatial resolution, a statistically complete sample of southern galaxies brighter than B = 12.0. The main objectives of this survey are to study: 1) the nuclear emission line properties; 2) the circumnuclear emission line properties; 3) the central stellar kinematics and 4) the central stellar archaeology. Preliminary results of individual or small groups of galaxies have been published in 18 papers.

Keywords. galaxies: active – galaxies: nuclei – galaxies: statistics – galaxies: kinematics and dynamics – galaxies: stellar content – galaxies: Seyfert – surveys

---

1. Introduction

Statistical surveys may provide important clues about the nature, spatial distribution and evolution of objects. Astronomy has seen a growing number of surveys of all kinds. The area of Active Galactic Nuclei (AGN) has profited significantly from surveys of various sizes.

One such important survey was the Palomar Survey (Filippenko & Sargent 1985): a spectroscopic survey of all northern galaxies brighter than B = 12.5 with $\delta > 0°$ and $|b| > 10°$. This survey was done on the Palomar Telescope and used a slit of $2'' \times 4''$. Most of the data and results were published many years after the survey started (Ho *et al.* 1995, 1997a,b,c). An important review with the statistical analysis of this survey was published 23 years after the first paper by L.C. Ho: Nuclear Activity in Nearby Galaxies (Ho 2008).

2. The DIVING^{3D} Survey

Inspired by the success of the Palomar survey, we started a few years ago a survey with optical integral field unit (IFU) spectroscopy. All 170 galaxies with $B < 12.0$, $\delta < 0°$ and $|b| > 15°$ are by now already observed. The 8 m Gemini telescopes were used to observe 150 galaxies with good seeing conditions only. The Gemini Multi-Object Spectrograph (GMOS) IFU field of view (FOV) of the observations are $3.5'' \times 5''$ and fibres covered $0.2''$; all observations were seeing limited. The other 20 objects were observed with the SOAR Integral Field Spectrograph (SIFS) on the SOAR telescope. The SIFS FOV is $15'' \times 7.8''$ with a $0.3''$ fibre. The spectral coverage is 4300–6800 Å with a typical spectral resolution of 1.3 Å and a seeing condition of $0.5''$–$0.8''$. The idea of the survey is to have the highest possible spatial resolution and signal/noise. Other important surveys using optical IFUs were done previously (SAURON – Bacon *et al.* 2001; ATLAS^{3D} – Cappellari *et al.* 2011; CALIFA – Sánchez *et al.* 2012; MaNGA – Bundy *et al.* 2015) among others.

The main DIVING^{3D} objectives are:

(a) *Nuclear emission line properties*: we intend to quantify with high signal/noise and high (seeing limited) spatial resolution the properties of the nuclei of all galaxies in the sample. Our objective is to detect the faintest AGN that can currently be observed in the optical. The BPT diagrams of these objects may show a decrease in "Transition Objects" (TO) population, as H II regions tend to be separated from the "true nuclei".

(b) *Circumnuclear emission line properties*: the circumnuclear emission may reveal important aspects about the ionizing sources in the central region of the galaxies. Is the nuclear AGN responsible for the circumnuclear emission or do we need further sources of excitation/ionization such as shock waves or Post-AGB stars?

In addition to these main objectives, as by-products, we will obtain important information about:

(c) *Central stellar kinematics*: not only information about stellar disc rotation and bulge velocity dispersion but also anomalies such as counter-rotation and decoupling.

(d) *Central stellar archaeology*: stellar population synthesis will always be attempted in order to remove the stellar "noise" from the emission lines. As a byproduct, one may obtain information about the star formation history of the nucleus and the circumnuclear environment.

The DIVING^{3D} galaxies morphological types are:

- E: 30
- S0: 32
- Sab: 59
- Scd: 49
- Total: 170 galaxies

3. Methods and Early results

Our group has developed a series of techniques that treat the raw data. This includes instrumental fingerprints removal by using Principal Component Analysis (PCA) Tomography; Butterworth filtering for removal of high-frequency spatial noise; deconvolution etc (see Menezes *et al.* 2019). Fig. 1 shows the effect of such treatments for an extreme case.

In addition we also used stellar spectral synthesis by using STARLIGHT (Cid Fernandes *et al.* 2005) and PPXF (Cappellari & Emsellem 2004). It is important to remove the stellar component from the original data cube in order to remain with the gas cube (basically showing emission lines only) - see Fig. 2.

Emission lines are analysed in terms of Gaussian decomposition. This procedure starts with fitting the [S II] doublet with a two kinematic component model. This is, then, fitted to the [N II] + Hα lines. In order to improve the fit, a broad Hα component is added when necessary (see Fig. 3).

To see the importance of high spatial resolution ($FWHM_{PSF} \sim 0.70''$) we show, in Fig. 4, the counter-rotation between the stellar and gaseous discs in the nucleus of the LINER galaxy IC 1459. Just for comparison, the fibre of the SDSS is $3''$. The critical information here happens on a significantly smaller scale. This shows one of the key differences of the DIVING^{3D} survey when compared to other optical IFU surveys.

As preliminary results, to date, a total of 18 papers have been published on individual or small groups of objects from the DIVING^{3D} Survey: Menezes *et al.* (2013); Ricci *et al.* (2014a,b); Menezes *et al.* (2014a,b); Ricci *et al.* (2015, 2016); Menezes *et al.* (2016); Menezes & Steiner (2017); Diniz *et al.* (2017); da Silva *et al.* (2017); Ricci *et al.* (2018);

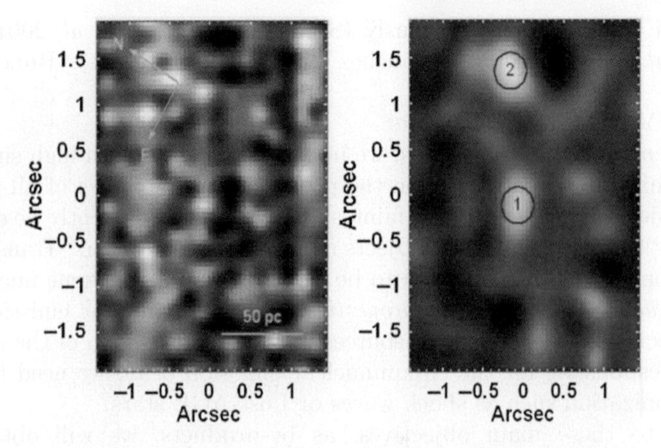

Figure 1. Typical galaxies have bright nuclei; some can be quite obscured as is the case of NGC 2835 (above). We have developed data treatment techniques that help in extracting information (see Menezes *et al.* 2019), transforming the raw data (left) into the treated data (right). Object 1 is the LINER nucleus and Object 2 is an H II region.

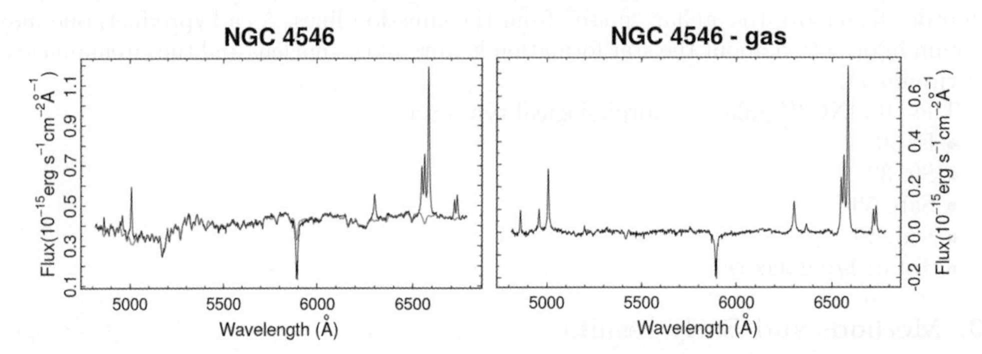

Figure 2. A typical spectrum of a S0 galaxy (left) and the same spectrum after removing the stellar component using spectral synthesis (right). This procedure is important, especially for studying weak emission lines.

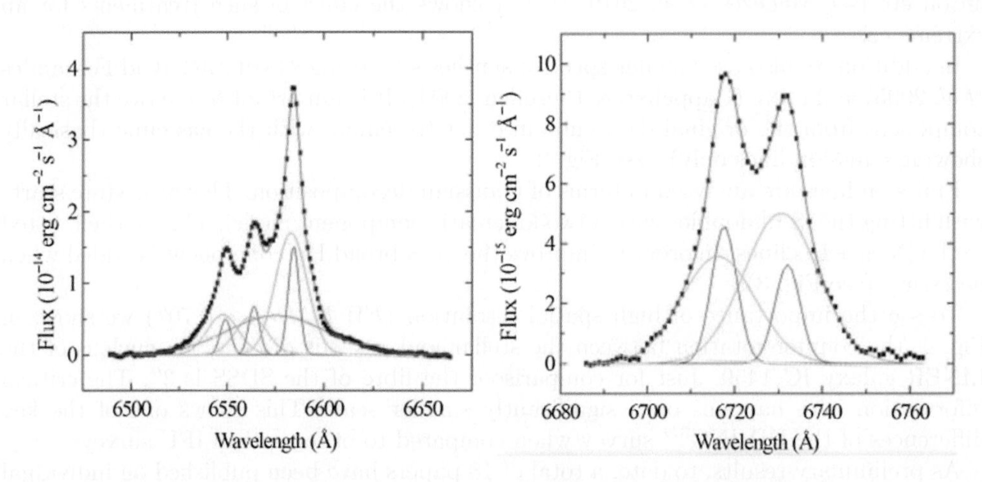

Figure 3. We modelled each blend by fitting the [S II] doublet (right) in two kinematic components. We, then, fitted the [N II]+Hα lines with the same kinematic components. A third broad Hα component was added when necessary.

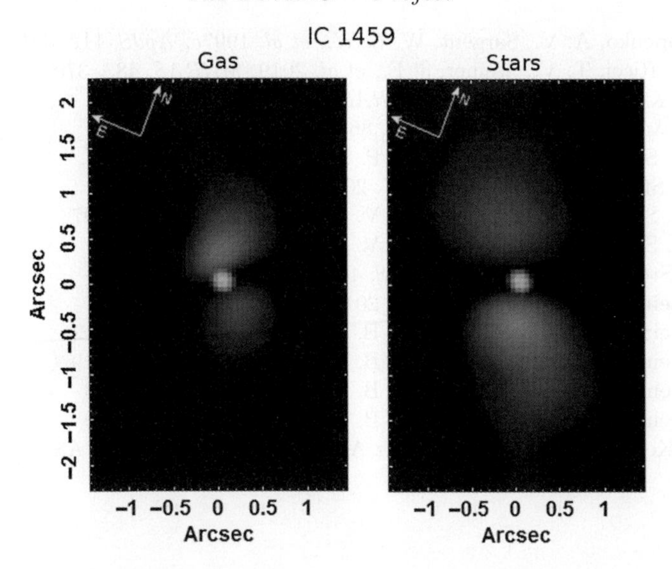

Figure 4. The counter-rotating gaseous and stellar discs (from Ricci *et al.* 2014b). This shows the spatial scale of such phenomena and the importance of high spatial resolution.

da Silva *et al.* (2018); Menezes & Steiner (2018); Dahmer-Hahn *et al.* (2019a,b); Ricci & Steiner (2019); da Silva *et al.* (2020).

4. Summary

The main early results of the survey can be seen in the contributions of Menezes *et al.* and Ricci *et al.* in this Proceedings. Below we summarize some of them:

- There seems to be a dichotomy in the BPT diagrams when using high spatial resolution data. The TO region seems to be depopulated, as a consequence of higher spatial resolution.
- There seem to be intrinsic TO objects, like superwind starbursting galaxies.
- 91 % of the early-type galaxies show nuclear emission lines.
- 50 % of the early-type galaxies show Seyfert or LINER type spectra.
- 31 % of early-type galaxies are AGN of Type 1 (as compared with 15 % in Palomar).

References

Bacon, R., Copin, Y., Monnet, G., *et al.* 2001, *MNRAS*, 326, 23
Bundy, K., *et al.* 2015, *ApJ*, 798, 7
Cappellari, M. & Emsellem, E. 2004, *PASP*, 116, 138
Cappellari, M., *et al.* 2011, *MNRAS*, 413, 813
Cid Fernandes, R., Mateus, A., Sodré, L., *et al.* 2005, *MNRAS*, 358, 363
da Silva, P., Menezes, R. B., & Steiner, J. E. 2020, *MNRAS*, 492, 5121
da Silva, P., Steiner, J. E., & Menezes, R. B. 2017, *MNRAS*, 470, 3850
da Silva, P., Steiner, J. E., & Menezes, R. B. 2018, *ApJ*, 861, 83
Dahmer-Hahn, L. G., Riffel, R., Steiner, J. E., *et al.* 2019a, *MNRAS*, 482, 5211
Dahmer-Hahn, L. G., Riffel, R., Ricci, T. V., *et al.* 2019b, *MNRAS*, 489, 5653
Diniz, S. I. F., Pastoriza, M. G., Hernand ez-Jimenez, J. A., *et al.* 2017, *MNRAS*, 470, 1703
Filippenko, A. V. & Sargent, W. L. W. 1985, *ApJS*, 57, 503
Ho, L. C. 2008, *ARA&A*, 46, 475
Ho, L. C., Filippenko, A. V., & Sargent, W. L. 1995, *ApJS*, 98, 477
Ho, L. C., Filippenko, A. V., & Sargent, W. L. W. 1997a, *ApJS*, 112, 315
Ho, L. C., Filippenko, A. V., & Sargent, W. L. W. 1997b, *ApJ*, 487, 568

Ho, L. C., Filippenko, A. V., Sargent, W. L. W., *et al.* 1997c, *ApJS*, 112, 391
Menezes, R. B., Ricci, T. V., Steiner, J. E., *et al.* 2019, *MNRAS*, 483, 3700
Menezes, R. B. & Steiner, J. E. 2017, *MNRAS*, 466, 749
Menezes, R. B. & Steiner, J. E. 2018, *ApJ*, 868, 67
Menezes, R. B., Steiner, J. E., & da Silva, P. 2016, *ApJ*, 817, 150
Menezes R. B., Steiner J. E., & Ricci T. V. 2013, *ApJ*, 765, L40
Menezes, R. B., Steiner, J. E., & Ricci, T. V. 2014a, *MNRAS*, 438, 2597
Menezes, R. B., Steiner, J. E., & Ricci, T. V. 2014b, *ApJ*, 796, L13
Ricci, T. V. & Steiner, J. E. 2019, *MNRAS*, 486, 1138
Ricci, T. V., Steiner, J. E., May, D., *et al.* 2018, *MNRAS*, 473, 5334
Ricci, T. V., Steiner, J. E., & Menezes, R. B. 2014a, *MNRAS*, 440, 2442
Ricci, T. V., Steiner, J. E., & Menezes, R. B. 2014b, *MNRAS*, 440, 2419
Ricci, T. V., Steiner, J. E., & Menezes, R. B. 2015, *MNRAS*, 451, 3728
Ricci, T. V., Steiner, J. E., & Menezes, R. B. 2016, *MNRAS*, 463, 3860
Sánchez S. F., Kennicutt, R. C., Gil de Paz A., *et al.* 2012, *A&A*, 538, A8

Galaxy Evolution and Feedback across Different Environments
Proceedings IAU Symposium No. 359, 2020
T. Storchi-Bergmann, W. Forman, R. Overzier & R. Riffel, eds.
doi:10.1017/S1743921320001702

Surface Brightness Fluctuations for constraining the chemical enrichment of massive galaxies

A. Vazdekis[1,2], P. Rodríguez-Beltrán[1,2], M. Cerviño[3], M. Montes[4], I. Martín-Navarro[5,6] and M. B. Beasley[1,2]

[1]Instituto de Astrofísica de Canarias, E-38200 La Laguna, Tenerife, Spain
email: vazdekis@iac.es

[2]Departamento de Astrofísica, Universidad de La Laguna, E-38205, Tenerife, Spain

[3]Centro de Astrobiología (CSIC/INTA), ESAC Campus, Camino Bajo del Castillo s/n, E-28692 Villanueva de la Cañada, Spain

[4]School of Physics, University of New South Wales, Sydney, NSW 2052, Australia

[5]Max-Planck Institut für Astronomie, Konigstuhl 17, D-69117 Heidelberg, Germany

[6]University of California Santa Cruz, 1156 High Street, Santa Cruz, CA 95064, USA

Abstract. Based on very deep photometry, Surface Brightness Fluctuations (SBF) have been traditionally used to determine galaxy distances. We have recently computed SBF spectra of stellar populations at moderately high resolution, which are fully based on empirical stars. We show that the SBF spectra provide an unprecedented potential for stellar population studies that, so far, have been tackled on the basis of the mean fluxes. We find that the SBFs are able to unveil metal-poor stellar components at the one percent level, which are not possible to disentangle with the standard analysis. As these metal-poor components correspond to the first stages of the chemical enrichment, the SBF analysis provides stringent constrains on the quenching epoch.

Keywords. galaxies: abundances, galaxies: elliptical and lenticular,cD, galaxies: stellar content, galaxies: distances and redshifts, galaxies: evolution

1. Introduction

Stellar populations are characterized by differences in the luminosity distribution of stars contributing to the flux in a given resolution element. By normalizing the variance, i.e. the second moment, of these fluctuations by the mean flux, i.e. the first moment, in that element we obtain the so called SBF (Tonry & Schneider 1988; Tonry, Ajhar & Luppino 1990). The SBF is an intrinsic property of an SSP, i.e. a simple, single-burst, stellar population characterized by a single-age and single-metallicity and, therefore, it depends on these parameters. For obtaining accurate fluctuation magnitudes we need very high quality photometry as well as subtracting a smooth galaxy model and it is often applied a Fourier Transform analysis to isolate the intrinsic fluctuations of the stellar populations.

So far the main application of the SBF method it has been its use for obtaining accurate galaxy distances in the nearby Universe, as the observed flux of the fluctuations depends on galaxy distance, with more distant galaxies appearing smoother. The SBFs provided distances to Virgo and Fornax with a precision of 2% (e.g., Blakeslee *et al.* 2010). SBFs have also been shown to provide additional constrains to relevant stellar population parameters (Liu, Charlot & Graham 2000; Blakeslee, Vazdekis & Ajhar 2001). In fact, by confronting theoretical SBF model predictions with the observational fluctuations it is possible to break the age/metallicity degeneracy (Worthey 1994; Cantiello *et al.* 2003). However such applications are very scarce in the literature, mostly due to the lack of SBF determinations in more than just a single band for a given object.

As it happens with the mean fluxes, the SBF method can be also used in spectroscopic studies, with increased abilities to break main stellar population degeneracies through key spectral features. Low resolution theoretical SBF spectra can be derived from the predictions of Buzzoni (1993) or at high resolution González Delgado *et al.* (2005) as shown in Cerviño (2013). These models are based on fully theoretical stellar spectral libraries. Recently were presented model SBF spectra at moderately high resolution (Mitzkus *et al.* 2018; Vazdekis *et al.* 2020), based on empirical stellar libraries. Very recently it has been presented the first observational SBF spectrum of a nearby S0 galaxy Mitzkus *et al.* (2018), using data from Multi Unit Spectroscopic Explorer (MUSE) Integral Field Spectrograph instrument. Although promising, much work is required to propose a robust treatment of the data to obtain the observational SBF spectra as well as to define an optimal methodology to extract the information contained in these spectra (Vazdekis *et al.* 2020). Here we employ fluctuation colour-colour plots from our recently computed E-MILES (Extended - MILes de EStrellas) SBF model spectra to study the first stages of the evolution of Early-Type Galaxies (ETGs). This allows us to constrain the very first stages of their chemical evolution.

2. Models

Here we employ the E-MILES SBF model spectra† presented in Vazdekis *et al.* 2020. Briefly, these models combine the isochrones of Girardi *et al.* (2000) and Pietrinferni *et al.* (2004) with fully empirical stellar spectral libraries. These libraries include the Hubble Space Telescope based Next Generation Stellar Library (NGSL, Gregg *et al.* 2006), Medium-resolution Isaac Newton telescope Library of Empirical Spectra (MILES, Sánchez-Blázquez *et al.* 2006), Indo-US (Valdes *et al.* 2004), Calcium Triplet (CaT, Cenarro *et al.* 2001a) and Infra-Red Telescope Facility library (IRTF, Cushing, Raynier & Vacca 2005; Raynier, Cushing & Vacca 2009). The models are computed for a suite of Initial Mass Function (IMF) shapes and slopes (Kroupa 2001; Chabrier 2001; Vazdekis *et al.* 1996). The various spectral ranges covered by these models are joined as described in Vazdekis *et al.* (2016) to build-up both the mean and the SBF extended E-MILES spectra at moderately high resolution.

Working with SBFs has some peculiarities to take into account as extensively described in Vazdekis *et al.* (2020). This concerns the modelling of more complex stellar populations and obtaining spectroscopic SBF magnitudes. In brief, due to the properties of the variance, we cannot combine the SBF spectra of the SSPs that contribute to a composite stellar population to obtain its SBF spectrum. The two, the SSP variance and the SSP mean spectra need to be combined separately and, only then, it is possible to divide them to obtain the composite SBF spectrum. Another important peculiarity to take into account concerns the obtention of SBF magnitudes from the SBF spectra. In this case we cannot simply convolve the SBF spectrum with the response of the desired filter, as there are correlations among the resolution elements. Several options were investigated in Vazdekis *et al.* (2020) to conclude that, under the very conservative hypothesis of full correlation among pixels, the spectroscopic SBF magnitude can be derived as

$$\bar{m}(\mathrm{spec}) = 2 \times m_{\sqrt{F_\lambda^{\mathrm{var}}}} - m_{F_\lambda^{\mathrm{mean}}} \tag{2.1}$$

† Models can be downloaded from the MILES website: http://miles.iac.es/

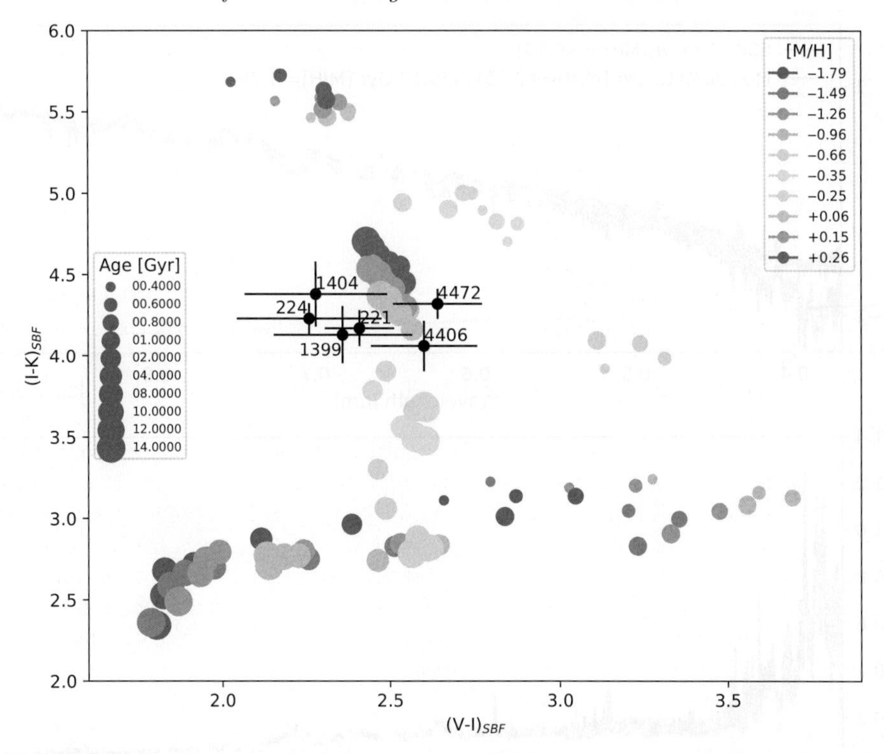

Figure 1. $V-I$ vs. $I-K$ fluctuation colour-colour diagram. The solid coloured circles represent the SBF colours corresponding to SSPs with varying metallicity [M/H] (from very metal-poor to metal-rich, i.e. from purple to red, respectively, as indicated in the top-right inset) and varying age (increasing from 0.4 to 14 Gyr, with increasing circle size as indicated in the bottom-left inset). The SBF data corresponding to a number of representative ETGs are shown in black solid circles, including their errorbars. Note that a fraction of the galaxies fall in a region that is not matched by the SSPs.

where $\bar{m}$ stands for the SBF magnitude, and $m_{\sqrt{F_\lambda^{\rm var}}}$ and $m_{F_\lambda^{\rm mean}}$ are the magnitudes obtained by convolving the filter response with the square root of the SSP variance and the mean SSP, respectively.

3. Results

According to the mean photometric and spectroscopic properties of massive ETGs these galaxies are found to be old and metal-rich (Renzini 2006). Fig. 1 illustrates the potential of the SBFs for further constraining the stellar populations properties of these galaxies. A number of representative ETGs from Cantiello *et al.* (2003) are shown in the $V-I$ vs. $I-K$ fluctuation colour-colour diagnostic diagram. We see that a fraction of these ETGs fall in a region that is not occupied by any of the old SSPs of varying metallicities. Only combinations of very metal-rich and very-metal poor SSPs are able to match these galaxies. As shown in Vazdekis *et al.* (2020), mass fractions of 1−5% of stellar populations with metallicitiies $[Fe/H] < -1$ are required on the top of the largely dominating old metal-rich stellar population to be able to match this set of galaxies.

4. Discussion

So far, these very small contributions from very metal-poor components have been elusive to the standard analysis based on the mean fluxes. Note that these contributions are not related to the well known age/metallicity degeneracy affecting the bulk of the

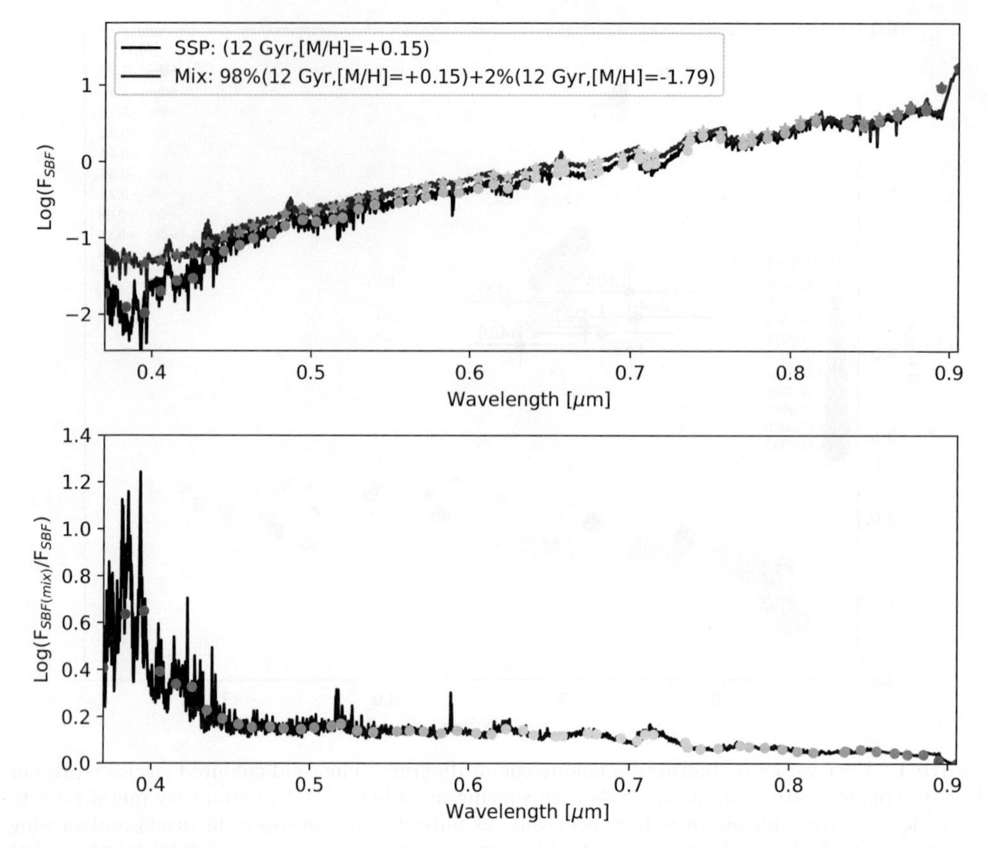

Figure 2. The upper panel shows two SBF spectra, one corresponding to a metal-rich old SSP (as indicated in the inset) and another one where this SSP is combined with a 2% (in mass fraction) of an equally old but very metal-poor SSP (see the inset). The spectroscopic fluctuation magnitudes corresponding to the narrow band filters of the J-PAS survey are indicated with circles of varying colours. The lower panel shows the resulting residuals between these two models (in magnitude). Note that such differences are significant and can be easily captured with the survey data.

population (Worthey 1994). These small components, which have been disentangled by the SBF diagnostic diagram, must be associated to the first stages of galaxy chemical enrichment. In fact, it is remarkable the agreement between these results and the full chemo-evolutionary modelling of Vazdekis *et al.* (1996). These models predicted a rapid (<200 Myr) enrichment for the innermost regions of massive ETGs, formed in-situ, leading to residual mass-fractions of very metal-poor stellar populations smaller than 5%. This light come from long-lived low-mass stars formed at $z > 2$.

The potential of the SBFs for the stellar population studies can be further optimised with the aid of narrow band observations such as those of the J-PAS survey (Javalambre Physics of the Accelerating Universe Astrophysical Survey) (Cenarro *et al.* 2010), which is composed of 56 narrow-band filters covering the optical spectral range. Such observations represent an intermediate step between imaging and spectroscopy, with the important advantage of achieving a high photometric precision (~0.05 mag). Fig. 2 illustrates this potential: tiny contributions from very metal-poor old components lead to magnitude differences above 0.1 mag in the optical range and much larger blueward ~4500 Å. Such differences can be easily detected with the survey data for nearby galaxies.

Urge investigating and proposing diagnostic SBF diagrams based on these filters, which could also help us to uncover the distribution of these metal-poor contributions within nearby galaxies.

References

Blakeslee, J. P., Vazdekis, A., Ajhar, E. A., *et al.* 2001, *MNRAS*, 320, 193

Blakeslee, J. P., Cantiello, M., Mei, S., *et al.* 2010, *ApJ*, 724, 657

Buzzoni, A. 1993, *A&A*, 275, 433

Cenarro, A. J., Cardiel, N., Gorgas, J., *et al.* 2001a, *MNRAS*, 326, 959

Cenarro, A. J., Moles, M., Cristóbal-Hornillos, D., *et al.* 2010, *SPIE*, 7738

Cantiello, M., Raimondo, G., Brocato, E., *et al.* 2003, *AJ*, 125, 2783

Cerviño, M. 2013, *New Astron. Revs.*, 57, 123

Chabrier, G. 2001, *ApJ*, 554, 1274

Cushing, M. C., Raynier, J. T., Vacca, W. D., *et al.* 2005, *ApJ*, 623, 1115

Girardi, L., Bressan, A., Bertelli, G., *et al.* 2000, *A&AS*, 141, 371

González Delgado, R. M., Cerviño, M., Martins, L. P., *et al.* 2005, *MNRAS*, 357, 945

Gregg, M. D., Silva, D., Rayner, J., *et al.* 2006, in The 2005 HST Calibration Workshop: Hubble After the Transition to Two-Gyro Mode, ed. A. M. Koekemoer, P. Goudfrooij, & L. L. Dressel, 209–215

Kroupa, P. 2001, *MNRAS*, 322, 231

Liu, M. C., Charlot, S., Graham, J. R., *et al.* 2000, *ApJ*, 543, 644

Mitzkus, M., Jakob Walcher, C. Roth, M. M., *et al.* 2018, *MNRAS*, 480, 629

Pietrinferni, A., Cassisi, S., Salaris, M., *et al.* 2004, *ApJ*, 612, 168

Raynier, J. T., Cushing, M. C., Vacca, W. D., *et al.* 2009, *ApJS*, 185, 289

Renzini, A. 2006, *ARAA*, 44, 141

Sánchez-Blázquez, P., *et al.* *MNRAS*, 371, 703

Tonry, J. & Schneider, D. P. 1988, *AJ*, 96, 807

Tonry, J. L., Ajhar, E. A., Luppino, G. A., *et al.* 1990, *AJ*, 100, 1416

Valdes, F., Gupta, R., Rose, J. A., *et al.* 2004, *ApJS*, 152, 251

Vazdekis, A., Casuso, E., Peletier, R. F., *et al.* 1996, *ApJS*, 106, 307

Vazdekis, A., Koleva, M., Ricciardelli, E., *et al.* 2016, *MNRAS*, 463, 3409

Vazdekis, A., Cerviño, M., Montes, M., *et al.* 2020, *MNRAS*, 493, 5131

Worthey, G. 1994, *ApJS*, 95, 107

Lucimara Martins

João Steiner

Galaxy Evolution and Feedback across Different Environments
Proceedings IAU Symposium No. 359, 2020
T. Storchi-Bergmann, W. Forman, R. Overzier & R. Riffel, eds.
doi:10.1017/S1743921320001489

POSTERS

Ionized gas kinematics and luminosity profiles of Low-z Lyman Alpha Blobs

María P. Agüero[1], Rubén Díaz[2] and Mischa Schirmer[3]

[1]Observatorio Astronómico de Córdoba, UNC, & CONICET, Argentina
email: mpaguero@unc.edu.ar

[2]Gemini Observatory, NSF's Optical Infrared Research Lab, USA

[3]MPIA, Heidelberg, Germany

Abstract. This work is focused on the characterization of the Seyfert-2 galaxies hosting very large, ultra-luminous narrow-line regions (NLRs) at redshifts z = 0.2−0.34. With a space density of 4.4 Gpc^{-3} at z ∼ 0.3, these "Low Redshift Lyman-α Blob" (LAB) host galaxies are amongst the rarest objects in the universe, and represent an exceptional and short-lived phenomenon in the life cycle of active galactic nuclei (AGNs). We present the study of GMOS spectra for 13 LAB galaxies covering the rest frame spectral range 3700–6700 Å. Predominantly, the [OIII]λ5007 emission line radial distribution is as widespread as that of the continuum one. The emission line profiles exhibit FWHM between 300–700 $\mathrm{Km\,s^{-1}}$. In 7 of 13 cases a broad kinematical component is detected with FWHM within the range 600–1100 $\mathrm{Km\,s^{-1}}$. The exceptionally high [OIII]λ5007 luminosity is responsible for very high equivalent width reaching 1500 Å at the nucleus.

Keywords. galaxies: Seyfert, galaxies: kinematics and dynamics, galaxies: stellar content

---

1. Background

Schirmer *et al.* (2013) serendipitously discovered J224024.1-092748 (hereafter J2240), a peculiar galaxy at z = 0.326 with "green" colors due to high [OIII] emission. Subsequently they systematically searched objects similar to J2240 in SDSS data, selecting objects with extremely green colors, and with resolved angular sizes. The authors obtained 29 galaxies that met these conditions and called them "Green Beans" (Schirmer *et al.* 2013). Later these objects were identified as Low redshift Lyman-α blob host galaxies (LABs, Schirmer *et al.* 2016; Kawamuro *et al.* 2017). The measured fluxes of the [OIII]λ5007 emission (hereafter [OIII]) of these galaxies are among the most energetic known, which result in very rare objects in the local universe (4.4 Gpc^{-3} at z ∼ 0.3). WISE 24 μm luminosities are 5–50 times lower than predicted by the [OIII] fluxes, and the X-ray and Radio emission are intrinsically low in comparison with such an energetic emission-line object. The NLRs seem to reflect earlier, very active quasar states that have strongly subsided in less than a galaxy's light-crossing time. These ionization echoes, are about 10–100 times more luminous than any other such echo known to date.

2. Ongoing work and Results

We obtained GMOS-Gemini Band-4 spectra for 13 LABs (seeing ∼0.6–1.3"). We used the R400 grating (spectral sampling of 1.02–1.14 Åpix^{-1}) covering a rest frame wavelength range 3700–6700 Å (redshift range 0.192–0.341). The observations were performed prioritizing the availability of a guide star and in many cases the slit position angle (PA)

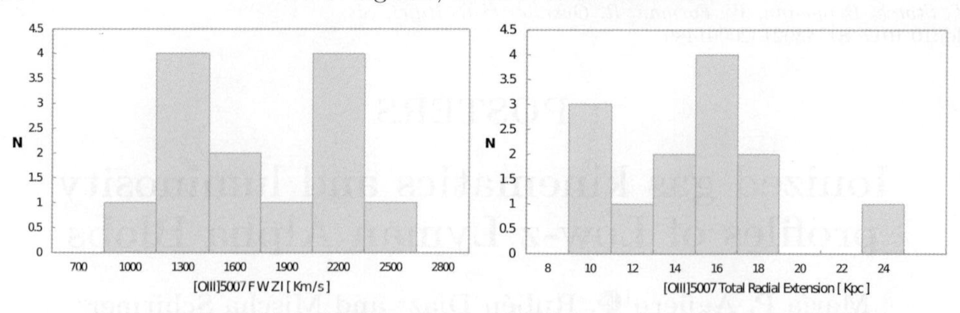

Figure 1. Full Width at Zero Intensity (Left) and Total Radial Extension (Right) distributions of the [OIII]5007 integrated emission line for the 13 GB sample.

does not match the galaxy major axis. The inclinations and major axis PAs used for deprojection were obtained from SDSS automatic surface brightness model fittings.

Several emission lines are detected in the LAB spectra. [OIII] is the most intense emission, with $\log(F([OIII])/F(H_\beta)) \geqslant 1$. Despite the large difference in flux, the [OIII] and hydrogen recombination line profiles are similar in shape. That is, both the radial and velocity FWHMs are comparable within the errors involved. However, the [OIII] high emission translates into extremely high values of the equivalent width (EW, range 160–1030Å). The radial distribution of EW([OIII]) has a steep rise at the nuclei up to 1500 Å. The integrated EW([OIII]) shows a positive correlation with the Concentration Index (ratio of Petrosian radii R_{90}/R_{50}). This would be consistent with the [OIII] being relatively more powerful in the earlier type galaxies of the sample. Although the integrated [OIII] FWHM is about $420\pm120\,\mathrm{Km\,s^{-1}}$, in a two component fit to the line profile, the widest kinematical component can reach FWHM $= 1100\ \mathrm{Km\,s^{-1}}$. The line profiles have asymmetries, and in four cases a broad blue component (FWHM $> 900\ \mathrm{Km\,s^{-1}}$) is detected, consistent with outflows.

The radial pure-continuum profile was determined at 5200 Å (green) and 4100 Å (blue). The average Half Light Diameter (HLD) of the sample is 4.1 ± 1.2 Kpc for the green continuum and 2.9 ± 1.4 Kpc for the [OIII]. The radial profile of [OIII] can be described with a Gaussian profile, with 7–15 % of the remaining flux in a weak and extended component. In contrast, the continuum profile has two more defined structures. The central region is characterised by a seeing-scale Gaussian, while the global component is best described by an exponential distribution.

From the SDSS magnitudes, the galaxy masses were estimated, which are of the order of $10^{11}\,M_\odot$. We found that the [OIII] extension is greater for galaxies with weak disks.

From the green and blue continuum profile, we set up the color radial distribution. Of the 13 LAB host galaxies studied, 4 have flat color profiles, 4 have symmetrical profiles with reddened centers and the rest have profiles with gradient from one side to the other of the nucleus (signs of dust in galactic disks). This last group is characterized by having steeper profiles showing a mixture of stellar populations or the presence of dust bands.

In conclusion, LAB host galaxy continuum emission radial profiles always exhibit a disk component, with some evidences of distorted morphology, emission line off-centering and possible dust lanes. The very large scale [OIII] emission structures with Seyfert characteristics have 13 Kpc and $1600\,\mathrm{Km\,s^{-1}}$ on average (Fig. 1), and seem relatively more powerful in the earlier type galaxies of the sample.

References

Kawamuro, T., Schirmer, M., Turner, J., *et al.* 2017, *ApJ*, 848, 1
Schirmer, M., Diaz, R., Holhjem, K., *et al.*2013, *ApJ*, 763, 60
Schirmer, M., Malhotra, s., Levenson, N., *et al.* 2016, *MNRAS*, 463, 1554

Galaxy Evolution and Feedback across Different Environments
Proceedings IAU Symposium No. 359, 2020
T. Storchi-Bergmann, W. Forman, R. Overzier & R. Riffel, eds.
doi:10.1017/S1743921320001787

Alternative classification diagrams for AGN-starburst galaxies

Catarina P. Aydar[1], J. E. Steiner[1] and Oli Dors Jr.[2]

[1]Instituto de Astronomia, Geofísica e Ciências Atmosférias, Universidade de São Paulo, 05508-090, São Paulo, SP, Brazil
email: catarina.aydar@usp.br

[2]Instituto de Pesquisa & Desenvolvimento, Universidade do Vale do Paraíba, 12244-390, São José dos Campos, SP, Brazil

Abstract. The aim of diagnostic diagrams is to classify galactic nuclei according to their photoionizing source using emission-line ratios, differentiating starburst regions from active galactic nuclei (AGN). However, the three traditional diagnostic diagrams can sometimes be ambiguous with regard to a single object. The main goal of the present work is to propose alternative diagnostic diagrams by using distinct combinations of emission lines ratios. We present these diagrams using data from the Sloan Digital Sky Survey. With these new diagrams, it is possible to better distinguish the ionizing source in nuclei of galaxies and also to study the parameters that are relevant when considering both kinds of objects, starbursts and AGN.

Keywords. galaxies: nuclei, galaxies: active, galaxies: abundances, galaxies: starburst

1. Introduction

Galactic nuclei can be ionized by stellar sources and by the accretion of gas onto a supermassive black hole; the difference between these mechanisms can be identified through the analysis of the spectrum of that region. The BPT diagnostic diagrams (Baldwin, Phillips & Terlevich 1981) are schemes by which one can recognize the physical mechanism behind the spectral ionization. Along the years, they have been improved to classify the nuclei of galaxies as starbursts, transition objects, Low-Ionization Nuclear Emission-line Regions (hereafter LINERs) or Seyferts (Kewley *et al.* 2001; Kauffmann *et al.* 2003; Schawinski *et al.* 2007), represented in Figures 1 and 2 as green stars, purple circles, black triangles, and pink squares, respectively. However, galaxy nuclei classifications based on the classical BPT diagrams are, in some cases, ambiguous. In this context, we propose alternative diagnostic diagrams to be used in gaseous nebulae studies.

2. Goals and Methodology

Although the BPT diagrams persist as a general prognostic to classify the ionization source, alternative axis considering the same emission line ratios can be useful to distinguish the ionization source in galactic nuclei. We aim to explore new combinations of the line ratios [O III] λ5007/Hβ, [O I] λ6300/Hα, [N II] λ6583/Hα and ([S II] λ6716+[S II] λ6731)/Hα to find new insights in the physics of galactic nuclei and their host galaxies.

We considered data from Sloan Digital Sky Survey DR7 (Abazajian *et al.* 2009) with redshift $z < 0.016$ and error bars smaller than 0.1 dex in all the considered emission line ratios, in order to select only high quality data.

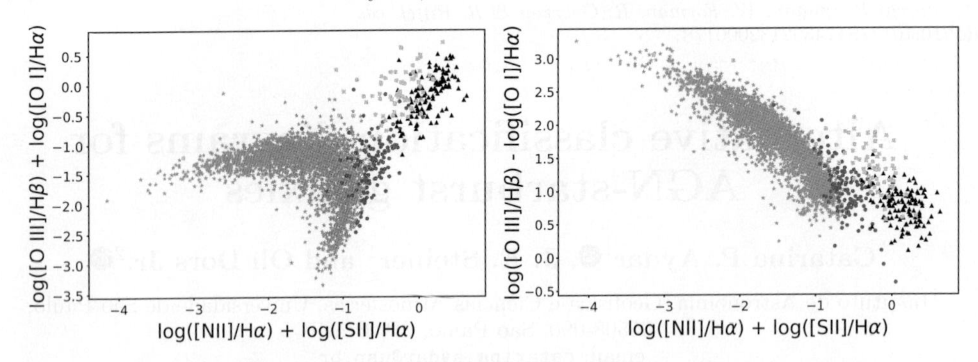

Figure 1. Alternative diagnostic diagrams with the sum of the logarithms of [N II]λ6584/Hα and [S II]λλ6716, 6731/Hα line ratios as abscissas. In the left panel, the ordinate is the sum of the logarithms of [O III]λ5007/Hβ and [O I]λ6300/Hα line ratios (that separates starbursts from AGN) and in the right panel the ordinate is the difference between such logarithms of line ratios (that distinguishes Seyferts and LINERs as in Heckman 1980).

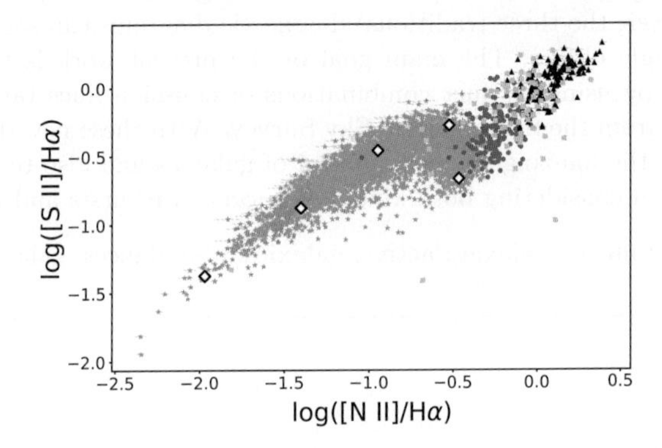

Figure 2. This alternative diagnostic diagram combines the line ratios of [N II]λ6584 and [S II]λλ6716, 6731 with respect to Hα. The wavelengths of such emission lines are similar, avoiding errors due to calibration or reddening effects. We can differentiate starbursts from AGN using only these two species. The white diamonds are the resulting models of starbursts simulated with Starburst99 and CLOUDY, considering equation (3.1).

3. Results and Conclusion

The relation between nitrogen and oxygen abundances commonly used in the literature (Vila-Costas & Edmunds 1993) is inadequate to describe the starbursts' distribution in Fig. 2. We propose the necessity of a second order term:

$$\log(\mathrm{N/H}) = \log(\mathrm{O/H}) + \log\left[0.039 + 20(\mathrm{O/H}) + 1.8 \times 10^5 (\mathrm{O/H})^2\right] \tag{3.1}$$

The alternative diagnostic diagrams proposed are successful to separate the different classes of objects - starbursts and AGN - and also to distinguish the two classes of AGN - Seyferts and LINERs. With these diagrams it is possible to distinguish uniquely the ionization source and also to study more accurately the model parameters that are relevant when considering both kinds of objects, starbursts and AGN.

References

Abazajian, K. N., Adelman-McCarthy, J. K., & The SDSS Colaboration *ApJS*, 182, 543
Baldwin, J. A., Phillips, M. M., & Terlevich, R. 1981, *PASP*, 93, 5

Heckman, T. M. 1980, *A&A*, 87, 152
Kauffmann, G., Heckman, T. M., *et al.* *MNRAS*, 346, 1055
Kewley, L. J., Dopita, M. A., Sutherland, R. S., *et al.* *ApJ*, 556, 121
Schawinski, K., Thomas, D., Sarzi, M., *et al.* *MNRAS*, 382, 1415
Vila-Costas, M. B. & Edmunds, M. G. 1993, *MNRAS*, 265, 199

Galaxy Evolution and Feedback across Different Environments
Proceedings IAU Symposium No. 359, 2020
T. Storchi-Bergmann, W. Forman, R. Overzier & R. Riffel, eds.
doi:10.1017/S1743921320001994

Unveiling the nuclear region of NGC 6868: Mapping the stellar population and ionized gas

João P. V. Benedetti[1], Rogério Riffel[1], Tiago V. Ricci[2], João E. Steiner[3], Rogemar A. Riffel[4], Miriani G. Pastoriza[1], Daniel Ruschel-Dutra[5] and Juliana C. Motter[1]

[1]Departamento de Astronomia, Universidade Federal do Rio Grande do Sul, 91501-970, Porto Alegre, RS, Brazil
email: jpvbene@gmail.com

[2]Universidade Federal da Fronteira Sul, Campus Cerro Largo, 97900-000, Cerro Largo, RS, Brazil

[3]Instituto de Astronomia, Geofísica e Ciências Atmosféricas, Universidade de São Paulo, 05508-900, São Paulo, Brazil

[4]Departamento de Física, Universidade Federal de Santa Maria 97105-900, Santa Maria, RS, Brazil

[5]Departamento de Física, Universidade Federal de Santa Catarina 88040-900, Florianópolis, SC, Brazil

Abstract. We mapped the stellar population and emission gas properties in the nuclear region of NGC 6868 using datacubes extracted with Gemini Multi-Object Spectrograph (GMOS) in the Integral Field Unit (IFU) mode. To obtain the star-formation history of this galaxy we used the STARLIGHT code together with the new generation of MILES simple stellar population models. The stellar population dominating (95% in light fraction) the central region of NGC 6868 is old and metal rich (~10 Gyr, 2.2 Z⊙). We also derived the kinematics and emission line fluxes of ionized gas with the IFSCUBE package. A rotation disk is clearly detected in the nuclear region of the galaxy and no broad components were detected. Also, there is a region where the emission lines disappear almost completely, probably due to diffuse ionized gas component. Channel maps, diagnostic diagrams and stellar kinematics are still under analysis.

Keywords. galaxies: individual (NGC 6868), galaxies: kinematics and dynamics, galaxies: stellar content

---

1. Introduction

Looking at the central part of nearby objects is important in order to map this region in detail and unveil what phenomena mediate the interaction between the central black hole and its host galaxy. The chosen galaxy for this study is NGC 6868 which is a cD galaxy and the central of the Telescopium group ~40 Mpc ($1'' \sim 180$ pc).

Observations were carried out with the Gemini telescope using GMOS in the IFU mode. The data cube has 5×3.5 arcsec^2 with a spectral coverage ranging from 4260 to 6795 Å. The spatial resolution is $0.6''$ (~100 pc). A Butterworth filter was applied.

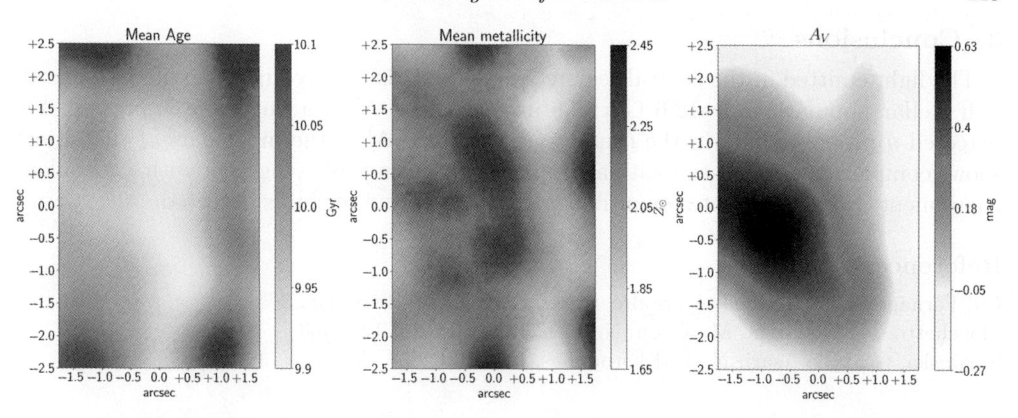

Figure 1. From left to right: the mean age, the mean metallicity and reddening in the V band.

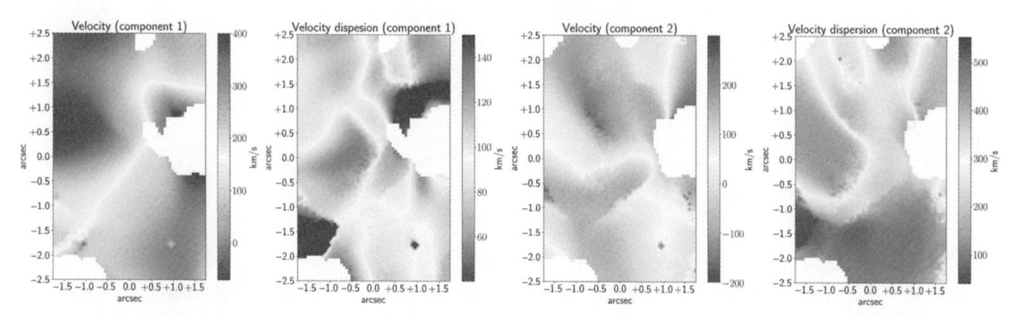

Figure 2. In the first two panels, the velocity and velocity dispersion of the rotation component, respectively. The last two show the same information for the second component.

2. Results

To derive the Star Formation History of our galaxy, we used the STARLIGHT code (Cid Fernandes *et al.* (2005)) to perform stellar population synthesis. Our Simple Stellar Population (SSP) base was composed with the latest MILES SSP models. As can be seen in figure 1, the populations that dominate are old (∼12.6 Gyr) and metal rich (2.6 and 1.8 Z⊙). The HOLMES (HOt Low-Mass Evolved Stars) could be the explanation for the LINER (Low-Ionization Nuclear Emission-line Region) emission observed (Stasińska *et al.* (2008)). The central region has a higher metallicity compared to the outer regions, in agreement with a successive enrichment of the interstellar medium. Also, no contribution of a power-law is found. As in Macchetto *et al.* (1996), the higher A_V towards the center probes a dust lane in the direction of the line of sight.

We analysed Hα λ 6563 Å and [NII] $\lambda\lambda$ 6548,6584 Å lines as they have a higher signal-to-noise ratio in our data. One of the most interesting regions is where the emission lines disappear or are below our detection limit, suggesting this is a Diffuse Ionized Gas region.

In order to understand the kinematics of the ionized gas, we fitted Gaussians to the emission lines using the IFSCUBE† package (figure 2). At the center of the galaxy one of the components has a rotation profile (probably a disc) whereas the other is a more turbulent one. Any broad component compatible with active galactic nuclei is identified.

† https://bitbucket.org/danielrd6/ifscube/src/master/

3. Conclusions

The light emitted in the central region of NGC 6868 is mostly due to an old and metal-rich stellar populations (∼12.6 Gyr; 2.6 and 1.8 Z⊙). Also, a dust lane is present. We detected a probable DIG in the center of the galaxy. Also, the inner part of the object shows complex gas kinematics with an apparent rotation profile together with a turbulent component. No component compatible with a broad line region is detected.

References

Cid Fernandes, R., Mateus, A., Sodré, L., *et al.*2005, *MNRAS*, 358, 363
Macchetto, F., Pastoriza, M., Caon, N., *et al.*1996, *A&AS*, 120, 463
Stasińska, G., Vale Asari, N., Cid Fernandes, R., *et al.*2008, *MNRAS*, 391, L29

Galaxy Evolution and Feedback across Different Environments
Proceedings IAU Symposium No. 359, 2020
T. Storchi-Bergmann, W. Forman, R. Overzier & R. Riffel, eds.
doi:10.1017/S1743921320004202

Recovering the origin of the lenticular galaxy NGC3115 using multi-band photometry

Maria Luísa Buzzo[1], Arianna Cortesi[2,1], Ariel Werle[3,1] and Claudia Mendes de Oliveira[1]

[1]Universidade de São Paulo, IAG, Rua do Matão 1226, São Paulo, Brazil
[2]Observatório do Valongo, Ladeira do Pedro Antônio 43, Rio de Janeiro, RJ, Brazil
[3]INAF - Osservatorio Astronomico di Padova, Vicolo dell'Osservatorio 5, I-35122 Padova, Italy

Abstract. We perform simultaneous multi-band fitting, using the routine GALFITM, of the galaxy NGC3115, in order to recover the stellar populations of its main components (a bulge, a thin disc and a thick disc). We model 11 bands, from ultraviolet to infrared, in order to take into account the galaxy younger stellar population and the presence of the Active Galactic Nuclei (AGN). We find that the majority of the galaxy baryonic mass belongs to the thick disc, which is also the oldest galaxy component, consistent with results from the literature. Differently from previous works, we find that the bulge has the bluest colour and it is younger than the thick disc, either as a result of recent star formation activity, or AGN feedback, or white dwarf emission in an old stellar population. Finally, we propose that NGC3115 was formed either through a two-phase formation scenario, or via an outside-in quenching of an isolated spiral galaxy, whose thick disc had been heated-up via minor mergers with dwarf satellites.

Keywords. galaxies: evolution, galaxies: formation, galaxies: elliptical and lenticular

1. Introduction

We try to answer the question of how the closest lenticular galaxy to the Milky Way, NGC3115, was formed. To perform this study, we use multi-band fitting techniques with data from ultraviolet to infrared, decomposing the galaxy into its main components to analyse their properties over a large wavelength range. In general, spectral data allows to precisely retrieve age and metallicity of a galaxy, but cover only a small wavelength range (usually the optical). Furthermore, spectroscopy is limited in aperture – even integral field units (IFU) data usually do not cover galaxies' outskirts. Spectral Energy Distributions (SEDs) obtained from broad-band photometry are far less restricted in spatial coverage. The ability of this technique to gather information from ultraviolet to infrared make up for the loss of detailed λ-by-λ constraints (Salim *et al.* 2014). Thus, multi-wavelength photometric SEDs are the right data sources to study the distribution of stellar populations from the inner parts to the outskirts of galaxies. By comparing the stellar population of the disk and the spheroid we can understand where the last formation episode occurred and from where the quenching has started, recovering the formation history of the galaxy.

2. Results

The final model created for NGC3115 using the routine GALFITM in 11 images, from ultraviolet (GALEX) to infrared (WISE), is shown in Fig. 1a and was constructed using

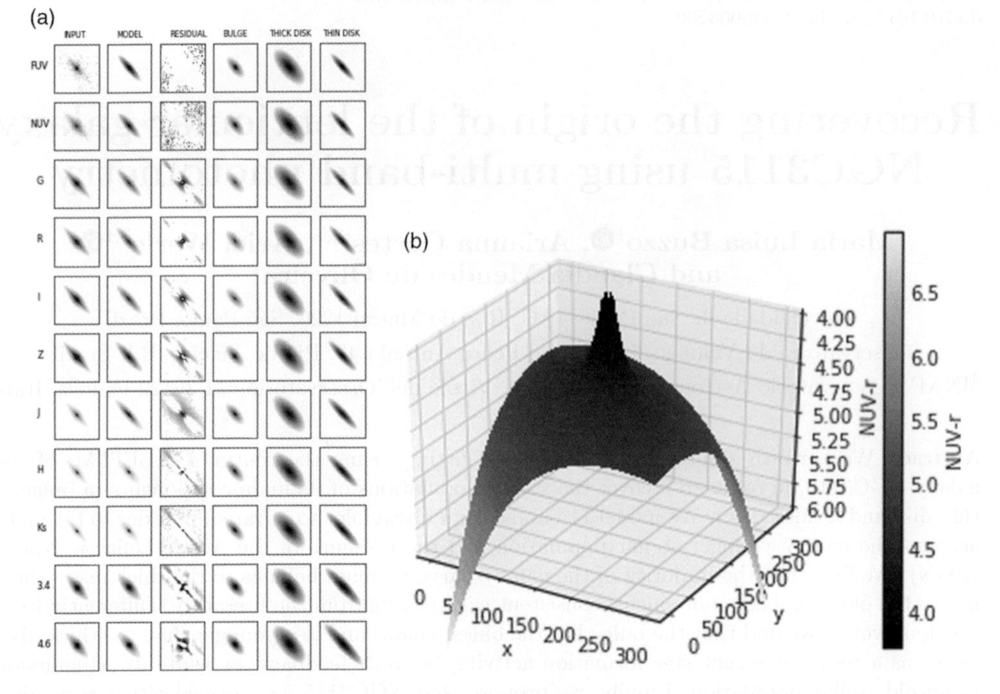

Figure 1. (a) Model of NGC3115 created using GALFITM with a bulge, a thick and a thin disc. (b) near-UV−r colour gradient of each pixel of the galaxy.

3 components: a bulge, a thick disc and a thin disc (also shown separately in the figure). With this model we are able to understand the stellar populations and ages of each component, deriving their SEDs and analysing the colour gradient across the galaxy. We have also calculated the near-UV (NUV)−r colour of each pixel, shown in Fig. 1b. From this figure, we identify a positive age gradient, i.e. a bulge bluer than the outskirts. The NUV−r colour of the bulge could be attributed to a young stellar population, the presence of the AGN (Wong *et al.* 2011) or the emission of white dwarf stars (Lisker *et al.* 2008). With these results and other properties retrieved from our model, we end up with two possible formation scenarios: the "two-phase scenario" or secular evolution (starvation induced by the AGN (Menezes *et al.* 2014)).

In the literature, the most likely formation scenario for NGC3115 is the so-called "two phase" scenario (Guérou *et al.* 2016), which consists in an in-situ formation at high redshift by dissipative gas collapse (creating the central stellar mass), followed by a significant merger event, responsible for the formation of the thick disc (or fast rotating oblate spheroid). Accretion of satellites or gas inflow would be responsible for the formation of the thin disc. Alternatively, NGC3115 could be a spiral galaxy whose thick disc has been heated up by several minor mergers. In a second moment, AGN and supernovae mechanical feedbacks would have been responsible for the quenching of the thick disc of the galaxy, and the continuing inflow of gas into the centre was capable of inducing star-formation events that rejuvenated the stellar population of the bulge. We reinforce, nevertheless, that all the components of this galaxy have, today, very low levels of star-formation and are all very old. Acquiring deeper images (surface brightness 28 mag/arcsec^2) would be a way to tell if the galaxy went through several accretions or not.

References

Guérou, A., Emsellem, E., & Krajnović, D. 2016, *A&A*, 591, A143
Lisker, T., Grebel, E. K., & Binggeli, B. 2008, *Aj*, 135, 380
Menezes, R. B., Steiner, J. E., & Ricci, T. V. 2014, *ApJ*, 796, L13
Salim, S. 2014, *Serbian Astronomical Journal*, 189, 1
Wong, K., Irwin, J. A., Yukita, M., *et al.* 2011, *ApJ*, 736, L23

Galaxy Evolution and Feedback across Different Environments
Proceedings IAU Symposium No. 359, 2020
T. Storchi-Bergmann, W. Forman, R. Overzier & R. Riffel, eds.
doi:10.1017/S174392132000188X

Direct measures of chemical abundances from stacked spectra of star-forming galaxies: Implications for the mass–metallicity–star formation rate relation

Katia Slodkowski Clerici and Natalia Vale Asari

Departamento de Física - CFM, Universidade Federal de Santa Catarina, C.P. 476, 88040-900, Florianópolis, SC, Brazil
email: katia@astro.ufsc.br

Abstract. The stellar mass–star formation rate–metallicity relation provides clues on the chemical evolution of galaxies. We revisit this relation by measuring the gas-phase metallicity using the direct method. For metal-rich galaxies this is not straightforward, because auroral emission lines sensitive the electron temperature are lost in spectral noise. In order to increase the spectral signal-to-noise ratio and detect faint auroral lines, we stack the spectra of similar galaxies. This allows us to use the direct method to obtain consistent metallicity measurements.

Keywords. galaxies: abundances, galaxies: ISM, ISM: HII regions, ISM: abundances

1. Introduction

Chemical abundances are the product of various processes within and interactions outwith galaxies, e.g. stellar activity, infall and outflow of chemically-poor or enriched gas. These processes are interlinked to physical properties of galaxies, such as their stellar mass and star formation rate. Empirical relations between the oxygen abundance and other physical properties of galaxies are thus informative for chemical evolution models.

Oxygen abundance measurements are hard to pin down, especially if using indirect methods based on strong emission lines (e.g. Kewley & Ellison 2008). The golden standard to measure gas-phase abundances is the direct method, which relies on measuring the electron temperature from colisionally-excited lines. In a ionized nebula, free electrons collide with ions and atoms and excite their bound electrons; the latter can de-excite radiatively emitting photons at specific wavelengths. Therefore, the electronic temperature can be measured from line ratios of the same ion with different excitation potentials, for example [O III]λ4363/[O III]λ5007. However, auroral lines such as [O III]λ4363 are very faint, and to detect them we need high signal-to-noise ratio (SNR) spectra.

2. Preliminary results from stacked spectra

To increase the spectral SNR, we stack of galaxies with similar properties. We select 111, 253 star-forming galaxies (below the Kauffmann *et al.* 2003 line on the [N II]λ6584/Hα versus [O III]λ3727/Hβ plane) from the Sloan Digital Sky Survey (SDSS) Data Release 7 (Abazajian *et al.* 2009). Before stacking galaxies we perform the following procedures to individual spectra. **<u>First</u>**, we model and subtract the stellar continuum spectra using STARLIGHT (Cid Fernandes*et al.* 2005). **<u>Second</u>**, we deredden galaxy spectra with a Cardelli *et al.* (1989) dust attenuation law assuming an intrinsic ratio for

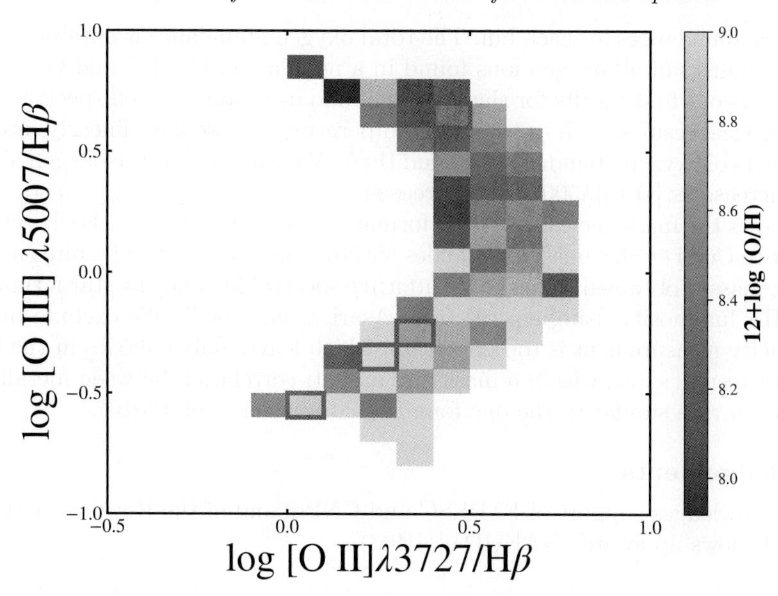

Figure 1. Oxygen abundances measured in SDSS stacked spectra using the direct method. O/H for the bins outlined in red have been calculated as a linear interpolation of the surrounding bins.

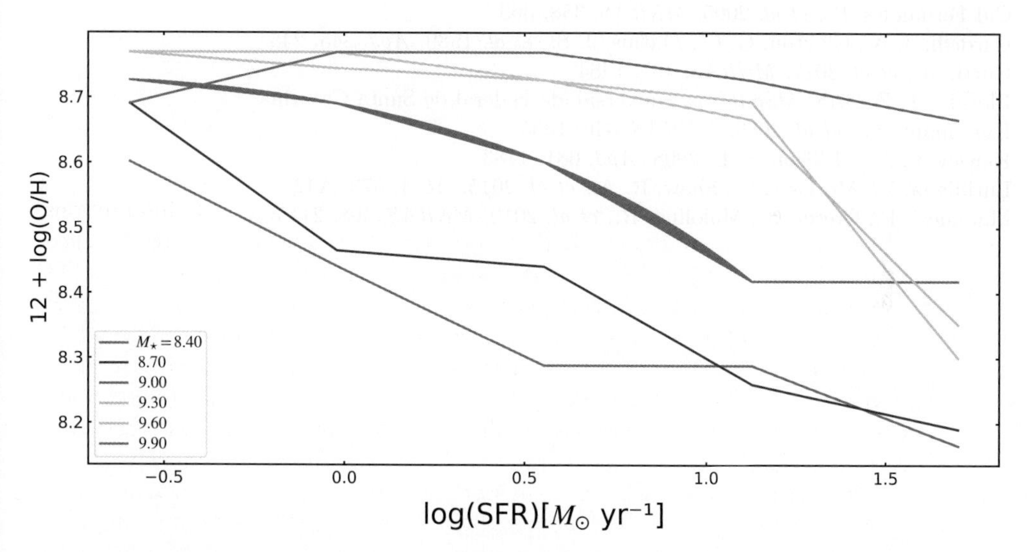

Figure 2. Preliminary results for the mass–metallicity–star formation relation. We show only the results for low-mass galaxies. There is an anti-correlation between O/H and star formation rate within a mass bin.

the Balmer lines $H\alpha/H\beta = 2.86$ (Case B), and normalise them by the dust-corrected $H\beta$ luminosity. **Third**, we shift the spectra to their rest-frame adopting as a reference the systemic velocity of the [O III]λ5007 emission line. Emission lines are modelled with Gaussian profiles using DOBBY (Florido 2018).

We stack galaxies in 0.1-dex-wide bins in the [O II]λ3727/Hβ versus [O III]λ5007/Hβ plane, following Curti *et al.* (2017). Bins contain 100 to 6442 spectra, and are expected to house galaxies with approximately the same oxygen abundance and ionization parameter. We use PyNeb (Luridiana *et al.* 2015) to measure the nebular electron temperature and ionic oxygen abundances for each bin. The total oxygen abundance is obtained by adding up the abundances of all oxygen ions found in a nebula, i.e. O^+/H^+ and O^{++}/H^+.

Fig. 1 shows our first results for the oxygen abundance from stacked spectra. Four bins had non-physical results, such as too-high temperatures, so we have linearly interpolated measurements of oxygen abundances around them. We can see that the oxygen abundance tends to increase as [O III]λ5007/Hβ decreases.

Fig. 2 shows the mass–metallicity–star formation rate relation with the direct method. In contrast to Curti et al.'s method, galaxies within a bin are assigned its bin's metallicity. The stellar mass is obtained from the STARLIGHT spectral fits and the star formation rate from the Hα luminosity using eq. (9) from Asari *et al.* (2007). We exclude bins where the metallicity measurement is too uncertain, which leaves only galaxies in the low-mass regime. Our results show, within a mass bin, an anti-correlation between metallicity and star formation rate similar to the one found by Mannucci *et al.* (2010).

Acknowledgements

NVA acknowledges support of FAPESC and CNPq, and of the Royal Society–Newton Advanced Fellowship award (NAF\R1\180403).

References

Abazajian, K. N., *et al.* 2009, *ApJS*, 182, 543
Asari, N. V., *et al.* 2007, *MNRAS*, 381, 263
Cid Fernandes, R., *et al.* 2005, *MNRAS*, 358, 363
Cardelli, J. A., Clayton, G. C., Mathis, J. S., *et al.* 1989, *ApJ*, 345, 245
Curti, M., *et al.* 2017, *MNRAS*, 465, 1384
Florido, T. Z. 2018, *MSc thesis*, Universidade Federal de Santa Catarina
Kauffmann, G., *et al.* 2003, *MNRAS*, 346, 1055
Kewley, L. J. & Ellison, S. L. 2008, *ApJ*, 681, 1183
Luridiana, V., Morisset, C., Shaw, R. A., *et al.* 2015, *A&A*, 573, A42
Mannucci, F., Cresci, G., Maiolino, R., *et al.* 2010, *MNRAS*, 408, 2115

Galaxy Evolution and Feedback across Different Environments
Proceedings IAU Symposium No. 359, 2020
T. Storchi-Bergmann, W. Forman, R. Overzier & R. Riffel, eds.
doi:10.1017/S1743921320002100

A panchromatic spatially resolved study of the inner 500 pc of NGC 1052

Luis G. Dahmer-Hahn[1], Rogério Riffel[2], Tiago V. Ricci[3], João E. Steiner[4], Thaisa Storchi-Bergmann[2], Rogemar A. Riffel[5,6], Roberto B. Menezes[7], Natacha Z. Dametto[8], Marlon R. Diniz[5], Juliana C. Motter[2] and Daniel Ruschel-Dutra[8]

[1]Laboratório Nacional de Astrofísica, Itajubá-MG, Brazil
email: lhahn@lna.br

[2]Departamento de Astronomia, Universidade Federal do Rio Grande do Sul

[3]Universidade Federal da Fronteira Sul, Campus Cerro Largo

[4]Instituto de Astronomia, Geofísica e Ciências Atmosféricas, Universidade de São Paulo

[5]Universidade Federal de Santa Maria, Departamento de Física

[6]Center for Astrophysical Sciences, Department of Physics and Astronomy

[7]Centro de Ciências Naturais e Humanas, Universidade Federal do ABC

[8]Departamento de Física - CFM - Universidade Federal de Santa Catarina

Abstract. We analyzed the inner $320 \times 535\,\mathrm{pc}^2$ of the elliptical galaxy NGC 1052 with integral field spectroscopy, both in the optical and in the near-infrared (NIR). The stellar population analysis revealed a dominance of old stellar populations from the optical data, and an intermediate-age ring from NIR data. When combining optical+NIR data, optical results were favoured. The emission-line analysis revealed five kinematic components, where two of them are unresolved and probably associated with the active galactic nucleus (AGN), one is associated with large-scale shocks, one with the radio jets, and the last could be explained by either a bipolar outflow, rotation in an eccentric disc or a combination of a disc and large-scale gas bubbles. Our results also indicate that the emission within the galaxy is caused by a combination of shocks and photoionization by the AGN.

Keywords. galaxies: individual: NGC 1052 - galaxies: jets - galaxies: nuclei - galaxies: elliptical and lenticular, cD

1. Introduction

A Low-ionization nuclear emission-line region (LINER, Heckman 1980) is a very well studied type of galaxy. The main conundrum behind LINERS is the fact that many mechanisms are capable of producing a low ionization spectrum, such as (i) shocks, (ii) photoionization by low luminosity active galactic nuclei, (iii) post-asymptotic giant branch stars, and (iv) starbursts dominated by Wolf-Rayet stars.

NGC 1052 is the prototypical LINER galaxy, located at a distance of 19.1 Mpc, subject of a long and intense debate surrounding its LINER emission. Whereas some authors (Koski & Osterbrock 1976; Fosbury *et al.* 1978, 1981; Dopita *et al.* 2015) suggest shocks as the main ionization source behind this galaxy, other authors found evidence of photoionization as the main driver behind its lines (Diaz *et al.* 1985; Gabel *et al.* 2000).

2. The data

In order to disentangle the various mechanisms which are capable of producing a LINER-like spectrum, we employed optical and NIR integral field specroscopy obtained with Gemini Multi-Object Spectrograph (GMOS) and Near-Infrared Integral-Field Spectrograph (NIFS), respectively. Both data sets were reduced using the standard reduction scripts distributed by the Gemini team. We also performed differential atmospheric refraction correction, Butterworth spatial filtering, and instrumental fingerprint removal. Lastly, we combined optical and NIR datacubes by slicing each NIR spaxel to 900 subspaxels (30×30) and adding their fluxes to the optical spaxel that matched the central position of the NIR subspaxel.

3. Stellar population

To study the stellar population of NGC 1052, we used the STARLIGHT code (Cid Fernandes *et al.* 2004, 2005) with E-MILES library of simple stellar populations (SSPs, Vazdekis *et al.* 2016). By fitting the optical datacube, we found only old stellar content in the entire datacube, dominated by ∼12Gyr SSPs. When fitting the NIR datacube, we found the same results in the nucleus of the galaxy, with the circumnuclear population dominated by ∼2.5 Gyr SSPs. By performing the synthesis using the panchromatic datacube, we found again a dominance of older stellar populations (∼12 Gyr). We attribute this difference between optical and NIR results to the lack of H-band data and the low signal-to-noise ratio of our J-band data.

4. Gas excitation and Kinematics

The emission-line fluxes were measured in a pure emission-line spectrum, free from the underlying stellar flux contributions. Close to the centre of our FoV, an unresolved broad component is present in Hα. Also in the nucleus, an unresolved blue wing is visible on [OIII] with $1380\,\mathrm{km\,s^{-1}}$. In order to reproduce the extended emission, two Gaussian functions were needed in order to fit each emission-line profile, one narrow ($100\,\mathrm{km\,s^{-1}} < \mathrm{FWHM} < 150\,\mathrm{km\,s^{-1}}$) and one with intermediate width (IW, $280\,\mathrm{km\,s^{-1}} < \mathrm{FWHM} < 450\,\mathrm{km\,s^{-1}}$). The narrow emission is compatible with two previously detected gas bubbles, which were attributed to large-scale shocks. The IW component, on the other hand, can be explained by an outflow, an eccentric disc, or a combination of a disc with large-scale shocks. When analysing density, temperature, and diagnostic diagrams, our results suggest that the ionization within the FoV of our data cannot be explained by one mechanism alone. Rather, our results suggest that photoionization is the dominant mechanism in the nucleus, with the extended regions being ionized by a combination of shocks and photoionization.

References

Cid Fernandes, *et al.* 2004, *MNRAS*, 355, 273
Cid Fernandes, R., Mateus, A., Sodré, L., *et al.* 2005, *MNRAS*, 358, 363
Dopita, M. A., *et al.* 2015, *ApJ*, 801, 42
Diaz, A. I., Terlevich, E., Pagel, B. E. J. *et al.* 1985, *MNRAS*, 214, 41P
Fosbury, R. A. E., Mebold, U., Goss, W. M., *et al.* 1978, *MNRAS*, 183, 549
Fosbury, R. A. E., Snijders, M. A. J., Boksenberg, A., *et al.* 1981, *MNRAS*, 197, 235
Gabel, *et al.* 2000, *ApJ*, 532, 883
Heckman, T. M. 1980, *A&A*, 87, 142
Koski, A. T. & Osterbrock, D. E. 1976, *ApJ*, 203, L49
Vazdekis, A., Koleva, M., Ricciardelli, E., *et al.* 2016, *MNRAS*, 463, 3409

Galaxy Evolution and Feedback across Different Environments
Proceedings IAU Symposium No. 359, 2020
T. Storchi-Bergmann, W. Forman, R. Overzier & R. Riffel, eds.
doi:10.1017/S1743921320002331

Hα plumes or arms associated with the nucleus of NGC 7020

H. Dottori[1], R. Díaz[2], G. Gimeno[2], A. Bianchi[1] and G. Gaspar[3]

[1]Instituto de Física, UFRGS, Av. B. Gonçalves, 9500 CEP: 91501-970, P. Alegre, Brasil
email: hdottori@gmail.com

[2]Gemini Observatory, NSF's Optical Infrared Research Lab, USA

[3]Observatorio Astronómico de Córdoba, UNC, Córdoba, Argentina

Abstract. We imaged the galaxy NGC 7020 with Gemini and GMOS-S interference filters centered on the Hα emission line and nearby continuum in order to detect and quantify the HII regions. Among about 200 HII regions, we detected two Hα emitting plumes or arms emerging from the galactic nucleus which, together with the nuclear emission, might indicate a process of feedback from the central region.

Keywords. Galaxy: disk, Galaxy: center, Galaxy: nucleus, ISM: HII regions

1. Ongoing work and Results

NGC 7020 is a weakly barred spiral with two symmetrical condensations or *ansae*, at the vertices of a regular hexagon, stretched along the line of the nodes and centered on the nucleus at a radius of ∼7.1 kpc (Fig. 1 Left).

This hexagon was explained by Patsis *et al.* (2003) as 6:1 resonant stellar orbits resulting from an almost square bar potential and finely tuned Jacobi energies. Orbits with 4:1 resonances also result in such conditions, which we also detected centered on the nucleus with a mean distance of ∼3.5 kpc. Stellar continuum isodensities of these structures are shown in green in Fig. 1 Right, superimposed on an image showing the pure Hα emission.

NGC 7020 Hα emission has already been studied by Buta (1990), Wozniak *et al.* (1995) and Crocker *et al.* (1996). This is located in an outer ring of 33 kpc mean diameter and also in the inner disc inside the hexagon including the nucleus itself. Curiously, NGC 7020 exhibits widespread star-forming activity in its inner disc, which originated in a highly perturbing phenomenon. The star formation is superposed on remarkable ring structures that are formed from highly constrained resonant stellar orbits in barred spiral galaxies as described by Patsis *et al.* (2003). In order to map the HII regions and determine their epoch of formation, we took images with the Gemini Multiobject Spectrograph (GMOS-S) attached to the Gemini South telescope with the G0136 and G0137 interference filters, centered on the Hα line and adjacent continuum. This allows us to study HII region ages from the equivalent width of the Hα emission line (Dottori 1981; Leitherer *et al.* 1999).

HII region detection was done with SExtractor (Bertin & Arnaut 1996). We detected approximately two hundred HII regions. The most intense regions in the internal disk are the nucleus itself and two plumes or arms that extend ∼1 kpc on both sides of the nucleus. Fig. 1 Right, shows that the emission in the inner disk is weaker than that of the outer ring.

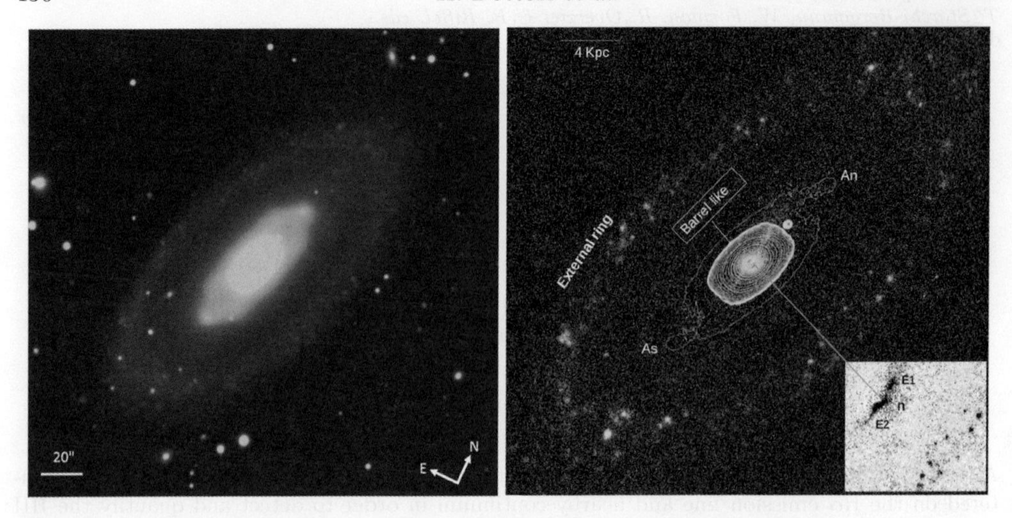

Figure 1. Left: NGC 7020 red continuum image observed with GMOS-S at Gemini, depicting the hexagonal and barrel like brightness levels of the bar, corresponding to 6:1 and 4:1 resonances respectively. Right: NGC 7020 continuum subtracted Hα emission. Green isophotes trace the continuum distribution, the fainter isophote is inside 6:1 resonance and embraces the ansae (An and As). Towards the nucleus the barrel-like 4:1 resonances appear. The inset shows the Hα ejecta (E1 and E2) originating at the nucleus (n). $1'' = 206$ pc

Concluding Remarks. We detected two Hα emitting plumes or arms with ~1 kpc extents. They seems to emerge from the nucleus approximately in the N-S direction. If the Hα emission is produced by ionizing stars, the age deduced for all of them including that from the nucleus is $\geqslant$10 Myr, but the figure would be different if the ionization of the plumes or arms is not of stellar origin. We are processing spectroscopic observations to study the kinematics and origin of these structures.

References

Bertin, E. & Arnouts, S. 1996, *AAS*, 117, 393B

Buta, R. 1990, *ApJ*, 356, 87

Crocker, D., Baugus, P. & Buta, R. 1996, *ApJS*, 105, 353

Dottori H. A. 1981, *ApSS*, 80, 267

Leitherer, C., Schaerer, D., Goldader, J. D., *et al.* 1995, *ApJS*, 123, 3

Patsis, P. A., Skokos, C., & Athanassoula E. 2003, *MNRAS*, 346, 1031

Wozniak, H., Friedli, D., Martinet, L., *et al.* 1995, *AAS*, 111, 115

Galaxy Evolution and Feedback across Different Environments
Proceedings IAU Symposium No. 359, 2020
T. Storchi-Bergmann, W. Forman, R. Overzier & R. Riffel, eds.
doi:10.1017/S1743921320002069

Structure and morphology of relic galaxies in the Local Universe

Rodrigo F. Freitas[1], Ana L. Chies-Santos[1], Cristina Furlanetto[1] and Fabricio Ferrari[2]

[1]Instituto de Física, Universidade Federal do Rio Grande do Sul, Porto Alegre, Brazil
[2]Instituto de Matemática, Estatística e Física, Universidade Federal do Rio Grande, Rio Grande, Brazil

Abstract. Red Nugget galaxies found at high-z have analogues in the Local Universe which are called relic galaxies. Because of their proximity to Earth, the relics allow a more detailed analysis of their properties and can help us understand the formation of massive early-type galaxies, since Red Nuggets could be their first phase of formation. The main goal of this work is to characterize the structure and morphology of candidates and confirmed relic galaxies in the Local Universe to further search for similar objects observationally and within cosmological simulations.

Keywords. galaxy: formation, galaxy: structure, relic, photometry

---

1. Introduction

Considering the two-phase galaxy formation scenario of massive early-type galaxies (Huang *et al.* 2016), the compact core forms first (at $z > 3$), followed by the growth of mass and size through mergers with other galaxies (from $z < 2$). Given the stochastic nature of the mergers, we can expect that some compact seeds of galaxies that formed in the beginning of the Universe are left almost untouched until the present epoch. Thus, we expect to find them in the Local Universe (Quilis & Trujillo 2013; Beasley *et al.* 2018). These compact ($R_e \leqslant 2$ kpc) massive galaxies ($M_\star \sim 10^{11} M_\odot$) when found at high redshift are called Red Nuggets and they have analogues in the Local Universe with old ($t \geqslant 10$ Gyr) stellar populations that are called relics (Ferré-Mateu *et al.* 2015; Yıldırım *et al.* 2017). It is important to study these relics because they are probes to understand the formation of massive early-type galaxies that we observe today.

2. Data & Method

The first sample studied in this work is composed of 16 galaxies with public images from the Hubble Space Telescope (HST) in H and I bands. The sample was first selected by Yıldırım *et al.* (2017) and 14 galaxies in the sample are very likely to be relics. The second sample studied was composed of 87 candidate relic galaxies (see Lohmann *et al.* 2020 in this IAU proceedings volume) with public images from the Sloan Digital Sky Survey (SDSS) in g, r and i bands (see Lohmann *et al.* in this volume). The objects were found in the MaNGA survey database and selected as outliers in terms of their radii from the galaxy mass-size distribution at $z = 0.25$ (van der Wel *et al.* 2014). A secondary SDSS control sample was built using galaxies with similar central dispersion velocities to those of the relic candidates.

To characterize the structure of the galaxies we used the Sérsic profile (Sérsic 1968; Graham & Driver 2005), performing single and double 2D profile fitting with the software IMFIT (Erwin 2015) for all the galaxies in both samples (Fig. 1).

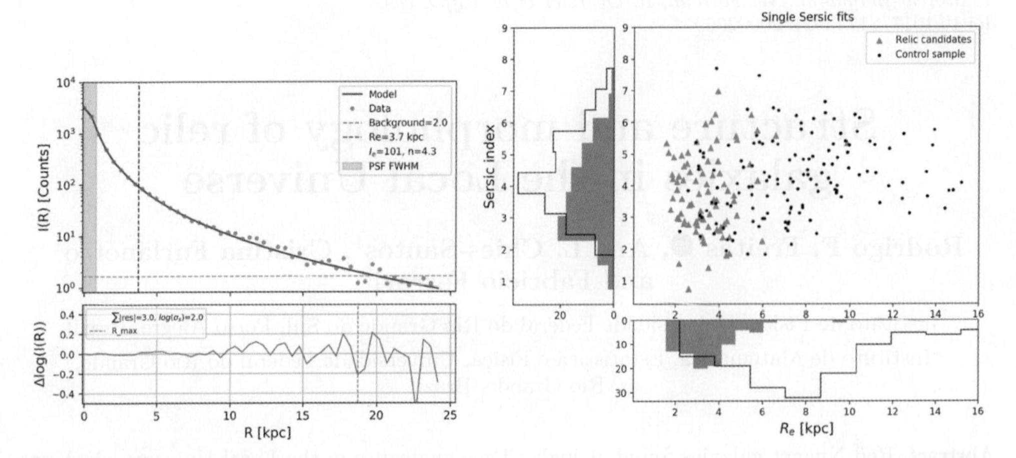

Figure 1. *Left:* Profile fitting (r-band) for the galaxy 1-48084 of the SDSS sample. *Right:* Results for relic candidates sample with single Sérsic profile fittings applied to r-band images for the SDSS sample. Red triangles are the candidates and black dots are the control sample.

3. Results & Discussion

HST sample. From the double Sérsic fitting of the HST sample we found that almost all of the galaxies can be well modeled with a small ($R_e < 2$ kpc) and concentrated ($n \geqslant 4$) component plus a larger ($R_e > 2.5$ kpc) component with low Sérsic index ($n \leqslant 1$), hinting that the structure of this objects is more complex than just central compact nuggets.

SDSS sample. From the single Sérsic fitting (Fig. 1) of the relic candidates in SDSS r-band images we noted that many objects have Sérsic index n $\sim$2.5, and effective radii $R_e \sim 3.0$ kpc, while the control sample seems to have different peaks in the distribution of the same parameters. We show in Fig. 1 a profile fitting example and the results for the structural analysis of the SDSS sample:

In addition, a two-sample Kolmogorov-Smirnov test at significance level of 0.05 shows that the Sérsic indices of the control and candidate samples probably come from different distributions, suggesting that the sample of relic candidates could be a population of compact objects in the Local Universe with structure that could resemble confirmed relic properties, such as disk-like morphology if $n \leqslant 2.5$ Buitrago *et al.* (2018).

The next step of this work is to search for similar objects in cosmological simulations in order to analyze their formation and evolution environments.

References

Beasley, M., Trujillo, I., Leaman, R., *et al.* 2018, *Nature*, 555, 483
Buitrago, F., Ferreras, I., Kelvin, L. S., *et al.* 2018, *A&A*, 619, 137
Erwin, P. 2015, *ApJ*, 799, 2
Ferré-Mateu, A., Mezcua M., Trujillo I., *et al.* 2015, *ApJ*, 808, 79
Graham, A. W. & Driver, S. P. 2005, *PASA*, 22, 2
Huang, S., Ho, L. C., Peng, C. Y., *et al.* 2016, *ApJ*, 821, 2
Quilis, V. & Trujillo I. 2013, *ApJ*, 773, L8
Sérsic, J. L. 1968, *Atlas de las galaxias australes*
van der Wel, A., *et al.* 2014, *ApJ*, 788, 28
Yıldırım, A., van den Bosch, R. C. E., van de Ven, G., *et al.* 2017, *MNRAS*, 468, 4

Galaxy Evolution and Feedback across Different Environments
Proceedings IAU Symposium No. 359, 2020
T. Storchi-Bergmann, W. Forman, R. Overzier & R. Riffel, eds.
doi:10.1017/S1743921320002173

Selection and characterisation of Red Geysers in the MaNGA survey

Gabriele S. Ilha[1,2], Rogemar Riffel[1,2], Sandro B. Rembold[1,2] and Jáderson S. Schimóia[1,2]

[1]Universidade Federal de Santa Maria, Departamento de Física,
Centro de Ciências Naturais e Exatas, 97105-900, Santa Maria, RS, Brazil
email: gabrieleilha1994@gmail.com

[2]Laboratório Interinstitucional de e-Astronomia - LIneA, Rua Gal. José Cristino 77,
Rio de Janeiro, RJ - 20921-400, Brazil

Abstract. Red Geysers are quiescent galaxies that show a bi-polar outflow, but the mechanism that produces this outflow is still unclear. Using MaNGA data, we find that Red Geysers correspond to ~1.6% of the sample of galaxies already observed by MaNGA. About ~16% of the Red Geysers show clear evidence of Active Galactic Nuclei, as revealed by emission-line ratios.

Keywords. galaxies: red geysers, galaxies: kinematics, galaxies: nuclei

---

1. Introduction

Based on the analysis of MaNGA (Mapping Nearby Galaxies at APO) data from SDSS-IV (Sloan Digital Sky Survey-IV), Cheung *et al.* (2016) revealed a new class of galaxies: the "Red Geysers". Red Geysers are quiescent galaxies and are characterised by a bipolar outflow. Cheung *et al.* (2016) showed for Akira – the prototype of this class – that this bipolar outflow is probably caused by a low-luminosity AGN (LLAGN). In addition, Roy *et al.* (2018) using MaNGA data from Data Release 13 (DR13), which are identical to the data of the DR14 (Abolfathi *et al.* 2018), and radio observations from VLA-FIRST found that most Red Geysers probably have a radio-mode AGN. Riffel *et al.* (2019) combined large scale data from MaNGA with observations from Gemini North Multi-Object Spectrograph (GMOS) Integral Field Unit (IFU) to constrain the gas and stellar kinematics of the galaxy Akira on scales of hundreds of pc and to better understand the mechanism which produces the outflow. Riffel *et al.* (2019) found that the orientation of outflow changes radially from the nuclear region to kpc scales and suggest that the outflow is produced by precession of the supermassive black hole accretion disk. In order to verify if the scenario proposed by Riffel *et al.* (2019) occurs in other Red Geysers, we have used the MaNGA data contained in SDSS-IV DR15 (Aguado *et al.* 2019) to select a sample of Red Geysers.

2. The sample and Results

Following Roy *et al.* (2018), we have defined the sample of Red Geysers in the MaNGA survey by adopting the following selection criteria: rest frame colour NUV$-r > 5$, star formation rate with log SFR$[M_\odot/\mathrm{yr}] < -2$, bi-symmetric pattern in Hα-EW maps aligned with the gas kinematic axis and misaligned with the stellar kinematic axis, velocity fields of Hα reaching values of ± 300 km/s and being at least twice as high as the values of the stellar velocity fields. To measure the orientation of the line of nodes of stellar and gas

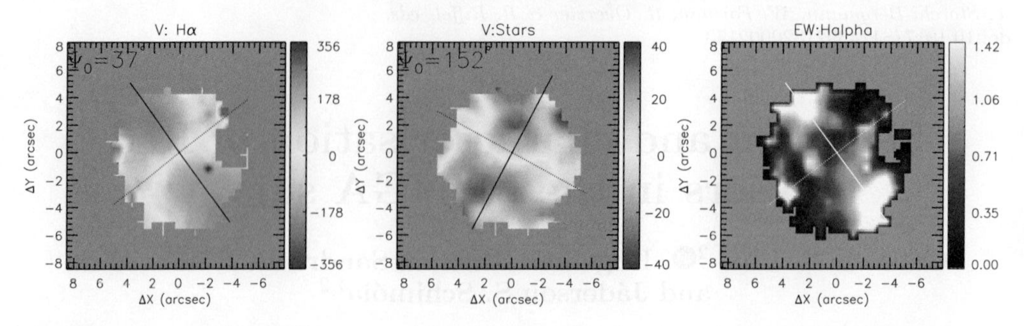

Figure 1. Kinematic structures and gas emission for galaxy MaNGA 1-523238. From left to right: Hα velocity field, stellar velocity field and Hα-EW map. The colour bars show the velocities in units of km s^{-1} and the EW values in Å. The black solid (first and second panel) and white solid (third panel) lines show the orientation of the line of nodes (Ψ_0) as measured by the kinemetry method.

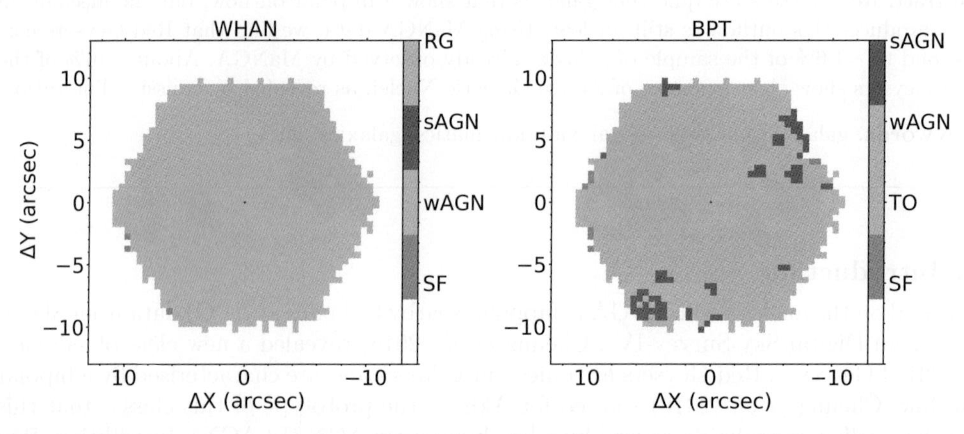

Figure 2. WHAN and BPT diagrams for galaxy MaNGA 1-217022. The following labels were used: SF or star-forming galaxies, TO (transition objects), wAGN or weak AGN (low-luminosity AGN), sAGN or strong AGN (high-luminosity AGN) and RG (retired galaxies–Hα-EW < 3 Å).

velocity fields we used the kinemetry method (Krajnović *et al.* 2006). We then added the following criteria: difference in the orientation of the line of nodes of stellar and gas velocity fields of $10° < \Delta PA < 170°$. Fig. 1 shows examples of the gas and stellar velocity fields and Hα-EW map for one target classified as a Red Geyser.

We find that ∼1.6% of the MaNGA galaxies in DR15 satisfy the selection criteria of the previous section and are thus classified by us as Red Geysers. In order to investigate if the objects of our sample present AGN activity, we constructed the BPT (Baldwin *et al.* 1981) and WHAN (Cid Fernandes *et al.* 2010) diagrams. Fig. 2 shows examples of such diagrams for MaNGA 1-217022, which present a weak AGN. We observe clear evidence of AGN (objects classified as AGN in both BPT and WHAN diagrams) in only 17 (16%) galaxies of our sample. However, the low spatial resolution of MaNGA (∼2.″5) does not allow the detection of weak AGN, whose emission may be diluted by those from circumnuclear regions. Thus, the AGN fraction determined here can be considered a lower limit. We are conducting follow-up observations with Gemini to obtain high resolution optical integral field spectroscopy for a sub-sample of Red Geysers in order to characterise the gas kinematics and excitation in the inner kpc (Ilha *et al.* in prep.).

References

Abolfathi, B., *et al.* 2018, *ApJS*, 235, 42
Aguado, *et al.* 2019, *ApJS*, 240, 23
Baldwin, J. A., Phillips, M. M., Terlevich, R., *et al.* 1981, *PASP*, 93, 5
Cheung, E., *et al.* 2016, *Nature*, 533, 504
Cid Fernandes, R., *et al.* 2010, *MNRAS*, 403, 1036
Krajnović, *et al.* 2006, *MNRAS*, 366, 787
Riffel, R. A., *et al.* 2019, *MNRAS*, 485, 5590
Roy, N., *et al.* 2018, *ApJ*, 869, 117

Galaxy Evolution and Feedback across Different Environments
Proceedings IAU Symposium No. 359, 2020
T. Storchi-Bergmann, W. Forman, R. Overzier & R. Riffel, eds.
doi:10.1017/S1743921320002148

Circumnuclear multi-phase gas around nearby AGNs investigated by ALMA

Takuma Izumi

National Astronomical Observatory of Japan, 2-21-1 Osawa, Mitaka, Tokyo 181-8588, Japan
email: takuma.izumi@nao.ac.jp

Abstract. Since the advent of the Atacama Large Millimeter/submillimeter Array (ALMA), more attention has been paid on the $\lesssim$100 pc scale circumnuclear disk (CND) to reveal feeding and feedback processes of active galactic nuclei (AGNs). By using cold molecular CO and atomic C^0 emission line observations, we have revealed that there are multi-component gas dynamical flows around the AGN of the Circinus galaxy, which may explain the physical origin of the AGN torus. In the luminous Seyfert galaxy NGC 7469, we found that [CI](1–0) line is extraordinary bright relative to CO lines (for example $J = 2 - 1$), manifesting the physical/chemical influence of the AGN on the surrounding gas in the form of X-ray dominated region (XDR).

Keywords. galaxies: active, galaxies: Seyfert, galaxies: evolution, ISM: evolution

---

1. Introduction

Almost thirty years have passed since the advent of the active galactic nucleus (AGN) torus paradigm (Antonucci 1993). In this scheme, AGN is surrounded by a dusty/gaseous donut-like structure, along the hole of which we can directly see the highly ionized broad line region, whereas our line-of-sight is blocked when we look at the core from its side. This scheme explains many observational features of AGNs (e.g., Ramos Almeida & Ricci 2017), however, its physical origin remains unclear. Wada (2012) attributes this origin to multi-phase gas dynamics in the circumnuclear disk (CND) and AGN radiation-driven winds, based on high resolution hydrodynamic simulations that also incorporate X-ray dominated region (XDR) chemistry (Maloney *et al.* 1996). In their simulations, dense molecular gas is confined in the mid-plane of the disk. Hot ionized gas is predominantly seen in AGN-driven outflows. Atomic gas is intermediate: a fraction of this phase gas is in warm outflows, part of which eventually falls back to the disk due to the disk gravity (= failed wind). The circulation of gas inflow (through the disk mid-plane), outflow, and failed wind (fountain) causes turbulence, which naturally forms a geometrically thick structure, i.e., torus. Although all of these occur at the central < 100 pc region of AGNs, now the Atacama Large Millimeter/submillimeter Array (ALMA) enables us to probe that compact region and test the above predictions. Here, we present two high resolution ALMA programs toward nearby Seyfert galaxies, namely the Circinus galaxy and NGC 7469. In these experiments, we observed both molecular CO lines and atomic [CI](1–0) line, so that we can probe both the dense molecular and diffuse atomic parts of the CND.

2. High resolution multi-phase gas dynamics in the Circinus galaxy

In our ALMA Cycle 4 observation, we successfully detected CO(3–2) and [CI](1–0) line emission from the central region of the Circinus galaxy ($D = 4.2$ Mpc) at the spatial resolutions of $\sim$6$-$15 pc, as presented in Izumi *et al.* (2018). The line emission was bright

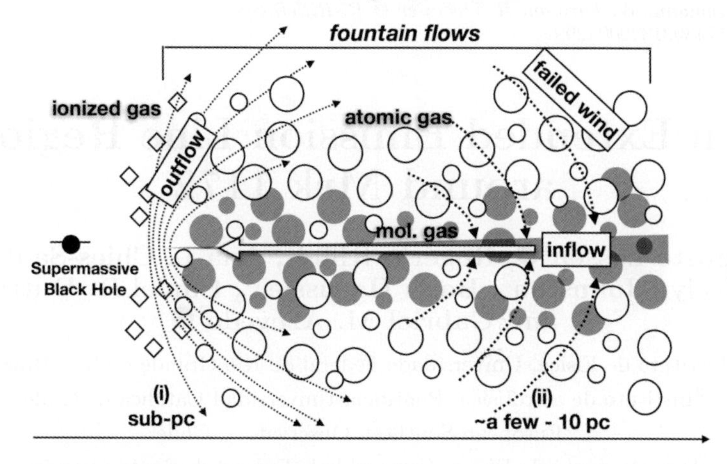

Figure 1. Our new picture of multi-phase AGN torus (modified from Izumi *et al.* 2018). Each phase gas preferentially traces different flows including inflows (molecular; filled circle), hot outflows (ionized; diamond), and warm outflows + failed winds (atomic; open circle), respectively.

at the innermost <100 pc region, manifesting the genuine existence of a massive CND. At the AGN position, we found that the spectral profiles of CO and [CI] are clearly different. The CO(3–2) shows a single Gaussian-like profile, which is usually seen in the CND-scale of nearby AGNs. On the other hand, the [CI](1–0) shows a triple Gaussian-like feature. The latter manifests the existence of atomic gas outflows. The estimated outflow velocity, however, does not exceed the escape velocity from the system. Hence at least part of the atomic outflow eventually falls back to the disk, i.e., failed wind. These are consistent with the predictions of the radiation-driven fountain torus model, which provides a key observational insight to understand the physical origin of AGN torus. A new dynamic picture of AGN torus based on our work is shown in Figure 1.

3. X-ray dominated region caught in the act in NGC 7469

To investigate potential AGN influence on the surrounding gas in the form of the XDR, we also observed multiple ^{12}CO lines (from $J = 1$–0 to 3–2), ^{13}CO(2–1), and [CI](1–0) emission toward the CND of the luminous type-1 Seyfert galaxy NGC 7469 ($D = 70.8$ Mpc) during ALMA Cycle 5. The CO lines are bright both in the CND and the surrounding starburst ring (SB ring; ∼1 kpc diameter), with two bright peaks on either side of the AGN. By contrast, the [CI](1–0) line is strongly peaked on the AGN. Consequently, the line brightness temperature ratio of [CI](1–0) to ^{13}CO(2–1) is ∼10 times higher around the AGN than at the SB ring. Our non local thermodynamic equilibrium simulations suggest that at least enhanced C^0/CO abundance ratio (∼10), as well as high gas temperature (>100−500 K) around the AGN relative to the starburst ring, are required to explain this line intensity enhancement. These physical and chemical features are consistent with the XDR model prediction that shows elevated C^0 abundance due to efficient dissociation of CO around AGNs (Maloney *et al.* 1996).

References

Antonucci, R. 1993, *ARAA*, 31, 473
Izumi, T., Wada, K., Fukushige, R., *et al.* 2018, *ApJ*, 867, 48
Maloney, P. R., Hollenbach, D. J., & Tielens, A. G. G. M. 1996, *ApJ*, 466, 561
Ramos Almeida, C. & Ricci, C. 2017, *Nature Astronomy*, 1, 679
Wada, K. 2012, *ApJ*, 758, 1

Galaxy Evolution and Feedback across Different Environments
Proceedings IAU Symposium No. 359, 2020
T. Storchi-Bergmann, W. Forman, R. Overzier & R. Riffel, eds.
doi:10.1017/S1743921320004299

An Extended Emission Line Region around Mrk 1172

Augusto Lassen[1], Rogério Riffel[1], Ana L. Chies-Santos[1], Evelyn Johnston[2], Boris Haeussler[3], Daniel R. Dutra[4] and Gabriel M. Azevedo[1]

[1]Instituto de Física, Universidade Federal do Rio Grande do Sul, Brazil

[2]Instituto de astrofísica, Pontificia Universidad Católica de Chile

[3]European Southern Observatory, Chile

[4]Departamento de Física, Universidade Federal de Santa Catarina

Abstract. We serendipitously found an intriguing Extended Emission Line Region (EELR) near the quiescent and massive early-type Mrk 1172, with a projected extension of approximately 14×14 kpc. Its irregular shape, high gas content, strong emission lines and proximity to an isolated possible faded quasar raise questions about the ionization of this gas and the nature of this object. Analyzing the stellar population in both objects we observe that the EELR has a dominance of young-intermediate and intermediate stellar populations (200 Myr $< t <$ 1 Gyr) with significant star formation activity, while Mrk 1172 is dominated by old stellar population ($t > 5$ Gyr). BPT diagnostic diagrams indicate that the gas in the EELR is photoionized by hot massive stars rather than by a hard radiation field or by shocks. Further analysis on abundances of the gas and its kinematics shall be performed to better comprehend the nature of this object and how it is interacting with Mrk 1172.

Keywords. ISM: structure, Galaxy: stellar content, Galaxy: evolution

---

1. A new extended emission line region

During the inspection of public data from the Multi Unit Spectroscopic Explorer (MUSE), we have serendipitously found an Extended Emission Line Region (EELR) near the Early-Type Galaxy (ETG) Mrk 1172 with similar redshift ($z = 0.04115$ for Mrk 1172 and $z \sim 0.0403$ for the EELR). In figure 1 we show the MUSE Field-of-View (FoV) for the system in the continuum and in H_α,[OIII]λ5007 and [SII] λ6716 + [SII] λ6731 wavelength range, where the red bars and the yellow parts of spectra represent the slices taken to produce the images. Spectra shown are from spaxels in Mrk 1172 (top right) and in the EELR (bottom right), showing that the EELR is very faint in the continuum, despite its emission lines luminosity, and was never reported in the literature, to the best of our knowledge. To estimate the projected extension of the EELR we used a square box of $\sim$14 kpc $\times$ 14 kpc that contains this region.

We performed a spatially resolved stellar population synthesis analysis on the system. This allowed us to remove the stellar contribution from the gas emission lines. To perform the stellar population synthesis we used MEGACUBE (Mallmann *et al.* 2018) with the templates of "GM base" introduced in Fernandes *et al.* (2014). The fit was performed in the $4800 \sim 6900$ Å range with normalization at $\lambda_0 = 5600$ Å. The analysis reveals that the EELR has a dominant stellar population with ages in the range of $0.2 \sim 1.0$ Gyr. Mrk 1172 is dominated by an old stellar population ($t > 5$ Gyr).

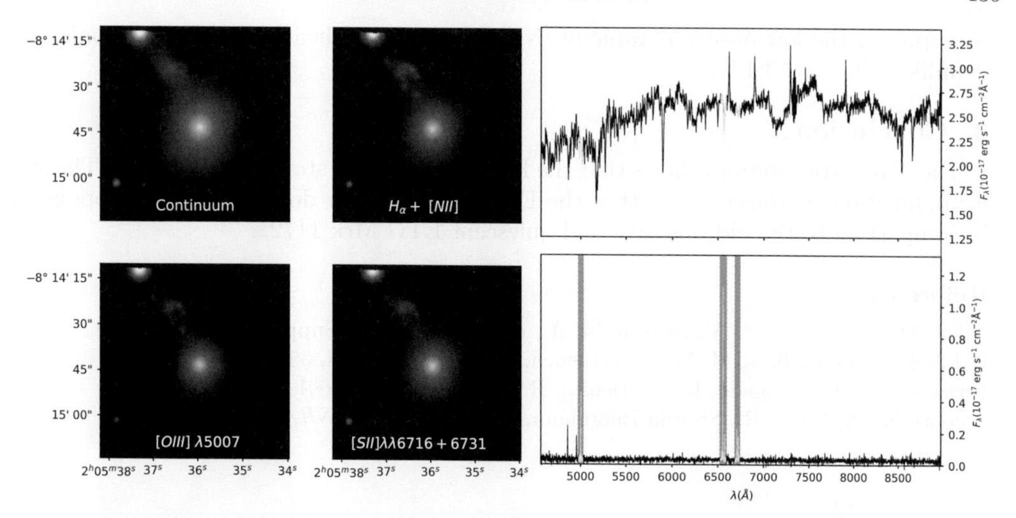

Figure 1. Mrk 1172 FoV in the continuum (top left) and H_α+[NII]$\lambda\lambda$6550 + 6585 (top central), [OIII]λ5007 (bottom left) and [SII]$\lambda\lambda$6716 + 6731 (bottom central) wavelength ranges. The wavelength windows used to produce these images is highlighted in yellow in top and bottom right panels. In right (top) the rest-frame spectrum of Mrk 1172 and (bottom) the rest-frame EELR spectrum. The bright object in the top of the figure is a star.

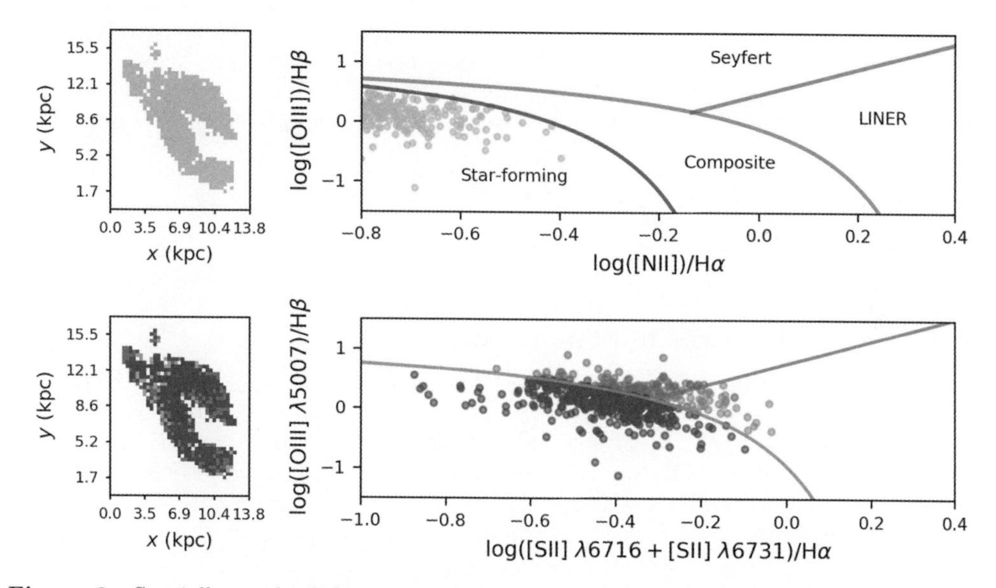

Figure 2. Spatially resolved diagnostic diagram for the EELR. *Right top/bottom panel:* The dots represent the position of each single spaxel in the diagnostic diagram, where the solid lines represent the curves that define the excitation regions. *Left top/bottom panel*: The EELR with each spaxel coloured corresponding to its position in the diagnostic diagram.

In figure 2 we present the spatially resolved diagnostic diagrams for the EELR (Baldwin, Phillips & Terlevich 1981). Both diagrams indicate that the gas seems to be ionized by young massive stars rather than by an Active Galactic Nuclei (AGN). Ionization by shocks was investigated using the fast radiative shock models adopting solar metallicity, $n_e = 1.0$ cm^{-3} and varying the values of magnetic field Allen *et al.* (2008). In any of these cases the curve of emission line ratio vs. shock velocity was able to approach the low observed value of [NII]/$H_\beta \sim 0.56$, indicating that shock ionization is unlikely in this EELR.

2. Conclusion

The diagnostic diagram shows that the EELR is an active star-forming region. The stellar population synthesis shows that the EELR has younger dominant stellar population in comparison to the old, massive and quiescent ETG Mrk 1172.

References

Allen, M. G., Groves, B. A., Dopita, M. A., *et al.* 2008, *ApJS*(Supplement Series) 178
Baldwin, J. A., Phillips, M. M., & Terlevich, R. 1981, *PASP* 93, 551
Fernandes, R. C., Delgado, R. G., Benito, R. G., *et al.* 2014, *A&A* 561, A130
Nicolas, N. D., Riffel, R., Storchi-Bergmann, T., *et al.* 2018, *MNRAS* 478, 4

Galaxy Evolution and Feedback across Different Environments
Proceedings IAU Symposium No. 359, 2020
T. Storchi-Bergmann, W. Forman, R. Overzier & R. Riffel, eds.
doi:10.1017/S1743921320002045

The Quest for Relics: Massive compact galaxies in the local Universe

F. S. Lohmann, A. Schnorr-Müller, M. Trevisan, R. Riffel, N. Mallmann, Ana L. Chies-Santos and C. Furlanetto

Departamento de Astronomia, Universidade Federal do Rio Grande do Sul,
PO Box 15051, 91501-970 Porto Alegre, Brazil
emails: felipe.s.lohmann@gmail.com, allan.schnorr@ufrgs.br,
marina.trevisan@ufrgs.br, riffel@ufrgs.br, nicolas.fisica@gmail.com,
ana.chies@ufrgs.br, cristina.furlanetto@ufrgs.br

Abstract. Observations at high redshift reveal that a population of massive, quiescent galaxies (called red nuggets) already existed 10 Gyr ago. These objects undergo a significant size evolution over time, likely due to minor mergers. In this work we present an analysis of local massive compact galaxies to assess if their properties are consistent with what is expected for unevolved red nuggets (relic galaxies). Using integral field spectroscopy (IFS) data from the MaNGA survey from the Sloan Digital Sky Survey (SDSS), we characterized the kinematics and properties of stellar populations of massive compact galaxies, and find that these objects exhibit, on average, a higher rotational support than a control sample of average sized early-type galaxies. This is in agreement with a scenario in which these objects have a quiet accretion history, rendering them candidates for relic galaxies.

Keywords. Galaxies: evolution, Galaxies: formation, Galaxies: kinematics and dynamics

1. Introduction

Observations of the high redshift Universe revealed that a population of massive quiescent galaxies (called red nuggets) already existed 10 billion years ago. These galaxies are thought to be the progenitors of local massive quiescent galaxies. However, compared to their local Universe counterparts, these objects show significant distinctions: they are very compact, more elongated and many have disks, suggesting that they undergo a significant size evolution over time (Trujillo *et al.* 2007). Simulations suggest that this evolution in size is mainly due to minor mergers (Naab *et al.* 2009). Thus, the study of these objects is crucial to understand the formation of massive galaxies. However, spatially resolved spectroscopic studies of large samples of red nuggets are not feasible with the current generation of telescopes. Since mergers occur in a stochastic manner, it is expected that there exists a population of galaxies in the local Universe that have not gone through these processes since they became quiescent. These galaxies, called relic galaxies, are local analogues of the high redshift red nuggets. Studying these objects can provide hints on the formation of massive quiescent galaxies at high redshift and how they transform into local early-type galaxies.

The aim of this work is to characterize the kinematics and the stellar populations of a sample of massive compact galaxies in the local Universe. The work focuses on the following items:

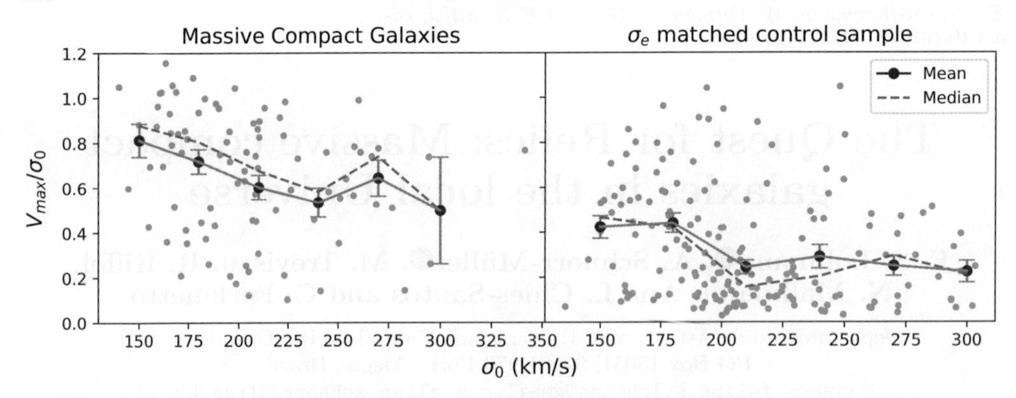

Figure 1. Distribution of $V_{\max}/\sigma_0$, the ratio between the maximum velocity along the major axis ($V_{\max}$) and the central velocity dispersion (σ_0, defined as the velocity dispersion measured in the central pixel), for both samples.

• Determine which galaxies have properties consistent with relic galaxies and understand their importance in the formation and evolution of massive galaxies;

• Investigate formation and evolution scenarios for these objects, as well as their relationship with local massive elliptical galaxies.

2. Methodology

We studied a sample of 87 galaxies selected from the MaNGA survey based on the following criteria:

(a) Stellar mass: $10^{10.5} M_{\odot} < M_{\star} < 10^{11.5} M_{\odot}$;

(b) The size of the semimajor axis of the half-light ellipse is at least 1σ below the value predicted by the local mass-size relation for early-type galaxies (van der Wel *et al.* 2014).

In order to verify if massive compact galaxies have different kinematics from massive average-sized galaxies, we defined a control sample with 174 objects with half-light radii within 1σ of the median size for a given mass. The samples were matched by σ_e, defined as the velocity dispersion measured in an effective radius aperture R_e, and by specific star formation rate. Using integral field spectroscopy data, velocity maps (V), velocity dispersion (σ) and Gauss-Hermite parameters h_3 and h_4 were obtained using the pPXF code (Cappellari 2017). The Gauss-Hermite h_3 parameter is of interest for this work as it is linked to the galaxy assembly history: simulations revealed that fast-rotators with a gas rich merger history show an anti-correlation between h_3 and V/σ, while fast-rotators with a gas poor merger history do not (Naab *et al.* 2014).

3. Conclusions

• The compact galaxy sample has, on average, a higher rotational support than the control sample (Fig. 1). This difference is driven mainly by a significant fraction of slow rotators in the control sample. This is in agreement with a scenario in which these objects did not go through gas-poor major mergers since their formation;

• Several objects in the compact galaxy sample (~80%) present a strong anti-correlation between h_3 and V/σ. This anti-correlation indicates that these objects have a gas-rich merger history;

• However, to confirm whether these objects are indeed relics, further studies of their stellar populations - to constrain the time of their formation - and of the stellar kinematics on their outskirts - to rule out a significant contribution of accreted stars - are necessary.

References

Cappellari, M. 2017, *MNRAS*, 466, 798
Naab, T., Johansson, P. H., & Ostriker, J. P. 2009, *ApJ* (Letters), 699, L178
Naab, T., Oser, L., Emsellem, E., *et al.* 2014, *MNRAS*, 444, 3357
Trujillo, I., Conselice, C. J., Bundy, K., *et al.* 2007, *MNRAS*, 382, 109
van der Wel, A., Franx, M., van Dokkum, P. G., *et al.* 2009, *ApJ*, 788, 28

Galaxy Evolution and Feedback across Different Environments
Proceedings IAU Symposium No. 359, 2020
T. Storchi-Bergmann, W. Forman, R. Overzier & R. Riffel, eds.
doi:10.1017/S1743921320002318

Curvature of galaxy brightness profiles

Geferson Lucatelli[1], Fabricio Ferrari[1], Arianna Cortesi[2], Ana L. Chies-Santos[3], Fernanda Roman de Oliveira[3], Claudia Mendes de Oliveira[4] and Lilianne M. Izuti Nakazono[4]

[1]Institudo de Matemática, Estatística e Física, Universidade Federal do Rio Grande, Rio Grande-RS, Brasil
email: gefersonlucatelli@furg.br

[2]Observatório do Valongo, Universidade Federal do Rio de Janeiro, Rio de Janeiro-RJ, Brasil

[3]Departamento de Astronomia, Universidade Federal do Rio Grande do Sul, Porto Alegre-RS, Brasil

[4]Instituto de Astronomia, Geofísica e Ciências Atmosféricas, Universidade de São Paulo, São Paulo-SP, Brasil

Abstract. Galaxy morphologies reflect the shapes of galaxies and their structural components, such as bulges, discs, bars, spiral arms, etc. The detailed knowledge of the morphology of a galaxy provides understanding of the physics behind its evolution, since the time of its formation, including interaction processes and influence of the environment. Thus, the more precisely we can describe a galaxy structure, the more we may understand about its formation and evolution. We present a method that measures curvature, using images, to describe galaxy structure and to infer the morphology of each component of a galaxy. We also include some preliminary results of curvature measurements for galaxies of the Southern Photometric Local Universe Survey (S-PLUS) DR1 data release and for jellyfish galaxies of the Omega Survey. We find that the median of the curvature parameter and the integrated area under the curvature give us clues on the morphology of a galaxy.

Keywords. galaxies:structure, galaxies:morphology, galaxies:morphometry, differential geometry: curvature

---

1. Introduction

One way to understand better how the morphology of a galaxy behaves is to quantify how many structural components there are in a galaxy. Each one of these components are directly or indirectly related to its dynamical properties. Thus, having a complete picture of the structure of a galaxy in terms of its components – how the light is distributed along different parts – might improve our knowledge regarding their morphology and evolution. There are many attempts to extract morphometric properties of galaxies (e.g. Conselice 2014). However, it is not certain how we can use them to study different parts of a galaxy. In (Lucatelli *et al.* 2019) we have introduced the method of curvature in order to discriminate these differences in galaxy morphologies. This enabled us to quantify approximately how many components there are in a galaxy in cases we have sufficient spatial resolution.

1.1. *Curvature* $\widetilde{\kappa}$

From differential geometry, the curvature κ of a uni-dimensional function $f(x)$ is defined as (e.g. Tenenblat 2008) $\kappa(x) \equiv f''/(1+f'^2)^{3/2}$ where $f' = \frac{\mathrm{d}f}{\mathrm{d}x}$. It measures how

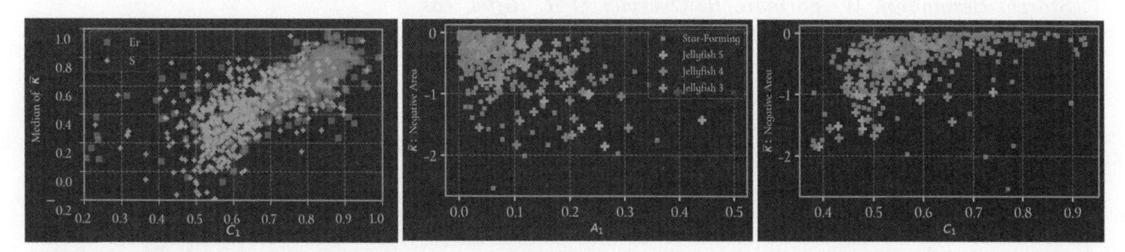

Figure 1. Left: Relation between the median of $\widetilde{\kappa}$ with the concentration C_1 index for a sample of galaxies from S-PLUS–DR1 data (Mendes de Oliveira *et al.* 2019). "S" means spirals and "Er" elliptical rounded, from GalaxyZoo. **Centre/Right:** Relation between area under $\widetilde{\kappa}$, C_1 and asymmetry A_1, for galaxies of the OMEGA survey. Jellyfish galaxies are indicated by "plus" labels and by their JClasses, 5, 4 and 3, see Roman-Oliveira *et al.* (submitted).

the tangent vector's curve changes as we move on the function and here it is applied to the light profile†

$$\widetilde{\kappa}(\chi) \equiv \frac{\mathrm{d}^2\nu(\chi)}{\mathrm{d}\chi^2}\left[1+\left(\frac{\mathrm{d}\nu(\chi)}{\mathrm{d}\chi}\right)^2\right]^{-3/2} \xrightarrow{\text{discrete}} \widetilde{\kappa} \equiv \frac{\delta\chi_i\,\delta^2\nu_i - \delta\nu_i\,\delta^2\chi_i}{(\delta\chi_i^2+\delta\nu_i^2)^{3/2}}, \tag{1.1}$$

where χ is the normalized projected radius of the galaxy and $\nu(\chi)$ is the normalized logarithm of the light profile I.

2. Results and Discussions

We present here some applications and preliminary results from curvature applied to S-PLUS and Omega survey galaxies. These first results are based on a simple approach which is to compute the area under the curvature profile and its median. This gives us a scalar value from it, being interpreted as a morphometric index.

In the left of Figure 1 it is shown the relation between the median of the curvature with the concentration index C_1 Conselice (2003) for a sample of S-PLUS-DR1 galaxies with classes from GalaxyZoo (Lintott *et al.* 2011). There is a distinct relation between both parameters. Spirals (green) have lower C_1 and curvature values than ellipticals (blue). In the centre and right of the same figure we show the relation between the total integrated area under the curvature with C_1 and the asymmetry A_1 parameter (Ferrari *et al.* 2015; Abraham *et al.* 1996) for galaxies of the Omega Survey (HST F606W). Star-forming galaxies are labelled in blue squares and jellyfish galaxies in cross symbols (purple, green and red), with colours indicating JClass 3, 4 and 5, Roman-Oliveira *et al.* (submitted). Despite some scatter, we can see a distinction between the Jellyfish candidates and other star forming galaxies in the $\widetilde{\kappa}$Area $\times\, C_1 \times A_1$ diagram.

References

Abraham, R., van den Bergh, S., Glazebrook, K. *et al.* 1996, *ApJS*, 107, 1
Conselice, C. 2003, *ApJS*, 147, 1
Conselice, C. 2014, *ARAA*, 52, 291
de Oliveira, C. M., Ribeiro, T., Schoenell, W., *et al.* 2019, *MNRAS*, 489, 241
Ferrari, F., de Carvalho, R. R., & Trevisan, M. 2015, *Apj*, 814, 55
Lintott, C., Schawinski, K., Bamford, S., *et al.* 2011, *MNRAS*, 410, 166
Lucatelli, G. & Ferrari, F. 2019, *MNRAS*, 489, 1161
Tenenblat, K. 2018, *Edgard Blucher*, 2nd

† https://gitlab.com/gefersonlucatelli/kurvature

Galaxy Evolution and Feedback across Different Environments
Proceedings IAU Symposium No. 359, 2020
T. Storchi-Bergmann, W. Forman, R. Overzier & R. Riffel, eds.
doi:10.1017/S1743921320001519

Dark matter bar evolution in triaxial spinning haloes

Daniel A. Marostica and Rubens E. G. Machado

Universidade Tecnológica Federal do Paraná, Brazil
email: marostica@alunos.utfpr.edu.br

Abstract. Dark matter bars are structures that may form inside dark matter haloes of barred galaxies. Haloes can depart from sphericity and also be subject to some spin. The latter is known to have profound impacts on the evolution of both stellar and DM bars, such as stronger dynamical instabilities, more violent vertical bucklings and dissolution or impairment of stellar bar growth. On the other hand, dark matter bars of spherical haloes become initially stronger in the presence of spin. In this study, we add spin to triaxial halos in order to quantify and compare the strength of their bars. Using N-body simulations, we find that spin accelerates main instabilities and strengthens the halo bars, although their final strength depends only on triaxiality. The most triaxial halo barely forms a halo bar, showing that flattening opposes to DM bar strengthening and indicating that there is a limit on how flattened the parent structure can be.

Keywords. galaxies: kinematics and dynamics, galaxies: halos, galaxies: evolution

---

1. Introduction

Dark matter (DM) bars are structures that may form inside dark matter haloes of barred galaxies. Haloes can depart from sphericity and also be subject to some spin, which is known as cosmological spin parameter $\lambda \equiv J_{\rm h}/\sqrt{2}M_{\rm vir}R_{\rm vir}v_{\rm c}$ – where $J_{\rm h}$ is the halo angular momentum, $M_{\rm vir}$ and $R_{\rm vir}$ are the virial mass and radius, and $v_{\rm c}$ is the circular velocity at $R_{\rm vir}$.

Spin has profound impacts on the evolution of both stellar and DM bars. For example, Collier *et al.* (2018) found that dynamical instabilities are accelerated when the disc is embedded in haloes with net rotation. This results in more violent vertical bucklings, which are followed by the dissolution of the stellar bar or impairment of its subsequent recovery. On the other hand, dark matter bars of spherical haloes are stronger in the presence of spin: while the strength before the vertical buckling of the stellar bars is somehow indifferent to λ, that of the DM bars increases with it Collier *et al.* (2019).

2. Methods and Results

In this study, we add spin to triaxial haloes of N-body simulations in order to compare the effects of both variables on the evolution of their DM bars. A set of simulations was run with the code GADGET-2, using the three haloes from Athanassoula *et al.* (2013): a spherical halo, a triaxial halo ($b/a = 0.8$ and $c/a = 0.6$) and a more triaxial halo ($b/a = 0.6$ and $c/a = 0.4$). For each halo, we ran simulations with three different spins: $\lambda \approx 0.00$, 0.03 and 0.07. The spinning haloes were created by inverting the tangential velocities of a certain fraction of retrograde halo particles.

We find that larger spin accelerates stellar bar formation, slightly decreasing its strength before buckling. The capacity of the stellar bar to recover its strength is inversely

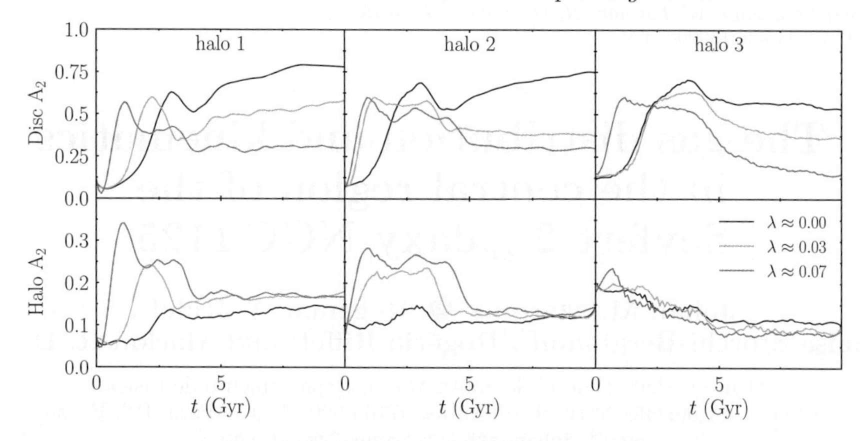

Figure 1. Bar strength (A_2) as a function of time. From left to right, galaxies with increasing halo triaxiality. Colours represent different spin parameters (λ).

proportional to λ. Increasing halo triaxiality does not greatly affect stellar bar strength, but negatively affects the ability of the bar to regrow, causing it to, sometimes, completely dissolve.

Larger spin accelerates the formation of the halo bar, which becomes approximately three times stronger for $\lambda \approx 0.07$. Spin does not affect the final strength of the DM bar in any of the simulated haloes. Increasing triaxiality slightly weakens the final halo bar strength (Figure 1). Spin alone is not capable of developing stronger bars for the most triaxial halo, drawing the line on how flattened haloes can be in order to host a stellar-bar-induced DM bar.

3. Conclusions

In this work, we have recovered the results of Collier *et al.* (2019) for spherical haloes and extended the analyses to fully triaxial spinning haloes. We found that the stellar bars of all the three haloes form earlier, according to their λ, and may eventually dissolve. The DM bars of haloes 1 and 2 form earlier, along with the stellar bar, and become much stronger before buckling, proportionally to λ. On the other hand, the stellar bar of halo 3 is not capable of inducing a halo bar, showing an inverse effect of triaxiality on DM bar strength.

Acknowledgements

The authors thank *Sociedade Astronômica Brasileira* and *Conselho Nacional de Desenvolvimento Cientfico e Tecnológico* (CNPq). The authors also acknowledge the National Laboratory for Scientific Computing (LNCC/MCTI, Brazil) for providing HPC resources of the SDumont supercomputer, which contributed to the research results in this paper.

References

Athanassoula, E., Machado, R. E. G., & Rodionov, S. A. 2013, *MNRAS*, 429, 1949
Collier, A., Shlosman, I., & Heller C. 2018, *MNRAS*, 476, 1331
Collier, A., Shlosman, I., & Heller, C. 2019, *MNRAS*, 488, 5788

Galaxy Evolution and Feedback across Different Environments
Proceedings IAU Symposium No. 359, 2020
T. Storchi-Bergmann, W. Forman, R. Overzier & R. Riffel, eds.
doi:10.1017/S1743921320002379

The gas distribution and kinematics in the central region of the Seyfert 2 galaxy NGC 1125

Johan M. Marques[1], Rogemar A. Riffel[1], Thaisa Storchi-Bergmann[2], Rogério Riffel[2] and Marlon R. Diniz[1]

[1]Universidade Federal de Santa Maria, Departamento de Física,
Centro de Ciências Naturais e Exatas, 97105-900, Santa Maria, RS, Brazil
email: johanmatheusmarques@gmail.com
[2]Instituto de Física, Universidade Federal do Rio Grande do Sul,
Av. Bento Gonçalves 9500, 91501-970 Porto Alegre, RS, Brazil

Abstract. We present Gemini Near-Infrared Integral-Field Spectrograph (NIFS) observations of the inner 660×660 pc^2 of the Seyfert 2 galaxy NGC 1125, which reveals that the emission-line profiles present two kinematic components: a narrow one ($\sigma < 150$ km s^{-1}) due to emission of the gas in the disk and a broad component ($\sigma > 150$ km s^{-1}) produced by a bipolar outflow, perpendicular to the galaxy's disk.

Keywords. galaxies: Seyfert, galaxies: kinematics and dynamics, galaxies: individual (NGC 1125)

It is now widely accepted that feeding and feedback processes of active galactic nuclei (AGN) play an important role in the evolution of galaxies (Harrison *et al.* 2018; Storchi-Bergmann & Schnorr-Müller 2019). Mapping the multiphase gas kinematics and distribution allow us to map and quantify the feeding and feedback processes of AGN. Recent studies, using near-IR integral field spectroscopy, have shown that the H_2 and ionised gas have distinct flux distributions and kinematics, with the former usually tracing gas in orbit in the disk of the galaxy and the latter tracing AGN winds usually extending to high latitudes relative to the galaxy disk (Riffel *et al.* 2015; Schönell *et al.* 2019).

We use J and K band integral field spectroscopy of the inner the 660×660 pc^2 of the Seyfert 2 galaxy NGC 1125 obtained with the Gemini NIFS instrument at a spatial resolution of ~25 pc to map the emission-line flux distributions and kinematics of both molecular and ionized gas. We use the the python library IFSCUBE (Ruschel-Dutra 2020)to fit the emission line profiles of both ionized and molecular gas by two Gaussian functions. Figure 1 shows an example of the fitting of the [Fe II]$\lambda 1.2570\,\mu$m and Paβ emission line profiles observed at the nucleus of the galaxy.

Figure 2 shows the flux, centroid velocity and σ maps for the Paβ narrow ($\sigma <$ 150 kms^{-1}) and broad ($\sigma > 150$ km s^{-1}) components. Similar maps are obtained for other near-infrared emission lines. For the narrow component, the flux distributions for all emission lines are more elongated along the major axis of the galaxy and the corresponding velocity fields show a similar rotation pattern as that of the stars. Thus, we interpret this component as being produced by emission from gas in the disk. On the other hand, most of the emission from the broad component is most extended perpendicular to the major axis of the galaxy, the velocity gradient is perpendicular to that of

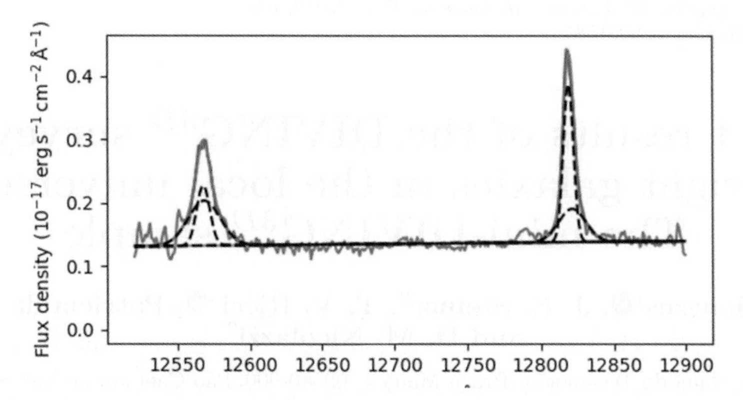

Figure 1. Example of fits of the [Fe II]$\lambda1.2570\,\mu$m and Paβ emission-line profiles by two Gaussian components (black solid + black dashed lines). The observations are shown in blue and the fit of the gaussian is shown as the red solid line. The narrow component is attributed to emisson from the disk and the broad component from an outflow.

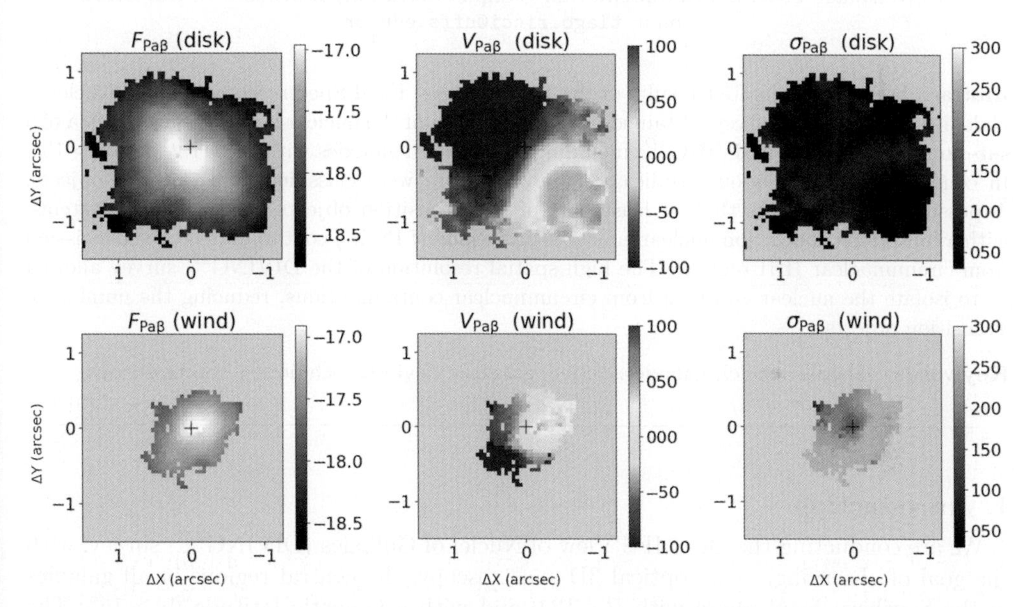

Figure 2. Flux (left), centroid velocity (middle) and velocity dispersion (right) maps for the disk (top) and wind (bottom) components of the Paβ emission line. The color bars show the fluxes in logarithmic scale of erg s^{-1} cm^{-2} spaxel^{-1} and the velocities (relative to the systemic velocity of the galaxy) and velocity dispersions (σ) in km s^{-1}.

the stars, and the velocity dispersion is higher than that observed for the stars and for the disk component. We interpret the observed broad component as a bipolar outflow.

References

Harrison, C., *et al.* 2018, *NatAs*, 2, 198
Riffel, R. A., *et al.* 2015, *MNRAS*, 451, 3587
Ruschel-Dutra, D. 2020, IFSCUBE, Retreived from https://github.com/danielrd6/ifscube
Schönell, A. J., *et al.* 2019, *MNRAS*, 485, 2054
Storchi-Bergmann, T. & Schnorr-Müller, A. 2019, *NatAs*, 3, 48

Galaxy Evolution and Feedback across Different Environments
Proceedings IAU Symposium No. 359, 2020
T. Storchi-Bergmann, W. Forman, R. Overzier & R. Riffel, eds.
doi:10.1017/S1743921320001581

First results of the DIVING^{3D} survey of bright galaxies in the local universe: The mini-DIVING^{3D} sample

R. B. Menezes[1], J. E. Steiner[2], T. V. Ricci[3], Patrícia da Silva[2] and D. M. Nicolazzi[2]

[1]Instituto Mauá de Tecnologia, Praça Mauá 1, 09580-900, São Caetano do Sul, SP, Brazil
email: roberto.menezes@maua.br

[2]Instituto de Astronomia, Geofísica e Ciências Atmosféricas, Departamento de Astronomia, Universidade de São Paulo, 05508-090, SP, Brazil
email: joao.steiner@iag.usp.br

[3]Universidade Federal da Fronteira Sul, Campus Cerro Largo, 97900-000, RS, Brazil
email: tiago.ricci@uffs.edu.br

Abstract. We present the first results of the Deep Integral Field Spectroscopy View of Nuclei of Galaxies (DIVING^{3D}) survey, obtained from the analysis of the nuclear emission-line spectra of a sub-sample we call mini-DIVING^{3D}, including all southern galaxies with $B < 11.2$ and $|b| > 15°$. In comparison with previous studies, very few galaxies were classified as Transition objects. A possible explanation is that at least part of the Transition objects are composite systems, with a central low-ionization nuclear emission-line region (LINER) contaminated by the emission from circumnuclear H II regions. The high spatial resolution of the DIVING^{3D} survey allowed us to isolate the nuclear emission from circumnuclear contaminations, reducing the number of Transition objects.

Keywords. galaxies: nuclei, galaxies: active, galaxies: Seyfert, techniques: spectroscopic

---

1. Introduction

We are conducting the Deep IFS View of Nuclei of Galaxies (DIVING^{3D}) survey, with the goal of observing, using optical 3D spectroscopy, the central regions of all galaxies in the Southern hemisphere with $B < 12.0$ and with a Galactic latitude $|b| > 15°$. The complete sample has a total of 170 objects. All the observations are being taken with the Integral Field Unit (IFU) of the Gemini Multi-Object Spectrograph (GMOS), on the Gemini South and Gemini North telescopes, and with the SOAR Integral Field Spectrograph (SIFS), on the SOAR telescope.

Here we present the first results of the analysis of the DIVING^{3D} sample, focused on the nuclear emission-line properties of all galaxies brighter than $B = 11.2$. We call this sub-sample the mini-DIVING^{3D}, which has a total of 57 objects.

2. Observations, Reduction and Data treatment

The GMOS/IFU data cubes were reduced using the Gemini package, in IRAF environment. On the other hand, the SIFS data reduction was performed using scripts in Interactive Data Language (IDL).

After the data reduction, a data treatment was applied to all data cubes. Such a procedure included: correction of the differential atmospheric refraction (DAR); combination

Table 1. Fractions of galaxies with the classifications determined from the diagnostic diagram analysis. The fractions obtained from a sub-sample of the PALOMAR survey, selected with the same criteria used for the selection of the mini-DIVING^{3D} sample, are also shown.

	mini-DIVING^{3D}(%)	PALOMAR with B $<$ 11.2(%)
H II regions	15.8 ± 2.1	25
Transition objects	7.0 ± 1.1	12
LINERs	23 ± 4	29
Seyferts	9 ± 3	7
LINERs/Seyferts	18 ± 7	10
Transition/LINERs/Seyferts	1.8 ± 1.8	0

of the data cubes of each galaxy into one in the form of a median; Butterworth spatial filtering; "instrumental fingerprint" removal; Richardson-Lucy deconvolution. For more detail about the treatment procedure, see Menezes *et al.* (2019).

3. Analysis and Results

We extracted the nuclear spectrum of each galaxy in the mini-DIVING^{3D} sample from a circular region, centered on the stellar nucleus of the object, with a radius equal to half of the FWHM of the PSF at the median wavelength of the data cube.

The subtraction of the stellar continuum from the extracted nuclear spectra was performed with the Penalized Pixel Fitting technique (pPXF - Cappellari & Emsellem 2004).

For the galaxies without blended emission lines, the integrated line fluxes were obtained via direct integration of the emission lines. On the other hand, for the objects with blended emission lines, the fluxes were determined by fitting the emission lines with a sum of Gaussian functions. After that, a dignostic diagram analysis was applied and the galaxies were classified as H II regions, Transion objects, Seyferts, and Low Ionization Nuclear Emission-Line Regions (LINERs). Table 1 shows the fractions of galaxies with different classifications. For a comparison, Table 1 also shows the corresponding fractions obtained from a sub-sample of the PALOMAR survey, selected with the same criteria used for the selection of the mini-DIVING^{3D} sample.

4. Conclusions

• By comparing our results with those obtained from a sub-sample of the PALOMAR survey (mini-PALOMAR), selected with the same criteria used for the mini-DIVING^{3D}, we verified that the fractions of objects in these two sub-samples classified as LINERs, Seyferts or with partial classifications of LINER/Seyfert and Transition/LINER/Seyfert are compatible, at the 1σ or 2σ levels

• The fractions of objects in the mini-DIVING^{3D} sample classified as H II regions and Transition objects are lower than the corresponding fractions in the mini-PALOMAR sample, not being compatible, even at the 3σ level

• Considering that PALOMAR long-slit spectra have a lower spatial resolution than DIVING^{3D} data cubes, the result obtained for Transition objects in the mini-DIVING^{3D} sample suggests that some Transition objects may be composite systems, with a central LINER emission contaminated by the emission from circumnuclear H II regions

References

Cappellari, M. & Emsellem, E. 2004, *PASP*, 116, 138
Menezes, R. B., Ricci, T. V., Steiner, J. E., *et al.* 2019, *MNRAS*, 483, 3700

Galaxy Evolution and Feedback across Different Environments
Proceedings IAU Symposium No. 359, 2020
T. Storchi-Bergmann, W. Forman, R. Overzier & R. Riffel, eds.
doi:10.1017/S1743921320004044

Gas kinematics and stellar archaeology of the Seyfert galaxy NGC 5643

P. H. Cezar[1], J. E. Steiner[1] and R. B. Menezes[2]

[1]IAG - Instituto de Astronomia, Geofísica e Ciências Atmosféricas,
Universidade de São Paulo, 05508-090, São Paulo, SP, Brazil
e-mail: pedrocezar@usp.br

[2]Instituto Mauá de Tecnologia, 09580-900, São Caetano do Sul, SP, Brazil

Abstract. In this work we derive stellar archaeology and kinematics of the central 400 pc of NGC 5643. The star formation history (SFH) reveals nuclear contribution of stellar populations older (20% older than 3.5 Gyr) and younger (60% younger than 320 Myr) as compared to the circumnuclear region. The [OIII] 5007 Å kinematics reveals the eastern ionization cone with an outflow (-60 km/s $\leqslant$ v $\leqslant$ 120 km/s).

Keywords. Active Galatic Nuclei, Outflows, Stellar Populations

---

1. Introduction

NGC 5643 is an almost face-on SAB(rs)c and Seyfert 2 galaxy. Some works obtained evidence of a double-sided ionization cone (Morris *et al.* 1985, Schmitt *et al.* 1994 and Cresci *et al.* 2015) almost in the bar direction (PA = 90°). Other works suggest that the cone region is tilted with respect to the galaxy plane (Schmitt *et al.* 1994, Simpson *et al.* 1997) and Cresci *et al.* (2015) found a western dust structure that may be connected to the dust lane of the bar.

The work of Cresci *et al.* (2015) suggests a blueshifted outflow of [OIII] 5007 Å on the East side of the nucleus. In contrast, Menezes *et al.* (2015), using Spectrograph for Integral Field Observations in the Near Infrared (SINFONI) data show that $Br\gamma$ presents a redshifted outflow in the central 200 pc.

This research is conducted in the context of the Deep IFS View of Nuclei of Galaxies (DIVING^{3D}) project, a survey of galactic nuclei of the southern hemisphere using optical integral field spectroscopy (IFS) and high spatial resolution (Steiner *et al.* this proceedings). The first goal of the present work is to map the distribution and kinematics of highly excited gas that can reveal the ionization cone of the active galactic nucleus (AGN) and possible outflowing components. We aim also to model the stellar populations present in the nuclear and circumnuclear regions, deriving the star formation history (SFH).

The observations were carried out in the Gemini South telescope with the Gemini Multi-Object Spectrographs (GMOS) in integral field spectroscopy with one-slit mode (5" × 3.5"). The data cube covers 4800−6800Å with a resolution of 1.3 Å at 5850 Å. The data treatment was conducted with the methods developed by our group (Menezes *et al.* 2019).

The stellar populations were modeled with the spectral synthesis code Starlight (Cid Fernandes*et al.* 2005) using 200 MILES (Vazdekis *et al.* 2010) simple stellar populations (SSPs) with 50 ages (63 Myr−17.78 Gyr) and 4 metalicities (Z = 0.19, 0.40, 1.00, 1.66). The stellar fit was subtracted to obtain the pure emission line cube.

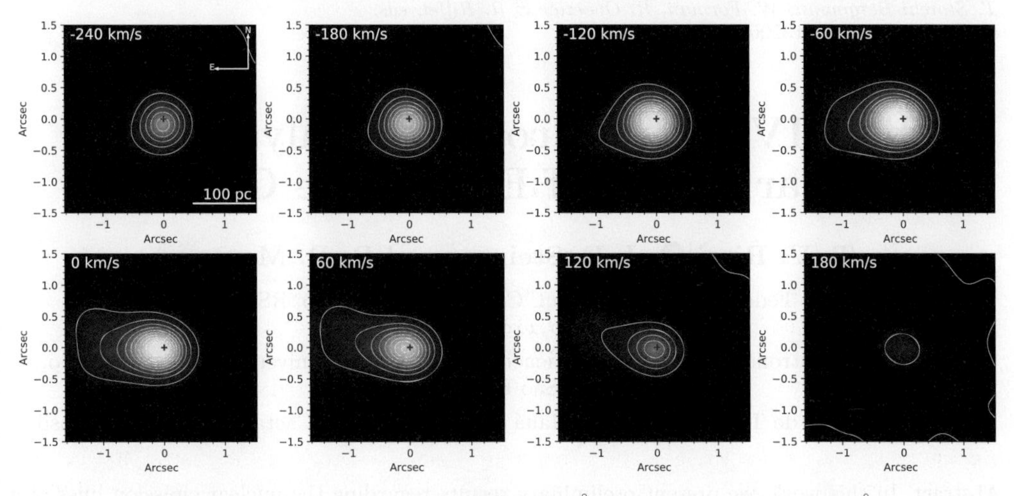

Figure 1. Channel maps of [OIII] 5007 Å in the range 5003–5010 Å.

To study the gas kinematics we constructed channel maps of the [OIII] 5007 Å emission line (Figure 1). We also compare the outflowing maps with the near infrared counterpart (Menezes *et al.* 2015).

2. Results

The derived SFH indicates that the circumnuclear region does not present significant star formation younger than 500 Myr and older than 3.5 Gyr. In the nuclear region, about 20% of the stellar populations are in the range 3.5–8 Gyr. Younger populations were found in the nuclear region, with a 60% contribution of SSPs younger than 320 Myr.

A one-sided ionization cone is revealed towards the East (close to PA $= 90^o$) - see Figure 1, but the West side cone probably is obscured by a dust structure clearly shown in Cresci *et al.* (2015). An extended emission of [OIII] is located close to the rest frame and in redder wavelengths, suggesting the existence of an outflow (-60 km/s $\leqslant$ v $\leqslant 120$ km/s). Menezes *et al.* (2015) found a similar outflow in $Br\gamma$ with velocities $v < 187 km/s$.

3. Conclusions

• The [OIII] 5007 Å reveals the East ionization cone; the West side is probably obscured by a dust structure; in the [OIII] emission line kinematics we found an outflow (-60 km/s $\leqslant$ v $\leqslant 120$ km/s) also seen in the kinematics of the $Br\gamma$ (Menezes *et al.* 2015);

• From the stellar spectral synthesis we show that the nuclear region presents older (20% older than 3.5 Gyr) and younger (60% younger than 320 Myr) populations when compared to the circumnuclear region.

References

Cid Fernandes, R., Mateus, A., Sodré, L., *et al.* 2005, *MNRAS*, 350, 363
Cresci, G., Marconi, A., Zibetti, S., *et al.* 2015, *A&A*, 582A, 63C
Menezes, R. B., da Silva, P., Ricci, T. V., *et al.* 2015, *MNRAS*, 450, 369
Menezes, R. B., Ricci, T. V., Steiner, J. E., *et al.* 2019, *MNRAS*, 483, 3700
Morris, S., Ward, M., Whittle, M., *et al.* 1985, *MNRAS*, 216, 193
Schmitt, H. R., Storchi-Bergmann, T., Baldwin, J. A., *et al.* 1994, *ApJ*, 423, 237
Simpson, C., Wilson, A. S., Bower, G., *et al.* 1997, *ApJ* 474, 121
Vazdekis, A., Sánchez-Blázquez, P., Falcón-Barroso, *et al.* 2010, *MNRAS*, 440, 1639

Galaxy Evolution and Feedback across Different Environments
Proceedings IAU Symposium No. 359, 2020
T. Storchi-Bergmann, W. Forman, R. Overzier & R. Riffel, eds.
doi:10.1017/S174392132000157X

The DIVING^{3D} Project: Analysis of the nuclear region of Early-type Galaxies

T. V. Ricci[1], J. E. Steiner[2] and R. B. Menezes[3]

[1]Universidade Federal da Fronteira Sul, Campus Cerro Largo, RS, 97900-000, Brasil
email: tiago.ricci@uffs.edu.br

[2]Instituto de Astronomia, Geofísica e Ciências Atmosféricas, Universidade de São Paulo, 05508-090, São Paulo, SP, Brasil

[3]Instituto Mauá de Tecnologia, Praça Mauá 1, 09580-900, São Caetano do Sul, SP, Brasil

Abstract. In this work, we present preliminary results regarding the nuclear emission lines of a statistically complete sample of 56 early-type galaxies that are part of the Deep Integral Field Spectroscopy View of Nuclei of Galaxies (DIVING^{3D}) Project. All early type galaxies (ETGs) were observed with the Gemini Multi-Object Spectrograph Integral Field Unit (GMOS-IFU) installed on the Gemini South Telescope. We detected emission lines in 93% of the sample, mostly low-ionization nuclear emission-line region galaxies (LINERs). We did not find Transition Objects nor H II regions in the sample. Type 1 objects are seen in ∼23% of the galaxies.

Keywords. galaxies: active – galaxies: elliptical and lenticular, cD – galaxies: nuclei – galaxies: statistics

1. Introduction

Statistical studies of the central regions of galaxies allow one to understand the main properties of active galactic nuclei (AGNs) in a given sample of objects. In the local Universe, the Palomar Survey (Filippenko & Sargent 1985; Ho *et al.* 1997a) is an example of such an analysis. If only the Early-type galaxies (ETGs, elliptical and lenticular galaxies) from this survey are considered, emission lines are seen in half of the objects, mostly LINERs (Ho *et al.* 1997b; Ho 2008).

The DIVING^{3D} Project (Deep IFS View of Nuclei of Galaxies) is a statistically complete sample of objects that contains all 170 galaxies of the Southern Hemisphere with B < 12.0 mag and galactic latitude $|b| < 15°$ from the RSA and the RC 3. All objects were observed with the Gemini South Telescope using the Gemini Multi-Object Spectrograph (GMOS) under the Integral Field Unit (IFU) mode (Allington-Smith *et al.* 2002) and with the SOAR Integral Field Spectrograph (SIFS). One of the main advantages of the DIVING^{3D} Project, when compared to the Palomar Survey, is the use of seeing-limited IFU data to analyse the very central region of galaxies.

The goal of this work is to present preliminary results of the nuclear region of a statistically complete sample of all 56 ETGs that belong to the DIVING^{3D} Project.

2. Results

Emission lines were detected in ∼93% of the nuclei of the sample ETGs after subtracting their stellar component with the PPXF technique (Cappellari & Emsellem 2004) and using the simple stellar population models described in Vazdekis *et al.* (2015). However,

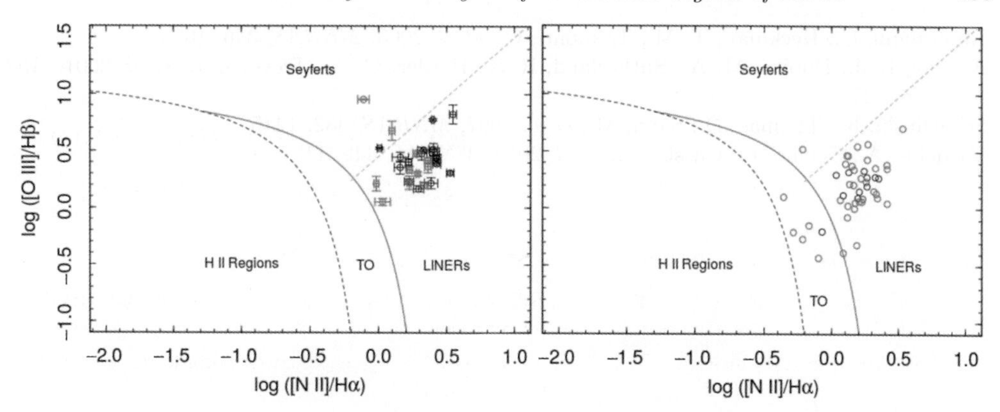

Figure 1. Diagnostic diagram containing only nuclei of ETGs. Left: data from the DIVING^{3D} Project. Right: data from the Palomar Survey (Ho *et al.* 1997a). The black circles correspond to Type 1 AGNs, while the red circles are related to objects with no broad component detected in the permitted lines. The dashed blue line is for the maximum starburst line (Kewley *et al.* 2001), the full red line is for the empirical division between H II regions and AGNs (Kauffmann *et al.* 2003), and the dashed green line is for the LINER/Seyfert division (Schawinski *et al.* 2007). TOs are Transition Objects, which correspond to LINER/H II region composite spectra.

for only half of the sample we were able to accurately measure the flux of the Hα, Hβ, [O III]λ5007 and [N II]λ6583 lines. For these objects, we used a BPT diagnostic diagram, shown in Fig. 1, to classify their nuclear emission.

3. Conclusions

• We detected emission lines in ~93% of the ETGs from the DIVING^{3D} Project, which is higher than the fraction seen in the Palomar sample (~ 60%).

• Also, the number of ETGs with a broad Hα component is higher in the objects from the DIVING^{3D} Project (~23%) than in the Palomar sample (~15%).

• Both the above results reflect the fact that the ETGs from the DIVING^{3D} Project were observed with a more modern instrument and a better spatial resolution than the Palomar Survey observations.

• Transition Objects are seen in the ETGs from the Palomar survey (9%), but not in the galaxies from the DIVING^{3D} Project. This is also a consequence of the better spatial resolution of the DIVING^{3D} observations, since the use of seeing limited IFU data allows one to isolate the nuclear region, thus avoiding contamination from H II regions.

• The fraction of ETG nuclei containing Seyferts + LINERs is the same in both samples (~50%).

• When only highly accurate measurements are used in BPT diagrams to determine an emission line nuclei, we noticed that all ETGs are consistent as being classified as LINERs with exception of only one coronal line Seyfert 2 galaxy.

References

Allington-Smith, J., Murray, G., Content, R., *et al.* 2002, *PASP*, 114, 892
Cappellari, M. & Emsellem E. 2004, *PASP*, 116, 138
Filippenko, A. V. & Sargent, W. L. W. 1985, *ApJS*, 57, 503
Ho, L. C. 2008, *ARA&A*, 46, 475
Ho, L. C., Filippenko, A. V., Sargent, W. L. W., *et al.* 1997a,*ApJS*, 112, 315
Ho, L. C., Filippenko, A. V., Sargent, W. L. W., *et al.* 1997b, *ApJ*, 487, 568

Kauffmann, G., Heckman, T. M., Tremonti, C., *et al.* 2003, *MNRAS*, 346, 1055
Kewley, L. J., Dopita, M. A., Sutherland, R. S., Heisler, C. A., Trevena, J., *et al.* 2001, *ApJ*, 556, 121
Schawinski, K., Thomas, D., Sarzi, M., *et al.* 2007, *MNRAS*, 382, 1415
Vazdekis, A., Coelho, P., Cassisi, S., *et al.* 2015, *MNRAS*, 449, 1177

Galaxy Evolution and Feedback across Different Environments
Proceedings IAU Symposium No. 359, 2020
T. Storchi-Bergmann, W. Forman, R. Overzier & R. Riffel, eds.
doi:10.1017/S1743921320001684

The radial acceleration relation and its emergent nature

Davi C. Rodrigues and Valerio Marra

Núcleo de Astrofísica e Cosmologia, PPGCosmo & Departamento de Física,
Universidade Federal do Espírito Santo, 29075-910, ES, Brazil
emails: davi.rodrigues@cosmo-ufes.org, marra@cosmo-ufes.org

Abstract. We review some of our recent results about the Radial Acceleration Relation (RAR) and its interpretation as either a fundamental or an emergent law. The former interpretation is in agreement with a class of modified gravity theories that dismiss the need for dark matter in galaxies (MOND in particular). Our most recent analysis, which includes refinements on the priors and the Bayesian test for compatibility between the posteriors, confirms that the hypothesis of a fundamental RAR is rejected at more than 5σ from the very same data that was used to infer the RAR.

Keywords. galaxies: kinematics and dynamics, dark matter, gravitation

1. Introduction

The Radial Acceleration Relation (RAR) (McGaugh *et al.* 2016) shows a sharp correlation between two accelerations associated to galaxy rotation curves. Since this correlation with its small dispersion is not an obvious outcome of the standard dark matter picture, several works interpreted the RAR as evidence for modified gravity such as the Modified Newtonian Dynamics (MOND) (e.g., Li *et al.* 2018). For the latter model, such correlation is a fundamental property of gravity, which is achieved by introducing a fundamental acceleration scale a_0, while removing dark matter.

Rodrigues *et al.* (2018a) have shown, using Bayesian inference and the SPARC data (Lelli *et al.* 2016), that the a_0 credible intervals for different galaxies are not compatible among themselves at more than 10σ. Hence, also considering that high-quality rotation curve data were used, this led to a re-interpretation of the RAR as the strongest evidence against MOND as a gravitational theory (Marra *et al.* 2020). Here we consider the approach of Rodrigues *et al.* (2018a) together with further recent refinements by Marra *et al.* (2020) on the statistical analysis.

To evaluate if MOND works as a dark matter replacement in galaxies, one has to address the issue of finding a_0. A common practice is to fit many galaxies and take the median of the best-fit values. Doing so is not optimal since it neglects the information from the a_0 posterior distributions of the individual galaxies (i.e., the "errors" on a_0 for each one of the galaxies). Moreover, and most importantly, from those posteriors one can test if the observational data are compatible with the existence of a common a_0 value. If they are not compatible, then a_0 is not fundamental and the RAR is necessarily an emergent correlation. Assuming standard dark matter, the RAR must be emergent (see e.g., Stone & Courteau 2019). If the RAR is emergent it can be useful (like many other emergent correlations), but it cannot directly reflect a fundamental property of gravity.

Table 1. The rejection level of the fundamental a_0 hypothesis.

Method	RAR sample (153 galaxies)	$\mathcal{S}_2$ subsample (91 galaxies)
Monte Carlo X^2 test	$> 5.7\sigma$	5.3σ

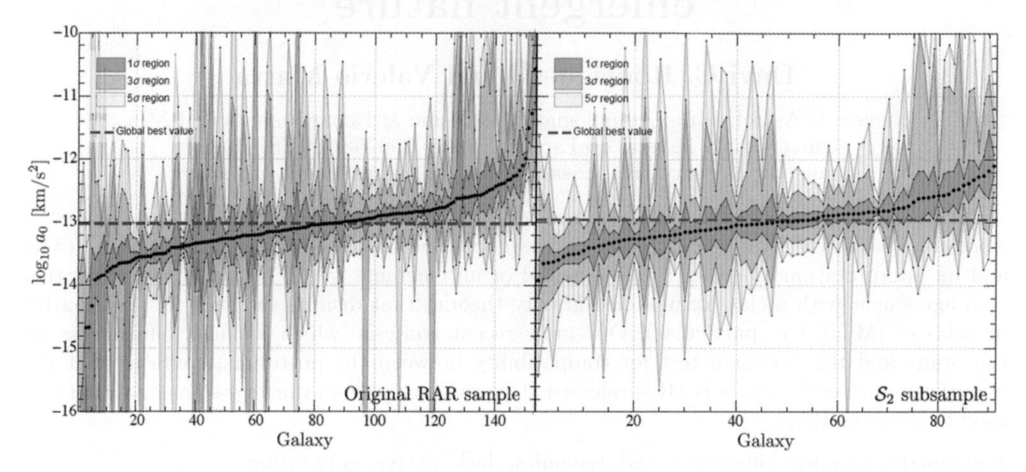

Figure 1. The a_0 modes (black dots) and the 1σ, 3σ and 5σ credible intervals for each one of the galaxies. The left panel shows all the 153 RAR galaxies, while the right panel shows a subsample ($\mathcal{S}_2$) with a stronger quality cut: galaxies with too high $\chi^2_{\rm min}$ values are also eliminated (91 galaxies are left). See Marra *et al.* (2020) for further details.

2. Methods and Results

Marra *et al.* (2020) improved the methodology of Rodrigues *et al.* (2018a) by considering priors more closely related to the observational uncertainties (see also Rodrigues *et al.* 2018b). A summary of the results from Marra *et al.* (2020) is shown in Table 1 and Fig. 1. Beyond the variation of a_0 from galaxy to galaxy, it is also considered the variation of mass-to-light ratios, distances and inclinations. Another improvement concerns the comparison of the credible intervals, since the method no longer employs Gaussian approximations (see also Cameron *et al.* 2020; Rodrigues *et al.* 2020). In particular, the existence of a common a_0 value is tested using the X^2 statistics, an extension of the tension estimator of Verde *et al.* (2013), which is based on the Bayes factor.

We conclude with our opinion on MOND: historically it has stimulated relevant developments for galaxy astrophysics. As an effective theory for galaxy dynamics valid on average, it is useful and it is the RAR (i.e., a correlation between accelerations). As a theory for gravity, it has many problems, even for galaxy rotation curves.

Acknowledgements

DCR thanks the organizers of the GALFEED Symposium. DCR and VM also thank CNPq (Brazil) and FAPES (Brazil) for partial financial support

References

Cameron, E., Angus, G. W., Burgess, J. M., *et al.* 2020, *Nat. Astron.*, 4, 132
Lelli, F., McGaugh, S. S., Schombert, J. M., *et al.* 2016, *AJ*, 152, 157
Li, P., Lelli, F., McGaugh, S., Schormbert, J., 2018, *Astron. Astrophys.*, 615, A3
Marra, V., Rodrigues, D. C., de Almeida, Á. O., 2020, *MNRAS*, 2002.03946

McGaugh, S., Lelli, F., Schombert, J., 2016, *Phys. Rev. Lett*, 117, 201101
Rodrigues, D. C., Marra, V., Del Popolo, A., Davari, Z., *et al.* 2018a, *Nat. Astron.*, 2, 668
Rodrigues, D. C., Marra, V., Del Popolo, A., Davari, Z., *et al.* 2018b, *Nat. Astron.*, 2, 927
Rodrigues, D. C., Marra, V., Del Popolo, A., Davari, Z., *et al.* 2020, *Nat. Astron.*, 4, 134
Stone, C. & Courteau, S. 2019, *ApJ*, 882, 6,
Verdem L., Protopapasm P., Jimenezm R. 2013, *Phys. Dark Univ.*, 2, 166

Galaxy Evolution and Feedback across Different Environments
Proceedings IAU Symposium No. 359, 2020
T. Storchi-Bergmann, W. Forman, R. Overzier & R. Riffel, eds.
doi:10.1017/S174392132000160X

Properties of AGN in NIR within the context of the Eigenvector 1

D. Dias dos Santos[1], A. Rodríguez-Ardila[1,2] and M. Marinello

[1]Instituto Nacional de Pesquisas Espaciais, Av. dos Astronautas, CEP 12227-010, São José dos Campos - SP, Brazil
email: denimaradds@id.uff.br

[2]Laboratório Nacional de Astrofísica, R. dos Estados Unidos, CEP 37504-364, Itajubá - MG, Brazil

Abstract. We present a spectral atlas of 70 type-I AGN with the wavelength ranging 0.4–2.5 μm. For 37 sources, this is the first report of NIR spectroscopy in literature. The sample was constructed to study narrow line Seyfert 1 and quasars, with a large range of line widths (800 km s^{-1} < FWHM < 4000 km s^{-1}) and Fe II intensities ($0.2 < R_{4570} < 2.8$). This work presents partial results of an ongoing project that has the objective of modeling the continuum emission and emission lines in order to derive the physics driven the Eigenvector 1 through a panchromatic spectral analysis, with emphasis on strong to super-strong Fe II emitters. Our results show that hot dust near the sublimation temperature is necessary to explain the 1μm break of the power law component of the continuum. We estimated the hot dust mass and found a weak or absent correlation with the Fe II intensity. Moreover, we found that low ionisation ions are formed in an outer region of the BLR.

Keywords. active galaxy nuclei, Seyfert, infrared, emission lines, spectroscopic

1. Introduction

An Active Galactic nuclei (AGN) is an energetic phenomenon which produces electromagnetic spectrum features in a broad range of wavelength. Narrow-line Seyfert 1 galaxies (NLS1) are a particular subclass of AGNs due to some specific characteristics. In these sources, the Balmer lines are narrower than a normal Seyfert 1 (FWHM < 2100 km s^{-1}), shows a stronger Fe II emission, and weaker [OIII] lines ([OIII/Hβ] < 3) (Shen & Ho (2014)). These objects are important in the physics context of the Eigenvector 1 (EV1) since they represent extreme objects in the quasar main sequence. The physical drivers of the EV1 are not completely understood yet, specially in the extreme cases such NLS1, and in order to improve our knowledge on these sources it is necessary to study how to distinguish these extreme objects.

In this work, we selected a sample of Type-I sources with extreme properties in the EV1 context. The 70 AGN in the samples covers the near infrared region from 0.4–2.5 μm. For 37 out of 70 sources, this is the first report of NIR spectroscopy in literature. We aim to model the continuum and measure the emission line properties, to derive its physical properties within the context of the Eigenvector 1.

2. Methodology and Database

We present for the first time near-infrared observations of 37 Type-I AGN optically identified as strong Fe II emitters. Additionally we complete the sample with previously published data of other 33 AGN from Riffel *et al.* (2006) and Marinello *et al.* (2016), achieving a broad range of Fe II emission ($0.2 < R_{4570} < 2.8$) and a redshift coverage of $0.002 < z < 0.2$.

In order to keep consistency in the methodology used to measure R_{4570}† we re-estimate this quantity using the optical spectral available for our sample. For the 46 out of 70 AGN we had the optical spectrum available. We modeled the continuum using a power law and the bump centered at $4570AA$ using the template from Boroson & Green (1992). For 33 out of 46 objects we measured $R_{4570} > 1$, which classifies them as strong Fe II emitters.

In order to analyse the continuum shape in the NIR we modeled this emission using to main components: (i) a power law, which extends from the UV to the NIR; and (ii) a black body, which is responsible for pump up the continuum at wavelengths redward 1.2 μm. The power law index is an important parameter in the SED modeling, and gives information of how much flat is the optical spectrum. On the other hand, the black body component gives information on how warm is the dust present in the outer region of the BLR (Rodriguez-Ardila *et al.* (2006)), and allow an estimative of the dust mass in this region, which it is derived from the 2.2 μm flux (Barvainis (1987)).

To probe the Fe II emitting region we use different low ionization emission lines, which are believed to be produced in the same region as Fe II, such as O Iλ8446 and Ca IIλ8663, and compare them with Paschen and H IIλ10830 lines. To model the profile of these lines we used the python library LMFIT. We fit the lines deblending the broad and narrow component using Lorentzian and Gaussian profiles, respectively, in order to obtain their FWHM, flux, and centroid.

3. Results and Conclusions

Our results shows the presence of dust emission heated by a AGN at T > 1000 K, likely representing the emission from the torus, a common characteristic in the NIR continuum of Type I AGN (Rodriguez-Ardila *et al.* (2006)). This dust could, hypothetically, carry iron grains towards the BLR via eccentric movement of dust cloud within the torus. In order to test this hypothesis we compare the dust mass with R_{4570} to verify a possible correlation. Our results show, however, a very week correlation between the dust mass and the Fe II intensity. We obtain a very flat correlation with a Spearman Rank Coefficient $S_r = 0.15$, with a $p - value = 0.35$. This suggests the absence of dust in the Fe II emitting region, which means that the hot dust observed in the continuum is probably the warm internal face of the torus.

We found similar values of FWHM for Ca II, O I and Fe II. Assuming a virialized movement for the BLR clouds, this result suggests a similar region of formation for these ions, in an outer part of the BLR. On the other hand, the FHWM of He II and Hydrogen are systematically larger than Fe II, suggesting a inner formation regions for these lines.

References

Barvainis, R. 1987, *ApJ*, 320, 537–544
Boroson & Green 2006, *ApJ, Supplement Series*, 80, 109B
Marinello, M., Rodriguez-Ardila, A., Garcia-Rissmann, A., *et al.* 2016, *ApJ*, 820, 116
Riffel, R., Rodriguez-Ardila, A., Pastoriza, M. G., *et al.* 2006, *A&A*, 457, 61–70
Rodriguez-Ardila, A. & Mazzalay, X. 2006, *MNRAS*, 367, L57–L61
Shen, Y. & Ho, L. C. 2014, *Nat.*, 513, 210–213

† The R_{4570} was defined as the flux ratio between Fe II complex centered at 4570 Å and the broad component of Hβ.

Galaxy Evolution and Feedback across Different Environments
Proceedings IAU Symposium No. 359, 2020
T. Storchi-Bergmann, W. Forman, R. Overzier & R. Riffel, eds.
doi:10.1017/S1743921320002367

Characterizing circumnuclear starbursts in the local universe with the VLA

Yiqing Song[1], Sean T. Linden[1], Aaron S. Evans[1,2], Loreto Barcos-Muñoz[2] and Eric J. Murphy[2]

[1]Department of Astronomy, University of Virginia, 530 McCormick Rd, Charlottesville, VA 22904, USA
email: ys7jf@virginia.edu

[2]National Radio Astronomy Observatory, 520 Edgemont Rd, Charlottesville, VA 22903, USA

Abstract. Nuclear rings are excellent laboratories to study star formation (SF) under extreme conditions. We compiled a sample of 9 galaxies that exhibit bright nuclear rings at 3-33 GHz radio continuum observed with the Jansky Very Large Array, of which 5 are normal star-forming galaxies and 4 are Luminous Infrared Galaxies (LIRGs). Using high frequency radio continuum as an extinction-free tracer of SF, we estimated the size and star formation rate of each nuclear ring and a total of 37 individual circumnuclear star-forming regions. Our results show that majority of the SF in the sample LIRGs take place in their nuclear rings, and circumnuclear SF in local LIRGs are much more spatially concentrated compared to those in the local normal galaxies and previously studied nuclear and extra-nuclear SF in normal galaxies at both low and high redshifts.

Keywords. radio continuum: galaxies, galaxies: starburst

---

1. Introduction

Nuclear star-forming rings occur in about 20% of nearby spiral galaxies (Knapen *et al.* 2004). Studies on nuclear rings in mostly local low-luminosity galaxies have shown that they are locations of active star formation (SF) (e.g. Buta & Combes 1996). In this work, we use extinction-free 3-33 GHz radio continuum observations from the Jansky Very Large Array (VLA) to study the physical properties of circumnuclear star formation (CNSF) along the nuclear rings of 4 local Luminous Infrared Galaxies (LIRGs; $L_{\rm IR}[8{-}1000\mu m] \geqslant 10^{11} L_\odot$), which are mostly gas-rich mergers (Sanders & Mirabel 1996). To provide context, we also performed the same analyses on 5 lower-luminosity normal star-forming galaxies with nuclear rings. We selected the sample LIRGs (NGC 1614, NGC 1797, NGC 7469, NGC 7591) from a subset of 68 LIRGs in the Great Observatories All-sky LIRG Survey (GOALS; Armus *et al.* 2009) that were observed with the VLA, and the low-luminosity galaxies (NGC 1097, NGC 3351, NGC 4321, NGC 4826, IC342) from the Star Formation in Radio Survey (SFRS; Murphy *et al.* 2012) of 56 nearby normal star-forming galaxies. The selected 9 galaxies are the only ones that exhibit distinct bright nuclear ring features, out of a combined total of 124 galaxies observed with the two surveys. VLA observations for all 9 sample galaxies share the same spectral setup as well as similar physical resolution (~100pc), which allows us to consistently measure and compare the radio properties of their nuclear rings and CNSF regions.

2. Circumnuclear Starbursts in different host environments

Through fitting the azimuthally averaged light profiles of the highest resolution observation, we estimated the radius of each ring and derived its star formation rate (SFR) based on recipes given by Murphy *et al.* (2012). We found that the nuclear rings in the sample LIRGs (median SFR $\sim 17 M_{\odot}/\mathrm{yr}$) produce stars at more than an order of magnitude higher rates than the ones in the sample normal galaxies (median SFR $\sim 0.3 M_{\odot}/\mathrm{yr}$). However, the overall sizes of the nuclear rings in the LIRGs (median $R \sim 250\mathrm{pc}$) are smaller than the ones in the normal galaxies (median $R \sim 360\mathrm{pc}$).

To better picture the roles played by CNSF in our sample galaxies, we compare our radio-derived SFR of each nuclear ring to the total SFR of each host galaxy derived from a combination of $L_{\mathrm{IR}}[8-1000\mu m]$ and L_{FUV} from the literature, accounting for significant AGN contribution to the bolometric luminosities of the LIRGs estimated by Díaz-Santos *et al.* (2017). The result shows that nuclear rings account for over 50% of the total SFRs in all 4 sample LIRGs. In the sample normal galaxies, the contribution from the nuclear rings is much lower (all below 50%), indicating that most of their SF takes place in the extra-nuclear regions. The high SFR contribution of the nuclear rings in the LIRGs demonstrates the crucial role of CNSF in the galaxy transformation process.

Taking advantage of our high resolution data, we performed structural analysis of our data with the *Astrodendro* Python package and further extracted a combined total of 37 individual CNSF regions in the nuclear rings of our sample galaxies. This was done to estimate sizes, SFRs and SFR surface densities (Σ_{SFR}) of individual star-forming knots. Our results show that CNSF in the sample of LIRGs (median $\Sigma_{\mathrm{SFR}} \sim 170 M_{\odot}\mathrm{kpc}^{-2}\mathrm{yr}^{-1}$) are around 100 times more spatially concentrated than for the sample of normal galaxies (median $\Sigma_{\mathrm{SFR}} \sim 2 M_{\odot}\mathrm{kpc}^{-2}\mathrm{yr}^{-1}$).

3. Circumnuclear Starbursts vs. Extra-nuclear star formation

To investigate the behaviour of our CNSF regions in contrast to general SF behaviours in systems represented by our sample, we gathered size and SFR measurements of SF regions in 48 local LIRGs from GOALS (Larson *et al.* 2020), 41 local normal galaxies from SINGS (Kennicutt *et al.* 2003) and 25 normal galaxies at $z \sim 1-4$ from Livermore *et al.* (2012) and Livermore *et al.* (2015). We found that CNSF measured for LIRGs in our present study is more spatially concentrated than SF previously probed in mostly extra-nuclear regions in both normal and IR-luminous galaxies. Furthermore, our comparison shows that CNSF regions in the sample LIRGs have higher Σ_{SFR} than SF regions in normal galaxies at $z > 1$, which means that the extreme environment of the central regions of LIRGs may be unique at both low and high redshifts.

References

Armus, L., Mazzarella, J. M., Evans, A. S., *et al.* 2009, *PASP*, 121, 559
Buta, R. & Combes, F. 1996, *FCPh*, 17, 95
Díaz-Santos, T., Armus, L., Charmandaris, V., *et al.* 2017, *ApJ*, 846, 32
Kennicutt, R. C., Armus, L., Bendo, G., *et al.* 2003, *PASP*, 115, 928
Knapen, J. H., Stedman, S., Bramich, D. M., *et al.* 2004, *A&A*, 426, 1135
Larson, K. L., Díaz-Santos, T., Armus, L., *et al.* 2020, *ApJ*, 888, 92
Livermore, R. C., Jones, T., Richard, J., *et al.* 2012, *MNRAS*, 427, 688
Livermore, R. C., Jones, T. A., Richard, J., *et al.* 2015, *MNRAS*, 450, 1812
Murphy, E. J., Bremseth, J., Mason, B. S., *et al.* 2012, *ApJ*, 761, 97
Sanders, D. B. & Mirabel, I. F. 1996, *ARAA*, 34, 749

Galaxy Evolution and Feedback across Different Environments
Proceedings IAU Symposium No. 359, 2020
T. Storchi-Bergmann, W. Forman, R. Overzier & R. Riffel, eds.
doi:10.1017/S1743921320002197

Turbulence/outflows perpendicular to low-power jets in Seyfert galaxies†

Giacomo Venturi[1,2], Alessandro Marconi[3,2], Matilde Mingozzi[4], Giovanni Cresci[2], Stefano Carniani[5,2] and Filippo Mannucci[2]

[1]Instituto de Astrofísica, Pontificia Universidad Católica de Chile, Avda. Vicuña Mackenna 4860, 8970117, Macul, Santiago, Chile
email: gventuri@astro.puc.cl

[2]INAF - Osservatorio Astrofisico di Arcetri, Largo E. Fermi 5, I-50125, Firenze, Italy

[3]Dipartimento di Fisica e Astronomia, Università degli Studi di Firenze, Via G. Sansone 1, I-50019, Sesto Fiorentino, Firenze, Italy

[4]INAF - Osservatorio Astronomico di Padova, Vicolo dell'Osservatorio 5, 35122, Padova, Italy

[5]Scuola Normale Superiore, Piazza dei Cavalieri 7, I-56126 Pisa, Italy

Abstract. We present recent results from our MAGNUM survey of nearby active galactic nuclei (AGN), which exploits observations from the optical/near-IR integral field spectrograph MUSE at VLT. We detect strongly enhanced line widths in emission line maps of four galaxies perpendicularly to their low-power jets and AGN ionisation cones, indicative of turbulent/outflowing material. The observation of a similar phenomenon in other works suggests that it originates from an interaction mechanism between the jet and the galaxy disc through which it propagates.

Keywords. galaxies: jets, galaxies: Seyfert, galaxies: individual (IC 5063, NGC 5643), galaxies: ISM, galaxies: kinematics and dynamics, techniques: spectroscopic

---

1. Introduction

Ionised outflows are routinely observed in active galactic nuclei (AGN), either powered by the strong AGN radiation pressure or by powerful jets (see e.g. Fabian 2012).

Here we focus on new results from our 10-to-100 pc spatially-resolved MAGNUM survey (Measuring Active Galactic Nuclei Under MUSE Microscope) of nearby AGN (e.g. Venturi *et al.* 2017, 2018; Mingozzi *et al.* 2019) observed with the optical and near-IR integral field spectrograph MUSE at VLT (Bacon *et al.* 2010), revealing a peculiar phenomenon in objects hosting low-power jets.

2. Turbulence/outflows perpendicular to low-power jets

In Fig. 1 we present MUSE maps of two Seyfert galaxies from our survey, IC 5063 and NGC 5643. Here, while the high bulk velocity gas (panels b and e) indicates an outflow in the direction of the AGN ionisation cones (traced by [O III], panels a and d), as normally expected in AGN (e.g. Fischer *et al.* 2013), a strongly enhanced line velocity width ($\gtrsim$800 km/s; panels c and f), indicating turbulent/outflowing motions with low bulk velocity, is observed instead perpendicularly to the ionisation cones and radio jets (black contours in panels c and f). We detect this phenomenon in two other galaxies of

† Based on observations made with ESO Telescopes at the La Silla Paranal Observatory under program IDs 094.B-0321(A), 60.A-9339(A), 095.B-0532(A).

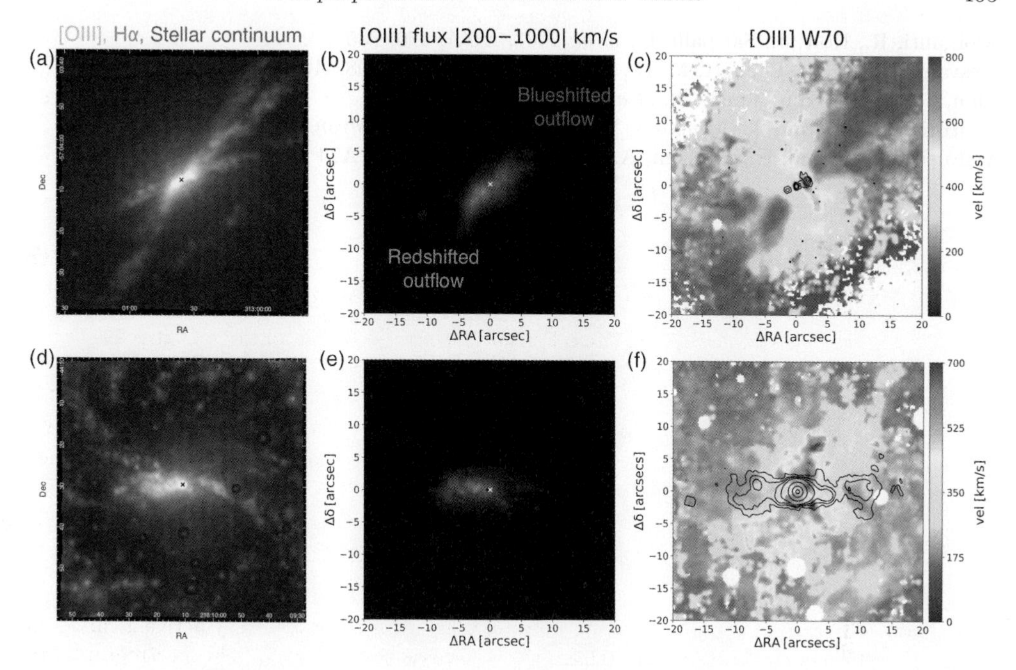

Figure 1. VLT/MUSE maps of IC 5063 (top) and NGC 5643 (bottom). a) and d) Flux maps of ionised gas, [O III]λ5007 (green) and Hα (red), and stellar continuum (blue). b) and e) Bulk of the high-velocity outflow, traced by the flux of [O III] line profile integrated in the velocity range $\pm|200-1000|$ km/s (blue if approaching, red if receding) with respect to the stellar velocity in each spaxel (to exclude contributions from gas rotating in the disc). c) and f) [O III] W70 line velocity width map (i.e difference between the 85th- and 15th-percentile velocities of the line profile), with radio jet contours superimposed, ATCA 17.8 GHz from Morganti *et al.* (2007) for IC 5063, 8.4 GHz VLA from Leipski *et al.* (2006) for NGC 5643.

our sample, NGC 1068 and NGC 1386. We note that all the four mentioned galaxies host a low-power radio jet with low inclinations to galaxy disc.

3. Discussion and Conclusions

We have shown MUSE emission-line maps of two nearby Seyfert galaxies from our MAGNUM survey, which - together with two other sources of the sample - exhibit enhanced line velocity widths perpendicularly to their low-power radio jets and ionisation cones, indicative of turbulent/outflowing motions. A similar phenomenon has been reported in other recent works also in galaxies hosting jets lying low onto the galaxy disc (e.g. Shin *et al.* 2019). This suggests that the jets, through the interaction with the gas in the disc, are responsible for generating the observed perpendicular turbulence/outflow. Moreover, our results confirm that not only powerful jets in "radio loud" objects (e.g. Nesvadba *et al.* 2008) are capable of affecting the gas in the host, but also low-power jets residing in "radio quiet" galaxies, as indicated by recent works (e.g. Jarvis *et al.* 2019).

References

Bacon, R., Accardo, M., Adjali, L., *et al.* 2010, *Proc. SPIE*, 7735, 773508
Fabian, A. C. 2012, *ARA&A*, 50, 455
Fischer, T. C., Crenshaw, D. M., Kraemer, S. B., *et al.* 2013, *ApJS*, 209, 1
Jarvis, M. E., Harrison, C. M., Thomson, A. P., *et al.* 2019, *MNRAS*, 485, 2710
Leipski, C., Falcke, H., Bennert, N., *et al.* 2006, *A&A*, 455, 161
Mingozzi, M., Cresci, G., Venturi, G., *et al.* 2019, *A&A*, 622, A146

Morganti, R., Holt, J., Saripalli, L., *et al.* 2007, *A&A*, 476, 735
Nesvadba, N. P. H., Lehnert, M. D., De Breuck, C., *et al.* 2008, *A&A*, 491, 407
Shin, J., Woo, J.-H., Chung, A., *et al.* 2019, *ApJ*, 881, 147
Venturi, G., Marconi, A., Mingozzi, M., *et al.* 2017, *Front. Astron. Space Sci.*, 4, 46
Venturi, G., Nardini, E., Marconi, A., *et al.* 2018, *A&A*, 619, A74

Galaxy Evolution and Feedback across Different Environments
Proceedings IAU Symposium No. 359, 2020
T. Storchi-Bergmann, W. Forman, R. Overzier & R. Riffel, eds.
doi:10.1017/S1743921320001611

Unveiling the afterlife of galaxies with ultraviolet data

Ariel Werle[1,2]

[1]INAF - Osservatorio Astronomico di Padova, Vicolo dell'Osservatorio 5, 35122 Padova, Italy
[2]Instituto de Astronomia, Geofísica e Ciências Atmosféricas, Universidade de São Paulo, R. do Matão 1226, 05508-090 São Paulo, Brazil
email: ariel.werle@inaf.it

Abstract. Recent works have shown that early-type galaxies (ETGs) are much more complex than early studies suggested. We present early results from a combined analysis of optical spectra and ultraviolet photometry for a sample of 3453 red sequence galaxies in at $z < 0.1$ that are classified as elliptical by Galaxy Zoo. By measuring the Gini index of the star-formation histories derived by STARLIGHT, we investigate the complexity of the mixture of stellar populations required to describe ETGs in our sample. When fitting only optical spectra, STARLIGHT assigns more or less the same mixture of stellar populations to all ETGs, while the addition of UV data unveils a bimodallity in the star-formation histories of these galaxies. We find evidence for stellar populations younger than 1 Gyr in 17 per cent of our sample, indicating that some galaxies do not stay permanently quenched after reaching the red sequence.

Keywords. galaxies: evolution, galaxies: elliptical and lenticular, cD, ultraviolet: galaxies

---

1. Introduction

Studies of the stellar populations of early-type galaxies (ETGs) based on their optical stellar continuum are unable to capture some nuances in their star-formation histories. Fortunately, this limitation can be surpassed with the addition of ultraviolet data, which allows the detection of even small amounts of young stellar populations, and also provides information on old hot stars (e.g. horizontal branch, extreme horizontal branch and post-asymptotic giant branch).

Here we present early results of the simultaneous analysis of optical spectra from the Sloan Digital Sky Survey (SDSS, York *et al.* 2000) Data Realease 7 (Abazajian *et al.* 2009) and photometry from the Galaxy Evolution Explorer (GALEX, Martin *et al.* 2005) in the NUV and FUV bands for a sample of 3453 galaxies with $NUV - r > 5$ at $z < 0.1$ that are classified as elliptical by Galaxy Zoo. For this, we use a new version of the STARLIGHT spectral synthesis code (Cid Fernandes *et al.* 2005), wich allows the combined analysis of spectroscopic and photometric data (Werle *et al.* 2019). Our spectral fitting is based on an updated version of the Bruzual & Charlot (2003) stellar population models (Charlot & Bruzual in prep, private communication).

2. Analysis

STARLIGHT decomposes galaxy spectra into a non-parametric combination of stellar population models. Here we use the fraction of stellar mass attributed by STARLIGHT to each stellar population model (μ) to compute mass-weighted mean stellar ages ($\langle \log\, t \rangle_M$) and also the Gini index of μ for ETGs in our sample. The lower the Gini index, the more STARLIGHT has to mix different stellar populations in order to match GALEX and SDSS

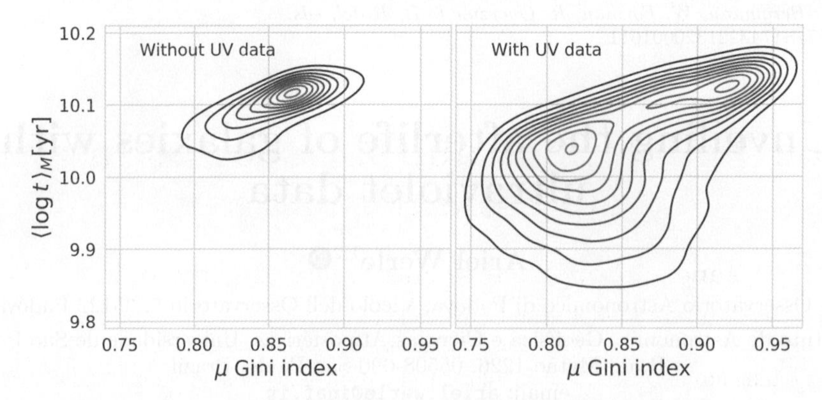

Figure 1. Gini index of the stellar populations against mass-weighted mean stellar ages for STARLIGHT fits of optical spectra (left) and for fits taking into account the combination of optical spectra and ultraviolet photometry (right).

data of a given galaxy. If the Gini index is equal to 1, the galaxy is fitted by one single stellar population model.

In Fig. 1, we plot the Gini index of μ against $\langle \log t \rangle_M$ for two different sets of STARLIGHT fits. When fitting only SDSS spectra (left panel), our sample covers a small range of values in both Gini index and age. With the addition of GALEX data (right panel), a bimodality appears, allowing us to distinguish between two sub-populations. While some galaxies are described solely by very old stellar populations (large Gini index and old age), others had more extended periods of star-formation (lower Gini index and younger age). Stellar populations younger than 1 Gyr are required to fit the GALEX photometry of about 17 per cent of galaxies in our sample.

3. Final remarks

The combined analysis of SDSS spectra and GALEX photometry reveals a bimodality in the star-formation histories of ETGs. Our results indicate that some galaxies do not stay permanently quenched after reaching the red sequence, as has been found by a number of studies in recent years. Instead, some early-type galaxies may undergo "rejuvenation events" triggered by minor mergers or the accretion of cold gas from the intergalactic medium. These and others results from our spectral synthesis will be explored in more detail in Werle *et al.*, in prep.

References

Abazajian, K. N., *et al.* 2009, *ApJS*, 182, 543
Bruzual, G. & Charlot S. 2003, *MNRAS*, 344, 1000
Cid Fernandes, R., Mateus, A., Sodré, L., *et al.* 2005, *MNRAS*, 358, 363
Martin, D. C., *et al.* 2005, *ApJ*, 619, L1
Werle, A., Cid Fernandes, R., Vale Asari, N., *et al.* 2019, *MNRAS*, 483, 2382
York, D. G., *et al.* 2000, *AJ*, 120, 1579

Galaxy Evolution and Feedback across Different Environments
Proceedings IAU Symposium No. 359, 2020
T. Storchi-Bergmann, W. Forman, R. Overzier & R. Riffel, eds.
doi:10.1017/S1743921320002410

Stellar populations and ionised gas in central spheroidal galaxies

Vanessa Lorenzoni[1] and Sandro B. Rembold[2]

[1]Departamento de Física, Universidade Federal de Santa Maria, Santa Maria, Brazil
email: vanessalorenzonii@gmail.com

[2]Departamento de Física, Universidade Federal de Santa Maria, Santa Maria, Brazil
email: sandro.rembold@ufsm.br

Abstract. We investigate the stellar populations and ionised gas properties of a sample of central spheroidal galaxies in order to better constrain their history of star formation and gas excitation mechanism. We select galaxies from Spheroids Panchromatic Investigation in Different Environmental Regions (SPIDER) catalogue and separate these galaxies in different regimes of halo and galaxy mass. To characterise the stellar population properties of these galaxies we use the stellar population synthesis method with the STARLIGHT code, and the presence of ionised gas is identified by measurements of the Hα equivalent width. We analyse how these properties behave as a function of the galaxy stellar mass and the parent halo mass. A trend is observed in the sense of increased ionised gas emission for low-mass centrals in high-mass halos. We interpret this trend in a scenario of intracluster medium (ICM) cooling versus active galactic nuclei (AGN) feedback in a Bondi accretion context.

Keywords. galaxies: clusters: general – galaxies: clusters: intracluster medium – galaxies: active

---

1. Introduction

Central galaxies of dark matter massive halos, usually of spheroidal morphology, have distinct physical properties relative to non-central spheroidal galaxies with comparable mass. Several studies suggest that, because they are in a privileged halo location, such objects are subject to evolutionary processes distinct from those that operate in non-central galaxies. Such mechanisms, however, are poorly understood, in particular regarding their effects on the barionic content evolution (Von Der Linden *et al.* 2007).

2. Methods

We use Spheroids Panchromatic Investigation in Different Environmental Regions (SPIDER, La Barbera *et al.* 2010) sample of early-type galaxies ($0.05 < z < 0.095$), selected from the sixth data release (DR6) of the Sloan Digital Sky Survey (SDSS). We characterise the environment where these galaxies are located using the Yang group catalogue (Yang *et al.* 2007), being 15,572 central galaxies. The stellar population parameters were derived from SDSS DR12 optical spectra (Alam *et al.* 2015). In order to increase the spectral signal-to-noise ratio, we stack the galaxy spectra in a grid of stellar velocity dispersion and parent halo mass by median-combining the flux-normalised individual spectra. Stellar population synthesis was performed with the STARLIGHT code (Cid Fernandes *et al.* 2005) using simple stellar populations (SSPs) drawn from the BC03 (Bruzual & Charlot 2003) and Vazdekis (Vazdekis *et al.* 2015) evolutionary models.

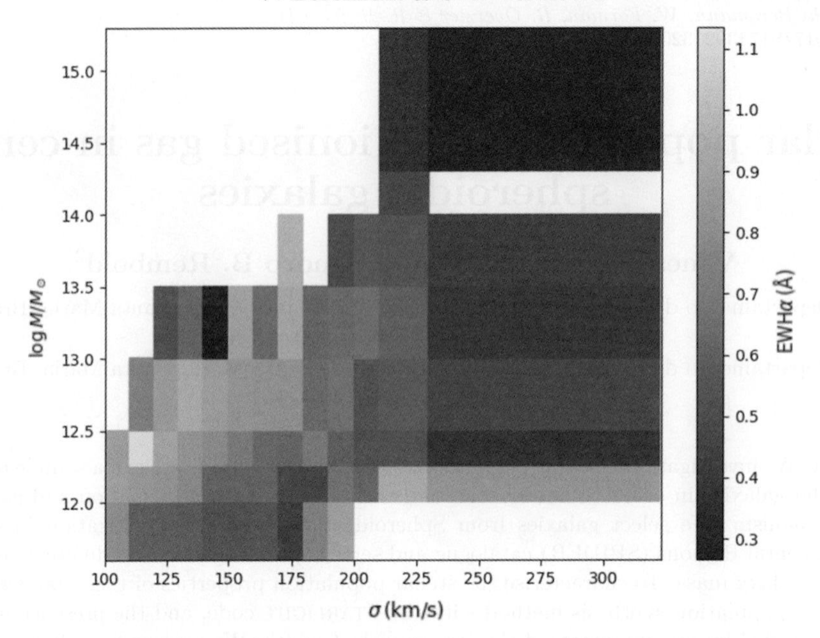

Figure 1. EWHα as a function of the galaxy velocity dispersion and halo mass.

The synthesis include a contribution of a power law ($F_\lambda = \lambda^{-1.5}$) in order to characterise a possible active galactic nuclei (AGN) signature.

We construct a model for the ionised gas content, explicitly in terms of galaxy stellar mass (σ) and halo mass (M_h), since these are the parameters that we use in the creation of our stacks and both are directly related to the AGN feedback and the intracluster medium (ICM) thermal energy (Stott *et al.* 2012). Using the Bondi accretion model (Bondi 1952) to describe the AGN feeding, we model the ratio between the AGN power and the internal energy of the ICM. We find the following relation for equivalent width of Hα (EWHα):

$$\log EW_{H\alpha} \propto \gamma(-10.74 \log \sigma + 1.67 \log M_h). \tag{2.1}$$

3. Results

We find that the extinction A_V, the EWHα and the contribution of a power law increases as the halo mass increases and the mass of the central galaxy decreases. The mean stellar ages present the opposite behaviour, while the mean stellar metallicity is higher for higher halo masses and galaxy velocity dispersion. In Figure 1, we present the EWHα as a function of the galaxy velocity dispersion and halo mass. For each observable A, we fit the function $\log(A) = a \log \sigma + b \log M_h + c$ by a least squares method.

We find that with $\gamma \sim 0.073$ the theoretical coefficients become very similar to those observed in our stacks (~ -0.78 and $\sim$0.121 for a and b, respectively). This suggests that the ionised gas emission is governed by the ratio of the thermal energy of the ICM to the instantaneous power of the AGN in a Bondi accretion scenario.

References

Alam, S., *et al.* 2015, *ApJS*, 219, 12
Bondi, H. 1952, *MNRAS*, 112, 195
Bruzual, G. & Charlot S. 2003, *MNRAS*, 344, 1000
Cid Fernandes, R., *et al.* 2005, *MNRAS*, 358, 363–378

La Barbera, F., *et al.* 2010, *MNRAS*, 408, 3, 1313–1334
Stott, J. P., *et al.* 2012, *MNRAS*, 422, 2213–2229
Vazdekis, A., *et al.* 2015, *MNRAS*, 449, 1177–1214
Von Der Linden, A., *et al.* 2007, *MNRAS*, 379, 867–893
Yang, X., *et al.* 2007, *ApJ*, 671, 153–170

Catarina Aydar

Session 7: Discussion - questions and answers

Galaxy Evolution and Feedback across Different Environments
Proceedings IAU Symposium No. 359, 2020
T. Storchi-Bergmann, W. Forman, R. Overzier & R. Riffel, eds.
doi:10.1017/S174392132000424X

Discussion Session

For this Session, Roderick Overzier invited everybody to submit questions relevant to the subject of the Symposium, that we then tried to answer during the Session. The questions and the corresponding answers - some discussed at the Symposium, and some sent by the participants after the Symposium - are listed below.

1. When and how do we expect to finally confirm what were the seed masses of supermassive black holes?

***Amirnezam Amiri*:** Understanding how super massive black holes' seeds distributed or evolved is one of the prominent keys in our knowledge in galaxy formation and evolution. There are different ways to create a super massive BH in the central galaxy region, for example: a) Direct matter collapsed on the halo when the baryonic and dark matter halo interact, gravitationally; b) Mass gathering (light elements, particularly hydrogen) as a function of time. Passing a criterion away, the mass gathered collapsed to a BH; c) Merging astrophysical BH together to make a SMBH. A good paper can be found here: https://iopscience.iop.org/article/10.1086/344675/pdf.

***Richard Bower*:** Ooooo! A tricky one: by looking for the remnants that failed to merge, but how would we find such objects? Possibly by extending the black hole mass function to lower masses, but the difficulty would be to distinguish low accretion rate objects from low mass objects. I think that the way we will make progress on this is through gravitational wave detections, particularly in the LISA era. That will be particularly well suited to answering this issue.

***Rainer Weinberger*:** Existence of specific channels: intermediate mass BHs (excluding: space based gravitational wave detectors). Excluding seed channels that produce higher mass seeds will be very difficult.

***Roderik Overzier*:** The combination of JWST and future X-ray telescopes Athena and Lynx should be able to detect the massive, direct-collapse seeds, if they exist. Shortly after their formation, accretion onto the massive seeds outshines any stellar emission from their hosts leading to strong infrared and X-ray emission. Not detecting any of such objects at very high redshifts should thus also be extremely insightful.

2. Should we believe SMBH masses at high redshift?

***Amirnezam Amiri*:** One of the highest SMBH is at redshift ∼7.5 with an 800 million solar mass (https://arxiv.org/abs/1712.01860). There are two scenarios, from my mind, to describe them: a) The available gas in the whole galaxy moves to central part as a function of short time (angular transportation or other mechanisms) and collapse to SMBH; b) Colliding or cannibalism of a satellite galaxy to central region. It seems possible we can detect more SMBHs based on their activities (∼quasars) at larger distances when our telescope facilities develop more.

***Thaisa Storchi-Bergmann*:** One major caveat is the method used to measure these masses: scaling relations and the assumed virial equilibrium for the clouds of the Broad Line Region (BLR). It is not clear that these properties – derived for nearby targets – apply also at the early phases of the Universe. In particular, in these earlier phases, non-circular motions could be dominating the BLR kinematics, resulting in overestimated SBMH masses.

***Rainer Weinberger*:** The precise number may be uncertain, but as an order of magnitude estimate.

3. What processes may explain overmassive black holes with respect to their bulges?

***Richard Bower*:** When we see such objects in our simulations, they are the result of a more massive galaxy being stripped of their stellar mass. It is possible, however, that the black hole enters the rapid growth phase "prematurely" in a relatively low mass galaxy. While the growth of the central black hole seems supressed in most galaxies until $M^* \sim 10^{10.5}$ Msol due to the interaction with supernova-driven feedback, this suppression may sometimes fail leading to premature rapid BH growth. We need to understand this interaction better, and spotting these "exceptions" may help us do exactly that.

***Benjamin Davis*:** Sahu, Graham, & Davis (2019, ApJ, 887, 10) presented a separate, steeper central black hole mass – stellar velocity dispersion (M–sigma) relation for core-Sérsic galaxies. This substructure in the M–sigma relation reveals that dry mergers do not increase the velocity dispersion, relative to the increased black hole mass, at the pace followed by Sérsic galaxies (built through either gas-rich mergers or accretion of gas from their surroundings). Such distinctive coevolution is typically manifest in brightest cluster galaxies (BCGs) because they undergo multiple gas-poor (dry) mergers, resulting in over-massive black holes with only mildly increased velocity dispersions.

***Storchi-Bergmann*:** What about loss of mass of the galaxy outskirts due to tidal striping or harassment?

***Weinberger*:** From a simulation perspective, the only thing that happens is stripping of the host halo (reduction of stellar mass). However, I could imagine that the coupling efficiency related to AGN feedback is low in some systems, which could allow over-massive BHs. In simulations, this efficiency is postulated to be constant.

4. Does AGN feedback play an important role on shaping the SMBH mass-galaxy (or halo) mass relation or not?

***Richard Bower*:** At high galaxy/halo masses this is certainly the case: the growth of the black hole will be curtailed when the energy injection rate becomes comparable to the cooling rate of the halo.

At lower masses, below the mass of the MilkyWay, I'd say no! the effect is the other way around and the growth of the black hole is controlled by the effectiveness of the supernova-driven feedback. Just above the mass of the Milky Way, the supernova regulation breaks down and the black hole grows rapidly until its growth is curtailed by the halo mass.

The thing I like about this scenario, is that it simulataneously explains the transition in galaxy properties from late to early type and the appearance of hot X-ray corona.

***Montserrat Villar Martin*:** We do not know. The scenario is very promising and attractive. It is successful from the theoretical point of view, but the observational evidence is controversial. The lack of solid alternative explanations for several open issues related to galaxy formation and evolution (including the relation between the SMBH mass and some large scale galaxy properties) strengthens the AGN feedback scenario even more. At the same time, I have the impression that this has led to the assumption that it is the only option (at least in massive galaxies) and that if we look hard enough, we will find definite evidence for the impact of AGN feedback in galaxy evolution. Currently, theory and observations leave sufficiently broad room for interpretation to reconcile the data with this scenario (and with the opposite!). Clearly, this does not mean that it has been corroborated (anyway the goal is not to accommodate the data to the scenario we like). It is essential to constrain much better the triggering mechanism and outflow properties (masses, energies, mass outflow rates, sizes, geometries...) and effects to evaluate the true impact of AGN feedback in galaxy evolution in general. It is also essential to advance on the treatment of feedback in numerical simulations so that they can be compared with the observations.

***Weinberger*:** It seems so, though how much is unclear.

5. What is the role of the hot halo in AGN feedback? Is it important in objects that are not part of groups/clusters?

***Richard Bower*:** In the group/cluster regime, AGN feedback plays a key role as the question suggests. At lower masses, there is no hot halo for the black hole to heat (and the black hole does not undergo a sustained period of (close-to) Eddington limited accretion). Is there no hot halo because the BH haloes has limited accretion though? I'd argue that it is the other way around, and that the longer cooling times in groups/clusters allow for the creation of the hot halo that leads to a reduction in the effectiveness of SN feedback and to sustained high BH accretion rates.

***Bianca Maria Poggianti*:** There is recent evidence that when a galaxy infalls into a cluster and moves at high speed within the hot intracluster medium filling the cluster, the ram pressure may cause the galaxy interstellar medium to flow towards the center, thus triggering the AGN activity. As a result, the fraction of AGN among galaxies with long extraplanar tails of stripped gas is unusually high, compared to both field and general cluster samples of galaxies. The ram pressure-triggered AGN, in turn, provides feedback that can impact a large fraction of the stellar disk: visible effects are a sharp decrease of the molecular gas content and the star formation activity, like a "cavity" that can extend for several kpc (see my contribution in these proceedings).

On the other hand, the large amount of energy injected by the AGN into the interstellar medium, decreasing its binding energy, can increase the efficiency of ram pressure stripping, possibly helping producing the striking tails of stripped gas that are observed in clusters. Of these two effects (the ram pressure causing the AGN, and the viceversa), the first one seems the dominant effect. This conclusion is reached observing that the triggering of the AGN is strongly linked with the velocity and position of the galaxy with respect to the cluster center, hence to the conditions for ram pressure stripping. This clearly does not exclude the possibility that the energy injected by the AGN contributes to efficient gas loss.

Weinberger: Many small-scale simulations show that the hot halo is the medium into which most of the AGN energy is transferred to. Thus, its existence might be crucial, not just for groups and clusters, but also in Milky-Way mass systems.

6. Does the environment play a role in shaping galaxies that are not part of a group or cluster?

Amirnezam Amiri: Role of environment on galaxy evolution is an exciting topic. We can generally divide galaxies in two environments: High-density regions (cluster or groups of galaxies) and low-density regions (Isolated/void galaxies). It is well known the stellar parameters (such as star formation rate, stellar age, surface gas density) in void galaxies (or isolated galaxies) are younger than the galaxies in the crowded regions (e.g. https://arxiv.org/abs/1601.08228, https://arxiv.org/abs/1601.04092). The secular evolution can go ahead quietly and in this case the evolution of galaxies can be different. We should be aware that all void galaxies were not isolated at all. From some aspects of galaxy structure and formation at high redshift (the structure formation based on non-linearity regions) most of galaxies can interact together in time. Their evolution as a function of redshift (from dark energy effect at $z \sim 1$ to fly-by galaxy evolution) can exclude them into a low-density region.

Richard Bower: An interesting question. In simulations, we certainly see galaxies that briefly interact with groups and clusters and then separate from them again. These are rare, however.

Francoise Combes: Yes, this is a domain in which much work has been done in the recent years, both from observations and simulations. Simulations for instance by Bahe *et al.* (2013) show that the environment plays a role, much farther the virial radius of a cluster. Galaxies are pre-processed in cosmic filaments, before entering a group or cluster. Spiral galaxies flow along filaments towards each other to merge and form an elliptical, and the orientation of their spin is specific (Codis *et al.* 2015). We have observed galaxies in filaments around intermediate redshift clusters with ALMA, and it appears that the gas content of these galaxies reveal their environment (Sperone-Longin *et al.* 2020), We have also observed with IRAM the molecular content of galaxies around Virgo, at about 5-7 Virial radii, and there exist already significant environmental effects (Castignani *et al.* 2020).

Weinberger: To some degree, though the underlying mechanism is gravity not hydrodynamics.

7. How much do you really believe the outflow rates (ionised, neutral, molecular) we estimate?

Santiago Garcia Burillo: Mass outflow rates as well as momentum and energy rates suffer from large uncertainties due to inaccurate estimates of the masses, sizes and velocities of the outflowing components in all tracers. In particular, conversion factors between luminosities and masses plague the estimates of the molecular gas masses in outflows. While we normally accept the use of standard (Milky-way or ULIRG-like) CO-to-H2 conversion factors, typical of optically thick gas, when deriving masses of molecular outflows, there is now growing evidence that lower conversion factors may be more appropriate in some systems where the outflowing molecular gas is optically thin in CO lines. If this is the rule, numbers may change dramatically.

Ric Davies: There are significant uncertainties in all aspects of the derivation of outflow rates, from the estimation of the speed itself (the maximum measured velocity in a line is taken to represent the outflow speed of all the gas), to the calculation of the mass (for ionised outflows the adopted density has a major impact; for molecular outflows the equivalent is the CO luminosity-to-mass conversion factor), to the outflow model itself (the two main classes of models differ by a factor 3 in outflow rate). Estimates of outflow rate that are based on integrated properties - as most are - have additional uncertainties, since one might expect, for example, outflow speed and gas density to vary with distance from the AGN. As such, one should treat any calculation of outflow rate as, at best, an order of magnitude estimation.

Steven Kraemer: Those based on spatially-resolved spectra are reliable, within a factor of a few. The global estimates have much larger uncertainties. However, the optical/resolved studies only sample a portion of the outflows. The largest reservoir for mass outflow may be in the form of X-ray emitting gas. But the lack of spatially resolved X-ray spectra make it difficult to determine the X-ray outflow rates. So, we have estimates of varying reliability and we are not getting a comprehensive view of the outflows.

Giacomo Venturi: I believe that they should be handled carefully, since the uncertainties on their estimation are several (arising e.g. from the density estimation, the CO-to-H2 conversion factor, the lack of spatial information) and the methods to obtain such quantities are really variable depending on the given work. For example, in Venturi *et al.* (2018) we found that by determining the mass outflow rate of the same object first from spatially resolved data and then by using the integrated information and a different recipe for calculating the mass outflow rate, the results changed by a factor of 20 between the two cases.

Weinberger: Not much.

8. What would it take to reduce outflow rate uncertainties by a factor of 10?

Santiago Garcia Burillo: High-spatial resolution combined with IFUs to build-up 3D models used to fit these observations. In molecular gas, use several lines to derive physical conditions and constrain conversion factors.

Ric Davies: Several things would help considerably: (i) using spectral and/or spatial information to constrain the outflow geometry, and hence how the measured maximum velocity relates to the actual outflow speed; (ii) for ionized outflows, a reliable way to estimate gas density is essential, and ideally one would know how the density varies with distance from the AGN; (iii) a better understanding of how outflows evolve over time (e.g. whether the outflow rate decreases with time or remains constant).

9. Is it important to have the exact feedback recipes (radiative, mechanical) in simulations of galaxy formation or only the global energy?

Richard Bower: The main challenge is to ensure that the energy that is put into the simulation is not immediately radiated. This happens because the limited resolution that we need to work at cannot reliably model the multiphase nature of the ISM. In reality, the cooling rate of hot (supernova-heated) gas can be much lower than you'd estimate from the average temperature density of a >100pc patch. Given this uncertainty, the exact recipe for heating is much less important than calibrating the net effective heating rate in some way.

Similar issues exist for the BH feedback, so I would optimistically think that the key is to heat the correct volume of gas by the correct change in entropy/energy. Hopefully the details of how this is done are not so important. Whether my optimism is well found is still an open question, however!

Santiago Garcia Burillo: I think the effects of the various feedback recipes is expected to be dissimilar on the different ISM phases. In particular, momentum and energy is injected differently in molecular gas and this can tested observationally. The net efficiency of feedback on molecular gas may change depending on the feedback recipe.

Weinberger: Energy + post-injection temperature and its timing are the key aspects to get global (integrated) galaxy properties correct; however, it is likely that internal galaxy properties are more sensitive to the exact mechanism. This needs to be explored further.

10. What quantities should be compared between observations and simulations? Wind velocity, outflow rate, kinetic energy?

Weinberger: The velocity PDF might be quite helpful to determine underlying mechanisms, but it is hard to see how this alone helps to determine its impact on the galaxy. In the latter case, an energy estimate might be required.

Storchi-Bergmann: From the physics point of view, the energy deposited on the galaxy should be the most important parameter. But observations are usually done for one gas phase, due to instrument constraints. Observations with multiple instruments, covering different wavelength ranges and gas phases are necessary to account for the energy deposited by outflows on the host galaxy.

11. How can <sub-pc scale simulations help to improve feedback prescriptions in large simulations?

Weinberger: The main problem with feedback models on kpc scales is that it is not even clear what exactly they should produce. Small scale simulation, capturing the necessary physics should ideally be able to provide this information.

12. What kind of AGN feedback has a bigger impact on galaxy evolution in general; radiative or mechanical?

Richard Bower: If my optimism truns out to be correct, this doesn't matter too much. The mechanical energy input will get converted into heat and the two schemes will look much the same, especially if they are both calibrated to match a subset of the observational data (eg., the entropy porofile of galaxy groups).

Montserrat Villar Martin: This question takes for granted that AGN feedback has a significant impact in galaxy evolution which remains to be demonstrated. On the other hand, it is of course very relevant to discern the precise action of the different modes of feedback. This question poses a major challenge to understand the connection between galaxy evolution and AGN physics. At the moment, there is little observational direct evidence for AGN driven outflows to affect star formation activity in galaxies (and, consequently, their evolution) even for the most extreme outflows. The lack of evidence, of course, is not evidence of lack. Multiple questions need to be answered: How do AGN (and starburst!!) driven outflows affect the different gas phases in galaxies? (ALMA has been wonderful on this regard to discover and parametrize molecular outflows!) What are the precise masses, mass outflow rates, energy injection rates? What are the geometries, sizes? What are the time scales in relation to the star formation episodes? How much gas can escape the galaxy? Are we missing most of the gas affected by the outflows due to its undetectability with current technology? How frequent are radio jets in active galaxies? Do these provide an efficient feedback mechanism ?

***Weinberger*:** Mechanical feedback seems to have a substantially higher coupling efficiency.

***Storchi-Bergmann*:** From the observational point-of-view, one measures higher velocities and mass-outflow rates in the vicinity of the AGN in radiation-driven outflows occurring in gas-rich galaxies. Thus radiative feedback seems to have a larger impact in the vicinity of the AGN in such objects. Mechanical feedback, that occurs in radio galaxies, seem to have higher impact outside the galaxy, as observed via X-ray cavities produced by radio jets in galaxy clusters.

13. Why does AGN feedback always have to be so negative? Could it also, in cases, promote star formation?

***Giacomo Venturi*:** Models actually predict that AGN can also promote star formation, indeed, and in two different forms:

a) AGN outflows or jets, by compressing the molecular gas in the interstellar medium (ISM), would be able to trigger star formation (so-called "positive" feedback; see e.g. Silk & Norman 2009, Nayakshin & Zubovas 2012, Silk 2013, Zubovas *et al.* 2013, Nayakshin 2014). Observational evidence for AGN jets triggering star formation in companion galaxies was found for instance by Croft *et al.* (2006), Feain *et al.* (2007), Elbaz *et al.* (2009). Jet-induced star formation has also been observed by Crockett *et al.* (2012) and Santoro *et al.* (2015, 2016), who found star-forming clumps triggered by jet compression around Centaurus A. Observational evidence of positive feedback by AGN outflows has been found both at high and low redshift by Cresci *et al.* (2015a) and (2015b), respectively. So far, evidence of positive feedback is episodic and further observational effort is needed to assess its importance in galaxy formation and evolution, mostly by taking advantage of spatially resolved observations to uncover such outflow-ISM interactions.

b) Models predict that stars may be able to form even within the outflowing gas itself, by the effect of gas cooling and fragmentation (e.g. Ishibashi & Fabian 2012, 2014, 2017, Ishibashi *et al.* 2013, Zubovas *et al.* 2013a, Zubovas & King 2014), with potentially important implication on structural and chemical formation and evolution of the spheroidal component of galaxies and on the chemically enrichment of the circum-galactic and inter-galactic medium as well as on its re-ionization during the early Universe. Large quantities of dense and clumpy molecular gas, having the physical conditions of giant molecular clouds in which stars normally form, are indeed observed in outflows (e.g. Cicone *et al.* 2014, Aalto *et al.* 2015, Pereira-Santaella *et al.* 2016, Fluetsch *et al.* 2019). Finally, Maiolino *et al.* (2017) found first observational evidence of this new in-outflow star formation mode, by detecting the presence of outflowing gas ionised in situ by young stars and of a young stellar population with kinematics consistent with stars formed at high velocity in the inner region of the outflowing gas and then decelerated by the gravitational potential of the galaxy. Further studies (Gallagher *et al.* 2019, Rodríguez del

Pino *et al.* 2019) have shown that such mechanism could be quite common in the local AGN population, but tricky to uncover. So, further focused observational effort through detailed observations is needed in this case as well.

Weinberger: There is certainly a possibility for enhanced star formation due to increased pressure. This does not mean that the net long term effect is an increase, though.

14. Are we focusing too much on outflows in bright and rare quasars? Are these common enough to relate to every quenched galaxy?

Montserrat Villar Martin: Regarding the first question, not really. The most extreme outflows are expected for AGN powers in the quasar regime. They often attract a lot of attention because sometimes they are indeed spectacular. Apart from this, I think that quasars are very relevant to understand the AGN feedback phenomenon in galaxies in general. Both theory and observations indicate that the more powerful the AGN is, the more extreme the outflows are (higher mas outflow rates, energy injection rates, etc). Thus, it is natural to expect that the most dramatic impact of AGN feedback on the host galaxies occurs in the most powerful AGN: QSOs. Let's imagine that observations contradict this and QSO outflows have little or no impact: this would question the efficiency of AGN feedback in less powerful active galaxies in general.

There are many exciting ongoing projects that cover a broad range of galaxy types, from statistical studies of very large samples to detailed studies of individual sources. This is, of course, essential since most galaxies will never go through an active phase as extreme as that of quasars. I find very interesting, for instance, the growing number of works focusing on the role of AGN feedback in low mass galaxies (<1e9 Msun), a regime that has been largely unexplored.

Giacomo Venturi: Powerful outflows in bright quasars are considered the best candidates to seek the "smoking gun" of AGN feedback and study its properties. However, eventually, statistical studies involving the most common types of objects are required to determine the impact on the bulk of the galaxy population.

Weinberger: They are certainly helpful to understand underlying mechanisms, but the question of abundance should always be considered when talking about galaxy populations.

15. How much should we believe the merger - ULIRG - blowout - quasar - elliptical cartoon? Are there other routes?

Ric Davies: This scenario is entirely believable, but it is certainly only one possible evolutionary route. There is plenty of evidence that in the local universe, secular fuelling of AGN is the dominant mode (even if not the most spectacular); and there is evidence that this is also the case at high redshift. It is important to take into acount the differences between, for example, what causes the most luminous AGN, what generates the strongest feedback, what is the most common mode of fuelling, and what leads to the most black hole growth.

Andrew Newman: I interpret this question to ask whether there are other routes to form ellipticals. While dry major mergers lead to formation of ellipticals, simulations show that this is not necessarily the outcome of a major merger when the merging galaxies' gas fraction is high. Thus, when considering massive galaxies that quenched early, there is reason to suspect that even if they experience a major merger that triggers a quasar phase and ends star formation, the end product may still be disk-dominated. There is support for this observationally: the ellipticity distribution of quiescent galaxies shifts to flatter values at high redshifts, suggesting a larger population of disks. The stellar-to-dynamical mass ratio in $z \sim 2$ quiescent galaxies correlates with ellipticity in the manner expected when viewing rotating disky galaxies at various inclinations. And the four $z > 2$ quiescent galaxies with spatially resolved kinematics are all disk-dominated "fast rotators" with V/sigma values 5-10x higher than classical ellipticals. All these observations suggest that quenching and elliptical formation are disconnected processes, at least in the gas-rich high-z universe, and that morphological transformation comes later. Most likely it occurs gradually through the same series of minor mergers that are commonly invoked to explain galaxies' evolution in size after quenching.

***Weinberger*:** It still is a possible idea, but I would consider it quite plausible that there are other routes to form quiescent galaxies. It is also likely that there exists quasar activity without quenching.

16. We have seen that outflows can originate from the accretion disk, or by the radiation pressure on the galaxy gas disk. How do we reconcile the two?

***Weinberger*:** These are not contradicting each other, but one has to be careful how to interpret outflows at different scales.

17. How do we reconcile massive molecular outflows observed on large scales with the less massive ionized flows on small scales?

***Santiago Garcia Burillo*:** The AGN winds are less massive than molecular outflows, but their typical velocities are much higher (by roughly an order of magnitude). The ionized winds are nevertheless the key actors by being powerful enough to launch more massive molecular outflows due to radiation and/or thermal pressure.

***Weinberger*:** A very dynamic multi-phase gas with a lot of entrainment. Maybe metal abundances can help with the interpretation.

18. Is AGN feedback sufficient to quench all star formation in massive galaxies at high redshift, or do we need other ingredients?

***Giacomo Venturi*:** So far, observational evidence from individual objects seems to show that outflow can be capable of shutting down star formation locally, but the overall SFR in the rest of the host galaxy remains still high. Clearest observational evidence for the impact of AGN feedback in single objects comes in the local Universe in the steadier "kinetic" mode through the action of jets which heat the gas halo around massive red galaxies at the centre of galaxy clusters, preventing gas cooling and re-accretion on the galaxy, and thus further star formation. Results from statistical studies are controversial, but clear and conclusive evidence that AGN feedback is able to shut down star formation is still missing. However, for instance, the feedback mechanism might act on different timescales than those on which we observe strong AGN activity and powerful outflows, with star formation resulting to be shut down only at later times, with a "delayed" feedback mechanism.

***Weinberger*:** It seems sufficient. At least I am not aware of any massive inconsistency in the simulated AGN population in simulations that rely on AGN feedback only.

19. How exactly do we define an "outflow" and how do we measure the effect it has on star formation?

***Giacomo Venturi*:** This is a debated topic. Some people define as outflow a motion which does not follow rotation in the galaxy disk and moves outward from the galaxy, and is not either in inflow or belonging to tidal tails or to stripped material (which can be tricky to distinguish observationally if good spatial information is lacking). Some define a velocity threshold with respect to systemic above which the gas is considered as in outflow and/or define a velocity dispersion threshold above which the emission in line profiles is considered as stemming from an outflow. Other even define as outflow only material moving outwards from the galaxy which has a velocity higher than the escape velocity from the galaxy (or galaxy halo) gravitational potential, which narrows down a lot the outflow candidates. The effect of outflows on star formation can be seek in different ways. Either in a direct way, e.g. by searching for regions of the galaxy where star formation appears damped in spatial correspondence with outflowing material, or through statistical studies investigating if star formation in the host (in the form of SFR, sSFR, SFE etc...) appears to be lowered in presence of outflows compared to cases without outflows. However, finding clear evidence may be tricky if the process of star formation quenching by outflows acts on different timescales than those in which outflows and intense star formation co-exist ("delayed" feedback) or if it operates on longer timescales through many outflow episodes with at a steadier and less intense pace, rather than through a single really intense blow-out phase.

***Weinberger*:** Not sure the effect on star formation is 'measurable' on an individual galaxy.

20. Is it necessary for feedback to expel the gas from a galaxy/halo for it to have an effect on the host galaxy?

***Richard Bower*:** I think the dominant effect is to cut off the supply of further cold gas by disrupting the cold stream as they pass through the hot halo of the galaxy. Supernova feedback is able to quickly heat/expel the gas that remains in the galaxy.

***Santiago Garcia Burillo*:** Not necessarily. Feedback can stabilise molecular gas by injecting kinetic energy that is eventually converted into turbulence. This will inhibit star formation for a while. A significant delay of the star formation process in the system, without having to resort to expelling the gas, is another form of feedback.

***Weinberger*:** Depends on mass of the galaxy/halo.

21. At what redshifts or in what environments does radio lobe-Xray cavity interaction become important?

***Weinberger*:** Might be as low as Milky-Way mass. Future soft X-ray observations will hopefully tell.

22. How common are the small scale radio jets in QSOs or galaxies in general?

***Montserrat Villar Martin*:** There is evidence for a compact relativistic jet in the Milky Way. This opens the possibility that SMBH driven jets exist in many galaxies, even with such low levels of nuclear activity that we consider them "inactive".

Different studies have shown since the 1980's that Seyfert galaxies and LINERs often host compact (∼10−100 pc, sometimes larger) jets. We do not know whether this is the case for QSOs also. The nature of the radio emission in radio quiet quasars (RQQ, ∼90% of all QSO) is still a matter of debate. It is not clear whether it is dominated by star formation in the host galaxy or by non thermal emission due to processes related to the nuclear activity. Some works suggest that the relative contribution of AGN related processes (including jets) to the radio emission in RQQ increases with radio luminosity. Even when the radio emission has a clear AGN origin, it is often difficult to discern the precise mechanism that produces it (e.g. jets/lobes/hot spots vs. relativistic electrons in wide angle quasar outflows). Moreover, the fact that the radio emission is consistent with star formation does not exclude the existence of jets.

***Giacomo Venturi*:** While jets in historically known as "radio quiet" sources were usually considered not being capable of affecting their host galaxies, as opposed to the much more powerful and extended jets in "radio loud" objects, recently this paradigm has been challenged, with the discovery of low power jets in radio quiet sources driving outflows and interacting with the ISM of the galaxy in more and more objects (e.g. Combes *et al.* 2013, García-Burillo *et al.* 2014, Morganti *et al.* 2015, Oosterloo *et al.* 2017, Jarvis *et al.* 2019). Discriminating if a "radio quiet" source actually hosts a small-scale jet or not requires high spatial resolution observations to resolve the small scales. Complete high-resolution studies of galaxy samples are thus necessary to determine how common these small-scale low-power jets are. Moreover, though synergies with observations in other bands revealed that they seem to be able to drive outflows or turbulence throughout the host galaxy, so far it is still not clear at which extent they are capable of affecting the gas reservoir in the host and its star formation processes. Being a recent topic, more of such multi-band detailed works are necessary to better investigate the phenomenon and assess its role.

***Weinberger*:** Seem very abundant.

23. Is the small-scale feedback they drive enough to disrupt growth?

***Weinberger*:** Might certainly limit it to a certain degree.

24. Do we worry enough about the importance of AGN variability?

***Santiago Garcia Burillo*:** We should worry. Outflows may sometimes mimic the AGN variability or the duration of a complete 'AGN cycle' to some extent. Some relic outflows are being discovered in systems with no current AGN activity. Different regions in spatially extended outflows do have different associated time-scales and they can reflect different stages of the AGN activity.

***Ric Davies*:** Not enough by far. There have been important papers published about this topic. But still any comparison of active and inactive galaxies needs to allow for the fact that an inactive galaxy might be in a short 'off' phase during a longer period of stronger activity, or it might have been in a low activity phase for a long period. In the former case, other aspects of the galaxy may still be very similar to an AGN; while in the latter case they may be quite different. One should also consider whether the fuelling path (via secular evolution or minor/major merger, etc) and environment (which might help or hinder these fuelling paths, or have little impact) has an impact on long vs short term variability.

Weinberger: In the context of interpretation of AGN luminosities, it is likely very important. With regard to AGN feedback and large-scale outflows not that much.

25. Do observations of luminous AGN with massive star formation imply that AGN feedback is not so important?

Giacomo Venturi: Not necessarily. First, more than focusing only on the absolute SFR, it is important to also consider for instance the specific SFR (sSFR), its efficiency (SFE) or the departure of the source from the star formation main sequence. In some case also, AGN triggering may be associated with large quantities of material being funnelled to the galaxy centre and gas mixing (e.g. by merging processes), leading in parallel to strong star formation processes as well. Moreover, the fact that current star formation seems to be unaffected by AGN does not necessarily mean that they are not important, as their feedback effect might arise at later times, through a "delayed" feedback mechanism, or on longer and steadier timescales rather than in a fast and intense phase, maybe through the cumulative effect of multiple AGN episodes and outflow events, thereby with mechanisms more difficult to detect observationally.

Weinberger: No.

Roderik Overzier

Tiago Ricci and Ana L. Chies-Santos

Travis Fischer

Montserrat Villar Martin Cristina Ramos Almeida

Poster-Sessions Photos

Santiago Garcia-Burillo, Francoise Combes and Keiichi Wada

Anelise Audibert, Miriani Pastoriza and Natacha Dametto

Ana L. Chies-Santos, Sandra Raimundo and Michael Beasley

Thaisa with Miriani Pastoriza and Eduardo Telles: Miriani was Thaisa's PhD advisor and to this date a guide and mentor for the Astronomy Department of IF-UFRGS.

Author index